COPPER-OXYGEN CHEMISTRY

Edited by

KENNETH D. KARLIN

SHINOBU ITOH

CW01506958

⊛WILEY

A JOHN WILEY & SONS, INC., PUBLICATION

Published by John Wiley & Sons, Inc., Hoboken, New Jersey
Published simultaneously in Canada

For general information on our other products and services or for technical support, please contact
our Customer Care Department within the United States at (800) 762-2974, outside the United States
at (317) 572-3993 or fax (317) 572-4002.

Wiley also publishes its books in a variety of electronic formats. Some content that appears in print may
not be available in electronic formats. For more information about Wiley products, visit our web site
at www.wiley.com.

Library of Congress Cataloging-in-Publication Data:

Copper-oxygen chemistry / edited by Kenneth D. Karlin, Shinobu Itoh.
    p. cm. – (Wiley series of reactive intermediates in chemistry and biology ; v. 8)
    Includes index.
  ISBN 978-0-470-52835-8 (hardback)
  1. Copper proteins. 2. Copper–Peroxidation. 3. Bioinorganic chemistry. I.
Karlin, Kenneth D., 1948- II. Itoh, Shinobu.
  QP535.C9C67 2011
  612.3'924–dc22

                                                          2011010606

Printed in the United States of America

oBook ISBN: 978-1-118-09436-5
ePDF ISBN: 978-1-118-09434-1
ePub ISBN: 978-1-118-09435-8

10 9 8 7 6 5 4 3 2 1

COPPER-OXYGEN
CHEMISTRY

Wiley Series of Reactive Intermediates in Chemistry and Biology

Steven E. Rokita, Series Editor

CONTENTS

PREFACE TO SERIES

Most stable compounds and functional groups have benefitted from numerous monographs and series devoted to their unique chemistry, and most biological materials and processes have received similar attention. Chemical and biological mechanisms have also been the subject of individual reviews and compilations. When reactive intermediates are given center stage, presentations often focus on the details and approaches of one discipline despite their common prominence in the primary literature of physical, theoretical, organic, inorganic and biological disciplines. The *Wiley Series on Reactive Intermediates in Chemistry and Biology* is designed to supply a complementary perspective from current publications by focusing each volume on a specific reactive intermediate and endowing it with the broadest possible context and outlook. Individual volumes may serve to supplement an advanced course, sustain a special topics course, and provide a ready resource for the research community. Readers should feel equally reassured by reviews in their speciality, inspired by helpful updates in allied areas and intrigued by topics not yet familiar.

This series revels in the diversity of its perspectives and expertise. Where some books draw strength from their focused details, this series draws strength from the breadth of its presentations. The goal is to illustrate the widest possible range of literature that covers the subject of each volume. When appropriate, topics may span theoretical approaches for predicting reactivity, physical methods of analysis, strategies for generating intermediates, utility for chemical synthesis, applications in biochemistry and medicine, impact on the environmental, occurrence in biology and more. Experimental systems used to explore these topics may be equally broad and range from simple models to complex arrays and mixtures such as those found in the final frontiers of cells, organisms, earth and space.

   Advances in chemistry and biology gain from a mutual synergy. As new methods are developed for one field, they are often rapidly adapted for application in the other. Biological transformations and pathways often inspire analogous development of new procedures in chemical synthesis, and likewise, chemical characterization and identification of transient intermediates often provide the foundation for understanding the biosynthesis and reactivity of many new biological materials. While individual chapters may draw from a single expertise, the range of contributions contained within each volume should collectively offer readers with a multi-disciplinary analysis and exposure to the full range of activities in the field. As this series grows, individualized compilations may also be created through electronic access to highlight a particular approach or application across many volumes that together cover a variety of different reactive intermediates.

   Interest in starting this series came easily, but the creation of each volume of this series required vision, hard work, enthusiasm and persistence. I thank all of the contributors and editors who graciously accepted the challenge.

STEVEN E. ROKITA

University of Maryland

INTRODUCTION

There is a great deal of current interest in the subject of oxidative processes mediated by copper ion. Not only have there been considerable recent advances in chemical applications, those useful to synthetic organic and pharmaceutical researchers, but there have been major advances in the clarification of biochemical oxidations that occur widely and critically in biological systems. In fact, the two areas have a synergistic relationship.

Copper-mediated biological oxidations include a diverse array of reaction types. The insertion of one or both oxygen atoms from molecular oxygen (O_2) into an organic substrate underscores mild and highly selective transformations. Such transformations have been and are highly worthy of careful attention by synthetic and catalytic chemists. Many other oxidation (i.e., dehydrogenation) reactions utilize the cheap, abundant, and energy containing O_2 molecule. Copper metalloproteins thus also mediate energetic processes.

In synthetic chemistry, many recent successes have in actual fact been bioinspired. The subfield of catalytic alcohol oxidation chemistry, using O_2, has seen explosive growth and following insights obtained from complementary copper protein studies. Also, there now exists a significantly increasing literature on oxygen-atom insertion, C−H bond activation, and C−C bond formation reactions, as well as very important enantioselective oxidative chemistries. Specific applications to the synthesis of sophisticated molecules important in natural product or pharmaceutical chemistries derive from information on copper enzymes or their synthetic models. Long known industrial examples include the oxidative coupling of phenols, requiring molecular oxygen and copper catalysis.

In addition, as mentioned, there has been an explosion of recent activity and significant insights obtained from biochemical and model compound chemistries.

These studies have highlighted the application of advanced physical–spectroscopic techniques, theory, and synthesis and study of copper coordination complex biomimics. The result has been the discovery of new types of interactions of molecular oxygen with copper ion, and most importantly the elucidation of reactivity patterns or oxidative capabilities, many previously unknown.

Thus, the current Volume (Vol. 4) on Copper–Oxygen Chemistry, within the Wiley Series on *Reactive Intermediates in Chemistry and Biology (S. R. Rokita, Ed.)*, will be an important addition. The authors have been selected for their international reputations and expertise. We choose to divide the topics and volume into logical areas, (A) Biological Systems, (B) Theory, and (C) Bioinorganic Models and Applications. The overlap seen between these areas will be very apparent in the final published volume, because of the synergism that exists. There will be considerable reference to other chapters and subjects covered in this volume.

The treatment here will be broad, including all the major and important areas and aspects of the field of Copper–Oxygen Chemistry. This volume will unquestionably appeal to a very broad audience. Biochemists, biophysicists, medical–pharmaceutical chemists, organic synthetic and (bio)inorganic chemists in academia and industry should find the volume to be highly interesting and useful, covering front-line areas and thorough in its coverage by prominent authors in the field.

KENNETH D. KARLIN

Johns Hopkins University

SHINOBU ITOH

Osaka University

CONTRIBUTORS

Doreen E. Brown, Department of Chemistry and Biochemistry, Montana State University, Bozeman, MT, USA

Simon de Vries, Department of Biotechnology, Delft University of Technology, Julianalaan 67, 2628 BC, Delft, The Netherlands

David M. Dooley, Department of Chemistry and Biochemistry, Montana State University, Bozeman, MT, *and* University of Rhode Island, Green Hall, 35 Campus Avenue, Kingston, RI, USA

Zakaria Halime, Department of Chemistry, Johns Hopkins University, 3400 N. Charles Street, Baltimore, MD 21218, USA

Shinobu Itoh, Department of Material and Life Science, Division of Advanced Science and Biotechnology, Graduate School of Engineering, Osaka University, 2-1 Yamada-oka, Suita, Osaka 565-0871, Japan

József Kaizer, Department of Chemistry, University of Pannonia, 8201 Veszprém, Hungary

Kenneth D. Karlin, Department of Chemistry, Johns Hopkins University, 3400 N. Charles Street, Baltimore, MD 21218, USA

Kunishige Kataoka, Graduate School of Natural Science and Technology, Kanazawa University, Kakuma, Kanazawa 920-1192, Japan

Judith P. Klinman, Department of Chemistry, Department of Molecular and Cellular Biology, and California Institute for Quantitative Biosciences, University of California - Berkeley, Berkeley, CA 94720, USA

Marisa C. Kozlowski, Department of Chemistry, Roy and Diana Vagelos Laboratories, University of Pennsylvania, Philadelphia, Pennsylvania, 19104, USA

Robert L. Osborne, Department of Chemistry and California Institute for Quantitative Biosciences, University of California - Berkeley, Berkeley, CA 94720, USA

József Sándor Pap, Department of Chemistry, University of Pannonia, 8201 Veszprém, Hungary

Angela Paulus, Department of Biotechnology, Delft University of Technology, Julianalaan 67, 2628 BC, Delft, The Netherlands

Jean-Noël Rebilly, Université Paris Descartes, UMR 8601, Laboratoire de Chimie et Biochimie, Pharmacologiques et Toxicologiques, 45 rue des Saints-Pères, 75006 Paris, France

Olivia Reinaud, Université Paris Descartes, UMR 8601 Laboratoire de Chimie et Biochimie, Pharmacologiques et Toxicologiques, 45 rue des Saints-Pères, 75006 Paris, France

Dalia Rokhsana, Department of Chemistry and Biochemistry, Montana State University, Bozeman, MT, USA

Justine P. Roth, Department of Chemistry, Johns Hopkins University, 3400 N. Charles Street, Baltimore, MD 21218, USA

Takeshi Sakurai, Graduate School of Natural Science and Technology, Kanazawa University, Kakuma, Kanazawa 920-1192, Japan

Eric M. Shepard, Department of Chemistry and Biochemistry, Montana State University, Bozeman, MT, USA

Gábor Speier, Department of Chemistry, University of Pannonia, 8201 Veszprém, Hungary

Kazunari Yoshizawa, Institute for Materials Chemistry and Engineering and International Research, Center for Molecular Systems, Kyushu University, Fukuoka 819-0395, Japan

1

INSIGHTS INTO THE PROPOSED COPPER–OXYGEN INTERMEDIATES THAT REGULATE THE MECHANISM OF REACTIONS CATALYZED BY DOPAMINE β-MONOOXYGENASE, PEPTIDYLGLYCINE α-HYDROXYLATING MONOOXYGENASE, AND TYRAMINE β-MONOOXYGENASE

ROBERT L. OSBORNE

Department of Chemistry and California Institute for Quantitative Biosciences, University of California, Berkeley, CA 94720

JUDITH P. KLINMAN

Department of Chemistry, Department of Molecular and Cellular Biology, and California Institute for Quantitative Biosciences, University of California, Berkeley, CA 94720

Copper-Oxygen Chemistry, First Edition. Edited by Kenneth D. Karlin and Shinobu Itoh.
© 2011 John Wiley & Sons, Inc. Published 2011 by John Wiley & Sons, Inc.

1.1. GENERAL INTRODUCTION

Oxidation–reduction reactions are essential to all biological systems and are involved in energy storage and countless biosynthetic reactions. Monooxygenases represent a large class of redox enzymes that incorporate one atom from dioxygen (O_2) into organic substrates while the other oxygen appears in water (Eq. 1.1).[1]

$$R_3C-H + O_2 + 2H^+ + 2e^- \rightarrow R_3C-OH + H_2O \tag{1.1}$$

Without monooxygenases, the spin-forbidden hydroxylation of innumerable ground-state singlet organic substrates by ground-state triplet O_2 would not occur at measurable rates. Nature has evolved to utilize organic cofactors and transition metal ions in order to bypass the spin-forbidden properties of these reactions. The total number of copper-containing monooxygenases is relatively small, but they catalyze physiologically essential reactions and are of great interest to medical science.[2] This chapter focuses on the chemical mechanism of dopamine β-monooxygenase (DβM), peptidylglycine α-hydroxylating monooxygenase (PHM), and tyramine β-monooxygenase (TβM), which belong to a small class of homologous, eukaryotic copper-, ascorbate-, and oxygen-dependent enzymes.[3–5] Herein, we aim to highlight parallel features that are unique to the members of this important class of enzymes. In particular, spectroscopic, structural, and detailed kinetic analyses are critically examined, discussed, and compared to the significant, recent contributions to the literature. A re-examination of previous findings, together with recent evidence, points toward a network of communication between the two copper-containing domains that are separated by an 11Å, solvent-filled cleft. The goal of this chapter is to provide a comprehensive examination of the complex mechanism of O_2 activation coupled to hydrogen atom transfer catalyzed by this fascinating class of enzymes.

1.2. COMPARATIVE PROPERTIES OF DOPAMINE β-MONOOXYGENASE (DβM), PEPTIDYLGLYCINE α-HYDROXYLATING MONOOXYGENASE (PHM), AND TYRAMINE β-MONOOXYGENASE (TβM)

Dopamine β-monooxygenase (DβM) is an essential enzyme in the catecholamine biosynthetic pathway that catalyzes the hydroxylation of dopamine to yield norepinephrine (Eq. 1.2), both of which are neurotransmitters in the sympathetic nervous system.[6–8]

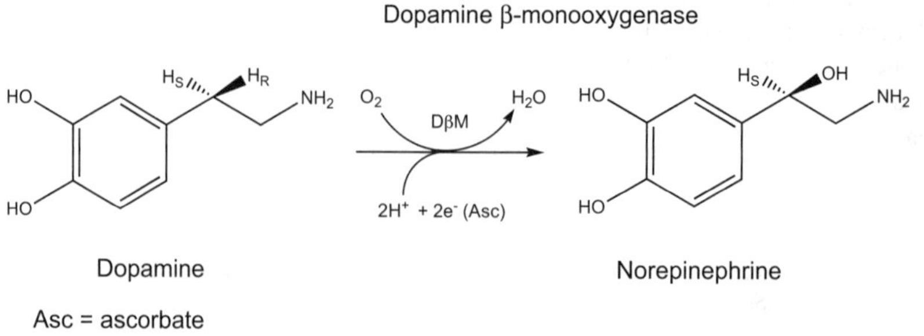

Dopamine β-monooxygenase

$$\tag{1.2}$$

Dopamine   Norepinephrine

Asc = ascorbate

Peptidylglycine α-amidating monooxygenase (PAM) is a bifunctional enzyme, which contains two independent enzymatic domains: a monooxygenase domain and a lyase domain. Peptidylglycine α-hydroxylating monooxygenase (PHM) is the monooxygenase domain and catalyzes the hydroxylation of peptidylglycine substrates (Eq. 1.3) en route to the biosynthesis of C-terminally carboxamidated peptide hormones.[4,9]

Peptidylglycine α-hydroxylating monooxygenase

2 Ascorbate   2 Semidehydroascorbate

$$(1.3)$$

Peptidylglycine   Peptidyl α-hydroxyglycine

Tyramine β-monooxygenase (TβM) is the insect homologue of DβM and plays a critical role in the biosynthesis of invertebrate neurotransmitters by catalyzing the hydroxylation of tyramine (Eq. 1.4).[10–12]

Tyramine β-monooxygenase

$$(1.4)$$

Tyramine   Octopamine

The enzymes DβM and PHM exist in both soluble and membrane-bound forms localized in neurosecretory vesicles of the adrenal gland or synaptic vesicles of the sympathetic nervous system (DβM) or secretory vesicles of the pituitary gland (PHM).[9,13–15] The location of TβM expression is not definitively known, but studies indicate that TβM is localized to octopaminergic neurons.[11,12] Initial studies on this important class of enzymes were focused on DβM. However, extensive experimental evidence, which will be discussed throughout this chapter, indicates a highly conserved chemical mechanism among all three enzymes.

The two-electron ($2e^-$) oxidation of dopamine to norepinephrine (Eq. 1.2), C-terminal glycine-extended peptides to their α-hydroxylated products (Eq. 1.3), and tyramine to octopamine (Eq. 1.4) involves the incorporation of an oxygen atom into the β-carbon of the ethylamine side chain (DβM and TβM) or the α-carbon of peptide hormone precursors (PHM). Early experiments revealed that DβM was a copper-dependent enzyme and that the incorporated oxygen was derived from molecular oxygen. However, understanding the relationship between copper stoichiometry and enzyme activity remained a challenge until rapid mixing techniques were used to demonstrate that optimal catalysis required two copper atoms per DβM subunit.[16] Electron paramagnetic resonance (EPR) spectra collected for PHM samples reconstituted with copper were found to contain ~ 2 equiv of copper per mole of enzyme.[4] Substrate oxidation is coupled to the $4e^-$ reduction of O_2 to water by using two electrons stored on copper cofactors of the enzyme, supplied by an exogenous donor, and two from the substrate undergoing oxidation. The preferred *in vitro* and *in vivo* electron donor is ascorbic acid, which is oxidized sequentially to yield 2 equiv of semidehydroascorbate with concomitant cyclical reduction of the copper sites.[17–21] The Cu(I) sites are recycled to Cu(II) in the presence of substrate and O_2 via a "ping–pong" mechanism.

Despite numerous efforts, a robust DβM expression system has not been developed. However, the availability of a fully active truncated PHM expression system[4] and the very recently developed facile TβM expression system[21] has made the pursuit of structure–function–dynamics relationships possible.

1.3. SEQUENCE, STRUCTURE, AND SPECTROSCOPY

Comparison of the primary sequence of PHM with a core of $\sim$300 amino acids from DβM indicates 27% identity and 40% similarity.[4,22,23] The sequence of TβM is 39% identical and 55% similar to its mammalian equivalent, DβM (Fig. 1.1).[21,24] Among these enzymes are six conserved copper ligands. Eight of the 10 cysteine residues in PHM are conserved.[4,15] The 12 intrasubunit cysteine residues of DβM are conserved in TβM, but the intersubunit cysteines of DβM are not conserved.[21] However, only five cysteine residues are strictly conserved when the primary sequences of DβM (rat), PHM (rat), and TβM (*Drosophila*) are aligned (Fig. 1.1).[24] In the last decade, a series of X-ray crystal structures of different forms of the catalytic core of PHM have been published that are fully consistent with previous indirect structural probes and spectroscopic studies. PHM is composed of two domains $\sim$150 residues in length, each of which contains one catalytic copper (Fig. 1.2).[25] The domain structures include an eight-stranded antiparallel jelly-roll motif. The solvent-exposed active-site coppers are separated by $\sim$11 Å, and the coordination environment of each copper is unique (Fig. 1.3).[25] The unusually long distance between copper sites without bridging ligands is consistent with earlier EPR studies in which no evidence of spin coupling between the paramagnetic copper centers was detected.[2] The EPR studies on PHM and TβM further support the structural homology between the copper centers among this class of enzymes.[4,21]

Extended X-ray absorption fine structure (EXAFS) spectroscopy has been used to probe the coordination environment of DβM and PHM and, most recently, TβM. The EXAFS data for oxidized Cu(II) proteins reveal an average of 2.5 N (histidine) and 1.5 O or N ligands per copper at 1.97 Å. [4,26–29] In addition, a weak sulfur ligand is detected at 2.71 Å in PHM[4] in agreement with the Cu_M site visualized in the crystal structure.[25] The EXAFS studies of reduced Cu(I) proteins detect a shorter sulfur ligand distance (2.27 Å in PHM, 2.25 Å in DβM, and 2.24 Å in TβM) than in the oxidized

```
DBM   -----------------------------------MQPHLSHQPCWSLP  14
TBM   MLKIPLQLSSQDGIWPARFARRLHHHHQLAYHHHKQEQQQQQQQQQQQQA  50
PHM   -------------------------------------------------

DBM   SPSVREAASMYG---TAVAIFLVILVAALQG------SEPPESPFPYHIP  55
TBM   KQKQKQNGVQQGRSPTFMPVMLLLLMATLLTRPLSAFSNRLSDTKLHEIY 100
PHM   -------------------------------------------------

DBM   LDPEGTLELSWNVSYDQEIIHFQLQ--VQGPRAGVLFGMSDRGEMENADL 103
TBM   LD-DKEIKLSWMVDWYKQEVLFHLQNAFNEQHRWFYLGFSKRGGLADADI 149
PHM   -------------------------------------------------

DBM   VMLWTDGDRTYFADAWSDQKGQIHLDTHQDYQLLQAQRVSNSLSLLFKRP 153
TBM   CFFENQNG--FFNAVTDTYTSPDGQWVRRDYQQDCEVFKMDEFTLAFRRK 197
PHM   -------------------------------------------------

DBM   FVTCDPKDYVIEDDTVHLVYG-----ILEEPFQSLEAINTSGLHTGLQQV 198
TBM   FDTCDPLDLRLHEGTMYVVWARGETELALEDHQFALPNVTAPHEAGVKML 247
PHM   --------------------------------------------N  45

DBM   QLLKPEVSTPAMPADVQTMDIRAPDVLIPSTETTYWC̲YITELPLHFPR-H 247
TBM   QLLRADK-ILIPETELDHMEITLQEAPIPSQETTYWC̲HVQRLEGNLRRRH 296
PHM   ECLGTIGPVTPLDASDFALDIRMPG-VTPKESDTYFC̲MSMRLP--VDEEA  92
                        *      *   **  *      *

DBM   HIIMYEAIVTEGNEALVH̲H̲MEVFQC̲TN-ESEAFPMFNGPCDSKMKPDRLN 296
TBM   HIVQFEPLIR--TPGIVH̲H̲MEVFHC̲EAGEHEEIPLYNG--DCEQLPPRAK 342
PHM   FVIDFKPRAS---MDTVH̲H̲MLLFGC̲NMPSSTGSYWFCDEGTCTDKAN--- 136
                      ****    *  *

DBM   YCRHVLAAWALGAKAFYYPEEAGVPLGSSGSSRFLRLEVH̲YHNPRNIQG- 345
TBM   ICSKVMVLWAMGAGTFTYPPEAGLPIGGPGFNPYVRLEVH̲FNNPEKQSG- 391
PHM   ----ILYAWARNAPPTRLPKGVGFRVGGETGSKYFVLQVH̲YGDISAFRDN 182
          **   *       *    *    *          * **

DBM   RRDSSGIRLHYTASLRPNEAGIMELGLVYTPLMAIPPQETTFVLTGYCTD 395
TBM   LVDNSGFRIKMSKTLRQYDAAVMELGLEYTDKMAIPPGQTAFPLSGYCVA 441
PHM   HKDCSGVSVHLTRVPQP---LIAGMYLMMSVDTVIPPGEKVVNAD----I 225
         * ***              *             ***

DBM   RC̲TQMALPKSGIRIFASQLH̲T̲H̲LTGRKVITVLARDGQQREVVNRDNHYSP 445
TBM   DC̲TRAALPATGIIIFGSQLH̲T̲H̲LRGVRVLTRHFRGEQELREVNRDDYYSN 491
PHM   SC̲QYKMYP---MHVFAYRVH̲T̲H̲HLGKVVSGYRVRNGQWTLIGRQNPQLP- 271
       *        *      ***  *   *       *     *

DBM   HFQEIRMLKNAVTVHQGDVLITSC̲TYNTENRTMATVGGFGILEEM̲CVNYV 495
TBM   HFQEMRTLHYKPRVLPGDALVTTC̲YYNTKDDKTAALGGFSISDEM̲CVNYI 541
PHM   --QAFYPVEHPVDVTFGDILAARC̲VFTGEGRTEATHIGGTSSDEM̲CNLYI 319
          *    *  ** *    * *       *        ***  *

DBM   HYYPKTELELCKSAVDDGFLQKYFHIVNRFGNEEVCTCPQASVPQQFASV 545
TBM   HYYPATKLEVCKSSVSEETLENYFIYMKRTEHQHGVHLNGARS-SNYRSI 590
PHM   MYYMEAKYALSFMTCTKNVAPDMFRTIPAEANIPI-------------- 354
       **                      *

DBM   PWNSFNRDMLKALYNYAPISVHCNKTSAVRFPGNWNLQPLPKITSAVEEP 595
TBM   EWTQPRIDQLYTMYMQEPLSMQCNRSDGTRFEGRSSWEGVAATPVQIRIP 640
PHM   -------------------------------------------------

DBM   DPRCPIRQTRGPAGPFVVITTEADTE----- 621
TBM   -IHRKLCPNYNPLWLKPLEKGDCDLLGECIY 670
PHM   -------------------------------
```

FIGURE 1.1. Sequence alignment of the mononuclear copper-containing monooxygenases using Clustal W 2.0.10.[24] Sequences for DβM and PHM are from rat and TβM from *Drosophila melanogaster*. Metal-binding ligands are shown in bold and underlined and conserved cysteine residues in italic and underlined. All conserved residues are indicated with a * below the sequences.

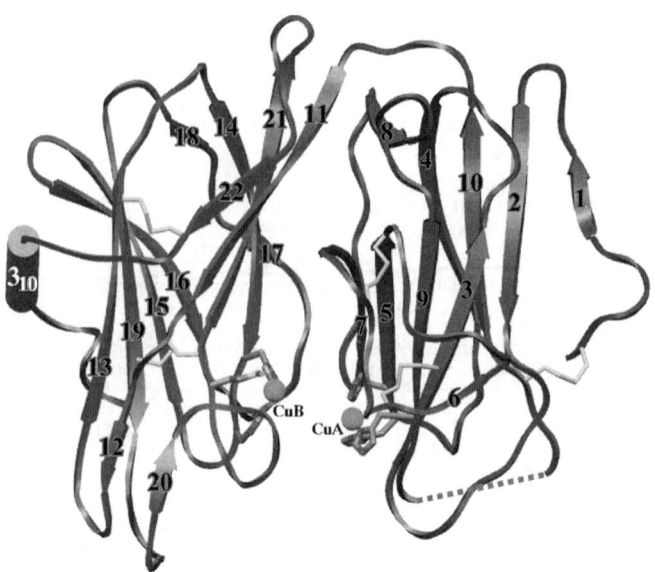

FIGURE 1.2. A representation of the catalytic core of PHM. The backbone is shown in gray with the catalytic coppers represented by green spheres. Strands are numbered arrows and the cylinder is a 3_{10} helix. Side chains of ligands to the two catalytic coppers (green spheres) are colored by atom type (carbon is gray, nitrogen is blue, sulfur is yellow). The dashed gray line indicates a six-residue loop (I176 to D181) not built into the final model. This figure[25] is from Prigge, S. T.; Eipper, B. A.; Mains, R. E.; Amzel, L. M. Amidation of bioactive peptides: The structure of peptidylglycine α-hydroxylating monooxygenase *Science* **1997**, *278*, 1300–1305. (Reprinted with permission from AAAS.) (See color insert.)

enzymes.[4,27–29] X-ray absorption spectroscopy (XAS) experiments indicate that the copper geometry of PHM in solution is predominately four-coordinate with some contribution of five-coordinate geometry.[26] As Figure 1.3 illustrates, these spectroscopic structural probes are in reasonable agreement with the copper sites observed crystallographically. Domain I binds one active site copper, designated Cu_H (Cu_A on the right), by three histidine residues (His107, His108, and His172) (Fig. 1.3). Domain II coordinates the second catalytic copper, Cu_M (Cu_B on the left), with two histidines and a methionine residue (His242, His244, and Met314).[25] Previous studies established that substrate hydroxylation is catalyzed at the Cu_M domain through a radical-initiated hydrogen-atom abstraction mechanism.[30–32] Spectroscopic, kinetic, and structural analyses support a mechanism whereby O_2 and substrate bind to the Cu_M site during catalysis.[27,32–35] The Cu_H site is assigned the role of an electron reservoir that supplies the second electron required for hydroxylation of the C–H bond by long-range intramolecular electron transfer (ET) from Cu_H to Cu_M.[33] An early structure of PHM in the presence of a substrate analogue, *N*-α-acetyl-3,5-diiodotyrosylglycine, bound at the Cu_M site, showed no change in the 11 Å distance between copper sites (i.e., the failure of the two copper sites to move closer together when substrate is present).[25]

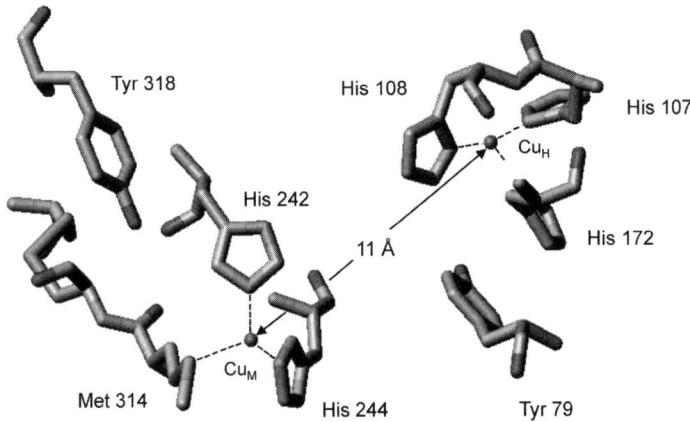

FIGURE 1.3. (Recreated from PDB: 1PHM).[25] The coordination environment of Cu_M (His242, His244, and Met314), Cu_H (His107, His108, and His172), and the relationship of Tyr79 and Tyr318 to the active site coppers of PHM.

1.4. ENZYME MECHANISMS DERIVED FROM KINETIC CHARACTERIZATION AND KINETIC ISOTOPE EFFECTS

One of the primary goals in the mechanistic characterization of the coupling of O_2 to substrate activation for DβM, PHM, and TβM has been to isolate the individual chemical steps of the reaction from substrate binding to product release. Klinman et al.[36] pioneered the use of kinetic isotope effects (KIEs) to delineate complex kinetic mechanisms, showing that the order of addition of reactants can be determined for multireactant enzymes from the magnitude of the isotope effects on the kinetic parameters for the various substrates.[36] Earlier studies with DβM had led to the proposal of an equilibrium-ordered mechanism in which O_2 binds prior to substrate.[37] In accord with this mechanism, the KIE on V_{max}/K_m for substrate would be independent of O_2 concentration. Contrarily, the primary tritium KIE was found to be dependent on O_2 concentration, providing the first unequivocal evidence for a random kinetic mechanism for DβM catalysis (Scheme 1.1).[36] The kinetic mechanism for PHM was subsequently determined using protiated and dideuterated hippuric acid as substrate. The initial rate patterns and double-reciprocal plots are indicative of an equilibrium-ordered mechanism with O_2 binding after hippuric acid to reduced PHM (Scheme 1.2).[38] Similarly, the kinetic characterization of TβM has recently been completed using protiated and dideuterated tyramine. Analogous to DβM, the kinetic data support a random mechanism where either tyramine or O_2 can bind first to reduced TβM (Scheme 1.1). The most notable difference between the kinetic mechanism of DβM and TβM is the substrate inhibition observed for the reaction of TβM with tyramine. The pattern of substrate inhibition indicates tyramine can interact with oxidized TβM to form an inhibitory complex, $E \cdot Cu(II) \cdot S$. The observed inhibition and kinetic profile of TβM suggest a tighter regulation of neurotransmitter levels in insects.[39]

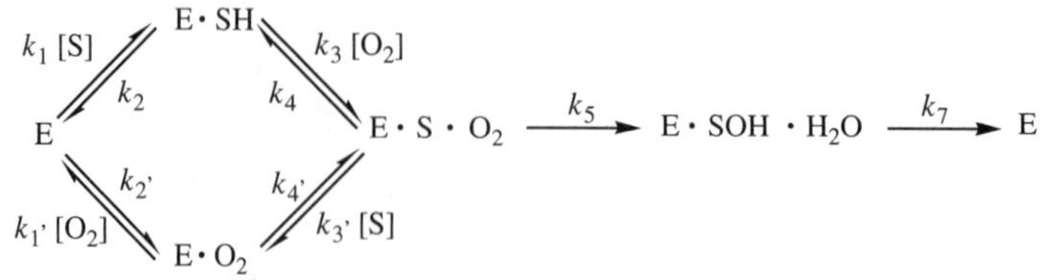

SCHEME 1.1. Kinetic mechanism for random-ordered binding of substrate and O_2.

The reactions catalyzed by this class of enzymes require the abstraction of a hydrogen atom from substrate by the activated oxygen species to yield hydroxylated product and water. Thus, the use of isotopically labeled substrates as mechanistic probes has served as a powerful tool for delineating complex chemical and kinetic mechanisms. The magnitude of primary KIEs for solution reactions can be equal to the intrinsic KIE because the chemical step is rate determining. However, the kinetic complexity of enzyme-catalyzed reactions often results in diminished primary KIEs given that substrate activation is only partially rate limiting. Such is the case when using deuterated dopamine, hippuric acid, and tyramine for reactions catalyzed by DβM, PHM, and TβM, respectively, resulting in comparatively small experimental KIEs under steady-state conditions.[38–40] Accordingly, isolation of the C—H bond cleavage step in enzyme-catalyzed reactions and the measurement of the intrinsic primary KIE is essential for accurate interpretation of steady-state KIEs. The method developed by Northrop[41] made it possible to extract the intrinsic primary KIE, and this method has been applied to DβM and PHM catalyzed reactions (Table 1.1).[38,40,42,43]

The intrinsic primary KIE for the reaction of DβM with dopamine yields a value of 10.9 ± 1.9.[40] Subsequently, this value was used in concert with a structure–reactivity study of a series of para-substituted phenethylamines to calculate the rate constant for the C—H bond cleavage step during DβM-catalyzed reactions.[32] Equation 1.5 demonstrates how the rate of C—H bond cleavage (k_{C-H}) can be calculated using the rate constant for protiated substrate (k_{cat}), the intrinsic KIE for dopamine (^{D}k), and the observed KIE on k_{cat} ($^{D}k_{cat}$).

$$k_{C-H} = k_{cat}(^{D}k - 1)/[^{D}k_{cat} - 1] \qquad (1.5)$$

The observed increase in rate with ring-donating substituents was initially interpreted to be the result of an electron-deficient transition state resulting from O_2. However,

SCHEME 1.2. Kinetic mechanism for equilibrium-ordered binding of substrate before O_2.

TABLE 1.1. Primary and Secondary Intrinsic Hydrogen and Oxygen Isotope Effects Demonstrating the High Conservation of Mechanism between DβM and PHM[a]

Enzyme	Intrinsic		Observed	
	k_H/k_D for C−H		k_{16}/k_{18} for O−O	
	1°	2°	C−H	C−D
DβM	10.9	1.19	1.0197	1.0256
	(1.9)	(0.06)	(0.0003)	(0.0003)
PHM	10.6	1.20	1.0167	1.0216
	(0.8)	(0.03)	(0.0032)	(0.0014)

[a] This table was derived from Refs. 38, 40, 42 and 43.

a modified Marcus model (i.e., for ground-state tunneling) whereby hydrogen transfer occurs quantum mechanically (Eq. 1.6), is now used to express the hydrogen-transfer reaction catalyzed by DβM.[42]

$$k_{tun} = \exp\left\{ -(\Delta G° + \lambda)^2 / (4\lambda RT) \right\} \int_{r_1}^{r_0} \exp^{-m_H \omega_H r^2_H / 2\hbar} \exp^{-E_X/k_B T} dX \quad (1.6)$$

According to Eq. 1.6, $\Delta G°$ is the reaction driving force; λ is the environmental reorganization; m_H, ω_H, and r_H are the mass, frequency and distance traveled for the transferred particle; E_x is the barrier for inter nuclear distance sampling with $\hbar$, k_B, R and T, Planck's constant over 2π, Boltzmann's constant, the gas constant, and temperature, respectively. According to the formalism in Eq. 1.6, the transition state for hydrogen transfer reflects the motions of the heavy atoms of the environment and not the cleavage of the C−H bond *per se*. In this context, the structure–reactivity correlations reported earlier for DβM are reflective of the changing $\Delta G°$ among substrates with varying structures. Although the ability to correlate the k_{tun} of Eq. 1.6 with $\Delta G°$ would be of considerable interest, there are currently neither computational nor experimental estimates for the size of $\Delta G°$ that represents the equilibration between phenethylamine substrates and the substrate-derived free radical.

Evidence for hydrogen transfer by a quantum mechanical mechanism for the PHM-catalyzed reaction has come from the kinetic isolation of the hydrogen-transfer step from other partially rate-limiting steps and determination of the intrinsic KIE as a function of temperature. The intrinsic primary KIE is nearly temperature-independent within experimental error. A fit of the intrinsic KIE to the Arrhenius equation for isotope effects (Eq. 1.7) yields $A_H/A_D = 5.9 \pm 3.2$

$$^D k_{int} = A_H/A_D \exp\{[E_a(D) - E_a(H)]/RT\} \quad (1.7)$$

and $E_a(D) - E_a(H) = 0.37 \pm 0.33$ kcal/mol.[42] These values lie significantly outside semiclassical limits [$0.7 < A_H/A_D < 1.4$ and $E_a(D) - E_a(H) = 1.2$ kcal/mol] and cannot be explained by a tunneling correction model.[44–46] Additionally, the combined magnitudes of the intrinsic primary and secondary KIEs (Table 1.1) cannot be rationalized by classical theory.

The relationship between hydrogen and O_2 activation in this class of enzymes can be further probed by measuring an oxygen KIE. The ability to detect small changes in ^{16}O and ^{18}O by mass spectrometry permits precise quantification of the oxygen KIE that can be related to the structure of the activated oxygen at Cu_M. Further, comparing the magnitude of the oxygen KIE using both protio and deuterio substrates allows discrimination among a range of possible mechanisms. At least six mutually exclusive mechanisms for O_2 activation by DβM, PHM, and TβM have been posited: (A) formation of a $2e^-$ reduced $Cu_M(II)$-hydroperoxo or (B) $Cu_M(II)$-peroxo species, (C) reductive cleavage of a $Cu_M(II)$-hydroperoxo intermediate by a conserved active-site tyrosine to form a $Cu_M(II)$-oxo species, (D) a superoxide-channeling mechanism, whereby O_2 initially binds at Cu_H to form a Cu_H-superoxide intermediate, which dissociates and migrates across a solvent interface to bind to and be further reduced at the Cu_M site,[47,48] (E) a mechanism where the function of the coppers is reversed with substrate hydroxylation occurring via a $Cu_H(II)$-superoxide intermediate,[49] and (F) a $1e^-$ reduced $Cu_M(II)$-superoxo intermediate.

The propensity and favorable energetics for hydrogen peroxide (H_2O_2) formation from O_2 led to an initial focus on mechanisms A and B above. To test for the formation of a $Cu_M(II)$-hydroperoxo species (mechanism A), it was reasoned that generation of such an intermediate at the highly solvent-exposed active site upon reaction could result in uncoupling between O_2 activation and substrate hydroxylation. However, full coupling of O_2 consumption and product formation was observed in studies with DβM using a range of substrates that varied by three-orders of magnitude in reactivity.[50] Similarly, tight coupling of O_2 uptake and product formation was observed for H172A PHM, a variant at Cu_H that retains its copper-binding and hydroxylase activity yet results in a three-order of magnitude decrease in k_{cat}.[51] Contrarily, formation of metal hydroperoxide intermediates in other enzymes (i.e., cytochrome P450s) often involves concomitant uncoupling of substrate hydroxylation and O_2 reduction.[52,53] Additionally, the experiments described above appear to rule out a superoxide channeling mechanism (mechanism D) based on the high probability of some degree of uncoupling between O_2 and substrate activation during migration of a highly reactive superoxide anion across a solvent-exposed active site.[47,48]

Generation of the Y318F PHM variant made it possible to test whether reductive cleavage of a $Cu_M(II)$-hydroperoxo intermediate by a proximal and conserved active-site tyrosine could lead to formation of the reactive $Cu_M(II)$-oxo species (mechanism C).[54] A full kinetic analysis of the Y318F PHM mutant resulted in only a four-fold reduction in the rate for C–H bond cleavage, and the intrinsic hydrogen and ^{18}O isotope effects are nearly identical to that of WT PHM. Based on this comprehensive kinetic analysis, mechanism C was ruled out as well. A mechanism whereby the roles of the copper sites are reversed (mechanism E) is intriguing, but not supported by experimental evidence. The compilation of kinetic, spectroscopic, and structural evidence supports a mechanism where substrate and O_2 activation take place at the Cu_M site.[27,32,33,35] The formation and utilization of a $Cu_M(II)$-peroxo (mechanism B) oxidizing intermediate is primarily based on the X-ray structure of PHM. The absence of any residues capable of general acid catalysis led to the proposal that a $Cu_M(II)$-peroxo species could be responsible for catalysis.[25,55] To test this hypothesis, reduced DβM was reacted with O_2 and an unreactive substrate analogue, β,β-difluorophenylethylamine.[50] Reduction of O_2 and the formation of a $Cu_M(II)$-peroxo intermediate is predicted to result in oxidation of both copper sites to $Cu(II)$, which should be detectable by EPR. However, this study did not result in any observable $Cu(II)$ EPR signal. Generation of a very low level of the $Cu_M(II)$-peroxo species, outside the sensitivity of the EPR experiments, could explain this result. However, the failure to observe any oxidized copper is best explained by the formation of a diamagnetic $Cu_M(II)$-superoxo intermediate (mechanism F), which is EPR silent.

The mechanism proposed by Klinman and co-workers[50] (Fig. 1.4) satisfies the extensive data available for DβM, PHM, and TβM. Substrate and O_2 bind to the reduced enzyme forming the ternary complex, which induces the initial O_2 activation involving ET from $Cu_M(I)$–dioxygen to form a $Cu_M(II)$–superoxo species.[50] The magnitude of ^{18}O KIEs using either protio- or deuterio-labeled substrates for DβM and PHM is consistent with this proposed mechanism (Table 1.1). For both enzymes, the magnitude of k_{16}/k_{18} increases with deuterated substrate, which reveals that the activation of substrate and O_2 must be coupled by a reversible chemical process.[38,43] The recently measured ^{18}O KIE for galactose oxidase, a mononuclear copper enzyme that undergoes hydrogen-atom transfer to a $Cu(II)$–superoxo species, is nearly identical to that determined for DβM/PHM.[56] The ^{18}O KIEs provide substantial evidence for a $Cu_M(II)$–superoxo species and establish a limiting KIE for hydrogen-atom abstraction by a $Cu_M(II)$–superoxo intermediate. Furthermore, density functional theory (DFT) calculations found the $Cu_M(II)$–superoxo species to be kinetically and thermodynamically more favorable than the $Cu_M(II)$–hydroperoxo intermediate.[57] Subsequently, the $Cu_M(II)$–superoxo intermediate abstracts a hydrogen atom from the substrate via a hydrogen tunneling mechanism.[38,42] A critical feature of this mechanism involves the timing and pathway for long-range transfer of the second electron from $Cu_H(I)$. If ET from $Cu_H(I)$ to $Cu_M(II)$–superoxo preceded C–H activation, it would have to happen significantly faster than C–H activation, which is estimated as $10^3\,s^{-1}$.[32,38] This scenario cannot be excluded in the event that the driving force ($\Delta G°$) and reorganization energy (λ) are both relatively small.[58] However, the implication of $Cu_M(II)$–superoxo as the oxygen species responsible for C–H activation, together with the clear demonstration that C–H abstraction is irreversible,[40] requires that k_{cat}/K_m for O_2 and substrate be independent of the second ET from Cu_H.

The consensus mechanism places the long-range ET from Cu_H to Cu_M within k_{cat}.[59] The route of long-range ET has been proposed to involve substrate and/or a protein network.[49,59] When the Q170A PAM variant was generated to investigate the role of an interdomain, hydrogen-bonded protein network only minor effects on

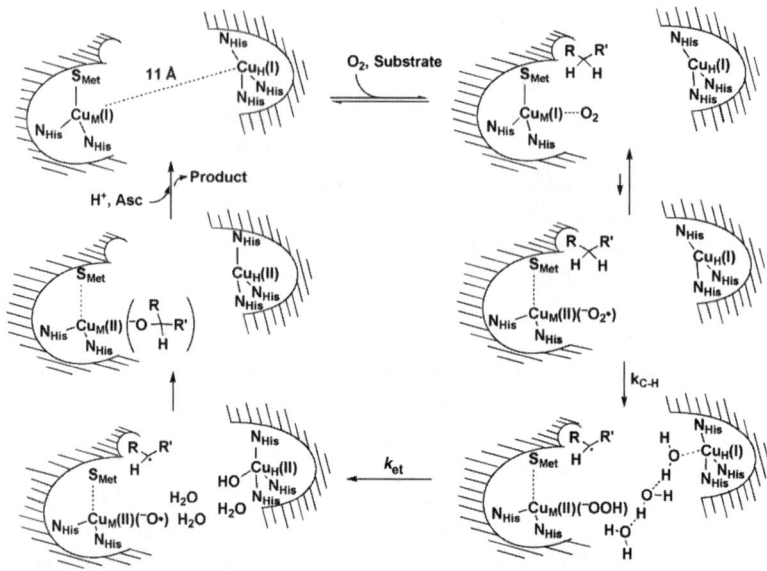

FIGURE 1.4. Copper-superoxo mechanism for DβM, PHM, and TβM. This research was originally published in the *Journal of Biological Chemistry.* Evans, J. P.; Ahn, K.; Klinman, J. P. Evidence that dioxygen and substrate activation are tightly coupled in dopamine β-monooxygenase. Implications for the reactive oxygen species[50] *J. Biol. Chem.* **2003**, *278*, 49691–49698. Copyright © The American Society for Biochemistry and Molecular Biology.

enzyme activity were observed, ruling out this pathway for ET from Cu_H to Cu_M.[49] Similarly, examination of the PHM-catalyzed reaction with two different substrates differing in length (Fig. 1.5) results in similar kinetic parameters (Table 1.2), indicating that the substrate backbone is an unlikely pathway for ET between copper sites.[59] The collective data on DβM and PHM are consistent with a mechanism in which long-range ET occurs within the solvent-exposed interface between Cu_M and Cu_H. As outlined in the proposed mechanism (Fig. 1.4), interdomain ET occurs after C–H activation, which reduces the transfer distance to $\sim 7\,\text{Å}$.[50] Water-mediated ET over this distance is thoroughly consistent with the experimentally measured values for k_{cat} for this class of enzymes (~ 3–$40\,\text{s}^{-1}$).[32,38,39] Electron transfer from Cu_H to Cu_M is proposed to induce reductive cleavage of the $Cu_M(II)$–hydroperoxo intermediate yielding water and a $Cu_M(II)$–oxo radical, which recombines with the substrate radical to form an

$$\phi\text{-}\overset{\text{O}}{\overset{\|}{\text{C}}}\text{-NH-CL}_2\text{-CO}_2^-$$

1

$$\text{DNS-GLY-GLY-SER-}\overset{\text{O}}{\overset{\|}{\text{C}}}\text{-NH-CL}_2\text{-CO}_2^-$$

2

FIGURE 1.5. The structures used in the study are from Ref. 59 and referred to in Table 1.2.

TABLE 1.2. Comparison of Kinetic Parameters for Truncated and Extended Peptide Substrates with PAM^a

Substrate	$k_{cat}/K_m(O_2)$ (μM^{-1}, s^{-1})	$^D[k_{cat}/K_m(O_2)]$	k_{cat} (s^{-1})	$^D k_{cat}$
1^b	0.14 (0.01)	3.1 (0.3)	37 (0.1)	1.0 (0.1)
2^b	0.23 (0.03)	2.3 (0.2)	17 (0.8)	1.0 (0.1)

^{a} This table was derived from Ref. 59.

^{b} Figure 1.5 illustrates the structure of the substrates used in this study denoted by "1" and "2" in the table.

inner-sphere alcohol product.[50] Alternatively, computational experiments suggest a thermodynamically favorable mechanism in which the substrate-derived radical abstracts a hydroxyl radical from the Cu$_M$(II)–hydroperoxo intermediate, and the resulting Cu$_M$(II)-oxo species is subsequently reduced via ET from Cu$_H$.[57] In reality, the net process is likely to be a concerted one, in which long-range ET occurs concomitant with the attack of substrate radical on the Cu$_M$(II)–hydroperoxo. However, the increase in k_{cat} for the oxidation of phenethylamine substrates with electron-withdrawing groups catalyzed by DβM has been interpreted to be the result of changes in the partially rate-limiting dissociation of the inner-sphere alcohol complex to form free product.[32] The intermediacy of an inner-sphere alcohol moiety is further supported by experimental evidence and QM/MM simulations for PAM catalysis with a non-natural substrate, benzaldehyde imino-oxy acetic acid.[60]

1.5. A NETWORK OF COMMUNICATION BETWEEN Cu$_M$ AND Cu$_H$

How DβM, PHM, and TβM catalyze such tightly coupled reactions using two mononuclear Cu atoms separated by ~ 11 Å and completely exposed to solvent is still not well understood. In this section, evidence will be presented that supports the idea that subtle, but essential communication occurs between the mononuclear Cu sites and throughout the entire protein structure, which plays a central role in ensuring efficient wave function overlap between the hydrogen donor and the metal-superoxo species at the Cu$_M$ site and in maintaining tight coupling between C−H and O$_2$ activation. A number of PHM variants have been generated, and the resulting data are generally analyzed in terms of how mutations near Cu$_M$ and Cu$_H$ impact the specific functions of each Cu site (i.e., C−H and O$_2$ activation and substrate hydroxylation at Cu$_M$ and long-range ET at Cu$_H$). However, a trend is starting to emerge illustrating how an individual mutation at one copper domain impacts the other copper domain.

Kinetic analyses of the Y79F[4] and Y79W[49] PHM variants provided early evidence for a network of communication between the two copper domains, although the data were not initially discussed in this context. Based on sequence conservation and prior to any knowledge that Tyr79 of PHM is located ~ 4 Å from Cu$_H$, this residue was one of the first mutated within recombinant PHM.[4] The Y79F PHM variant has a significant impact on k_{cat}/K_m(substrate), giving a 15-fold decrease in rate (k_{cat})

TABLE 1.3. Relative Decrease in Kinetic Parameters for Variants of PHM Compared to WT PHM[a]

PHM/Mutants	k_{cat}	$k_{cat}/K_m(S)$	$k_{cat}/K_m(O_2)$	k_{C-H}
Y79F[b]	~ 15-fold	~ 60-fold		
Y79W[c]	~ 200-fold	~ 300-fold		
H172A[d]	~ 3000-fold		~ 300-fold	~ 12,000-fold

[a] The data in this table are given as relative decreases compared to WT PHM because exact values were not always provided for the kinetic parameters, and the substrates were not always the same from study to study. However, the same substrate was used when comparing the impact on kinetic parameters of each individual PHM variant to WT PHM.
[b] Data taken from Ref. 4.
[c] Data taken from Ref. 49.
[d] Data taken from Ref. 51.

relative to a WT PHM construct for the oxidation of α-N-acetyl-Tyr-Val-Gly (Table 1.3). Y79F PHM also exhibited an ~ four-fold increase in the K_m(substrate).[4] Based on this simple analysis, Y79F PHM results in an ~60-fold decrease for k_{cat}/K_m (substrate), which is the kinetic parameter that reports on C−H activation and substrate hydroxylation at the Cu_M site (Table 1.3). Y79W PHM was generated after a number of PHM crystal structures had been solved. The rationale for generating this variant was to use Trp79 as a fluorescent indicator that could be used to monitor the active site.[49] Additionally, a general kinetic examination of Y79W PHM was completed, and the largest effect observed (~ 300-fold reduction) was on the second-order rate constant, k_{cat}/K_m (substrate) (Table 1.3).[49] The impact of Y79F PHM on K_m (substrate) led to the suggestion that Tyr79 might interact with the glycyl moiety of glycine extended peptides[4]; little discussion was provided in regards to the impact of Y79F PHM on k_{cat}/K_m (substrate). The effect of Y79W PHM on the measured kinetic parameters in concert with the fluorescence measurements led to a proposed mechanism in which the roles of Cu_M and Cu_H are reversed.[49] Based on the current working mechanism (Fig. 1.4), we now propose that the two Tyr79 mutants disrupt a network of hydrogen-bonded water molecules connecting the separate copper domains. The Tyr79 residue may be critical for long-range ET based on the decrease in k_{cat}, but the largest impact for both variants is on the kinetic parameter k_{cat}/K_m (substrate), which measures C−H activation efficiency at Cu_M. The Tyr79 residue forms a pi–pi interaction with His172, a ligand to Cu_H, and replacement of this tyrosine with any amino acid likely invokes subtle changes in the coordination environment at Cu_H, which we now propose disrupts a network of connectivity that significantly impacts chemistry taking place at Cu_M.

A number of studies with Met314 PHM variants have been carried out to probe the importance of this residue during catalysis. Initial reports showed that substitution of Met314 with Ile, His, or Cys did not restore catalysis.[4,61] However, a recent study with TβM demonstrates that mutation of the parallel residue, Met471, to Cys retains activity while the Asp471 and His471 variants are inactive. Interestingly, M471C TβM undergoes a secondary inactivation pathway during turnover, and full conversion of tyramine to octopamine is not observed under any conditions.[62] The coordination environment of the reduced Met471 variants was investigated by EXAFS, and slight, subtle differences for M471C TβM compared to the Asp471 and His471 mutants indicate the possibility of a small contribution from cysteine ligation.[29] This observation could account for the activity of M471C TβM. A crystallographic study with M314I PHM revealed a number of interesting results. Predictably, structural alignment between oxidized M314I PHM and the published structure of oxidized PHM showed significant changes close to the Cu$_M$ site. Displacement of the mutated residue, Ile314, from Cu$_M$ resulted in disorder of the flanking loop (299–313) and caused loop 212–218 to move as far as 5 Å away from Cu$_M$. Unexpectedly, replacement of the Met314 ligand does not prohibit copper binding at the Cu$_M$ site. The Met314 residue is replaced with a water molecule, and the remaining coordinating ligands shift slightly to retain the distorted tetrahedral geometry. In fact, mutation of Met314 appears to have a more dramatic impact on the Cu$_H$ domain, even though this domain resides 11 Å away from Cu$_M$. Specifically, His107 and Cu$_H$ are highly disordered in the oxidized M314I PHM structure. At the same time, significant structural changes in the protein regions connecting the two copper domains were not observed.[63] These findings reinforce a previous EXAFS study, in which the occupancy of the Cu$_H$ appears to be coupled to the movement of Met314 at Cu$_M$.[64] The significant impact on the coordination environment at Cu$_H$ as a result of mutation of the methionine ligand to Cu$_M$ further demonstrates the interconnectivity between the two domains and the sensitivity of each copper site to long-range perturbations.

Recently, an extensive kinetic and deuterium isotope study was completed for H172A PHM.[51] Histidine at position 172 is a ligand to the ET copper, Cu$_H$, and the H172A PHM variant retains copper binding and hydroxylase activity.[48] Changing the coordination environment at Cu$_H$ was predicted to impact the long-range ET step without affecting other steps in the kinetic mechanism. Unexpectedly, both k_{cat} and $k_{cat}/K_m(O_2)$ for H172A PHM decrease significantly (Table 1.3), and the deuterium isotope effects on these parameters increase. For H172A PHM, C–H abstraction has become more rate-limiting, indicating that altering the coordination and the redox properties at Cu$_H$ impacts the rate of chemistry taking place at Cu$_M$.[51] The impact of mutating His172 was further examined by determining the intrinsic rate constants (Scheme 1.1), which can be calculated from the measured kinetic parameters, their isotope effects, and the intrinsic isotope effect. The intrinsic isotope effect for H172A PHM was not determined, but the value was determined previously for WT PHM[38] and assumed to be similar for H172A PHM. Remarkably, this approach leads to an estimated ~12,000-fold decrease in the rate of the hydrogen-transfer step for H172A PHM (Table 1.3). Now we propose that the dramatic impact on the rate of C–H cleavage is due to subtle, but critical, changes to a water-mediated hydrogen-bonding network linking the two mononuclear copper domains.

Re-examination of the existing family of PHM crystal structures shows a conserved network of waters from His108, bound to Cu$_H$, to the peptide backbone carbonyl of His244 or Met314, a ligand to Cu$_M$ (Fig. 1.6). This network is present in all of the structures, and does not appear to be dependent on the presence of substrate or the oxidation state of the coppers. Herein, we suggest that this network could be one of the primary structural features that links the two copper domains. The observation that substitution of His108 with alanine eliminated detectable activity for PHM, and the fact that this residue is the only Cu_H ligand observed to move when comparing the oxidized and reduced structures is consistent with our hypothesis.[4,63] If this network couples the two copper sites, substitution of any active-site residue at Cu_H might disrupt substrate binding at Cu_M such that its orientation becomes nonoptimal for C–H cleavage. As discussed above, the transfer of hydrogen from substrate to the activated oxygen species in WT PHM transpires via hydrogen tunneling.[38,42] We believe the PHM data can be fully understood in the context of a model for hydrogen tunneling in enzyme-catalyzed reactions that requires the contribution of two types of protein motions to achieve effective wave function overlap between the hydrogen

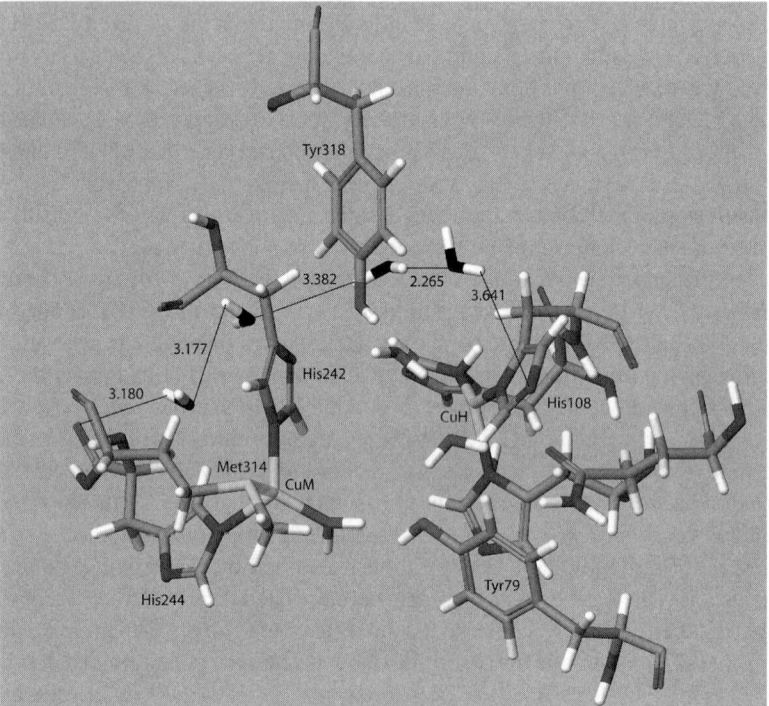

FIGURE 1.6. Active-site structure of reduced PHM recreated from PDB: 3PHM.[55] The copper atoms (Cu_M to the left and Cu_H to the right), coordinating ligands, Tyr79 (bottom), and Tyr318 (top) are colored by atom type (carbon is gray, nitrogen is blue, sulfur is yellow, oxygen is red, copper is aqua, and hydrogen is white). The proposed network of connectivity between Cu_H and Cu_M through hydrogen-bonded water molecules is shown (oxygen atoms of networked waters are shown in black for clarity). Other active-site water molecules that might stabilize the active site and/or contribute to a pathway of long-range ET are shown (oxygen atoms of water molecules are red). The active-site structure was recreated using the program Maestro. (See color insert.)

donor and acceptor.[65] The first type of motion has been labeled preorganization and reflects the sampling of a large number of protein conformational substates, with only a small subset of conformers optimized for tunneling to occur in the WT enzyme. Once the family of preorganized conformations has been achieved, reorganization invokes heavy atom motions to achieve barrier crossing. Disturbing the network of connectivity by making mutations at Cu_H could have a significant impact on the ability of the enzyme to find catalytically relevant conformers resulting in dramatic impacts on the rate of C−H activation taking place at Cu_M. A structured solvent-exposed active site, comprised of ordered water molecules, is likely a required feature for this small family of enzymes orchestrating the intricate and tightly coupled chemistry that transpires.

1.6. CONCLUDING REMARKS AND FUTURE PROSPECTS

Herein, we present a review of the literature relating to the mechanism for reactions catalyzed by DβM, PHM, and TβM. Specifically, experimental evidence for the structure of the activated oxygen intermediate, the nature of hydrogen transfer, and how C−H and O_2 activation are coupled have been discussed. Based on the multiple X-ray structures available for PHM, we propose a network of connectivity between the two mononuclear copper domains (Fig. 1.6). Researchers have spent an abundant amount of energy trying to explain the anomalous nature of the solvent-exposed active site that is characteristic of this small class of enzymes. We suggest that nature has engineered these enzymes to utilize water as a specific medium for communication between the separate active-site domains from Cu_H to Cu_M in a way that may prevent deleterious uncoupling reactions. For the future, appropriately designed experiments and computations are needed to clarify the precise pathway of long-range ET and to provide more insight into the nature of backbone connectivity between the fully solvent-exposed copper domains that characterize this unique family of enzymes.

ABBREVIATIONS

Asc	Ascorbate
Asp	Aspartate
Cys	Cysteine
DβM	Dopamine β-monooxygenase
DFT	Density functional theory
EPR	Electron paramagnetic resonance
ET	Electron transfer
EXAFS	Extended X-ray absorption fine structure
His	Histidine
Ile	Isoleucine
KIE(s)	Kinetic isotope effect(s)

Met	Methionine
PAM	Peptidylglycine α-amidating monooxygenase
PHM	Peptidylglycine α-hydroxylating monooxygenase
QM/MM	Quantum mechanical/molecular mechanics
TβM	Tyramine β-monooxygenase
Tyr	Tyrosine
WT	Wild type
XAS	X-ray absorption spectroscopy

REFERENCES

1. Malmstrom, B. G. Enzymology of oxygen *Ann. Rev. Biochem.* **1982**, *51*, 21–59.

2. Klinman, J. P. Mechanisms whereby mononuclear copper proteins functionalize organic substrates *Chem. Rev.* **1996**, *96*, 2541–2561.

3. Klinman, J. P. The copper-enzyme family of dopamine β-monooxygenase and peptidylglycine α-hydroxylating monooxygenase: Resolving the chemical pathway for substrate hydroxylation *J. Biol. Chem.* **2006**, *281*, 3013–3016.

4. Eipper, B. A.; Quon, A. S. W.; Mains, R. E.; Boswell, J. S.; Blackburn, N. J. The catalytic core of peptidylglycine α-hydroxylating monooxygenase: Investigation by site-directed mutagenesis, Cu X-ray absorption spectroscopy, and electron paramagnetic resonance *Biochemistry* **1995**, *34*, 2857–2865.

5. Orchard, I.; Ramirez, J.; Lange, A. B. A multifunctional role for octopamine in locus flight *Ann. Rev. Entomol.* **1993**, *38*, 227–249.

6. Kim, C-H.; Zabetian, C. P.; Cubells, J. F.; Cho, S.; Biaggioni, I.; Cohen, B. M.; Robertson, D.; Kim, K-S. Familial paraganglioma and gastric stromal sarcoma: A new syndrome distinct from the Carney triad *Am J. Med. Genet.* **2002**, *108*, 140–147.

7. Timmers, H. J. L. M.; Deinum, J.; Wevers, R. A.; Lenders, J. W. M. Congenital dopamine-β-hydroxylase deficiency in humans *Ann. N. Y. Acad. Sci.* **2004**, *1018*, 520–523.

8. Cubells, J. F.; Zabetian, C. P. Human genetics of plasma dopamine β-hydroxylase activity: Applications to research in psychiatry and neurology *Psychopharmacology* **2004**, *174*, 463–476.

9. Prigge, S. T.; Mains, R. E.; Eipper, B. A.; Amzel, L. M. New insights into copper monooxygenases and peptide amidation: Structure, mechanism and function *Cell Mol. Life Sci.* **2000**, *57*, 1236–1259.

10. Roeder, T. Tyramine and octopamine: Ruling behavior and metabolism *Annu. Rev. Entomol.* **2005**, *50*, 447–477.

11. Lehman, H. K.; Schulz, D. J.; Barron, A. B.; Wraight, L.; Hardison, C.; Whitney, S.; Takeuchi, H.; Paul, R. K.; Robinson, G. E. Division of labor in the honey bee (*Apis mellifera*): The role of tyramine β-hydroxylase *J. Exp. Biol.* **2006**, *209*, 2774–2784.

12. Monastirioti, M. Distinct octopamine cell population residing in the CNS abdominal ganglion controls ovulation in *Drosophila melanogaster Dev. Biol.* **2003**, *264*, 38–49.

13. Winkler, H.; Carmichael, S. W. *The Secretory Granule*, A. M. Poisner and J. M. Trifaro, Eds., Elsevier Biomedical Press, Amsterdam, The Netherlands, **1982**.

14. Stewart, L. C.; Klinman, J. P. Dopamine beta-hydroxylase of adrenal chromaffin granules: Structure and function *Annu. Rev. Biochem.* **1988**, *57*, 551–592.

15. Eipper, B. A.; Stoffers, P. A.; Mains, R. E. The biosynthesis of neuropeptides: Peptide α-amidation *Annu. Rev. Neurosci.* **1992**, *15*, 57–85.

16. Klinman, J. P.; Krueger, M.; Brenner, M.; Edmondson, D. E. Evidence for two copper atoms/subunit in dopamine beta-monooxygenase catalysis *J. Biol. Chem.* **1984**, *259*, 3399–3402.

17. Skotland, T.; Ljones, T. Direct spectrophotometric detection of ascorbate free radical formed by dopamine β-monooxygenase and by ascorbate oxidase *Biochim. Biophys. Acta* **1980**, *630*, 30–35.

18. Diliberto, E. J., Jr.; Allen, P. L. Mechanism of dopamine-β-hydroxylation. Semidehydroascorbate as the enzymic oxidation product of ascorbate *J. Biol. Chem.* **1981**, *256*, 3385–3393.

19. Brenner, M. C.; Klinman, J.P. Correlation of copper valency with product formation in single turnovers of dopamine beta-monooxygenase *Biochemistry* **1989**, *28*, 4664–4670.

20. Freeman, J. C.; Villafranca, J. J.; Merkler, D. J. Redox cycling of enzyme-bound copper during peptide amidation *J. Am. Chem. Soc.* **1993**, *115*, 4923–4924.

21. Gray, E. E.; Small, S. N.; McGuirl, M. A. Expression and characterization of recombinant tyramine β-monooxygenase from *Drosophila*: A monomeric copper-containing hydroxylase *Prot. Exp. Purif.* **2006**, *47*, 162–170.

22. Lamouroux, A.; Vigny, A.; Faucon Biguet, V.; Darmon, M. C.; Frank, R.; Henry, J. P.; Mallet, J. The primary structure of human dopamine-beta-hydroxylase: Insights into the relationship between the soluble and the membrane-bound forms of the enzyme *EMBO J.* **1987**, *6*, 3931–3937.

23. Southan, C.; Kruse, L. I. Sequence similarity between dopamine β-hydroxylase and peptide α-amidating enzyme: Evidence for a conserved catalytic domain *FEBS Lett.* **1989**, *255*, 116–120.

24. Chenna, R.; Sugawara, H.; Koike, T.; Lopez, R.; Gibson, T. J.; Higgins, D. G.; Thompson, J. D. Multiple sequence alignment with the Clustal series of programs *Nucleic Acids Res.* **2003**, *31*, 3497–3500.

25. Prigge, S. T.; Eipper, B. A.; Mains, R. E.; Amzel, L. M. Amidation of bioactive peptides: The structure of peptidylglycine α-hydroxylating monooxygenase *Science* **1997**, *278*, 1300–1305.

26. Boswell, J. S.; Reedy, B. J.; Kulathila, R.; Merkler, D.; Blackburn, N. J. Structural investigations on the coordination environment of the active-site copper centers of recombinant bifunctional peptidylglycine α-amidating enzyme *Biochemistry* **1996**, *35*, 12241–12250.

27. Reedy, B. J.; Blackburn, N. J. Preparation and characterization of half-apo dopamine-β-hydroxylase by selective removal of CuA. Identification of a sulfur ligand at the dioxygen binding site by EXAFS and FTIR spectroscopy *J. Am. Chem. Soc.* **1994**, *116*, 1924–1931.

28. Scott, R. A.; Sullivan, R. J.; DeWolf, W. E.; Dolle, R. E.; Kruse, L. I. The copper sites of dopamine β-hydroxylase: An x-ray absorption spectroscopic study *Biochemistry* **1988**, *27*, 5411–5417.

29. Hess, C. R.; Klinman, J. P.; Blackburn, N. J. The copper centers of tyramine β-monooxygenase and its catalytic-site methionine variants. An x-ray absorption study. *J. Biol. Inorg. Chem.* **2010**, *15*, 1195–1207.

30. Fitzpatrick, P. F.; Villafranca, J. J. Mechanism-based inhibitors of dopamine beta-hydroxylase containing acetylenic or cyclopropyl groups *J. Am. Chem. Soc.* **1985**, *107*, 5022–5023.

31. Fitzpatrick, P. F.; Flory, D. R., Jr.; Villafranca, J. J. 3-Phenylpropenes as mechanism-based inhibitors of dopamine beta-hydroxylase: Evidence for a radical mechanism *Biochemistry* **1985**, *24*, 2108–2114.

32. Miller, S. M.; Klinman, J. P. Secondary isotope effects and structure–reactivity correlations in the dopamine beta-monooxygenase reaction: Evidence for a chemical mechanism *Biochemistry* **1985**, *24*, 2114–2127.

33. Blackburn, N. J.; Pettingill, T. M.; Seagraves, K. S.; Shigeta, R. T. Characterization of a carbon monoxide complex of reduced dopamine β-hydroxylase. Evidence for inequivalence of the Cu(I) centers *J. Biol. Chem.* **1990**, *265*, 15383–15386.

34. Cook, P. F.; Cleland, W. W. Mechanistic deductions from isotope effects in multireactant enzyme mechanisms *Biochemistry* **1981**, *20*, 1790–1796.

35. Prigge, S. T.; Eipper, B. A.; Mains, R. E.; Amzel, L. M. Dioxygen binds end-on to mononuclear copper in a precatalytic complex *Science* **2004**, *304*, 864–867.

36. Klinman, J. P.; Humphries, H.; Voet, J. G. Deduction of kinetic mechanism in Multisubstrate enzyme reactions from tritium isotope effects *J. Biol. Chem.* **1980**, *255*, 11648–11651.

37. Friedman, S.; Kaufman, S. 3,4-Dihydroxyphenylethylamine beta-hydroxylase. Physical properties, copper content, and role of copper in the catalytic activity. *J. Biol. Chem.* **1965**, *240*, 4763–4773.

38. Francisco, W. A.; Merkler, D. J.; Blackburn, N. J.; Klinman, J. P. Kinetic mechanism and intrinsic isotope effects for the peptidylglycine α-amidating enzyme reaction *Biochemistry* **1998**, *37*, 8244–8252.

39. Hess, C. R.; McGuirl, M. M.; Klinman, J. P. Mechanism of the insect enzyme, tyramine β-monooxygenase, reveals differences from the mammalian enzyme, dopamine β-monooxygenase *J. Biol. Chem.* **2008**, *283*, 3042–3049.

40. Miller, S. M.; Klinman, J. P. Magnitude of intrinsic isotope effects in the dopamine β-monooxygenase reaction *Biochemistry* **1983**, *22*, 3091–3096.

41. Northrop, D. B. Steady state analysis of kinetic isotope effects in enzymatic reactions *Biochemistry* **1975**, *14*, 2644–2651.

42. Francisco, W. A.; Knapp, M. J.; Blackburn, N. J.; Klinman, J. P. Hydrogen tunneling in peptidylglycine α-hydroxylating monooxygenase *J. Am. Chem. Soc.* **2002**, *124*, 8194–8195.

43. Tian, G.; Berry, J. A.; Klinman, J. P. Oxygen-18 kinetic isotope effects in the dopamine β-monooxygenase reaction: Evidence for a new chemical mechanism in non-heme metallomonooxygenases *Biochemistry*, **1994**, *33*, 226–234.

44. Schneider, M. E.; Stern, M. J. Arrhenius preexponential factors for primary hydrogen kinetic isotope effects *J. Am. Chem. Soc.* **1972**, *94*, 1517–1522.

45. Bell, R. P. *The Tunnel Effect in Chemistry*; Chapman & Hall: New York, 1980.

46. Melander, L.; Saunders, W. H. J. *Reaction Rates of Isotopic Molecules*; John Wiley & Sons, Inc.: New York, 1980.

47. Jaron, S.; Blackburn, N. J. Does superoxide channel between the copper centers in peptidylglycine monooxygenase? A new mechanism based on carbon monoxide reactivity *Biochemistry* **1999**, *38*, 15086–15096.

48. Jaron, S.; Mains, R. E.; Eipper, B. A.; Blackburn, N. J. The catalytic role of the copper ligand H172 of peptidylglycine alpha-hydroxylating monooxygenase (PHM): A spectroscopic study of the H172A mutant *Biochemistry* **2002**, *41*, 13274–13282.

49. Bell, J.; El Meskini, R.; D'Amato, D.; Mains, R. E.; Eipper, B. A. Mechanistic investigation of peptidylglycine α-hydroxylating monooxygenase via intrinsic tryptophan fluorescence and mutagenesis *Biochemistry* **2003**, *42*, 7133–7142.

50. Evans, J. P.; Ahn, K.; Klinman, J. P. Evidence that dioxygen and substrate activation are tightly coupled in dopamine beta-monooxygenase. Implications for the reactive oxygen species *J. Biol. Chem.* **2003**, *278*, 49691–49698.

51. Evans, J. P.; Blackburn, N. J.; Klinman, J. P. The catalytic role of the copper ligand H172 of peptidylglycine α-hydroxylating monooxygenase: A kinetic study of the H172A mutant *Biochemistry*, **2006**, *45*, 15419–15429.

52. Solomon, E. I.; Brunold, T. C.; Davis, M. I.; Kemsley, J. N.; Lee, S. –K.; Lehnert, N.; Neese, R.; Skulan, A. J.; Yang, Y. –S.; Zhou, J. Geometric and electronic structure/function correlations in non-heme iron enzymes *Chem. Rev.* **2000**, *100*, 235–349.

53. Gorsky, L. D.; Koop, D. R.; Coon, M. J. On the stoichiometry of the oxidase and monooxygenase reactions catalyzed by liver microsomal cytochrome P-450. Products of oxygen reduction *J. Biol. Chem.* **1984**, *259*, 6812–6817.

54. Francisco, W. A.; Blackburn, N. J.; Klinman, J. P. Oxygen and hydrogen isotope effects in an active site tyrosine to phenylalanine mutant of peptidylglycine α-hydroxylating monooxygenase: Mechanistic implications *Biochemistry* **2003**, *42*, 1813–1819.

55. Prigge, S. T.; Kolhekar, A. S.; Eipper, B. A.; Mains, R. E.; Amzel, L. M. Substrate-mediated electron transfer in peptidylglycine α-hydroxylating monooxygenase *Nature (London)* **1999**, *6*, 976–983.

56. Humphreys, K. J.; Mirica, L. M.; Wang, Y.; Klinman, J. P. Galactose oxidase as a model for reactivity at a copper superoxide center *J. Am. Chem. Soc.* **2009**, *131*, 4657–4663.

57. Chen, P.; Solomon, E. I. Oxygen activation by the noncoupled binuclear copper site in peptidylglycine alpha-hydroxylating monooxygenase. Reaction mechanism and role of the noncoupled nature of the active site *J. Am. Chem. Soc.* **2004**, *126*, 4991–5000.

58. Personal communication with Harry B. Gray.

59. Francisco, W. A.; Wille, G.; Smith, A. J.; Merkler, D. J.; Klinman, J. P. Investigation of the pathway for inter-copper electron transfer in peptidylglycine α-amidating monooxygenase *J. Am. Chem. Soc.* **2004**, *126*, 13168–13169.

60. McIntyre, N. R.; Lowe, Jr., E. W.; Merkler, D. J. Imino-oxy acetic acid dealkylation as evidence for an inner-sphere alcohol intermediate in the reaction catalyzed by peptidylglycine α-hydroxylating monooxygenase *J. Am. Chem. Soc.* **2009**, *131*, 10308–10319.

61. Kolhekar, A. S.; Keutmann, H. T.; Mains, R. E.; Quon, A. S.; Eipper, B. A. Peptidylglycine alpha-hydroxylating monooxygenase: active site residues, disulfide linkages, and a two domain model of the catalytic core *Biochemistry* **1997**, *36*, 10901–10909.

62. Hess, C. R.; Wu, Z.; Ng, A.; Gray, E. E.; McGuirl, M. M.; Klinman, J. P. Hydroxylase activity of Met471Cys tyramine β-monooxygenase *J. Am. Chem. Soc.* **2008**, *130*, 11939–11944.

63. Siebert, X.; Eipper, B. A.; Mains, R. E.; Prigge, S. T.; Blackburn, N. J.; Amzel, L. M. The catalytic copper of peptidylglycine α-hydroxylating monooxygenase also plays a critical structural role *Biophys. J.* **2005**, *89*, 3312–3319.

64. Jaron, S.; Blackburn, N. J. Characterization of the half-apo derivative of peptidylglycine monooxygenase. Insight into the reactivity of each active site copper *Biochemistry* **2001**, *40*, 6867–6875.

65. Nagel, Z. D.; Klinman, J. P. Tunneling and dynamics in enzymatic hydride transfer *Chem. Rev.* **2006**, *106*, 3095–3118.

2

COPPER DIOXYGENASES

József Kaizer, József Sándor Pap, and Gábor Speier

Department of Chemistry, University of Pannonia, 8201 Veszprém, Hungary

2.1. INTRODUCTION

Dioxygen (O_2) is an essential element for life. For example, it plays key roles in the metabolism of essential substances for vital functions, (e.g., amino acids, lipids,

Copper-Oxygen Chemistry, First Edition. Edited by Kenneth D. Karlin and Shinobu Itoh.
© 2011 John Wiley & Sons, Inc. Published 2011 by John Wiley & Sons, Inc.

sugars, vitamins, hormones, and various poisons). The metabolism generates energy and controls vital functions. However, owing to its triplet ground state, O_2 cannot directly react with singlet ground-state substrates to produce singlet-state products because that would imply a violation of the conservation of the total angular momentum. Some way to overcome the spin forbiddenness of the process is therefore required to make O_2 react with organic molecules. Often, metal cofactor-containing enzymes (e.g., oxygenases, oxidases, or peroxidases) are used to circumvent the spin barrier.[1]

Oxygenases are a class of enzymes that catalyze the incorporation of oxygen atoms from O_2 into a substrate. They can be divided into two subclasses depending on whether both atoms of O_2 are incorporated in the substrate (*dioxygenases*) or only one of them (*monooxygenases*). The main question in the enzymatic dioxygenations is the mode of the formation and structure of active oxygen species. Dioxygenases usually employ a transition metal, or an organic cofactor, to mediate O_2 activation. Iron, copper, and manganese are the metals most commonly used because, in their lower oxidation states, they can form complexes with O_2, organic substrates, or both, and affect the electronic structure of the bound compound to alter its reactivity.[1,2]

Flavonol 2,4-dioxygenase (FDO) is the only dioxygenase unambiguously known to contain copper.[3–11] Dioxygenase enzymes, as well as their synthetic models containing other metal centers (Fe, Mn, Co, Ni), have also been reported.[12–22] They have, however, not been studied as extensively as the copper-containing enzymes. Flavonol 2,4-dioxygenase facilitates the incorporation of O_2 into flavonols, thereby cleaving their heterocyclic ring to produce the corresponding depside (phenolic carboxylic acid ester) and carbon monoxide (CO). This enzyme is unique, because CO forming enzymes are rare. To date, four prokaryotic dioxygenases are known, which catalyze oxidative C−C bond cleavage with incorporation of two oxygen atoms into, and release of CO from the substrate; flavonol 2,4-dioxygenase (2.1), 3-hydroxyquinaldin-4(1*H*)-one 2,4-dioxygenase (2.2, R=CH$_3$), 3-hydroxyquinolin-4-(1*H*)-one 2,4-dioxygenase (2.2, R=H), and 1,2-dihydroxy-5-(methylthio)pent-1-en-3-one anion 1,3-dioxygenase (2.3).[23–28]

$$(2.1)$$

$$(2.2)$$

$$\text{(structure)} \xrightarrow[]{O_2 \quad CO} \quad /^S\diagdown\diagup CO_2^- \; + HCO_2H \qquad (2.3)$$

Flavonoids are polyphenolic pigments found in higher plants and some fungi.[29] Numerous representatives [e.g., quercetin (3,5,7,3′,4′-pentahydroxyflavone)] are well known for their antioxidant and antimicrobial properties. Being present in many fruits and vegetables, these species provide an important source of antioxidant and antibacterial dietary supplement for humans.[30,31] With the degradation of the plant material, however, microbiocidal flavonoids get into the soil, too, where bacteria and fungi are exposed to these compounds. In response, soil microorganisms have developed effective catabolic systems utilizing flavonoids as a carbon source with the help of FDO enzymes, which can transform flavonoids under aerobic conditions. Not surprisingly, these metalloenzymes have drawn much attention in the last few decades.

Flavonol 2,4-dioxygenases were first isolated and most extensively studied from *Aspergillus* species, *A. flavus*,[3,6,7] *A. niger*,[5] and *A. japonicus*.[8–10] These were all characterized as highly glycosylated metal-dependent enzymes belonging to the bicupin proteins. Fungal FDOs contain a mononuclear Cu^{2+} center, whereas the enzyme from *Bacillus subtilis* is presumed to prefer Mn^{2+} as cofactor,[14–17] despite having been purified as an iron enzyme from a recombinant *Escherichia coli* clone. In contrast to the enzymes from *B. subtilis*, *Aspergillus* spp., and *Penicillium olsonii*,[11] the FDO protein of *Streptomyces* sp. FLA is most active with Ni^{2+} ions as the cofactor, followed by Co^{2+}.[12,13]

Model chemistry of relevance to dioxygenases has progressed remarkably in recent years, and contributed greatly to clarification of structures and mechanisms of dioxygenases.[22] This chapter focuses on the functional and structural model systems related to the fungal copper-containing native FDOs.[32,33] First, we give a brief overview of the studies on the enzymatic reaction and the structure of the enzymatic active site from *A. japonicus*. A more detailed insight will be presented into the mechanistic details of the oxygenation reactions including the mode of the O_2 activation carried out with model substrates. We provide a thorough understanding of the influence of metal and ligand moieties, as well as those of the substrate coordination modes on the reaction mechanism.

2.2. FUNGAL FLAVONOL 2,4-DIOXYGENASE ENZYMES

2.2.1. Structure of the Enzymatic Active Site

Studies on the FDOs from *A. flavus* and *A. niger* DSM 821 have shown that the enzyme does not require any additional organic cofactors for catalysis. The *A. flavus* FDO has a molecular mass of 111 kDa, with a sugar content of 27.5% and a metal content of 2 mol Cu per mole of enzyme.[3] The FDO isolated from *A. niger* DSM 821 is composed of three different subunits, with molecular masses of 63–67, 53–57, and 31–35 kDa,

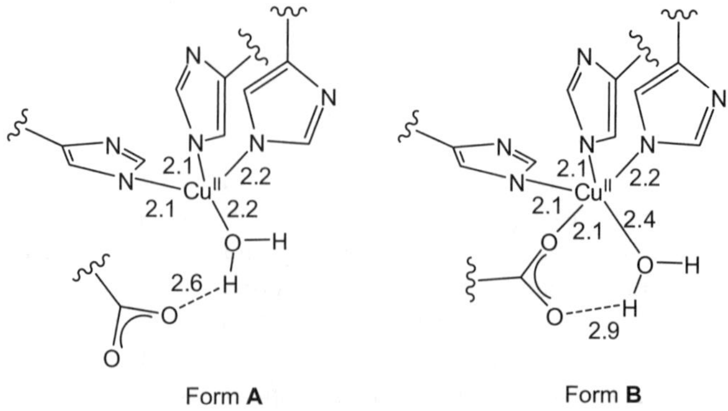

Form **A**   Form **B**

FIGURE 2.1. Coordination modes of Cu in the FDO active site from *A. japonicus*. Distances indicated in the scheme are in angstroms (Å). (See Ref. 33.)

respectively. It has a carbohydrate content of 46–54% and contains 1.0–1.6 mol of Cu per mole of enzyme.[5] Although FDO enzymes had been spectroscopically investigated before, the real breakthrough was the structural characterization of the enzyme from *A. japonicus*.[8] The enzyme is described as a homodimer containing one Cu(II) ion per monomer unit, as revealed by atomic absorption spectroscopy. The 1.6 Å resolution crystal structure of this ~ 100 kDa molecular weight dimeric glycoprotein reveals that the active site is found in a cavity ~ 10 Å from the protein surface, and it is solvent exposed. In the apoenzyme, there are two alternative coordination forms for the Cu ion in accordance with electron paramagnetic resonance (EPR) observations. In form **A** (Fig. 2.1), the Cu is ligated by three histidine (His) residues and by a water molecule in a distorted tetrahedral geometry. This form is present in 70% in the native enzyme and gives the major $g_{\parallel}$ signal at 2.330 and with $A_{\parallel} = 13.7$ mT in the X-band EPR spectrum.[9] Form **B** (Fig. 2.1) represents a trigonal bipyramidal (TBPY) coordination mode and is present in nearly 30% giving the minor EPR signal ($g_{\parallel} = 2.290$ and $A_{\parallel} = 12.5$ mT). In this case, an additional glutamate is bound to the Cu that puts the water molecule farther away from the Cu site. Carboxylate ligation has never before been observed in natural Cu proteins. It has been proposed that, for some iron redox centers, coordination by a carboxylate residue might be responsible for modulating small energy barriers. So, carboxylate coordination to a native copper-containing active site is unique and, as will be discussed in Section 2.2.3, it has a crucial role in the catalytic activity.

2.2.2. Structure of the Enzyme–Substrate Complex

X-ray absorption and EPR spectroscopic data indicate that, on anaerobically binding of flavonol substrates, the heterogeneous distorted tetrahedral (major)/distorted TBPY (minor) native Cu gains structural order, changing to a square pyramidal (SPY) geometry represented by Figure 2.2.[9,34]

It has been clear since early studies that the number and distribution of the hydroxyl functions on the substrates has a great influence on the reactivity[7] resulting in varying

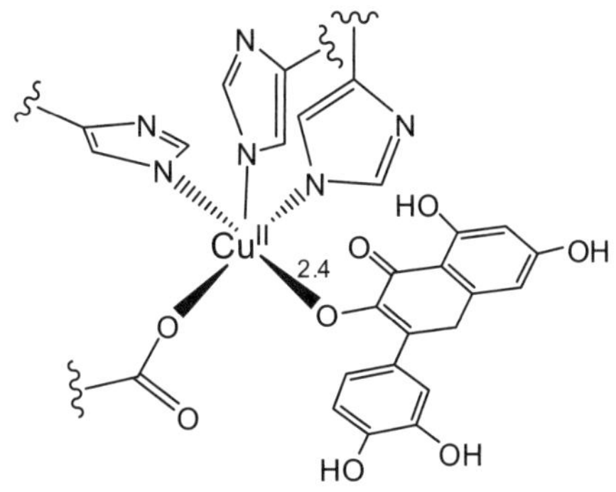

FIGURE 2.2. X-ray crystal structure (at 1.6-Å resolution) of the quercetin–enzyme complex. (The Cu−O distance is in angstroms.) (See Ref. 33.)

reaction rates for their oxygenation. Apart from the effect of the hydroxyl functions on the intrinsic reactivity of the substrate,[35] hydrogen bonding between these functions (in the 3′, 4′, 5, or 7 positions, see Eq. 2.1) and the peptide residues (tyrosine and threonine) found nearby the active site, which positions the flavonol derivatives to the copper site with different orientation, also plays an important role in the substrate-dependent reactivity.[34] For example, hydroxyl groups on the C7 and C3′ atoms of quercetin result in a faster oxidation rate, whereas hydroxyl substitution on the C8 atom slows down the oxidation dramatically. Manual modeling of the substrate coordination at the active site revealed that the most reactive quercetin, partly due to the above-mentioned interactions, is very likely bound as a monodentate ligand via the 3-hydroxyl group displacing the water molecule from the coordination sphere. Further structural information on the enzyme-substrate (ES) complex was deduced from the structural analysis of FDO crystallized with kojic acid as ligand on the metal. This study showed that similar to form **B** of the enzyme the glutamate remains bound to the copper site after the formation of the ES complex (Fig. 2.2) and probably plays a role in substrate deprotonation. However, in contrast to flavonol derivatives, which are bound in a monodentate fashion, kojic acid chelates the Cu ion in FDO through its hydroxyl (O3) and carbonyl oxygen (O4) atoms. Further evidence on the coordination mode of the ES complex was provided based on extended X-ray absorption fine structure (EXAFS) experiments on anaerobic complexes of FDO with quercetin and myricetin (5′-hydroxyquercetin).[34] The copper coordination environment is five-coordinate, and is best modeled by a single shell of N/O scatterers, ~ 2.00 Å from the metal, also an indication of the monodentate substrate coordination.

2.2.3. Mechanism of the Enzymatic Reaction

General questions regarding the function of dioxygenases are the number of O_2 molecules involved in the reaction and the order and mechanism of activation of the reactants (i.e., substrate and O_2) at the active site. The oxidation of flavonol derivatives is often paralleled with the catabolism of heme (degradation of heme into biliverdin via hydroxyheme formation). A major difference, however, is that while heme undergoes a two-molecule mechanism, during which the oxygen atoms that are incorporated into the substrate originate from two O_2 molecules, FDO utilizes only one O_2 molecule per substrate (one-molecule mechanism). This finding was proven by mass spectral analysis of the reaction product by FDO in a mixture of $^{18}O_2$ and $^{16}O_2$.[36] The order of activation was understood via studying the anaerobic ES complexes and some mechanistic information was obtained from EPR spectroscopic experiments.[9,34] The results indicate that the first step in the enzymatic process is the binding of the flavonol derivative as a monodentate ligand to the copper site. This step does not require the presence of O_2 (i.e., the substrate activation step precedes the oxidation of the metal site). After formation of the ternary ESO_2 complex, oxygenation of the substrate takes place and CO is released before the depside leaves the site. Some of these steps are further supported by studies on model reactions (in Section 2.3).

Another intriguing question is the place of O_2 attack on the ES complex. The generally accepted presumption is that an **ES_{rad}** valence isomer is formed that can react with $^3O_2$ in two ways, as is shown in Figure 2.3 (paths *a* and *b*). Although some hybrid density functional theory (DFT) calculations indicate that the energetically preferred one is path *b*, others do not exclude path *a* as an alternative.[35,37,38] Geometrical considerations that take into account the X-ray crystal structure findings on ES complexes also support the latter pathway.[34]

Although the experiments on native enzymes answered some fundamental questions, some remained unresolved. These include the activation mode of the coordinated flavonolate (coordinated anion vs. radical formation) and the possible role of the glutamate function coordinated to copper. Model reactions and complexes that are discussed in the other chapters aim to elucidate such problems.

2.3. MODEL SYSTEMS

Model oxygenation reactions on quercetin and the parent compound flavonol have been carried out in order to understand the enzymatic reaction. Early studies involved simple base-catalyzed reactions of quercetin and related compounds with O_2, as well as direct reactions of flavonol with superoxide as models for the enzymatic process.[39–41] Photosensitized oxygenation of flavonols causes a similar type of degradation, suggesting that singlet excited O_2 may participate in the reaction. The necessity of singlet O_2 is, however, excluded by the fact that the base-catalyzed oxygenation of flavonols with the normal (triplet) O_2 leads to the highly selective enzyme-type oxygenolysis to give the corresponding depsides and CO quantitatively. Metal complexes of cobalt and copper have been found to act as catalysts for the oxygenation reaction.[22,42] Many reports have appeared for the formation and structures of ES from both enzyme and model studies, but until now there has been little mechanistic evidence for the formation of ESO_2 species. There are two types of discussions on the structure of ES. One concerns whether flavonol coordinates to the metallic center as a mono *versus* bidentate ligand or dissociates from the metallic ion

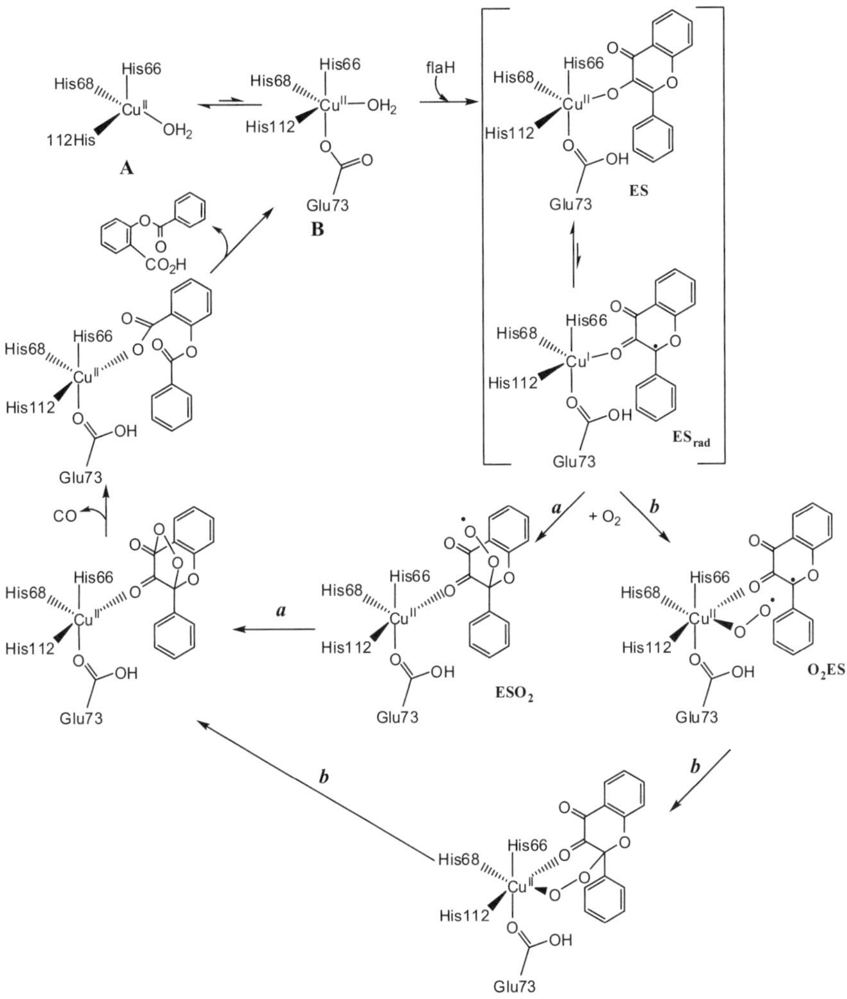

FIGURE 2.3. Proposed mechanism for the enzymatic oxygenation of flavonols. (See Ref. 16.)

forming free flavonolate ion in the activation step. The other concerns the change of the character of the flavonolate ligand (i.e., how the flavonolate ligand is activated for oxygenation): whether the flavonolate ligand can be regarded as a radical, as shown in **A** or has an anionic character as in **B** (Fig. 2.4).

On the other hand, the fact, that the FDO enzymes employ a Cu cofactor early shifted attention to elucidate the possible role of this redox-active metal during the enzymatic oxygenation process with the help of Cu(I)- and Cu(II)-containing flavonolate complexes that have been directly oxygenated, or tried as catalysts in the oxygenation of the substrate. Conclusions based on such model systems and non-redox metal assisted systems have been discussed in a recent review, and will be mentioned only briefly.[32] In the following sections, we present an overview of model

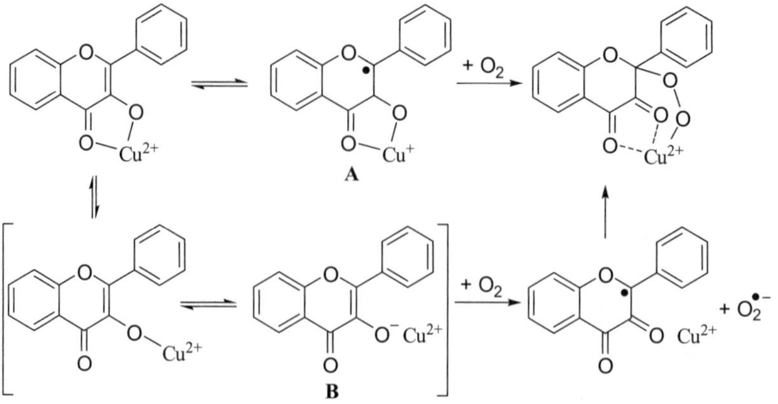

FIGURE 2.4. Probable modes of activation of flavonol and/or dioxygen.

complexes and reactions involving N- and O-donor ligands, focusing on the spec-troscopic and structural properties of the former in connection with the mechanistic aspects of the latter.

2.3.1. Base-Catalyzed and Non-Redox Metal Assisted Dioxygenation of Flavonols

These reactions are carried out either under aqueous or nonaqueous conditions.[39,41] In both cases, O_2 reacts with the substrate analogously to the enzymatic reaction and in addition to the enzymatic product depside, hydrolyzed secondary products were identified. Formation of the enzymatic products was rationalized by assuming that 2-hydroperoxyflavan-3,4-dionate anion is formed first according to Figure 2.5.

A more detailed mechanistic study was carried out with flavonol and its 4′-substituted derivatives (see also Fig. 2.5) in a 50% dimethyl sulfoxide (DMSO)–water solvent mixture.[43] The reaction showed specific base catalysis and fits a Hammett linear free energy relationship for the 4′-substituted derivatives

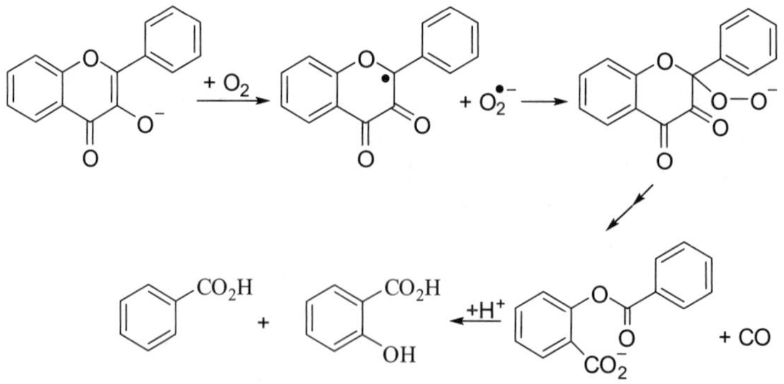

FIGURE 2.5. Base-catalyzed dioxygenation of flavonol.

($\rho = -0.50$). The latter indicates higher electron density on the C ring of the flavonol (i.e., more electron-donating substituents) that makes the electrophilic attack of O_2 easier, an indication for the potential of the deprotonated substrate to reduce triplet O_2 directly. Reactivity of potassium flavonolate derivatives in aprotic medium were also studied.[41] The presence of a flavonoxy radical presumes a single electron transfer (SET) mechanism in which again, the anionic substrate reduces the triplet O_2, which then provides the enzymatic product via recombination and formation of a putative 2-hydroperoxyflavan-3,4-dionate intermediate species as seen in Figure 2.5. A similar mechanism is proposed for the oxygenation of a zinc-containing model complex with the 3N-donor 3,3'-iminobis(N,N-dimethylpropylamine) (idpa, see Fig. 5.6).[44] In this case, the formation of superoxide radical anion is observed [using the nitroblue tetrazolium (NBT) test] suggesting an SET mechanism.

Recombination of the presumed zinc-bound flavonoxy radical and the superoxide radical anion results in a transient peroxide species, which then decomposes into the enzymatic product. The bimolecular rate equation and the negative activation entropy (-96 ± 13 J mol K) support the above associative mechanism. As in the case of the base-catalyzed oxygenations, the influence of the 4'-substituted groups on the reaction rate showed a linear Hammett plot with a reaction constant of $\rho = -0.83$, indicating that electron-releasing groups lead to a remarkable increase (~ 2.5-fold) in the reaction rates (Fig. 2.7). In the oxygenation reaction of the flavonolate in various systems the reaction rates are in the following order: n-Bu$_4$Nfla > flaK > [Zn(fla) (idpa)]ClO$_4$. Beside electronic factors, steric effects were also investigated for the dioxygenation reaction. We have found that the rate of dioxygenolysis is dramatically enhanced by various coligands, [e.g., acetate (CH$_3$CO$_2{}^-$), phenyl- (PhCH$_2$CO$_2{}^-$), diphenyl- (Ph$_2$CHCO$_2{}^-$), or triphenylacetate (Ph$_3$CCO$_2{}^-$)]. For example, addition of 10 equiv of the bulky Ph$_2$CHCO$_2{}^-$ to [Zn(fla)(idpa)]ClO$_4$ accelerated its decay by almost two orders of magnitude ($V_r = 66$). In summary, non-redox metal-assisted oxygenolysis of flavonol derivatives in aprotic solvents possess an SET mechanism in which the substrate with increased basicity reduces molecular oxygen, initiating further steps that lead to the enzyme-like product. It can be said also that bulky carboxylates as coligands dramatically enhance the reaction rate, which can be explained by the formation of more reactive monodentate flavonolatozinc complexes, which is necessary for the intermolecular ET.

2.3.2. Copper-Containing Models for the Enzyme–Substrate and Enzyme–Product Complexes

The reactivity of the Cu(II) flavonolate binary complex and the Cu(I) triphenylphosphine flavonolate ternary complex were studied and discussed in earlier reviews.[32,45–49] Both complexes mimic the enzyme action and provide the corresponding O-benzoylsalicylate complexes as oxygenation products, but with a surprisingly low reaction rate. Since key features of the Cu(II) flavonolate complexes are that the flavonols coordinate as a chelate to the copper center, this is thought to be the main reason for the lower reaction rate compared to that of the enzyme. It is further evidenced by the accelerating effect of added pyridine that competes with the

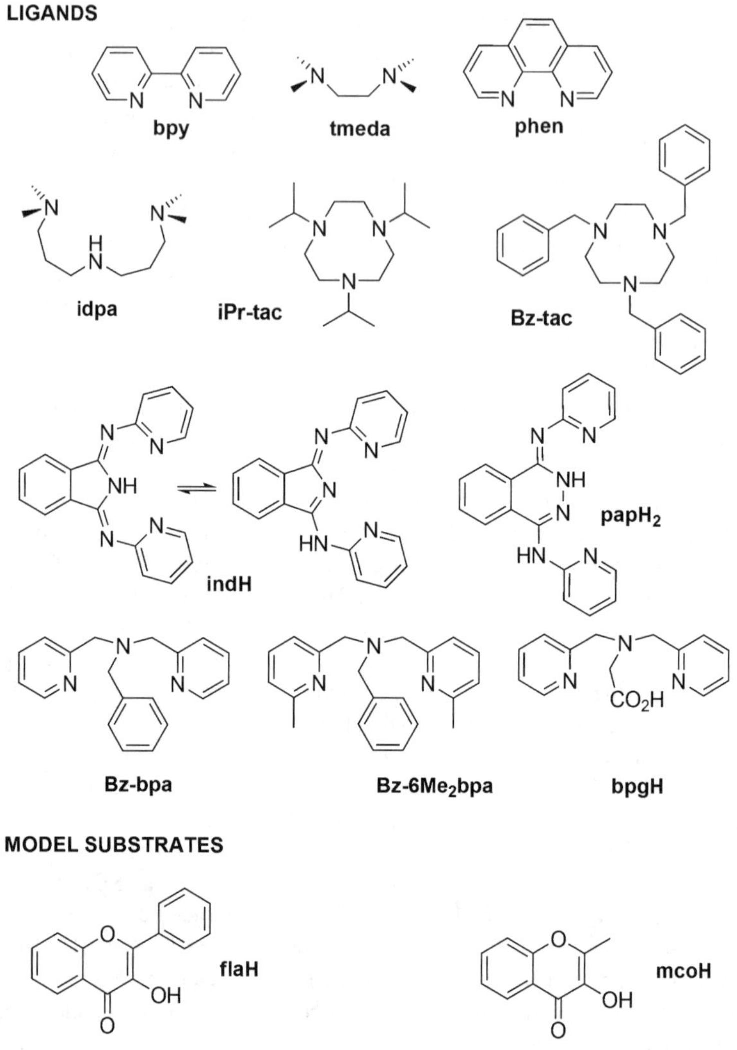

LIGANDS

bpy   tmeda   phen

idpa   iPr-tac   Bz-tac

indH   papH$_2$

Bz-bpa   Bz-6Me$_2$bpa   bpgH

MODEL SUBSTRATES

flaH   mcoH

FIGURE 2.6. Applied ligands and model substrates with their abbreviated names. (See Ref. 33.) [Abbreviations: flaH (3-hydroxyflavone), idpa [3,3′-iminobis(*N,N*-dimethylpropylamine)], *i*-Pr-tac (*N,N,N*-triisopropyl-1,4,7-triazacyclononane), Bz-tac (*N,N,N*-tribenzyl-1,4,7-triazacyclononane), Bz-bpa [*N*-benzyl-*N,N′*-bis(2-pyridylmethy)amine], Bz-6Me$_2$bpa [*N*-benzy]-*N,N′*-bis(6-methyl-2-pyridylmethyl)amine], bpgH (*N,N*-bispicolylglicine), indH [1,3-bis(2′pyridylimino)isoindoline], mcoH (3-hydroxy-(4*H*)-benzopyran-4-one), phen (1,10-phenantroline), bpy (2,2′-bipyridine), tmeda (*N,N,N′,N′*-tetramethylethylenediamine), *O*-bsH (*O*-benzoylsalicylic acid), aspH (acetylsalicylic acid; aspirine), salH [salicylic acid.])

flavonolate ligand for the sites thus forcing monodentate flavonolate coordination.[45] Thus, we may presume that application of ligands in model complexes that are bound to the copper center, and model the enzymatic metal-binding site, might result in better structural models for the ES complex, as well as better functional models. We

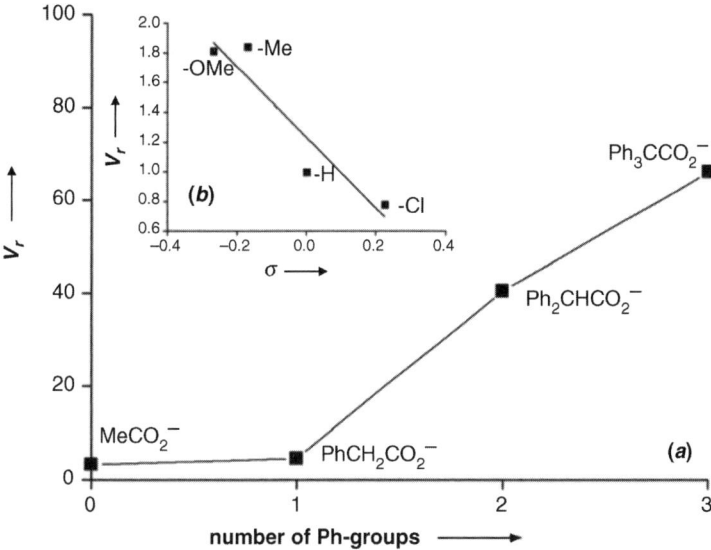

FIGURE 2.7. (*a*) Steric effects on the reaction rate for the dioxygenation of [Zn(fla)(idpa)] ClO_4 in the presence of 10 equiv acetates in dimethylformamide (DMF) at 100°C (correlation between the number of phenyl substituents of acetates and the relative rates). (*b*) Substituent effects on the rate constants for the dioxygenation of [Zn(4′R-fla)(idpa)]ClO_4 (R = −OMe; −Me; −H; −Cl), where Me = methyl and Ph = phenyl, in DMF (solvent) at 100°C.

have prepared several complexes using these ideas, employing the ligands seen in Fig. 2.6, which made it possible to draw conclusions from reactivity–structure relationships.

2.3.2.1. Structural Features of Enzyme–Substrate Model Complexes.

All complexes have been characterized by infrared (IR) and ultraviolet–visible (UV–vis) spectroscopy and some with X-ray crystallography. Coordination of the substrates flaH and mcoH (Fig. 2.3) to the copper site is indicated by the characteristic ν_{CO} band between 1540 and 1580 cm^{-1} (Table 2.1, see also references cited therein). Compared to that of the ν_{CO} vibration at 1610 cm^{-1} of free flavonol, this band is shifted by 30–70 cm^{-1} to lower energies.[50] This finding can be interpreted by the formation of a stable five-membered chelate that is formed upon the coordination of the 3-OH and 4-CO oxygen atoms of the flavonol and partially populates the π^* antibonding orbital of the carbonyl group via back-donation from copper d orbitals. In the UV–vis absorption spectrum, the bathochromic shift of the flavonol $\pi–\pi^*$ transition from ~340 to 420–440 nm shows unambiguously the presence of the coordinated substrate (Table 2.1). Bands at lower energies (600–700 nm) originate from d–d transitions of the Cu(II) ions.

Crystal structures of the complexes (Fig. 2.8) further support the bidentate coordination of flavonol that leads to the formation of a very stable five-membered

TABLE 2.1. The IR and UV–vis Spectroscopy Data for the Studied ES Complexes^a

Complex^b	IR (KBr), ν_{CO} (cm^{-1})	UV–vis (DMF), λ(nm)(log ε)	Reference
[Cu(fla)(idpa)]ClO$_4$	1559	732 (2.31); 427 (3.99)	50
[Cu(fla)(i-Pr-tac)]ClO$_4$	1554	1065 (1.89); 653 (2.15); 434 (3.52)	51
[Cu(Bz-tac)(fla)]ClO$_4$	1545	1040 (1.73); 615 (2.90); 432 (2.23)	51
[Cu(Bz-bpa)(fla)]ClO$_4$	1541	421 (4.13)	52
[Cu(Bz-6Me$_2$bpa)(fla)]ClO$_4$	1545	429 (4.09)	52
[Cu(bpg)(fla)]	1543	421 (4.32)	52
[Cu(ind)(mco)]	1577	618 (1.87); 450 (4.43) 423 (4.49)	53
[Cu(fla)(ind)]	1574	631 (2.18); 444 (3.96) 415 (4.17)	54
[Cu(fla)(phen)$_2$]ClO$_4$	1581	702 (1.71); 419 (3.92)	55
[Cu(fla)$_2$(phen)]	1564	674 (2.58); 426 (4.75)	55
[Cu(bpy)(fla)$_2$]	1578	431 (4.77)	55
[Cu(fla)$_2$(tmeda)]	1573	674 (2.58); 430 (4.78)	55
[Cu(fla)(MeCN)(tmeda)]ClO$_4$	1545	623 (1.50); 427 (4.27)	33
[Cu(bpy)(fla)ClO$_4$]	1540	637 (2.30); 427 (4.26)	56
[Cu(fla)$_2$]	1536	426 (4.56) 410 (4.53)	45

^{a}See Ref. 33.
^{b}See Figure 2.6 for abbreviations.

chelate ring in each case. All the applied coligands are 3N-donors. Thus the copper site is five coordinated with distorted geometries. The extent of distortion from an ideal TBPY or SPY arrangement ($\tau = 0$ in an ideal TBPY complex and $\tau = 1$, in case of an ideal SPY arrangement) is highly influenced by the 3N-donor ligand. As seen from Table 2.2, the τ parameter embraces a wide range ($\tau = 0.12$–0.71).[58]

Structures of the Cu(fla)(L) model complexes studied differ from that of the enzymatic active site regarding the substrate coordination mode: Bidentate coordination is experienced in the models in all cases, whereas in the enzyme only the 3-enolate function is bound to the metal. This difference can be explained by means of van der Waals forces and hydrogen bonding that ban the formation of the otherwise thermodynamically favored five-membered chelate ring. These secondary interactions are ruled out in the case of our synthetic models.

On the other hand, changes in the geometry of the five-coordinate copper complexes will influence the electron delocalization, and thus the bond distance between the copper center and the flavonolate oxygens. While in SPY geometries the Cu–O$_{enolate}$ and the Cu–O$_{keto}$ bond lengths are nearly equidistant, in complexes with a TBPY geometry this Δ value approaches 0.3 Å. If the difference between the Cu–O bond distances is plotted against the τ value, a clear correlation is observed (Fig. 2.9). As will be seen in Section 2.3.3.1 and Table 2.5, distortion of the Cu–fla chelate will considerably change the reaction rate for oxygenation in model complexes.

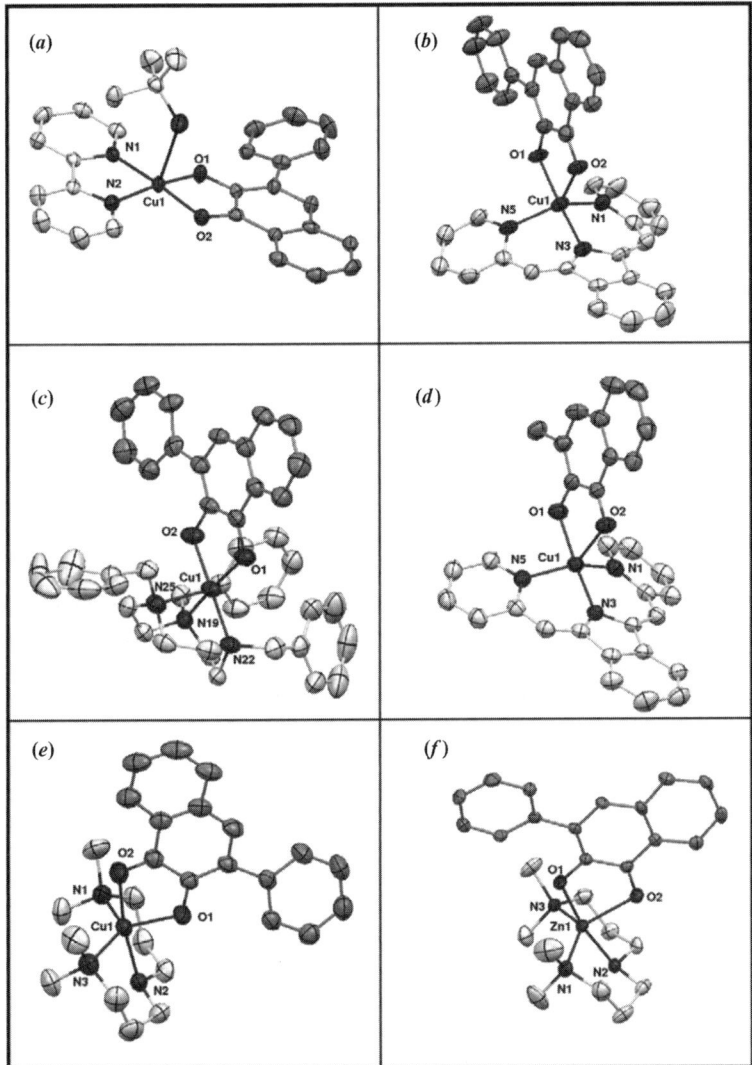

FIGURE 2.8. X-ray structure of some synthetic ES complexes: [Cu(bpy)(fla)ClO$_4$] (*a*); [Cu(fla)(ind)] (*b*); [Cu(Bz-tac)(fla)]ClO$_4$ (*c*); [Cu(ind)(mco)] (*d*); [Cu(fla)(idpa)]ClO$_4$ (*e*); [Zn(fla)(idpa)]ClO$_4$ (*f*);. Thermal ellipsoids are at the 50% probability level. (See Ref. 33.)

2.3.2.2. Structural Features of Enzyme–Product Model Complexes. Synthetic enzyme-product (EP) model complexes can be prepared directly, or by reacting the corresponding ES complex with O$_2$. The EP complexes have been structurally characterized (X-ray diffraction, IR, and UV–vis spectroscopy). The relevant spectroscopic and structural data are given in Table 2.3 (IR, UV–vis; Table 2.3) and Table 2.4 (X-ray).

TABLE 2.2. Structural Data for Synthetic ES Type Complexes[a]

Complex		Bond Distance (Å)	Δ(Å) |(Cu1−O1) − (Cu1−O2)|	τ	Reference
[Cu(fla)(idpa)]ClO$_4$	Cu1−O1	2.210(3)	0.292	0.61	50
	Cu1−O2	1.918(3)			
[Cu(ind)(mco)]	Cu1−O1	1.9506(14)	0.347	0.38	53
	Cu1−O2	2.2966(14)			
[Cu(Bz-tac)(fla)]ClO$_4$	Cu−O1	1.917(3)	0.095	0.12	51
	Cu−O2	2.012(3)			
[Cu(fla)(ind)]	Cu1−O1	1.942(5)	0.264	0.50	54
	Cu1−O2	2.206(6)			
[Cu(bpy)(fla)ClO$_4$]	Cu1−O1	1.971(3)	0.074	0.10	56
	Cu1−O2	2.897(3)			

[a]See Ref. 33.

The ν(CO) bands for the ester bond in the IR, spectrum of the EP complexes remain unchanged compared to the corresponding value for the free O-benzoylsalicylic acid (O-bsH). Consequently, this moiety does not take part in the ligation. The value of the ν_{as}(CO) and ν_s(CO) bands for the O-bs⁻ carboxylate function on the other hand provides us information on the coordination mode. Differences between the energy of the symmetric and asymmetric stretching modes allows us to predict the mono-, or bidentate coordination mode of the carboxylate moiety to the copper. X-ray diffraction structures (where available) are in good accordance with this IR based estimation.

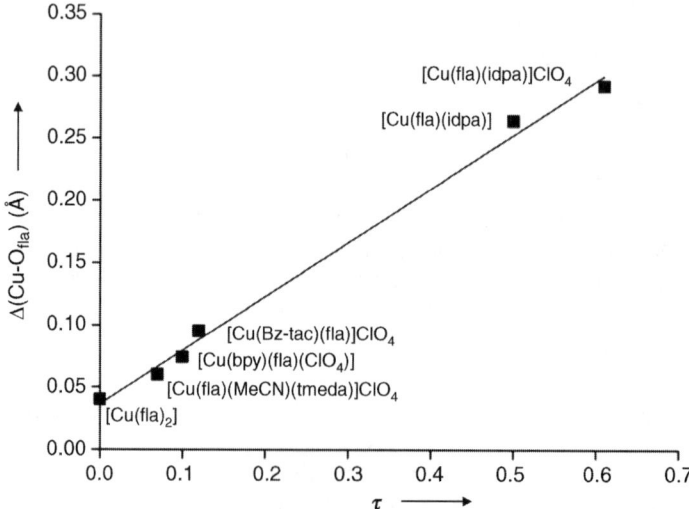

FIGURE 2.9. The π-electron delocalization in flavonolate ligand in correlation with the geometry of the Cu(L)(fla) complexes. (See Ref. 33.)

TABLE 2.3. The IR and UV–vis Data for EP Complexes

Complex	IR (KBr) (cm^{-1}) ν_{as}(CO), ν_s(CO)	UV–vis (DMF) λ(nm) (log ε)	Reference
[Cu(idpa)(O-bs)]ClO$_4$	1572, 1375	716 (2.45); 276 (3.98)	50
[Cu(i-Pr-tac)(O-bs)]ClO$_4$	1584, 1428	1060 (1.73); 670 (1.43); 278 (3.23)	51
[Cu(Bz-tac)(O-bs)]ClO$_4$	1586, 1432	1065 (1.62); 650 (1.96); 276 (3.81)	51
[Cu(Bz-bpa)(O-bs)]ClO$_4$	1559, 1374	682 (2.41); 270 (3.85)	52
[Cu(Bz-6Me$_2$bpa)(O-bs)]ClO$_4$	1570, 1375	736 (2.16); 271 (3.93)	52
[Cu(bpg)(O-bs)]	1571, 1344	914 (1.50); 270 (3.72)	52
[Cu(ind)(O-bs)]	1578, 1380	816 (1.17); 447 (3.97); 272 (4.04) 421 (4.04) 334 (4.02) 309 (4.02)	54
[Cu(asp)(H$_2$O)(ind)]	1531, 1357	662 (2.13); 446 (4.29); 272 (4.30) 421 (4.35) 335 (4.27) 312 (4.28)	59
[Cu(O-bs)(phen)$_2$]ClO$_4$	1572, 1388	694 (2.06); 331 (3.31); 294 (4.25)	55

[a]See Ref. 33.

Formation of the EP complexes can be followed by UV–vis spectrophotometry in case we oxygenate the precursor ES complexes, since the π–π^* charge-transfer band that is characteristic for the flavonolate ligand disappears. Thus UV–vis spectroscopy becomes a useful tool in the kinetic investigations of the ES complex oxygenation reactions.

TABLE 2.4. Selected Structural Data for the EP Complexes

Complex		Bond length (Å)	τ	Reference
[Cu(idpa)(O-bs)]ClO$_4$	Cu1–O1	1.995(5)	0.16	50
	Cu1–O2	2.344(9)		
[Cu(Bz-tac)(O-bs)]ClO$_4$	Cu–O1A	1.9772(17)	0.16	51
	Cu–O2A	2.0685(19)		
[Cu(i-Pr-tac)(O-bs)]ClO$_4$	Cu1–O1	1.981(6)	0.10	51
	Cu1–O2	2.020(7)		
[Cu(Bz-6Me$_2$bpa)(O-bs)]ClO$_4$	Cu1–O1	1.9835(19)	–	52
[Cu(bpg)(O-bs)]	Cu1–O1	1.972(2)	0.51	52
	Cu1–O3	2.015(2)		
[Cu(O-bs)(phen)$_2$]ClO$_4$	Cu1–O1	1.993(4)	0.47	55
[Cu(H$_2$O)(ind)(sal)]	Cu1–O24	2.0053	0.25	59
	Cu1–O1	2.4344		

[a]See Ref. 33.

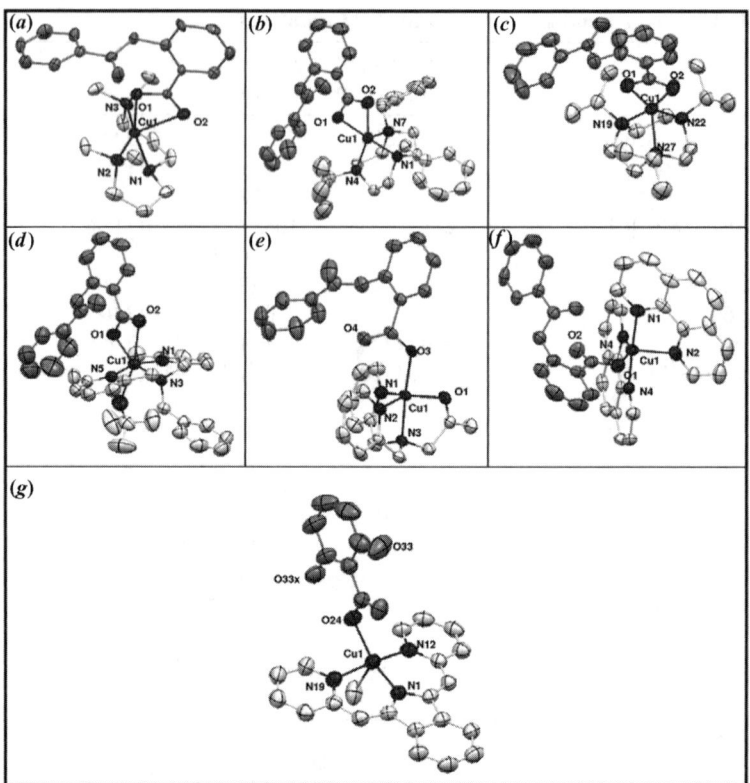

FIGURE 2.10. X-ray structure of the EP complexes: [Cu(idpa)(*O*-bs)]ClO$_4$ (*a*); [Cu(Bz-tac)(*O*-bs)]ClO$_4$ (*b*); [Cu(*i*-Pr-tac)(*O*-bs)]ClO$_4$ (*c*); [Cu(Bz-6Me$_2$bpa)(*O*-bs)]ClO$_4$ (*d*); [Cu(bpg)(*O*-bs)] (*e*); [Cu(*O*-bs)(phen)$_2$]ClO$_4$ (*f*); [Cu(asp)(H$_2$O)(ind)] (*g*). (See Ref. 33.)

Among the X-ray crystallographically characterized compounds (Fig. 2.10), we can find examples for the bidentate *O*-bs$^-$ coordination (*a*, *b*, *c*, and *d*) and for the monodentate (*e*, *f*, and *g*) as well. Bond distances and angles are found in Table 2.4. Increasing the ν_{as}(CO)–ν_{s}(CO) difference with the Δ(Cu–O$_{carboxylate}$) value shows a linear increase with a good correlation (Fig. 2.11). In other words, a relatively big difference between the ν_{as}(CO) and ν_{s}(CO) indicates a monodentate coordination mode for the *O*-bs$^-$ ligand. Of the EP models, compounds with a TBPY geometry correctly structurally describe the native enzyme-product complex. Based on the available data, [Cu(bpg)(*O*-bs)] ($\tau = 0.51$) is the best structural model so far.[52]

2.3.3. Functional Models for Flavonol 2,4-Dioxygenase

Studies on the FDO enzymes imply that the substrate coordination and concomitant activation at the active site makes the oxidative decarbonylation of quercetin viable. Early model reactions involving ES type complexes, however, took place typically at a reasonable rate, but only at higher temperatures (80–120°C). At such high thermal

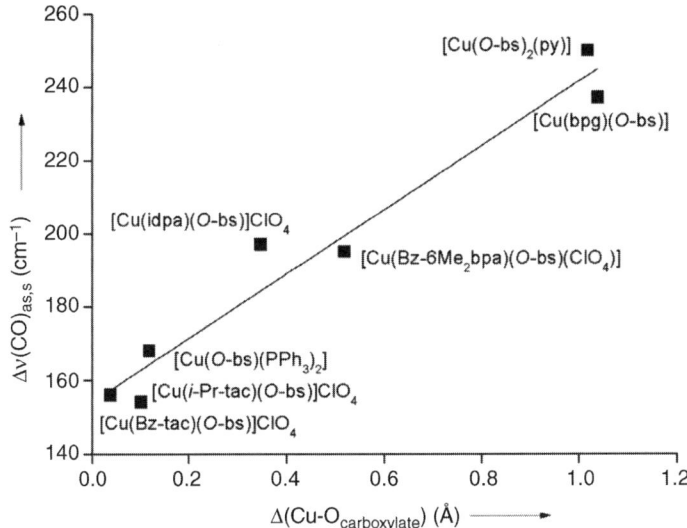

FIGURE 2.11. Carboxylate coordination mode for O-bs$^-$. Correlation between the energy difference of the IR stretching bands of the carboxylate function and the copper carboxylate oxygen bond distance differences. (See Ref. 33.)

energy levels the initial activation of O_2 instead of that of the substrate remains an alternative mechanism. To elucidate the order of substrate versus dioxygen activation, $EO_2 + S$ type reactions were carried out with the known Cu(II)– and Cu(III)–O_2 adducts, $[\{Cu^{III}(Bz\text{-}tac)\}_2(\mu\text{-}O)_2](ClO_4)_2$ and $[\{Cu^{II}(i\text{-}Pr\text{-}tac)\}_2(\mu\text{-}\eta^2:\eta^2\text{-}O)_2]$ $(ClO_4)_2$, and flavonol.[60–63] The reaction yields only the corresponding ES type complexes, excluding O_2 activation.[51]

Mechanistic details for some model systems have been comprehended in earlier reviews.[32,33,63] Studies summarized therein highlight the redox role of the copper center, the electronic effect of the substituents on the flaH moiety, and the influence of the N-ligands on the chemoselectivity. While some complexes follow the enzymatic reaction pathway, such as ([Cu(fla)$_2$] and [Cu(fla)(PPh$_3$)$_2$]) during oxygenation, others deviate from it producing no CO, but different oxygenation products; these include ([Cu(fla)$_2$(tmeda)], [Cu(bpy)(fla)$_2$], and [Cu(fla)$_2$(phen)]).[55] In the latter reaction, Cu(II) 2-hydroxyphenylglyoxylate is formed, possibly via a 1,2-dioxethane intermediate. On the other hand, complexes containing only one fla$^-$ ligand with formula [Cu(fla)(L)$_2$]ClO$_4$, follow the enzymatic route, but typically with a much lower rate. Kinetic studies, ^{18}O-labeling experiments and product analyses support the mechanistic pathways depicted in Figure 2.12.[64] Available kinetic data are summarized in Table 2.5. Apart from major differences, a common feature is that both involve a pre-equilibrium between the Cu(II) flavonolate and its proposed valence tautomer, a Cu(I) flavonoxy radical complex.

The flavonoxy radical is thought to be very unstable, which is supported by the observation of its difficult preparation, and detection in EPR spectroscopy.[65] One electron oxidation of flavonol derivatives leads to the formation of the unstable

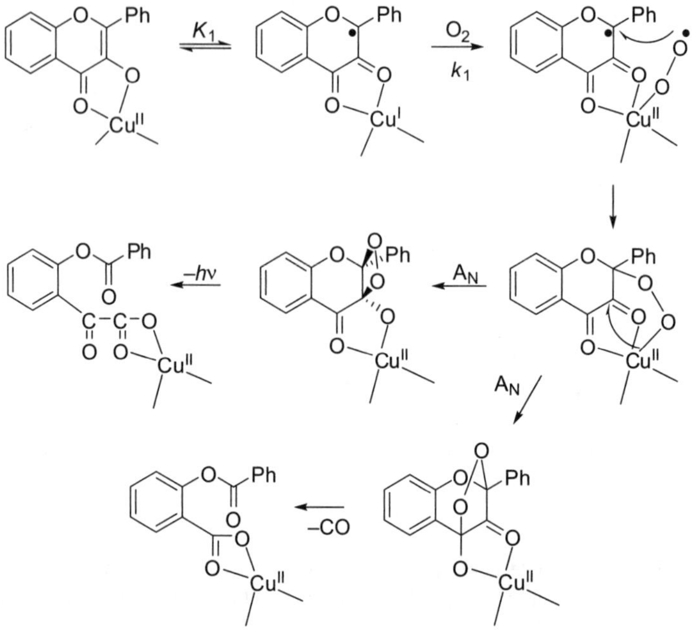

FIGURE 2.12. Proposed mechanisms for the copper-containing flavonol 2,4-dioxygenase models.

flavonoxy radical (Fig. 2.13). This reaction can be facilitated by using the outer-sphere $1e^-$ oxidant [Ce(IV) ammonium nitrate, CAN], or a stable radical initiator like TEMPO (2,2,6,6-tetramethylpiperidinyloxy free radical). Under inert atmosphere, the radical generated forms a dehydro dimer (Fig. 2.13) that is bound via the 3-O (Fig. 2.13, **A**) of one flavonoxy and the C2 (Fig. 2.13, **B**) atom of the other flavonoxy radical, demonstrating spin delocalization in the radical.[65] The weak EPR signal

TABLE 2.5. Kinetic Parameters for the Oxygenation Reaction of Copper Flavonolate Complexes in DMF[a]

Complex	Temperature (°C)	k ($10^3 M^{-1}s^{-1}$)	$\Delta H^{\ddagger}$ (kJ/mol)	$\Delta S^{\ddagger}$ (J/mol K)	Reference
[Cu(fla)(idpa)]ClO$_4$	100	6.13 ± 0.16	75 ± 2	-92 ± 3	50
	120	20.86 ± 0.94			
[Cu(Bz-tac)(fla)]ClO$_4$	120	3.31 ± 0.01			51
[Cu(fla)(i-Pr-tac)]ClO$_4$	120	1.51 ± 0.01			51
[Cu(fla)$_2$(tmeda)]	80	24.0 ± 1.0			55
[Cu(bpy)(fla)$_2$]	80	20.0 ± 1.0			55
[Cu(fla)$_2$(phen)]	80	95.0 ± 5.0	79 ± 16	-40 ± 44	55
[Cu(fla)(phen)$_2$]ClO$_4$	120	183.0 ± 8.0			55
[Cu(fla)$_2$]	80	15.7 ± 0.8	53 ± 6	-138 ± 11	45

[a]See Ref. 33.

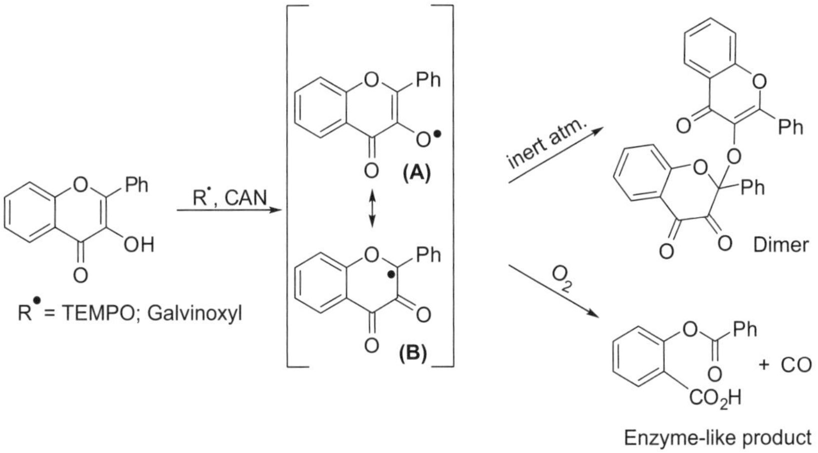

FIGURE 2.13. Cerium(IV) and free radical mediated oxidation of flavonol.

$(a_{H(1)} = 8.1; a_{H(1)'} = 6.4; a_{H(2)} = 1.08$ G; $g = 2.0067)$ of the sample from the MeflaH reaction with CAN is a resolved triplet where the shape of lines indicates two slightly different major and two nonresolved additional couplings. This finding can be unambiguously attributed to an oxygen-centered radical, whereas no signal could be assigned to the C2 centered radical. The DFT calculations showed that from the three possible radical–radical coupling combinations, this one is the most thermodynamically and sterically favored (bond energies are $C-O = 79$ kcal/mol and $O-O = 34$ kcal/mol).[65]

Oxidation of the radical follows the enzymatic pathway producing O-benzoylsalicylic acid and CO.[66,67] It can be concluded that in the enzyme and in the model systems (1) the $^3O_2$ attack can take place on the C2 carbon atom of the flavonoxy radical, besides its rather likely attack on the Cu(I) center, as is shown in Figure 2.3, (2) the flavonolate ligand is rich enough in energy to be able to give up one electron to $^3O_2$ yielding a superoxide anion representing an alternative pathway for the enzyme-like reaction. The latter point is further supported by the mechanism found in case of $[Zn(fla)(idpa)]ClO_4$, where the valence tautomerism does not function so the coordinated flavonolate anion directly reduces the O_2 in an intermolecular step yielding the superoxide anion and flavonoxy radical.[44,68]

Here we focus on the nature and reactivity of the flavonoxy radical and further factors that influence the reaction rate, namely, the geometry of the ES complexes and the application of carboxylate coligands. Substrate specificity will be also demonstrated.

2.3.3.1. Dioxygenation Reactions of Enzyme–Substrate Model Complexes. The bimolecular rate constants for all the oxygenation reactions of the model complexes are orders of magnitude lower than the enzymatic rate constant. The reason is that in the enzyme active site the quercetin is coordinated as a monodentate ligand. On the other hand, in the model systems studied, only bidentate coordination of the

flavonolate moiety was observed. This finding leads to an extended delocalization of π-electrons in the C ring and a rapid equilibrium between the Cu(II) flavonolate and a proposed Cu(I) flavonoxy radical valence tautomer that stabilizes the chelate-bound substrate anion (or radical) and makes the oxygenation less favored than in the case of the singly coordinated substrate.[69] The extent of delocalization correlates well with the τ value of the five-coordinate model complexes. This result can be rationalized if we realize that the delocalization decreases changing from SPY to TBPY geometries (i.e., the $\Delta Cu-O_{fla}$ increases). Consequently, one observes higher rates for the oxygenation of the models bearing TBPY geometry (Fig. 2.9, Table 2.5).

Note that model complexes cannot approximate the efficiency of the FDO enzymes ($K_M = 0.0052$ M for FDO from *Asp. flavus*,[7] or 0.019 M from *Penicillium oslonii*).[11] However, when oxygenation of [Cu(fla)(idpa)]ClO₄ is performed in the presence of excess acetate coligand, the reaction rate is increased dramatically (Fig. 2.14).[70,71]

Cathodic reduction of [Cu(fla)(idpa)]⁺ in DMF solution with 0.1 M tetrabutylammonium perchlorate (TBAP) electrolyte occurs at -1.23 and -1.51 V versus the Fc^+/Fc couple, indicating the presence of a putative [Cu(dmf)(fla)(idpa)]⁺ species. The processes are irreversible within the time scale of the cyclic voltammetry experiment. Upon addition of acetate coligand to the inert solution in a 1:10 ratio with the complex, the -1.23 V current peak shifts by 80 mV to the positive direction, and the -1.51 V peak disappears. Addition of the bulkier phenyl-, diphenyl-, and triphenylacetate coligands causes only minor shifts (0, 10, and 25 mV, respectively) compared to the acetate. These shifts can be explained by the formation and exclusive

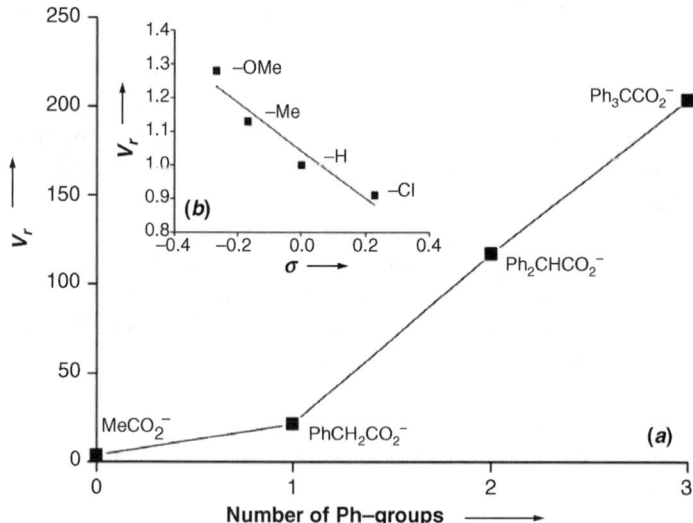

FIGURE 2.14. (*a*) Steric effects on the reaction rate for the dioxygenation of [Cu(fla)(idpa)] ClO₄ in the presence of 10 equiv acetates in DMF (solvent) at 100°C (correlation between the number of phenyl substituents of acetates and the relative rates). (*b*) Substituent effects on the rate constants for the dioxygenation of [Cu(4'R-fla)(idpa)]ClO₄ (R = −OMe; −Me; −H; −Cl) in DMF (solvent) at 100°C.

presence of [Cu(fla)(idpa)(RCO$_2$)] complexes. The acetate is thought to compete with the 4-CO group of the flavolate ligand for the coordination site, that is, [Cu(fla)(idpa)]ClO$_4$ is transformed to [Cu(fla)(idpa)(OAc)]. This finding results in the weakening of the chelate effect regarding the substrate, thus accelerating its oxidation. Moreover, when bulky triphenylacetate is added to the complex, the effect is one order of magnitude greater (Fig. 2.14).[33] Although electronic releasing substituents on the flavonolate ligand were shown to somewhat enhance the reaction rate, considering only the above electronic effect of the added carboxylates, one would expect somewhat lower reaction rate for the oxygenation of the [Cu(fla)(idpa)]$^+$ complex when these coligands are present. Experiments, on the other hand, show that the rate of dioxygenolysis is dramatically enhanced by various carboxylate coligands, especially by the bulky triphenylacetate ($V_r = 200$, at 100°C). The main mechanistic difference between the direct and carboxylate-enhanced dioxygenation of [Cu(fla)(idpa)]$^+$ is that in the latter case there is an ET from the [Cu(fla)(idpa)]$^+$ to O$_2$ resulting in the formation of free superoxide anion (Fig. 2.15). In conclusion, steric hindrance of the added coligand is thought to be the governing factor in the oxygenation of flavonolate complexes that renders monodentate coordination of the flavonol to the copper, which is less stable than the bidentate coordination. Since the electron density on the C2 atom of the flavonolate moiety is higher when it is only weakly coordinated to the metal, direct ET from the activated flavonolate to O$_2$ becomes viable resulting in free superoxide radical production that reacts in further steps with the Cu(II) flavonoxy intermediate to yield the enzyme-like product (Fig. 2.16). Formation of

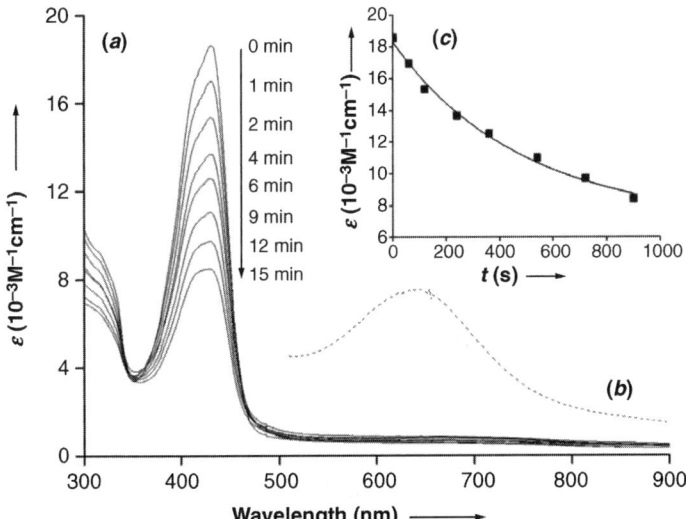

FIGURE 2.15. (*a*) Visible spectral change for the decay of [Cu(fla)(idpa)]ClO$_4$ in the presence of 10 equiv Ph$_3$CCO$_2^-$ in DMF (solvent) at 100 °C, (*b*) in the presence of NBT, and (*c*) time-dependent conversion of [Cu(fla)(idpa)]ClO$_4$ under the condition described above monitored at 407 nm.

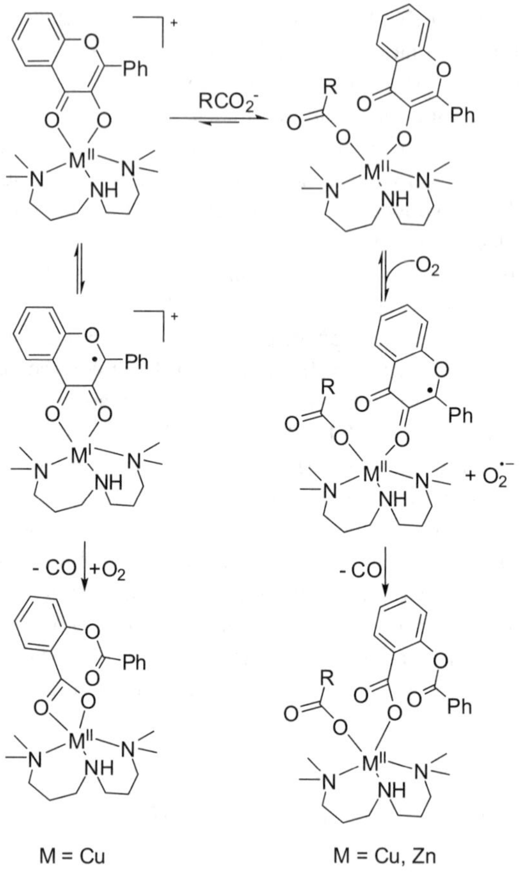

FIGURE 2.16. Mechanistic differences between the oxygenation of [M(fla)(idpa)]ClO$_4$ (M = Cu, Zn) in the presence and absence of carboxylate coligands.

superoxide was proved by adding nitroblue tetrazolium (NBT) to the reaction mixture, where the reduction of the added NBT by O$_2$·$^-$ took place by the enzyme-like oxygenation of [Cu(fla)(idpa)]ClO$_4$. An analogous reaction pathway without the need for redox cycling of the metal was suggested for the nickel- and copper-containing flavonol 2,4-dioxygenases,[12] and for the iron- and zinc-containing model systems.[18] The added carboxylate in these model systems helps the understanding of the role of the glutamate ligand at the active site of the FDO enzyme, which plays a role in positioning the quercetin as a monodentate ligand.

Presumably, the coordination mode and the ligand environment around the copper center, and the Lewis acidity of the copper ion all influence the chemoselectivity of the oxygenation reaction. Since all complexes are sterically very crowded, discrimination of the intramolecular nucleophilic peroxide group attack between the 4-CO and 3-CO carbon atoms seems to be governed by the different electron densities on the indicated carbon atoms, which partially depends on the ligand environment.

TABLE 2.6. Oxygenation of flaH in the Presence of ES and EP Complexes in DMF[a]

Complex	T (°C)	t (h)	Conversion (%)	Reference
[Cu(fla)(idpa)]ClO$_4$	100	8	53	50
[Cu(bpg)(O-bs)]	100	20	90	52
[Cu(Bz-bpa)(O-bs)]ClO$_4$	100	20	91	52
[Cu(Bz-6Me$_2$bpa)(O-bs)]ClO$_4$	100	20	69	52
[Cu(fla)(phen)$_2$]ClO$_4$	100	10	90	55
[Cu(fla)$_2$(phen)]	100	8	100	55
[Cu(fla)(ind)]	100	8	17	55
[Cu(fla)$_2$(pap)]	100	8	100	51
[Cu$_2$(ClO$_4$)$_2$(fla)$_2$(papH$_2$)]	100	8	87	51
[Cu(fla)$_2$]	80	10	78	74

[a]See Ref. 33.

2.3.4. Dioxygenation of Flavonols Catalyzed by Model Complexes

Several ES and EP models (see in Table 2.6) function as catalysts in the oxygenation reaction of flaH. The catalyst/substrate ratio ranges from 1:5 to 1:20.[72-74] All the studied systems produce the enzymatic product O-bsH (see Table 2.6) with satisfactory catalytic activity. The $3N,O$-donor set bearing [Cu(bpg)(O-bs)], which is a good model for the enzyme form **B** (Fig. 2.1) did not exhibit outstanding activity. This finding is not necessarily contradictory if we note that all the reactions were carried out in DMF (solvent), which is a good coordinating solvent itself, and which possibly plays a role similar to that of the carboxylate function (i.e., it competes with the substrate for the coordination site and constrains the substrate's monodentate coordination). Another explanation can be the analogous effect of the product O-bsH. Consequently, a built-in carboxylate arm in the ligand may play only a minor role in determining the structure of the transient species, and the reaction rate. Interestingly, [Zn(fla)(idpa)]ClO$_4$ does not catalyze the reaction,[44] even at extreme conditions, probably due to stable product–zinc complex formation after the stoichiometric oxygenation of the catalyst. Detailed kinetic studies were carried out on the most efficient catalyst among the $3N$-donor ligand containing complexes that aim to model form **A** of the active site of the FDO enzyme (Fig. 2.1). These studies on the [Cu(fla)(idpa)]ClO$_4$–flaH–O$_2$ system reveal that the rate-determining step is the reaction of the copper flavonolate complex with O$_2$ and the free flaH does not play a role in the slow step. Instead, it exchanges the formed O-bs$^-$ ligand from the complex in a fast step of the cycle. Thus mechanistically, one cycle corresponds to the oxygenation of the ES model. This mechanism is presumed to apply to the other studied systems, too.

2.4. CONCLUDING REMARKS

Flavonol 2,4-dioxygenases from *Aspergillus* species and *B. subtilis* are metal cofactor dependent enzymes that catalyze the spin-forbidden oxidative decarbonylation of flavonol derivatives into the corresponding depsides using one O_2 molecule per substrate. Fungal FDOs contain Cu^{2+} centers, whereas the iron-containing enzyme from *B. subtilis* is able of incorporating different metal ions while retaining some dioxygenase activity, being most active with Mn^{2+}. In contrast to these enzymes, the FDO from *Streptomyces* species FLA is most active with Ni^{2+} ions as cofactor, followed by Co^{2+}. Taken together, the biochemical and spectroscopic data on cobalt- and nickel-containing FDOs, and the chemical mechanism of non-redox metal-containing model reactions suggest the possibility of a non-redox role of the metal cofactor of FDO. However, lots of problems remain in the O_2 activation mechanism. For example, the structures of the peroxo intermediate and the active species have not yet been determined. Furthermore, there is no information on how the peroxo intermediate is converted to the active species. The structures of these unstable intermediates may be clarified from studies using model compounds.

In the case of the copper-containing enzyme, it is thought to circumvent the high energy need of the spin-forbidden process by the $1e^-$ oxidation of the bound deprotonated substrate yielding a Cu(I) flavonoxy radical valence isomer that can react with triplet O_2 in a spin allowed step. However, the primary activation of O_2 stands as an alternative pathway. Studies on $[Cu_2(\mu\text{-}O)_2]$, and $[Cu_2(\mu\text{-}\eta^2{:}\eta^2O)_2]$ models credibly excluded this alternative as they are inert in their reaction with flaH.

X-ray structural, EXAFS, and EPR spectroscopic investigations on the copper-containing FDO from *A. japonicus* have clarified that the Cu(II) ion is five coordinate involving a glutamate function as ligand, a unique situation among copper utilizing oxygenases. Formation of a substrate-metal binding that is important for activation of the substrate, has been suggested for these enzymes. Manual modeling of the substrate coordination at the active site, and the EXAFS experiments on anaerobic complexes of FDO revealed that the most reactive quercetin is bound as a monodentate ligand. Model systems involving multidentate ligands (aimed to mimic the enzymatic environment) and flavonol coordinated to the copper center yield enzyme-like products upon oxidation, but only at high temperatures. Structural characterization of ES type complexes revealed that the substrate forms a stable chelate with the copper center that explains its sluggish behavior. Geometric analyses of the ES complexes show that in cases where the $(Cu-O_{fla})$ bond lengths differ significantly (those with more 5-TBPY character) exhibit higher reactivity. It was demonstrated that addition of acetate in high excess further increases the reaction rate significantly. A similar result was found for the non-redox zinc flavonolate complexes. These results support the role of glutamate not merely in deprotonation of the docked substrate, but also in directly preventing its chelation. In conclusion, a significant steric effect on the oxidative decay of ES complexes has been found. Addition of bulky carboxylates as coligands dramatically enhance the reaction rate, which can be explained by two different mechanisms (redox vs. non-redox), caused by the formation of more reactive monodentate copper flavonolate complexes.

The EP type models provide useful information about the possible coordination modes of the product depside to the copper, since there is no available information about such transient species from the enzymatic systems. The depside is coordinated exclusively via the carboxylate function, the ester carbonyl does not interact with the metal. Mono- and bidentate modes alter in the models characterized by X-ray crystallography, steric effects seem to have only a minor effect, and supposedly donor properties of the ligands determine the structure.

Both ES and EP complexes that have been investigated catalyze the enzyme-like oxygenation of flaH that, in contrast with the early models, makes the ligand–Cu (II)–substrate (or product) complexes useful both as structural and as functional models. On the basis of the similarity of redox and non-redox metal-containing ES model studies discussed above, an analogous reaction pathway, that is, direct ET from the activated flavonol to dioxygen without the need for redox cycling of the metal, may be envisaged for the reaction catalyzed by Ni-, Mn-, Fe-, Co-, and Cu-FDO. The major role of the divalent metal ion in the active site of FDO could be to control the orientation of bound substrates, to contribute to modulating the reduction potential of the bound flavonol, and to provide electrostatic stabilization of anionic intermediates, rather than to participate directly in redox chemistry.

As seen in this chapter, it is not unreasonable to expect that recent great progress in the chemical and enzymatic analysis of dioxygenases will in the near future lead not only to the clarification of reaction mechanisms, but also to the development of efficient tailored oxygenase-type catalysts by transition metal complexes.

ACKNOWLEDGMENTS

Financial support of the Hungarcian National Research Fund (OTKA K67871, OTKA K75783, and OTKA PD75360) and COST are gratefully acknowledged.

REFERENCES

1. Funabiki, T., *Oxygenases and Model Sytems*, T. Funabiki, Ed., Kluwer Academic Publishers, Dordrecht, The Netherlands, 1997, pp 1–89.
2. Karlin, K. D. Metalloenzymes, structural motifs, and inorganic models. *Science* **1993**, *261*, 701–708.
3. Oka, T.; Simpson, F. J. Quercetinase, a dioxygenase containing copper. *Biochem. Biophys. Res. Commun.* **1971**, *43*, 1–5.
4. Sharma, H. K.; Vaidyanathan, C. S. A new mode of ring cleavage of 2,3-dihydroxybenzoic acid in *Tecoma stans* (L.). Partial purification and properties of 2,3-dihydroxybenzoate 2, 3-oxygenase. *Eur. J. Biochem.* **1975**, *56*, 163–171.
5. Hund, H. K.; Breuer, J.; Lingens, F.; Huttermann, J.; Kappl, R.; Fetzner, S. Flavonol 2,4-dioxygenase from *Aspergillus niger DSM 821*, a type 2 Cu^{II}-containing glycoprotein. *Eur. J. Biochem.* **1999**, *263*, 871–878.
6. Oka, T.; Simpson, F. J.; Child, J. J.; Mills, C. Degradation of rutin by *Aspergillus flavus*. Purification of the dioxygenase quercetinase. *Can. J. Microbiol.* **1971**, *17*, 111–118.

7. Oka, T.; Simpson, F. J.; Krishnamurty, H. G.; Degradation of rutin by *Aspergillus flavus*. Studies on specificity, inhibition, and possible reaction mechanism of quercetinase. *Can. J. Microbiol.* **1972**, *18*, 493–508.

8. Fusetti, F.; Schröter, K. H.; Steiner, R. A.; van Noort, P. I.; Pijning, T.; Rozeboom, H. J.; Kalk, K. H.; Egmond, M. R.; Dijkstra, B. W. Crystal structure of the copper-containing quercetin 2,3-dioxygenase from *Aspergillus japonicus*. *Structure* **2002**, *10*, 259–268.

9. Kooter, I. M.; Steiner, R. A.; Dijkstra, B. W.; van Noort, P. I.; Egmond, M. R.; Huber, M. EPR characterization of the mononuclear Cu-containing *Aspergillus japonicus* quercetin 2,3-dioxygenase reveals dramatic changes upon anaerobic binding of substrates. *Eur. J. Biochem.* **2002**, *269*, 2971–2979.

10. Steiner, R. A.; Kooter, I. M.; Dijkstra, B. W. Functional analysis of the copper-dependent coordination changes probed by X-ray crystallography, ordering effect, and mechanistic insights. *Biochemistry* **2002**, *41*, 7955–7962.

11. Tranchimand, S.; Ertel, G.; Gaydou, V.; Gaudin, C.; Tron, T.; Iacazio, G. Biochemical and molecular characterization of a quercetinase from *Penicillium olsonii*. *Biochimie* **2008**, *90*, 781–789.

12. Merkens, H.; Kappl, R.; Jakob, R. P.; Schmid, F. X.; Fetzner, S. Quercetinase QueD of *Streptomyces sp* FLA, a monocupin dioxygenase with a preference for nickel and cobalt. *Biochemistry* **2008**, *47*, 12185–12196.

13. Merkens, H.; Fetzner, S. Transcriptional analysis of the queD gene coding for quercetinase of Streptomyces sp FLA. *FEMS Microbiol. Lett.* **2008**, *287*, 100–107.

14. Bowater, L.; Fairhurst, S. A.; Just, V. J.; Bornemann, S. *Bacillus subtilis* YxaG is a novel Fe-containing quercetin 2,3-dioxygenase. *FEBS Lett.* **2004**, *557*, 45–48.

15. Barney, B. M.; Schaab M. R.; LoBrutto, R.; Francisco, W. A. Evidence for a new metal in a known active site: Purification and characterization of an iron-containing quercetin 2,3-dioxygenase from *Bacillus subtilis*. *Protein Expression Purif.* **2004**, *35*, 131–141.

16. Gopal, B.; Madan, L. L.; Betz, S. F.; Kossiakoff, A. A. The crystal structure of a quercetin 2,3-dioxygenase from *Bacillus subtilis* suggests modulation of enzyme activity by a change in the metal ion at the active site(s). *Biochemistry* **2005**, *44*, 193–201.

17. Schaab, M. R.; Barney, B. M.; Francisco, W. A. Kinetic and spectroscopic studies on the quercetin 2,3-dioxygenase from *Bacillus subtilis*. *Biochemistry* **2006**, *45*, 1009–1016.

18. Baráth, G.; Kaizer, J.; Speier, G.; Párkányi, L.; Kuzmann, E.; Vértes, A. One metal-two pathways to the carboxylate-enhanced, iron-containing quercetinase mimics. *Chem. Commun.* **2009**, 3630–3632.

19. Kaizer, J.; Baráth, G.; Pap, J.; Speier, G.; Giorgi, M.; Réglier, M. Manganese and iron flavonolates as flavonol 2,4-dioxygenase mimics *Chem. Commun.* **2007**, 5235–5237.

20. Hiller, W.; Nishinaga, A.; Rieker, A. A simple model for the enzyme-substrate-complex of the quercetinase reaction—crystal structure of flavonolato-cobalt(III)-(salen). *Z. Naturforsch.* **1992**, *47b*, 1185–1188.

21. Nishinaga, A.; Numada, N.; Maruyama, K. Substrate anion cobalt(III) complex intermediate in model quercetinase reaction using cobalt schiff-base complex. *Tetrahedron Lett.* **1989**, *30*, 2257–2258.

22. Nishinaga, A., *Oxygenases and Model Sytems*, T. Funabiki,Ed., Kluwer Academic Publishers, Dordrecht, The Netherlands, 1997, pp 157–194.

23. Qi, R.; Fetzner, S.; Oakley, A. J. Crystallization and diffraction data of 1*H*-3-hydroxy-4-oxoquinoline 2,4-dioxygenase: a cofactor-free oxygenase of the alpha/beta-hydrolase family. *Acta Crystallogr., Sect. F.* **2007**, *63*, 378–381.

24. Steiner, R. A.; Frerichs-Deeken, U.; Fetzner, S. Crystallization and preliminary X-ray analysis of 1*H*-3-hydroxy-4-oxoquinaldine 2,4-dioxygenase from *Arthrobacter nitroguajacolicus Ru61a*: a cofactor-devoid dioxygenase of the alpha/beta-hydrolase-fold superfamily. *Acta Crystallogr., Sect. F.* **2007**, *63*, 382–385.

25. Fetzner, S. Oxygenases without requirement for cofactors or metal ions. *Appl. Microbiol. Biotechnol.* **2002**, *60*, 243–257.

26. Fischer, F.; Fetzner, S. Site-directed mutagenesis of potential catalytic residues in 1*H*-3-hydroxy-4-oxoquinoline 2,4-dioxygenase, and hypothesis on the catalytic mechanism of 2,4-dioxygenolytic ring cleavage. *FEMS Microbiol. Lett.* **2000**, *190*, 21–27.

27. Dai, Y.; Pochapsky, T. C.; Abeles, R. H. Mechanistic studies of two dioxygenases in the methionine salvage pathway of *Klebsiella pneumoniae*. *Biochemistry* **2001**, *40*, 6379–6387.

28. Frerichs-deeken, U.; Ranguelova, K.; Kappl, R.; Hüttermann, J.; Fetzner, S. Dioxygenases without requirement for cofactors, and their chemical model reaction: Compulsory order ternary complex mechanism of 1*H*-3-hydroxy-4-oxoquinaldine 2,4-dioxygenase involving general base catalysis by histidine 251 and single-electron oxidation of the substrate dianion. *Biochemistry* **2004**, *43*, 14485–14499.

29. Iwashina, T. The structure and distribution of the flavonoids in plants. *J. Plant Res.* **2000**, *113*, 287–299.

30. Pietta, P.-G. Flavonoids as antioxidants. *J. Nat. Prod.* **2000**, *63*, 1035–1042.

31. Cushine, T. P. T.; Lamb, A. J. Antimicrobal activity of flavonoids. *Int. J. Antimicrob. Agents* **2005**, *26*, 343–356.

32. Kaizer, J.; Balogh-Hergovich, É.; Czaun, M.; Csay, T.; Speier, G. Redox and nonredox metal assisted model systems with relevance to flavonol and 3-hydroxyquinolin-4(1*H*)-one 2,4-dioxygenase. *Coord. Chem. Rev.* **2006**, *250*, 2222–2233.

33. Pap, J. S.; Kaizer, J.; Speier, G. Model Systems for the CO-Releasing Flavonol 2,4-Dioxygenase Enzyme. *Coord. Chem. Rev.* **2010**, *254*, 781–793.

34. Steiner, R. A.; Kalk, K. H.; Dijkstra, B. W. Anaerobic enzyme.substrate structures provide insight into the reaction mechanism of the copper-dependent quercetin 2,3-dioxygenase. *Proc. Natl. Acad. Sci. USA* **2002**, *99*, 16625–16630.

35. Antonczak, S.; Fiorucci, S.; Golebiowski, J.; Cabrol-Bass, D. Theoretical investigations of the role played by quercetinase enzymes upon the flavonoids oxygenolysis mechanism. *Phys. Chem. Chem. Phys.* **2009**, *11*, 1491–1501.

36. Brown, S. B.; Rajananda, V.; Holroyd, J. A.; Evans, E. G. V. A study of the mechanism of quercetin oxygenation by *O*-18 labeling a comparison of the mechanism with that of heme degradation. *Biochem. J.* **1982**, *205*, 239–244.

37. Siegbahn, P. E. M. Hybrid DFT study of the mechanism of quercetin 2,3-dioxygenase. *Inorg. Chem.* **2004**, *43*, 5944–5953.

38. Fiorucci, S.; Golebiowski, J.; Cabrol-Bass, D.; Antonczak, S. Molecular simulations bring new insights into flavonoid/quercetinase interaction modes. *Proteins* **2007**, *67*, 961–970.

39. Nishinaga, A.; Tojo, T.; Tomita, H.; Matsuura, T. Base-catalyzed oxygenolysis of 3-hydroxyflavones. *J. Chem Soc., Perkin Trans. I.* **1979**, 2511–2516.

40. Nishinaga, A.; Matsuura, T. Base-catalyzed autoxidation of 3,4'-dihydroxyflavone. *J. Chem. Soc., Chem. Commun.*, **1973**, 9–10.

41. Barhács, L.; Kaizer, J.; Speier, G. Kinetics and mechanism of the oxygenation of potassium flavonolate. Evidence for an electron transfer mechanism. *J. Org. Chem.* **2000**, *65*, 3449–3452.

42. Balogh-Hergovich, É.; Speier, G. Oxidation of 3-hydroxyflavones in the presence of copper(I) and copper(II) chlorides. *J. Mol. Catal.* **1992**, *71*, 1–5.

43. Balogh-Hergovich, É.; Speier, G. Kinetics and mechanism of the base-catalyzed oxygenation of flavonol in DMSO–H₂O solution. *J. Org. Chem.* **2001**, *66*, 7974–7978.

44. Barhács, L.; Kaizer, J.; Speier, G. Kinetics and mechanism of the stoichiometric oxygenation of the zinc(II) flavonolate complex [Zn(fla)(idpa)]ClO₄ (fla = flavonolate; idpa = 3,3'- iminobis(N,N-dimethylpropylamine)). *J. Mol. Catal. A: Chem.* **2001**, *172*, 117–125.

45. Balogh-Hergovich, É.; Kaizer, J.; Speier, G.; Argay, G.; Párkányi, L. Kinetic studies on the copper(II)-mediated oxygenolysis of the flavonolate ligand. Crystal structures of [Cu (fla)₂] (fla = flavonolate) and [Cu(O-bs)₂(py)₃] (O-bs = O-benzoylsalicylate). *J. Chem. Soc., Dalton Trans.* **1999**, 3847–3854.

46. Balogh-Hergovich, É.; Kaizer, J.; Speier, G.; Fülöp, V.; Párkányi, L. Quercetin 2,3-dioxygenase mimicking ring cleavage of the flavonolate ligand assisted by copper. Synthesis and characterization of copper(I) complexes [Cu(PPh₃)₂(fla)] (fla = flavonolate) and [Cu(PPh₃)₂(O-bs)] (O-bs = O-benzoylsalicylate). *Inorg. Chem.* **1999**, *38*, 3787–3795.

47. Balogh-Hergovich, É.; Speier, G.; Argay, G. The oxygenation of flavonol by copper(I) and copper(II) flavonolate complexes - the crystal and molecular-structure of bis(flavonolato) copper(II). *J. Chem. Soc., Chem. Commun.* **1991**, 551–552.

48. Speier, G.; Fülöp, V.; Párkányi, L. Chelated flavonol coordination in flavonolatobis (triphenylphosphine)copper(I). *J. Chem. Soc., Chem. Commun.* **1990**, 512–513.

49. Speier, G., *Bioinorganic Chemistry of Copper*, K. D. Karlin and Z. Tyeklár, Eds., Chapman & Hall, New York, **1993**, pp 382–394.

50. Balogh-Hergovich, É.; Kaizer, J.; Speier, G.; Huttner, G.; Zsolnai, L. Copper-mediated oxygenation of flavonolate int he presence of a tridentate N-ligand. Synthesis and crystal structures of [Cu(fla)(idpaH)]ClO₄ and [Cu(idpaH)(O-bs)]ClO₄, [fla = flavonolate, idpaH = 3,3'- iminobis(N,N-dimethylpropylamine)]. *Inorg. Chim. Acta* **2000**, *304*, 72–77.

51. Kaizer, J.; Pap, J.; Speier, G.; Párkányi, L. The reaction of $\mu-\eta^2$: η^2-peroxo- and bis (μ-oxo)dicopper complexes with flavonol. *Eur. J. Inorg. Chem.* **2004**, 2253–2259.

52. Kaizer, J.; Goger, S.; Réglier, M.; Giorgi, M. (O-benzoylsalicylato)copper(II) complexes as synthetic enzyme-product models for flavonol 2,4-dioxygenase. *Inorg. Chem. Commun.* **2006**, *9*, 251–254.

53. Kaizer, J.; Pap, J.; Speier, G.; Réglier, M.; Giorgi, M. Synthesis, properties and crystal structure of a novel 3-hydroxy-(4H)-benzopyran-4-one containing copper(II) complex, and its oxygenation and relevance to quercetinase. *Trans. Met. Chem.* **2004**, *29*, 630–633.

54. Balogh-Hergovich, É.; Kaizer, J.; Speier, G.; Huttner, G.; Jacobi, A. Preparation and oxygenation of (flavonolato)copper isoindoline complexes with relevance to quercetin dioxygenase. *Inorg. Chem.* **2000**, *39*, 4224–4229.

55. Balogh-Hergovich, É.; Kaizer, J.; Pap, J.; Speier, G.; Huttner, G.; Zsolnai, L. Copper-mediated oxygenolysis of flavonols via endoperoxide and dioxetan intermediates;

Synthesis and oxygenation of [Cu^{II}(phen)$_2$(fla)]ClO$_4$ and [Cu^{II}(L)(fla)$_2$] [flaH = flavonol; L = 1,10-phenanthroline (phen), 2,2'-bipyridine (bpy), N,N,N',N'-tetramethylethylenediamine (tmeda)] complexes. *Eur. J. Inorg. Chem.* **2002**, 2287–2295.

56. Lippai, I.; Speier, G.; Huttner, G.; Zsolnai, L. (2,2'-bipyridine)(flavonolato)copper(II) perchlorate, [Cu(bpy)(fla)]ClO$_4$. *Acta Crystallogr., Sect. C* **1997**, *53*, 1547–1549.

57. Bellamy, L. M. *Ultrarot-Spektrum und chemische Konstitution*, Dr. D. Steinkopff Verlag, Darmstadt, Germany, 1966, p 12.

58. Addison, A. W.; Rao, T. N.; Reedijk, J.; van Rijn, J.; Verschoor, G. C. Syntheses, structure, and spectroscopic properties of copper(II) compounds containing nitrogen sulfur donor ligands- the crystal and molecular-structure of aqua[1,7-bis(N-methylbenzimidazol-2'-yl)-2,6-dithiaheptane]copper(II) perchlorate. *J. Chem. Soc., Dalton Trans.* **1984**, 1349.

59. Kaizer, J.; Pap, J.; Speier, G.; Párkányi, L. Crystal structure of aqua[1,3-bis(2-pyridylimino)isoindolinato](salicylato)copper(II). *Z. Kristallogr. NCS.* **2004**, *219*, 141–142.

60. Halfen, J. A.; Mahapatra, S.; Wilkinson, E. C.; Kaderli, S.; Young, V. G.; Que, L.; Zuberbuehler, A. D.; Tolman, W. B. Reversible cleavage and formation of the O−O bond within a dicopper complex. *Science* **1996**, *271*, 1397–1400.

61. Mahapatra, S.; Halfen, J. A.; Wilkinson, E. C.; Que, L., Jr., ; Tolman, W. B. Modeling copper-dioxygen reactivity in proteins - aliphatic C−H bond activation by a new dicopper (II)−peroxo complex. *J. Am. Chem. Soc.* **1994**, *115*, 9785–9786.

62. Tolman, W. B. Making and breaking the dioxygen O−O bond: New insight from studies of synthetic copper complexes. *Acc. Chem. Res.* **1997**, *30*, 227–237.

63. Lewis, A.; Tolman, W. B. Reactivity of dioxygen–copper systems. *Chem. Rev.* **2004**, *104*, 1047–1076.

64. Lippai, I.; Speier, G.; Huttner, G.; Zsolnai, L. Crystal and molecular structure of a ketocarboxylatocopper(II) intermediate in the oxygenation of a copper(I) flavonolate complex. *Chem. Commun.* **1997**, 741–742.

65. Kaizer, J.; Ganszky, I.; Speier, G.; Rockenbauer, A.; Korecz, L.; Giorgi, M.; Réglier, M.; Antonczak, S. Cerium(IV)-mediated oxidation of flavonol with relevance to flavonol 2,4-dioxygenase. Direct evidence for spin delocalization in the flavonoxy radical. *J. Inorg. Biochem.* **2007**, *101*, 893–899.

66. Pap, J.; Kaizer, J.; Speier, G. DPPH-initiated oxygenation of 3-hydroxyflavone to *O*-benzoylsalicylic acid. *React. Kinet. Catal. Lett.* **2005**, *85*, 115–121.

67. Kaizer, J.; Speier, G. Radical-initiated oxygenation of flavonols by dioxygen. *J. Mol. Catal. A: Chem.* **2001**, *171*, 33–36.

68. Kaizer, J.; Kupan, A.; Pap, J.; Speier, G.; Michel, G. Crystal structures of [3,3'-iminobis (N,N-dimethylpropylamine)](flavonolato)zinc(II) perchlorate. *Z. Kristallogr. NCS.* **2000**, *215*, 571–572.

69. Speier, G.; Tyeklár, Z.; Toth, P.; Speier, E.; Tisza, S.; Rockenbauer, A.; Whalen, A. M.; Alkire, N.; Pierpont, C. G. Valence tautomerism and metal-mediated catechol oxidation for complexes of copper with 9,10-phenanthrenequinone. *Inorg. Chem.* **2001**, *40*, 5653–5659.

70. Barhács, L.; Kaizer, J.; Pap, J.; Speier, G. Kinetics and mechanism of the stoichiometric oxygenation of [Cu^{II}(fla)(idpa)]ClO$_4$ [fla = flavonolate, idpa = 3,3'-iminobis(N,N-

dimethylpropylamine)] and the $[Cu^{II}(fla)(idpa)]ClO_4$-catalyzed oxygenation of flavonol. *Inorg. Chim. Acta* **2001**, *320*, 83–91.

71. Balogh-Hergovich, É.; Kaizer, J.; Speier, G. Carboxylate enhanced reactivity in the oxygenation of copper flavonolate complexes. *J. Mol. Catal. A: Chem.* **2003**, *206*, 83–87.

72. Lippai, I.; Speier, G. Quercetinase model studies. The oxygenation of flavonol catalyzed by a cationic 2,2′-bipyridine copper(II) flavonolate complex. *J. Mol. Catal. A: Chem.* **1998**, *130*, 139–148.

73. Balogh-Hergovich, É.; Kaizer, J.; Speier, G. Synthesis and characterization of copper(I) and copper(II) flavonolate complexes with phthalazine ligand, and their oxygenation and relevance to quercetinase. *Inorg. Chim. Acta* **1997**, *256*, 9–14.

74. Balogh-Hergovich, É., Kaizer, J.; Speier, G. Kinetics and mechanism of the Cu(I) and Cu(II) flavonolate-catalyzed oxygenation of flavonol. Functional quercetin 2,3-dioxygenase models. *J. Mol. Catal. A: Chem.* **2000**, *159*, 215–224.

3

AMINE OXIDASE AND GALACTOSE OXIDASE

DALIA ROKHSANA, ERIC M. SHEPARD, AND DOREEN E. BROWN

Department of Chemistry and Biochemistry, Montana State University, Bozeman, MT

DAVID M. DOOLEY

University of Rhode Island, Green Hall, 35 Campus Avenue, Kingston, RI

Copper-Oxygen Chemistry, First Edition. Edited by Kenneth D. Karlin and Shinobu Itoh.
© 2011 John Wiley & Sons, Inc. Published 2011 by John Wiley & Sons, Inc.

3.1. INTRODUCTION

Mononuclear copper-containing oxidases are ubiquitous in nature and have diverse and critical roles in many organisms.[1-3] These enzymes often employ both copper ions and reactive organic cofactors, derived by post-translational modification of amino acids in the polypeptide chain of the enzyme. It is well established that such enzymes are activated by metal ions and molecular oxygen to generate these organic redox cofactors.[1,4] Copper appears to play multiple roles in the biogenesis of these novel cofactors and in the catalytic cycles of the oxidases. Substantial and impressive progress has been made in understanding the structures, mechanisms, and biological roles of copper-containing oxidases. However, there are still some underlying questions as to how these enzymes activate molecular oxygen for use as an oxidant. Herein, we describe two exemplary mononuclear copper proteins, galactose oxidase (GO) and amine oxidase (AO), which utilize and activate dioxygen (O_2) during both post-translational modification and catalysis.

Clear similarities between these enzymes have been documented in terms of their active-site structures and functions. First, both enzymes contain a copper ion and a protein derived cofactor: 2,4,5-trihydroxyphenylalanine quinone (TPQ) in amine oxidases and a Tyr-Cys cross-link (a tyrosine covalently cross-linked to a cysteine residue) in galactose oxidase.[5,6] Second, the protein derived cofactors result from post-translational modification of endogenous tyrosine residues, and following chemical modification the cofactors are involved in redox chemistry during catalysis.[5,7-9] Third, both enzymes perform similar $2e^-$ chemistry during the oxidation of substrate amine or alcohol to product aldehyde with the concomitant reduction of O_2 to hydrogen peroxide (H_2O_2). The H_2O_2 is believed to have vital physiological roles, particularly in cell signaling, and in inflammation and diseases of tissue in mammals.[10,11] Finally, the substrate oxidation mechanisms employed by AOs and GO appear to proceed through a Cu(I) intermediate, which will be discussed in Section 3.4.2.2 in great detail. In this chapter, a current understanding of O_2 activation, as well as the structure and function of AOs and GO will be discussed together with comparisons among them.

3.2. BASIC STRUCTURES

Structures of different copper-containing amine oxidases (CuAOs) and GO have been solved by X-ray crystallography.[12-20] These enzymes contain a mononuclear copper center with N/O donor ligands. Copper(II) is coordinated to three histidines and two water molecules (one axial, one equatorial) at the active site of AO and two histidines (H496 and H581), a unique Tyr-Cys cross-link, tyrosine (Y495), and an exogenous water molecule at the active site of GO (Fig. 3.1).

As observed in Figure 3.1, both enzymes contain water molecules in their active sites. An important aspect of these coordinated waters may be to accommodate flexibility during redox activation, allowing the metal atom to change coordination number without major effects on the protein structure.[21,22] For example, GO displays

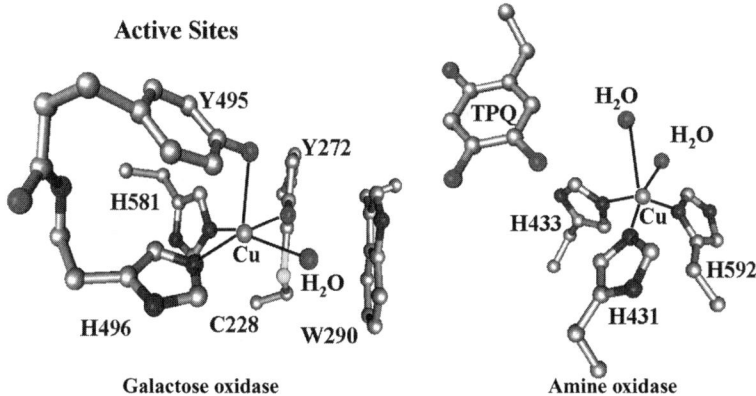

Active Sites

Galactose oxidase   Amine oxidase

FIGURE 3.1. Active-site structures of GO from *Hypomyces rosellus* (PDB ID: 1GOG) and AO from *Arthrobacter globiformis* (PDB ID: 1AV4). (See color insert.)

three distinct oxidation states: oxidized [Cu(II)−Tyr•-Cys] (green); semireduced [Cu(II)–Tyr-Cys] (blue); and reduced [Cu(I)–Tyr-Cys] (colorless). It is apparent that the coordination geometry is altered as a consequence of copper [Cu(II) to Cu(I)] and tyrosine (Tyr• to Tyr) redox state changes (see section 3.4.1.1), which is evident in their distinct color transformations (green → blue → colorless).

3.2.1. Galactose Oxidase Structure

Galactose oxidase (EC 1.1.3.9) is a 68-kDa monomeric fungal enzyme secreted, for example, by *Fusarium graminearum* into the extracellular environment.[23,24] The crystal structure of GO has been solved at 1.7-Å resolution (PDB ID: 1GOG).[14] Mature-GO is comprised of three distinct domains: (a) the N-terminal (domain I, 155 residues); (b) the catalytic (domain II, residues 155–552); (c) and the C-terminal (domain III, residues 553–639). The N-terminal domain forms a globular structure containing a divalent metal-binding site and a noncatalytic carbohydrate-binding site that may be involved in protein–protein interactions.[14] The catalytic domain is comprised of a sevenfold β-propeller structure that is capped on one face by the C-terminal domain III. Domain III donates a loop strand that passes through the axis of domain II and contributes the His581 ligand to the metal-binding site.[14]

The copper site of GO has five ligands arranged in a distorted square-pyramidal coordination (Fig. 3.1). The four equatorial ligands are Tyr272, His496, His581, and either H_2O (PDB ID: 1GOG) or an acetate ion (PDB ID: 1GOF). An unusual feature at the metal site is the presence of a Tyr-Cys cross-link, where the Cys228 sulfur is covalently linked to Tyr272. The side chain of Trp290 stacks over the Tyr-Cys cross-link with the aromatic ring of Trp290 positioned above the sulfur atom, resulting in equidistant carbon-to-sulfur distances of ∼ 3.8 Å.

Two partially processed forms of GO (identified as precursor GO (pro-GO, ∼ 70.2 kDa) and premature GO (premat-GO, ∼ 68.5 kDa) have been discovered in the growth media from a heterologous expression of the complete *Fusarium* goa gene

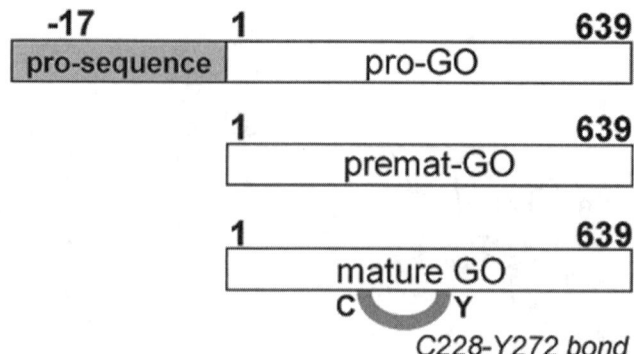

FIGURE 3.2. Schematic of the isolated forms of galactose oxidase. (Reprinted with permission from Ref. 26.)

under reduced copper conditions. As shown in Fig. 3.2, the pro-GO contains a 17-amino acid pro-sequence and no Tyr-Cys cross-link, while the premat-GO lacks both the pro-sequence and the Tyr-Cys cross-link. The presence of different forms of the GO protein indicates that it goes through a multistep post-translational modification process in order to yield the mature GO enzyme that contains both Cu and the Tyr-Cys cross-link (see the discussion in Section 3.3.2.1).

Crystal structures for the pro-GO (PDB ID: 1K3I)[25] and premat-GO (PDB ID: 2VZ1)[26] enzyme forms show remarkable overall structural similarities. However, there are local structural differences between the precursor forms and the mature GO (Fig. 3.3), which delineate the changes that occur during protein maturation. Only the main differences between pro-GO and mature-GO will be discussed in this section. In pro-GO, some active-site residues show significant rearrangements of side chains and of adjacent main-chain regions relative to mature GO. Between the two structures, one of the main-chain regions (residues 216–227) differs significantly by up to 6.1 Å (Fig. 3.3). At the end of this strand is Cys228, the cysteine of the Tyr-Cys cross-link.

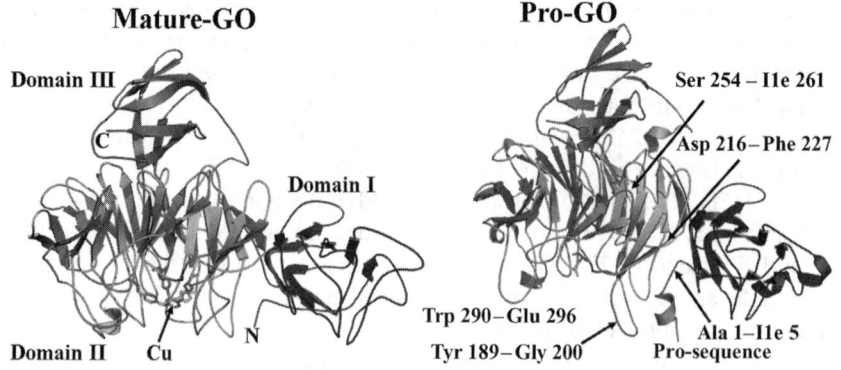

FIGURE 3.3. Structures of the mature form and the precursor (pro-GO) form of GO. Arrows in pro-GO delineate loop regions that differ between the two structures.

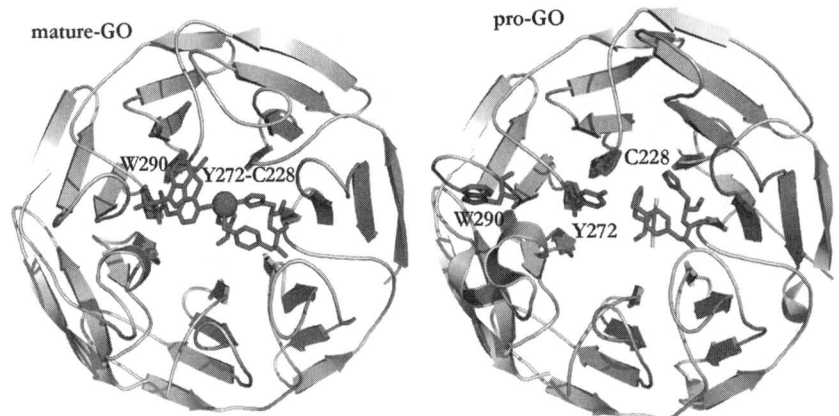

FIGURE 3.4. Orientation of the active-site residues in mature-GO and pro-GO. Protein backbone is colored in gray. The active-site residues W290, Y272, and C228 are represented as sticks. Copper is shown as a dark sphere (PDB ID: IGOG and 1K3I for mature-GO and pro-GO, respectively).

Another region affected is the loop containing the active-site residue Trp290, which in the mature protein stacks over the Tyr272 and the Tyr-Cys cross-link (Figs. 3.1 and 3.4). The Cα of Trp290 has moved by 6.3 Å in the pro-GO structure relative to its location in the mature protein. The backbone of Trp290 and Glu296 differs by up to 8.0 Å between these two structures and the electron density in this region is quite poor, suggesting high mobility or disorder.

At the copper-binding site, the loop rearrangement that affects Trp290 leads to a much more open conformation of the active site in pro-GO. The two residues forming the Tyr–Cys cross-link (Cys228 and Tyr272) in this structure are more exposed to solvent and are positioned differently from those in the mature form (Fig. 3.4). Three copper ligands (Tyr495, His496, and His581) have main-chain Cα movements of < 0.4 Å. The side chains of the other two copper ligands (Tyr495 and His581) are not significantly different from their mature positions, but the side chain of His496 shows larger deviation by rotating ∼ χ_1 by 37°. The electron density at Cys228 in pro-GO shows additional electron density, suggestive of oxidation to an S−OH group. It is uncertain whether the oxidation is an artifact resulting from crystallization or radiation damage or represents an intermediate in Tyr-Cys bond formation.[25]

3.2.2. Amine Oxidase Structure

Structures of CuAOs have been solved from *Arthrobacter globiformis* (AGAO),[20] pea seedlings (PSAO),[16] *Hansenula polymorpha* (HPAO),[17] *Escherichia coli* (ECAO),[19] *Pichia pastoris* (PPLO),[13] bovine serum (BSAO),[18] human vascular adhesion protein-1 (VAP-1),[15] and recombinant human kidney diamine oxidase (rhDAO).[12] All known amine oxidases are dimeric with molecular masses ranging from 140 to 240 kDa and contain one active site per monomer; the active site is comprised of a single copper ion and the 2,4,5-trihydroxyphenylalanine quinone (TPQ) cofactor

(a)   (b)

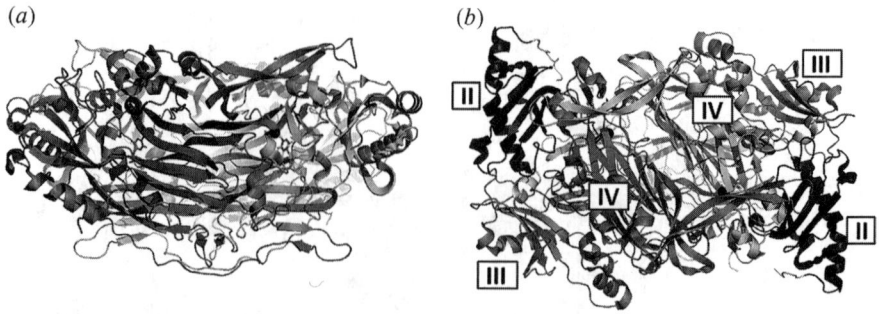

FIGURE 3.5. Secondary structure of rhDAO (PDB ID: 3HI7). The "front" view (a) shows each monomer (dark gray and gray). The "top" view (b) shows each domain.

(Fig. 3.5). Copper amine oxidases obtained from different sources share considerable structural homology and the Cu(II) site is highly conserved.[3] Each monomeric unit is composed of three or four domains with a large, C-terminal β-sandwich catalytic domain, designated as domain IV (Fig. 3.5). A β-ribbon from one monomer extends to the active-site channel in the opposite monomer to varying extents. Residues at the end of these β-strands form part of the active-site channel entrance and vary considerably in size and polarity among AOs (Fig. 3.6).[12,13,17] The presence of aspartate and/or glutamate residues in the active-site channel may be specifically linked to the recruitment of positively charged amine substrates to the active-site channel.[16,17,19]

In addition to the copper active sites of these enzymes, CuAOs also have peripheral second metal sites located ~ 30 Å from the copper site. These secondary metal sites are believed to be occupied by Ca(II) or Mn(II) under physiological conditions.[27] The function of the second metal-binding site has remained somewhat elusive; however, the second metal site has long been thought to play a role in either proper protein folding or dimer integrity.[27] Intriguingly, there is a direct structural link between the more highly conserved, solvent inaccessible second metal site and the active-site copper center. The ligands to the peripheral metal site lie on one end of an antiparallel β-strand that contains two of the three histidine ligands to copper on the opposite end of the β-strand. Recent results exploring the role of peripheral metal sites on k_{cat} and k_{cat}/K_M(amine) in ECAO clearly show that binding of certain divalent cations in the secondary sites increases catalytic activity.[28] Smith et al.[28] propose that these effects

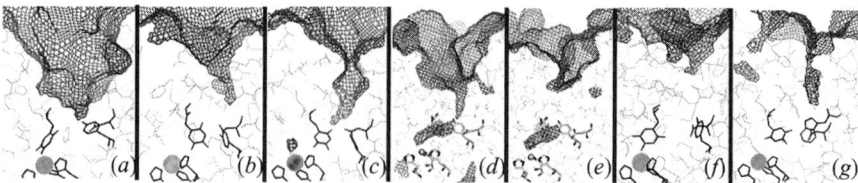

FIGURE 3.6. Solvent accessibility calculations (gray mesh) of CuAO crystal structures showing the amine substrate channel. Here (a) is PPLO, (b) is AGAO, (c) is PSAO, (d) is BPAO, (e) is HVAP, (f) is ECAO, and (g) is HPAO. (Adapted from Ref. 13.)

may be attributed to long-range conformational control of TPQ mobility via tweaking of β-strand residues resulting in alteration of active-site interactions. Another possibility relates to the efficient delivery of O_2 to the copper site, such that alterations in the divalent cation sites on the periphery of the enzyme could perturb the dynamics of O_2 entry through the hydrophobic, funnel-shaped channel believed to be responsible for O_2 delivery to the active site.[28]

With regard to the Cu(II) environment and the overall protein topology, crystallographically characterized CuAOs are remarkably similar. In spite of the close three-dimensional (3D) structural similarities, a sequence alignment of the catalytic domains from bacterial and plant sources showed sequence identity of $< 35\%$.[29] A consequence of the overall low-sequence identity among all CuAOs is reflected in differences in TPQ solvent accessibility (Fig. 3.6) and in the amino acid composition and dimensions of the active-site channel, which likely contribute to substrate and inhibitor specificity.[13,29–32] The copper site is located ~ 10–$12\,\text{Å}$ from the protein surface in CuAOs.[5,13,33,34] As mentioned above, the Cu(II) ion is five coordinate with three equatorial histidine ligands and two water molecules, one bound equatorially and the other bound axially, to form a distorted square-pyramidal coordination geometry.[35] The TPQ cofactor is in close proximity to the Cu(II) ion and has been isolated in both "off"-copper (Fig. 3.7a) and "on"-copper (Fig. 3.7b) conformations, which differ by a $\sim 180°$ rotation about the $C\alpha$–$C\beta$ bond of TPQ.[3,20,36] These two conformations are believed to exist in equilibrium in solution.[37] The "off"-copper conformation (Fig. 3.7a) binds amine substrate for oxidation, optimally positioning the enzyme–substrate complex toward a catalytic-base residue.[20,38] In the "on"-copper conformation, Cu(II) is coordinated by the same three histidines and by O4 of the TPQ ring in a quasi-trigonal pyramidal geometry.[20] Originally, the "on"-copper structure was hypothesized to be inactive, a designation that may be misleading in light of the possibility that the "on"- and "off"-copper forms may rapidly equilibrate in solution at physiological temperatures.

A crystal structure of apo-AGAO also has been resolved at 2.2-Å resolution and shows the three histidine ligands and the precursor tyrosine residue primed for Cu(II)

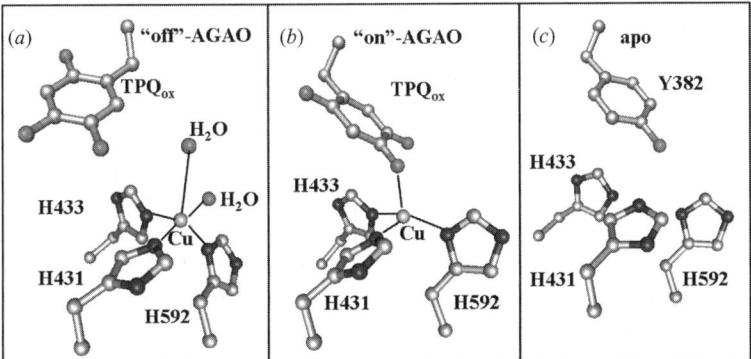

FIGURE 3.7. Active-site representations of AGAO. Here (a) is "off"-copper, (b) is "on"-copper, and (c) are the apo states.

coordination (Fig. 3.7c).[20] All three of these AO structures (Fig. 3.7), possess an overall similar structural fold within the limit of precision; the only significant differences among these forms are limited to the active-site residues. The apo-AGAO structure reveals four endogenous ligands, His431, His433, His592, and an unmodified Tyr382, arranged around the empty metal site in a way that metal can bind to these residues in a tetrahedral fashion resembling the active-site geometry of the "on"-copper" form. The Tyr382 is located at a position close to that of the TPQ in the "on"-copper holo-AGAO and is converted to TPQ during biogenesis (Fig. 3.7c).

3.3. POST-TRANSLATIONAL MODIFICATION

3.3.1. Determination of Self-Processing

The post-translational modification of endogenous amino acid residues enormously expands the functional versatility of proteins. More specifically, post-translational modification may be required for enzymatic activity and may either be a self-processing event or require the participation of chaperone proteins or additional enzymes.[26] Many enzymes, including CuAOs, GO, lysyl oxidase (LOX), methylamine dehydrogenase, cytochrome oxidase, and catechol oxidase are known to contain post-translationally modified amino acids in their active sites.[39,40] In many of these cases, the cross-linked amino acid side chains generate redox active cofactors, such as those identified in galactose oxidase (Tyr-Cys), lysyl oxidase (Tyr-Lys), amine dehydrogenase (Trp-Trp), and catechol oxidase (His-Cys).[39] Another subset of post-translationally modified amino acids are quinones, derived from the oxidation of either tyrosine or tryptophan residues.[41,42] Quinone cofactors, such as TPQ in CuAOs, lysine tyrosyl quinone (LTQ) in copper lysyl oxidases, tryptophan tryptophylquinone (TTQ) in bacterial methylamine dehydrogenase, and cysteine tryptophyl quinone (CTQ) of bacterial quinohemoprotein amine dehydrogenase have been identified.[39]

3.3.2. Mechanistic and Structural Data on Biogenesis

3.3.2.1. Tyr-Cys Biogenesis in GO. As discussed earlier, GO can be isolated in various forms, which suggests that the formation of mature-GO proceeds through a series of post-translational modifications: the removal of both the signal sequence and the N-terminal pro-sequence, the formation of the Tyr-Cys cross-link followed by generation of the Tyr-Cys radical by a 1e⁻ oxidation.[26,43] Several investigations have been conducted to help elucidate the mechanism of GO biogenesis by trapping the kinetic and structural intermediates during the reaction.[26,43,44] The maturation of pro-GO, as first reported by Rogers et al.,[43] requires only copper and molecular oxygen. When copper was added to pro-GO under aerobic conditions, two absorption bands were initially observed in the visible region at 410 and 750 nm. These intermediate transitions subsequently shifted to the characteristic bands ($\lambda_{max} = 445$ and 800 nm) of the mature protein [Cu(II)–Tyr•-Cys], and protein maturation was

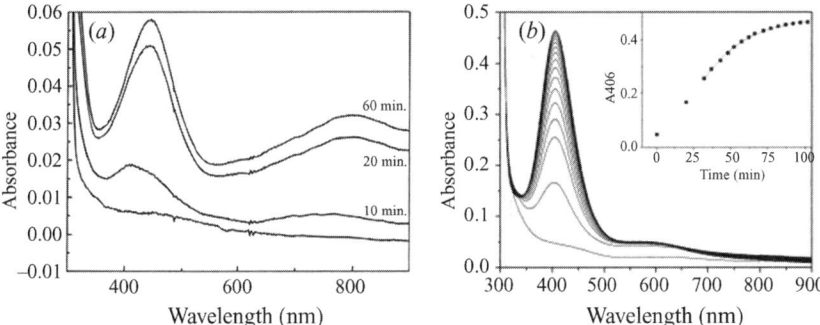

FIGURE 3.8. (*a*) Absorbance spectral changes accompanying the aerobic addition of copper to "metal-free" unprocessed pro-GO. The 60-min spectrum corresponds to fully processed GO. (*b*) Formation of the anaerobic Cu(II)–premat-GO complex. The formation of the 406-nm species over time is shown in the inset. (Reprinted with permission from Refs. 26, 43.)

completed within 1 h (Fig. 3.8*a*).[43,45] Rogers et al.[43] also observed that anaerobic conditions or other divalent metals, such as Ca(II), Mg(II), Mn(II), Zn(II), Co(II), and Mn(II), did not support the cleavage of the pro-sequence or cross-link formation. These results led to a series of investigations regarding the possible roles of the pro-sequence, Cu and O_2.[26,44] However, expression and isolation of substantial amounts of unprocessed, metal-free pro-GO has been a major challenge, resulting in the use of premat-GO (lacking the N-terminal 17-aa pro-sequence) for biogenesis studies. Whittaker et al.[44] demonstrated that formation of mature GO from premat-GO upon exposure of precursor protein to Cu and O_2 readily occurred, indicating that the N-terminal pro-sequence is not absolutely required for biogenesis of the Tyr-Cys cross-link.[44]

The biogenesis reaction has also been investigated with Cu(I) and Cu(II) to assess dependence on the metal oxidation state. Interestingly, it was reported that Tyr-Cys cross-link formation was 4700 times faster with Cu(I) ($t_{1/2} = 3.9$ s) than with Cu(II) ($t_{1/2} = 5.1$ h) under aerobic conditions and at a copper/protein ratio of 0.8:1.[44] Addition of excess copper did not affect the yield of the processed enzyme. However, the excess was reported to cause protein precipitation during maturation. The upper limit of copper required for maximal processing varied slightly from $\sim$ 0.5 to 0.8 equiv depending on the different enzyme preparations.[44] The Cu(I) initiated biogenesis required 1.8 mol of O_2 per mol of enzyme for full maturation and the resulting enzyme had a specific activity comparable to the wild-type enzyme.[44] Whittaker et al.[44] also reported that the rate of cross-link formation was also enhanced as a function of pH over the range 5.0–8.5. The increased rate observed under basic conditions was attributed to the existence of protonated–deprotonated states of the Cys228 thiol.[44]

Rogers et al.[26] reported the rapid maturation of Cu(II)–premat-GO under aerobic conditions with a 3.5-fold excess of copper. Rates of Tyr-Cys cross-link and Tyr•-Cys radical formation were 0.3 ± 0.05 min^{-1} [SDS–PAGE] and 0.21 ± 0.03 min^{-1} [ultraviolet–visible (UV–vis) based on A_{445}], respectively. These results suggested that the rate-limiting step during the formation of the oxidized, mature GO is the formation of the Tyr-Cys cross-link. Mature oxidized enzyme was obtained with a specific activity of $\sim 70\%$ of the wild-type enzyme. A stoichiometry of ~ 1.3 mol of O_2 reduced per mol of premat-GO was also reported. The lower amount of O_2 reduction with Cu(II) processing could be attributed to $1e^-$ reduction of Cu(II) to Cu(I) by a protein residue. It is interesting to note that the rate of aerobic maturation of premat-GO with Cu(II) appears to be slow ($t_{1/2} = 5.1$ h).[44] However, the maturation of pro-GO with Cu(II) can be fully achieved within 1 h.[43] This later result indicates that the N-terminal 17 amino acid residue pro-sequence may facilitate some step or steps in the biogenesis reaction. This hypothesis has not been fully explored.

The data clearly show that the precursor protein can be processed utilizing either Cu(I) or Cu(II) under aerobic conditions. However, based on the rapid Cu(I) dependent biogenesis, it was concluded that *in vivo* processing likely utilizes Cu(I) and a proposed biogenesis mechanism was formulated based on the kinetics, the O_2 stoichiometry, and the pH sensitivity of the reaction (Scheme 3.1).[44] Copper(I)

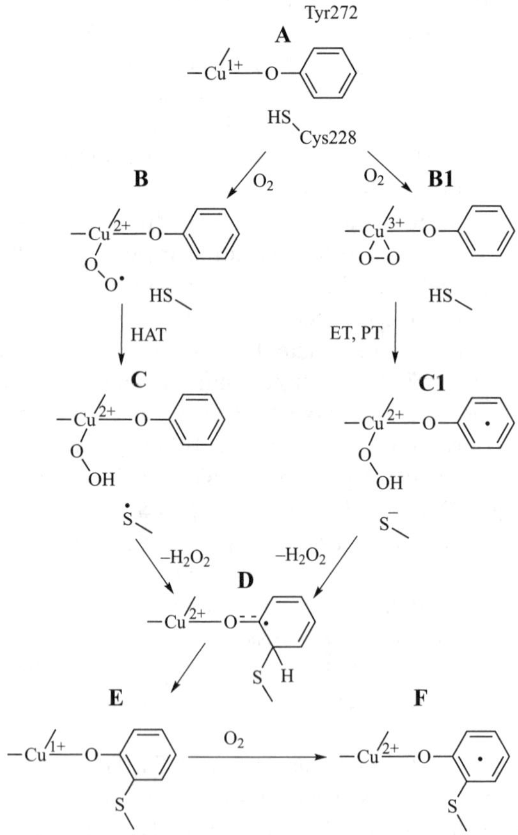

SCHEME 3.1. Proposed mechanism for the formation of the Tyr-Cys cross-link with Cu(I) in GO. (Adapted from Ref. 44.).

was proposed to bind at the active site of premat-GO with a three-coordinate geometry, and the binding was reported to be relatively strong as the processing was not affected by the presence of ethylenediaminetetraacetic acid (EDTA).[44] As discussed, Cu(I)–premat-GO utilizes two molecules of O_2 to generate the Tyr-Cys cross-link.[44] Dioxygen can preferentially bind to Cu(I), which has been observed in various model complexes,[46–51] and is proposed to bind to copper either in end-on or side-on fashion following exposure of the Cu(I)–premat-GO complex to O_2. End-on binding of O_2 to Cu(I)–premat-GO would yield a Cu(II)-superoxo species (Scheme 3.1**B**), whereas side-on binding of O_2 would generate a Cu(III)-peroxo species (Scheme 3.1**B1**). Variations in pH during the reaction can control the mechanism for the conversion of **B** → **C** and **B1** → **C1** in Scheme 3.1. For example, conversion of the Cu(II)-superoxo to the thiyl radical intermediate (**B** → **C**) can occur either through an electron transfer (ET) (at basic pH) or hydrogen-atom transfer (at neutral pH) mechanism. The phenoxyl pathway, conversion of **B1** → **C1** via an ET event followed by proton transfer, was hypothesized to be less likely due to the lack of an intense absorption at 410 nm that would be characteristic of a phenoxyl radical species.[52] Regardless, loss of H_2O_2 from either species **C** or **C1** results in a Cu(II)–tyrosyl based radical moiety (Scheme 3.1**D**). The Cu(II)–tyrosyl radical is assumed to convert to the fully reduced [Cu(I)–Tyr-Cys] species via proton transfer, which would then utilize a second O_2 molecule to generate the fully processed, catalytically competent [Cu(II)–Tyr•-Cys] cofactor.[44]

Anaerobic spectroscopic and crystallographic freeze-trap experiments were also performed to trap the initial biogenesis intermediates of premat-GO with Cu(II) prior to introduction with O_2.[26,44] Anaerobic addition of Cu(II) to premat-GO resulted in several distinct intermediates over time.[44] The most dominant, later-stage species was attributed to a Cu(II)–thiolate complex ($\lambda_{max} = 406$ nm, and $k_{406} = 3.2 \times 10^{-4}$ s^{-1}), as supported by resonance Raman (RR) assignments.[44] The formation of the yellow Cu(II)–premat-GO complex was also confirmed by Rogers et al.[26] with a similar rate of formation ($k_{406} = 6.0 \times 10^{-4}$ s^{-1}) (Fig. 3.8b). The spectroscopic evidence [circular dichroism (CD) and (RR)] of the yellow, 406 nm Cu(II)–premat-GO complex suggests that copper is coordinated by nitrogen and sulfur of His and Cys, respectively.[26,44,53] Similar optical absorption features have been observed in four helix bundle proteins containing a copper center coordinated by these ligands,[54,55] as well as in variants of blue copper proteins.[56] A three-coordinate copper site in the Cu(II)–premat-GO complex is further supported by the determination of a freeze-trap structure obtained by anaerobic soaking of the apo premat-GO crystal (Fig. 3.9a) with Cu(II) for ~ 3 min (Fig. 3.9b).[26] Comparison of the apo-GO and 3-min copper-soaked Cu(II)–premat-GO structures revealed that at the initial stage of the biogenesis reaction, Cys228 has moved from its premat-GO position to coordinate to copper and the plane of the Tyr272 ring is rotated by $\sim 90°$. The rotation of the Tyr272 ring allows the π-system of tyrosine to overlap with the S orbital of Cys228, which is essential for the cross-linking reaction to progress. This complex was believed to be the yellow, 406 nm Cu(II)–premat-GO species observed during the solution studies.[26,44]

Intriguingly, the yellow Cu(II)–premat-GO complex was initially reported to be stable by Whittaker and Whittaker,[44] however, Rogers et al.[26] observed the decay of

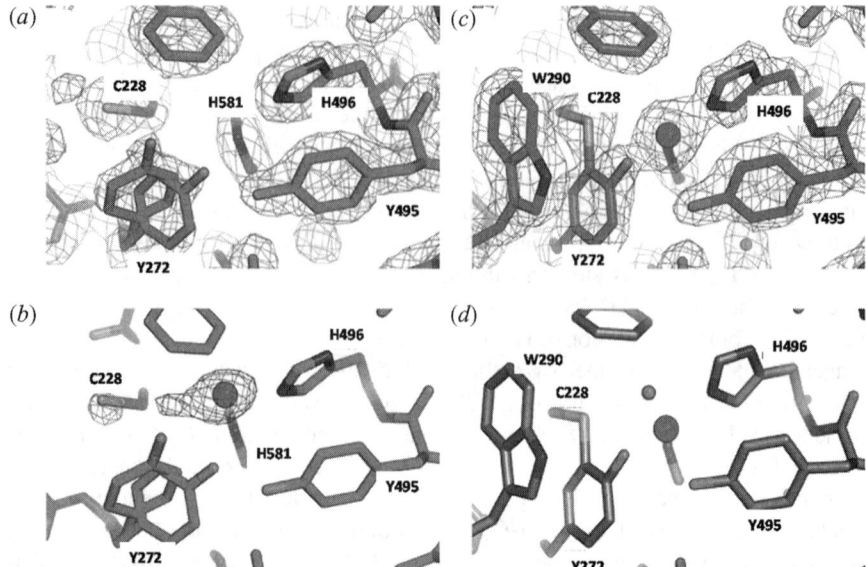

FIGURE 3.9. Crystal structures of the active site of GO processing forms. (*a*) Premat-GO prepared and crystallized under copper-free conditions. (*b*) After a 3-min Cu(II) anaerobic soak. (*c*) After a 24-h Cu(II) anaerobic soak. (*d*) Fully processed mature GO (PDB entry 1GOG). (Reprinted with permission from Ref. 26.)

this complex over time ($k_{406\text{decay}} = 0.003 \pm 0.0003$ min^{-1}). The presence of excess Cu (3.5-fold) was discovered to enhance the decay rate of the Cu$^{(II)}$–premat-GO complex ($k_{406\text{decay}} = 0.245 \pm 0.022$ min^{-1}). A bleached species with weak absorption bands at ~ 430 and 625 nm were observed as the product of the decay reaction. In the presence of 3.5 equiv of Cu(II), $\sim 80\%$ of Tyr-Cys cross-link formation occurred during the bleaching of the Cu(II)–premat-GO complex and EPR copper spin quantification showed that only 1 M equiv of Cu(II) was reduced in this sample.[26] This observation implied that only 1 equiv of Cu(II) is required for Tyr-Cys cross-link formation that is bound at the active site and Cu can act as an electron acceptor in the absence of O$_2$.[26] This conclusion was further supported by the structural analysis, which revealed that allowing longer exposure to copper (from 3 min to 24 h) enabled the initial 3-min structure representing the Cu(II)–premat-GO complex (Fig. 3.9*b*) to process into a structure resembling holo-GO, where the sulfur of Cys228 no longer coordinates to Cu and the Tyr-Cys cross-link is present (Fig. 3.9*c*) (PDB ID: 2VZ3).[26] Additionally, Tyr-Cys cross-link formation between Cu(II) coordinated cysteine and exogenously added tyrosine has been observed under anaerobic conditions in a model system.[57] This observation led to the conclusion that O$_2$ is not absolutely required for the formation of the Tyr-Cys cross-link, although the rate of Tyr-Cys formation is clearly significantly faster in the presence of O$_2$.

A mechanism for Tyr-Cys cross-link formation utilizing Cu(II) was proposed by Rogers et al.[26] (Scheme 3.2). In general, Tyr–Cys bond formation requires removal of

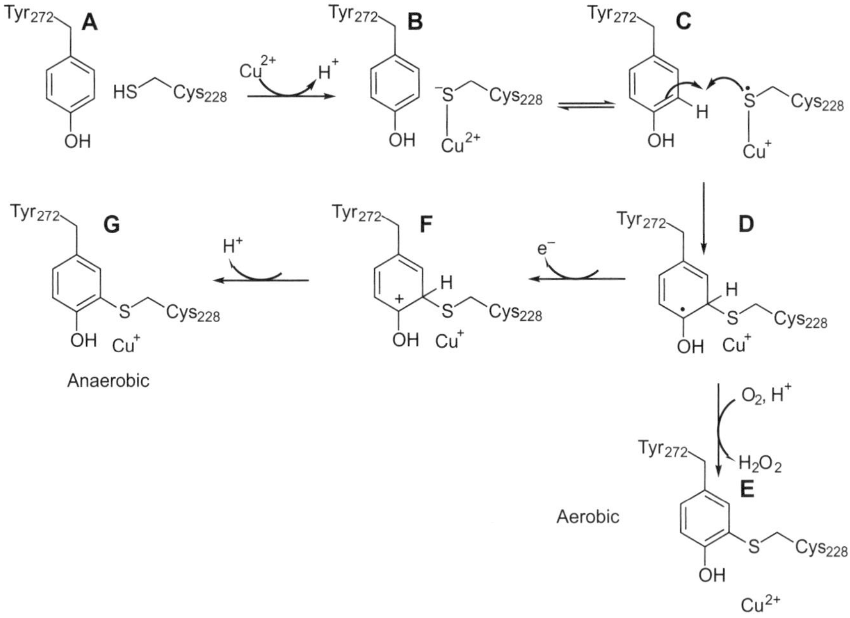

SCHEME 3.2. Proposed mechanism for the aerobic and anaerobic formation of the Tyr-Cys cross-link with Cu(II) in GO. (Adapted from Ref. 26.)

two electrons and two protons. Under both aerobic and anaerobic conditions, Cu(II) was proposed to bind to the deprotonated Cys228, forming a Cu(II)-thiolate species (Scheme 3.2**B**). An electron is transferred from thiolate to Cu(II), resulting in a Cu(I)–CysS• species (Scheme 3.2**C**). Subsequently, the cysteine radical can couple with an electron from the aromatic ring of Tyr272, resulting in a tyrosyl-based radical (Scheme 3.2**D**). Under aerobic conditions, species **D** can reduce O_2 while forming H_2O_2 with reoxidation of Cu(I) to Cu(II) (Scheme 3.2**E**). However, it is still unclear how O_2 binds to species **D** during the conversion of **D** to **E**, as there is no concrete evidence of any trapped copper-O_2 reactive intermediates to date. Under anaerobic conditions, species **D** can convert to species **F** by releasing a proton and an electron. However, the identity of the electron acceptor in this step is not resolved. Potential candidates for the electron acceptor, including C515 and C518, were ruled out by Rogers et al.[26] as the disulfide bond between these cysteine residues was intact in the crystal structure obtained after 24 h.

In summary, multiple biogenesis pathways of Tyr-Cys formation in GO exist depending, at least in part, on the selection of the metal oxidation state [Cu(II) vs Cu(I) pathways]. As suggested by Whittaker and Whittaker,[44] Cu(I) mediated biogenesis may represent the preferred *in vivo* mechanism based on substantially faster reaction kinetics and the intracellular availability of Cu(I) compared to Cu(II). It is interesting that processing also occurs with Cu(II), which may permit the formation of active GO under a variety of conditions. Under anaerobic conditions, Cu(II) binds with the premat-GO in a three-coordinate geometry with sulfur and nitrogen ligation. This result was confirmed from the crystal structure, as well as by CD and RR spectroscopy. Anaerobically soaked apo-GO crystals with Cu(II) for a long time period resulted in the formation of the Tyr-Cys cross-link.[26] Based on this evidence, Rogers et al.[26] proposed that excess Cu(II) can act as an oxidant in an anaerobic environment (and also under aerobic conditions) during biogenesis. However, convincing evidence of other intermediates during processing of the cross-link is still lacking in both Cu(II) and Cu(I) mediated pathways and further investigation is warranted to sort out all the important mechanistic details in the biogenesis pathways.

3.3.2.2. TPQ Biogenesis in AO. There are several features that are directly relevant to understanding the mechanism of cofactor biogenesis in CuAOs. First, the TPQ group and the Cu(II) ion are in close proximity, and in fact, TPQ can coordinate Cu(II), as shown in the "on"-copper form (Fig. 3.7*b*). Comparisons among the known AO structures indicate that the TPQ ring can adopt various conformations within the active site.[20] The presence of active-site water molecules (Fig. 3.7) implies that the copper ion can adopt different coordination geometries upon reduction, and this has been confirmed by X-ray absorption spectroscopy (XAS) [copper K-edge and extended X-ray absorption fine structure (EXAFS)] studies of the Cu(I) state.[58] Furthermore, histidine–water coordination is frequently observed for copper sites that react with O_2.[59]

The TPQ cofactor was first identified in 1990 by Klinman and co-workers[41] through mass spectral analysis of the BPAO phenylhydrazine adduct with subsequent confirmation for a variety of CuAOs through RR spectroscopy.[60] The TPQ is derived from a conserved tyrosine in the consensus sequence T/S-X-X-N-Y(TPQ)-D/E-Y/N and biogenesis is a self-processing reaction only requiring Cu and O_2. The mechanism of TPQ biogenesis has been studied extensively for AGAO[20,61–66] and HPAO.[67–74] The modification can proceed under aerobic conditions by addition of Cu(II) to apo-protein or exposing the anaerobic Cu(II) bound precursor protein to O_2. Conversion of tyrosine to TPQ is overall a $6e^-$ oxidation

$$E-Tyr + 2O_2 + H_2O \rightarrow E-TPQ + H_2O_2 \qquad (3.1)$$

process, and during the biogenesis reaction, 2 equiv of O_2 are consumed while producing 1 equiv of TPQ and H_2O_2 (Eq. 3.1).[66]

Evidence of intermediates during TPQ biogenesis have been observed using spectroscopic[67–70,73,74] and crystallographic[20,38,61,64,75] techniques. The colorless apo-AGAO is converted into pink holo-AGAO ($\lambda_{max} = 480$ nm for TPQ_{OX}) during the biogenesis reaction. Distinct spectral features of the initial and final states of the biogenesis reaction provide an excellent handle to monitor the reaction by trapping intermediates during crystallographic and solution studies, and analysis of these species has provided us with our current understanding of TPQ biogenesis.

The crystal structure of apo-AGAO revealed that the precursor Tyr382 is located within bonding distance from the vacant metal site and is presumably primed to bind Cu(II) and serve as an axial ligand (Fig. 3.7*c*).[20] This hypothesis was substantiated by two subsequent reports of Zn(II) substituted HPAO[75] and Co(II) substituted AGAO[64] structures in which the precursor tyrosine is coordinated to the respective metal centers. These results corroborate the initial hypothesis that the precursor tyrosine may be activated by Cu(II) during biogenesis (Scheme 3.3).[20,65] Concrete evidence of precursor tyrosine coordination to Cu(II) was provided by Kim et al.[61] in the freeze-trapped crystal structure of anaerobically soaked apo-AGAO with Cu(II). The anaerobic copper-bound AGAO structure is nearly identical to that of apo-AGAO, clearly showing copper bound to the vacant metal site with axial coordination from O4 of Tyr382 to the copper center (Fig. 3.10 and Scheme 3.3).[20,61] Interestingly, the structure showed a long Cu(II)-O4 of Tyr282 bond distance (2.5 Å) in this intermediate, a distance not expected to yield an intense ligand-to-metal charge transfer (LMCT) transition. This observation suggested that the precursor tyrosine remained protonated in the crystal and that deprotonation occurred producing the Cu(II)-tyrosinate species only after exposure to oxygen, supported by studies on both AGAO and HPAO (Scheme 3.3**B** → 3.3**C**).[65,67,73,74] Following exposure of the Cu(II) bound AGAO complex to O_2, rapid formation of mature TPQ (TPQ$_{OX}$; $\lambda_{max} = 480$ nm) was observed, and the reaction was complete within 2 min.[65]

Subsequent biogenesis studies in HPAO showed the formation of a spectral feature ($\lambda_{max} = 350$ nm) in the Cu(II) bound apo-HPAO under anaerobic conditions that decayed with isosbestic formation of TPQ$_{OX}$ following exposure to O_2.[67] The 350-nm species was assigned to that of a Cu$^{(II)}$–tyrosinate complex proposed to exist in equilibrium with a Cu(I)–tyrosyl radical.[67,68,73,74] Available data suggests that the precursor tyrosine and O_2 form a "collision complex", which is linearly dependent on the dissolved O_2 concentration.[64,68] The "collision complex" is proposed to subsequently isomerize to form the 350-nm Cu(II)–tyrosinate complex. This O_2 dependent process indicates that dioxygen binding promotes a conformational change in the active site that supports formation of the tyrosinate-copper species, although the detailed basis of this structural change is currently unknown. Two O_2 binding sites have been proposed to exist: either the vacant equatorial site of the copper center[20,66] or at an off-metal site adjacent to the Tyr(TPQ) precursor involving Met634.[74] Density

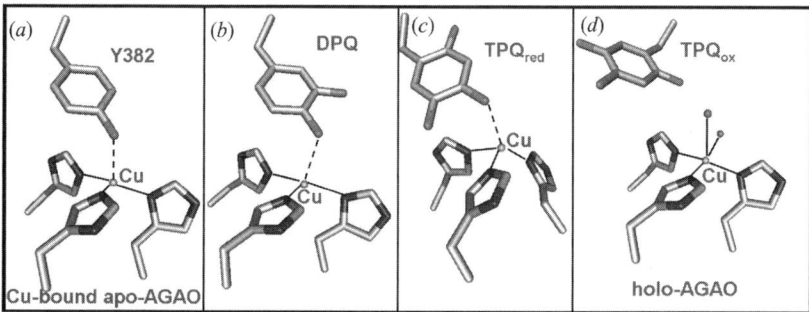

FIGURE 3.10. Crystal snapshots of intermediates during TPQ biogenesis. (*a*) The Tyr382 precursor (PDB ID: 1IVU), (*b*) dopaquinone (DPQ) (PDB ID: 1IVV), (*c*) 2,4,5-trihydroxyl-benzene (TPQred) form of the reduced TPQ (PDB ID:1IVW), and (*d*) the oxidized form of TPQ (TPQox) (PDB ID:1IVX).

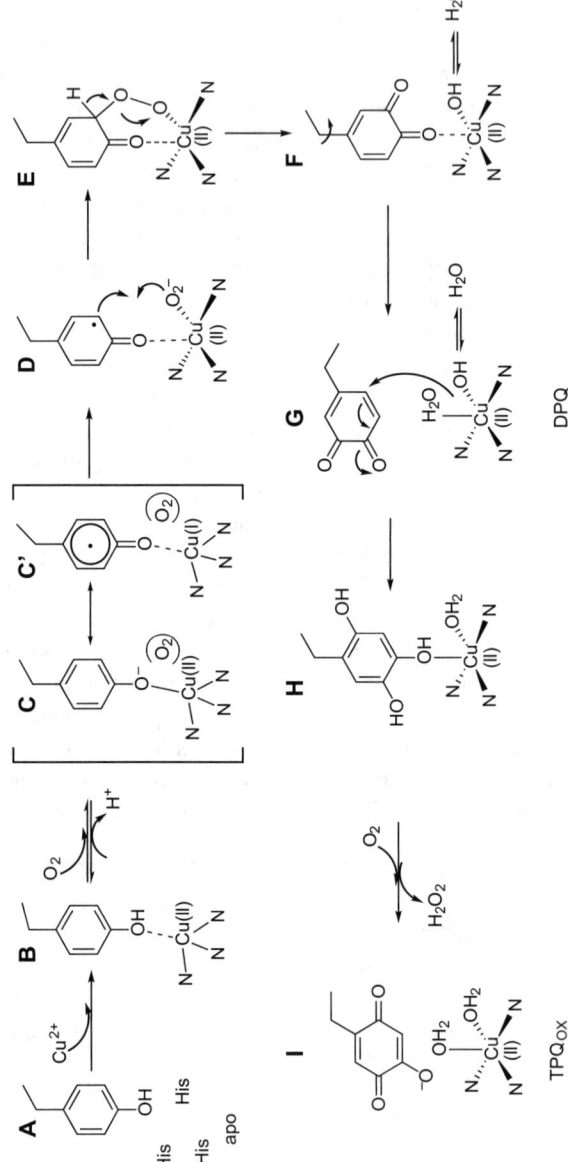

SCHEME 3.3. Proposed mechanism of TPQ biogenesis. (Adapted from Ref. 20.)

functional calculations concluded that the dioxygen-binding energy to the equatorial copper position is more favorable by 11.7 kcal/mol.[76]

Regardless of where O_2 actually binds, the event triggers activation of the phenol ring toward nucleophilic attack by O_2, where either full (HPAO) or partial (AGAO) charge transfer from the tyrosinate to copper may occur resulting in activation of the tyrosine carbon ring atoms.[20,62,67] It is then believed that O_2 is reduced to $O_2^{\bullet-}$ through the facile reaction with Cu(I), [20,66] although given the existence of the Cu(I)−Y$^{\bullet}$ resonance form, it should be stated that Cu(I) may not be the reactant species, and outer-sphere mechanisms for O_2 reduction from the phenolate complex have been proposed by Samuels and Klinman.[72] In either scenario, the superoxide species is then proposed to form a bond at the C3 position of the tyrosine ring by radical coupling, resulting in Cu(II)–aryl–peroxide intermediate formation (Scheme 3.3**D** → **E**). The observation that the rate of TPQ biogenesis in HPAO was identical when using ring-3,5-[$^2H_2$]-tyrosine substituted protein indicated that there was no isotope effect on C3−H bond cleavage, providing support that breakdown of the Cu$^{(II)}$–aryl-peroxide intermediate (Scheme 3.3**E**) into dopaquinone was not rate limiting and that the rate-determining step was associated solely with formation of the 350-nm intermediate species.[68]

An early intermediate trapped by Kim et al.[61] in the biogenesis of TPQ when apo-AGAO was exposed to an aerobic Cu(II) solution for a short period of time (~10 min) clearly showed that Tyr382 had been modified to either 3,4-dihydroxyphenylalanine or the oxidized quinone form (3,4-dihydroxyphenylalanine quinone or DPQ) (Fig. 3.10). In addition, a water–hydroxide molecule is found in this structure located ~2.1 Å from the Cu center in a similar position to the equatorial water position observed in holo-AGAO. This water–hydroxide species would be expected to readily exchange with solvent,[20] thereby permitting solvent oxygen incorporation into TPQ, as consistent with RR.[77]

Additional evidence for the DPQ intermediate was also observed from a site-directed mutagenesis study of AGAO.[78] Two residues in AGAO close to the C6 (C2 in TPQ$_{OX}$) position of dopaquinone (Fig. 3.10b) were each mutated to lysine residues in an attempt to trap a dopaquinone species in one of two ring-flipped orientations (Scheme 3.3**F** and 3.3**G**) and to additionally observe if a lysine tyrosylquinone-like species could be trapped. Lysine tyrosylquinone (LTQ) is the cofactor found in copper-containing (LOX) enzymes[79] Formation of the LTQ cofactor from the 1,4-addition of the ε-amino side chain of lysine to dopaquinone in a preorganized active site was shown to be possible through model studies.[80] The copper-free precursor form of recombinant LOX (apo-LOX) was isolated recently by Bollinger et al.[81] and the formation of LTQ was observed by extensive dialysis of apo-LOX against copper under aerobic conditions. From their studies, it was proposed that LTQ biogenesis is a self-catalyzed reaction requiring only Cu(II) and molecular oxygen, just as observed for TPQ biogenesis in CuAOs. Additionally, Moore et al.[78] showed that a LTQ-like cofactor formed in the AGAO D298K mutant. This result not only provides substantive support for the intermediacy of dopaquinone during TPQ biogenesis, but also indicates that the dopaquinone species is mobile and can clearly dissociate from the Cu(II) site (Scheme 3.3**F** → **G**). Collectively, these results support proposals that both TPQ and LTQ biogenesis proceed through a dopaquinone (DPQ) type intermediate.

Following rotation of the Cu(II) coordinated DPQ species away from the Cu center, C6 becomes positioned for nucleophilic attack by copper-bound hydroxide, a step that has been shown to come from solvent and not from molecular oxygen (Scheme 3.3**G**).[63] Kim et al.[61] also trapped the late intermediate during biogenesis by flash freezing anaerobic Cu(II)–AGAO exposed to an air-saturated solution for 100 min. This structure shows clear evidence for the formation of the reduced 2,4,5-trihydroxyphenylalanine species (Fig. 3.10c and Scheme 3.3**H**), where the C2 oxygen of the cofactor forms a hydrogen bond with the backbone carbonyl of Thr403. The reduced 2,4,5-trihydroxyphenylalanine moiety is oxidized to form the mature TPQ$_{OX}$ cofactor in holo-protein in the final stage of biogenesis (Scheme 3.3**H** → **I**) in a step requiring O$_2$ and producing H$_2$O$_2$.[66] The final product of biogenesis has also been obtained by exposing the anaerobic apo-AGAO crystal to an aerobic solution containing Cu(II) for 1 week.[61] The crystal turned pink, a visual indicator of the presence of the fully oxidized TPQ cofactor. The final stage structure clearly shows the oxidized cofactor in the productive conformation (dissociated away from Cu), with the Cu center in five-coordinate distorted square-pyramidal geometry (Fig. 3.10d, Scheme 3.3**I**).

For a number of years, Cu(II) was believed to be the only metal capable of supporting TPQ biogenesis, as a number of different divalent ions were found not to promote biogenesis.[82] However, a fortuitous observation by Tanizawa and co-workers[64] led to a reinvestigation of the ability of certain metals to support biogenesis in AGAO. Okajima et al.[64] showed that Co(II) and Ni(II) ions were also capable of supporting TPQ formation, although at rates that were 1200-fold slower than with Cu(II). The drastically reduced k_{obs} rates for biogenesis with Co(II) and Ni(II) substituted AGAO most likely explain why these ions were not identified in the initial study as being able to support TPQ formation. Furthermore, this report showed that out of several tested metals, only Cu(II), Co(II), Ni(II), and Zn(II) can tightly bind to the enzyme active site. Similar studies were performed with HPAO and it was shown that Ni(II), but not Co(II), supported biogenesis at a reported rate that was 120-fold slower than with wild-type, Cu(II) containing enzyme.[72]

These results raise some interesting issues in terms of the proposed biogenesis mechanism (Scheme 3.3), as these four metals share distinct Lewis acid and redox properties. At neutral pH, Cu(II) coordinated H$_2$O is 100-fold more dissociated than H$_2$O coordinated to either Co(II), Ni(II), or Zn(II), given the relative pK_a values for coordinated H$_2$O (7.5 for Cu(II)–H$_2$O).[64] The higher Lewis acidity for Cu may result in either more efficient tyrosine precursor ring activation (Scheme 3.3**C**) or the addition of metal-bound hydroxide to dopaquinone (Scheme 3.3**G**) and may therefore explain the rate enhancement observed for this metal. Along these lines, substitution of Cu(II) with Ni(II) in HPAO was discovered to affect the rate of the Michael addition step of the metal-bound hydroxide to DPQ in a manner that made this step partially rate limiting.[72]

Another possibility that may explain rate enhancement with Cu(II) relates to the ability of transition metals to promote O–O bond heterolysis of coordinated alkyl peroxides. This chemistry is promoted by Lewis acids and thus the replacement of

Cu(II) with metals that are weaker Lewis acids could be expected to have a detrimental effect on this metal-assisted cleavage step (Scheme 3.3**E**).[72] However, these issues alone do not explain the kinetic behavior observed with these metals, as Zn(II) fails to support biogenesis to any measurable extent. In terms of redox chemistry, if full charge transfer from the tyrosinate to Cu(II) indeed occurs (Scheme 3.3**C**′), then the facile reaction of Cu(I) with O_2 would be expected to occur readily. This chemistry, however, is only feasible with Cu(II), as Zn(II) is not capable of redox chemistry, and the $1e^-$ reduction of either Co(II) or Ni(II) is unlikely given the respective low reduction potentials of these metals. It therefore seems that a much less efficient and completely different mechanism may be operative in the cases of Co(II) and Ni(II) supported TPQ biogenesis. The mechanism put forth for Ni(II) substituted HPAO provides further insight into this alternate chemistry, whereby the principal rate-limiting step in Ni(II) supported biogenesis was proposed to be the initial outer-sphere ET from a Ni(II)–phenolate complex to O_2, forming a transient Ni(II)–tyrosyl radical species.[72]

These results collectively indicate that CuAO enzymes have tuned their active sites toward chemistry associated with Cu(II)/Cu(I). This finding may relate to the Lewis acidity of Cu, and to the ability of Cu to form a Cu(I)–tyrosyl radical intermediate, although direct detection of this intermediate has eluded investigators. Intriguingly, biogenesis rates in Cu(I) substituted HPAO were 17-fold slower than the wild-type Cu(II) enzyme and lacked the characteristic 350-nm species assigned to the Cu(II)–tyrosinate complex.[83] Electron paramagnetic resonance analysis of Cu(I)–HPAO samples following exposure to O_2 supported a mechanism in which the Cu(I) must first be oxidized by oxygen to yield a Cu(II)–superoxide species, with rate-limiting dissociation of the superoxide species then enabling tyrosine(O4) coordination to the copper center, thereby representing a point where the pathways for Cu(I) and Cu(II) supported biogenesis merge.[83] This work underscores two important ideas: (1) that O_2 can and does readily react with Cu(I) species in the amine oxidase active site, a topic that gains significance during the oxidative half-reaction of amine catalysis (see Section 3.4.2.3), and (2) that a critical component of rapid TPQ biogenesis is the activation of the tyrosine ring via coordination to the Cu center. While it appears that an outer-sphere mechanism can operate to support TPQ biogenesis in these enzymes, the wild-type Cu(II) enzymes from both HPAO and AGAO generate TPQ at rates that are two or three orders of magnitude faster, respectively, than their metal-substituted counterparts. While concrete evidence for O_2 reduction to superoxide through an inner-sphere mechanism during TPQ biogenesis (Scheme 3.3**C**′ → **D**) is still lacking, this mechanism may be the dominant factor contributing to copper's ability to rapidly oxidize tyrosine into TPQ.

3.4. CATALYTIC MECHANISM

3.4.1. Galactose Oxidase

Galactose oxidases contain a five-coordinate active site, utilizing both copper and a unique Tyr-Cys cross-link to carry out the $2e^-$ oxidation of a broad range of alcohols to aldehydes, with the concomitant reduction of O_2 to H_2O_2. The turnover reaction in GO occurs in two distinct half-reactions: substrate oxidation (Eq. 3.2) and O_2 reduction (Eq. 3.3).[84] These half-reactions are resolved in practice by eliminating one or the other reactants (substrate alcohol, O_2), allowing the enzyme to be reduced by its alcohol substrate anaerobically and subsequently reoxidized by O_2. This behavior is consistent with a ping–pong-type mechanism,[84,85] similar to CuAOs.

$$E_{ox} + RCH_2OH \rightarrow E_{red} + RCHO \qquad (3.2)$$

$$E_{red} + O_2 \rightarrow E_{ox} + H_2O_2 \qquad (3.3)$$

A detailed, proposed mechanism of GO catalysis has been established on the basis of extensive spectroscopic and computational work (Scheme 3.4).[9,85,86] However, none of the catalytic intermediates have been structurally characterized. The absence of detectable intermediates may be a consequence of the character of the overall reaction. Briefly, substrate binds at the equatorial position, replacing the coordinated H_2O from the oxidized [Cu(II)–Tyr•-Cys] form (Scheme 3.4A). Associated with this ligand-substitution reaction, a proton is transferred from the alcohol to the axial Y495 resulting in an equatorially coordinated, deprotonated alcohol (Scheme 3.4A → B). Next, a hydrogen atom is transferred from the substrate alcohol to the Tyr-Cys cross-link (Scheme 3.4B → C). The resulting substrate-derived ketyl radical is then oxidized through ET to the Cu center yielding Cu(I) and aldehyde (Scheme 3.4D).[9] During the oxidative half-reaction, Cu(I) and tyrosine are reoxidized by molecular oxygen, regenerating Cu(II) and tyrosyl radical and liberating H_2O_2 as product (Scheme 3.4D → A).[86]

3.4.1.1. Mechanistic Issues. There are several mechanistic issues still unresolved. First, the underlying question during the reductive half-reaction (Scheme 3.4A → D) is whether the mechanism proceeds by hydrogen-atom transfer (HAT) followed by ET, or proton-coupled electron transfer (PCET) during the transformations (Scheme 3.4B → D). In addition, it is unknown whether HAT and PCET proceed in a stepwise or concerted manner. Second, the detailed mechanistic steps involved in the oxidative half-reaction, the conversion of substrate reduced Cu(I) to the oxidized active state in the presence of O_2 (Scheme 3.4D → A), have not been fully elucidated.

As isolated, mature-GO is a mixture of two oxidation states: oxidized [Cu(II)–Tyr•-Cys] and semireduced [Cu(II)–Tyr-Cys]. The oxidized [Cu(II)–Tyr•-Cys] (Scheme 3.4A) form is the only form that reacts with substrate alcohol. The presence of a mixture of two oxidation states complicates the understanding of the kinetics of the first half-reaction. Computational investigations have provided some insight into both half-reactions, which to some extent correlate with the experimental observations such as proton transfer to the axial Tyr495 (Scheme 3.4B) and the rate-limiting HAT step (Scheme 3.4B → C).[87] However, the structure of the oxidized model was not reported (Scheme 3.4A), and the substrate-bound active-site model (Scheme 3.4B) was used to map the potential energy surface of the catalytic cycle.[87] In their model, one of the histidine residues was deprotonated to obtain the correct

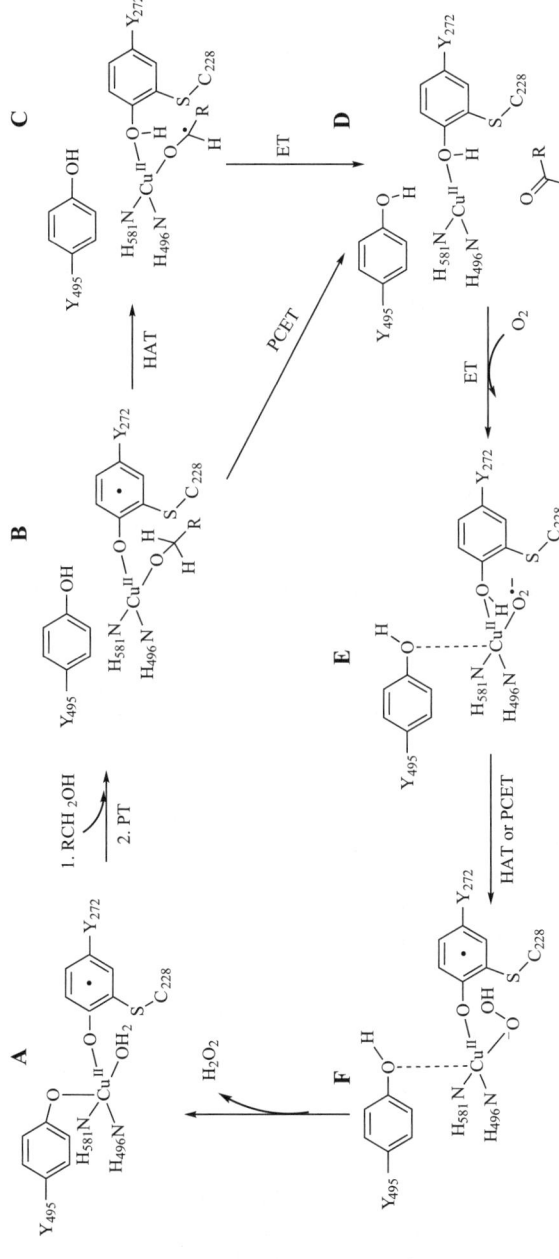

SCHEME 3.4. Proposed catalytic cycle in GO (PT = proton transfer, HAT = hydrogen-atom transfer, ET = electron transfer and PCET = proton-coupled electron transfer).

oxidation state of copper $(2+)$, which resulted in an overall neutral charge for the model. However, oxidized GO with two histidines and tyrosines at the active site contains an overall positive charge, and this has been supported by the observation that negatively charged ligands (e.g., N_3^- and CN^-) have higher binding affinity.[88] Further, the radical in the substrate-bound computational model (Scheme 3.4**B**) was proposed to localize at the axial Tyr495, instead of the equatorial Tyr-Cys cross-link,[87] but recent experimental and computational work provides strong support for the location of this radical to be on the Tyr-Cys cross-link in the oxidized form.[84,89–91] A brief overview of both half-reactions based on our current understanding follows.

3.4.1.2. Reductive Half-Reaction. The proposed catalytic cycle is shown in Scheme 3.4**A** → **D** starting with the oxidized active GO form (Scheme 3.4**A**). Evidence for the structure of the substrate-bound GO complex (Scheme 3.4**B**) is indirect as there is no crystal structure available for substrate-bound GO.[6] However, the substrate-binding site is well established based on ligand-binding studies to the copper center.[88,92] Azide binds at the equatorial position replacing the copper-coordinated water as evident by the X-band EPR data of water and azide bound as-isolated GO [$g_z = 2.27$ and $a_{cu} = 127$ G (H_2O) and $g_z = 2.24$ and $a_{cu} = 171$G (N_3^-)].[88] Interestingly, azide binding leads to the uptake of 0.8 protons per active site, which is attributed to the conversion of tyrosinate 495 to tyrosine.[88] It was shown that mutation of the Y495F resulted in an 1100-fold reduction of k_{cat}/K_M and, importantly, a loss of proton uptake on azide binding.[93] Resonance Raman studies of active and azide inhibited GO by McGlashen et al.[94] have supported the conclusion that contributions from the bound tyrosinate are lost on azide binding. Knowles et al.[92] also implicated a loss of a rapidly exchanging equatorial water in the presence of azide from nuclear magnetic relaxation dispersion (NMRD) measurements. These experiments, in the presence of the substrate dihydroxyacetone, look very similar to those with azide. Axial Y495 activates the bound substrate by deprotonation, setting the pathway for further substrate oxidation in the subsequent reactions.[88,95]

Substrate alcohol binding is also influenced by the stacking W290 residue.[96] Site directed mutants of W290X (X = G, H, F) have been investigated using kinetics and structural data. All three mutants show a similar coordination environment in the crystal structure.[96] However, a lower K_M for alcohol substrate for both W290G and W290F mutants coupled with a similar K_M for W290H relative to wild type indicate that the indole ring of W290 is critical for proper substrate positioning.[96] The stability of the Tyr-Cys radical is also influenced in the W290 variants, which is reflected in their redox potentials. Both W290H and W290G showed higher redox potentials compared to W290F. The effect of W290 in the radical stabilization of the Tyr-Cys cross-link also has been observed in the computational modeling of the active site.[91]

Once substrate alcohol is bound to the active site and deprotonated, subsequent oxidation of the copper-bound deprotonated alcohol is proposed to occur either via a hydrogen-atom transfer to the modified tyrosine radical followed by ET to copper or proton coupled ET.[85,86] On the basis of experiments with various inhibitors,

Branchaud and co-workers[97–99] suggested the possibility of a concerted mechanism in which the hydrogen atom and ET from substrate undergo simultaneous transfer to the Tyr-Cys radical and Cu, respectively. In support of this, a large kinetic isotopic effect (KIE) ($k_H/k_D = 22$) ascribed to hydrogen atom tunneling from substrate to the Tyr-Cys radical was observed in steady-state measurements using 1-O-methyl-α-D-galactopyranoside (rate of enzyme reduction, $k_{red} = 1.59 \times 10^4 \, M^{-1} \, s^{-1}$ for protio-substrate and $k_{red} = 7.5 \times 10^2 \, M^{-1} \, s^{-1}$ for deuterated substrate).[85,100]

Whittaker and Whittaker[86] compared the solvent (k_{H_2O}/k_{D_2O}) and substrate (k_H/k_D) KIE values for the reductive half-reaction using meta or para substituents of benzyl alcohol and concluded that the reaction proceeds in a single-step PCET. It was proposed that initial proton abstraction from a coordinated substrate activates the alcohol toward inner-sphere ET to the Cu(II) metal center in an unfavorable redox equilibrium, forming a substrate radical that undergoes hydrogen-atom abstraction by the Tyr-Cys cross-link to form the aldehyde product.[86] The presence of a substrate-derived radical intermediate has been supported by the demonstration of highly efficient inactivation of GO using a radical probe molecule that undergoes ring opening subsequent to reduction of the tyrosyl radical to tyrosine.[101]

When alcohol substrate was added to active GO under anaerobic conditions, the formation of a reduced intermediate was observed (Scheme 3.4D).[102,103] Following release of aldehyde product, the reduced enzyme has been suggested to exist as a three-coordinate Cu(I) center with N/O ligands. This structure has been supported by computational models and structural studies of GO.[91,102,103] A reduced Cu(I) site with this ligation environment is primed for direct reaction with O$_2$ during the oxidative half-reaction.

3.4.1.3. Oxidative Half-Reaction.

The oxidative half-reaction in the catalytic cycle of GO is less well understood (Scheme 3.4D → A). According to the current mechanism proposed in Scheme 3.4, dioxygen is proposed to bind the Cu(I) in the reduced enzyme, supposedly a three-coordinate Cu(I) center.[103] An electron is transferred from Cu(I) to bound O$_2$ through an inner-sphere mechanism, generating Cu(II)–superoxo (Scheme 3.4E). At this stage, Cu(II) bound superoxide abstracts a hydrogen atom from the phenolic hydroxyl group of the Tyr-Cys cofactor to produce hydroperoxide (Scheme 3.4F) and the Tyr-Cys radical. Finally, a proton is transferred from the axial tyrosine to metal-bound hydroperoxide to produce H$_2$O$_2$ and the fully reoxidized enzyme (Scheme 3.4A).[100] Overall, O$_2$ accepts two protons and two electrons from the active-site copper and tyrosines to produce H$_2$O$_2$ and the oxidized GO cofactor [Cu(II)–Tyr$^\bullet$-Cys].[100]

Molecular oxygen can preferentially bind to Cu(I), which has been observed in various model complexes.[46–51] As mentioned previously, the O$_2$ reduction chemistry in the GO active site is exclusively a two-electron reaction.[104] The structural basis for this controlled reactivity is still unclear, but the crystal structure of the resting enzyme provides a hint in the organization of solvent within the active site.[14] In GO, the active site contains a pair of solvent molecules, one coordinated to the copper ion and the other hydrogen bonded to the active-site tyrosine.[14] The position of the two water molecules could mimic the dioxygen-binding site. In the absence of other structural information, this suggests a model for a coordinated hydroperoxide product complex, which during turnover would subsequently be displaced by the protonation of the coordinated oxygen atom. Tyrosine residues (Y272 and Y495) at the active site are believed to be the proton donors for H_2O_2 production given their optimal position close to the copper center.[100,104] Dioxygen is also proven not to be essential for the oxidative half-reaction in GO.[105] Alternative oxidants, including hexacyanoferrate(III), can replace O_2 and support turnover under anaerobic conditions, although at a much lower rate than is found for the oxygen-driven reaction.[105]

Kinetic studies have been performed to gain additional insight into the oxidative half-reaction in GO.[85] A significant substrate KIE on the O_2 reduction step has been observed ($k_H/k_D = 8.7$) for 1-O-methyl-6,6′-di-[^2H]-β-D-galactopyranoside from V/K_M measurements as monitored by O_2 uptake.[85] This data implies that hydrogen transfer is fully rate limiting in the oxygen-driven half-cycle.[85] If substrate oxidation is complete when O_2 reduction occurs, there is no obvious reason for the latter reaction to be sensitive to substrate isotopic labeling. However, the two electrons and two protons involved in O_2 reduction ultimately originate in the substrate molecule, and deuterium abstracted from the substrate alcohol during the oxidative half-reaction may be retained in the active site between the two half-reactions.[85] The KIE observed on O_2 reduction could then arise as a transferred isotope effect.

Recently, Klinman and co-workers[100] investigated the oxidative half-reaction (Scheme 3.4**D** → **A**) under steady-state conditions by monitoring the O_2 consumption using 1-O-methyl-α-D-galactopyranoside as a substrate. The kinetic parameter for the oxidative half-reaction [$k_{cat}/K_M(O_2)$] was reported to be $2.5 \times 10^6 \, M^{-1} \, s^{-1}$. No statistically significant solvent isotope effect was observed for the reoxidation reaction. The absence of a solvent isotope effect led to the conclusion that hydrogen-atom transfer from the reduced cofactor to a Cu(II)-superoxo intermediate is fully rate determining for $k_{cat}/K_M(O_2)$ and that the active-site tyrosines are the source for the protons in H_2O_2 formation. Further, the ^{18}O KIE was reported to be 1.0185 with either protonated or deuterated sugar, providing evidence for the reactive superoxo species.[100] Similar isotope effects have been obtained for the dopamine β-monooxygenase (^{18}O KIE = 1.0214) and peptidylglycine α-hydroxylating monooxygenase (^{18}O KIE = 1.0212) family that utilize a similar copper-superoxo-type species.[100]

3.4.2. Amine Oxidase

Copper amine oxidases utilize the active-site TPQ cofactor to bind and oxidize a wide range of primary amines, from short- to long-chain aliphatic mono- and diamines, including multiple arylalkylamines.[17,106,107] The $2e^-$ oxidation of amines, with the subsequent reduction of O_2 to H_2O_2, proceeds through a ping–pong mechanism. Numerous biochemical studies have identified that AO catalysis proceeds through a reductive half-reaction wherein primary amines are oxidized to their corresponding aldehydes (Eq. 3.4, Scheme 3.5), and an oxidative half-reaction, in which reoxidation of the reduced TPQ cofactor by molecular O_2 yields production of H_2O_2 (Eq. 3.5, Scheme 3.6).[5,20,36,108,109]

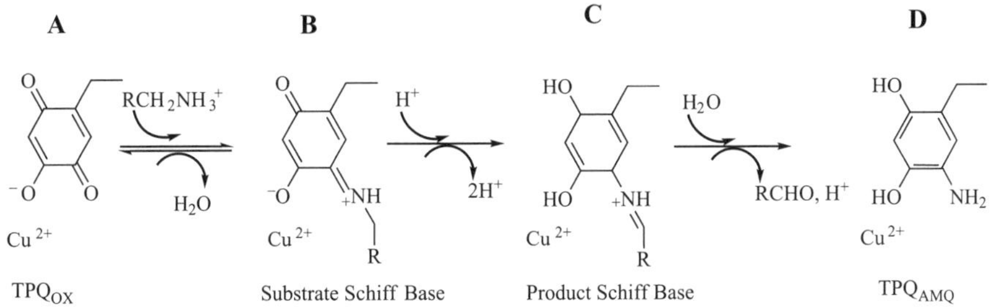

SCHEME 3.5. Proposed catalytic mechanism of the reductive half-reaction in amine oxidases.

$$E_{ox} + RCH_2NH_2 \leftrightarrow E{-}RCH_2NH_2 \rightarrow E_{red} + RCHO \tag{3.4}$$

$$E_{red} + O_2 + H_2O \rightarrow E_{ox} + H_2O_2 + NH_3 \tag{3.5}$$

Given the complicated, multistep mechanism (Schemes 3.4 and 3.5), it is not surprising that the overall catalytic rate may be limited by several different steps. Indeed, the catalytic rate-limiting step (r.l.s.) depends on the enzyme source. In BSAO, for example, ^{18}O isotope effects on catalysis indicated that ~60% of the r.l.s. could be attributed to proton abstraction during the reductive half-reaction, with the remaining ~40% ascribed to the first ET event to O_2 during the oxidative half-reaction.[111] The r.l.s. in HPAO has been proposed to be formation of cationic TPQ_{SQ} and metal-bound superoxide.[71,112] This result contrasts to lentil seedling AO (LSAO), where formation of TPQ_{AMQ} and release of aldehyde product is rate limiting.[113] In the closely related pea seedlings AO, the r.l.s. has been shown to be substrate dependent; the poor substrate benzylamine shows accumulation of the product Schiff-base species, while the preferred substrate putrescine shows contributions from both substrate Schiff-base formation and ammonia dissociation from TPQ_{IMQ}.[114] Similarly, TPQ_{IMQ} was observed to accumulate under single turnover conditions in AGAO, suggesting that either hydrolysis of this species or dissociation of metal-bound peroxide is rate limiting in this enzyme.[35,110]

3.4.2.1. Mechanistic Issues. A comprehensive understanding has been established from numerous studies regarding the reductive half-reaction in terms of the role for both Cu and TPQ.[5,36,109,115] However, the mechanism of the oxidative half-reaction has proved more difficult to conclusively define, particularly with regard to the roles of Cu and TPQ in oxygen reduction, and might vary somewhat among CuAOs. It is well established that upon substrate reduction under anaerobic conditions the Cu(II)–TPQ_{AMQ} (aminoquinol form of TPQ) couple exists in equilibrium with a Cu(I)–TPQ_{SQ} (semiquinone form of TPQ) couple (Scheme 3.6**B** $\leftrightarrow$ **C**), although the magnitude of K_{eq} for this interconversion varies among CuAOs from different sources.[108] Due to this internal redox equilibrium, it has proven to be a challenge to discern which species is directly responsible for reducing molecular oxygen. Along these lines, it has been suggested that O_2 binds to the Cu(I) center of the Cu(I)–TPQ_{sq}

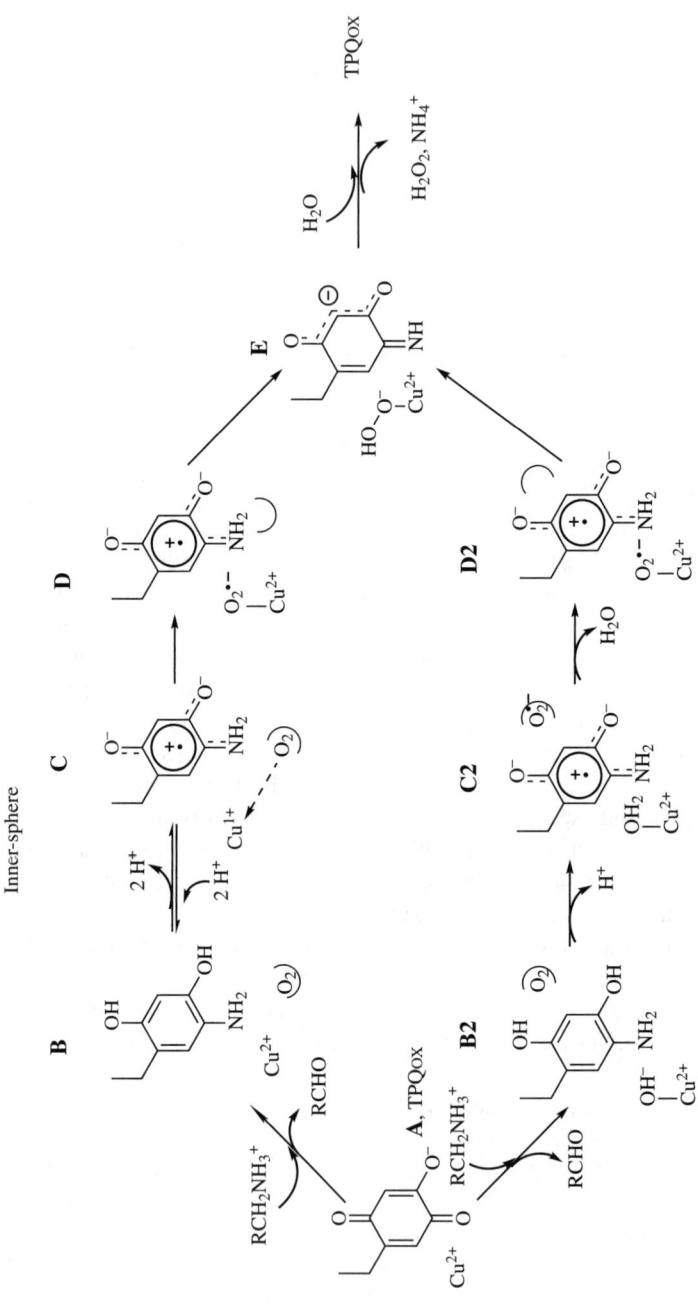

SCHEME 3.6. Catalytic mechanism illustrating the two proposed pathways for oxidative half-reaction in CuAOs. (Adapted from Ref. 110.)

moiety and is reduced to $O_2^{\bullet-}$ by an inner-sphere ET process in nonrate-limiting fashion.[108] However, other studies have suggested that O_2 binds in a hydrophobic pocket near the quinone cofactor and is reduced to $O_2^{\bullet-}$ by TPQ_{AMQ} in the rate-limiting step by outer-sphere ET (Scheme 3.6**B2** ↔ **C2**), with the role of Cu(II) being relegated to electrostatic stabilization of reduced O_2 species.[115,116] Despite these fundamental mechanistic differences, both inner- and outer-sphere pathways converge at the copper-bound hydroperoxide, TPQ_{IMQ} (iminoquinone form of TPQ), stage of the oxidative half-reaction (Scheme 3.6**E**).[110]

To explore these fundamental mechanistic issues, several lines of research have been conducted, including the examination of potential binding sites for O_2 and reduced oxygen species, the redox role of the copper center, the rate of ET from TPQ_{AMQ} to Cu(II), and competitive oxygen isotope effects. Small molecule copper ligands (e.g., azide and cyanide) have been used to probe copper-centered O_2 binding,[117–119] and several amine oxidases have been crystallized under high pressures of Xe gas to examine protein-based hydrophobic pockets that could represent sites for O_2 binding.[120,121] Additional insight has been provided by crystallographic characterization of an AO under turnover conditions.[122] Metal-substitution studies in several CuAO enzymes have been conducted in attempts to explore the function of the copper center in terms of its redox role during the oxidative half-reaction.[35,112,123] Electron transfer processes have been investigated using temperature jump relaxation methods,[124–126] stopped-flow kinetics[110] and ^{18}O kinetic isotope experiments together with density functional theory (DFT).[114] A summary of both the reductive and oxidative half-reactions and our current understanding of the catalytic mechanism in CuAOs follows.

3.4.2.2. Reductive Half-Reaction. The reductive half-reaction, involving the oxidation of an amine to an aldehyde, is depicted in Scheme 3.5**A** → **D**. The oxidized enzyme Cu(II)–TPQ_{ox} (species **A**) reacts with substrate amine to form the substrate Schiff base (species **B**), which is converted to a product Schiff base (species **C**), in a reaction facilitated by abstraction of a proton from the α-carbon of substrate by a conserved aspartate residue acting as a catalytic base.[35,109] The aldehyde product is then released by hydrolysis, generating the reduced cofactor Cu(II)–TPQ_{AMQ} (species **D**).

Evidence for substrate binding has come from a substrate analogue, 2-hydrazinopyridine (2-HP), that contains a nitrogen (as opposed to a carbon) adjacent to the primary amine group thereby preventing proton abstraction and trapping the enzyme in a complex that mimics the substrate Schiff-base moiety (species **B**).[36] The 2-HP bound ECAO structure provides a snapshot of the initial enzyme–substrate complex and notably shows that TPQ must be in an "off"-copper or productive conformation with the C5 carbonyl of TPQ positioned toward the catalytic aspartic acid residue to facilitate stereospecific proton abstraction.[36,115,127–129] In addition, the absolutely conserved Tyr369 forms a hydrogen bond with O4 of TPQ, and the gate residue Tyr381 is seen to rotate out of the substrate channel, permitting access to the active site.[36] Additional substrate analogues (benzylhydrazine and tranylcypromine) have also been trapped as substrate Schiff-base moieties in AGAO and ECAO.[130,131]

Collectively, these structures provide useful information regarding initial substrate binding and the residues that affect the various steps in the reductive half-reaction.

Site-directed mutagenesis has also provided critical information regarding the proton abstraction steps by the conserved aspartate residue.[109,132,133] Mutations of this aspartate residue in ECAO, HPAO, and AGAO have conclusively shown that the catalytic base performs multiple roles: facilitates the key step of proton abstraction during catalysis; assists substrate binding; positions intermediate species for optimal reactivity; and functions as a general acid–base catalyst for multiple proton-transfer steps during the reductive half-reaction.[109,132,133] Underlying the ability of the aspartic acid residue to carry out its function as an acid–base catalyst, studies have shown that this residue contains an unusually high pK_a value ranging between 7.5 and 8.1 for CuAOs from different sources,[18,71,132,134] which is likely a result of charge repulsion between the C4 oxoanion of TPQ and the carboxylate of the aspartic acid.[115]

Another residue implicated in positioning TPQ derived intermediates during turnover is the conserved tyrosine residue that hydrogen bonds with O4 of TPQ in the ECAO-2-HP structure.[36] A mutation of this tyrosine to a phenylalanine residue in ECAO resulted in a dramatic reduction in catalytic efficiency, and TPQ was found to predominantly occupy a "nonproductive" conformation, substantiating the hypothesis that this residue positions TPQ for optimal interaction with substrate.[135] Furthermore, mutagenesis experiments of this residue in HPAO showed an 800-fold decrease in k_{obs} for the rate of TPQ biogenesis relative to wild-type enzyme, indicating that the conserved tyrosine residue also coordinates intermediate species during biogenesis in a manner more conducive to the generation of TPQ.[71]

3.4.2.3. Oxidative Half-Reaction. The oxidative half-reaction involving the reoxidation of reduced TPQ is shown in Scheme 3.6. As described above, following release of product aldehyde, the reduced enzyme exists as an equilibrium between a Cu(II)–TPQ$_{AMQ}$ (Scheme 3.6B) and a Cu(I)–TPQ$_{SQ}$ (Scheme 3.6C) state. Notably, both Cu(I) and TPQ$_{SQ}$ may react with O$_2$. In the proposed inner-sphere mechanism (**B → D**), O$_2$ binds to the Cu(I) center of the Cu(I)–TPQ$_{SQ}$ intermediate and is reduced to O$_2^{\bullet-}$ (Scheme 3.6D). In essence, this pathway would constitute the preferred route to product if the Cu(I) center reacts more rapidly with O$_2$ than the TPQ$_{SQ}$ species, which is quite reasonable given the very well documented reactivity of three-coordinate Cu(I) complexes toward O$_2$.[46–51,108,119,136] Further reduction of the Cu(II)–superoxide from the semiquinone results in an iminoquinone moiety, Cu(II)–TPQ$_{IMQ}$ (Scheme 3.6E), which hydrolyzes to liberate NH$_4^+$ and the resting cofactor Cu(II)–TPQ$_{OX}$ (Scheme 3.6A), while the reduced oxygen species is converted to H$_2$O$_2$.[20] The release of ammonia may also occur through a transimination reaction between amine substrate and TPQ$_{IMQ}$ (Scheme 3.6E), generating the substrate Schiff base (Scheme 3.5B).[137]

Central to the proposal for a Cu(I)-mediated inner-sphere mechanism for O$_2$ reduction is that such a mechanism circumvents the spin conversion problem associated with two electron O$_2$ reduction reactions[138] and that there is ample precedent for the reactivity of Cu(I) sites with O$_2$ in copper-containing metalloproteins[4,139–141] and copper model complexes.[136,142] However, results obtained with HPAO and BPAO led to the proposal that Cu(II) reduction is not essential for enzymatic reoxidation, giving rise to the suggestion of an alternative reaction pathway (Scheme 3.6**B2** → **D2**) based on outer-sphere ET.[74,111,112,116] This mechanism essentially posits that TPQ$_{AMQ}$ is more reactive toward O$_2$ than the Cu(I) center.

To further explore these two potential mechanisms of cofactor reoxidation, several studies have examined the effects of the small molecule copper ligands on catalysis.[117–119,143,144] Given the ability of cyanide to coordinate both Cu(II) and Cu(I), this ligand has traditionally been used to probe the electronic environment in both oxidized- and substrate-reduced CuAO enzyme forms.[118,119,145–148] These studies were primarily concerned with determining cyanide's affects on amine oxidation and O$_2$ reduction. The ligand substitution chemistry of cyanide is comparable among CuAO enzymes from different sources, preferentially displacing an equatorial water molecule.[144,145,149] Consequences are reflected in the reactivity of the quinone cofactor toward substrate amines, substrate amine analogues, and the copper ion.[118,119,145–147] For example, cyanide inhibition has been reported previously as competitive,[146] noncompetitive,[145] and mixed[147] with respect to O$_2$. McGuirl et al.[118] reported that cyanide inhibition in several AO enzymes occurs by binding to Cu(I) and preventing the Cu(I)–TPQ$_{SQ}$ intermediate from further oxidation with O$_2$, which is also substantiated by Greenaway and He[147] and Shepard and co-workers[119] Subsequent work showed that cyanide also bound to the C5 carbonyl of TPQ forming a cyanohydrin adduct in both AGAO and PSAO.[119] Cyanide inhibition against amine substrate was found to be uncompetitive for both PSAO and AGAO, and under the conditions of the kinetics assays the inhibition exclusively arose from binding of cyanide to Cu(I) in the Cu(I)–TPQ$_{SQ}$ state in spite of cyanohydrin adduct formation.[119] These studies provide support for an inner-sphere mechanism, where binding of cyanide to Cu(I) presumably acts in direct competition with O$_2$ binding to the copper center, preventing reoxidation of the reduced enzyme.[118,119,147]

The results with cyanide showed that this ligand preferentially binds to Cu(I) over Cu(II).[119] Similar to cyanide, azide also replaces the equatorial H$_2$O ligand of Cu, although it preferentially binds to Cu(II), as opposed to Cu(I).[117,143,144] Therefore, inhibition by these ligands can be rationalized largely in terms of the effects they exert on the internal redox equilibrium in the reduced enzyme (Scheme 3.6**B** ↔ **C**).[118] Inhibition by azide versus substrate amine varies according to enzyme source, where PPAO displays uncompetitive inhibition,[150,151] while inhibition patterns for BSAO have been reported as mixed at low amine concentrations with a shift to uncompetitive at higher substrate concentrations.[144] In terms of the oxidative half-reaction, azide has been described as a competitive inhibitor in pig plasma AO with inhibition constants consistent with K_d values for the Cu(II)–N$_3^-$ complex[145,146] and a noncompetitive inhibitor[74] with respect to O$_2$. This latter result in HPAO was proposed to provide support for the hypothesis that O$_2$ is prebound in a hydrophobic pocket near TPQ and is then reduced to superoxide by TPQ$_{AMQ}$.[74]

In order to shed light on some of these differences, Juda et al.[117] carried out an extensive kinetic comparison of the effects of azide on both the reductive and oxidative half-reactions among four AO enzymes to specifically examine if azide competes with O_2 during cofactor reoxidation. Given the ping–pong bi–bi nature of AO catalysis, the two half-reactions (Schemes 3.5 and 3.6) are kinetically independent. To examine the effects of azide on each half-reaction independently, experiments were performed with variable concentrations of one substrate (either amine or O_2) while the other substrate was kept at saturating concentrations, in the presence of varying amounts of azide. It was discovered that azide exhibited assorted modes of inhibition during the reductive half-reaction (saturating O_2, varied amine concentrations) from noncompetitive to competitive with K_i values that varied by an order of magnitude, ranging from 18 ± 2 to 250 ± 18 mM. The differences in modes of inhibition indicate that azide binds to Cu(II) randomly and reversibly at multiple stages during catalysis (Schemes 3.5 and 3.6) in a manner that is independent of substrate amine binding. For the case of competitive inhibition in rhDAO, it seems that azide complexation may distort the orientation of TPQ, thereby preventing amine binding.[117]

With regard to the oxidative half-reaction (saturating amine, varied O_2 concentrations), different modes of inhibition were observed, from pure competitive, to partial competitive, to noncompetitive. For certain AO enzymes, like AGAO, pure noncompetitive inhibition is observed for both reductive and oxidative half-reactions with nearly identical K_i values (18 ± 2 and 19 ± 1 mM). Our results indicate that for an enzyme like AGAO the inhibition effects on the reductive half-reaction can never be fully overcome when examining the oxidative half-reaction (despite saturating amine concentrations), such that the two half-reactions cannot be kinetically separated. This event complicates the interpretation of azide's apparent noncompetitive inhibition toward O_2. However, for other AO enzymes (e.g., PSAO) azide is a poor noncompetitive inhibitor of the reductive half-reaction (with a K_i value of 250 ± 18 mM) and a more potent competitive inhibitor of the oxidative half-reaction (with a K_i value of 31 ± 2 mM, Fig. 3.11). The relatively good agreement between the latter K_i value and the experimentally determined K_d value for the Cu(II)–N_3^- complex in both oxidized and reduced PSAO enzyme forms (53 ± 5 and 59 ± 10 mM) suggests that azide coordination to the TPQ_{AMQ}–Cu(II) state shifts the TPQ_{AMQ}–Cu(II) $\leftrightarrows$ TPQ_{SQ}–Cu(I) equilibrium toward the former species, effectively blocking ET from the aminoquinol to Cu(II). This effect is experimentally manifested as azide directly competing with O_2. Competition between azide and O_2 arises from the rapid ET between TPQ_{AMQ} and Cu(II) in this enzyme and the facile ligand-substitution chemistry displayed by tetragonal Cu(II) complexes with solvent-derived ligands, such that reversible coordination by N_3^- is fast.[144,152] Specifically, NMRD measurements performed on CAOs have demonstrated the presence of a rapidly exchanging Cu bound H_2O molecule,[152] and electron spin echo envelope modulation (ESEEM) studies revealed that azide (and cyanide) displace the equatorial, not the axial, H_2O ligand to the Cu center.[149] In summary, studies of the ligand-substitution chemistry of CuAOs and the effects of ligand substitution on the enzyme kinetics are generally consistent with an inner-sphere mechanism for the reoxidation of reduced TPQ among CuAOs from various sources.[117,119,145,146]

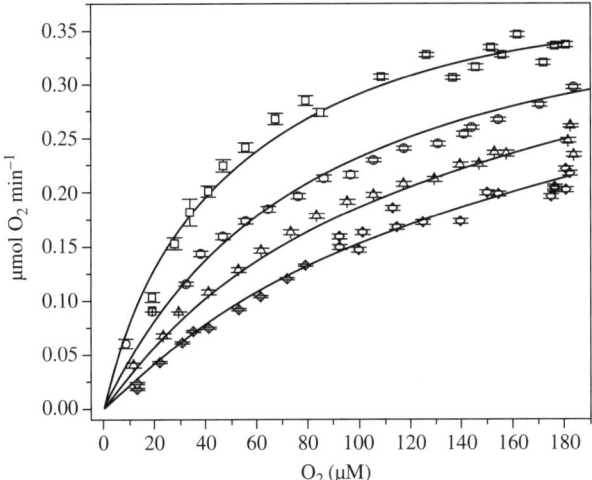

FIGURE 3.11. The competitive nature of azide inhibition of PSAO during the oxidative half-reaction. Azide concentrations were $0 = \square$, 25 mM $= \bigcirc$, 55 mM $= \triangle$, and 90 mM $= \diamondsuit$.

Oxygen binding to the copper center has additionally been investigated by trapping and structurally characterizing intermediate species during the oxidative half-reaction. Various ECAO intermediates were trapped crystallographically both under anaerobic and aerobic conditions, and the anaerobic substrate-reduced enzyme was additionally exposed to the presumed O_2 mimic nitric oxide.[122] The anaerobically substrate-reduced structure clearly shows an ordered TPQ_{AMQ} species with a copper center ligated by three histidines and the axial H_2O ligand; the equatorial H_2O ligand is not resolved in this structure.[122] The observation that the copper ligation environment in the reduced enzyme may be four coordinate as opposed to five coordinate in the oxidized enzyme was not surprising, as EXAFS studies have demonstrated a decrease in coordination number in the active sites of several reduced CuAO enzymes, resulting in three-coordinate Cu(I) centers,[58] providing a convenient path for O_2 binding and inner-sphere ET. The anaerobic substrate-reduced structure also revealed the presence of the product aldehyde (phenylacetaldehyde in this case) bound in the active-site pocket. The presence of this product was surprising given the ping–pong bi–bi kinetics of CuAO enzymes in solution, and has been explained to occur as a result of crystal contacts, which could also be responsible for the retarded rate of the oxidative half-reaction in the crystals.[122]

Addition of NO to the substrate-reduced enzyme revealed that NO replaced the axial H_2O ligand to copper and was bound at a distance of 2.4 Å, with the TPQ_{AMQ} rotating 20° from the previous structure.[122] The oxygen atom of NO was hydrogen bonded to O2 of the TPQ cofactor and was suggested to be strongly interacting with the cofactor, as opposed to the copper center. Wilmot and co-workers[38] suggest that this structure is consistent with the first ET event coming either from Cu(I) or TPQ_{AMQ}. To further explore this, an equilibrium turnover species was also trapped in ECAO crystals aerobically exposed to excess substrate. This structure shows a

O_2 species occupying the position analogous to NO, but this species was bound in a different conformation, being bound side-on to the copper ion at distances of 2.8 and 3.0 Å.[122] This trapped intermediate was proposed to represent a $2e^-$ reduced form of O_2 (potentially OOH^- or H_2O_2) and the TPQ_{IMQ} moiety. This structure was also suggested to support the simple electrostatic role of copper during the oxidative half-reaction, although the authors point out that due to the bridging nature of the bound ligands, neither structure rules out a redox role for copper in oxygen activation.[38] Also note that no evidence of O_2 binding in the off-metal, hydrophobic site proposed by Klinman and co-workers[74] was observed in either structure.

In an effort to probe potential cavities suitable for oxygen binding, several studies have carried out solution-phase nuclear magnetic resonance (NMR) and X-ray structural determination of amine oxidases following exposure to high pressures of xenon gas.[18,120,121,153,154] Solution interactions between Xe and LSAO have been examined by NMR and have concluded that Xe binds near the active site and induces perturbations to neighboring residues that, in the absence of amine substrate, result in generation of the TPQ_{SQ}–Cu(I) species.[154] The mechanistic significance of this result is not clear, but suggests that the active-site environment in this enzyme is sensitive to binding of small hydrophobic molecules and that the Cu(I)–semiquinone intermediate readily forms under these conditions. The X-ray crystal structures of Xe bound AGAO, PPLO, PSAO, and BSAO show multiple Xe atoms bound to each structure with only a single Xe site in common among these four enzymes; this site is the closest to the Cu/TPQ centers, being $\sim$ 7.4–7.7 Å from copper and $\sim$ 9.3–9.8 Å from the C2 carbonyl of TPQ (Fig. 3.12).[18,120] Electron transfer from either Cu or TPQ would

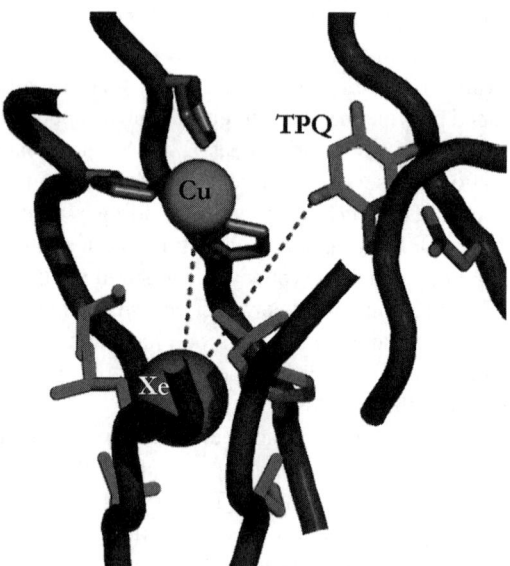

FIGURE 3.12. Xenon-binding site in PSAO. The pink sphere represents the consensus Xe binding site found in AGAO, PSAO, PPLO, and BSAO and the cyan sphere represents the copper site. The hydrophobic residues are colored as green. (See color insert.)

require migration of the putative O_2 molecule bound in the Xe site over a distance of 4–6 Å, but these structures clearly show that any movement toward the active-site center would require an initial close approach to Cu.[18,120]

Model building using the native HPAO crystal structure coordinates demonstrated that a Xe atom could be accommodated in a cavity corresponding to the Xe site in AGAO, PPLO, PSAO, and BSAO with minor movement of two side chains.[120] Additionally, modeling also confirmed that a Xe atom could be placed in the hydrophobic pocket initially proposed by Klinman and co-workers[75,155] for off-metal O_2 activation by TPQ_{AMQ}. This hypothetical second Xe binding site is much closer to TPQ and Cu(II), at distances of ~3.5 Å. Despite the attractive locality of this second presumed site, the placement of a Xe atom in this locale required the deletion of a solvent water molecule in the native structure, which was observed to be part of a chain of H_2O molecules.[120] Given the facts that Xe was not observed to be present in this alternate site in the AGAO, PPLO, PSAO, or BSAO xenon structures, or the Xe bound HPAO structure,[121] the experimental data suggest that this site may have greater affinity for hydrophilic molecules.[120] The Xe bound HPAO structure did demonstrate four xenon-binding sites per monomer and showed that the strongest anomalous Xe peak was located at the center of the large β-sandwich of the catalytic domain.[121] An overlay of all Xe−CuAO structures, including the recently determined ECAO Xe structure,[153] demonstrates that Xe atoms are positioned throughout the catalytic β-sandwich domain, suggesting that this region, as well as the substrate amine channel, are suitable for transient dioxygen-binding with subsequent delivery to the active site. Johnson et al.[121] also calculated the O_2 free energy maps for PPLO, PSAO, AGAO, and HPAO, which supported the different occupancies of the experimentally observed xenon-binding sites in these systems and concluded that the migration routes for O_2 among these proteins may be different.

Another critical component for understanding the involvement of Cu during the oxidative half-reaction is the rate of electron transfer (k_{ET}) from TPQ_{AMQ} to Cu(II) in the Cu(II)−TPQ_{AMQ} ⇄ Cu(I)−TPQ_{SQ} equilibrium. Electron transfer during the redox equilibrium has been investigated by temperature jump relaxation studies.[124–126] Our laboratory has measured the ET rate to Cu(II) from TPQ_{AMQ} in PSAO, APAO, and AGAO and calculated this rate to be ~20,000, ~ 60–75, and ~ 145 s^{-1}, respectively. The ET rate is significantly faster in the case of PSAO than that of APAO and AGAO, and this difference may reflect disorder in the ligation environment of the Cu(I) site that influences reorganization energy, thereby affecting k_{ET}.[58] Intriguingly, PSAO was the lone CuAO examined that showed no increase in disorder of the Cu(I) site upon treatment of the oxidized enzyme with dithionite, suggesting the reorganization energy is lower in this enzyme.[58] Despite the differences in the magnitude of the k_{ET} values, all three measured rates are faster than the respective k_{cat} values.[124–126] This result indicates that the Cu(I)–TPQ_{SQ} moiety could certainly be a kinetically competent intermediate during catalytic turnover, where the electron is transferred via an inner-sphere mechanism upon O_2 binding to the Cu(I) center in the transition state.

Studies of HPAO and BSAO, however, have led to the proposal that copper reduction is not essential for reoxidation of the TPQ_{AMQ} cofactor

(Scheme 3.6**B2** → **D2**).[74,111,112,116] Specifically, metal-replacement studies in HPAO provide evidence for an outer-sphere mechanism of O_2 reduction. Given the observation that Co(II) substituted HPAO was active and displayed a k_{cat} close to that of the wild-type Cu(II) enzyme near pH 7.0, Klinman and co-workers[116] argued that the first ET to O_2 occurred directly from TPQ_{AMQ}, with the Cu(II) center serving as a binding site for reduced oxygen species. The Co(II) reconstituted HPAO enzyme displayed a maximum k_{cat} value of 2.1 ± 0.5 S^{-1} versus $7.8 \pm 1.2\,s^{-1}$ for the native, copper-containing enzyme.[116] Differences in the kinetics of the Co(II) substituted enzyme compared to the Cu(II) enzyme were attributed to a drastically increased value for $K_M(O_2)$.[112,116] Molecular oxygen was proposed to initially bind in an off-metal site where rate-limiting ET from TPQ_{AMQ} (this event contributes $\sim 29\%$ to k_{cat})[156] would take place, with the superoxide species subsequently migrating to Cu(II).[115] The difference in oxygen-binding affinity for the nonmetal site was believed to arise from an increase in the pKa value for the axial H_2O ligand in cobalt-substituted HPAO relative to the native copper enzyme. This finding would directly result in an active-site net charge increase from $+1$ [with Cu(II)$-OH^-$, Scheme 3.6**B2**] to $+2$ [with Co(II)$-H_2O$].[112,116] Following the first ET to O_2 from TPQ_{AMQ}, it is believed the pKa of the resulting TPQ_{SQ} is perturbed in a way to allow for rapid proton transfer from the cofactor back to the metal-bound hydroxide species (C2), thereby forming H_2O, which would then be expected to undergo substitution with superoxide anion (C2 → **D2**).[115]

Metal substitution in BSAO shows that replacement of Cu(II) with Co(II) results in a significant reduction in activity. The Cu stripped form of the enzyme, which still contained a fraction of residual copper, displayed $\sim 5\%$ amine oxidation activity; addition of Co(II) resulted in recovery of activity to $\sim 20\%$ of wild-type levels.[157] These results were argued to support a mechanism where the TPQ_{SQ}–Cu(I) species was off-pathway. Additional support for a non-redox role for Cu comes from a theoretical study utilizing DFT, which implicates the paramagnetic copper center in the spin transitions required to produce singlet TPQ from the reaction of singlet TPQ_{AMQ} with triplet O_2.[158] It is difficult to apply these latter findings to all CuAO enzymes, as this study was premised on the rate-limiting step being the first ET to O_2, which is clearly not the case for either PSAO, LSAO, or AGAO.

The results for metal substitution in HPAO are in stark contrast to what is observed in AGAO and other amine oxidases. As mentioned previously, it has been definitively established that the rate-determining step during the oxidative half-reaction of AGAO is the formation of the charge delocalized, oxidized TPQ cofactor.[110,159] Both the Co(II) and Ni(II) substituted forms of AGAO, however, display remarkably different kinetics for the O_2 reaction than the wild-type, copper enzyme. Importantly, Tanizawa and co-workers[35] showed that Co(II) and Ni(II) reconstituted forms of AGAO had K_M values for amine and O_2 substrates that were essentially identical to native AGAO, however, k_{cat} values for Co(II)– and Ni(II)–AGAO were $\sim 1\,s^{-1}$ relative to $110\,s^{-1}$ for Cu(II)–AGAO. By evaluating, the crystal structures of Co(II)– and Ni(II)–AGAO, it was found that both Co(II) and Ni(II) bind octahedrally with the addition of another H_2O molecule as a ligand. The results indicate that the low catalytic activities of Co(II)– and Ni(II)–AGAO were associated with impaired efficiency of the oxidative half-reaction, with the rate-determining step likely being the reaction between TPQ_{AMQ} and O_2. This result considered, together with the low observed levels of catalysis ($\sim 2\%$ of wild-type levels) for Co(II)– and Ni(II)–AGAO, indicate that this enzyme may have the ability to reduce O_2 through an outer-sphere mechanism, consistent with the hypothesis suggested by Klinman and co-workers.[116] Moreover, Kishishita et al.[35] addressed the relatively low levels of catalytic turnover with Co(II) and Ni(II) by proposing that Cu plays an essential role in catalyzing ET between TPQ_{AMQ} and O_2, at least in part by acting as a binding site for reduced O_2 species to be efficiently protonated and released. The disparities in the catalytic rates for the Co(II) and Ni(II) substituted forms, compared to the native Cu(II) containing enzyme, may reflect different mechanisms for O_2 reduction. That is, the substantially higher rate for the reoxidation reaction of native AGAO with O_2 could be a consequence of the participation of the more reactive Cu(I)–TPQ_{SQ} intermediate, which is not possible in the metal-substituted enzymes given the low redox potentials of the alternate metals tested.

Furthermore, detailed kinetics studies of LSAO strongly support the participation of the Cu(I)–TPQ_{SQ} intermediate in the reduction of O_2 with a rate constant of $1.56 \times 10^7 \, M^{-1} \, s^{-1}$.[160,161] Neither Ni(II) nor Zn(II) substituted LSAO displayed any catalytic activity, while Co(II) substitution displayed minor (7%) activity relative to the wild-type enzyme.[162] In results that are remarkably similar to those reported for AGAO. Metal substitution in ECAO shows that the Co(II) substituted enzyme displays the ability to oxidize amines, although at a k_{cat} value that is 2.4% of the wild-type, copper enzyme. When corrected for the measured 1% activity in the metal-depleted enzyme, the activity of Co(II)–ECAO is quite low.[123] In order to test the hypothesis that $K_M(O_2)$ increased upon replacement of copper with cobalt, as was reported for HPAO,[116] Smith et al.[123] tested the rate of amine oxidation under both air-saturated and O_2 saturated conditions. Their results perhaps reveal a slight increase in k_{cat} for the Co(II) enzyme, from 0.27 ± 0.08 to $0.43 \pm 0.1 \, s^{-1}$ (relative to the 9.89 ± 0.6-s^{-1} value measured for copper–ECAO), but the reported values for the Co(II) enzyme are practically within error of one another.[123] This result indicates that ECAO behaves similarly to AGAO in that metal substitution does not alter the $K_M(O_2)$ value. Analogous to AGAO, it appears that the metal-substituted form of ECAO has the ability to reduce O_2 via an outer-sphere process, although the turnover rates are nearly that of background levels. Collectively, these data are consistent with the hypothesis that the active site has been tuned to exploit the reactivity of copper by taking advantage of an inner-sphere Cu(I) to O_2 electron-transfer mechanism.

Stopped-flow studies of substrate reduced AGAO have been performed in attempts to more definitively characterize the intermediate species that reacts with O_2.[110,159] Previous stopped-flow studies in AGAO demonstrated that the reduced enzyme reacted quite rapidly with O_2 (< 1 ms), forming a Cu-peroxy species.[159] The results were consistent with either TPQ_{SQ}–Cu(I) or TPQ_{AMQ}–Cu(II) reacting rapidly with O_2 or with rapid $1e^-$ transfer from TPQ_{AMQ} to Cu(II). Subsequent to this report, we showed that k_{ET} between TPQ_{AMQ} and Cu(II) in substrate reduced AGAO at 30°C and pH 7.2 was $\sim 145 \, s^{-1}$ (~ 3 times k_{cat} for the preferred amine substrate

β-phenylethylamine).[125] Given this result, we felt that the stopped-flow kinetics of enzyme reoxidation warranted further investigation.

The measured k_{ET} value permits stopped-flow methods to resolve the reactivity of either the TPQ_{SQ}–Cu(I) or TPQ_{AMQ}–Cu(II) with O_2. Accordingly, reoxidation rates of substrate reduced AGAO were examined as a function of temperature, pH, and O_2 concentration.[110] Reduced-enzyme spectra that were collected under the assay conditions, but in the absence of O_2, were deconvoluted to component spectra (TPQ_{AMQ} and TPQ_{SQ}), as each of these two quinone species has very distinct absorption profiles (TPQ_{SQ}; $\sim$ 360, 440, and 470 nm and TPQ_{AMQ}; $\sim$315 nm), and the reactivity of these species following introduction to O_2 could then directly be probed by global data analysis.

As shown in Figure 3.13, the initial spectra of the substrate-reduced enzyme show a significant amount of the TPQ_{SQ} species given its characteristic absorption at 436 and 466 nm. During the first few milliseconds of reaction with 700 μM dioxygen, the TPQ_{SQ} spectral features are observed to rapidly disappear (Fig. 3.13a).

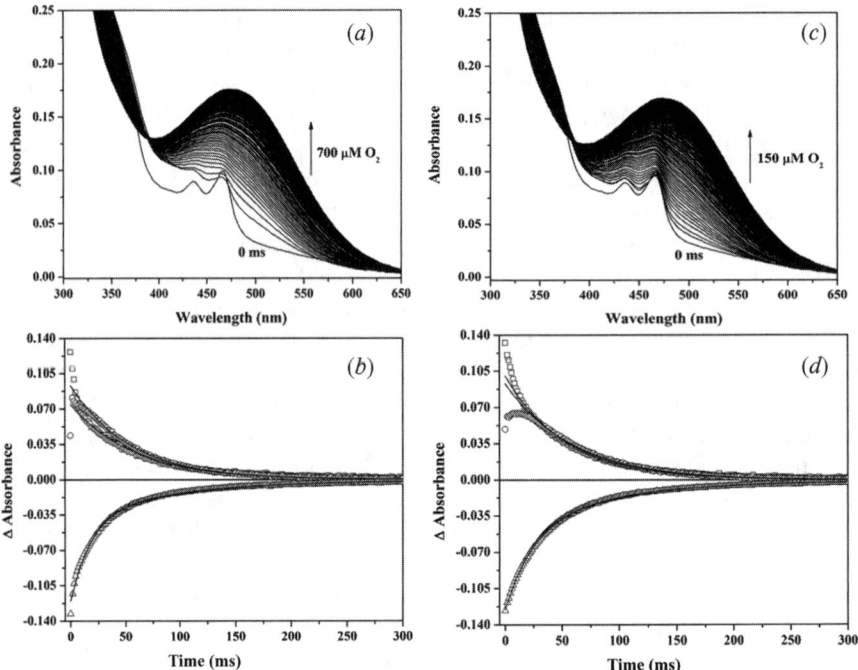

FIGURE 3.13. Stopped-flow spectral measurements of the reoxidation of reduced AGAO at 5°C and pH 7.2. Panels a (700 μM O_2) and c (150 μM O_2) are experimental spectral changes. The initial fully reduced enzyme spectrum is shown at 0 ms followed by absorption spectra following introduction to O_2 over the course of 500 ms. Panels b (700 μM O_2) and d (150 μM O_2) show absorbance changes at fixed wavelengths as created from difference spectra with associated single exponential fits. The following single wavelengths are shown: 314 ($\square$), 342 ($\bigcirc$), and 499 nm ($\triangle$). (Adapted from Ref. 110.)

Comparatively, the addition of 150 μM dioxygen, shows that the TPQ_{SQ} spectral features are more prevalent for a longer time period (Fig. 3.13c). Despite these qualitative differences, the calculated reaction rates for both 700 and 150 μM O_2 under all conditions tested were found to be independent of O_2 concentration, showing that the oxidative half-reaction was not limited by O_2 (Fig. 3.13b and d).[110] The near immediate disappearance of the TPQ_{SQ} moiety at saturating O_2 [700 μM O_2 is ~ 21 times $K_M(O_2)$] is fully consistent with the nonrate-limiting reaction of O_2 with the majority of the resting (prior to O_2 introduction) concentration of TPQ_{SQ}-Cu(I) during the dead time of the stopped-flow instrument. The greater prevalence of the TPQ_{SQ} spectral features at nonsaturating O_2 [150 μM O_2 is ~4.5 times $K_M(O_2)$], demonstrate that a significantly greater concentration of TPQ_{SQ}–Cu(I) is present in the initial spectrum following the initiation of the reaction with O_2, which provides qualitative evidence that O_2 binding to Cu(I) is reversible, as has been experimentally shown for PSAO.[110,114]

Collectively, on the basis of the difference spectra for the data collected as a function of temperature, O_2, and pH, the oxidative half-reaction appears to proceed in two phases as evidenced by two isosbestic points; the rate-limiting aspect of the oxidative half-reaction is the formation of TPQ_{OX} from TPQ_{IMQ} at rates that are ~1/4 to ~1/2 k_{ET} values (Scheme 3.6$\mathbf{E} \rightarrow TPQ_{OX}$).[110] Inspection of experimental difference spectra show the disappearance of a feature absorbing at ~310 nm during the initial phase of the oxidative half-reaction, which is interpreted as the conversion of the TPQ_{AMQ}–Cu(II) couple to the TPQ_{SQ}–Cu(I) intermediate, which then subsequently reacts rapidly with O_2. Moreover, at the higher O_2 concentration, the presence of a Cu–hydroperoxy intermediate ($\lambda_{max} = $ ~400 nm) is found to be more prominent in a D_2O solvent under basic conditions. Similar observations have been reported by Hirota et al.[110,159] for comparable reaction conditions, and both studies reported similar rates of decay for this species.

A global-fitting software program was utilized to fit the experimental data initially employing a three-state model with appropriate k_{ET} values to specifically probe the reactivity of the TPQ_{SQ} versus the TPQ_{AMQ}. The global analysis fully supported an inner-sphere mechanism of O_2 reduction with the TPQ_{SQ}–Cu(I) species, with the observed rapid decrease in the TPQ_{SQ} absorbance features occurring during the dead time of the stopped flow being indicative of a burst phase involving reaction of available TPQ_{SQ}–Cu(I) with O_2, followed by conversion of unreactive TPQ_{AMQ}–Cu(II) to reactive TPQ_{SQ}–Cu(I).[110] Moreover, this analysis showed that an outer-sphere mechanism for the reaction of TPQ_{AMQ} with O_2 was completely inconsistent with the experimental absorbance traces, as the rate of ET from the Cu(I) back to the TPQ_{SQ} to generate Cu(II) and TPQ_{AMQ} is simply too slow to adequately account for the spectral data.[110,125] The simplified three-state global data-fitting model was expanded to a complete reaction scheme with rate constants for all intermediates involved in the oxidative half-reaction. Second-order rate constants ranging between 1.4×10^6 and $1.0 \times 10^7 \, M^{-1} \, s^{-1}$ for the reaction of O_2 with TPQ_{SQ}–Cu(I) showed that the inner-sphere mechanism was nonrate-limiting; rates of $\geq 1000 \, s^{-1}$ provided acceptable concentration profiles and theoretical absorbance spectra that satisfactorily described the experimental data (Fig. 3.14).

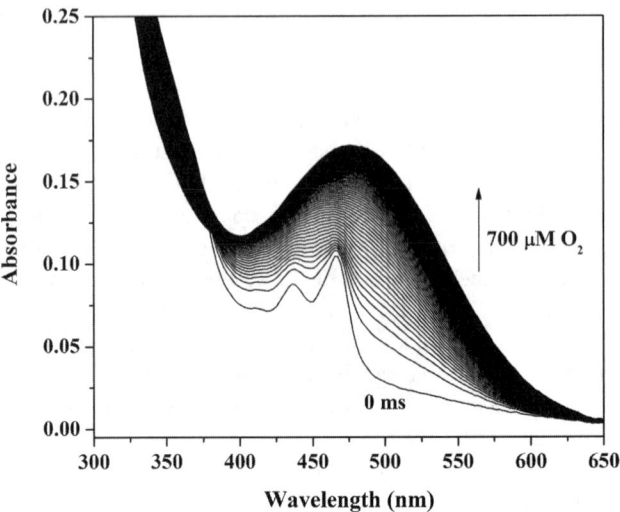

FIGURE 3.14. Best-fit modeled spectra calculated during global analysis for the 10°C and pH 7.2 experimental data set ($O_2 = 700 \, \mu M$). Experimentally obtained spectrum of the reduced enzyme is shown at 0 ms followed by theoretical spectra over the course of 500 ms. (Adapted from Ref. 110.)

Compelling evidence for an inner-sphere mechanism for O_2 reduction has been provided by a detailed experimental and computational examination of competitive oxygen kinetic isotope effects $k_{cat}/K_M(^{16,16}O_2)/k_{cat}/K_M(^{16,18}O_2)$ for the reaction between PSAO and O_2.[114] Thorough analysis of the internal redox equilibrium revealed that at 22°C and neutral pH, $\sim 80\%$ of the reduced enzyme exists in the TPQ_{SQ}–Cu(I) state. There was no discernable solvent viscosity effect on $k_{cat}/K_M(O_2)$, although k_{cat}/K_M(amine) displayed a small positive viscosity effect consistent with rate limitation by substrate binding. The $k_{cat}/K_M(O_2)$ parameter at 22°C is unchanged over the pH range from 6.0 to 9.5. A similar profile is observed in D_2O, although the average $k_{cat}/K_M(O_2)$ value is slightly decreased indicating that a small isotope effect is present, but this effect is too small for rate-determining proton transfer (Fig. 3.15a).[114] The pH and pD profiles for k_{cat}(amine) yield pK_a values of 5.2 (H_2O) and 6.0 (D_2O). The measured k_{cat} values in H_2O and D_2O are indistinguishable in the basic region, which suggests that ionization of the catalytic aspartate residue at the substrate Schiff-base stage is responsible for the deviation in the acidic region (Fig. 3.15b). From these pH profile studies, Mukherjee et al.[114] concluded that $k_{cat}/K_M(O_2)$ and k_{cat} do not share a common step. Moreover, it was shown that the rate-limiting step in PSAO is substrate dependent. For the slower substrate benzylamine, steady-state kinetics analysis established that a small contribution from C–H deprotonation and large contribution from the formation of the product Schiff base comprise k_{cat}. For the preferred substrate putrescine, significant contribution from C–H bond cleavage, resulting in accumulation of the substrate Schiff-base species, and ammonia dissociation from TPQ_{IMQ} comprise k_{cat}. Importantly, it should be noted that for neither benzylamine nor putrescine were spectral features due to the TPQ_{SQ} species observed

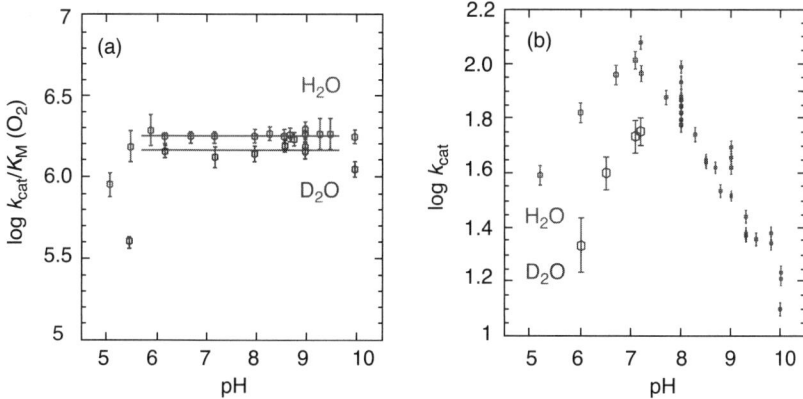

FIGURE 3.15. Profiles of (a) $k_{cat}/K_M(O_2)$ and (b) k_{cat} at 22°C (H_2O, top circles), (D_2O, bottom squares). (Reprinted with permission from Ref. 114.)

in the steady state, indicating that there are no rate-limiting ET steps during cofactor reoxidation.[114]

Figure 3.16 shows the competitive isotope effects during PSAO catalysis sampled over time to examine changes in the ratio of ^{18}O to ^{16}O. The curve clearly shows the accumulation of the heavier ^{18}O isotope during the course of the oxidative half-reaction, given the faster reaction of $^{16}O-^{16}O$ versus $^{18}O-^{16}O$. The ^{18}O measurements reveal a calculated KIE value of 1.0136 ± 0.0014 at 22°C (Fig. 3.16), and additional experiments demonstrated that this value is independent of the temperature and pH indicating that there is no change in the step that limits $k_{cat}/K_M(O_2)$ and

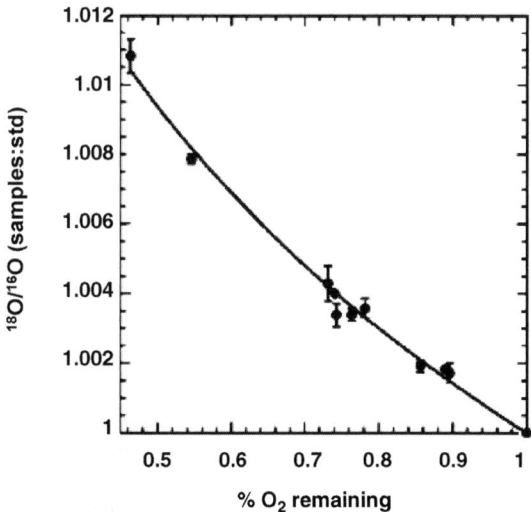

FIGURE 3.16. Oxygen isotope fractionation during PSAO turnover at pH 7.2 and 22°C, $\mu = 0.1$ M. (Reprinted with permission from Ref. 114.)

that kinetic complexity can be excluded.[114] The value for the ^{18}O KIE was subsequently shown through DFT methods to be fully consistent with the preequilibrium binding of O_2 to Cu(I), forming a $[Cu^{II}(\eta^1\text{-}O_2{}^{-1})]^+$ intermediate, followed by rate-determining ET from TPQ_{SQ}. The ^{18}O KIE values obtained for HPAO and BSAO $(1.0101 \pm 0.0029$ and $1.0097 \pm 0.0010)^{[111,116]}$ are inconsistent with the magnitude of outer-sphere ^{18}O KIE values that typically range between 1.025 and 1.030.[141,163–166] Intriguingly, a D630N HPAO mutant that forms 36% more TPQ_{SQ}–Cu(I) relative to the wild-type enzyme has an even smaller ^{18}O KIE value of 1.0074 ± 0.0008.[167] Despite this repressed value and that the D630N mutant showed a $k_{cat}/K_M(O_2)$ value an order of magnitude greater than for wild-type enzyme at pH 7, Klinman and co-workers[167] have suggested that kinetic complexity linked to either partially rate-limiting O_2 binding in a non-metal hydrophobic pocket or $O_2{}^{\bullet-}$ binding to Cu(II) explain the low ^{18}O KIE values during the outer-sphere oxidative half-reaction of HPAO (Scheme 3.6B2 $\rightarrow$ D2).[167] The results by Mukherjee et al.[114] clearly show that no such kinetic complexity exists during cofactor reoxidation in PSAO. Moreover, a positive value for the entropy of activation for $k_{cat}/K_M(O_2)$ in PSAO suggests the presence of a multistep process whereby ET follows prebinding of O_2 to Cu. This phenomena can be explained by release of bound H_2O from a metal center and EXAFS studies for PSAO, and several other CuAOs, demonstrated that the Cu center undergoes a decrease in coordination number from 5 in the oxidized enzyme to 3 upon reduction with the release of the copper-bound H_2O molecules.[58] This finding may provide a generic mechanism whereby CuAOs bind O_2 and propagate inner-sphere ET.[114]

3.5. CONCLUSIONS AND FUTURE PROSPECTS

This chapter has reviewed the roles of copper in the self-processing biogenesis reactions responsible for formation of the post-translationally modified active-site cofactors in galactose oxidase and amine oxidase. In fact, the active-site copper has two different, but equally critical roles: to aid assembly of the active site, and to act as an essential part of the final catalytic machinery. Biogenesis of both TPQ and the [Tyr$^{\bullet}$–Cys] cofactor requires only the addition of Cu and O_2; the Cu center has been shown to be involved in critical redox steps during both biogenesis mechanisms by coordinating a cysteine thiolate (GO) or by tyrosine coordination (AO). For galactose oxidase, few intermediates have been characterized and some issues relating to the *in vivo* preference for Cu(I) versus Cu(II) remain to be elucidated. It is clear that Cu in both oxidation states can support biogenesis, however, rates of biogenesis differ substantially. In amine oxidases, several intermediates have been thoroughly characterized and the biogenesis mechanism is well established. Lack of experimental observation of the highly reactive Cu(I)–tyrosyl radical species, however, at this time precludes the direct confirmation of inner-sphere reduction of O_2 from the metal center; it has also been proposed that an outer-sphere mechanism for ET from the Cu(II)–phenolate moiety to O_2 is operative for metal-substituted enzymes. Clarification of this issue is needed, although it may be that AO active sites display some degree of plasticity toward metal-assisted O_2 activation and slight variances in second sphere metal ligands dictate inner-sphere versus outer-sphere pathways. The observation that rates of biogenesis in wild-type copper-containing enzymes are faster by two or three orders of magnitude relative to their metal-substituted counterparts should be addressed and ultimately must be accounted for by proposed mechanisms for TPQ formation.

During catalysis, the Cu centers of GO and AO again play intimate roles in the chemical reactions and intermediate steps involved in product aldehyde formation and cofactor reoxidation. In both enzymes, inner-sphere ET from a reduced, three-coordinate Cu(I) center to O_2 has been proposed to be a critical step in the reoxidation of the reduced active site. The mechanism for GO is well agreed upon, whereby substrate alcohol coordinates directly to the copper center, subsequently undergoing oxidation via the copper and tyrosine–cysteine cross-link system. Upon introduction of molecular oxygen, an electron is transferred from Cu(I) to O_2 forming superoxide, which is then involved in rate-limiting hydrogen-atom transfer in H_2O_2 release. The reductive half-reaction of AO catalysis is also well understood, with the binding and oxidation of primary amines to aldehydes via covalent bond formation between the TPQ cofactor and the amine compound being thoroughly characterized. Yet, some debate remains for proposed mechanisms of cofactor reoxidation during the oxidative half-reaction. Experimental data for HPAO and for BSAO supports an outer-sphere mechanism for O_2 reduction directly from the reduced quinone cofactor, with copper acting as a binding site for reduced-oxygen species. Conversely, substantial experimental data for AGAO and PSAO strongly support the participation of a Cu(I)–semiquinone intermediate during cofactor reoxidation, where Cu(I) reduces O_2 to superoxide through an inner-sphere mechanism.

The similar active-site architecture of all structurally characterized amine oxidases at first glance provides the expectation that similar mechanisms for O_2 reduction will be operative for enzymes across this class. Metal substitution studies have proven to be a highly informative means to probe the role of Cu during O_2 reduction. Turnover numbers of Co(II) substituted enzymes like AGAO are a fraction of levels observed for wild-type enzymes, with the exception of Co(II)–HPAO, which shows a k_{cat} nearly equivalent to the Cu(II) enzyme near neutral pH. This single result is the strongest evidence for an outer-sphere ET process in the latter enzyme. Conversely, a broad range of studies for PSAO and AGAO provide compelling evidence that an inner-sphere mechanism is operative in these enzymes. It is clear that in order to delineate the role of Cu, several types of experimental approaches must be used; no single set of experimental data provides clear cut answers as to the role of Cu during O_2 activation. Some of the most critical experimental approaches used to-date include metal-substitution, X-ray crystallography, studies of the effects of ligand substitution, temperature jump relaxation and stopped-flow kinetics, and competitive ^{18}O KIE measurements coupled with DFT studies.

Taken collectively, it seems that the AO active site has some plasticity associated with it in terms of how the coordination environment of the copper site tunes the metal for either inner- or outer-sphere ET processes. While it is certainly plausible that all amine oxidases utilize a similar mechanism for O_2 reduction, the most convincing evidence for the hypothesis of active-site plasticity comes from AGAO; stopped-flow studies clearly show the Cu(I)–semiquinone species to be reacting with O_2 in nonrate-limiting fashion, with an experimentally determined ^{18}O KIE value (AGAO $= 1.0175$, Roth and Dooley manuscript in preparation)[168] that is fully consistent with inner-sphere ET from Cu(I) to O_2, while the replacement of Cu with Co results in very low levels of catalytic activity associated with rate-limiting, outer-sphere ET to O_2. Given these results, in light of those mentioned in this chapter for HPAO and PSAO, it appears that amine oxidases from various sources have likely utilized first- and second-sphere bonding interactions to fine-tune hydrogen bonding, redox potentials, and reorganization energies that alter reactivity toward O_2. Only further studies will resolve whether additional amine oxidases employ an outer-sphere mechanism for O_2 reduction, or if the majority of these enzymes have evolved to exploit the reactivity of the TPQ_{SQ}–Cu(I) moiety.

ABBREVIATIONS

AGAO	*Arthrobacter globiformis* amine oxidase
AO	Amine oxidase
BSAO	Bovine serum amine oxidase
CD	Circular dichroism
CTQ	Cysteine tryptophyl quinine
CuAOs	Copper-containing amine oxidases
DFT	Density functional theory
DPQ	3,4-Dihydroxyphenylalanine quinone
ECAO	*Escherichia coli* amine oxidase
EDTA	Ethylenediaminetetraacetic acid
EPR	Electron paramagnetic resonance
ESEEM	Electron spin echo envelope modulation
EXAFS	Extended X-ray absorption fine structure
ET	Electron transfer
GO	Galactose oxidase
HAT	Hydrogen atom transfer
HPAO	*Hansenula polymorpha* amine oxidase
KIE	Kinetic isotopic effect
LMCT	Ligand-to-metal charge transfer
LSAO	Lentil seedling AO
LOX	Lysyl oxidase
LTQ	Lysine tyrosyl quinine
NMRD	Nuclear magnetic relaxation dispersion
PAGE	Polyacrylamide gel electrophoresis
PCET	Proton-coupled electron transfer
PPLO	*Pichia pastoris* amine oxidase
premat-GO	Premature GO
pro-GO	Precursor GO

PSAO	Pea seedling amine oxidase
r.l.s.	Rate-limiting step
rhDAO	Recombinant human kidney diamine oxidase
RR	Resonance Raman
SDS	Sodium dodecylsulfate
TPQ	2,4,5-Trihydroxyphenylalanine quinone
TPQ_{AMQ}	TPQ Aminoquinol
TPQ_{IMQ}	TPQ Iminoquinone
TPQ_{OX}	Oxidized TPQ
TPQ_{SQ}	TPQ Semiquinone
TTQ	Tryptophan tryptophylquinone
UV–vis	Ultraviolet–visible
VAP-1	Human vascular adhesion protein-1

REFERENCES

1. Halcrow, M.; Knowles, P.; Phillips, S. Copper Proteins in the Transport and Activation of Dioxygen, and the Reduction of Inorganic Molecules, in *Handbook on Metalloproteins*, Bertini, I., Sigel, A., Sigel, H., Eds., Marcel Dekker, Inc., New York, 2001, pp 709–762.

2. Knowles, P. F.; Yadav, K. D. S. Amine Oxidases, R. Lontie, Ed, *Copper Proteins and Copper Enzymes*, 2, CRC Press, Boca Raton, FL, 1982, pp 103–129.

3. Knowles, P. F.; Dooley, D. M. Amine Oxidases. *Metal Ions in Biological Systems*, Singel, H., Singel, A.Eds., Marcel Dekker, Inc., New York, 1994, pp 361–403.

4. Whittaker, J. W. Oxygen reactions of the copper oxidases. *Essays Biochem.* **1999**, *34*, 155–172.

5. Halcrow, M.; Phillips, S.; Knowles, P. Amine Oxidases and Galactose Oxidase, in *Subcellular Biochemistry*, Holzenburg, A., Scrutton, N.,Eds., Kluwer Academic/ Plenum Publishers, New York, 2000, pp 183–231.

6. Knowles, P. F.; Ito, N. Galactose oxidase. *Perspectives Bioinorg. Chem.* **1993**, *2*, 207–244.

7. McPherson, M. J.; Mark R. Parsons; Spooner, R. K.; Wilmot, C. M. Galactose Oxidase. *Handbook of Metalloproteins*, John Wiley & Sons, Inc. Chichester, New York, 2001

8. Mure, M. Tyrosine-derived quinone cofactors. *Acc. Chem. Res.* **2004**, *37*, 131–139.

9. Whittaker, J. W. Galactose oxidase. *Adv. Protein Chem.* **2002**, *60*, 1–49.

10. Csiszar, A.; Labinskyy, N.; Zhao, X.; Hu, F.; Serpillon, S.; Huang, Z.; Ballabh, P.; Levy, R. J.; Hintze, T. H.; Wolin, M. S.; Austad, S. N.; Podlutsky, A.; Ungvari, Z. Vascular superoxide and hydrogen peroxide production and oxidative stress resistance in two closely related rodent species with disparate longevity. *Aging Cell* **2007**, *6*, 783–797.

11. Rhee, S. G. Cell signaling. H_2O_2, a necessary evil for cell signaling. *Science* **2006**, *312*, 1882–1883.

12. McGrath, A. P.; Hilmer, K. M.; Collyer, C. A.; Shepard, E. M.; Elmore, B. O.; Brown, D. E.; Dooley, D. M.; Guss, J. M. Structure and inhibition of human diamine oxidase. *Biochemistry* **2009**, *48*, 9810–9822.

13. Duff, A. P.; Cohen, A. E.; Ellis, P. J.; Kuchar J.A.; Langley, D. B.; Shepard, E. M.; Dooley, D. M.; Freeman, H. C.; Guss, J. M. The crystal structure of Pichia pastoris lysyl oxidase. *Biochemistry* **2003**, *42*, 15148–15157.

14. Ito, N.; Phillips, S. E.; Stevens, C.; Ogel, Z. B.; McPherson, M. J.; Keen, J. N.; Yadav, K. D.; Knowles, P. F. Novel thioether bond revealed by a 1.7 Å crystal structure of galactose oxidase. *Nature (London)* **1991**, *350*, 87–90.

15. Jakobsson, E.; Nilsson, J.; Ogg, D.; Kleywegt, G. J. Structure of human semicarbazide-sensitive amine oxidase/vascular adhesion protein-1. *Acta Crystallogr. D. Biol. Crystallogr.* **2005**, *61*, 1550–1562.

16. Kumar, V.; Dooley, D. M.; Freeman, H. C.; Guss, J. M.; Harvey, I.; McGuirl, M. A.; Wilce, M. C. J.; Zubak, V. M. Crystal structure of a eukaryotic (pea seedling) copper-containing amine oxidase at 2.2 Å resolution. *Structure* **1996**, *4*, 943–955.

17. Li, R. B.; Klinman, J. P.; Mathews, F. S. Copper amine oxidase from *Hansenula Polymorpha*: the crystal structure determined at 2.4 Å resolution reveals the active conformation. *Structure* **1998**, *6*, 293–307.

18. Lunelli, M.; Di Paolo, M. L.; Biadene, M.; Calderone, V.; Battistutta, R.; Scarpa, M.; Rigo, A.; Zanotti, G. Crystal structure of amine oxidase from bovine serum. *J. Mol. Biol.* **2005**, *346*, 991–1004.

19. Parsons, M. R.; Convery, M. A.; Wilmot, C. M.; Yadav, K. D. S.; Blakeley, V.; Corner, A. S.; Phillips, S. E. V.; McPherson, M. J.; Knowles, P. F. Crystal structure of a quinoenzyme: copper amine oxidase of *Escherichia Coli* at 2 Å resolution. *Structure* **1995**, *3*, 1171–1184.

20. Wilce, M. C. J.; Dooley, D. M.; Freeman, H. C.; Guss, J. M.; Matsunami, H.; McIntire, W. S.; Ruggiero, C. E.; Tanizawa, K.; Yamaguchi, H. Crystal structures of the copper-containing amine oxidase from *arthrobacter globiformis* in the holo and apo forms: implications for the biogenesis of topaquinone. *Biochemistry* **1997**, *36*, 16116–16133.

21. Jameson, R. F. Coordination Chemistry of Copper With Regard to Biological Systems. in *Properties of Copper*, Sigel, H., Ed., Dekker, New York, 2000, pp 1–30.

22. Adman, E. T. Copper protein structures. *Adv. Protein Chem.* **1991**, *42*, 145–198.

23. McPherson, M. J.; Ogel, Z. B.; Stevens, C.; Yadav, K. D. S.; Keen, J. N. Galactose oxidase of *Dactylium dendroides*. *J. Biol. Chem.* **1992**, *267*, 8146–8152.

24. Ogel, Z. B.; Ozilgen, M. Regulation and kinetic modeling of galactose oxidase secretion. *Enzyme Microb. Technol.* **1995**, *17*, 870–876.

25. Firbank, S. J.; Rogers, M. S.; Wilmot, C. M.; Dooley, D. M.; Halcrow, M. A.; Knowles, P. F.; McPherson, M. J.; Phillips, S. E. Crystal structure of the precursor of galactose oxidase: an unusual self-processing enzyme. *Proc. Natl. Acad. Sci. USA* **2001**, *98*, 12932–12937.

26. Rogers, M. S.; Hurtado-Guerrero, R.; Firbank, S. J.; Halcrow, M. A.; Dooley, D. M.; Phillips, S. E.; Knowles, P. F.; McPherson, M. J. Cross-link formation of the cysteine 228-tyrosine 272 catalytic cofactor of galactose oxidase does not require dioxygen. *Biochemistry* **2008**, *47*, 10428–10439.

27. Knowles, P. F.; Singh, I.; Yadav, K. D. S.; Mabbs, F. E.; Collison, D.; Cote, C. E.; Dooley, D. M.; McGuirl, M. A. Active Site Structures of Copper-Containing Oxidases, in *PQQ and Quinoproteins*, Jonegan, J. A. D., Duine, J. A.,Eds., Kluwer Academic Publishers, Dordrecht, The Netherlands, 1989, pp 283–288.

28. Smith, M. A.; Pirrat, P.; Pearson, A. R.; Kurtis, C. R. P.; Trinh, C. H.; Gaule, T. G.; Knowles, P. F.; Phillips, S. E. V.; McPherson, M. J. Exploring the roles of the metal ions in *Escherichia coli* copper amine oxidase. *Biochemistry* **2010**, *49*, 1268–1280.

29. Zhang, X. P.; McIntire, W. S. Cloning and sequencing of a copper-containing, topa quinone-containing monoamine oxidase from human placenta. *Gene* **1996**, *179*, 279–286.

30. Chang, C. M.; Klema, V. J.; Johnson, B. J.; Mure, M.; Klinman, J. P.; Wilmot, C. M. Kinetic and structural analysis of substrate specificity in two copper amine oxidases from *hansenula polymorpha*. *Biochemistry* **2010**, *49*, 2540–2550.

31. O'Connell, K. M.; Langley, D. B.; Shepard, E. M.; Duff, A. P.; Jeon, H. B.; Sun, G.; Freeman, H. C.; Guss, J. M.; Sayre, L. M.; Dooley, D. M. Differential inhibition of six copper amine oxidases by a family of 4-(aryloxy)-2-butynamines: evidence for a new mode of inactivation. *Biochemistry* **2004**, *43*, 10965–10978.

32. Shepard, E. M.; Smith, J.; Elmore, B.; Kuchar J.A.; Sayre, L. M.; Dooley, D. M. Towards the development of selective amine oxidase inhibitors. mechanism-based inhibition of six copper containing amine oxidases. *Eur. J. Biochem.* **2002**, *269*, 3645–3658.

33. Dawkes, H. C.; Phillips, S. E. V. Copper amine oxidase: cunning cofactor and controversial copper. *Curr. Opin. Struct. Biol.* **2001**, *11*, 666–673.

34. Dove, J. E.; Klinman, J. P. Trihydroxyphenylalanine quinone (TPQ) from copper amine oxidases and lysyl tyrosylquinone (LTQ) from lysyl oxidase. *Adv. Protein Chem.* **2001**, *58*, 141–174.

35. Kishishita, S.; Okajima, T.; Kim, M.; Yamaguchi, H.; Hirota, S.; Suzuki, S.; Kuroda, S.; Tanizawa, K.; Mure, M. Role of copper ion in bacterial copper amine oxidase: spectroscopic and crystallographic studies of metal-substituted enzymes. *J. Am. Chem. Soc.* **2003**, *125*, 1041–1055.

36. Wilmot, C. M.; Murray, J. M.; Alton, G.; Parsons, M. R.; Convery, M. A.; Blakeley, V.; Corner, A. S.; Palcic, M. M.; Knowles, P. F.; McPherson, M. J.; Phillips, S. E. V. Catalytic mechanism of the quinoenzyme amine oxidase from *Escherichia coli*: exploring the reductive half-reaction. *Biochemistry* **1997**, *36*, 1608–1620.

37. Green, E. L.; Nakamura, N.; Dooley, D. M.; Klinman, J. P.; Sanders-Loehr, J. Rates of oxygen and hydrogen exchange as indicators of TPQ cofactor orientation in amine oxidases. *Biochemistry* **2002**, *41*, 687–696.

38. Brazeau, B. J.; Johnson, B. J.; Wilmot, C. M. Copper-containing amine oxidases. biogenesis and catalysis; a structural perspective. *Arch. Biochem. Biophys.* **2004**, *428*, 22–31.

39. Okeley, N. M.; van der Donk, W. A. Novel cofactors via post-translational modifications of enzyme active sites. *Chem. Biol.* **2000**, *7*, 159–171.

40. Ostermeier, C.; Harrenga, A.; Ermler, U.; Michel, H. Structure at 2.7 Å resolution of the *paracoccus denitrificans* two-subunit cytochrome c oxidase complexed with an antibody fv fragment. *Proc. Natl. Acad. Sci. USA* **1997**, *94*, 10547–10553.

41. Janes, S. M.; Mu, D.; Wemmer, D.; Smith, A. J.; Kaur, S.; Maltby, D.; Burlingame, A. L.; Klinman, J. P. A new redox cofactor in eukaryotic enzymes: 6-hydroxydopa at the active site of bovine serum amine oxidase. *Science* **1990**, *248*, 981–987.

42. Janes, S. M.; Palcic, M. M.; Scaman, C. H.; Smith, A. J.; Brown, D. E.; Dooley, D. M.; Mure, M.; Klinman, J. P. Identification of topaquinone and its consensus sequence in copper amine oxidases. *Biochemistry* **1992**, *31*, 12147–12154.

43. Rogers, M. S.; Baron, A. J.; McPherson, M. J.; Knowles, P. F.; Dooley, D. M. Galactose oxidase pro-sequence cleavage and cofactor assembly are self-processing reactions. *J. Am. Chem. Soc.* **2000**, *122*, 990–991.

44. Whittaker, M. M.; Whittaker, J. W. Cu(I)-dependent biogenesis of the galactose oxidase redox cofactor. *J. Biol. Chem.* **2003**, *278*, 22090–22101.

45. Whittaker, J. W. Spectroscopic Studies of Galactose Oxidase. *Redox-Active Amino Acids in Biology*, Klinman, J. P., Ed., Academic Press, San Diego, CA, 1995, pp 262–278.

46. Karlin, K. D.; Tyeklar, Z.; Farooq, A.; Haka, M. S.; Ghosh, P.; Cruse, R. W.; Gultneh, Y.; Hayes, J. C.; Toscano, P. J.; Zubieta, J. Dioxygen-copper reactivity and functional modeling of hemocyanins-reversible binding of O_2 and CO to dicopper(I) complexes $[Cu(I)_2(L)]^{2+}$ (L = dinucleating ligand) and the structure of a bis(carbonyl) adduct, $[Cu(I)_2(L)(CO)_2]^{2+}$. *Inorg. Chem.* **1992**, *31*, 1436–1451.

47. Karlin, K. D.; Wei, N.; Jung, B.; Kaderli, S.; Niklaus, P.; Zuberbuehler, A. D. Kinetics and thermodynamics of formation of copper-dioxygen adducts: oxygenation of mononuclear copper(I) complexes containing tripodal tetradentate ligands. *J. Am. Chem. Soc.* **1993**, *115*, 9506–9514.

48. Karlin, K. D.; Tyekl'ar, Z. Functional biomimics for copper proteins involved in reversible O_2-binding, substrate oxidation/reduction and nitrite reduction. *Adv. Inorg. Biochem.* **1994**, *9*, 123–172.

49. Karlin, K. D.; Nasir, M. S.; Cohen, B. I.; Cruse, R. W.; Kaderli, S.; Zuberbuehler, A. D. Reversible dioxygen binding and aromatic hydroxylation in O_2-reactions with substituted xylyl dinuclear copper(I) complexes: syntheses and low-temperature kinetic/thermodynamic and spectroscopic investigations of a copper monooxygenase model system. *J. Am. Chem. Soc.* **1994**, *116*, 1324–1336.

50. Karlin, K. D.; Kaderli, S.; Zuberbuehler, A. D. Kinetics and thermodynamics of copper(I)/dioxygen interaction. *Acc.Chem. Res.* **1997**, *30*, 139–147.

51. Tolman, W. B. Making and breaking the dioxygen O–O bond: new insights from studies of synthetic copper complexes. *Acc. Chem. Res.* **1997**, *30*, 227–237.

52. Petersson, L.; Graslund, A.; Ehrenberg, A.; Sjoberg, B. M.; Reichard, P. The iron center in ribonucleotide reductase from *Escherichia-coli. J. Biol. Chem.* **1980**, *255*, 6706–6712.

53. Rogers, M. S.; Knowles, P. F.; Baron, A. J.; McPherson, M. J.; Dooley, D. M. Characterization of the active site of galactose oxidase and its active site mutational variants Y495F/H/K and W290H by circular dichroism spectroscopy. *Inorg. Chim. Acta* **1998**, *275–276*, 175–181.

54. Schnepf, R.; Horth, P.; Bill, E.; Wieghardt, K.; Hildebrandt, P.; Haehnel, W. De novo design and characterization of copper centers in synthetic four-helix-bundle proteins. *J. Am. Chem. Soc.* **2001**, *123*, 2186–2195.

55. Schnepf, R.; Haehnel, W.; Wieghardt, K.; Hildebrandt, P. Spectroscopic identification of different types of copper centers generated in synthetic four-helix bundle proteins. *J. Am. Chem. Soc.* **2004**, *126*, 14389–14399.

56. Hellinga, H. W. Construction of a blue copper analogue through iterative rational protein design cycles demonstrates principles of molecular recognition in metal center formation. *J. Am. Chem. Soc.* **1998**, *120*, 10055–10066.

57. Lee, Y.; Lee, D.-H.; Narducci Sarjeant, A. A.; Karlin, K. D. Thiol-copper(I) and disulfide-dicopper(I) complex O_2-reactivity leading to sulfonate-copper(II) complex or the

formation of a cross-linked thioether-phenol product with phenol addition. *J. Inorg. Biochem.* **2007**, *101*, 1845–1858.

58. Dooley, D. M.; Scott, R. A.; Knowles, P. F.; Colangelo, C. M.; McGuirl, M. A.; Brown, D. E. Structures of the Cu(I) and Cu(II) forms of amine oxidases from X-ray absorption spectroscopy. *J. Am. Chem. Soc.* **1998**, *120*, 2599–2605.

59. Dooley, D. M. Structure and biogenesis of topaquinone and related cofactors. *J. Biol. Inorg. Chem.* **1999**, *4*, 1–11.

60. Brown, D. E.; McGuirl, M. A.; Dooley, D. M.; Janes, S. M.; Mu, D.; Klinman, J. P. The organic functional group in copper-containing amine oxidases. Resonance raman spectra are consistent with the presence of topa quinone (6-hydroxydopa quinone) in the active site. *J. Biol. Chem.* **1991**, *266*, 4049–4051.

61. Kim, M.; Okajima, T.; Kishishita, S.; Yoshimura, M.; Kawamori, A.; Tanizawa, K.; Yamaguchi, H. X-ray snapshots of quinone cofactor biogenesis in bacterial copper amine oxidase. *Nature Struct. Biol.* **2002**, *9*, 591–596.

62. Matsunami, H.; Okajima, T.; Hirota, S.; Yamaguchi, H.; Hori, H.; Kuroda, S.; Tanizawa, K. Chemical rescue of a site-specific mutant of bacterial copper amine oxidase for generation of the topa quinone cofactor. *Biochemistry* **2004**, *43*, 2178–2187.

63. Nakamura, N.; Matsuzaki, R.; Choi, Y.-H.; Tanizawa, K.; Sanders-Loehr, J. Biosynthesis of topa quinone cofactor in bacterial amine oxidases. *J. Biol. Chem.* **1996**, *271*, 4718–4724.

64. Okajima, T.; Kishishita, S.; Chiu, Y. C.; Murakawa, T.; Kim, M.; Yamaguchi, H.; Hirota, S.; Kuroda, S.; Tanizawa, K. Reinvestigation of metal ion specificity for quinone cofactor biogenesis in bacterial copper amine oxidase. *Biochemistry* **2005**, *44*, 12041–12048.

65. Ruggiero, C. E.; Smith, J. A.; Tanizawa, K.; Dooley, D. M. Mechanistic studies of topa quinone biogenesis in phenylethylamine oxidase. *Biochemistry* **1997**, *36*, 1953–1959.

66. Ruggiero, C. E.; Dooley, D. M. Stoichiometry of the topa quinone biogenesis reaction in copper amine oxidases. *Biochemistry* **1999**, *38*, 2892–2898.

67. Dove, J. E.; Schwartz, B.; Williams, N. K.; Klinman, J. P. Investigation of spectroscopic intermediates during copper-binding and TPQ formation in wild-type and active-site mutants of a copper-containing amine oxidase from yeast. *Biochemistry* **2000**, *39*, 3690–3698.

68. DuBois, J. L.; Klinman, J. P. The nature of O_2 reactivity leading to topa quinone in the copper amine oxidase from *hansenula polymorpha* and its relationship to catalytic turnover. *Biochemistry* **2005**, *44*, 11381–11388.

69. DuBois, J. L.; Klinman, J. P. Mechanism of post-translational quinone formation in copper amine oxidases and its relationship to the catalytic turnover. *Arch. Biochem. Biophys.* **2005**, *433*, 255–265.

70. DuBois, J. L.; Klinman, J. P. Role of a strictly conserved active site tyrosine in cofactor genesis in the copper amine oxidase from *hansenula polymorpha*. *Biochemistry* **2006**, *45*, 3178–3188.

71. Hevel, J. M.; Mills, S. A.; Klinman, J. P. Mutation of a strictly conserved, active-site residue alters substrate specificity and cofactor biogenesis in a copper amine oxidase. *Biochemistry* **1999**, *38*, 3683–3693.

72. Samuels, N. M.; Klinman, J. P. 2,4,5-trihydroxyphenylalanine quinone biogenesis in the copper amine oxidase from *hansenula polymorpha* with the alternate metal nickel. *Biochemistry* **2005**, *44*, 14308–14317.

73. Schwartz, B.; Dove, J. E.; Klinman, J. P. Kinetic analysis of oxygen utilization during cofactor biogenesis in a copper-containing amine oxidase from yeast. *Biochemistry* **2000**, *39*, 3699–3707.

74. Schwartz, B.; Olgin, A. K.; Klinman, J. P. The role of copper in topa quinone biogenesis and catalysis, as probed by azide inhibition of a copper amine oxidase from yeast. *Biochemistry* **2001**, *40*, 2954–2963.

75. Chen, Z. W.; Schwartz, B.; Williams, N. K.; Li, R. B.; Klinman, J. P.; Mathews, F. S. Crystal structure at 2.5 Å resolution of zinc-substituted copper amine oxidase of *Hansenula polymorpha* expressed in *Escherichia coli*. *Biochemistry* **2000**, *39*, 9709–9717.

76. Prabhakar, R.; Siegbahn, P. E. M. A theoretical study of the mechanism for the biogenesis of cofactor topaquinone in copper amine oxidases. *J. Am. Chem. Soc.* **2004**, *126*, 3996–4006.

77. Nakamura, N.; Matsuzaki, R.; Choi, Y. H.; Tanizawa, K.; Sanders-Loehr, J. Biosynthesis of topa quinone cofactor in bacterial amine oxidases-solvent origin of C-2 oxygen determined by raman spectroscopy. *J. Biol. Chem.* **1996**, *271*, 4718–4724.

78. Moore, R. H.; Spies, M. A.; Culpepper, M. B.; Murakawa, T.; Hirota, S.; Okajima, T.; Tanizawa, K.; Mure, M. Trapping of a dopaquinone intermediate in the TPQ cofactor biogenesis in a copper-containing amine oxidase from *arthrobacter globiformis*. *J. Am. Chem. Soc.* **2007**, *129*, 11524–11534.

79. Wang, S. X.; Mure, M.; Medzihradszky, K. F.; Burlingame, A. L.; Brown, D. E.; Dooley, D. M.; Smith, A. J.; Kagan, H. M.; Klinman, J. P. A crosslinked cofactor in lysyl oxidase: redox function for amino acid side chains. *Science* **1996**, *273*, 1078–1084.

80. Mure, M.; Wang, S. X.; Klinman, J. P. Synthesis and characterization of model compounds of the lysine tyrosyl quinone cofactor of lysyl oxidase. *J. Am. Chem. Soc.* **2003**, *125*, 6113–6125.

81. Bollinger, J. A.; Brown, D. E.; Dooley, D. M. The formation of lysyltyrosine quinone (LTQ) is a self-processing reaction. Expression and characterization of a drosophila lysyl oxidase. *Biochemistry* **2005**, *44*, 11708–11714.

82. Matsuzaki, R.; Fukui, T.; Sato, H.; Ozaki, Y.; Tanizawa, K. Generation of the topa quinone cofactor in bacterial monoamine oxidase by cupric ion-dependent autooxidation of a specific tyrosyl residue. *FEBS Lett.* **1994**, *351*, 360–364.

83. Samuels, N. M.; Klinman, J. P. Investigation of Cu(I)-dependent 2,4,5-trihydroxyphenylalanine quinone biogenesis in *hansenula polymorpha* amine oxidase. *J. Biol. Chem.* **2006**, *281*, 21114–21118.

84. Whittaker, M. M.; Whittaker, J. W. The active site of galactose oxidase. *J. Biol. Chem.* **1988**, *263*, 6074–6080.

85. Whittaker, M. M.; Ballou, D. P.; Whittaker, J. W. Kinetic isotope effects as probes of the mechanism of galactose oxidase. *Biochemistry* **1998**, *37*, 8426–8436.

86. Whittaker, M. M.; Whittaker, J. W. Catalytic reaction profile for alcohol oxidation by galactose oxidase. *Biochemistry* **2001**, *40*, 7140–7148.

87. Himo, F.; Eriksson, L. A.; Maseras, F.; Siegbahn, P. E. M. Catalytic mechanism of galactose oxidase: a theoretical study. *J. Am. Chem. Soc.* **2000**, *122*, 8031–8036.

88. Whittaker, M. M.; Whittaker, J. W. Ligand interactions with galactose oxidase: mechanistic insights. *Biophys. J.* **1993**, *64*, 762–772.

89. Lee, Y. K.; Whittaker, M. M.; Whittaker, J. W. The electronic structure of the Cys-Tyr(•) free radical in galactose oxidase determined by EPR spectroscopy. *Biochemistry* **2008**, *47*, 6637–6649.

90. Rokhsana, D.; Dooley, D. M.; Szilagyi, R. K. Structure of the oxidized active site of galactose oxidase from realistic in silico models. *J. Am. Chem. Soc.* **2006**, *128*, 15550–15551.

91. Rokhsana, D.; Dooley, D. M.; Szilagyi, R. K. Systematic development of computational models for the catalytic site in galactose oxidase: impact of outer-sphere residues on the geometric and electronic structures. *J. Biol. Inorg. Chem.* **2008**, *13*, 371–383.

92. Knowles, P. F.; Brown, R. D.; Koenig, S. H.; Wang, S.; Scott, R. A.; McGuirl, M. A.; Brown, D. E.; Dooley, D. M. Spectroscopic studies of the active site of galactose oxidase. *Inorg. Chem.* **1995**, *34*, 3895–3902.

93. Reynolds, M. P.; Baron, A. J.; Wilmot, C. M.; Philips, S. E. V.; Knowles, P. F.; McPherson, M. J. Tyrosine 495 is a key residue in the active site of galactose oxidase. *Biochem. Soc. Trans.* **1995**, *23*, 5105.

94. McGlashen, M. L.; Eads, D. D.; Spiro, T. G.; Whittaker, J. W. Resonance Raman Spectroscopy of Galactose Oxidase. A New Interpretation Based on Model Compound Free Radical Spectra. *J. Phys. Chem.* **1995**, *99*(14), 4918–4922.

95. Reynolds, M. P.; Baron, A. J.; Wilmot, C. W.; Vinecombe, E.; Stevens, C.; Phillips, S. E. V.; Knowles, P. F.; McPherson, M. J. Structure and mechanism of galactose oxidase: catalytic role of tyrosine 495. *J. Biol. Inorg. Chem.* **1997**, *2*, 327–335.

96. Rogers, M. S.; Tyler, E. M.; Akyumani, N.; Kurtis, C. R.; Spooner, R. K.; Deacon, S. E.; Tamber, S.; Firbank, S. J.; Mahmoud, K.; Knowles, P. F.; Phillips, S. E.; McPherson, M. J.; Dooley, D. M. The stacking tryptophan of galactose oxidase: a second-coordination sphere residue that has profound effects on tyrosyl radical behavior and enzyme catalysis. *Biochemistry* **2007**, *46*, 4606–4618.

97. Wachter, R. M.; Montague-Smith, M. P.; Branchaud, B. P. Beta-haloethanol substrates as probes for radical mechanisms for galactose oxidase. *J. Am. Chem Soc.* **1997**, *119*, 7743–7749.

98. Wachter, R. M.; Branchaud, B. P. Molecular modeling studies on oxidation of hexopyranoses by galactose oxidase. an active site topology apparently designed to catalyze radical reactions, either concerted or stepwise. *J. Am. Chem Soc.* **1996**, *118*, 2782–2789.

99. Wachter, R. M.; Branchaud, B. P. Thiols as mechanistic probes for catalysis by the free radical enzyme galactose oxidase. *Biochemistry* **1996**, *35*, 14425–14435.

100. Humphreys, K. J.; Mirica, L. M.; Wang, Y.; Klinman, J. P. Galactose oxidase as a model for reactivity at a copper superoxide center. *J. Am. Chem. Soc.* **2009**, *131*, 4657–4663.

101. Branchaud, B. P.; Montague-Smith, M. P.; Kosman, D. J.; McLaren, F. R. Mechanism-based inactivation of galactose oxidase: evidence for a radical mechanism. *J. Am. Chem. Soc.* **1993**, *115*, 798–800.

102. Clark, K.; Penner-Hahn, J. E.; Whittaker, M. M.; Whittaker, J. W. Oxidation-state assignments for galactose oxidase complexes from X-ray absorption spectroscopy. evidence for Cu(II) in the active enzyme. *J. Am. Chem. Soc.* **1990**, *112*, 6433–6434.

103. Clark, K.; Penner-Hahn, J. E.; Whittaker, M.; Whittaker, J. W. Structural characterization of the copper site in galactose oxidase using X-ray absorption spectroscopy. *Biochemistry* **1994**, *33*, 12553–12557.

104. Whittaker, J. W. Free radical catalysis by galactose oxidase. *Chem. Rev.* **2003**, *103*, 2347–2363.

105. Hamilton, G. A.; Adolf, P. K.; de Jersey, J.; DuBois, G. C.; Dyrkacz, G. R.; Libby, R. D. Trivalent copper, superoxide, and galactose oxidase. *J. Am. Chem. Soc.* **1978**, *100*, 1899–1911.

106. Elmore, B.; Bollinger, J. A.; Dooley, D. M. Human kidney diamine oxidase: heterologous expression, purification, and characterization. *J. Biol. Inorg. Chem.* **2002**, *7*, 565–579.

107. Kuchar, J.; Dooley, D. Cloning, sequence analysis, and characterization of the "lysyl oxidase" from *pichia pastoris*. *J. Inorg. Biochem.* **2001**, *83*, 193–204.

108. Dooley, D. M.; McGuirl, M. A.; Brown, D. E.; Turowski, P. N.; McIntire, W. S.; Knowles, P. F. A Cu(I)-semiquinone state in substrate-reduced amine oxidases. *Nature (London)* **1991**, *349*, 262–264.

109. Murray, J. M.; Saysell, C. G.; Wilmot, C. M.; Tambyrajah, W. S.; Jaeger, J.; Knowles, P. F.; Phillips, S. E. V.; McPherson, M. J. The active site base controls cofactor reactivity in *escherichia coli* amine oxidase: X-ray crystallographic studies with mutational variants. *Biochemistry* **1999**, *38*, 8217–8227.

110. Shepard, E. M.; Okonski, K. M.; Dooley, D. M. Kinetics and spectroscopic evidence that the Cu(I)-semiquinone intermediate reduces molecular oxygen in the oxidative half-reaction of *arthrobacter globiformis* amine oxidase. *Biochemistry* **2008**, *47*, 13907–13920.

111. Su, Q. J.; Klinman, J. P. Probing the mechanism of proton coupled electron transfer to dioxygen: the oxidative half-reaction of bovine serum amine oxidase. *Biochemistry* **1998**, *37*, 12513–12525.

112. Mills, S. A.; Klinman, J. P. Evidence against reduction of Cu^{2+} to Cu^+ during dioxygen activation in a copper amine oxidase from yeast. *J. Am. Chem. Soc.* **2000**, *122*, 9897–9904.

113. Medda, R.; Padiglia, A.; Pedersen, J. Z.; Rotilio, G.; Finazzi-Agro, A.; Floris, G. The reaction mechanism of copper amine oxidase: detection of intermediates by the use of substrates and inhibitors. *Biochemistry* **1995**, *34*, 16375–16381.

114. Mukherjee, A.; Smirnov, V. V.; Lanci, M. P.; Brown, D. E.; Shepard, E. M.; Dooley, D. M.; Roth, J. P. Inner-sphere mechanism for molecular oxygen reduction catalyzed by copper amine oxidases. *J. Am. Chem. Soc.* **2008**, *130*, 9459–9473.

115. Mure, M.; Mills, S. A.; Klinman, J. P. Catalytic mechanism of the topa quinone containing copper amine oxidases. *Biochemistry* **2002**, *41*, 9269–9278.

116. Mills, S. A.; Goto, Y.; Su, Q. J.; Plastino, J.; Klinman, J. P. Mechanistic comparison of the cobalt-substituted and wild-type copper amine oxidase from *hansenula polymorpha*. *Biochemistry* **2002**, *41*, 10577–10584.

117. Juda, G. A.; Shepard, E. M.; Elmore, B. O.; Dooley, D. M. A comparative study of the binding and inhibition of four copper-containing amine oxidases by azide: implications for the role of copper during the oxidative half-reaction. *Biochemistry* **2006**, *45*, 8788–8800.

118. McGuirl, M. A.; Brown, D. E.; Dooley, D. M. Cyanide as a copper-directed inhibitor of amine oxidases:implications for the mechanism of amine oxidation. *J. Biol. Inorg. Chem.* **1997**, *2*, 336–343.

119. Shepard, E. M.; Juda, G. A.; Ling, K. Q.; Sayre, L. M.; Dooley, D. M. Cyanide as a copper and quinone-directed inhibitor of amine oxidases from pea-seedlings (*Pisum savitum*) and *Arthrobacter globiformis*: evidence for both copper coordination and cyanohydrin derivatization of the quinone cofactor. *J. Biol. Inorg. Chem.* **2004**, *9*, 256–268.

120. Duff, A. P.; Trambaiolo, D. M.; Cohen, A. E.; Ellis, P. J.; Juda, G. A.; Shepard, E. M.; Langley, D. B.; Dooley, D. M.; Freeman, H. C.; Guss, J. M. Using xenon as a probe for dioxygen-binding sites in copper amine oxidases. *J. Mol. Biol.* **2004**, *344*, 599–607.

121. Johnson, B. J.; Cohen, J.; Welford, R. W.; Pearson, A. R.; Schulten, K.; Klinman, J. P.; Wilmot, C. M. Exploring molecular oxygen pathways in *hansenula polymorpha* copper-containing amine oxidase. *J. Biol. Chem.* **2007**, *282*, 17767–17776.

122. Wilmot, C. M.; Hajdu, J.; McPherson, M. J.; Knowles, P. F.; Phillips, S. E. V. Visualization of dioxygen bound to copper during enzyme catalysis. *Science* **1999**, *286*, 1724–1728.

123. Smith, M. A.; Pirrat, P.; Pearson, A. R.; Kurtis, C. R. P.; Trinh, C. H.; Gaule, T. G.; Knowles, P. F.; Phillips, S. E. V.; McPherson, M. J. Exploring the roles of the metal ions in *Escherichia coli* copper amine oxidase. *Biochemistry* **2010**, *49*, 1268–1280.

124. Dooley, D. M.; Brown, D. E. Intramolecular electron transfer in the oxidation of amines by methylamine oxidase from *Arthrobacter P1*. *J. Bio. Inorg. Chem.* **1996**, *1*, 205–209.

125. Shepard, E. M.; Dooley, D. M. Intramolecular electron transfer rate between active-site copper and TPQ in *arthrobacter globiformis* amine oxidase. *J Biol Inorg. Chem.* **2006**, *11*, 1039–1048.

126. Turowski, P. N.; McGuirl, M. A.; Dooley, D. M. Intramolecular electron transfer rate between active-site copper and topa quinone in pea seedling amine oxidase. *J. Biol. Chem.* **1993**, *268*, 17680–17682.

127. Alton, G.; Taher, T. H.; Beever, R. J.; Palcic, M. M. Stereochemistry of benzylamine oxidation by copper amine oxidases. *Arch. Biochem. Biophys.* **1995**, *316*, 353–361.

128. Coleman, A. A.; Scaman, C. H.; Kang, Y. J.; Palcic, M. M. Stereochemical trends in copper amine oxidase reactions. *J. Biol. Chem.* **1991**, *266*, 6795–6800.

129. Uchida, M.; Ohtani, A.; Kohyama, N.; Okajima, T.; Tanizawa, K.; Yamamoto, Y. Stereochemistry of 2-phenylethylamine oxidation catalyzed by bacterial copper amine oxidase. *Biosci. Biotechnol. Biochem.* **2003**, *67*, 2664–2667.

130. Langley, D. B.; Trambaiolo, D. M.; Duff, A. P.; Dooley, D. M.; Freeman, H. C.; Guss, J. M. Complexes of the copper-containing amine oxidase from *arthrobacter globiformis* with the inhibitors benzylhydrazine and tranylcypromine. *Acta Crystallogr., Sect. F* **2008**, *64*, 577–583.

131. Wilmot, C. M.; Saysell, C. G.; Blessington, A.; Conn, D. A.; Kurtis, C. R.; McPherson, M. J.; Knowles, P. F.; Phillips, S. E. Medical implications from the crystal structure of a copper-containing amine oxidase complexed with the antidepressant drug tranylcypromine. *FEBS Lett.* **2004**, *576*, 301–305.

132. Chiu, Y. C.; Okajima, T.; Murakawa, T.; Uchida, M.; Taki, M.; Hirota, S.; Kim, M.; Yamaguchi, H.; Kawano, Y.; Kamiya, N.; Kuroda, S.; Hayashi, H.; Yamamoto, Y.; Tanizawa, K. Kinetic and structural studies on the catalytic role of the aspartic acid residue conserved in copper amine oxidase. *Biochemistry* **2006**, *45*, 4105–4120.

133. Plastino, J.; Green, E. L.; Sanders-Loehr, J.; Klinman, J. P. An unexpected role for the active site base in cofactor orientation and flexibility in the copper amine oxidase from *hansenula polymorpha*. *Biochemistry* **1999**, *38*, 8204–8216.

134. Farnum, M. F.; Klinman, J. P. Stereochemical probes of bovine plasma amine oxidase: evidence for mirror image processing and a syn abstraction of hydrogens from C-1 and C-2 of dopamine. *Biochemistry* **1986**, *25*, 6028–6036.

135. Murray, J. M.; Kurtis, C. R.; Tambyrajah, W.; Saysell, C. G.; Wilmot, C. M.; Parsons, M. R.; Phillips, S. E. V.; Knowles, P. F.; McPherson, M. J. Conserved tyrosine-369 in the active site of *escherichia coli* copper amine oxidase is not essential. *Biochemistry* **2001**, *40*, 12808–12818.

136. Cramer, C. J.; Tolman, W. B. Mononuclear Cu-O$_2$ complexes: geometries, spectroscopic properties, electronic structures, and reactivity. *Acc. Chem. Res.* **2007**, *40*, 601–608.

137. Mure, M.; Klinman, J. P. Model studies of topaquinone-dependent amine oxidases: characterization of reaction intermediates and mechanism. *J. Am. Chem. Soc.* **1995**, *117*, 8707–8718.

138. Ho, R. Y. N.; Liebman, J. F.; Valentine, J. S. Overview of the Energetics and Reactivity of Oxygen. in *Active Oxygen in Chemistry*, Foote, C. S., Valentine, J. S., Greenberg, A., Liebman, J. F.,Eds., Blackie Academic and Professional, New York, **1995**, pp 1–23.

139. Murthy, N. N.; Karlin, K. D. *Mechanistic Bioinorganic Chemistry*, Thorp, H. H., Pecoraro, V. L.,Eds., American Chemical Society, Washington DC, **1995**, pp 165–193.

140. Karlin, K. D.; Tyekl'ar, Z. *Bioinorganic Chemistry of Copper*, Chapman & Hall, New York, **1993**; pp 506.

141. Roth, J. P. Advances in studying bioinorganic reactions mechanisms: isotopic probes of activated oxygen intermediates in metalloenzymes. *Curr. Opin. Chem. Biol.* **2007**, *11*, 142–150.

142. Fry, H. C.; Scaltrito, D. V.; Karlin, K. D.; Meyer, G. J. The rate of O$_2$ and CO binding to a copper complex, determined by a "flash-and-trap" technique, exceeds that for hemes. *J. Am. Chem. Soc.* **2003**, *125*, 11866–11871.

143. Dooley, D. M.; Golnik, K. C. Spectroscopic and kinetic studies of the inhibition of pig kidney diamine oxidase by anions. *J. Biol. Chem.* **1983**, *258*, 4245–4248.

144. Dooley, D. M.; Cote, C. E. Copper(II) coordination chemistry in bovine plasma amine oxidase: azide and thiocyanate binding. *Inorg. Chem.* **1985**, *24*, 3996–4000.

145. Barker, R.; Boden, N.; Cayley, G.; Charleton, S. C.; Henson, R.; Holmes, M. C.; Kelly, I. D.; Knowles, P. F. Properties of cupric ions in benzylamine oxidase from pig plasma as studied by magnetic-resonance and kinetic methods. *Biochem. J.* **1979**, *177*, 289–302.

146. Olsson, B.; Olsson, J.; Pettersson, G. Effects on enzyme activity of ligand binding to copper in pig plasma benzylamine oxidase. *Eur. J. Biochem.* **1978**, *87*, 1–8.

147. He, Z.; Zou, Y.; Greenaway, F. T. Cyanide inhibition of porcine kidney diamine oxidase and bovine plasma amine oxidase: evidence for multiple interaction sites. *Arch. Biochem. Biophys.* **1995**, *319*, 185–195.

148. Finazzi-Agro, A.; Rinaldi, A.; Floris, G.; Rotilio, G. A free-radical intermediate in the reduction of plant Cu-amine oxidases. *FEBS Lett.* **1984**, *176*, 378–380.

149. McCracken, J.; Peisach, J.; Dooley, D. M. Cu(II) coordination of amine oxidases: pulsed EPR studies of histidine imidazole, water and exogenous ligand coordination. *J. Am. Chem. Soc.* **1987**, *109*, 4064–4072.

150. Lindstrom, A.; Olsson, B.; Pettersson, G. Effect of azide on some spectral and kinetic properties of pig plasma benzylamine oxidase. *Eur. J. Biochem.* **1974**, *48*, 237–243.

151. Dooley, D. M.; Cote, C. E.; Golnik, K. C. Inhibition of copper-containing amine oxidases by Cu(II) complexes and anions. *J. Mol. Catal.* **1984**, *23*, 243–253.

152. Dooley, D. M.; McGuirl, M. A.; Cote, C. E.; Knowles, P. F.; Singh, I.; Spiller, M.; Brown, R. D., III; Koenig, S. H. Coordination chemistry of copper-containing amine oxidases: nuclear magnetic relaxation dispersion studies of copper binding, solvent-water exchange, substrate and inhibitor binding, and protein aggregation. *J. Am. Chem. Soc.* **1991**, *113*, 754–761.

153. Pirrat, P.; Smith, M. A.; Pearson, A. R.; McPherson, M. J.; Phillips, S. E. V. Structure of a xenon derivative of *escherichia coli* copper amine oxidase: confirmation of the proposed oxygen-entry pathway. *Acta Crystallogr., Sect. F: Struct. Biol. Cryst. Commun.* **2008**, *64*, 1105–1109.

154. Medda, R.; Mura, A.; Longu, S.; Anedda, R.; Padiglia, A.; Casu, M.; Floris, G. An unexpected formation of the spectroscopic Cu(I)-semiquinone radical by xenon-induced self-catalysis of a copper quinoprotein. *Biochimie* **2006**, *88*, 827–835.

155. Goto, Y.; Klinman, J. P. Binding of dioxygen to non-metal sites in proteins: exploration of the importance of binding site size versus hydrophobicity in the copper amine oxidase from *hansenula polymorpha*. *Biochemistry* **2002**, *41*, 13637–13643.

156. Takahashi, K.; Klinman, J. P. Relationship of stopped flow to steady state parameters in the dimeric copper amine oxidase from *hansenula polymorpha* and the role of zinc in inhibiting activity at alternate copper-containing subunits. *Biochemistry* **2006**, *45*, 4683–4694.

157. Agostinelli, E.; De Matteis, G.; Mondovi, B.; Morpurgo, L. Reconstitution of Cu^{2+}-depleted bovine serum amine oxidase with Co^{2+}. *Biochem. J.* **1998**, *330*, 383–387.

158. Prabhakar, R.; Siegbahn, P. E. M.; Minaev, B. F. A theoretical study of the dioxygen activation by glucose oxidase and copper amine oxidase. *Biochim. Biophys. Acta Protein Struct. Mol. Enzymol.* **2003**, *1647*, 173–178.

159. Hirota, S.; Iwamoto, T.; Kishishita, S.; Okajima, T.; Yamauchi, O.; Tanizawa, K. Spectroscopic observation of intermediates formed during the oxidative half-reaction of copper/topa quinone-containing phenylethylamine oxidase. *Biochemistry* **2001**, *40*, 15789–15796.

160. Medda, R.; Padiglia, A.; Bellelli, A.; Sarti, P.; Santanche, S.; Agro, A. F.; Floris, G. Intermediates in the catalytic cycle of lentil (lens esculenta) seedling copper-containing amine oxidase. *Biochem. J.* **1998**, *332*, 431–437.

161. Padiglia, A.; Medda, R.; Bellelli, A.; Agostinelli, E.; Morpurgo, L.; Mondovi, B.; Finazzi-Agrò, A.; Floris, G. The reductive and oxidative half-reaction and the role of copper ions in plant and mammalian copper-amine oxidases. *Eur. J. Inorg. Chem.* **2001**, *1*, 35–42.

162. Padiglia, A.; Medda, R.; Pedersen, J. Z.; Finazzi-Agro, A.; Lorrai, A.; Murgia, B.; Floris, G. Effect of metal substitution in copper amine oxidase from lentil seedlings. *J. Biol. Inorg. Chem.* **1999**, *4*, 608–613.

163. Roth, J. P.; Wincek, R.; Nodet, G.; Edmondson, D. E.; McIntire, W. S.; Klinman, J. P. Oxygen isotope effects on electron transfer to O-2 probed using chemically modified flavins bound to glucose oxidase. *J. Am. Chem. Soc.* **2004**, *126*, 15120–15131.

164. Roth, J. P.; Klinman, J. P. *Isotope Effects in Chemistry and Biology*, CRC Press, Boca Raton FL, **2006**.

165. Smirnov, V. V.; Roth, J. P. Mechanisms of electron transfer in catalysis by copper zinc superoxide dismutase. *J. Am. Chem. Soc.* **2006**, *128*, 16424–16425.

166. Roth, J. P. Oxygen isotope effects as probes of electron transfer mechanisms and structures of activated O-2. *Acc. Chem. Res.* **2009**, *42*, 399–408.

167. Welford, R. W. D.; Lam, A.; Mirica, L. M.; Klinman, J. P. Partial conversion of *hansenula polymorpha* amine oxidase into a "plant" amine oxidase: implications for copper chemistry and mechanism. *Biochemistry* **2007**, *46*, 10817–10827.

168. Roth, J. P.; Dooley, D. M., in preperation.

4

ENERGY CONVERSION
AND CONSERVATION
BY CYTOCHROME OXIDASES

ANGELA PAULUS AND SIMON DE VRIES

Department of Biotechnology, Delft University of Technology, Delft, The Netherlands

Copper-Oxygen Chemistry, First Edition. Edited by Kenneth D. Karlin and Shinobu Itoh.
© 2011 John Wiley & Sons, Inc. Published 2011 by John Wiley & Sons, Inc.

4.1. INTRODUCTION

Aerobic organisms are found in all three Domains of Life: archaea, bacteria, and eukarya. In the aerobic organisms, oxygen serves as the terminal electron acceptor gathering the excess reducing equivalents produced in cellular metabolism, thereby making the overall chemistry of the cell electroneutral. The reduction of oxygen by membrane-bound terminal oxidases is usually coupled to the generation of a proton motive force, which enables the cell, for example, to produce adenosine triphosphate (ATP) and to import nutrients. Three evolutionary independent classes of terminal oxidases occur in nature: The heme–copper, the cytochrome *bd*, and the di-iron oxidases. The heme–copper oxidases are highly efficient proton pumps with a stoichiometry of eight protons translocated per oxygen reduced. The cytochrome *bd* oxidases, quinol-linked terminal oxidases harboring hemes *b* and *d*, are not proton pumps, but the reaction is electrogenic due to the uptake of four protons per water molecule formed from one side of the membrane. The cytochrome *bd* oxidases thus have half the bioenergetic efficiency of the heme–copper oxidases. The membrane-bound di-iron oxidases occur mainly in plant mitochondria and do not translocate protons. All free energy of the reduction of oxygen to water is released in the form of heat, which is used as a signal for flowering.

The superfamily of heme–copper oxidases comprises not only the cytochrome oxidases, which catalyze the reduction of molecular oxygen to water, but also the NO reductases, which reduce NO to N_2O.[1–6] The close relation between oxidases and NO reductases is illustrated in studies showing that the latter enzymes quite efficiently catalyze the reduction of oxygen to water,[7] while the cytochrome oxidases can reduce NO to N_2O, albeit at a low rate.[8] These observations pertaining to function suggest a high similarity between the structures of oxidases and NO reductases, as is also borne out by the sequence similarities between the two types of enzymes in the superfamily and their comparable biophysical properties.[7,9,10] The reduction of oxygen (Eq. 4.1) by cytochrome oxidases (but not by NO reductases) generates a proton electrochemical gradient across the cytoplasmic or mitochondrial membrane. Four protons are used for the formation of water and four are pumped across the membrane according to

$$4 \, \text{cyt} \, c^{2+} + O_2 + 8 \, H^+_c \rightarrow 4 \, \text{cyt} \, c^{3+} + 2 \, H_2O + 4 \, H^+_p \qquad (4.1)$$

where H^+_c are protons taken up from the cytoplasm (or mitochondrial matrix) and H^+_p are protons ejected to the periplasm (or mitochondrial intermembrane space).[11–15] Instead of cytochrome *c*, azurin, 4Fe–4S ferredoxins, ubiquinol, or menaquinol may serve as an electron donor. The ubiquinols, for example, are the reductants to the cytochrome *bo₃* of *Escherichia coli*.[16]

Bioinformatic analysis combined with three-dimensional (3D)-structural information has led to the classification of the superfamily of heme–copper oxidases into three oxidase families (Types A–C)[17]; a more recent bioinformatic search has identified eight oxidase families (Types A–C mentioned above and Types D–H) and five NO reductase families.[18] The Type D–H oxidases occur predominantly in the archaea and structural and kinetic characterization of these enzymes is presently very limited.

Crystal structures have been determined for Type A oxidases (e.g., cytochrome aa_3 oxidase from *Paracoccus denitrificans* [1QLE],[19] *Rhodobacter sphaeroides* cytochrome oxidase [2GSM],[20] *E. coli* cytochrome bo_3 oxidase [1FFT],[21] mitochondrial cytochrome aa_3 oxidase [2DYR][22]) and for Type B oxidases (*Thermus thermophilus* cytochrome ba_3 oxidase [1XME][23]). Recently Type C, the cytochrome cbb_3 oxidases [3MK7] were also included.[24] All these oxidases are named according to the types of heme groups that they contain.

This chapter discusses the details of the catalytic mechanism of mainly the Types A and B oxidases highlighting electron transfer (ET)-coupled proton-pumping. For these oxidases, a wealth of kinetic data is available that can be related to their known structures. Recent reviews covering cytochrome bd oxidases and archaeal oxidases are found in Refs. 25 and 18,26, respectively.

4.2. STRUCTURAL FEATURES OF HEME–COPPER OXIDASES

Cytochrome c oxidases are multisubunit enzymes having three subunits in common. Subunit I harbors the entire binuclear active site (heme Fe–Cu_B) and the heme center that acts as the direct electron donor to the binuclear center. The ET pathway extends to the Cu_A center in subunit II, near the cytochrome c binding site. The precise location of the quinol binding site in, for example, the bo_3 oxidases, is not known, but it may reside at the interface of subunits I and II. The proton-transfer pathways are mainly located in subunit I, a few residues being contributed by subunit II. Purified bacterial enzyme preparations consisting of subunits I and II only are capable of reducing oxygen at high rates, and they display full proton-pumping capacity.[27] The third subunit, an integral membrane subunit, binds phospholipids; however, the function of subunit III is not known. In some oxidases, subunits I and III are fused to a single subunit.[28] Bacterial oxidases usually contain a small fourth subunit of unknown function that consists of a single transmembrane alpha helix. Mitochondrial cytochrome oxidases may contain up to 13 subunits. The mitochondrial subunits I, II, and III are encoded by the mitochondrial deoxyribonucleic acid (DNA), pointing to their bacterial origin. The additional subunits are encoded by the nuclear DNA and appear not essential for catalytic activity *per se*. Some subunits are expressed in a tissue-specific or oxygen-dependent way.[29]

The active site where molecular oxygen binds, consists of a mono-histidine bound five-coordinate high-spin heme (heme a_3, o_3, or b_3) and a mono-copper site (Cu_B) ligated by three histidine residues (Fig. 4.1). The direct electron donor to the binuclear center is a low-spin bis(histidine) heme center (heme a or b), which receives its electrons from Cu_A or directly from the substrate quinol. The six histidine residues that serve as ligands to the metal centers are strictly conserved. The Cu_A is a mixed Cu^{2+}–Cu^{1+}/Cu^{1+}–Cu^{1+} mixed-valence center in which the two copper atoms are 2.5 Å apart[30] and that shuttles between the two redox states Cu^{2+}–Cu^{1+}

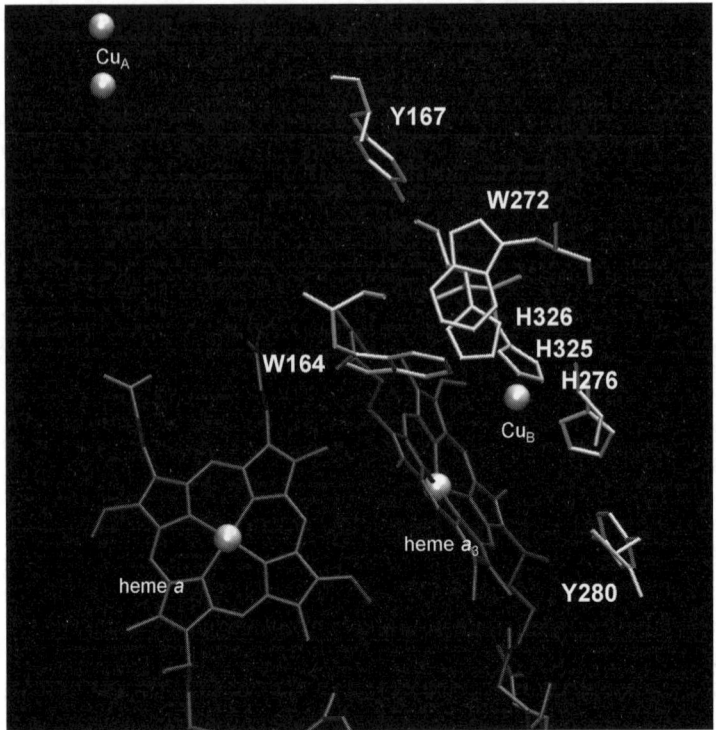

FIGURE 4.1. The active site of cytochrome c oxidase (CcO) highlighting the conserved aromatic residues in Type A and B oxidases (with the exception of W164, which occurs only in Type A oxidases) and all four redox-active metal centers. (See color insert.)

(oxidized) and Cu^{1+}–Cu^{1+} (reduced). These states are denoted as Cu_A^{2+} and Cu_A^{1+}, respectively.

4.3. ELECTRON TRANSFER

Rates of electron tunneling between the cofactors can be predicted within a factor 5–10 with the square-barrier approximation, which neglects any influence of the tunneling medium.[31] For every ~1.7 Å increase in tunneling distance, the rate of ET decreases by an order of magnitude. The best estimations for tunneling rates are obtained when the tunneling distance is determined as the shortest edge-to-edge distance between the cofactors rather than the metal-to-metal distance. The Cu_A center is located at a distance of 16.1 Å from the low-spin heme a and at 18.9 Å from heme a_3.[32] The edge-to-edge distance between hemes a and a_3 is 4.7 Å (metal-to-metal 13 Å).[33] The two metals of the binuclear center are separated by 5.3 Å in bacterial and 4.5 Å in mammalian CcO oxidase and they are antiferromagnetically coupled in the resting enzyme through a bridging peroxide residue.[34,35]

The rates of ET from Cu_A to heme a and from heme a to the binuclear center are calculated as 8.7×10^4 and $6.5 \times 10^8 \, s^{-1}$.[32] Direct measurements of ET rates yield values close to the calculations (2×10^4 [36,37] and $7 \times 10^8 \, s^{-1}$,[38] respectively). The rate of direct ET from Cu_A to heme a_3 is calculated at $\sim 250 \, s^{-1}$, which is not insignificant with respect to the turnover rate of $\sim 1000 \, e^- \, s^{-1}$.[32] The short-circuit between Cu_A and heme a_3 is generally assumed not to be functional and might be suppressed to physiologically insignificant rates in case this reaction path would have a high reorganization energy.

In addition to the metal-containing cofactors, aromatic residues might participate directly in ET. The binuclear center is surrounded by a number of conserved aromatic residues (Fig. 4.1), which can form neutral radicals during the reaction with oxygen and/or hydrogen peroxide (H_2O_2).[39–41] Furthermore, CcO from each of the three subtypes harbor a covalent cross-link of a histidine Cu_B ligand (H276) to its neighboring tyrosine (Y280).[42] The cross-link is formed post-translationally, presumably during the first turnover. Its strict conservation among Type A, B, and C oxidases suggest a vital role in the mechanism of CcO. Due to its unique electronic properties, the His-Tyr cross-link can be regarded as a cofactor that might mediate electron and proton transfer.

4.4. PROTON-CONDUCTING PATHWAYS

The protons that are taken up by cytochrome oxidase are conducted along proton pathways that consist of chains of hydrogen-bonded functional groups and water-filled cavities, forming effective proton relay networks. The proton-conducting chains also appear to contain "gaps", so that conformational changes are necessary for delivery of protons to the binuclear center or to the proton loading site. Proton-transfer rates may be modulated through these conformational changes.

At least two proton-conducting pathways were revealed by the X-ray crystal structure of the bovine enzyme.[43,44] The shorter pathway, which starts from the N-side of the membrane near a lysine residue (K354), is made up of residues from both helix VI and VIII of subunit I and is called the K-pathway (Fig. 4.2b). The K-pathway extends to the strictly conserved Y280 near the active site via T351 and S291 and conducts only two of the total of eight protons taken up during turnover. The K-pathway is used specifically in the reductive phase of the cycle. Six out of the eight protons (including all four pumped protons) are conducted by the much longer D-pathway (Fig. 4.2a). The D-pathway is located between D124 on the N-side of the membrane and the E278 near the binuclear center. It comprises the side chains of residues in helices III and IV of subunit I. Residues contributing to the D-pathway in *P. denitrificans* are D124, N199, N131, Y35, S134, S193, and E278. The E278 at the end of the pathway in Type A1 oxidases is lacking in Type A2 oxidases. Type A2 oxidases may have a tyrosine residue at a similar position, sometimes with a neighboring serine residue to form a YS-motif. The K- and the D-pathway are otherwise very well conserved among Types A1 and A2 oxidases.

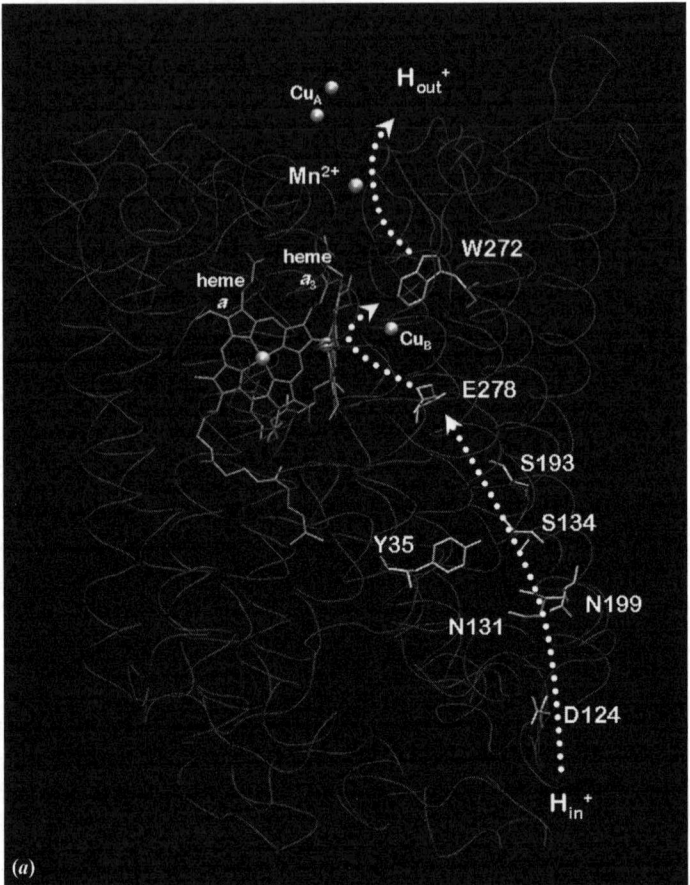

FIGURE 4.2. Amino acid residues, which contribute to the proton-conducting pathways in the type A1 cytochrome c oxidase (cytochrome aa_3 from $P.\ denitrificans$). (a) D-pathway: Protons taken up through this pathway are transferred to W272 and released to the P-side via the Mn^{2+} ion, as in the $F_{W\cdot}$ to O_H transition. (See Ref. 40.) Alternatively, the protons taken up through the D-pathway may be transferred to the binuclear center to bind to metallo-oxo/hydroxo reaction intermediates. (b) K-pathway: All protons that are taken up through this pathway are transferred to the binuclear center and play a role only in the reduction of oxygen to water, not in proton-pumping.

Sequence comparison indicates that the residues of the D- or K-pathway are not at all conserved in Type B oxidases. The crystal structure of the cytochrome ba_3 (Type B) oxidase, indicates the presence of K- and D-like proton-conducting pathways.[45] Recent mutational analyses of the Type B ba_3 oxidase have, however, shown that the four protons required for the formation of water and the pumped protons are delivered via the K-like pathway.[46] The K-pathway in Type B oxidases is formed by E15 (subunit II), T312, S309, Y248, S261, and Y237 (all located in subunit I). Mutations in the K-pathway severely diminish proton-pumping and ET rates in both the reductive

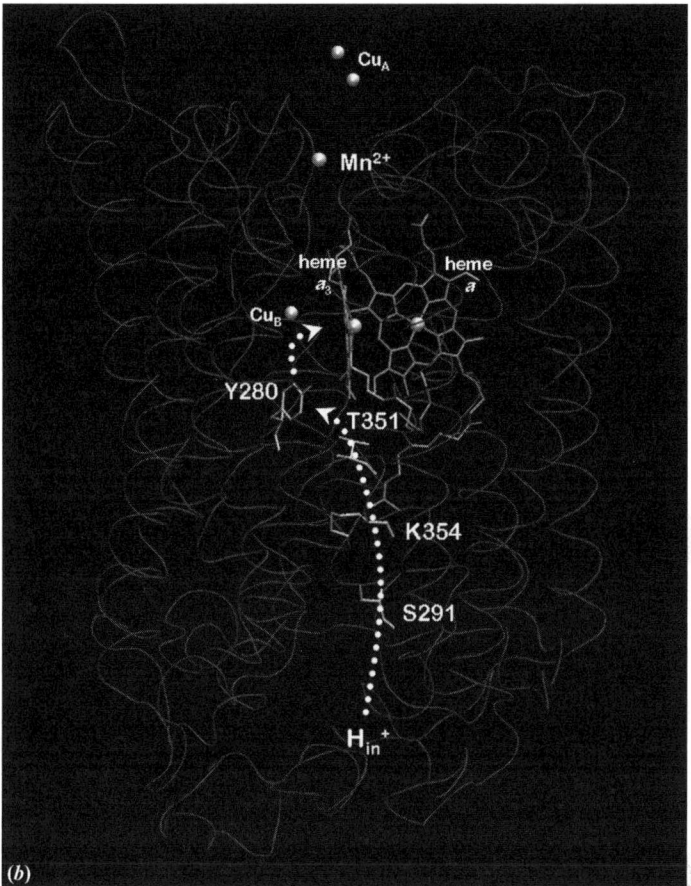

FIGURE 4.2. (*Continued*)

and the oxidative phases of the reaction. Mutations to the putative D-like proton pathway were shown to have much smaller effects or none at all. This behavior is completely different from that of Type A oxidases, where mutations blocking the D-pathway result in selective inhibition of the oxidative phase and abolished proton-pumping. Type C oxidases are suggested to use only one proton-transfer pathway, similar to Type B oxidases.[47] According to the crystal structure, this pathway is positioned equivalently to the K-pathway in Type A oxidases, although it is lined by different residues.[24]

Based on the crystal structures, additional putative proton-conducting pathways have been suggested for Type A and B oxidases. In Type A oxidases of mammalian[44,48] and bacterial origin,[43] there is a chain of protonatable residues starting at an aspartate (D407, *B. taurus*) or a glutamate residue (E442, *P. denitrificans*), leading toward heme *a*. The pathway is named after the first residue from the N-side that is conserved between the bacterial and the mammalian enzymes (H448).

The H-pathway in *P. denitrificans* may also be referred to as an E-pathway.[49] The residues of the H-pathway are only weakly conserved between mammalian and *P. denitrificans* oxidase and are not conserved in cytochrome bo_3 from *E. coli*. In addition to that, H-pathway mutants display high residual activity. The H-pathway may influence ET from heme *a* to the binuclear center through the many hydrogen-bond interactions between heme *a* and its surroundings, but it is probably not a proton-transfer pathway.

The third putative proton pathway in Type B oxidases is the Q-pathway,[50,51] which leads from a glutamine residue to the histidine ligand of heme a_3. Residues of the Q-pathway are not conserved among Type B oxidases and mutations blocking the pathway do not affect activity or proton-pumping.[46] Therefore, the Q-pathway does not appear to function as a proton entry pathway. It was suggested that the Q-pathway is involved in the release of protons to the proton exit pathway.[51]

4.5. FUNCTIONAL ASPECTS OF CYTOCHROME *c* OXIDASE

4.5.1. Proton-Pumping

Studies on C*c*O in whole cells, coupled mitochondria, or incorporated in lipid membrane vesicles have been carried out to determine the stoichiometry of proton-pumping (Table 4.1). For Type A and C oxidases it is firmly established that eight

TABLE 4.1. Overview of the Proton-Pumping Stoichiometries Determined for C*c*Os from Various Sources

Species	C*c*O Type	Hemes	Pumping Ratio ($H^+/1e^-$)	Reference	Complex
B. taurus[a]	A	aa_3	1.1	27	aa_3
P. denitrificans[b]	A	aa_3	1.04	52	aa_3
R. marinus[b]	A	caa_3	1	53	caa_3
Synechocystis sp PCC6803[b]	A	aa_3	0.9	54	aa_3
T. thermophilus[b]	B	ba_3	0.4–0.5	55	ba_3
R. sphaeroides[b]	C	cbb_3	0.6–1.1	56	cbb_3
S. acidocaldarius[c]	A	bb_3	3[d]		SoxM
	B	aa_3	1.2	57	SoxABCD
A. ambivalens[c]	B	aa_3	1	58	DoxBC
P. oguniense[c]	A	aa_3	Unknown		PoxC
	B	aa_3	Unknown		PoxI
A. pernix[c]	A	aa_3	Unknown		CO-2
	B	ba_3	Unknown		CO-1

[a] Eukaryote.
[b] Bacterium.
[c] Archaeon.
[d] Sum of the $H^+/1e^-$ ratios for the bb_3-type oxidase proton pump and for the electrogenic reaction of the bc_1-like complex.

protons are taken up during turnover, yielding an overall stoichiometry of $2 H^+/1e^-$. The stoichiometry of proton-pumping alone is $1 H^+/1e^-$, as four protons that are taken up are consumed in the chemical reduction of oxygen to water. It is tacitly assumed that all CcOs have the same total stoichiometry, although there are indications that the overall $H^+/1e^-$ ratio for Type B oxidases is smaller ($\sim$1.5 $H^+/1e^-$), yielding a stoichiometry of $0.5 H^+/1e^-$ for proton-pumping (Table 4.1). A significant difference in the stoichiometry of proton-pumping may be a reflection of the different structures of the proton pathways in Type A and B oxidases. In particular, the glutamate residue E278 in Type A oxidases that is involved in (gating) proton transfer appears to lack a protonatable equivalent in Type B oxidases. In *T. thermophilus* ba_3 oxidase, there is an isoleucine residue in this position, which may compromise the efficiency of the proton translocation process. On the other hand, the relatively low-proton-pumping stoichiometry of the Type B ba_3 oxidase may simply be due to suboptimal experimental conditions. *Thermus thermophilus* thrives at 95°C, whereas proton-pumping experiments are performed at temperatures <65°C. The observation that the Type B oxidase from *Acidianus ambivalens*[58] pumps protons at $H^+/1e^- = 1$ strengthens the belief that under the right circumstances all cytochrome oxidases translocate protons at the same maximum stoichiometry of $2 H^+/1e^-$. The effect of the specific experimental conditions on proton-pumping efficiency has been observed also in Type A[59,60] and in Type C oxidases.[61]

Gating of proton-pumping is necessary in order to prevent backflow of protons across the membrane. This backflow results in a lowered net $H^+/1e^-$ ratio and in dissipation of the membrane potential and/or proton motive force. The gating residue in Type A CcOs, acting as a valve or switch, is proposed to be E278.[62] This residue is not conserved in Type A2 and B oxidases, but might be functionally replaced by the YS motif. The pK_a value for E278 is high ($pK_a = 9.4$),[63] but in fact close to that expected for the tyrosine of the YS motif. Another possible candidate to gate proton-pumping is the strictly conserved W272, which was suggested to direct proton transfer through redox linked activity.[40,64]

4.5.2. Uncoupled and Decoupled Mutants

Point mutations in CcOs can affect the kinetics of proton uptake and thereby the stoichiometry of translocation. These mutant enzymes may reveal clues to the principles underlying proton-pumping. Most of the amino acid residues mentioned in the following paragraph are part of one of the proton-conducting pathways depicted in Figure 4.2a and b. Particularly interesting are the uncoupled mutant oxidases, which show significant or even increased oxygen reductase activity, with a severely decreased coupling of proton-pumping. The energy conservation efficiency for these enzymes is low compared to the wild-type enzyme. In fully uncoupled mutants, proton-pumping is lacking and the only contribution to the membrane potential comes from the uptake of protons and electrons from different sides of the membrane for the reduction of oxygen to water. In so-called decoupled mutants, proton translocation, proton-pumping, and respiration are all inhibited.

Several studies on site-directed mutants of the aa_3 terminal oxidase from *R. sphaeroides* have been reported.[65–69] Some of these mutants show a decreased coupling between ET and proton translocation.[66] Substitution of the *R. sphaeroides* asparagine residues N139 or N207 (N131 and N199 in *P. denitrificans*) by aspartate, for example, abolishes proton-pumping while ET activity is retained. The asparagines are located in the neck of the D-pathway and proposed to be involved in the organization of the water chain along the pathway. The mutants display a modified pH dependency for the P_R to F transition, which was interpreted as an increase in the pK_a of E278 compared to the wild-type value (10.9 instead of 9.4).[66] The decrease in coupling efficiency might therefore be explained by an increased proton affinity of E278. This result would prevent facile proton release toward the P-side, while proton transfer to the binuclear site is maintained due to the even higher proton affinity of the specific heme-Fe and Cu_B ligand states in the binuclear site. On the other hand, electrostatic calculations on the equivalent *P. denitrificans* mutant do not support this view. The influence of a mutation in the N131/N199 region (20 Å away) on the pK_a of E278 was predicted to be only minor.[70]

In crystallized samples of the N131D mutant, the water chain that connects the residues of the D-pathway appears structurally perturbed.[52] Four water molecules that are readily distinguished in the wild-type crystal structure appear missing in the mutant, suggesting higher mobility of the molecules in the water chain, which may disrupt a Grotthuss-type proton transfer along the chain. The direct environment of E278 was also altered in the mutant. The E278 side chain was found to adopt two distinct orientations (pointing toward and away from the D-pathway) in the same sample, indicating that the function of E278 as a proton "switch" or "valve" may be impeded by the mutation. In view of the above, the observation that the S189A mutation was without effect on oxygen reduction activity or proton-pumping efficiency, is unexpected.[71] In the S189A mutant, the serine residue, which is normally hydrogen bonded to water and located only 7 Å from E278 in the D-pathway, is replaced by the nonpolar alanine. Such a mutation might easily distort or disrupt the direct environment of E278 and the precise structure of the chain of water molecules. Apparently, the structure of the water chain in the D-pathway is rather robust and is not significantly affected by the hydrogen-bonding state of S189.

Mutation of the uncharged S189 residue to a negatively charged aspartate (S189D) results in uncoupling of the proton pump, which led to the suggestion that it is the alteration of the electrostatic potential within the pathway that uncouples ET from proton translocation.[71] However, that is not plausible, because the N131T mutation, which does not affect the electrostatic potential, also results in uncoupling.[72] The activity and stoichiometry of proton-pumping in D-pathway double mutants (D124N/N131D) are surprisingly unaffected,[73] that is, the two mutations appear compensatory. At present, there is no single explanation for the effect of mutations on D-pathway residues.

Some mutants of CcO are only partially decoupled (i.e., they do pump protons), but with lower stoichiometry. An example of such a mutant is the W164F mutant, which maintains 40% of wild-type activity and pumps protons at one-half of the stoichiometry.[74] The mutated tryptophan residue is located in the region above the hemes and is hydrogen bonded to the δ-propionates of both hemes a and a_3. Mutation of the tryptophan to a phenylalanine creates more space for the D-pathway E278 residue to adopt the conformation pointing away from the proton pathway. This result might prevent fast reprotonation of E278 by the D-pathway and/or increase the rate of proton back-flow by reprotonation from the P-side. In both cases, the apparent proton-pumping efficiency decreases. The pH dependence of the proton-pumping reaction in the wild-type and mutant enzymes suggests that there is a pH sensitive proton back-flow that becomes more significant at high pH.[74]

4.6. CATALYTIC CYCLE

4.6.1. Overview

The reduction of oxygen to water is a complex reaction, involving at least 13 components (8 H^+, 4 e^-, O_2) and perhaps as many steps, some of which have yet to be uncovered. The catalytic cycle of CcO can be divided into two parts (Fig. 4.3): The reductive phase ($\mathbf{O_H} \rightarrow \mathbf{MV}$) and the oxidative phase ($\mathbf{MV} \rightarrow \mathbf{O_H}$). The reductive phase starts with the fully oxidized enzyme ($\mathbf{O_H}$) and proceeds via the 1e^- reduced state ($\mathbf{E}$) to the 2e^- reduced form ($\mathbf{MV}$) or further to the artifactual 4e^- reduced state ($\mathbf{R}$), where all four redox-active metal centers are in their reduced state ($Cu_A{}^{1+}$, $Fe_a{}^{2+}$, $Fe_{a3}{}^{2+}$, $Cu_B{}^{1+}$). In the natural cycle of CcO, molecular oxygen binds as soon as the enzyme is 2e^- reduced ($\mathbf{MV}$), a 4e^- reduced state ($\mathbf{R}$) does not accumulate in the cell under aerobic conditions. However, the $\mathbf{R}$ state is easy to produce experimentally and a suitable state to study properties of initial catalytic intermediates. The oxidative phase constitutes the reaction between oxygen and the 2e^- or 4e^- reduced enzyme ($\mathbf{MV}$ or $\mathbf{R}$, respectively) and proceeds to the fully oxidized enzyme $\mathbf{O_H}$, (but not $\mathbf{O}$) via a number of partially oxidized intermediates. The oxidative phase includes heterolytic scission of the O$-$O bond in a reaction that requires four electrons and a proton.

4.6.2. The Individual Steps of the Catalytic Cycle

The reaction scheme that contains all known kinetically competent intermediates and the suggested proton-pumping steps (omitting the uptake of the four chemical protons) is depicted in Figure 4.3. The redox states of the metal centers in the fully oxidized enzyme $\mathbf{O_H}$ are $Cu_A{}^{2+}$, a^{3+}, $a_3{}^{3+}$, and $Cu_B{}^{2+}$. Electrons from substrates like cytochrome c or ubiquinol enter the enzyme via Cu_A and are further transferred, via heme a, to the binuclear center. The first ET yields $\mathbf{E}$. The formation of $\mathbf{E}$ occurs in two steps. The electron is initially shared between the Cu_A and heme a metallo-centers that are in electronic equilibrium[75] and in a subsequent proton-dependent step transferred to Cu_B.[76] The second electron arriving via Cu_A produces the mixed valence ($\mathbf{MV}$) state of the enzyme. In $\mathbf{MV}$, the reduced active-site heme a_3 binds oxygen forming $\mathbf{A}$,

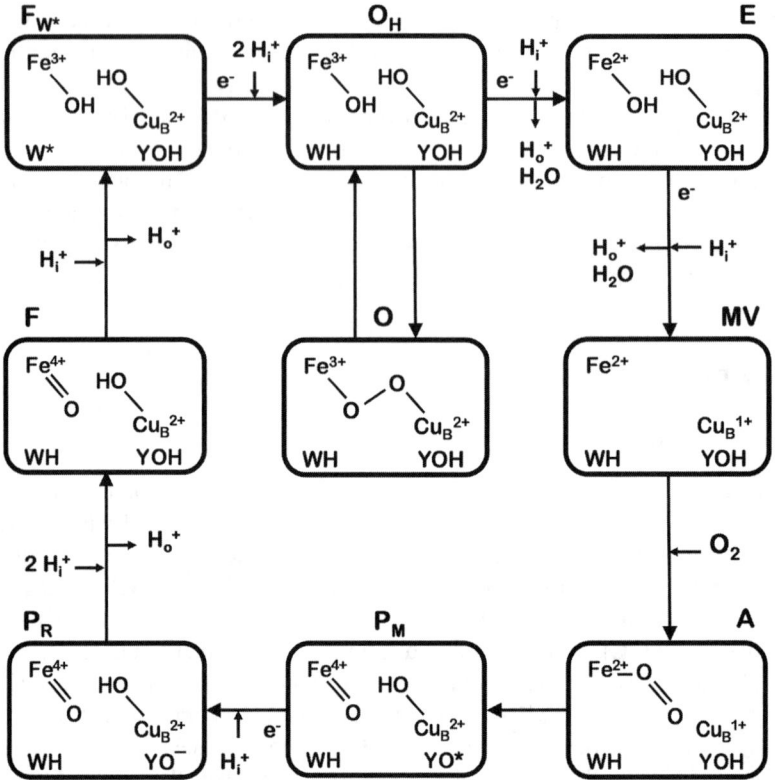

FIGURE 4.3. The protons taken up from the proton-conducting pathways to be pumped or to be used in the active-site chemistry are H_i^+, H_o^+ are the protons released to the periplasmic space, e^- are the electrons destined for the binuclear site, YOH is tyrosine280 and WH is tryptophan272. The electron donors to the binuclear site (Cu_A and heme a) and **R**, the artifactual $4e^-$ fully reduced enzyme that is not formed under physiological conditions, are omitted from the drawing. (See Ref. 35.)

the ferrous-oxy complex. Oxygen binds transiently to Cu_B^{1+} before binding to heme a_3. The intermediate **A** appears to decay to P_M with a half-life of only $\sim 32\ \mu s$,[40] a rate that is dependent on [O_2]. It was established by resonance Raman (RR) spectroscopy that the O—O bond is broken in the **P** state.[77] In the mixed-valence state, only three of the required four electrons are available for O—O bond splitting ($Fe_{a3}^{2+} \rightarrow Fe_{a3}^{4+}$; $Cu_B^{1+} \rightarrow Cu_B^{2+}$). The fourth electron and the proton are donated by a nearby amino acid residue, suggested as Y280.[78] As a result of the heterolytic O—O bond splitting, the heme a_3 has reacted to the oxoferryl form ($Fe_{a3}^{4+}=O^{2-}$) while the other oxygen atom ends up at Cu_B^{2+} as a hydroxo ligand. The subsequent third substrate electron reduces the YO' radical (Y280*) to its anion YO⁻. The P_R state thus formed no longer harbors a radical. The P_R state thus formed no longer harbors a radical and is indistinguishable from P_M by its optical spectrum, but not by electron spin resonance

(EPR).[79,80] From $\mathbf{P_R}$ to $\mathbf{F}$ ($t_{1/2} \sim 27$ μs), two protons are taken up from the N-side of the membrane through the proton-conducting pathways. One proton protonates the YO⁻ to Y280, and the other proton is released to the P-side of the membrane. The transition from the $\mathbf{F}$ (oxoferryl) to the recently described[40] $\mathbf{F_{W^{\cdot}}}$ state forms the rate-limiting step of the cycle with $t_{1/2} \sim 1.2$ ms. It involves the uptake of one proton from the N-side and the release of one proton to the P-side, as well as the transient formation of a neutral radical (W272*).[40] This radical is proposed to facilitate and gate proton-pumping in a redox-dependent manner.[64] In the $\mathbf{F} \rightarrow \mathbf{F_{W^{\cdot}}}$ transition, the oxoferryl heme a_3 is reduced to Fe(III)−OH⁻ by an electron from the nearby W272, yielding W272*, and protonated by a proton from E278. Since the W272* was determined to be a tryptophan *neutral* radical it was proposed that the W272 proton is ejected to the periplasm upon oxidation of W272 to W272*.[40] The E278 is replenished with a proton from the D-pathway and in the $\mathbf{F_{W^{\cdot}}} \rightarrow \mathbf{O_H}$ transition the W272* is reduced to W272⁻ by an electron from the Cu$_A$/heme a redox pair with a $t_{1/2}$ of 30–60 μs.[40] In its anionic form, the W272⁻ is a very strong base ($pK_a > 15$) and is able to attract the newly arrived proton at E278. The deprotonated glutamate is once again protonated by a proton from the D-pathway. In total, two protons are taken up via E278, one from W272 is ejected to the P-side, while the W272 electron reduces the oxo-ferryl intermediate, which brings the overall stoichiometry of the $\mathbf{F} \rightarrow \mathbf{O_H}$ transition to 2 H⁺/1e⁻, as found experimentally. According to this model[64] proton-pumping is directly coupled to the redox activity of the strictly conserved residue W272, which provides both the thermodynamic driving force and the unidirectionality required for proton-pumping.

We consider it very likely that the O−O bond breaking proceeds in the same way in either the fully reduced or mixed-valence enzyme.[5,40] The contrasting view[12,81] is that in the 2e⁻ reduced enzyme O−O splitting occurs as described above, involving Y280, but in the 4e⁻ reduced enzyme, the fourth electron to split the O−O bond is donated by heme a.

The $\mathbf{O_H}$ state, depicted as part of the cycle in Figure 4.3, is the activated (also pulsed or fast, see, e.g. Ref. 82) oxidized form of CcO that has just finished a turnover cycle. This state is known to pump protons during the reductive phase,[83] whereas the oxidized "resting" enzyme does not (although some reports disagree).[13] The states $\mathbf{O_H}$ and $\mathbf{O}$ can also be distinguished by their cyanide-binding kinetics.[82] The $\mathbf{O_H}$ binds CN⁻ by at least a factor of 100 more rapidly than $\mathbf{O}$. The ligands to the active-site metals in the $\mathbf{O}$ state have been proposed as H_2O for heme a_3 and OH⁻ for Cu$_B^{2+}$.[43] However, more recent crystallographic and extended x-ray absorption fine structure (EXAFS) studies and careful redox titrations by Mochizuki et al.[35] provide evidence for the presence of peroxide as a bridging ligand between heme a_3 and Cu$_B$ in the resting enzyme.[84–86] The catalytically relevant "high-energy" $\mathbf{O_H}$ might have OH⁻ ligands at both metal centers (e.g., Ref. 81). In the absence of substrates the active form of CcO relaxes back to the resting state on a time scale of minutes.

A branched, rather than a linear sequential, mechanism for CcO was suggested by Szundi et al. (see Ref. 87, or the review with Einarsdóttir and Szundi 88). This mechanism is linear between $\mathbf{F}$ and $\mathbf{A}$, but branches into two parallel pathways for $\mathbf{A} \rightarrow \mathbf{F}$. Equilibrium exists between the $\mathbf{P}$- and $\mathbf{F}$-states in this model and the interconversion between **P** and **F**, involving the uptake of a proton by the binuclear center, is pH dependent (**P** being the more stable species at high pH). It is possible that the kinetics of oxidation of reduced CcO appear as a series of unidirectional sequential transitions under regular experimental conditions, even if the mechanism is in reality branched or contains rapid protonic equilibria.

4.6.3. Radical Formation

Over the past decade, several groups have reported the formation of radicals in cytochrome oxidases from various organisms. Radicals have been observed both in the reaction with oxygen or with H_2O_2. The reaction between H_2O_2 and oxidized CcO is very slow compared to the oxidation of reduced (or mixed-valence) CcO by molecular oxygen. The **P$_M$** and **F'** states that are formed after addition of H_2O_2 are relatively stable enabling analysis by EPR spectroscopy. However, the actual reaction that occurs upon mixing H_2O_2 with *oxidized* oxidase is different from the reaction of *reduced* oxidase with molecular oxygen. The H_2O_2 acts as a $2e^-$ oxidant to the oxidized oxidase, converting the heme a_3 iron to the $Fe^{4+}=O^{2-}$ oxoferryl and abstracting $1e^-$ from the enzyme. This electron is donated by a nearby amino acid residue, which results in formation of an organic radical. The **P$_M$/F'** generated in this way is obtained from **O$_{(H)}$** by moving in reversed direction through the catalytic cycle, which is essentially irreversible. In addition, a true peroxy intermediate has never been detected in cytochrome oxidases. The role in the catalytic cycle of radicals formed by H_2O_2 is therefore questionable.

4.6.3.1. Radical Formation in the Reaction with Hydrogen Peroxide. Addition of H_2O_2 to oxidized CcO from *P. denitrificans* leads to formation of a tyrosine radical, Y167*.[89,90] The Y167* is not formed in the W272F mutant, suggesting that Y167* is a secondary radical produced via radical migration in the enzyme involving a W272 radical. While a neutral W272* is formed in the reaction with oxygen[40] the Y167* has so far been detected only in the presence of H_2O_2. The Y167F mutant retains 62% of the activity of the wild-type oxidase, which renders a specific role in catalysis as yet unclear.[90]

The broad radical signal observed in the **F'** state of the enzyme from bovine heart mitochondria treated with H_2O_2 was assigned to a tryptophan radical cation, based on electron nuclear double resonance (ENDOR) studies, and not to a tyrosine radical. A narrow EPR signal that was not linked to the kinetics of turnover, but was probably the result of a peroxide side reaction, was attributed to a porphyrin cation radical.[39] The **F'** state may be converted to **P$_M$** simply by changing the pH suggesting a change in electronic equilibrium between the two states. Since the **P$_M$** state did not appear to have an EPR detectable radical signal, the authors proposed that the radical in **P$_M$** resides wholly on the His276–Tyr280 pair and is EPR silent due to spin coupling with Cu_B^{2+}.[39] The radical W164 was put forward as the most likely candidate for generation of the EPR signal,[39] although involvement of W272 or W323 was not excluded. All three tryptophan residues are highly conserved (W272 even strictly) and located close to the binuclear center.

Svistunenko et al.[91] addressed the apparent discrepancy between the proposed radical origins in bovine and bacterial CcO treated with H_2O_2.[91] These authors show by EPR simulation studies that the signal detected in the bovine oxidase can be nicely simulated as originating from Y129, the equivalent of the Y167 residue in *P. denitrificans*. This study resolves the discrepancy that appeared to exist between the bacterial and bovine oxidases.

4.6.3.2. Radical Formation in the Reaction with Molecular Oxygen.

Studies on the rapid reaction of reduced CcO with oxygen to detect kinetically competent radicals present an extra challenge. One approach is to use E278 mutants, in which the reaction does not proceed noticeably beyond $\mathbf{P_R}$. Gorbikova et al.[92] analyzed the E278Q mutant during oxidation of its fully reduced form directly after photodissociation of CO. The Fourier transform infrared (FTIR) spectrum observed for $\mathbf{P_R}$ contained a contribution from the H276 cross-linked Y280 in its deprotonated form. Furthermore, the development of an electric potential during formation of $\mathbf{P}$ was consistent with proton transfer across the distance between Y280 and the bound oxygen. Formation of a Y280 radical could not be resolved in these experiments, but the observation of the deprotonated Y280 may suggest its involvement in O−O bond splitting in the $\mathbf{P_M}$ state.

Although the direct detection of a radical by electron paramagnetic resonance (EPR) of CcO in the $\mathbf{P_M}$ state has so far failed, radicals have been observed in the His-Tyr model compound in the absence or presence of copper[88,93] and in a CcO biomimetic compound brought to the $\mathbf{P_M}$ state containing Cu_B^{2+}.[93] The apparent EPR silence of $\mathbf{P_M}$ in CcO, as prepared in Ref. 39, should therefore not be taken for granted. The $\mathbf{P_M}$ prepared by the triple-trapping method is not EPR silent, but shows the EPR signal of Cu_B^{2+},[77,78] which was apparently not observed in Ref. 39. In addition, there are several ways of preparing $\mathbf{P_M}$, each yielding a state with slightly different properties.[94] Because a Y280* has not been detected, it remains undecided whether Y280 is indeed the primary donor of the proton and the electron needed for O−O bond scission. Calculations[95] may point to another residue as proton and/or electron donor to break the O−O bond, possibly W272.

A neutral W272* was recently observed in the reaction between CcO and molecular oxygen studied by means of microsecond freeze-hyperquenching combined with optical and EPR spectroscopy.[40,41] The observed W272* was proposed as an intermediate in the second part of the catalytic cycle, but not in the initial O−O bond breaking reaction. The W272* is formed in the $\mathbf{F} \rightarrow \mathbf{O_H}$ transition with a $t_{1/2} \sim 1.2$ ms, which led to the proposal of the intermediate state $\mathbf{F_{W^*}}$ (Fig. 4.3) between $\mathbf{F}$ and $\mathbf{O_H}$. The rate of W272* formation is close to the turnover rate of the CcO and represents the rate-limiting step of the reaction.[40] There is no disagreement in the literature that the rate-limiting step occurs in the transition from $\mathbf{F}$ to $\mathbf{O_H}$ (e.g., Refs. 64 and 88), and this reaction appears experimentally as a relatively slow ($\sim$1 ms) ET from heme a to heme a_3. However, the rate of $\sim$1 ms for this step is an apparent rate. It is the application of optical techniques like ultraviolet–visible (UV–vis) and RR spectroscopy in the majority of the work on CcOs that has led to the above assignment for the rate-limiting step. However, the optical techniques did not enable detection of the W272*, for which "time-resolved" EPR spectroscopy is needed. Electron transfer between hemes a and a_3 proceeds on the nanosecond time scale[32,38,83] and is thus definitely not "intrinsically slow". We can now ask and try to answer the question as to why reduction of the oxoferryl by W272 (Fig. 4.3) is so slow, or to put it slightly differently: Why does the $A \rightarrow F$ transition take $\sim$60 μs and the $F \rightarrow F_{W^*} \rightarrow O_H$ transitions $\sim$1.2 ms, a 20-fold difference in rates; or if one takes the view that it is really heme a to heme a_3 electron transfer that is limiting, why is it a million-fold slower? After all, the two reaction sequences above include similar steps like proton translocation, proton-pumping, and ET in which metallo-centers and aromatic amino acid residues forming radicals play a role, thus a large difference in rates is not to be expected *a priori*. To change the rate of ET from heme a to heme a_3 by a factor of 1 million ($\sim$1 ns to $\sim$1 ms) would require a change in relative reduction potentials or reorganization energies of 0.35–0.4 eV during the reaction, which seems unlikely. To explain the low rate of reduction of heme a_3 by W272, the reduction potential of W272 must be higher than that of the metal centers in the binuclear site by a similar amount (0.35–0.4 eV), thus in the range of $\sim$0.7–0.8 V. This value is not far from that of "free" tryptophan ($\sim$1 V) at pH $\sim$7. The actual reduction potential of Y280 ($\sim$0.9 V at pH 7 for "free" tyrosine) may quite well match the value needed for the $O-O$ bond splitting in the $A \rightarrow F$ transition. The reduction potentials of W272 and the heme a_3 iron may not match perfectly, explaining the 20-fold difference in rate between the $A \rightarrow F$ and $F \rightarrow O_H$ transitions. On the other hand, the actual reduction potential of W272 might be the result of a compromise because W272 might provide the electron and the proton to break the $O-O$ bond in the $A \rightarrow F$ transition and serve as the reductant for heme a_3 in the $F \rightarrow F_{W^*}$ transition.

4.7. CONCLUSION AND FUTURE PROSPECTS

Type A, B, and C oxidases are all proton pumps. With the possible exception of the Type B thermophilic cytochrome ba_3 oxidase, the stoichiometry of proton-pumping is $1\ H^+/1e^-$. Crystal structures that are available for Type A and B oxidases indicate a high 3D-structural conservation of the metallo-electron-transfer centers. Sequence comparison of subunit I of the Type C cbb_3 oxidases suggests a topology for the redox centers that is similar to that in Type A and B oxidases leading to the conclusion that ET pathways are highly similar in all cytochrome oxidases. This finding is perhaps unexpected in view of the quite loose distance constraint for rapid ET (i.e., metallo-centers must be located within $\sim$15 Å). Proton-transfer pathways appear to be far less conserved than ET pathways. Type A oxidases have two proton-transfer pathways (D and K), Type B has only one (the K proton pathway). This result is also surprising, since the Grotthuss mechanism for proton transfer requires formation of hydrogen bonds between many strategically placed acid–base groups, polar side chains, and/or water molecules in a hydrophobic environment for which a distance constraint of <2.5 Å applies. Nature has apparently found many solutions to cope with these latter restrictions. Recent experiments have implicated tyrosine and tryptophan radicals as competent reaction intermediates in $O-O$ bond cleaving and proton-pumping, respectively. In these reactions, the aromatic residues act both as electron and as proton donors, a capacity exceeding that of the metallo-centers. To complete their complex catalytic cycle, the cytochrome oxidases need the participation of six rather than four redox centers. Future studies on the precise roles of aromatic residues in O−O bond breaking and proton-pumping will help advance our knowledge on the link between ET and proton-pumping, or energy conversion and energy conservation.

ABBREVIATIONS

ATP	Adenosine triphosphate
CcO	Cytochrome c oxidase
DNA	Deoxyribonucleic acid
ENDOR	Electron nuclear double resonance
EPR	Electron paramagnetic resonance
ET	Electron transfer
EXAFS	Extended X-ray absorption fine structure
FTIR	Fourier transform infrared spectroscopy
H_2O_2	Hydrogen peroxide
MV	Mixed valence
NO	Nitric oxide
N_2O	Nitrous oxide
RR	Resonance Raman
UV–visible	Ultraviolet–visible
1FFT	*E. coli* cytochrome bo_3 oxidase
1QLE	*Paracoccus denitrificans* cytochrome aa_3 oxidase
1XME	*Thermus thermophilus* cytochrome ba_3 oxidase
2DYR	Mitochondrial cytochrome aa_3 oxidase
2GSM	*Rhodobacter sphaeroides* cytochrome oxidase
3D	Three dimensional

REFERENCES

1. Wasser, I. M.; de Vries, S.; Moenne-Loccoz, P.; Schroder, I.; Karlin, K. D. Nitric oxide in biological denitrification: Fe/Cu metalloenzyme and metal complex NO(x) redox chemistry. *Chem. Rev.* **2002**, *102*, 1201–1234.

2. Richter, O. M.; Ludwig, B. Cytochrome c oxidase—structure, function, and physiology of a redox-driven molecular machine. *Rev. Physiol. Biochem. Pharmacol.* **2003**, *147*, 47–74.

3. Michel, H.; Behr, J.; Harrenga, A.; Kannt, A. Cytochrome c oxidase: structure and spectroscopy. *Annu. Rev. Biophys. Biomol. Struct.* **1998**, *27*, 329–356.

4. Hendriks, J.; Oubrie, A.; Castresana, J.; Urbani, A.; Gemeinhardt, S.; Saraste, M. Nitric oxide reductases in bacteria. *Biochim. Biophys. Acta* **2000**, *1459*, 266–273.

5. Cherepanov, A. V.; De Vries, S. Microsecond freeze-hyperquenching: development of a new ultrafast micro-mixing and sampling technology and application to enzyme catalysis. *Biochim. Biophys. Acta* **2004**, *1656*, 1–31.

6. Babcock, G. T.; Wikstrom, M. Oxygen activation and the conservation of energy in cell respiration. *Nature (London)* **1992**, *356*, 301–309.

7. Girsch, P.; deVries, S. Purification and initial kinetic and spectroscopic characterization of NO reductase from Paracoccus denitrificans. *Biochim. Biophys. Acta Bioenerg.* **1997**, *1318*, 202–216.

8. Giuffre, A.; Stubauer, G.; Sarti, P.; Brunori, M.; Zumft, W. G.; Buse, G.; Soulimane, T. The heme-copper oxidases of Thermus thermophilus catalyze the reduction of nitric oxide: evolutionary implications. *Proc. Natl. Acad. Sci. USA* **1999**, *96*, 14718–14723.

9. Hendriks, J. H.; Jasaitis, A.; Saraste, M.; Verkhovsky, M. I. Proton and electron pathways in the bacterial nitric oxide reductase. *Biochemistry* **2002**, *41*, 2331–2340.

10. Hayashi, T.; Lin, M. T.; Ganesan, K.; Chen, Y.; Fee, J. A.; Gennis, R. B.; Moenne-Loccoz, P. Accommodation of Two Diatomic Molecules in Cytochrome *bo*(3): Insights into NO Reductase Activity in Terminal Oxidases. *Biochemistry* **2009**, *48*, 883–890.

11. Wikstrom, M. Cytochrome *c* oxidase: 25 years of the elusive proton pump. *Biochim. Biophys. Acta* **2004**, *1655*, 241–247.

12. Verkhovsky, M. I.; Belevich, I.; Bloch, D. A.; Wikstrom, M. Elementary steps of proton translocation in the catalytic cycle of cytochrome oxidase. *Biochim. Biophys. Acta* **2006**, *1757*, 401–407.

13. Ruitenberg, M.; Kannt, A.; Bamberg, E.; Fendler, K.; Michel, H. Reduction of cytochrome *c* oxidase by a second electron leads to proton translocation. *Nature (London)* **2002**, *417*, 99–102.

14. Faxen, K.; Gilderson, G.; Adelroth, P.; Brzezinski, P. A mechanistic principle for proton-pumping by cytochrome *c* oxidase. *Nature (London)* **2005**, *437*, 286–289.

15. Belevich, I.; Verkhovsky, M. I.; Wikstrom, M. Proton-coupled electron transfer drives the proton pump of cytochrome *c* oxidase. *Nature (London)* **2006**, *440*, 829–832.

16. Puustinen, A.; Verkhovsky, M. I.; Morgan, J. E.; Belevich, N. P.; Wikstrom, M. Reaction of the *Escherichia coli* quinol oxidase cytochrome *bo*3 with dioxygen: the role of a bound ubiquinone molecule. *Proc. Natl. Acad. Sci. USA* **1996**, *93*, 1545–1548.

17. van der Oost, J.; de Boer, A. P.; de Gier, J. W.; Zumft, W. G.; Stouthamer, A. H.; van Spanning, R. J. The heme-copper oxidase family consists of three distinct types of terminal oxidases and is related to nitric oxide reductase. *FEMS Microbiol. Lett.* **1994**, *121*, 1–9.

18. Hemp, J.; Gennis, R. B. Diversity of the heme-copper superfamily in archaea: insights from genomics and structural modeling. *Results Probl. Cell Differ.* **2008**, *45*, 1–31.

19. Harrenga, A.; Michel, H. The cytochrome *c* oxidase from Paracoccus denitrificans does not change the metal center ligation upon reduction. *J. Biol. Chem.* **1999**, *274*, 33296–33299.

20. Qin, L.; Hiser, C.; Mulichak, A.; Garavito, R. M.; Ferguson-Miller, S. Identification of conserved lipid/detergent-binding sites in a high-resolution structure of the membrane protein cytochrome *c* oxidase. *Proc. Natl. Acad. Sci. USA* **2006**, *103*, 16117–16122.

21. Abramson, J.; Riistama, S.; Larsson, G.; Jasaitis, A.; Svensson-Ek, M.; Laakkonen, L.; Puustinen, A.; Iwata, S.; Wikstrom, M. The structure of the ubiquinol oxidase from Escherichia coli and its ubiquinone binding site. *Nat. Struct. Biol.* **2000**, *7*, 910–917.

22. Shinzawa-Itoh, K.; Aoyama, H.; Muramoto, K.; Terada, H.; Kurauchi, T.; Tadehara, Y.; Yamasaki, A.; Sugimura, T.; Kurono, S.; Tsujimoto, K.; Mizushima, T.; Yamashita, E.; Tsukihara, T.; Yoshikawa, S. Structures and physiological roles of 13 integral lipids of bovine heart cytochrome *c* oxidase. *EMBO J.* **2007**, *26*, 1713–1725.

23. Hunsicker-Wang, L. M.; Pacoma, R. L.; Chen, Y.; Fee, J. A.; Stout, C. D. A novel cryoprotection scheme for enhancing the diffraction of crystals of recombinant cytochrome *ba3* oxidase from Thermus thermophilus. *Acta Crystallogr.* **2005**, *61*, 340–343.

24. Buschmann, S.; Warkentin, E.; Xie, H.; Langer, J. D.; Ermler, U.; Michel, H. The structure of *cbb3* cytochrome oxidase provides insights into proton-pumping. *Science* **2010**, *329*, 327–330.

25. Junemann, S. Cytochrome *bd* terminal oxidase. *Biochim. Biophys. Acta* **1997**, *1321*, 107–127.

26. Schroder, I.; De Vries, S. *Archaea: New Models for Prokaryotic Biology*, P. Blum, Ed., Caister Academic Press, Norfolk, UK, **2008**, pp. 1–26.

27. Solioz, M.; Carafoli, E.; Ludwig, B. The cytochrome *c* oxidase of Paracoccus denitrificans pumps protons in a reconstituted system. *J. Biol. Chem.* **1982**, *257*, 1579–1582.

28. Mather, M. W.; Springer, P.; Hensel, S.; Buse, G.; Fee, J. A. Cytochrome oxidase genes from *Thermus thermophilus*. Nucleotide sequence of the fused gene and analysis of the deduced primary structures for subunits I and III of cytochrome *caa*3. *J. Biol. Chem.* **1993**, *268*, 5395–5408.

29. Castello, P. R.; Woo, D. K.; Ball, K.; Wojcik, J.; Liu, L.; Poyton, R. O. Oxygen-regulated isoforms of cytochrome *c* oxidase have differential effects on its nitric oxide production and on hypoxic signaling. *Proc. Natl. Acad. Sci. USA* **2008**, *105*, 8203–8208.

30. Blackburn, N. J.; Barr, M. E.; Woodruff, W. H.; Vanderooost, J.; Devries, S. Metal–Metal Bonding in Biology - Exafs Evidence for a 2.5-Angstrom Copper–Copper Bond in the Cu-*a* Center of Cytochrome-Oxidase. *Biochemistry* **1994**, *33*, 10401–10407.

31. Moser, C. C.; Chobot, S. E.; Page, C. C.; Dutton, P. L. Distance metrics for heme protein electron tunneling. *Biochim. Biophys. Acta* **2008**, *1777*, 1032–1037.

32. Moser, C. C.; Page, C. C.; Dutton, P. L. Darwin at the molecular scale: selection and variance in electron tunnelling proteins including cytochrome *c* oxidase. *Philos. Trans. R. Soc. Lond* **2006**, *361*, 1295–1305.

33. Solomon, E. I.; Sundaram, U. M.; Machonkin, T. E. Multicopper Oxidases and Oxygenases. *Chem. Rev.* **1996**, *96*, 2563–2606.

34. Van Gelder, B. F.; Beinert, H. Studies of the heme components of cytochrome *c* oxidase by EPR spectroscopy. *Biochim. Biophys. Acta* **1969**, *189*, 1–24.

35. Mochizuki, M.; Aoyama, H.; Shinzawa-Itoh, K.; Usui, T.; Tsukihara, T.; Yoshikawa, S. Quantitative reevaluation of the redox active sites of crystalline bovine heart cytochrome *c* oxidase. *J. Biol. Chem.* **1999**, *274*, 33403–33411.

36. Adelroth, P.; Brzezinski, P.; Malmstrom, B. G. Internal electron transfer in cytochrome *c* oxidase from Rhodobacter sphaeroides. *Biochemistry* **1995**, *34*, 2844–2849.

37. Farver, O.; Grell, E.; Ludwig, B.; Michel, H.; Pecht, I. Rates and Equilibrium of CuA to heme a electron transfer in Paracoccus denitrificans cytochrome *c* oxidase. *Biophys. J.* **2006**, *90*, 2131–2137.

38. Pilet, E.; Jasaitis, A.; Liebl, U.; Vos, M. H. Electron transfer between hemes in mammalian cytochrome *c* oxidase. *Proc. Natl. Acad. Sci. USA* **2004**, *101*, 16198–16203.

39. Rich, P. R.; Rigby, S. E.; Heathcote, P. Radicals associated with the catalytic intermediates of bovine cytochrome *c* oxidase. *Biochim. Biophys. Acta* **2002**, *1554*, 137–146.

40. Wiertz, F. G.; Richter, O. M.; Ludwig, B.; de Vries, S. Kinetic resolution of a tryptophan-radical intermediate in the reaction cycle of *Paracoccus denitrificans* cytochrome *c* oxidase. *J. Biol. Chem.* **2007**, *282*, 31580–31591.

41. Wiertz, F. G.; Richter, O. M.; Cherepanov, A. V.; MacMillan, F.; Ludwig, B.; de Vries, S. An oxo-ferryl tryptophan radical catalytic intermediate in cytochrome *c* and quinol oxidases trapped by microsecond freeze-hyperquenching (MHQ). *FEBS Lett.* **2004**, *575*, 127–130.

42. Buse, G.; Soulimane, T.; Dewor, M.; Meyer, H. E.; Bluggel, M. Evidence for a copper-coordinated histidine-tyrosine cross-link in the active site of cytochrome oxidase. *Prot. Sci.* **1999**, *8*, 985–990.

43. Ostermeier, C.; Harrenga, A.; Ermler, U.; Michel, H. Structure at 2.7 angstrom resolution of the *Paracoccus denitrificans* two-subunit cytochrome *c* oxidase complexed with an antibody F-V fragment. *Proc. Natl. Acad. Sci. USA* **1997**, *94*, 10547–10553.

44. Tsukihara, T.; Aoyama, H.; Yamashita, E.; Tomizaki, T.; Yamaguchi, H.; Shinzawaltoh, K.; Nakashima, R.; Yaono, R.; Yoshikawa, S. The whole structure of the 13-subunit oxidized cytochrome *c* oxidase at 2.8 angstrom. *Science* **1996**, *272*, 1136–1144.

45. Soulimane, T.; Buse, G.; Bourenkov, G. P.; Bartunik, H. D.; Huber, R.; Than, M. E. Structure and mechanism of the aberrant *ba*(3)-cytochrome *c* oxidase from thermus thermophilus. *EMBO J.* **2000**, *19*, 1766–1776.

46. Chang, H. Y.; Hemp, J.; Chen, Y.; Fee, J. A.; Gennis, R. B. The cytochrome *ba*3 oxygen reductase from Thermus thermophilus uses a single input channel for proton delivery to the active site and for proton-pumping. *Proc. Natl. Acad. Sci. USA* **2009**, *106*, 16169–16173.

47. Hemp, J.; Han, H.; Roh, J. H.; Kaplan, S.; Martinez, T. J.; Gennis, R. B. Comparative genomics and site-directed mutagenesis support the existence of only one input channel for protons in the C-family (*cbb*3 oxidase) of heme-copper oxygen reductases. *Biochemistry* **2007**, *46*, 9963–9972.

48. Yoshikawa, S.; Shinzawa-Itoh, K.; Nakashima, R.; Yaono, R.; Yamashita, E.; Inoue, N.; Yao, M.; Fei, M. J.; Libeu, C. P.; Mizushima, T.; Yamaguchi, H.; Tomizaki, T.; Tsukihara, T. Redox-coupled crystal structural changes in bovine heart cytochrome *c* oxidase. *Science* **1998**, *280*, 1723–1729.

49. Pfitzner, U.; Odenwald, A.; Ostermann, T.; Weingard, L.; Ludwig, B.; Richter, O. M. Cytochrome *c* oxidase (heme *aa*3) from Paracoccus denitrificans: analysis of mutations in putative proton channels of subunit I. *J. Bioenerg. Biomembr.* **1998**, *30*, 89–97.

50. Pereira, M. M.; Santana, M.; Teixeira, M. A novel scenario for the evolution of haem-copper oxygen reductases. *Biochim. Biophys. Acta* **2001**, *1505*, 185–208.

51. Koutsoupakis, C.; Soulimane, T.; Varotsis, C. Probing the Q-proton pathway of *ba*3-cytochrome *c* oxidase by time-resolved Fourier transform infrared spectroscopy. *Biophys. J.* **2004**, *86*, 2438–2444.

52. Durr, K. L.; Koepke, J.; Hellwig, P.; Muller, H.; Angerer, H.; Peng, G.; Olkhova, E.; Richter, O. M.; Ludwig, B.; Michel, H. A D-pathway mutation decouples the *Paracoccus denitrificans* cytochrome *c* oxidase by altering the side-chain orientation of a distant conserved glutamate. *J. Mol. Biol.* **2008**, *384*, 865–877.

53. Pereira, M. M.; Verkhovskaya, M. L.; Teixeira, M.; Verkhovsky, M. I. The *caa*(3) terminal oxidase of *Rhodothermus marinus* lacking the key glutamate of the D-channel is a proton pump. *Biochemistry* **2000**, *39*, 6336–6340.

54. Alge, D.; Wastyn, M.; Mayer, C.; Jungwirth, C.; Zimmermann, U.; Zoder, R.; Fromwald, S.; Peschek, G. A. Allosteric properties of cyanobacterial cytochrome *c* oxidase: inhibition of the coupled enzyme by ATP and stimulation by ADP. *IUBMB Life* **1999**, *48*, 187–197.

55. Kannt, A.; Soulimane, T.; Buse, G.; Becker, A.; Bamberg, E.; Michel, H. Electrical current generation and proton-pumping catalyzed by the *ba*3-type cytochrome *c* oxidase from Thermus thermophilus. *FEBS Lett.* **1998**, *434*, 17–22.

56. Toledo-Cuevas, M.; Barquera, B.; Gennis, R. B.; Wikstrom, M.; Garcia-Horsman, J. A. The cbb3-type cytochrome *c* oxidase from Rhodobacter sphaeroides, a proton-pumping heme-copper oxidase. *Biochim. Biophys. Acta* **1998**, *1365*, 421–434.

57. Gleissner, M.; Kaiser, U.; Antonopoulos, E.; Schafer, G. The archaeal SoxABCD complex is a proton pump in Sulfolobus acidocaldarius. *J. Biol. Chem.* **1997**, *272*, 8417–8426.

58. Gomes, C. M.; Backgren, C.; Teixeira, M.; Puustinen, A.; Verkhovskaya, M. L.; Wikstrom, M.; Verkhovsky, M. I. Heme-copper oxidases with modified D- and K-pathways are yet efficient proton pumps. *FEBS Lett.* **2001**, *497*, 159–164.

59. Lepp, H.; Brzezinski, P. Internal charge transfer in cytochrome *c* oxidase at a limited proton supply: proton-pumping ceases at high pH. *Biochim. Biophys. Acta* **2009**, *1790*, 552–557.

60. Capitanio, N.; Capitanio, G.; De Nitto, E.; Boffoli, D.; Papa, S. Proton transfer reactions associated with the reaction of the fully reduced, purified cytochrome *c* oxidase with molecular oxygen and ferricyanide. *Biochemistry* **2003**, *42*, 4607–4612.

61. Raitio, M.; Wikstrom, M. An Alternative Cytochrome-Oxidase of Paracoccus-Denitrificans Functions as a Proton Pump. *Biochim. Biophys. Acta Bioenerg.* **1994**, *1186*, 100–106.

62. Kaila, V. R.; Verkhovsky, M. I.; Hummer, G.; Wikstrom, M. Glutamic acid 242 is a valve in the proton pump of cytochrome *c* oxidase. *Proc. Natl. Acad. Sci. USA* **2008**, *105*, 6255–6259.

63. Namslauer, A.; Aagaard, A.; Katsonouri, A.; Brzezinski, P. Intramolecular proton-transfer reactions in a membrane-bound proton pump: the effect of pH on the peroxy to ferryl transition in cytochrome *c* oxidase. *Biochemistry* **2003**, *42*, 1488–1498.

64. de Vries, S. The role of the conserved tryptophan272 of the Paracoccus denitrificans cytochrome *c* oxidase in proton-pumping. *Biochim. Biophys. Acta* **2008**, *1777*, 925–928.

65. Busenlehner, L. S.; Branden, G.; Namslauer, I.; Brzezinski, P.; Armstrong, R. N. Structural elements involved in proton translocation by cytochrome *c* oxidase as revealed by backbone amide hydrogen–deuterium exchange of the E286H mutant. *Biochemistry* **2008**, *47*, 73–83.

66. Han, D.; Namslauer, A.; Pawate, A.; Morgan, J. E.; Nagy, S.; Vakkasoglu, A. S.; Brzezinski, P.; Gennis, R. B. Replacing Asn207 by aspartate at the neck of the D channel in the *aa*3-type cytochrome *c* oxidase from Rhodobacter sphaeroides results in decoupling the proton pump. *Biochemistry* **2006**, *45*, 14064–14074.

67. Lee, H. J.; Ojemyr, L.; Vakkasoglu, A.; Brzezinski, P.; Gennis, R. B. Properties of Arg481 mutants of the *aa*3-type cytochrome *c* oxidase from Rhodobacter sphaeroides suggest that neither R481 nor the nearby D-propionate of heme *a3* is likely to be the proton loading site of the proton pump. *Biochemistry* **2009**, *48*, 7123–7131.

68. Mitchell, D. M.; Adelroth, P.; Hosler, J. P.; Fetter, J. R.; Brzezinski, P.; Pressler, M. A.; Aasa, R.; Malmstrom, B. G.; Alben, J. O.; Babcock, G. T.; Gennis, R. B.; Ferguson-Miller, S. A ligand-exchange mechanism of proton-pumping involving tyrosine-422 of subunit I of cytochrome oxidase is ruled out. *Biochemistry* **1996**, *35*, 824–828.

69. Namslauer, I.; Brzezinski, P. A mitochondrial DNA mutation linked to colon cancer results in proton leaks in cytochrome *c* oxidase. *Proc. Natl. Acad. Sci. USA* **2009**, *106*, 3402–3407.

70. Olkhova, E.; Helms, V.; Michel, H. Titration behavior of residues at the entrance of the D-pathway of cytochrome *c* oxidase from paracoccus denitrificans investigated by continuum electrostatic calculations. *Biophys. J.* **2005**, *89*, 2324–2331.

71. Namslauer, A.; Lepp, H.; Branden, M.; Jasaitis, A.; Verkhovsky, M. I.; Brzezinski, P. Plasticity of proton pathway structure and water coordination in cytochrome *c* oxidase. *J. Biol. Chem.* **2007**, *282*, 15148–15158.

72. Lepp, H.; Salomonsson, L.; Zhu, J. P.; Gennis, R. B.; Brzezinski, P. Impaired proton-pumping in cytochrome *c* oxidase upon structural alteration of the D pathway. *Biochim. Biophys. Acta* **2008**, *1777*, 897–903.

73. Branden, G.; Pawate, A. S.; Gennis, R. B.; Brzezinski, P. Controlled uncoupling and recoupling of proton-pumping in cytochrome *c* oxidase. *Proc. Natl. Acad. Sci. USA* **2006**, *103*, 317–322.

74. Ribacka, C.; Verkhovsky, M. I.; Belevich, I.; Bloch, D. A.; Puustinen, A.; Wikstrom, M. An elementary reaction step of the proton pump is revealed by mutation of tryptophan-164 to phenylalanine in cytochrome *c* oxidase from Paracoccus denitrificans. *Biochemistry* **2005**, *44*, 16502–16512.

75. Farver, O.; Grell, E.; Ludwig, B.; Michel, H.; Pecht, I. Rates and Equilibrium of CuA to heme a electron transfer in Paracoccus denitrificans cytochrome *c* oxidase. *Biophys. J.* **2006**, *90*, 2131–2137.

76. Belevich, I.; Bloch, D. A.; Belevich, N.; Wikstrom, M.; Verkhovsky, M. I. Exploring the proton pump mechanism of cytochrome *c* oxidase in real time. *Proc. Natl. Acad. Sci. USA* **2007**, *104*, 2685–2690.

77. Proshlyakov, D. A.; Ogura, T.; Shinzawa-Itoh, K.; Yoshikawa, S.; Appelman, E. H.; Kitagawa, T. Selective resonance Raman observation of the "607 nm" form generated in the reaction of oxidized cytochrome *c* oxidase with hydrogen peroxide. *J. Biol. Chem.* **1994**, *269*, 29385–29388.

78. Proshlyakov, D. A.; Pressler, M. A.; DeMaso, C.; Leykam, J. F.; DeWitt, D. L.; Babcock, G. T. Oxygen activation and reduction in respiration: Involvement of redox-active tyrosine 244. *Science* **2000**, *290*, 1588–1591.

79. Hansson, O.; Karlsson, B.; Aasa, R.; Vanngard, T.; Malmstrom, B. G. The structure of the paramagnetic oxygen intermediate in the cytochrome *c* oxidase reaction. *EMBO J.* **1982**, *1*, 1295–1297.

80. Morgan, J. E.; Verkhovsky, M. I.; Palmer, G.; Wikstrom, M. Role of the PR intermediate in the reaction of cytochrome *c* oxidase with O_2. *Biochemistry* **2001**, *40*, 6882–6892.

81. Wikstrom, M. Mechanism of proton translocation by cytochrome *c* oxidase: a new four-stroke histidine cycle. *Biochim. Biophys. Acta* **2000**, *1458*, 188–198.

82. Moody, A. J. 'As prepared' forms of fully oxidised haem/Cu terminal oxidases. *Biochim. Biophys. Acta* **1996**, *1276*, 6–20.

83. Verkhovsky, M. I.; Jasaitis, A.; Verkhovskaya, M. L.; Morgan, J. E.; Wikstrom, M. Proton translocation by cytochrome *c* oxidase. *Nature (London)* **1999**, *400*, 480–483.

84. Powers, L.; Chance, B.; Ching, Y.; Angiolillo, P. Structural features and the reaction mechanism of cytochrome oxidase: iron and copper X-ray absorption fine structure. *Biophys. J.* **1981**, *34*, 465–498.

85. Scott, R. A.; Schwartz, J. R.; Cramer, S. P. Structural aspects of the copper sites in cytochrome *c* oxidase. An X-ray absorption spectroscopic investigation of the resting-state enzyme. *Biochemistry* **1986**, *25*, 5546–5555.

86. Koepke, J.; Olkhova, E.; Angerer, H.; Muller, H.; Peng, G.; Michel, H. High resolution crystal structure of Paracoccus denitrificans cytochrome *c* oxidase: new insights into the active site and the proton transfer pathways. *Biochim. Biophys. Acta* **2009**, *1787*, 635–645.

87. Szundi, I.; Van Eps, N.; Einarsdottir, O. pH dependence of the reduction of dioxygen to water by cytochrome *c* oxidase. 2. Branched electron transfer pathways linked by proton transfer. *Biochemistry* **2003**, *42*, 5074–5090.

88. Einarsdottir, O.; Szundi, I. Time-resolved optical absorption studies of cytochrome oxidase dynamics. *Biochim. Biophys. Acta* **2004**, *1655*, 263–273.

89. MacMillan, F.; Kannt, A.; Behr, J.; Prisner, T.; Michel, H. Direct evidence for a tyrosine radical in the reaction of cytochrome *c* oxidase with hydrogen peroxide. *Biochemistry* **1999**, *38*, 9179–9184.

90. Budiman, K.; Kannt, A.; Lyubenova, S.; Richter, O. M.; Ludwig, B.; Michel, H.; MacMillan, F. Tyrosine 167: the origin of the radical species observed in the reaction of cytochrome *c* oxidase with hydrogen peroxide in Paracoccus denitrificans. *Biochemistry* **2004**, *43*, 11709–11716.

91. Svistunenko, D. A.; Wilson, M. T.; Cooper, C. E. Tryptophan or tyrosine? On the nature of the amino acid radical formed following hydrogen peroxide treatment of cytochrome *c* oxidase. *Biochim. Biophys. Acta* **2004**, *1655*, 372–380.

92. Gorbikova, E. A.; Belevich, I.; Wikstrom, M.; Verkhovsky, M. I. The proton donor for $O-O$ bond scission by cytochrome *c* oxidase. *Proc. Natl. Acad. Sci. USA* **2008**, *105*, 10733–10737.

93. Collman, J. P.; Devaraj, N. K.; Decreau, R. A.; Yang, Y.; Yan, Y. L.; Ebina, W.; Eberspacher, T. A.; Chidsey, C. E. A cytochrome *c* oxidase model catalyzes oxygen to water reduction under rate-limiting electron flux. *Science* **2007**, *315*, 1565–1568.

94. Ji, H.; Yeh, S. R.; Rousseau, D. L. Structural characterization of the Pco/o(2). compound of cytochrome *c* oxidase. *FEBS Lett.* **2005**, *579*, 6361–6364.

95. Blomberg, M. R.; Siegbahn, P. E.; Wikstrom, M. Metal-bridging mechanism for $O-O$ bond cleavage in cytochrome *c* oxidase. *Inorg. Chem.* **2003**, *42*, 5231–5243.

5

MULTICOPPER PROTEINS

TAKESHI SAKURAI AND KUNISHIGE KATAOKA

Graduate School of Natural Science and Technology, Kanazawa University, Kakuma, Kanazawa, 920-1192, Japan

Copper-Oxygen Chemistry, First Edition. Edited by Kenneth D. Karlin and Shinobu Itoh.
© 2011 John Wiley & Sons, Inc. Published 2011 by John Wiley & Sons, Inc.

5.1. INTRODUCTION

Multicopper protein refers, in a narrow sense, to multicopper oxidase (MCO) containing a type I copper (blue copper, abbreviated as T1), a type II copper (non-blue copper, abbreviated as T2), and a pair of type III coppers [electron paramagnetic resonance (EPR)-nondetectable coppers, abbreviated as T3].[1] Copper-containing nitrite reductase (NIR) occasionally has been included with multicopper proteins or MCOs due to its possessing T1 and T2, and more reasonably due to having evolved from common ancestors. Nitrite reductase utilizes NO_2^- as its final electron acceptor, which differs from MCOs use of dioxygen (O_2). Therefore, NIR is not included as a part of MCOs in this chapter in spite of an early study of the $2e^-$ reduction activity of O_2.[2] Cytochrome c oxidase (CcO), as involved in the aerobic respiratory chain and nitrous oxide reductase (N$_2$OR), which is involved in denitrification, may also be classified into multicopper proteins because they both possess the binuclear Cu$_A$ center and further the latter the tetranuclear Cu$_Z$ center. However, we focus only on MCOs and their $4e^-$ reduction of O_2 in this chapter, since CcO is reviewed in Chapter 4. Peptidylglycine α-amidating monooxygenase, dopamine β-monooxygenase, tyramine β-monooxygenase, particulate methane monooxygenase, and tyrosinase contain more than two isolated or clustered copper centers, but these copper-containing oxygenases activate O_2 and are also excluded from coverage here.

Until recently, MCOs had been regarded as a minor class of enzymes. Only laccase (Lc), ascorbate oxidase (AOase), and ceruloplasmin (Cp) have been studied in some detail, and studies on other MCOs [e.g., bilirubin oxidase (BOD) and phenoxazinone synthase]remained limited. However, according to recent genomics and proteomics studies, it has become apparent that MCOs are distributed from microorganisms to mammals and more widely than supposed before; profound structural and functional studies on a variety of novel MCOs have begun in the last 10 years.

Laccase was discovered by Yoshida,[3] from the sap of plant lacquer as early as 1883, and later from a variety of fungi, microorganisms,[4] and insects.[5] Sometimes Lc has been confused with phenol oxidases including tyrosinase, and confusions are still found even in databases. As judged from the amino acid sequence, structure, and properties of T1, Lcs from plants exhibit higher homologies with AOase rather than Lcs from microorganisms.[6] Further, the substrate specificities of Lcs from plants and microorganisms are considerably different from each other. Occasional confusion on Lcs arises from the fact that the same name was given to MCOs derived from plants and microorganisms. In addition to Lc, each MCO shows unique properties and activities, and accordingly, combined discussion on the structure and function of MCOs from different origins has also been confusing.

The T1 center in MCO is adjacent to the substrate-binding site on the surface of the MCO molecules, and mediates the electron transfer (ET) from substrate to the trinuclear copper center comprised of T2 and T3. The T1 center in MCOs is coordinated by 1Cys2His and additionally by 1Met or none. The ligand set of 1Cys2His1Met is the same with that of the blue copper center, which is the only one copper center to be present in many blue copper proteins (cupredoxins). The ligand set of 1Cys2His1Gln in a class of blue copper proteins, phytocyanin, with a comparatively low redox potential, has not been found in MCOs. The MCOs having a noncoordinating amino acid (Leu, Ile, or Phe for Met) show, in turn, high redox potentials compared to MCOs having a Met residue at the axial site of the T1 center.[7] Since the redox potential of T1 in MCO is quite high (0.4–0.8 V), diphenols (e.g., as urushiol), diaminophenols, and metal ions (e.g., Cu^+, Fe^{2+}, and Mn^{2+}) are natural substrates. Substances with comparatively high redox potentials [e.g., bilirubin, 2,2′-azino-bis(3-ethylbenzothiazoline-6-sulfonic acid) (ABTS), p-phenylenediamine, and 2,6-dimethoxyphenol] are also effectively oxidized by MCOs. The oxidase activity of MCOs is governed by how specifically the substrate is bound at the substrate-binding site, how efficiently electrons are transferred from substrate to the trinuclear copper center via T1, and how efficiently O_2 is converted into H_2O at the trinuclear copper center.

In order to treat the electron withdrawn from the substrate by T1, MCOs have the trinuclear copper center composed of T2 and coupled T3 centers, where four electrons are transferred to the final electron acceptor, O_2. No other protein is conjugated with MCO as an electron acceptor. Thus MCOs are superior due to this independency. Another superior behavior of MCOs is that activated oxygen species (e.g., superoxide and peroxide) are not formed or released from the protein molecule during the $4e^-$ reduction of O_2 at the trinuclear copper center. As for the coordination structure at the trinuclear copper center needed to realize the $4e^-$ reduction of O_2, the ligands selected to construct the T2 and T3 centers are $2His1H_2O/OH^-$ and $3His1OH^-$ (bridge), respectively. These sets of ligands are conserved in all MCOs in order to effectively bind O_2 and reduce it to H_2O, while the bridged OH^- is eliminated from the T3 centers in the reduced form. The redox potentials of the trinuclear copper center are high in order to effectively reduce O_2 with the reduction potential of 1.23 V at pH 0 and 0.82 V at pH 7.

The copper-binding sites in a MCO, CueO, is shown in Figure 5.1 together with those of a blue copper protein, pseudoazurin, and NIR from *Alcaligenes xylosoxidans*. Of various oxidoreductases that utilize O_2, only MCOs and CcO are able to perform the $4e^-$ reduction of O_2 to H_2O. Since most MCOs are soluble monomeric proteins, MCOs have been applied as cathodic catalysts in biofuell cells, in contrast to the normally membrane-bound protein CcO.

Table 5.1 lists MCOs with their source, function, and studies relevant to O_2 reduction.[1,7–10] The focus of this chapter is on the novel aspects of reactivity and related features of structure and the properties of MCOs.

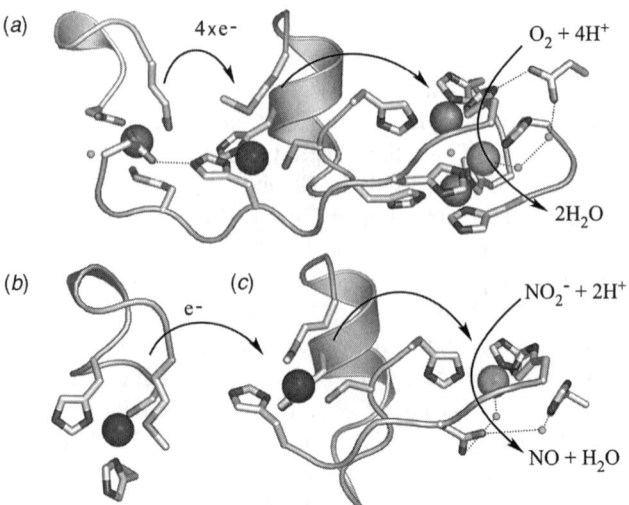

FIGURE 5.1. The copper-binding sites of CueO (*a*), pseudoazurin (*b*), and nitrite reductase (*c*). The T1, T2, T3, and the substrate copper ions are colored blue, green, orange, and red, respectively. (See color insert.)

5.2. MOLECULAR ARCHITECTURE OF MULTICOPPER OXIDASES

In addition to approaches from enzymology, kinetics, and biophysics, the current dramatic progress in genomics and molecular biology allow us a wider and more profound understanding of the structure and function of MCOs. In this section, the molecular architecture of MCOs is discussed based on amino acid sequences and molecular structures. The amino acid sequence of MCO was first determined for human Cp (hCp) in 1984.[11] On the other hand, the crystal structure of MCO was first solved for AOase in 1989.[12] Curuloplasmin and AOase are not necessarily the most typical MCOs with respect to structure and properties, but these have functioned as landmarks for the present more profound understanding of the structure–function relationship of MCOs. New findings continue to show structural variations in MCOs.

Most MCOs possess three domains, each of which has a β-barrel fold also found in cupredoxins.[1] Therefore, it is considered that MCOs evolved from triplication of the cupredoxin domains, the introduction of the crowded His ligands in the domains 1 and 3, and formation of the peculiar sequence found in MCOs, His-Cys-His. Homologies in the amino acid sequences among MCOs, blue copper proteins, and NIR (Fig. 5.2), strongly suggest that these copper proteins have common ancestors, while some other proteins, such as the blood coagulation factors and Stufl (FtsP) in the cell division apparatus, also belong to branches of the cupredoxin superfamily.[13–15] In addition to the His-Cys-His sequence, the T1 and T3 and the T2 and T3 centers are closely connected to each other with the ligand loops, His-X_{1-3}-His.[7,10]

Crystal structures of the three domain MCOs (Fig. 5.3*a* for AOase and *b* for CueO) indicate that the trinuclear copper center are located at the interface of domains 1 and 3,

TABLE 5.1. Origin, Function, and Dioxygen-Redcution-Relevant Study of MCOs

MCO	Origin	Function	Dioxygen-Reduction Relevant Study (Studied Form)
Lc	Rhus vernicifera	Oxidation of urushiol	Int. 1(T1Hg and T1redT2/3ox, DFT[a]), Int. 2(wild, DFT)
	Coprinus cinereus	Degradation of lignin	T2D
	Coriolus hirsatus		
	Melanocarpus albomyces		Str. of recombinant enzyme, 2,6-Dimethoxylphenol-soaked
	Rigioporus lignosus		Str. of resting enzyme partially reduced
	Cerrena maxima		Str. of resting enzyme
	Trametes versicolor		Arylamine bound near T1
	Trametes trogii		p-Toluate bound near T1
	Trametes hirsuta		Str. of resting enzyme
	Lentinus tigrinus		Str. of resting enzyme
CotA	Bacillus subtilis	Endospore coat formation	Str. of recombinant enzyme and mutants for T1
BOD	Myrothecium verrucaria	Oxidation of bilirubin	Int. 1(T1D), Int. 2(Glu mutant, Met mutant for T1)
CueO	Escherichia coli	Oxidation of Cu(I)	Int. 1(T1D, Asp mutant), Int. 2(Glu mutant)
PcoA	Escherichia coli	Oxidation of Cu(I)	
Fet3p	Saccharomyces cerevisiae	Oxidation of Fe(II)	Int. 1(T1D, DFT), Int. 2(DFT)
CumA, MoxA, MnxG	Pseudomonas putida, etc.	Oxidation of Mn(II)	
Phenoxazinone synthase	Streptomyces antibioticus	Biosynthesis of antibiotic	Str. of resting enzyme
Dihydrogeodin oxidase and sulochrin oxidase	Aspergillus terreus	Biosynthesis of grisans	
AOase	Higher plant	Cell division	Str. of resting and reduced enzyme, O_2^{2-}, N_3^- soaked
Cp	Vertebrate	Oxidation of Fe(II)	Str. of resting enzyme
Hephaestin	Vertebrate	Transport of Fe(II)	
SLAC	Streptomyces coelicilor		Radical intermediate, Str. of resting enzyme
EpoA	Streptomyces griseus	Oxidation of phenols	
BCO	Nitrosomonas europaea	Oxidation of p-phenylenediamine	Str. of resting enzyme

[a] Characterized by density functional theory (DFT) calculations.

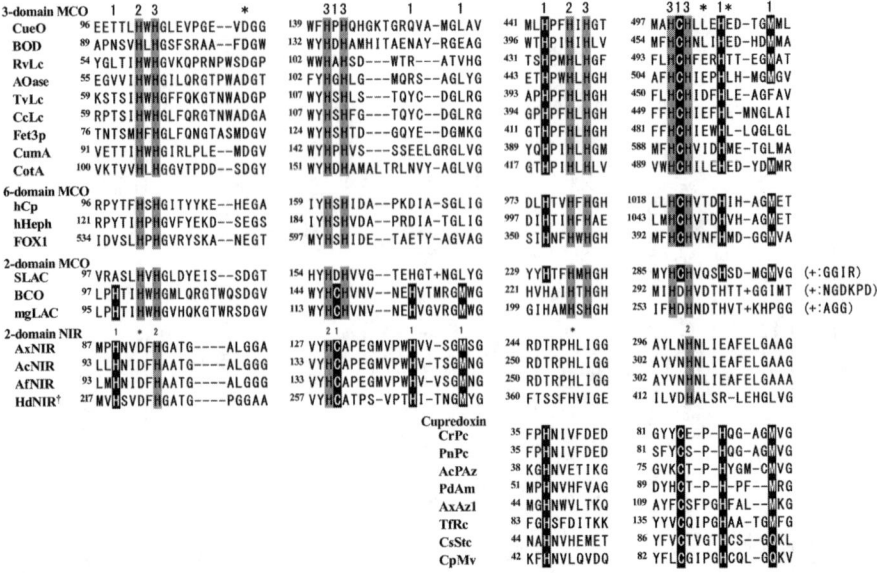

FIGURE 5.2. Homology of amino acid sequences around the copper-binding sites of MCO, NIR, and cupredoxines. The 1, 2, 3, and asterisk represent T1, T2, T3, and acidic amino acid concerned in O_2 reduction, respectively. Plus represents the three to six amino acids shown in the parentheses. Here TvLc, *Trametes versicolor* Lc; CcLc, *Coprinus cinrrius* Lc; CumA, manganese oxidase from *Pseudomonas putida*; hCp, human Cp; hHeph, human hephaestin; FOX1, *Chlamydomonas reinhardtii* ferroxidase; BCO, *Nitrosomonas europaea* blue copper oxidase; mgLAC, two-domain Lc from a metagenome; AxNIR, *Alcaligenes xylosoxidans* NIR; AcNIR, *Achromobacter cycloclastes* NIR; AfNIR; *Alcaligens faecalis* NIR; HdNIR, *Hyphomicrobium denitrificans* NIR; CrPC, *Chlamydomonas reinhardtii* plastocyanin; PnPC, *Populus nigra* plastocyanin; AcPAz, *Achromobacter cycloclastes* pseudoazurin; PdAm, *Paracoccus denitrificans* amicyanin; AxAZ1, *Alcaligenes xylosoxidans* azurin1; TfRc, *Thiobacillus ferroxidans* rusticyanin; CsStc, *Cucumis sativus* stellacyanin; CpMv, *Cucurbita pepo* mavicyanin.

while the His-Cys-His is located in the latter domain. Domain 2 functions as a structural factor. Differing from MCOs (e.g., AOase), CueO has an extra segment to cover the substrate-binding site, and, accordingly, is slightly bulky among the three-domain MCOs.

As for variants of MCOs, the two-domain MCOs, such as *Streptomyces* 2-domain Lc (SLAC), metagenome Lc (mgLAC), and *Nitrosomonas* Lc (BCO), have recently been discovered.[16–19] These two-domain MCOs have a hexagonal-like quaternary structure (Fig. 5.3c and d) analogous to NIR (Fig. 5.3e) (see below), while they are further classified into three subgroups on the basis of the location of the T1 center.[17]

The hexagonal-like structure is also found in a monomeric Cp comprised of the six domains (Fig. 5.3f). Only one complete trinuclear copper center is found at the interface of domains 1 and 6, and three mononuclear copper centers are in domains 2, 4, and 6, of which the last one harbors the T1 center coupled with the trinuclear copper center.[20] The fact that Cp is present in the serum of vertebrates and has the six-domain

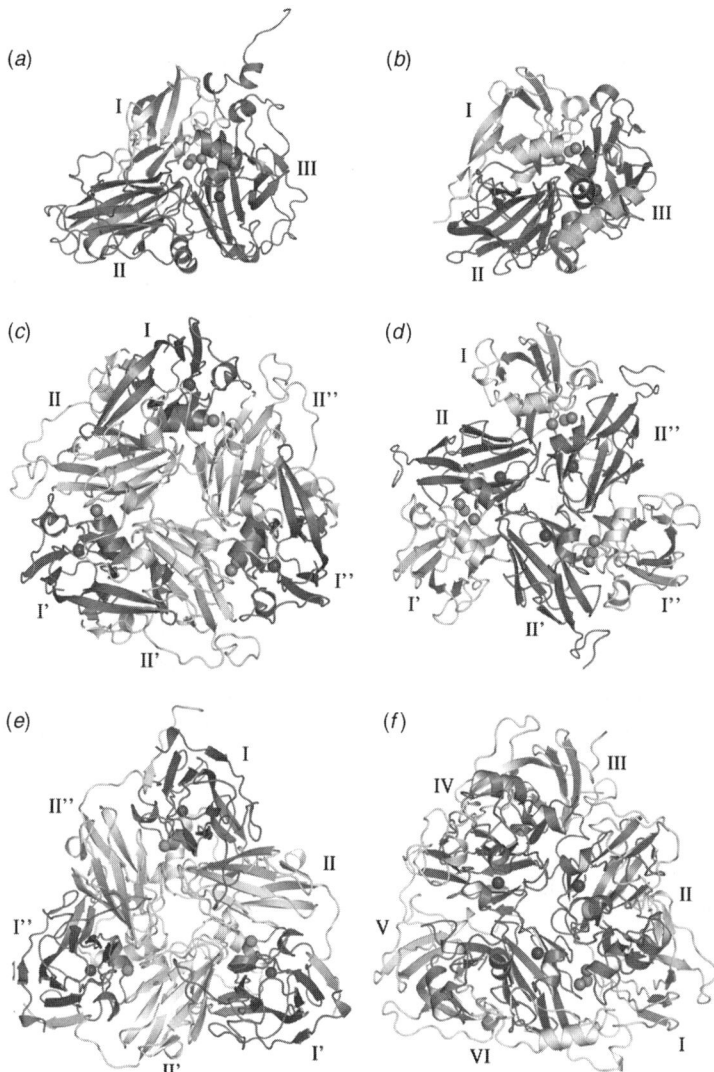

FIGURE 5.3. Crystal structures of two-, three-, and six-domain MCOs and NIR: (*a*) AOase, (*b*) CueO, (*c*) BCO, (*d*) SLAC, (*e*) AxNIR, and (*f*) hCp. The domain with T1 is colored blue and others are colored gray and red. Here T1, T2, and T3 are colored blue, green, and orange, respectively. The I, II, and III indicate the numbers of domains. (See color insert.)

structure as in NIR present in denitrifiers (see below), might represent a puzzle of molecular evolution of MCO families, while the three-domain MCOs are widely distributed in prokaryotes and eukaryotes excluding vertebrates. A variant of the six-domain MCO, (Fox1 from *Chlamydomonas*) has also been found.[21]

Nitrite reductase also forms a hexagonal-like quaternary structure as a homotrimer of the two-domain proteins (Fig. 5.3*e*). Figures 5.1–5.3 indicate that the T2 center in

NIR corresponds to one of the T3 centers in MCOs. From a molecular evolution point of view, the molecular architecture of NIR with T1 and the mononuclear NO_2^- reduction site is positioned between blue copper protein with only T1 and MCO with T1 and the trinuclear O_2 reduction site. The T2 center is located at the interface of domains belonging to different subunits, sharing His ligands. The three-domain MCO and NIR are structurally quite different in that the former is completed as a monomer, but the latter should have an oligomeric structure to exert activity. Future findings of novel MCOs and NIRs are expected to embed blanks of the undiscovered or lost proteins, displaying total features of the molecular evolution of cupredoxin super-family. As for NIR, variants with two T1 domains or the heme c binding domain have been discovered.[22–24]

Apart from the domain structure, metal oxidases, Cp, Fet3p, and Fox1 for Fe^{2+}, CueO for Cu^+, and MnxG, MoxA, and CumA for Mn^{2+}[25] harbor extended segment(s) covering the substrate-binding site so as to interfere with access of organic substrates to the T1 center (Fig. 5.3b). The substrate-binding site is constructed with an appropriate set of ligand groups to bind a metal ion in the reduced form and to release a metal ion in the oxidized form with moderate affinities (see Section 5.5 and 5.7.1).

In spite of the long history of studies on the *Rhus vernicifera* Lc (RvLc), only the amino acid sequence for structural information has been reported.[6] One reason why the plant Lc has not been crystallized is that its high sugar content coming up to ~50% did not allow the formation of crystals. Thirteen-to-fourteen N-linked glycosylation sites, Asn-X-Thr/Ser, are present in the isozymes of RvLc. Enzymatic and non-enzymatic attempts to hydrolyze off the sugars have been unsuccessful. The heterologous overexpression of holo-RvLc in *E. coil* also has not been successful and we must wait to obtain a recombinant plant Lc without sugars that is suitable for crystallography. In contrast, sugar contents in fungal Lc's are low, so they do not interfere with crystallizations, and an increasing number of crystal structure data is now available.

5.3. STRUCTURE OF THE COPPER-BINDING CENTERS

Crystal structure analyses of MCOs have shown that a specific sequence in MCO, His-Cys-His, connects the T1 and T3 centers. The copper centers that perform the ET and O_2 reduction are directly connected, making the long-range ET (12–13 Å) easy. Further, peptide backbones closely connect each copper center with the His-spacer-His sequence, and accordingly, any change induced at a copper site is propagated to the other copper centers.

There are two types of MCO T1 centers that concern ligand groups, 1Cys2His1Met and 1Cys2His.[7] The use of the fourth ligand appears to have been selected to adapt to the redox potential of a given substrate (see Section 5.7.2). The coordination geometry of the T1 center barely changes during the catalytic cycle, which is important in order to minimize the reorganization energy attended to by the change in the redox state of the T1 center.

The T2 center in the resting form is a T-shaped coordination geometry with the three coordination comprised of 2His and H_2O or OH^- depending on the pH. This favorable structure, for cuprous ion, has not been realized in the small molecule studies of Cu(II) complexes. Related to the facile reduction, T2 might be partly reduced in many resting MCOs. Therefore, it should be guaranteed that the hydrated electron did not reduce T2 in the X-ray crystallographic studies of MCOs.

Each T3 center is coordinated by the three His residues at the end of the His-spacer-His sequences. In addition, the T3 centers are bridged by an OH^- group, and are antiferromagnetically coupled in the resting form of MCOs. We refer to this structure as the classical one because a variety of resting structures other than this have been discovered, and the number of variations continues to increase. The actual structure of the resting MCO is now somewhat unclear except for the conserved coordination of the three His residues to each T3 center (see below). The steric structure of the T3 center is hindered tetrahedral as reflected in the EPR spectrum: An EPR signal with an intermediate magnitude of the A_z value between those given by T1 and T2 becomes observable, but only when the T3 centers become magnetically uncoupled.[26–28]

5.4. SPECTRAL PROPERTIES

Absorption bands at ~600 and ~330 nm are characteristic of T1 and the trinuclear copper center, respectively, and the band at ~750–800 nm is characteristic of the d–d transitions due to T1 (Fig. 5.4).[8] The absorption at ~330 nm is usually recognized as a shoulder in many MCOs, since it is located at the slope of the strong absorption band at ~280 nm derived from the protein molecule. While the band at ~330 nm has been

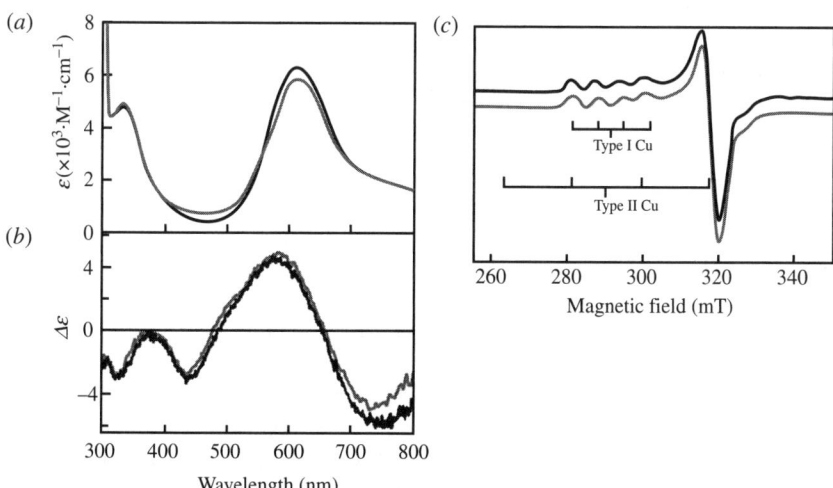

FIGURE 5.4. Absorption (*a*), CD (*b*), and EPR (*c*) spectra of the recombinant CueO (black) and the deletion mutant, $\Delta\alpha$5–7 CueO (gray).

assigned to coming from the charge transfer from OH^- to T3, the resting forms other than the classical one also afford a band at $\sim$330 nm (see below). When the absorption intensity at 330 nm is prominently higher or lower compared to that of the absorption at $\sim$600 nm, however, a mixing of other forms should be doubted. In the resting form of *Melanocarpus* Lc[29] and the double mutant of CueO, Cys500Ser/Glu506Gln,[30] a pronounced absorption at $\sim$360 nm (two bands) and a shoulder at $\sim$450 nm have been observed. These spectral features are concerned with an intermediate form of the O_2 reduction (see Section 5.6.4). On the other hand, the T3 center might be partly reduced or not properly formed, when the absorption in the near-UV region is not very peculiar.

Circular dichroism (CD) spectra of MCOs are sometimes more informative than corresponding absorption spectra because more than five bands are observable in the near-UV to visible region. However, curve analysis has not necessarily been successful for all MCOs, since many bands originated from the four copper centers are overlapped.[31] Magnetic circular dichroism (MCD) spectroscopy has also been applied to MCOs at high magnetic fields and cryogenic temperatures to characterize the electronic and magnetic properties of the trinuclear copper center and reaction intermediates.[32]

In addition to absorption and CD spectroscopies, EPR spectroscopy has also been frequently applied to MCOs. Both T1 and T2 signals are detectable in the resting state of MCOs. The T1 center affords the peculiar Cu^{2+}–EPR signal characteristic of its high covalency due to the binding of Cys, although the anisotropy is not as high as that given by phytocyanins with the coordination of Gln at the apical site. A pair of T3 centers is EPR nondetectable even at low temperatures due to the strong antiferromagnetic interaction, which is in accordance with the superconducting quantum interference device (SQUID) data.[33] However, in cryogenic measurements of the N_3^- ligated Lc and AOase, a broad signal assigned to the dipolar coupled Cu^{2+} pair in an $S = 1$ state was observed at $g \sim 1.9$,[34] similar to that seen for reaction intermediate II (see Section 5.6.5).

Vibrational spectroscopy, such as resonance Raman (RR) spectroscopy, has not necessarily been informative concerning the nature of the electronic state of the copper centers in MCOs except for T1.[35] The RR spectra of MCOs excited at the 600-nm band are analogous to those of blue copper proteins. Excitation at the 330-nm band has not given any information about the electronic state of the trinuclear copper center due to strong fluorescence,[36] in contrast to RR spectroscopy for the proteins containing the binuclear copper center (e.g., hemocyanin and tyrosinase[37]. Thus, vibrational spectroscopies have not provided much information about O_2 reduction by MCOs.

K-edge X-ray absorption spectra (XAS) have shown that the T3 center in the T2 depleted plant Lc[32] and AOase[38] remain in the cuprous state even under aerobic conditions, indicating that all three copper centers are indispensable for reduced MCOs to react with O_2. However, due to the high redox potential, reduction of the copper centers in MCO may readily take place with the hydrated electron formed by synchrotron radiation.[39,40] Therefore, data acquisition conditions are especially important in the performance of K-edge XAS spectroscopy, as they also are in the application of X-ray crystallography and RR spectroscopy.

5.5. SUBSTRATE BINDING AND SPECIFICITY

Multicopper oxidases are potentially able to oxidize substances for which the redox potential is more negative than that of T1. Note that the redox potential of substrates may differ in solution and on the protein surface. The MCO exhibits high oxidizing activity toward phenols, such as urushiol (phenol lipid), o- and p-diphenol, 2,6-dimethoxyphenol, diamines (e.g., p-phenylenediamine), ascorbate, bilirubin, and ABTS in addition to metal ions (e.g., Fe^{2+}, Cu^{+}, and Mn^{2+}). Among MCOs, AOase, BOD, and phenoxazinone synthase, [41] these are named for their substrates, but others are names, for their origins and recently, for their gene names.

The substrate-binding site of MCO is located in the cavity on the protein surface, being adjacent to the T1 center.[42] In the case of the docked model of AOase with L-ascorbate, the two O atoms of L-ascorbate are within hydrogen-bonding distances of the imidazole nitrogen of His512 coordinated to T1 and the indole nitrogen of Trp362 (figure not shown).[43] Both aromatic rings of His512 and Trp362 are approximately parallel to each other. Furthermore, Trp163 forms the wall of the substrate-binding pocket, facilitating the binding of L-ascorbate. According to a docking study of *Bacillus* Lc (CotA) with ABTS, the half-moiety of the bulky substrate in the U-shape binds in the pocket mainly formed with apolar residues, being close to the His497 coordinated to T1; the other half-moiety of the ABTS molecule is exposed to solvent (Fig. 5.5a).[44] One of the sulfonate groups in the ABTS molecule approaches the S−S bond of Cys229–Cys322 and forms a hydrogen bond with Gly323. The imidazole ring of His497 and the thiazoline ring of the partly buried half of the ABTS molecule were nearly perpendicular to each other, with the closest approach of 3.3 Å between the C2 and NE2 atom of His497. One can conclude that the ET from ABTS to T1 takes place through His497. Another striking feature to suggest the ET pathway from substrate to T1 has been shown in the crystal structure of *Trametes versicolor* Lc (TvLc), in which 2,5-xylidine used as an inducer in the culture medium was retained and docked in the cavity close to T1 (Fig. 5.5b).[45] The amino group in this arylamine formed hydrogen bonds with His458 coordinated to T1 and the uncoordinated Asp206 that

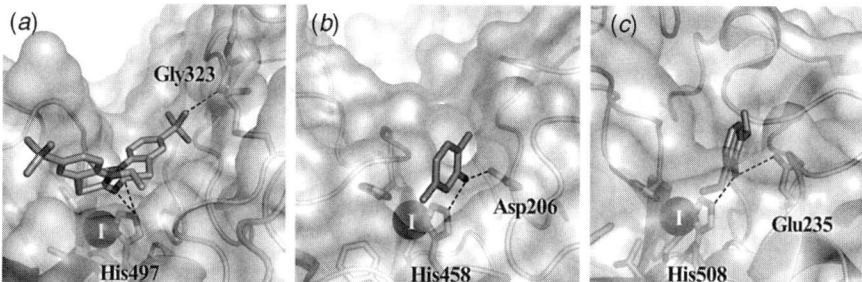

FIGURE 5.5. Substrate-binding pockets of MCOs. (a) The ABTS-bound CotA,[44] (b) 2,5-xylidine-bound TvLc,[45] and (c) 2,6-dimethoxyphenol-bound *M. albomyces* Lc[46] looked down from the same direction. It may not be easy to recognize that the substrate-binding pocket in (c) is considerably narrow and deep from the monochrome figure.

is conserved in Lcs (Fig. 5.2). Furthermore, 2,6-dimethoxyphenol-bound *Melano-carpus albomyces* Lc (Fig. 5.5c)[46] and *p*-toluate-bound *Trametes trogii* Lc[47] also show that the substrate or inhibitor forms hydrogen bonds directly with the His coordinated to T1 and indirectly with an Asp or Glu via a water molecule. The acidic amino acid that is located on the wall of the substrate-binding cavity may play a role in assisting the deprotonation of substrate phenolic OH groups and the succeeding ET from substrate to T1, while this does not seem to be the exclusive origin of high activities for these Lcs operating at acidic pH values. Nevertheless, the effective ET pathway leading from substrate to T1 is constructed in a common manner in many MCOs.

Metal oxidases differ from MCOs acting on organic substrates by possessing molecular architectures designed to specifically bind a metal ion in the neighborhood of the T1 site. The CueO, which is involved in the copper efflux system in *E. coli,* specifically binds Cu^+ at the site constructed with 2Asp2Met located under the Met-rich region involving helices 5–7 and a very flexible region. This set of amino acids is adapted to bind a cuprous ion and to release the oxidized cupric ion. The Met-rich segment comprised of 50 amino acids not only plays a role in blocking the access of organic substrates to the T1 center, but also assists the access of Cu^+ to the substrate-binding site with its combination of the rigid helices and a flexible loop.[48] We depleted this region wholly[49] and partly,[50] connecting the ends of the residual chains with short linkers. By these protein engineering efforts, the specificity to Cu^+ could be prominently decreased and the reactivity to organic substrates could be promoted or even newly emerged (Fig. 5.6). Deletion has also been performed for *Shigella dysenteriae* Lc.[51] According to the amino acid sequences of the bacterial PcoA and homologues,[52] the multiple repeats of the Met-rich domains are docked, and

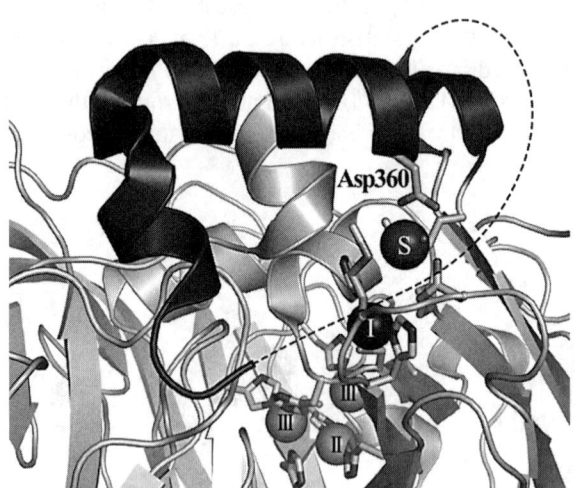

FIGURE 5.6. The substrate-binding region of CueO. The deleted segment rich in Met residues, which include helices 5–7 and the unresolved region (dotted line), is depicted in dark gray. Here T1, T2, T3, and substrate(product) copper are represented as I, II, III, and S, respectively.

accordingly, the molecular architecture found in CueO is not exceptional to this cuprous oxidase.

Manganese oxidases (e.g., MnxG, CumA, and MoxA)[53–55] are also known to contain the substrate-binding region as found in CueO. On the other hand, ferroxidases (e.g., Cp and Fet3p) have molecular architectures different from CueO in order to specifically bind an Fe^{2+} ion. The negatively charged pocket, constructed with 2Glu1Asp1His located near the T1 center in domains 4 and 6 in Cp, is the putative Fe^{2+} binding site.[56] It is assumed that the electron withdrawn from the Fe^{2+} ion bound at domain 4 is transferred to the T1 center in domain 6 via a through-bond process occurring > 18 Å. A p-phenylenediamine-binding site is also present at the bottom of domain 4, remote from the T1 center in this domain. Hephaestin, a ferroxidase in the epithelial cells in the intestine, differs from Cp since it is coupled with a transporter, Ireg1.[57] The Fet3p in *Saccharomyces cerevisiae* is a typical three domain MCO having a membrane anchor at its C-terminal end. It forms a complex with the permease, Ftr1p, in order to transport Fe through the membrane as Fe^{3+}.[58] The imidazole edge of a His coordinating to T1 is not directly exposed to solvent. Residues Glu185, Tyr354, and possibly Asp278 have been identified as the amino acid residues for the specific binding of Fe^{2+}. However, a MCO that shows high ferroxidase activity from basidiomycete *Phanerochaete chrysosoporium* has an Arg residue instead of a Tyr residue according to the amino acid sequence, although another Tyr might serve the function.[59] As for the ET pathway from substrate to T1, a hydrogen bond connects a His residue coordinating T1 and an acidic amino acid, which constitutes the binding site for the metallic substrate in CueO, Cp, and Fet3p.[49,60,61]

5.6. FOUR-ELECTRON REDUCTION OF DIOXYGEN BY MULTICOPPER OXIDASES

5.6.1. Diverse Resting Form

In the typical resting form of MCOs, the T3 centers bridged with an OH^- group are antiferromagnetically coupled. In contrast, T2 is magnetically isolated from the T3 centers, and is EPR detectable. These characteristics of the trinuclear copper center in MCO were first substantiated in the crystal structure of AOase,[12] and later in many other MCOs (*a* in Fig. 5.7). In this resting form (*a*), the bridging OH^- is slightly outside the triangle formed with the T2 and T3 centers.[42,43,62] A variant of this structure is that the OH^- is asymmetrically located between T3 centers, although it is not certain whether the likely partial reduction of the T3 center was examined or not.[63,64] The resting structure (*a*) is referred to as a classical resting one because a variety of other resting structures have been discovered (*b–f* in Fig. 5.7). Part *b* is a variant of *a*: the angle between Cu−O−Cu is near to 180° or even the bridging O atom is slightly inside the triangle.[41,45,49] In c[29,44] and d,[65,66] an O_2 molecule is bound at T3 centers in the end-on and side-on fashions, respectively. Here, the side-on binding includes the form that O_2 is not equally shared by the two T3 centers. On

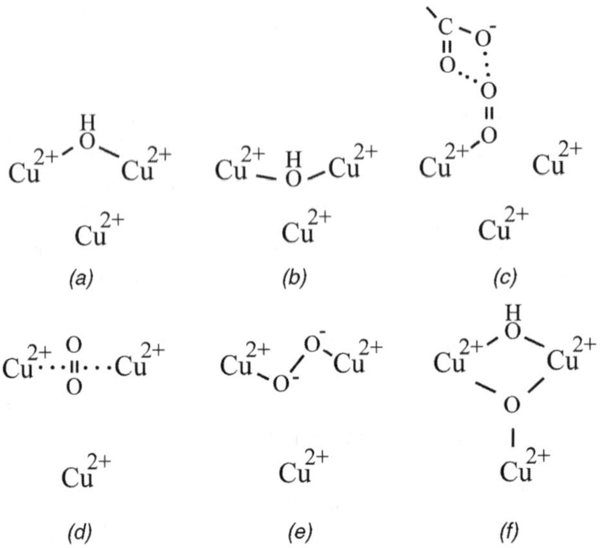

(a)   (b)   (c)

(d)   (e)   (f)

FIGURE 5.7. Diverse structures of the trinuclear copper center in the resting MCOs, parts a–f, which have been shown by X-ray crystallography hitherto.

the other hand, a peroxide ion was bound between the T3 centers in the trans-1,2 peroxide fashion in e.[65,67] The binding of the peroxide between the dinuclear center is the same as with the resting CcO.[68] In f, an O^{2-} anion is located at the center of the triangle[18,40,67,69] and an OH^- is bridged between the T3 centers or is bound to one of the T3 centers. This structure is practically the same as that proposed for an intermediate (intermediate II or native intermediate), to which MCO reaches the final stage of the single turnover process of the $4e^-$ reduction of O_2.

In addition, the resting state of MCOs is not necessarily in an exclusive state among the a–f state; that is, apparently its in a mixed b and d state in the deletion mutant of CueO.[40] One reason why more than two forms are observed in the crystal structures of MCOs is presumably because the copper centers are highly reducible by hydrated electrons, especially when synchrotron radiation is used as the source of X-ray. When we performed time-resolved measurements of the X-ray crystallography of the recombinant CueO, the structure f corresponding to intermediate II was exclusively present at 6 s. However, reduction of one of the T3 centers began with the progress of data acquisition, and finally the structure without the bridged OH^- became dominant. The stepwise reduction of Cu^{2+} ions with hydrated electrons was evidenced by the *in situ* Cu K-edge measurements. Therefore, conditions to acquire diffraction data are especially important in the crystal-structure analyses of MCO, in addition to resolution and modeling procedures.[70] The facile reduction of Cu^{2+} ions can lead to confusion about the nature of copper centers, especially those in the trinuclear copper center in MCO that have high redox potentials. Differing from the X-ray crystal structure analysis of the recombinant CueO, that for the truncated mutant of CueO, $\Delta\alpha$5–7CueO, in which 50 amino acids covering the substrate-binding site were removed, gave different results. The resting $\Delta\alpha5$–7CueO had structure b, but structure d with the side-on bound O_2 between the T3 centers was also found as a minor one, indicating that the resting form is in a mixed state. Regardless of these complexities about the resting structure of MCOs, it was ascertained that the authentic and recombinant BODs are in different resting forms and related by one cycling of the enzyme reaction.[71]

The presence of an O_2 molecule or O_2^{2-} anion between the T3 centers in the resting form of *Melanocarpus* Lc, CotA, Cp, and CueO is controversial (Fig. 5.7). The O_2 molecule might have been captured between the T3 centers due to a certain cage effect. However, the fact that T2 is required for the plant Lc to exert reactivity to O_2, is contradictory to the side-on binding of O_2 between the T3 centers. It is not certain whether the binding mode of peroxide found in the resting MCOs is the same with that in the intermediate I (see Section 5.6.4). In order to avoid confusions, electron densities in crystal structure analyses should guarantee full occupancies of O and copper atoms.

Ascomycete Lcs have a plug comprised of Asp-Ser-Gly-Leu for the O_2 channel at the C-terminus. The role of this plug is supposed to be a switch that transforms the pre-protein to an activated state. However, the fact that enzyme activities were lost by deleting this plug is controversial, while the expression system of Lc enzymes has not been fully revealed.[72]

5.6.2. Reduced Form

The X-ray crystallography of the reduced MCO has been performed for AOase.[73] On reduction of the resting enzyme, the bridged OH^- among the T3 centers was released, and the T3 centers moved toward their respective His residues, which has become three coordinate. The distances between copper atoms increased from an average of 3.7 to 4.1, 4.4, and 5.1 Å. The size of the copper triangle became apparently large compared with that of the resting form, becoming favorable to accommodate O_2. The size of the triangle will be the smallest when O_2 is bound, and will be larger as the $4e^-$ reduction of O_2 proceeds. Crystal structures of other reduced MCOs, CotA,[74] and CueO[40] have been obtained by exposing crystals of the resting enzymes to X-rays, affording analogous structures with the reduced AOase. Note that the facile reduction of the trinuclear copper center by X-rays took place during acquisition of diffraction data for resting Fet3p and Lcs.[39,67,75,76]

5.6.3. Multicopper Oxidases Require the Fully Reduced Trinuclear Copper Center to React with Dioxygen

The electron withdrawn from the substrate by T1 is transferred to the trinuclear copper center that is 12–13 Å away. The neighboring Cys and His residues coordinate T1 and T3 centers, respectively, and function as the through-bond pathway for this long-range ET. The two branches of the pathway may not be equivalent due to a hydrogen-bond shunt connecting the imidazole ring coordinated to one of T3s and the backbone carbonyl O atom of the Cys residue.[12,43]

The trinuclear copper center in MCOs, which differ from the proteins containing only a pair of T3 centers, should be fully reduced to bind and reduce O_2. The selective removal of T2 has been successfully carried out only for the plant Lc[77] and AOase,[78] in which the T3 centers were observed to be in the reduced state even in air.[38,79] Reactivity of the T2 depleted Lc to O_2 was restored by adding a cuprous ion under anaerobic conditions.[80] The reaction with O_2 was also possible by selectively substituting the T1 Cu with Hg.[81] These results indicate that the reactivity with O_2 is different in a paired T3 center in MCOs and the proteins containing only the T3 center.[82] This result might be due to a difference in the spatial arrangement of His residues coordinated to T3s and/or to a different size of the cavity to accommodate the copper centers, in addition to the contribution from T2. It is not known how the resting forms of MCOs bound with O_2, c–e, were formed, although the oxidation state of T2 had been considered to be cupric. In the resting form of CcO, O_2 is bound between the heme a_3 and Cu_B.[68] This resting form of CcO, which was found by utilizing diffraction data under conditions where the metal centers were not reduced, apparently differs from the putative O_2 bound form in the catalytic cycle.

5.6.4. Intermediate I Captured by the Reaction of Dioxygen under the One-Electron Deficient Conditions

No reaction intermediate of the 4e$^-$ reduction process of O_2 has ever been detected during turnover conditions of MCOs. However, the reaction intermediate I could be trapped by performing the reaction with O_2 at 1e$^-$ deficient conditions by using the plant Lc, where T1 was cupric[80] or the reaction with a plant Lc derivative where T1 Cu was substituted with Hg.[81] Although direct evidence was never obtained, intermediate I has also been referred to as the peroxide intermediate.[83]

Recently, it become possible to form reaction intermediate I of CueO,[30] BOD,[28] and Fet3p[84,85] by substituting the Cys residue coordinated to T1 with Ser. In these mutants, the T1 center was vacant. Thus it became possible to study the 1e$^-$ deficient reaction process of MCOs, whose expression systems was established. Absorption, CD, and EPR spectra of intermediate I, obtained from the Cys500Ser mutant of CueO, are shown in Figure 5.8. Intermediate I gave a characteristic absorption spectrum with maxima at $\sim$340, 470–475, and $\sim$680 nm. The absorption at $\sim$340 nm is comprised of the two oppositely signed intense bands according to the corresponding CD spectrum. The band at >700 nm originate from d–d transitions of the copper centers. Intermediate I was EPR silent, but the T2 EPR signal became observable with the decay of intermediate I. Therefore, it has been considered that T2 is cuprous in the intermediate I, and the decay of it involved the ET from T2 to the O_2 reduced species. The low-temperature MCD studies of Fet3p have also been performed to characterize intermediate I, suggesting that T2 or one of the T3 centers is cuprous in intermediate I.[86] Contrary to this finding, the T2 copper was cupric and paramagnetic in the peroxide adduct of the plant Lc.[87] Therefore, special precautions are required to discuss the reaction mechanism of MCOs based on characterizations of an artificially formed intermediate analogue.

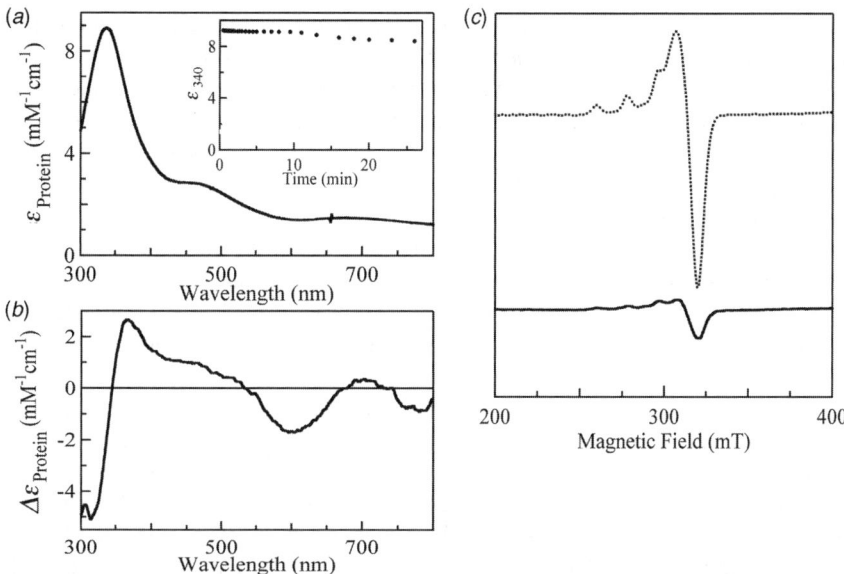

FIGURE 5.8. Intermediate I obtained from the reaction of Cys500Ser CueO: (*a*) absorption spectrum (Inset, decay curve), (*b*) CD spectrum, and (*c*) EPR spectrum (solid line) (dotted line is for the resting Cys500Ser mutant of CueO to give only the T2 signal).

Various structures, in many of which a peroxide or hydroperoxide is bound, have been proposed for intermediate I, taking into consideration that this form is EPR silent and gives strong absorptions in the near-UV to visible regions, as shown in Figure 5.8. In this regard, T2 is not involved in the side-on binding of O_2 between the T3 centers as in the resting CotA, fungal Lc, and Cp.[56,65] The end-on binding of a hydroperoxide at one of the T3 centers was realized by soaking peroxide into AOase, although the possibility that an artifact was obtained was not excluded due to the same reason mentioned above.[73]

The structure of intermediate I is still controversial in spite of its own designation, the peroxide intermediate. No RR spectrum to provide direct evidence for the presence of peroxide has ever been obtained for intermediate I of CueO, BOD, and plant Lc, in spite of the employment of relatively weak laser excitations for the shoulder band at $\sim$470 nm in order to avoid facile breaching of chromphores.[36] The absorption at $\sim$340 nm, which is more intense than that of the resting MCOs, seems to be typical of the peroxide to Cu^{2+} charge transfer according to small molecule studies.[88] However, it might be undeniable that intermediate I is a metastable state reached after forming the very short-lived peroxide bound form. To fix the oxidation state of T2 is mandatory to solve the puzzle of the structure of intermediate I. In connection with this, a double mutant of CueO, Cys500Ser/ Glu506Gln, as isolated gave the absorption, CD, and EPR spectra typical of intermediate I.[30] The X-ray crystal structure analysis of this mutant, however, indicated that the O—O bond had already been cleaved, giving the structure anticipated for intermediate II, in which an O atom is at the center of the copper triangle.[40] Diffraction data obtained within 6 s with synchrotron radiation and with a conventional diffractmeter, both avoided the reduction of the copper centers with hydrated electrons at low temperatures, and resulted in the same findings. This fact indicates that intermediate I is quite readily transformed into a subsequent form by accepting a solvated electron; or the O—O bond has not been cleaved by homolysis, but rather by heterolysis. Otherwise, intermediate I was not in a $2e^-$ reduced state, or the structure of the trinuclear copper center was changed during crystallization. Further studies to directly give information about the structure of intermediate I are required. In the case of CcO, rapid reduction of O_2 is performed by preventing leakage of deleterious intermediates in the conversion steps of the initially formed O_2 adduct to $Fe^{IV}=O$ and H_2O.[89]

In the decay process, intermediate I of RvLc, Fet3p, and possibly CueO accelerates with decreasing pH, suggesting the involvement of an acidic amino acid to function as the proton donor to intermediate I.[32,80,84,90,91] Thus, it is evident that the reduction of O_2 is governed by the noncoordinating amino acid(s) located near the trinuclear copper center (see Section 5.6.6). In the formation and decay of intermediates I and II, the acidic amino acids adjacent to the trinuclear copper center are indispensable. The roles of these acidic amino acids in the $4e^-$ reduction of O_2 by MCOs are discussed in Section 5.7.6.

5.6.5. Intermediate II Captured at the Single Turnover Conditions of Multicopper Oxidases

Another reaction intermediate has been trapped at the end of a single-turnover process in plant Lc,[92–95] BOD,[28] and CueO,[30] although the structure proposed for this intermediate has also been found in the resting TvLc,[69] SLAC,[18] and CueO,[40] as Figure 5.7f. Therefore, we do not refer to this intermediate as the native intermediate, but to avoid confusion, call it intermediate II. The blue color of T1 was already recovered, and accordingly, the ET from T1 to O_2 was finished when MCO reached this intermediate (Fig. 5.9). The EPR spectrum of intermediate II is characteristic: A broad signal with $g_\perp \ll 2$ is observed at <40 K, in addition to the T1 signal. This novel signal was not easily saturated with increasing microwave power, indicating it originated in a species in which all three Cu^{2+} ions were highly magnetically coupled. The T2 EPR signal was not detected when this signal was fully detected, but became observable with the decay of the $g_\perp \ll 2$ signal.

The lifetime of intermediate II was prominently lengthened by mutating an acidic amino acid located near the trinuclear copper center, Glu506 in CueO[30] and Glu486 in BOD.[96] In the case of BOD, intermediate II could also have been trapped in the single-turnover reaction of a T1 mutant, Met467Gln, in spite of the shift in the redox potential of T1 toward a negative potential. Intermediate II thus trapped showed its absorption spectrum features to be superficially analogous to that of intermediate I in the 300–500-nm regions, although the absorption intensities of the bands at $\sim$350 nm and the shoulder at $\sim$420 nm were not as intense as those given by intermediate I in addition to the contributions from the reoxidized T1.

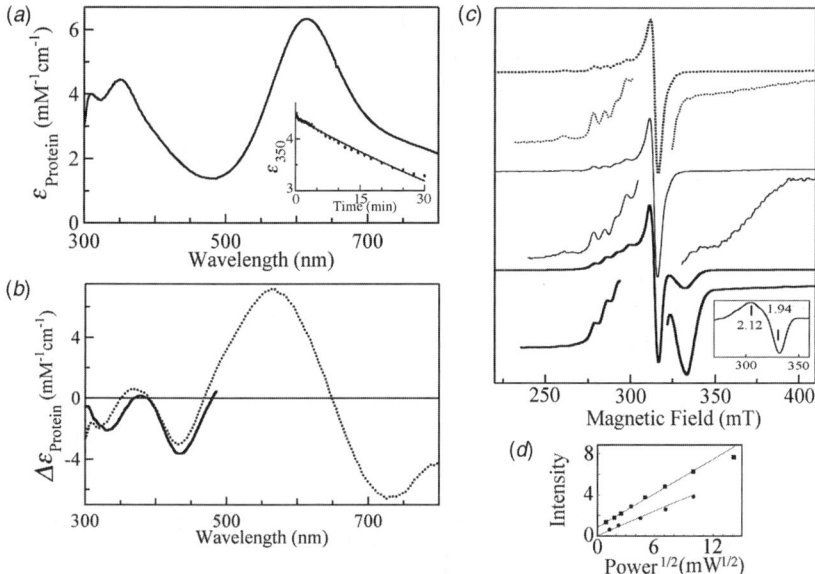

FIGURE 5.9. Intermediate II obtained from the reaction of Glu506Gln CueO; (*a*) absorption spectrum (Inset, decay curve), (*b*) CD spectrum (solid line) (dotted line is for the resting Glu506Gln), (*c*) EPR spectrum at 3.5 K (from top to bottom: the resting Glu506Gln, O_2-reacted recombinant CueO, and O_2-reacted Glu506Gln) (inset shows the novel signal derived from the intermediate II detected at <40 K obtained by subtracting the T1 EPR signal), and (*d*) power saturation behavior of the EPR signal of intermediate II derived from the O_2 reaction of CueO and Glu506Gln mutant.

Intermediate II has been considered to be in a fully oxidized form, in which the O atom is located at the center of the three Cu^{2+} ions and T3 centers are bridged with an OH^- group. With the assistance of an acidic amino acid to protonate the O atom, intermediate II is transformed into the classical resting form (Fig. 5.7*a*) by releasing a water molecule. Therefore, the decay process of intermediate II is accelerated with decreasing pH.[30,31,94,95] The decay process exhibited a large positive activation of enthalpy and a large negative activation of entropy, suggesting that a considerable deformation takes place on the skeletal structure of the trinuclear copper center.[96] With the decay of intermediate II to reach the resting form, the T2 center is magnetically isolated and changes so as to be EPR detectable.

In contrast to the reactions of the three-domain MCOs, a T1-depleted mutant of SLAC, Cys288Ser, showed a unique reaction.[97] This mutant did not give rise to intermediate I, but gave an intermediate having the transient absorptions at 330 and 420 nm and a magnetically coupled EPR signal assigned as coming from the exchange-coupled T2 and a tyrosyl radical. The same intermediate has also been observed at single turnover conditions of the recombinant SLAC. The $1e^-$ deficient reaction of SLAC did not stop at intermediate I, but reached a kind of intermediate II by receiving an electron presumably from Tyr108, which is separated by 4.6 Å from T2. Under turnover conditions, the fourth electron to fully reduce O_2 will be rapidly supplied from T1, and it is not known whether the possible contribution of the Tyr residue is limited to SLAC or it occurs more widely. Figure 5.2 demonstrates that this Tyr residue is not conserved in the three-domain MCOs, although a Tyr next to one of the His ligands to T3 is more highly conserved in MCOs. If the fourth electron is supplied from an aromatic amino acid residue in every MCO, it is during the decay of intermediate II that the radical source abstracts an electron from T1. This story about the $4e^-$ reduction of O_2 by MCOs is more similar to that by CcO utilizing the condensed His-Tyr cofactor that formed by a post-translational modification, as an electron donor. In any event, the $4e^-$ reduction of O_2 by MCO does not seem as simple as is often described, as proceeding in two sequential $2e^-$ steps.

5.6.6. Contributions from Acidic Amino Acids Located near the Trinuclear Copper Center in the Binding and Reduction of Dioxygen

For the copper-binding sites in CueO (Fig. 5.10), it is clear that adjacent acidic amino acid residues, Asp112 and Glu506, interact with the T2 and T3 centers: Asp112 is hydrogen bonded with His448 coordinated to one of the T3 centers and is also hydrogen bonded with a water molecule that is hydrogen bonded to the OH^- or H_2O coordinated to T2. Also, Glu506 is hydrogen bonded with His143 and His499 coordinated to a T3 center and is hydrogen bonded with a water molecule that is nearest to the bridged OH^- between T3 centers. Analogous acidic amino acids, Asp94 and Glu487, are also located at the trinuclear copper center in Fet3p, while Glu is not rigorously conserved in the amino acid sequences (Fig. 5.2).

When Glu506 in CueO was mutated to Gln, the catalytic activity was profoundly lowered, leading to the detection of intermediate II.[30] This fact strongly suggests that Glu506 is involved in the proton donation to intermediate II and may also donate to intermediate I. On the other hand, when Asp112 was mutated to Glu, the enzyme

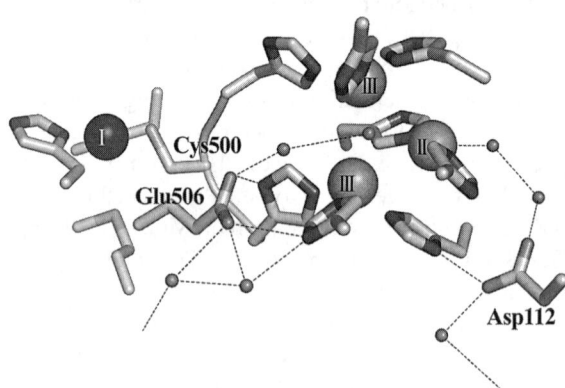

FIGURE 5.10. Residues Asp112 and Glu506Gln adjacent to the trinuclear copper centers in CueO, which are indispensable in the binding and $4e^-$ reduction of O_2.

activity reduced to about one-half, and when Asp112 was changed to amino acids having aliphatic side chains, the enzyme activities were reduced to $\sim$10%.[98] Analogous results have also been observed in BOD,[28] while reductions in the enzyme activities were not so significant as in the Fet3p mutants.[86] The affinity of the trinuclear copper center to O_2 did not change in the Glu mutants, but were reduced in other mutants. Therefore, it is apparent that Asp112 is involved in the binding of O_2 at the trinuclear copper center. Supporting this conclusion, the formation of intermediate I was diminished and slower in the reaction of the double mutant of CueO, Asp112Asn/Cys500Ser. It has been proposed for Fet3p that the deprotonation from the coordinated water to the reduced T2 is favored by the formation of the hydrogen bound with Asp94, stabilizing the bound peroxide in intermediate I,[31,86,91,99] while the oxidation state of T2 is unclear (see above). To fully reveal the O_2 reduction mechanism by MCOs, information concerning whether acidic amino acids involved in catalysis are protonated or deprotonated is required at each stage in the catalytic cycle. According to the attenuated total reflectance–Fourier transfer infrared (ATR-FTIR) study on CueO and BOD, Glu506 and Glu463, respectively, are deprotonated in the resting form, but are protonated upon reduction, being ready to assist the donation of protons to the intermediates.[100]

5.6.7. Reaction Mechanism of the Four-Electron Reduction of Dioxygen

Although activated oxygen species (e.g., superoxide, peroxide, and hydroxyl radical) are formed in the non-enzymatic reduction of O_2 to H_2O, no such activated oxygen species has been detected during the catalytic reactions of MCOs. This presumably occurs because MCOs avoid forming meta-stabilizing reactive activated-oxygen species. Differing from oxygenases, which utilize activated oxygen species, MCOs convert O_2 into water efficiently in order to minimize the risk of giving rise to species that may result in protein damage.

The presently most accepted reaction mechanism for MCOs is shown in Figure 5.11; direct support for this mechanism remains lacking, as described above. Possible hydrogen bonds involving the noncoordinating amino acids are also included in this figure. Reviews to describe the reactions by MCOs are available.[7,10,32,87] The reaction mechanisms delineated by combining individual results on different MCOs are confusing or even misleading. Under turnover conditions, MCO may vary between the reduced, intermediate I, and intermediate II forms, while the lifetime of intermediate I is short and undetectable. On the other hand, under single-turnover conditions, intermediate II has a long lifetime and reaches the classically known resting form after decay; the intermediate II state itself is more stable than the classical resting form in many MCOs. Since a variety of resting structures have been observed in MCOs and might be even mixed, it may now not make sense to fix and discuss the resting form of MCOs as only being the classical one, and accordingly, even altering the reaction mechanism. In the resting CotA, an O_2 molecule is simply caged between T3 centers, and this form occurs at the start of the reaction cycle.[101] It has been proposed that the peroxide-bound intermediate I, in which T2 is in the cupric state, is formed when the resting CotA is $2e^-$ reduced.

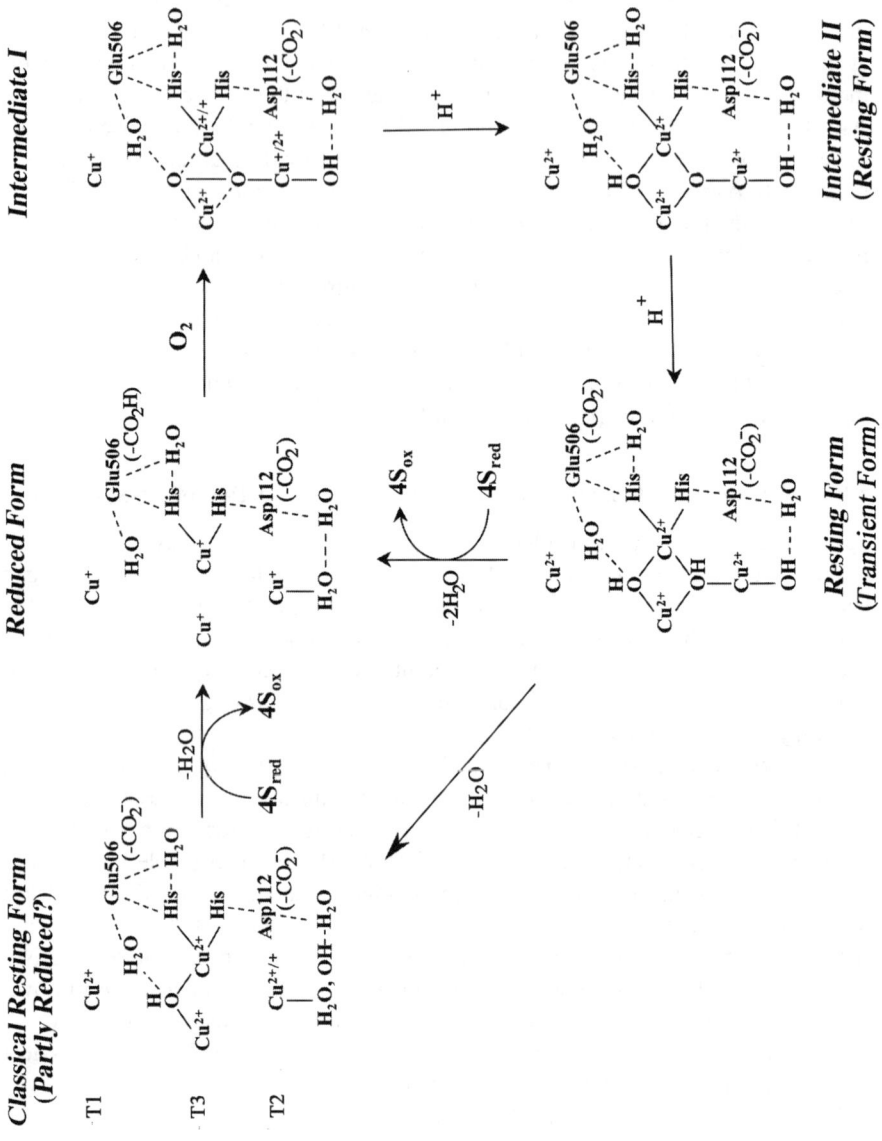

Intermediate I has never been detected during the reaction of MCOs, but this intermediate becomes detectable when ET from T1 is impossible. Therefore, intermediate I is considered to be a real intermediate, although its lifetime is undetectably short in the presence of the cuprous T1. The Asp112 residue in CueO is apparently indispensable for the trinuclear copper center to exert a high affinity to O_2, and therefore, is involved in the formation of intermediate I. This intermediate readily accepts electron(s), and is converted into the resting form. Our preliminary X-ray crystallographic study on Cys500Ser/Glu506Gln, which gives a spectrum identical with that known for intermediate I, showed the structure proposed for intermediate II, indicating that hydrated electrons formed by X-ray irradiation readily converted intermediate I to intermediate II. Although spectral properties of intermediate I are possibly those of a peroxide bound form, more efforts are required to obtain definite evidence for its structure.

The pH dependence observed during the decay of intermediate I suggests the involvement of an acidic amino acid, leading to the deduction that at least one proton is abstracted by one of the O atoms. The OH^- group bridged between T3 centers is a candidate for such a species; it may be exchangeable with water molecules present in the cavity leading to the exterior of the protein molecule.

The structure of intermediate II is considered to be O centered with respect to the trinuclear copper center, and an OH^- is bridged between T3s. Thus, all T2 and T3 cupric ions are magnetically interacting to give the $g \ll 2$ EPR signal. The pH dependent decay of intermediate II is the process where donation of a proton to the O atom in the inside of the trinuclear copper center occurs and the resulting OH^- flips around the T2 center. However, there left the puzzling fact that many resting MCOs have the structure in Figure 5.7f, the same one proposed for intermediate II (see Section 5.6.5). The striking difference in the magnetic properties of the resting MCO and intermediate II remains unrevealed if both forms have the same structure. This conflict might be accounted for if intermediate II is in fact a preceding undefined form just before it reaches structure 5.7f.

Thus, the $4e^-$ reduction of O_2 by MCO is not as simple as being composed of the two sequential $2e^-$ steps, involving the deformation steps coupled with transfers of electrons and protons. Therefore, we should not exclude any possible structures for the intermediates until convincing direct evidence is obtained. In addition, it might not make sense to discuss whether the reduction of the trinuclear copper center or the introduction of an O_2 molecule at this center takes place earlier in the turnover

---

FIGURE 5.11. An O_2 reduction mechanism by MCO. Of the various resting structures shown in Figure 5.7, structure of f continues to increase. Intermediate I, also called the peroxy intermediate, has been thought to be an EPR undetectable peroxide bound form. Intermediate II, also called the native intermediate, gives the $g \ll 2$ EPR signal at cryogenic temperatures, and has been considered to have the same skeletal structure shown for the resting MCO in this figure. A striking difference in properties of intermediate II and resting MCO is that the EPR signal of T2 is not detected in the former, but is detected in the latter. Preference of these forms differs from MCOs. At turnover conditions, the resting state may be skipped.

condition. Another reaction mechanism, where an O_2 molecule is captured among T3 centers in the resting form and the reaction is initiated by the subsequent reduction of the T3 centers,[101] arises based on the observed resting structures of CotA and some fungal Lc enzymes, c–e in Figure 5.7.

To put an end to confusion concerning the reaction mechanism of MCOs, properly modeled X-ray crystal structures of high-quality samples using diffraction data, where the trinuclear copper center is never deformed, are required in addition to structures including protons by neutron diffraction.[102] Thus, the $4e^-$ reduction of O_2 by MCOs is still in an undetermined state, and is far from having been fully revealed.

5.7. MODIFICATION OF MULTICOPPER OXIDASES

5.7.1. Mutation at the Substrate-Binding Site

Substrate specificity of MCOs is, in the first place, governed by the size of the cavity having an edge to a wall formed by the imidazole group coordinated to T1, when it is exposed, in order to mediate ET. The docking study of ABTS to CotA showed that a half-moiety of the substrate was bound at the cavity on the protein surface leading to T1 via the imidazole group. Meanwhile, the other half-moiety of the bulky ABTS molecule was directed outward to the solvent (Fig. 5.5a).[44] Besides the size of the cavity that accommodates an organic substrate, charge, polarity, and the location of specific amino acids in the substrate-binding cavity may also be factors that determine the specificity of a MCO. According to the crystal structure of TvLc, 2,5-xylidine when utilized as an inducer and docked at the substrate cavity, forms hydrogen bonds with the His458 coordinated to T1 and the neighboring Asp206 with its amino group (Fig. 5.5b). High activities shown by many fungal Lcs for phenolic substrates at acidic pH values might arise from the presence of this highly conserved acidic amino acid, while an induced fit may also take place upon binding of the substrate.

On the other hand, substrate specificities for metal oxidases depend on their unique molecular architectures toward each of the Cu^+, Fe^{2+}, and Mn^{2+} ions. In order to interfere with the access of organic substrates, cuprous oxidases (e.g., CueO) have an extra bulky segment comprised of three α-helices and a flexible region, which is usually a short loop in other MCOs. The substrate-binding site is located beneath this bulky segment and interferes with the access of organic substrates to the T1 center. Furthermore, the region covering the substrate-binding site is rich in Met residues in the case of CueO and related MCOs, as found in a copper transporter, Ctr1.[103] This arrangement of many Met residues in the extra region may guide the access of Cu^+ ion to its binding site. In addition to the steric hindrance against potential organic substrates, the binding site for a specific metal ion is constructed with an appropriate set of ligand groups. Amino acids selected to form the binding sites for these transition metal ions contrast to those used to construct the catalytic sites: 2Asp2Met for Cu^+ in CueO,[48] 2Glu1Asp1His for Fe^{2+} in Cp, and 1Glu2Asp for Fe^{2+} in Fet3p.[61] Metal ions in low oxidation states should be bound at these sites, and are released when oxidized. Therefore, ligand groups are selected in order to afford an appropriate lability for each kind of metal ion.

When we deleted the Met-rich region from the CueO molecule ($\Delta\alpha5$–7 CueO), which is comprised of 50 amino acids, oxidizing activities for organic substrates either newly emerged and/or increased.[49,50] Since one of the Asp residues to construct the substrate-binding site was in the deleted helix 5, the binding affinity for the Cu^+ ion became slightly reduced.[49] Although mutations were performed on the residual amino acid ligands for substrate, which were now exposed to solvent, the specificity for cuprous ion was not greatly modified.[104] In connection with this, the enzyme ferroxidase activity decreased by $\sim$50% due to the mutation at Glu935, which is involved in the binding of substrate Fe^{2+} in Cp; however, the diamine oxidase activity remained unaltered.[105]

5.7.2. Mutation at the Axial Ligand-to-Type I Copper

The T1 center is classified into three groups based on the set of ligand amino acids present; 1Cys2His1Gln, 1Cys2His1Met, or 1Cys2His1Phe/Leu/Ile. Redox potentials of these type T1 centers are in the range, 0.18–0.26, 0.29–0.68, and 0.55–0.79 V, respectively.[7] Only the second and third groups are found in MCOs. The axial ligand is Met in CotA, CueO, and BOD and can change to Gln and other noncoordinating amino acids. Shifts in the redox potential occurred in the expected directions toward negative potential for Gln and positive potentials for noncoordinating amino acids, except for the compensating coordination of an O atom, although the extents are not very large, ($\sim$.50–150 mV).[27,28,65,106–109] On the other hand, the proximal mutation at the T1 center in CotA, Ile494Ala, resulted in the coordination of a water molecule as the fifth ligand leading to an approximate trigonal bipyramidal (tbp) geometry.[110] Analogous occupation of a solvent water at the vacant site might have taken place in the situation where Met510 in CueO is mutated to Ala.[106]

Mutations of Met to bulky noncoordinating amino acids resulted in increases in enzyme activities of CueO, but resulted in reductions in the activities of CotA and BOD.[111] The increases in enzyme activities would have been brought about from an increased rate of ET from substrate to T1 in spite of the unfavorable ET between T1 and the trinuclear copper center. Decreases in enzyme activities for CotA and BOD might have occurred due to be a contrasting situation having to do with the driving force of the ETs involving the mutated T1 center. In the case of CotA, decreased copper incorporation was also suggested. Mutations at the T1 center to change Met to Gln in BOD and CueO led to reductions in enzymatic activities because the abstraction of a substrate electron became unfavorable due to negative shifts in the redox potential of T1. Furthermore, the compensatory binding of Asn459 to T1 was suggested for the negative shift in the redox potential of BOD.[111] This result was confirmed by the recent crystal structure analysis with the access of Asn459 to the opposite axial position of Met467.[112] Modifications induced by conservative mutations at the axial ligand of T1 in CueO, BOD, Fet3p, and CotA are expected to be universally applicable to other MCOs.

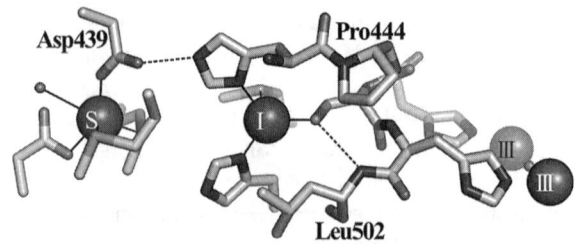

FIGURE 5.12. Hydrogen bonds involving the coordinating groups to T1 and the absence of a hydrogen bond due to the location of Pro444.

5.7.3. Mutation to Form or to Break Hydrogen Bonds Concerned with the Coordinating Groups to Type I Copper

Modifications to form or break down a hydrogen bond involving a ligand to T1 have been performed on blue copper proteins, leading to changes in the redox potential.[113–115] However, these mild modifications have not been performed on the T1 center in MCOs. According to the crystal structures of MCOs, the imidazole groups of the His residue coordinating to T1 are hydrogen bonded with the amino acids located in the outer-sphere region. The presence of these hydrogen bonds may induce an effect to favor the cupric state, in addition to an effect to stabilize the T1 center. Therefore, it is expected that the breaking down of these hydrogen bonds favors T1 to be in the cuprous state. We verified this hypothesis by mutating Asp439, which is hydrogen bonded to the imidazole group of His143 coordinated to T1 in CueO (Fig. 5.12). The redox potential shifted ~50 mV more positive, leading to a 17-fold increase in enzyme activities (Fig. 5.13).[107]

Figure 5.12 shows that the S atom of the coordinated Cys in CueO is hydrogen bonded with an amide NH in the main chain. However, another hydrogen bond is not formed due to the location of Pro444, which is located in a spacer to connect the T1 and the trinuclear copper center. If the S atom is doubly hydrogen bonded, the anionic character of the S atom is shared, and the redox potential of T1 may shift toward positive potentials due to a greater favoring of the cuprous state. Mutations of Pro444 to Ala, Leu, or Ile in CueO showed shifts in the redox potential of T1 as expected, leading to a 10-fold increase in enzyme activities.[116] The double mutation both at Asp439 and Pro444 reached a 40-fold increase in activity due to an cumulative effect.

5.7.4. Mutations around the Trinuclear Copper Center

Mutations at ligands to the trinuclear copper center have not been widely performed. When one or two His ligands to T2 or T3 are substituted to noncoordinating amino acid(s) (i.e., His94Val, His134,136Val, and His456,458Val in BOD), three copper ions were not fully incorporated into the trinuclear copper center.[27] These incompleted trinuclear copper centers led to the cross-ligation of the His residues to copper ions. On the other hand, when a His ligand to each T3 center is substituted by a coordinating Lys or Asp (His456,458Lys/Asp in BOD), the reducing activity of O_2

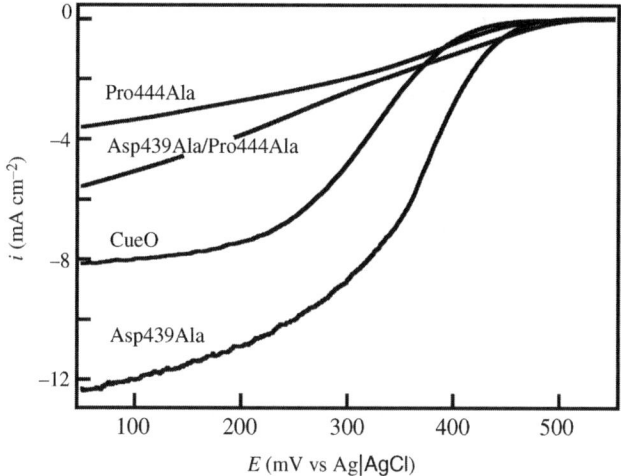

FIGURE 5.13. Cyclic voltammograms of the recombinant CueO, Asp439Ala, Pro444Ala, and Asp439Ala/Pro444Ala to show the effective O_2 reduction and the shift in the redox potential toward positive potential by the breaking down and/or forming hydrogen bond.

was completely lost in spite of retaining an ability to bind O_2.[35] In the case of Fet3p, a His ligand to one T3s was substituted, also resulting in the enzyme's reduced activity to convert O_2 to H_2O.[85]

5.7.5. Mutations to the Acidic Amino Acids Adjacent to the Trinuclear Copper Center

When studying the mutations of the highly conserved Asp residue in CueO and Fet3p, which is directly hydrogen bonded with a His ligand to T3 and indirectly with a coordinated water molecule to T2 via a water molecule, we find that the binding affinity of the trinuclear copper center for O_2 is lowered. This finding indicates that the role of this amino acid is to assist the binding of O_2 (see Section 5.6.6).[30,86,98] In most MCOs, this amino acid is Asp, and the enzyme activities are reduced to almost one-half by substitutions to Glu in CueO and BOD. Therefore, the active site of most MCOs involving this intrinsic Asp residue located in the outer sphere of the trinuclear copper center is not as flexible enough to allow substitution by Glu. In contrast, Cp is adapted to accommodate Glu instead of Asp (Fig. 5.2) to assist binding of O_2 at its trinuclear copper center.

The Glu residue that forms hydrogen bonds directly with His residues coordinated to T3 and an OH^- bridged between T3s via a water molecule, is a key amino acid that assists in proton donation to O_2. The decay rate of intermediate II was significantly slowed by mutations at this Glu residue (see Section 5.6.6).[30,31]

Any mutation on these amino acids led to decreases in enzyme activities due to slowing down of the turnover rate. At present, acceleration of the O_2 reduction process has not been reported following any mutations occurring around the trinuclear copper center.

5.7.6. Applications of Multicopper Oxidases

The MCOs have been widely utilized in the diagnosis of disease states, as food additives, in dye formation, in bleaching of dye and pulp, in detergents, and so on. All this is in addition to uses in traditional lacquer work and adhesives in East Asia.[117] These industrial applications of MCOs are due to their substrate specificities and the fact that these are comparatively stable and soluble proteins. The use of O_2 as the final electron acceptor without being coupled to another protein is also a prominent feature of MCOs and in their value for practical uses, especially as an electrode catalyst of biofuel cells and sensors.[107,108,118–120] Since O_2 is converted into water without forming activated oxygen species by MCOs, the electrode catalyst systems is clean. A large voltage difference between anode and cathode is attained in the biofuel cell if the MCO T1 center has a redox potential near 0.82 V at pH 7, which is the particular thermodynamic driving force for which O_2 is converted into H_2O. Although Cu ions may function as structural factors to stabilize the protein molecule in addition to being the centers of catalysis, most MCOs require mild conditions. In this regard, the highly thermostable CotA, which is located in the endospore coat of *Bacillus subtilis*, has a great potential, while this MCO is now commercially available as BOD2 because of the high oxidizing activity of bilirubin.[121] In order to meet wide demands, naturally occurring MCOs have limitations intrinsic to their specificity, optimum pH, stability, and so on. Accordingly, novel MCOs, especially those from thermophiles, are desirable and sought out. It may be that modifications of known MCOs with point to random mutations, will be effective as a means to obtain useful and/or more stable MCOs.

5.8. CONCLUSIONS AND FUTURE PROSPECTS

The MCOs have received special attention because of their oxidizing activities toward organic substances and possession of metal ions that facilitate the $4e^-$ reduction of O_2 to H_2O. Both of these properties make MCOs applicable to a variety of fields. Studies on the increasing numbers of MCOs have been performed from the perspective of enzymology, biophysics, kinetics, structural biology, molecular biology, and so on. However, we are so to speak in a state of chaos concerning the structures of MCOs, although the available numbers of the molecular structures continue to increase in the last 20 years. In addition to the classical resting structure of MCO, in which an OH^- group bridges among T3 centers and T2 is magnetically isolated, many other resting structures have been provided. It is not known whether every MCO is in the same resting form or each MCO possesses its own unique resting form. Not in every case, but in many, copper ions in the trinuclear copper center might be partly reduced by hydrated electrons formed from X-ray radiation, and this may be one of the causes for the observation of many resting forms.

The mechanism of the $4e^-$ reduction of O_2 by MCO has been in a black box for a long time. However, we can now trap the two reaction intermediates by virtue of molecular biology techniques, as applied to MCOs whose effective expression systems are available. Intermediates I and II, which have also been called the peroxide and native intermediates, have been trapped and characterized. However, direct evidence has not yet been obtained to support their proposed structures, and questions arrive even for the structures of resting forms. The peroxide binding has not been directly evidenced in spite of its biophysical characterizations and spectral similarities with small molecule $Cu-O_2$ complexes. The O centered structure proposed for intermediate II is now found in many resting MCOs. To understand this puzzle would constitute a breakthrough in understanding an enzyme reaction mechanism. The whole $MCO-O_2$ reduction process will in the future be revealed and characterized in more detail by use of time-resolved measurements, or by slowing down the steps in the catalytic cycle for mutant enzyme forms where amino acids located in the inner or outer sphere of the copper centers are modified.

Some MCOs are apparently in a mixed resting form. This finding may be the intrinsic property of MCOs, and accordingly, one is always required to carry out studies that include checking whether the structure under discussion is the main one or not, and if the local structure is unaltered or not. Since results on a given MCO may not be necessarily applicable to other MCOs, combined discussions using results on different MCOs might lead to misunderstanding or confusion. Nevertheless, now we are informed more than before concerning the structures and reactions of MCOs. A full understanding of how O_2 is converted into H_2O by MCOs will be extremely valuable in the construction of functional small molecules. In the future, tailor-made modifications may also become feasible for each MCO by virtue of research leading to more profound structural and functional insights.

ACKNOWLEDGMENTS

This work was supported by Grants-in-aid for Scientific Research 19350081 and 21655061 from the Ministry of Education, Culture, Sports, Science and Technology of Japan and NEDO.

ABBREVIATIONS

ABTS	2,2′-Azino-bis(3-ethylbenzothiazoline-6-sulfonic acid)
AOase	Ascorbate oxidase
ATR	Attenuated total reflectance
BCO	*Nitrosomonas* laccase
BOD	Bilirubin oxidase
C*c*O	Cytochrome *c* oxidase
CD	Circular dichroism
Cp	Curuloplasmin
DFT	Density functional theory
EPR	Electron paramagnetic resonance
ET	Electron transfer

FTIR   Fourier transform infrared
hCp   human ceruloplasmin
Lc   Laccase
MCD   Magnetic circular dichroism
MCO   Multicopper oxidase
mgLAC   metagenome laccase
NIR   Nitrite reductase
N_2OR Nitrous oxide reductase
O_2 Dioxygen
RR   Resonance Raman
RvLc   *Rhus vernicifera* laccase
SLAC   *Streptomyces* 2-domain laccase
SQUID   Superconducting quantum interference device
T1   Type I copper (Blue copper)
T2   Type II copper (Non-blue copper)
T3   Type III copper (EPR nondetectable coppers)
tbp   Trigonal bipyrmidal
TvLc   *Trametes versicolor* laccase
UV   Ultraviolet
XAS   X-ray absorption spect.

REFERENCES

1. *Multicopper Oxidases*, A. Messershmidt, Ed., World Scientific Singapore, Singapore, 1977.
2. Kakutani, T.; Watanabe, H.; Arima, K.; Beppu, T. Purification and properties of a copper-containing nitrite reductase from a denitrifying bacterium, *Alcaligenes faecalis* strain S 6. *J. Biochem.* **1981**, *89*, 463–472.
3. Yoshida, H. Zur chemie des urushi-fuirniss. *J. Chem. Soc. (Tokyo)* **1883**, *43*, 472–486.
4. Reinhammer B. *Copper Proteins and Copper Enzymes* Vol. III, R. Lontie, Ed., CRS Press Boca Raton, Fl., 1984, pp 1–35.
5. Yatsu, J.; Asano, T. Cuticle laccase of the silkworm, Bombyx mori: purification and presence of its inactive precursor in the cuticle. *Insect Biochem. Mol. Biol.* **2009**, *39*, 254–262.
6. Nitta, K.; Kataoka, K.; Sakurai, T. Primary structure of a Japanese lacquer tree laccase as a prototype enzyme of multicopper oxidase. *J. Inorg. Biochem.* **2002**, *91*, 125–131.
7. Sakurai, T.; Kataoka, K. Basic and Applied features of multicopper oxidases, CueO, bilirubin oxidase, and laccase. *Chem. Rec.* **2007**, *7*, 220–229.
8. Solomon, E. I.; Sundaram, U. M.; Machonkin, T. E. Multicopper oxidases and oxygenases. *Chem, Rev.* **1996**, *96*, 2563–2605.
9. Stoj, C. S.; Kosman, D. J. *Encyclopedia of Inorganic Chemistry*, 2nd ed., R. B. King, Ed., John Wiley & Sons, Inc., New York, **2005**, pp 1134–1159.

10. Sakurai, T.; Kataoka, K. Structure and function of type I copper in multicopper oxidase. *Cell. Mol. Life Sci.* **2007**, *64*, 2642–2656.

11. Takahashi, H,; Ortel, T. L.; Putnam, F. W. Single-chain structure of human ceruloplasmin: the complete amino acid sequence of the whole molecule. *Proc. Natl. Acad. Sci. USA* **1984**, *81*, 390–394.

12. Messerschmidt, A.; Rossi, A.; Ladenstein, R.; Huber, R.; Bolognesi, M.; Gatti, G.; Marchesini, A.; Petruzzelli, R.; Finazzi-Agro, A. X-ray structure of the blue oxidase ascorbate oxidase from zucchini. Analysis of the polypeptide fold and a model of the copper sites and ligands. *J. Mol. Biol.* **1989**, *206*, 513–529.

13. Nakamura, K.; Go, N. Function and molecular evolution of multicopper blue proteins. *Cell. Mol. Life Sci.* **2005**, *62*, 2050–2066.

14. Hoegger, P. J.; Kilaru, S.; James, T. Y.; Thacker, J. R.; Kues, U. Phylogenetic comparison and characterization of laccase and related multicopper oxidase protein sequence. *FEBS J.* **2006**, *273*, 2308–2326.

15. Tarry, M.; Arends, S. R. J.; Roversi, P.; Sargent, F.; Berks, B. C.; Weiss, D.; Lea, S. M. The *Escherichia coli* cell division protein and model Tat substrate StufI (FtsP) localizes to the septal ring and has a multicopper oxidase-like structure. *J. Mol. Biol.* **2009**, *386*, 504–516.

16. Machczynski, M. C.; Vijgenboom, E.; Samyn, B.; Canters, G. W. Characterization of SLAC; a small laccase from *Streptomyces coelicolor* with unprecedented activity. *Protein Sci.* **2004**, *13*, 2388–2397.

17. Lawton, T. J.; Sayavedra-Soto, L. A.; Arp, D. J.; Rosenberg, A. C. Crystal structure of a two-domain multicopper oxidase: Implications for the evolution of multicopper blue proteins. *J. Biol. Chem.* **2009**, *284*, 10174–10180.

18. Skalova, T.; Dohnalek, J.; Ostergaard, L. H.; Ostergaard, P. R.; Kolenko, P.; Duslova, J.; Stepankova, A.; Hasek, J. The Structure of the small laccase from *Streptomyces coelicolor* reveals a link between laccases and nitrite reductase. *J. Mol. Biol.* **2009**, *385*, 1165–1178.

19. Komori, H.; Miyazaki, K.; Higuchi, Y. X-ray structure of a two-domain type laccase: a missing link in the evolution of multi-copper proteins. *FEBS Lett.* **2009**, *583*, 1189–1195.

20. Zaitseva, I.; Zaitsev. V.; Card, G.; Moshkov, K.; Bax, B.; Ralph, A.; Lindley, P. The X-ray structure of human serum ceruloplasmin at 3.1 Å: nature of the copper centres. *J. Biol. Inorg. Chem.* **1996**, *1*, 15–23.

21. Terzulli, A. J.; Kosman, D. J. The Fox1 ferroxidase of *Chlamydomonas reinhardtii*: a new multicopper oxidase structural paradigm. *J. Biol. Inorg. Chem.* **2009**, *14*, 315–325.

22. Yamaguchi, K.; Kataoka, K.; Kobayashi, M.; Itoh, K.; Fukui, A.; Suzuki, S. Characterization of two type 1 Cu sites of *Hyphomicrobium denitrificans* nitrite reductase: a new class of copper-containing nitrite reductases. *Biochemistry* **2004**, *43*, 14180–14188.

23. Nojiri, N.; Xie, Y.; Inoue, T.; Yamamoto, T.; Matsumura, H.; Kataoka, K.; Deligeer, Yamaguchi, K.; Kai, Y.; Suzuki, S. Structure and function of a hexameric copper-containing nitrite reductase: *Proc. Natl. Acad. Sci. USA*, **2007**, *104*, 4315–4320.

24. Nojiri, N.; Koteishi, H.; Nakagami, T.; Kobayashi, K.; Inoue, T.; Yamaguchi, K.; Suzuki, S. Structural basis of inter-protein electron transfer for nitrile reduction in denitrification. *Nature (London)* **2009**, *462*, 117–120.

25. Ridge, J. P.; Lin, M.; Larsen, E. I.; Fegan, M.; McEwan, A. G.; Sly, L. I. A multicopper oxidase is essential for manganese oxidation and laccase-like activity in *Pedomicrobium* sp. ACM 3067. *Environm. Microbiol.* **2007**, *9*, 944–953.

26. Sakurai, T.; Takahashi, J. EPR spectra of type 3 copper centers in *Rhus vernicifera* lacacse and *Cucumis sativus* ascorbate oxidase. *Biochim. Biophys. Acta* **1995**, *1248*, 143–148.

27. Shimizu, A.; Kwon, J.-H.; Sasaki, T.; Satoh, T.; Sakurai, N.; Sakurai, T.; Yamaguchi, S.; Samejima, T. *Myrothecium verrucaria* bilirubin oxidase and its mutants for potential copper ligands. *Biochemistry* **1999**, *38*, 3034–3042.

28. Kataoka, K.; Kitagawa, R.; Inoue, M.; Naruse, D.; Sakurai, T.; Huang, H.-W. Point mutations at the type I ligands, Cys457 and Met467, and the putative proton donor, Asp105, in *Myrothecium verrucaria* bilirubin oxidase and reactions with dioxygen. *Biochemistry* **2005**, *44*, 7004–7012.

29. Hakulinen, N.; Kiiskinen, L.-L.; Kruus, K.; Saloheimo, M.; Paananen, A.; Koivula, A.; Rouvinen, J. Crystal structure of a laccase from *Melanocarpus albomyces* with an intact trinuclear copper site. *Nat. Str. Biol.* **2002**, *9*, 601–605.

30. Kataoka, K.; Sugiyama, R.; Hirota, S.; Inoue, M.; Urata, K.; Minagawa, Y.; Seo, D.; Sakurai, T. Four-electron reduction of dioxygen by a multicopper oxidase, CueO, and roles of Asp112 and Glu506 located adjacent to the trinuclear copper center. *J. Biol. Chem.* **2009**, *284*, 14405–14413.

31. Augustine, A. J.; Quintanar, L.; Stoj, C. S.; Kosman, D. J.; Solomon, E. I. Spectroscopic and kinetic studies of the perturbed trinuclear copper clusters: the role of protons in reductive cleavage of the O–O bond in the multicopper oxidase Fet3p. *J. Am. Chem. Soc.* **2007**, *129*, 13118–13126.

32. Solomon, E. I.; Augustine A.; Yoon, J. O_2 reduction to H_2O by the multicopper oxidases. *Dalton Trans.* **2008**, 3921–3932.

33. Huang, H.-W.; Sakurai, T.; Monjushiro, H.; Takeda, S. Magnetic studies of the trinuclear center in laccase and ascorbate oxidase approached by EPR spectroscopy and magnetic susceptibility measurements. *Biochim. Biophys. Acta* **1998**, *1384*, 160–170.

34. Gromov, I.; Marchesini, A.; Farver, O.; Pecht, I.; Goldfarb, D. Azide binding to the trinuclear copper center in laccase and ascorbate oxidase. *Eur. J. Biochem.* **1999**, *266*, 820–830.

35. Shimizu, A.; Samejima, T.; Hirota, S.; Yamaguchi, S.; Sakurai, N.; Sakurai, T. Type III Cu mutants of *Myrothecium verrucaria* bilirubin oxidase. *J. Biochem.* **2003**, *133*, 767–772.

36. Hirota, S.; Kataoka, K.; Sakurai, T. unpublished data.

37. Holt, B. O. T.; Vance, M. A.; Mirica, L. M.; Heppner, D. E.; Stack, T. D.; Solomon, E. I. Reaction coordinate of a functional model of tyrosinase; spectroscopic and computational characterization. *J. Am. Chem. Soc.* **2009**, *131*, 6421–6438.

38. Sakurai, T.; Suzuki, S.; Sano, M. X-ray absorption study on the type II copper-depleted cucumber ascorbate oxidase. *Inorg. Chim. Acta* **1988**, *152*, 3–4.

39. Hakulinen, N.; Kruus, K.; Koivula, A.; Rouvinen, J. A. crystallographic and spectroscopic study on the effect of X-ray radiation on the crystal structure of *Melanocarpus albomyces* laccase. *Biochem. Biophys. Res. Commun.* **2006**, *350*, 929–934.

40. Komori, H.; Higuchi, Y.; Sugiyama, R.; Moriguchi, Y.; Kataoka, K.; Sakurai, T. unpublished data.

41. Smith, A. W.; Camara-Artigas, A.; Wang, M.; Allen, J. P.; Francisco, W. A. Structure of phenoxazinone synthase from *Streptomyces antibioticus* reveals a new type 2 copper center. *Biochemistry* **2006**, *45*, 4378–4387.

42. Enguita, F.; Martins, L. O.; Henriques, A. O.; Carrondo, M. A. Crystal structure of a bacterial endospore coat component. A laccase with enhanced thermostability properties. *J. Biol. Chem.* **2003**, *278*, 19416–19425.

43. Messerschmidt, A.; Ladenstein, R.; Huber, R.; Bolognesi, M.; Avigliano, L.; Petruzzelli, R.; Rossi, A.; Finazzi-Agro, A. Refined crystal structure of ascorbate oxidase at 1.9 Å resolution. *J. Mol. Biol.* **1992**, *224*, 179–205.

44. Enguita, F.; Marcal, D.; Martins, L. O.; Grenha, R.; Henriques, A. O.; Lindley, P. F.; Corrondo, M. A. Substrate and dioxygen binding to the endospore coat laccase from *Bacillus subtilis*. *J. Biol. Chem.* **2004**, *279*, 23472–23476.

45. Bertrand, T.; Jolivalt, C.; Briozzo, P.; Caminade, E.; Joly, N.; Madzak, C.; Mougin, C. Crystal structure of a four-copper laccase complexed with an arylamine: insights into substrate recognition and correlation kinetics. *Biochemistry* **2002**, *41*, 7325–7333.

46. Kallio, J. P.; Auer, S.; Jänis, J.; Andberg, M.; Kruus, K.; Rouvinen, J.; Koivula, A.; Hakulinen, N. Structure-function studies of a *Melanocarpus albomyces* laccase suggest pathway for oxidation of phenolic compounds. *J. Mol. Biol.* **2009**, *392*, 895–909.

47. Matera, I.; Gullotto, A.; Tilli, S.; Ferraroni, M.; Scozzafava, A.; Briganti, F. Crystal structure of the blue multicopper oxidase from the white-rot fungus *Trametes troggi* complexes with *p*-toluate. *Inorg. Chim. Acta* **2008**, *361*, 4129–4137.

48. Roberts, S. A.; Weichsel, A.; Grass, G.; Thakali, K.; Hazzard, J. T.; Tollin, G.; Rensing, C.; Montfort, W. R. Crystal structure and electron transfer kinetics of CueO, a multicopper oxidase required for copper homeostasis in *Escherichia coli*. *Proc. Natl. Acad. Sci. USA* **2002**, *99*, 2776–2771.

49. Kataoka, K.; Komori, H.; Ueki, Y.; Konno, Y.; Kamitaka, Y.; Kurose, S.; Tsujimura, S.; Higuchi, Y.; Kano, K.; Seo, D.; Sakurai, T. Structure and function of the engineered multicopper oxidases CueO from *Escherichia coli*—deletion of the methionine-rich helical region covering the substrate binding site. *J. Mol. Biol.* **2007**, *373*, 141–152.

50. Kurose, S.; Kataoka, K.; Otsuka, K.; Tsujino, Y.; Sakurai, T. Promotion of laccase activities of *Escherichia coli* cuprous oxidase, CueO by deleting the segment covering the substrate binding site. *Chem. Lett.* **2007**, *36*, 232–233.

51. Shao. X.; Gao, Y.; Jiang, M.; Li, L. Deletion and site-directed mutagenesis of laccase from *Shigella dysenteriae* results in enhaced enzymatic activity and thermostability. *Enzyme Microbiol. Technol.* **2009**, *44*, 274–280.

52. Huffman, D. L.; Huyett, J.; Outten, F. W.; Doan, P. E.; Finney, L. A.; Hoffman, B. M.; O'Halloran, T. V. Spectroscopy of Cu(II)–PcoA and multicopper oxidase function of PcoA, two essential components of *Escherichia coli pco* copper resistance operon. *Biochemistry* **2002**, *41*, 10046–10055.

53. Francis, C. A.; Casciotti, K. L.; Tebo, B. M. Localization of Mn(II)-oxidizing activity and the putative multicopper oxidase, MnxG, to the exosporium of the marine *Bacillus* sp. strain SG-1. *Arch. Microbiol.* **2002**, *178*, 450–456.

54. Brouwers, G.; de Vrind, J. P. M.; Corstjens, P. L. A. M.; Cornelis, P.; Baysse, C.; de Vrind-de Jong E. W. *cumA*, a Gene encoding a multicopper oxidase, is involved in Mn^{2+} oxidation in *Pseudomonas putida* GB-1. *Appl. Environ. Microbiol.* **1999**, *65*, 1762–1768.

55. Ridge, J. P.; Lin, M.; Larsen, E. I.; Fegan, M.; McEwan, A. G.; Sly, L. I. A multicopper oxidase is essential for manganese oxidation and laccase-like activity in *Pedomicrobium* sp. ACM 3067. *Environ. Microbiol.* **2007**, *9*, 944–953.

56. Bento, I.; Peixoto, C.; Zaitsev, V. N.; Lindley, P. F. Ceruloplasmin revised: structural and functional roles of various metal-binding sites. *Acta Crystallogr.* **2007**, *D63*, 240–248.

57. Grifiths, T. A. M.; Mauk, A. G.; MacGilliveray, T. A. Recombinant expression and functional characterization of human hephaestin: a multicopper oxidase with ferroxidase activity. *Biochemistry* **2005**, *44*, 14725–14731.

58. Kwok, E. Y.; Severance, S.; Kosman, D. J. Evidence for iron channeling in the Fet3p-Ftr1p high-affinity iron uptake complex in the yeast plasma membrane. *Biochemistry* **2006**, *45*, 6317–6327.

59. Larrondo, L. F.; Salas, L.; Melo. F. Vicuna, R.; Cullen, D. A novel extracellular multicoper oxidase from *Phanerochaete chrysosoporium* with ferroxidase activity. *Appl. Environ. Microbiol.* **2003**, *69*, 6257–6263.

60. Roberts, S. A.; Wildner, G. F.; Grass, G.; Weichsel, A.; Ambrus, A.; Rensing, C.; Montfort, W. R. A labile regulatory copper ion lies near the T1 copper site in the multicopper oxidase CueO. *J. Biol. Chem.* **2003**, *278*, 31958–31963.

61. Quintanar, L.; Gebhard, M.; Wang, T.-P.; Kosman, D. J.; Solomon, E. I. Ferrous binding to the multicopper oxidases *Saccharomyces cerevisiae* Fet3p and human ceruloplasmin: contribution to ferroxidase activity. *J. Am. Chem. Soc.* **2004**, *126*, 6579–6589.

62. Piontek, K.; Antorini, M.; Choinowski, T. Crystal structure of a laccase from the fungus *Trametes versicor* at 1.90-Å resolution containing a full complement of coppers. *J. Biol. Chem.* **2002**, *277*, 37633–37699.

63. Ducros, V.; Brzowsli, A. M.; Wilson, K. S.; Brown, S. H.; Osteregaard, P.; Schneider, P.; Yaver, D. S.; Pedersen A. H.; Davies, G. J. Crystal structure of the type-2 Cu depleted laccase from *Coprinus cinereus* at 2.2 Å resolution. *Nat. Str. Biol.* **1998**, *5*, 310–316.

64. Lyashenko, A. V.; Bento, S.; Zaitsev, V. N.; Zhuklistova, N. E.; Zhukova, Y. N.; Gabdoulkhakov, A. G.; Morgunova, E. Y.; Voelter, W.; Kachalova, G. S.; Stepanova, E. V.; Koreleva, O. V.; Lamzin, V. S.; Tishkov, V. I.; Betzel, C.; Lindley, P. F.; Mikhailov, A. M. X-ray structural studies of the fungal laccase from *Cerrena maxima*. *J. Biol. Inorg. Chem.* **2006**, *11*, 963–973.

65. Durao, P.; Bento, I.; Fernandes, A. T.; Melo, E. P.; Lindley, P. F.; Martins, L. O. Perturbations of the T1 copper site in the CotA laccase from *Bacillus subtilis*: structural, biochemical, enzymatic and stability studies. *J. Biol. Inorg. Chem.* **2006**, *11*, 514–526.

66. Hakulinen, N.; Andberg, M.; Kallio, J.; Koivula, A.; Kruus, K.; Rouvinen, J. A near atomic resolution structure of a *Melanocarpus albomyces* laccase. *J. Str. Biol.* **2008**, *162*, 29–39.

67. Ferraroni, M.; Myasoedova, M. N.; Schmatechenko, V.; Leontievsky, A. A.; Golvolvela, L. A.; Scozzafava, A.; Briganti, F. Crystal structure of a blue laccase from *Lentinus tigrinus*: evidences for intermediates in the molecular oxygen reductive splitting by multicopper oxidases. *BMC Str. Biol.* **2007**, *7*, 60–73.

68. Aoyama, H.; Muramoto, K.; Shinzawa-Ito, K.; Hirata, K.; Yamashita, E.; Tsukihara, T.; Ogura, T.; Yoshikawa, S. A peroxide bridge between Fe and Cu ions in the O_2 reduction site of fully oxidized cytochrome *c* oxidase could suppress the proton pump. *Proc. Natl. Acad. Sci. USA* **2009**, *106*, 2165–2169.

69. Polyakov, K.; Fedorova, T.; Stepanova, E. V., Cherlashin, E. A.; Kurteev, S. A.; Strokopytov, B. V.; Lamzin, V. S.; Koroleva, O. V. Structure of native laccase from *Trametes hirsute* at 1.8 Å resolution. *Acta Crystallogr.* **2009**, *D65*, 611–617.

70. Wlodawer, A.; Minor, W.; Dauter, Z.; Jaskolski, M. Protein crystallography for non-crystallographers, or how to get the best (but not more) from published macromolecular structures. *FEBS J.* **2008**, *275*, 1–21.

71. Sakurai, T.; Zhan, L.; Fujita, T.; Kataoka, K.; Shimizu, A.; Samejima, T.; Yamaguchi, S. Authentic and recombinant bilirubin oxidase are in different resting forms. *Biosci. Biotechnol. Biochem.* **2003**, *67*, 1157–1159.

72. Andberg, M.; Hakulinen, N.; Auer, S.; Saloheimo, M.; Koivula, A.; Rouvinen, J.; Kruus, K. Essential role of the C-terminus in *Melanocarpus albomyces* laccase for enzyme production, catalytic properties, and structure. *FEBS J.* **2009**, *276*, 6285–6300.

73. Messerschmidt, A.; Luecke, H.; Huber, R. X-ray structures and mechanistic implications of three functional derivatives of ascorbate oxidase from zucchini. *J. Mol. Biol.* **1993**, *230*, 997–1014.

74. Bento, L.; Martins, L. O.; Lopes, G. G.; Carrondo, M. A.; Lindley, P. F. Dioxygen reduction by multi-copper oxidases; a structural perspective. *Dalton Trans.* **2005**, 3507–3513.

75. Garavaglia, S.; Cambria, M. T.; Miglio, M.; Ragusa, S.; Iacobazzi, V.; Palmieri, F.; D'Ambrosio, C.; Scaloni, A.; Rizzi, M. The structure of *Rigidoporus lignosus* laccase containing a full complement of copper ions, reveals an asymmetrical arrangement for the T3 copper pair. *J. Biol. Chem.* **2004**, *342*, 1515–1531.

76. Taylor, A. B.; Stoj, C. S. : Ziegler, L.; Kosman, D. J.; Hart, P. J. The copper–iron connection in biology: structure of the metallo-oxidase Fet3p. *Proc. Natl. Acad. Sci. USA* **2005**, *102*, 15459–15464.

77. Morpurgo, L.; Graziani, T.; Desideri, A.; Rotilio, G. Titrations with ferrocyanide of Japanese-lacquer-tree (*Rhus vernicifera*) laccase and of the type 2 copper-depleted enzyme. *Biochem. J.* **1980**, *187*, 367–370.

78. Sakurai, T.; Sawada, S.; Suzuki, S.; Nakahara, A. Properties of the type II copper-depleted cucumber ascorbate oxidase and its reaction with azide. *Biochim. Biophys. Acta* **1987**, *915*, 238–245.

79. Woolery, G. L.; Powers, L.; Peisach, J.; Spiro, T. G. X-ray absorption study of *Rhus vernicifera* laccase: evidence for a copper-copper interaction, which disappears on type 2 copper removal. *Biochemistry* **1984**, *23*, 3428–3434.

80. Zoppellaro, G.; Sakurai. T.; Huang, H.-W. A novel mixed valence form of *Rhus vernicifera* laccase and its reaction with dioxygen to give a peroxide intermediate bound to the trinuclear center. *J. Biochem.* **2001**, *129*, 949–953.

81. Cole, J. L.; Tan, G. O.; Yang. E. K.; Hodgson, K.O.; Solomon, E. I. Reactivity of the laccase trinuclear copper center active site with dioxygen: an X-ray absorption edge study. *J. Am. Chem. Soc.* **1990**, *112*, 2243–2249.

82. Spira-Solomon, D. J., Solomon, E. I. Chemical and Spectroscopic studies of the coupled binuclear copper site in the type 2 depleted *Rhus* laccase: comparison to the hemocyanin and tyrosinase. *J. Am. Chem. Soc.* **1987**, *109*, 6421–6432.

83. Shin, W.; Sundaram, U. M.; Dole, J. L.; Zhang, H. H.; Hedman, B.; Hodgson, K. O.; Solomon, E. I. Chemical and spectroscopic definition of the peroxide-level intermediate

in the multicopper oxidases: relevance to the catalytic mechanism of dioxygen reduction to water. *J. Am. Chem. Soc.* **1996**, *118*, 3202–3215.

84. Palmer, A. E.; Quintanar, L.; Severance, S.; Wang, T.-P., Kosman, D. J.; Solomon, E. I. Spectroscopic characterization and O_2 reactivity of the trinuclear Cu cluster of mutants of the multicopper oxidase Fet3p. *Biochemistry*, **2002**, *41*, 6438–6448.

85. Augustine, A. J.; Kragh, M. E.; Sarangi, R.; Fujii, S.; Liboiron, B. D.; Stoj, C. S.; Kosman, D. J.; Hodgson, K. O.; Dedman, B. Solomon, E. I. Spectroscopic studies of perturbed T1 Cu sites in the multicopper oxidases, *Saccharomyces cerevisiae* Fet3p and *Rhus vernicifera* laccase: allosteric coupling between the T1 and trinuclear Cu sites. *Biochemistry* **2008**, *47*, 2036–2045.

86. Quintanar, L.; Stoj, C.; Wang, T.-P.; Kosman, D. J.; Solomon, E. I. Role of aspartate 94 in the decay of the peroxide intermediate in the multicopper oxidase Fet3p. *Biochemistry* **2005**, *44*, 6081–6091.

87. Solomon, E. I.; Chen, P.; Metz, M.; Lee, S.-K.; Palmer, A. E. Oxygen binding, activation, and reduction to water by copper proteins. *Angew. Chem. Int. Ed.* **2001**, *40*, 4570–4590.

88. Mirica, L. M.; Ottenwaelder, X.; Stack, T. D. P. Structure and spectroscopy of copper-dioxygen complexes. *Chem. Rev.* **2004**, *104*, 1013–1045.

89. Ogura, T.; Hirota, S.; Proshllyakov, D. A.; Shinkawa-Itoh K.; Yoshikawa, S.; Kitagawa, T. Time-resolved resonance Raman evidence for tight coupling between electron transfer and proton pumping of cytochrome *c* oxidase upon the change from the Fe^V oxidation level to the Fe^{IV} oxidation level. *J. Am. Chem. Soc.* **1996**, *118*, 5443–5449.

90. Palmer, A. E.; Lee, S. Y.; Solomon, E. I. Decay of the peroxide intermediate in laccase: reductive cleavage of the O–O bond. *J. Am. Chem. Soc.* **2001**, *123*, 6591–6599.

91. Yoon, J.; Solomon, E. I. Electronic structure of the peroxy intermediate and its correlation to the native intermediate in the multicopper oxidases: insights into the reductive cleavage of the O–O bond. *J. Am. Chem. Soc.* **2007**, *129*, 13127–13136.

92. Reinhammer, B.; Malkin, R.; Jensen, P.; Karlsson, B.; Andreasson, L.-E.; Aasa, R.; Vanngaard, T.; Malmstrom, B. G. A new copper(II) electron resonance paramagnetic signal in two laccases and in cytochrome *c* oxidase. *J. Biol. Chem.* **1980**, *255*, 5000–5003.

93. Goldberg, M.; Farver, O.; Pecht, I. Interaction of *Rhus* laccase with dioxygen and its reduction intermediates. *J. Biol. Chem.* **1980**, *255*, 7353–7361.

94. Lee, S.-K.; George, S. D.; Antholine, W. E.; Hedman, B.; Hodgson, K. O.; Solomon, E. I. Nature of the intermediate formed in the reduction of O_2 to H_2O at the trinuclear copper cluster active site in native laccase. *J. Am. Chem. Soc.* **2002**, *124*, 6180–6193.

95. Huang, H.-W.; Zoppelllaro, G.; Sakurai, T. Spectroscopic and kinetic studies on the oxygen-centered radical formed during the four-electron reduction process of dioxygen by *Rhus verinicifera* laccase. *J. Biol. Chem.* **1999**, *274*, 32718–32724.

96. Morishita, H.; Kataoka, K.; Sakurai, T. unpublished data.

97. Tepper, A. W. J. W.; Miliksyants, S.; Scottini, S.; Vijgenboom, E.; Groenen, E. J. J.; Canters, G. W. Identification of a radical intermediate in the enzymatic reduction of oxygen by a small laccase. *J. Am. Chem. Soc.* **2009**, *131*, 11680–11682.

98. Ueki, Y.; Inoue, M.; Kurose, S.; Kataoka, K.; Sakurai, T. Mutations at Asp112 adjacent to the trinuclear Cu center in CueO as the proton donor in the four-electron reduction of dioxygen. *FEBS Lett.* **2006**, *580*, 4069–4072.

99. Augustine, A.; Kjaergaard, C.; Qayyum, M.; Ziegier, L.; Kosman, D. J.; Hodgson, K. O.; Hedman, B.; Solomon, E. I. Systematic perturbation of the trinuclear copper center in the

multicopper oxidases: The Role of active site symmetry in its reduction of O_2 to H_2O. *J. Am. Chem. Soc.* **2010**, *132*, 6057–6067.

100. Iwaki, M.; Kataoka, K.; Kajino, T.; Sugiyama, R.; Morishita, H.; K.; Sakurai, T. ATR–FTIR study of the protonation states of the Glu residue in the multicopper oxidases, CueO and bilirubin oxidase. *FEBS Lett.* **2010**, *584*, 4027–4031.

101. Bento, S.; Carrondo, M. A.; Lindley, P. F. Reduction of dioxygen by enzymes containing copper. *J. Biol. Inorg. Chem.* **2006**, *11*, 539–547.

102. Sukmar, N.; Mathews, F. S.; Langan, P.; Davidson, D. V. A joint X-ray and neutron study on amicyanin reveals the role of protein dynamics in electron transfer. *Proc. Natl. Acad. Sci. USA* **2010**, *107*, 6817–6822.

103. Boal, A. K.; Rosenberg, A. C. Structural biology of copper trafficking. *Chem. Rev.* **2009**, *109*, 4760–4779.

104. Ueki, Y.; Kataoka. K.; Sakurai, T. unpublished results.

105. Brown, M. A.; Stenberg, L. M.; Mauk, A. G. Identification of catalytically important amino acids in human ceruloplasmin by site-directed mutagenesis. *FEBS Lett.* **2002**, *520*, 8–12.

106. Kurose, S.; Kataoka, K.; Shinohara, N.; Miura, Y.; Tsutsumi, M.; Tsujimura, S.; Kano, K.; Sakurai, T. Modification of Spectroscopic properties and catalytic activity of *Escherichia coli* CueO by mutations of methionine 510, the axial ligand to the type I Cu. *Bull. Chem. Soc. Jpn.* **2009**, *82*, 504–508.

107. Miura, Y.; Tsujimura, S.; Kamitaka, Y.; Kataoka, K.; Sakurai, T.; Kano, K. Direct electrochemistry of CueO and its mutants at residues to and near type I Cu for oxygen-reducing biocathode. *Fuel Cells*, **2009**, *9*, 70–78.

108. Kamitaka, Y.; Tsujimura, S.; Kataoka, K.; Sakurai, T.; Ikeda, T.; Kano, K. Effects of axial ligand mutation of the type I copper site in bilirubin oxidase on direct electron transfer-type biocatalytic reduction of dioxygen. *J. Electroanal. Chem.* **2007**, *601*, 119–124.

109. Shimizu, A.; Sasaki, T.; Kwon, J.-H. Odaka, A.; Satoh, T.; Sakurai, N.; Sakurai, T.; Yamaguchi, S.; Samejima, T. Site-directed mutagenesis of a possible type I copper ligand of bilirubin oxidase; a Met467Gln mutant shows stellacyanin-like properties. *J. Biochem.* **1999**, *125*, 662–668.

110. Durao, P.; Chen, Z.; Silva, C. S.; Soares, C. M.; Pereira, M. M.; Todorovic, S.; Hiderbandt, P.; Bento, I.; Lindley, P. F.; Martins, L. O. Proximal mutations at the type 1 copper site of CotA laccase: spectroscopic. redox, kinetic and structural characterization of I494A and L386A mutants. *Biochem. J.* **2008**, *412*, 339–346.

111. Kataoka, K.; Tsukamoto, K.; Kitagawa, R.; Ito, T.; Sakurai, T. Compensatory binding of an asparagine residue to the coordination of unsaturated type I Cu center in bulirubin oxidase mutants. *Biochem. Biophys. Res. Commun.* **2008**, *371*, 416–419.

112. Mizutani, K.; Toyoda, M.; Sagara, K.; Takahashi, N.; Sato, A.; Kamitaka, Y.; Tsujimura, S.; Nakanishi, Y.; Sugiura, T.; Yamaguchi, S.; Kano, K.; Mikami, B. X-ray analysis of bilirubin oxidase from *Myrothecium verrucaria* at 2.3 Å resolution using a twinned crystal. *Acta Crystallogr.* **2010**, *F66*, 765–770.

113. Machczynski, M. C.; Gray, H. B.; Richards, J. H. An outer-sphere hydrogen bond network constrains copper coordination in blue proteins. *J. Inorg. Biochem.* **2002**, *88*, 375–380.

114. Carrell C. J.; Sun, D.; Jiang, S.; Davidson, V. L.; Mathews, F. S. Structural studies of two mutants of amicyanin from *Paracoccus denitrificans* that stabilize the reduced state of the copper. *Biochemistry* **2004**, *43*, 9372–9380.

115. Yanagisawa, S.; Banfield, M. J.; Dennison, C. The role of hydrogen bonding at the active site of a cupredoxin; the Phe113Pro azurin variant. *Biochemistry* **2006**, *45*, 8812–8822.

116. Kataoka, K.; Hirota, S.; Maeda, Y.; Kogi, H.; Shinohara, N.; Sekimoto, M.; Sakurai, T. Enhancement of laccase activity through the construction and breakdown of a hydrogen bond at the type I copper center in *Escherichia coli* CueO and the deletion mutant Δα5-7 CueO. *Biochemistry* **2011**, *50*, 558–565.

117. Couto, S. R.; Herrera, J. L. T. Industrial and biological applications of laccases: a review *Biotechnol. Adv.* **2006**, *24*, 500–513.

118. Shleev, S.; Tkac, J.; Christenson, A.; Ruzgas, T., Yaropolov, A. I.; Whittaker, J. W., Gorton, L. Direct electron transfer between copper-containing proteins and electrodes. *Biosens. Bioelectro.* **2005**, *20*, 2517–2554.

119. Miura, Y.; Tsujimura, S.; Kamitaka, Y.; Kurose, S.; Kataoka, K.; Sakurai, T.; Kano, K. Biocatalytic reduction of O_2 catalyzed by CueO from *Escherichia coli* adsorbed on a highly oriented pyrolytic graphite electrode. *Chem. Lett.* **2007**, *36*, 1323–133.

120. Tsujimura, S.; Miura, Y.; Kano. K. CueO-immobilized porous carbon electrode exhibiting improved performance of electrochemical reduction of dioxygen to water. *Electrochim. Acta* **2008**, *53*, 5716–5720.

121. Sakasegawa, S.; Ishikawa, H.; Imamura, S.; Sakuraba, H.; Goda, S.; Ohshima, T. Bilirubin oxidase activity of *Bacillus subtilis* CotA. *Appl. Environm. Microbiol.* **2006**, *72*, 972–975.

6

STRUCTURE AND REACTIVITY OF COPPER–OXYGEN SPECIES REVEALED BY COMPETITIVE OXYGEN-18 ISOTOPE EFFECTS

JUSTINE P. ROTH

Department of Chemistry, Johns Hopkins University, Baltimore, MD 21218

6.1. INTRODUCTION

Competitive oxygen-18 (^{18}O) isotope effects can be used to identify reactive copper–oxygen species (CuO_2) that are typically elusive due to their kinetic and thermodynamic properties.[1,2] Importantly, the isotope effects can probe reactive intermediates under catalytic conditions.[3,4] Measurements provide the means to relate the CuO_2 formed in an enzyme active site to model complexes that can be synthesized and characterized, usually at low temperatures in organic solvents.[5] This chapter describes efforts to this end, where competitive oxygen-18 kinetic isotope effects (^{18}O KIEs) and equilibrium isotope effects (^{18}O EIEs) are used together with density functional theory (DFT) computational methods to expose structures and mechanisms.[6,7]

The physical and chemical properties that influence CuO_2 reactivity in Nature have not been understood.[8,9] These fundamental issues notwithstanding, copper-mediated redox catalysis continues to inspire the preparation of new classes of bio-inspired inorganic compounds.[10] Though there are few success stories, synthetic modeling efforts have forged new approaches to understanding small molecule activation, as illustrated by the isotope fractionation technique described here.[11]

The competitive fractionation of natural abundance molecular oxygen ($^{16}O^{16}O$ vs $^{16}O^{18}O$) is a well-established phenomenon in the atmospheric and geological sciences.[12,13] It has more recently been adapted for the determination of highly precise isotope effects on reactions involving enzymes,[3,4,14–17] inorganic compounds,[5–7,18] and even environmental contaminants in complex media.[19,20] This chapter focuses on CuO_2 complexes, which are transformed into CuO_2^- and CuO_2H by electron transfer (ET) and proton-coupled electron transfer, (PCET) respectively. The competitive ^{18}O isotope fractionation technique described here has been applied to elucidating such mechanisms.[21] The relevant experimental and computational methodology developed over the past decade will be described, together with advances in the field to which this monograph/series is dedicated.

6.2. INSTRUMENTATION

Competitive oxygen isotope effects are typically measured using a stable isotope mass spectrometer[22] and a high-vacuum apparatus for sample preparation, which will be described.[4,5,23] The apparatus consists of two manifolds. One manifold is used to isolate samples of gas from a reaction solution while the other is used to manipulate the gas for determination of its pressure and isotope composition, as described below.

A reaction chamber is connected to the primary manifold. This chamber can be a glass vessel with a movable plunger or a collapsible, solvent impervious bag, equipped with an injection port and magnetic stirrer. A sample tube built into the manifold is used to remove fixed-volume aliquots from the reaction chamber. Turning a cross-over stopcock fills an evacuated section of the vacuum line with a solution that is subsequently dispensed into a gas bubbler. Helium is then flowed through the entire system while under dynamic vacuum.

Noncondensable gases are entrained in the helium while condensable gases (e.g., H_2O and CO_2) are removed by a series of cold traps. The O_2 is eventually collected in a liquid-nitrogen-cooled trap containing 5-Å molecular sieves. Removal of helium, by placing the cold sieves under dynamic vacuum, is followed by heating to release the O_2. At this point, the purified O_2 is transferred into the second manifold.

In the second manifold, O_2 is quantitatively combusted to CO_2. Recirculation over a Pt/graphite rod at $\sim$800°C produces CO_2, which is condensed onto a cold finger. The conversion of O_2 to CO_2 is monitored by the accompanying pressure changes. Once conversion is complete, the CO_2 is transferred to a second cold finger where its pressure is determined. The isotope content and quantity of CO_2 is identical to the O_2 remaining in solution at a specific reaction yield. The $^{18}O/^{16}O$ is then determined by isotope ratio mass spectrometry (IRMS).

6.3. METHODOLOGY

The precision of the competitive isotope effect measurements can partly be attributed to the use of natural abundance O_2, which makes the technique relatively insensitive to leakage or atmospheric contamination. Isotope effects are determined from the two most abundant isotopologues: $^{16}O-^{16}O$ and $^{16}O-^{18}O$ at concentrations similar to that of O_2 in ambient air. Analogously, the isotope compositions of nonenriched KO_2 and H_2O_2, as well as H_2O, can be analyzed.

One restriction of the technique is sample size. While typically not a problem, the high effective concentration of H_2O can make fractionation measurements on this substrate difficult. Since natural abundance ^{18}O is low ($\sim$0.2%), samples must consist of at least 2 µm of gas. While often easier for measurements involving inorganic compounds, it can be a limitation for measurements on enzymes with low activities.

Experiments typically involve determining the changes in $^{18}O/^{16}O$ at varying conversions during a catalytic or stoichiometric reaction. The fixed-field IRMS used for the analysis is equipped with a mechanism for internal referencing against a standard sample gas. The isotope composition of the sample is reported relative to an internal standard, which is referenced to standard mean ocean water (SMOW = standard mean ocean water). Errors associated with the IRMS measurements are ±0.1 parts per thousand (ppt), however, error in the experimental precision can be up to 10 times larger. For example, the repeated analysis of ambient air under typical experimental conditions has indicated 1.0235 ± 0.0015 versus SMOW.

Two types of competitive isotope fractionation measurements have been performed to determine oxygen (^{18}O): ^{18}O EIEs and ^{18}O KIEs.[9] The ^{18}O EIEs correspond to the ratio of equilibrium constants, $K(^{16}O-^{16}O)/K(^{16}O-^{18}O)$ and the ^{18}O KIEs correspond to a ratio of second order rate constants, $k(^{16}O-^{16}O)/k(^{16}O-^{18}O)$. Both isotope effects are determined from a ratio of ratios, minimizing the error such that it is only 5–10% of the ^{18}O EIE or KIE.

The approaches used to determine ^{18}O EIEs and ^{18}O KIEs differ with respect to the experimental method used. The ^{18}O EIEs are measured in an open reaction vessel that allows rapid exchange between the atmosphere and the total gas dissolved in solution. Experiments are performed to determine the change in total O_2 concentration (i.e., $O_2^{unbound} + O_2^{bound}$) and the accompanying isotope composition upon coordination of O_2. Only the bound O_2 is fractionated in a manner that reflects its reduction and/or interaction with a redox-active metal center. Importantly, the bound O_2 must be released from the metal center so that it can be quantified and analyzed. The pressure and isotope composition of the unbound O_2 is determined independently and then used to determine the ^{18}O EIE according to Eq. 6.1, where terms include R_u and R_t for the $^{18}O/^{16}O$ of the unbound and total O_2, respectively, and f for the fraction of unbound O_2 remaining in solution at equilibrium.

$$^{18}O \text{ EIE} = \frac{1 - f}{(R_t/R_u) - f} = \frac{K(^{16}O^{16}O)}{K(^{16}O^{18}O)} \tag{6.1}$$

The determination of ^{18}O KIEs requires a sealed reaction vessel that keeps the dissolved O_2 in solution as the reaction progresses. Moles of O_2 consumed are accurately quantified. At varying stages of the reaction, the O_2 is collected and the $^{18}O/^{16}O$ at a given percent conversion is determined. The ^{18}O KIE is then calculated from Eq. 6.2, where terms include R_0 for the initial $^{18}O/^{16}O$ of the O_2 at 0% conversion and R_f for the $^{18}O/^{16}O$ of O_2 at the conversion designated f.

$$^{18}O \text{ KIE} = \left[1 + \frac{\ln(R_f/R_0)}{\ln(1 - f)}\right]^{-1} = \frac{k(^{16}O^{16}O)}{k(^{16}O^{18}O)} \tag{6.2}$$

6.4. OXYGEN EQUILIBRIUM ISOTOPE EFFECTS

Calculations performed using the Bigeleisen and Goeppert-Mayer[24] formalism provide a frame of reference for interpreting ^{18}O EIEs on reversible reactions involving CuO_2 complexes as stable intermediates and proposed transients. In accord with the theoretical formalism, the ^{18}O EIE is defined for an isotope exchange reaction as depicted in Eq. 6.3. The ^{18}O EIE is expressed in terms of the reduced gas-phase partition functions corresponding to isotope effects on the zero-point energy (ZPE), excited-state vibrational energy (EXC) and mass and moments of inertia (MMI) (Eq. 6.4). It is generally observed that isotope effects computed in the gas phase agree with solution measurements.[25,26]

$$O_2 + CuO_2^* \rightleftharpoons O_2^* + CuO_2 \tag{6.3}$$

$$^{18}O \text{ EIE} = ZPE \times EXC \times MMI \tag{6.4}$$

The isotopic partition functions in Eq. 6.4 are computed from the vibrational frequencies of the heavy and light isotopologues of the reactant and product states. No scaling of frequencies or correction for anharmonicity is typically included.[6] The ZPE and EXC terms represent the isotope effects deriving from quantized vibrational energy levels, whereas MMI represents the isotope effect due to changes in the rotational and translational modes.[27] The formulas for the partition functions are presented in Eqs. (6.5–6.8), where v is a vibrational frequency, T is the temperature, h is Planck's constant, k is Boltzmann's constant and N is the number of atoms. As in Eq. 6.3, the asterisk designates the site containing ^{18}O. The MMI can be expressed as a vibrational product (VP) according to the Redlich–Teller product rule (Eq. 6.7) or in terms of masses and rotational constants following the classical Newtonian equations of motion (Eq. 6.8).

$$ZPE = \frac{\left[\prod_{j}^{3N-6} \frac{\exp^{(hv_j^{B*}/2kT)}}{\exp^{(hv_j^{B}/2kT)}} \right]}{\left[\prod_{i}^{3N-5} \frac{\exp^{(hv_i^{A*}/2kT)}}{\exp^{(hv_i^{A}/2kT)}} \right]} \tag{6.5}$$

$$EXC = \frac{\left[\prod_{j}^{3N-6} \frac{1-\exp^{-\left(hv_j^{B*}/kT\right)}}{1-\exp^{-\left(hv_j^{B}/kT\right)}} \right]}{\left[\prod_{i}^{3N-5} \frac{1-\exp^{-\left(hv_i^{A*}/kT\right)}}{1-\exp^{-\left(hv_i^{A}/kT\right)}} \right]} \tag{6.6}$$

$$MMI = VP = \frac{\prod_{j}^{3N-6} (v_j^{B}/v_j^{B*})}{\prod_{i}^{3N-5} (v_i^{A}/v_i^{A*})} \tag{6.7}$$

$$MMI = \frac{\left[\left(\frac{M_{16-16}}{M_{16-18}}\right)^{3/2} \prod_{i}^{n_{rot}} \left(\frac{I_{i,16-16}}{I_{i,16-18}}\right)^{1/2} \right]_{B}}{\left[\left(\frac{M_{16-16}}{M_{16-18}}\right)^{3/2} \prod_{i}^{n_{rot}} \left(\frac{I_{i,16-16}}{I_{i,16-18}}\right)^{1/2} \right]_{A}} \tag{6.8}$$

Calculations of ^{18}O EIEs upon reactions of O_2 or other small molecules require accurate vibrational frequencies for the $^{16}O-^{16}O$ and $^{18}O-^{16}O$ isotopologues ($^{16,16}v$ and $^{18,16}v$). For most stable molecules, the frequencies reported in the literature have been estimated from published force constants.[5,28] Density functional theory calculations provide full sets of vibrational frequencies for complex molecules, as well as the principal rotational moments of inertia, which together with the masses, factor into the MMI calculation (Eq. 6.8). Agreement of Eqs. 6.7 and 6.8 is used to verify the quality of the energy minimized structures due to the applicability of the Redlich–Teller product rule.[29] In the event that "cut-off" models are used and the full set of frequencies is not considered in the calculations, isotope effects on the partition functions are redistributed in a manner that is not physically meaningful, although the calculated ^{18}O EIE may in some cases be comparable to that derived from full-frequency analysis.[5–7,2] More often, the ^{18}O EIE is overestimated because the net isotope shift upon converting the reactant to the product is underestimated.

6.4.1. Computational Studies

Computational studies of copper-containing proteins and synthetic Cu(I) and Cu(II) compounds suggest that ^{18}O EIEs can be used to characterize the bonding within the CuO_2 adduct. Klinman and Tian[28] were the first to experimentally determine ^{18}O EIEs on reversible reactions, where O_2 binds to metalloproteins such as hemocyanin (Hc). In this case, the oxygenated protein has been characterized as a μ-η^2,η^2-peroxide bonded to two antiferromagnetically coupled Cu(II) centers[30] using spectroscopy and DFT calculations.[31–33]

Studies of structurally defined compounds deriving from transition metals in groups VIII–X have demonstrated that side-on (η^2) bound metal peroxide (O_2^{-II}) complexes exhibit much larger ^{18}O EIEs than end-on (η^1) superoxide complexes.[34] This observation is reproduced by DFT calculations performed using the local functional, mPWPW91,[35] and the following atomic orbital basis functions: Cu (the compact relativistic effective core potential basis CEP-31G),[36] N and O (6-311G*), C (6-31G), and H (STO-3G).[37]

A number of calculations have been performed on the μ-η^2,η^2-peroxo structure and bis-μ-oxo model structures for oxyHc.[38] Surprisingly, the best agreement to the experimentally determined ^{18}O $EIE_{exp} = 1.0184 \pm 0.0023$ is observed for calculations on the bis-μ-oxo complex, where ^{18}O $EIE_{calc} = 1.0200$. The ^{18}O EIE_{calc} varies little upon changing imidazole to methylimidazole and inducing structural asymmetry by coordinating two rather than three ligands to one copper site. A larger ^{18}O $EIE_{calc} = 1.0284$ is computed for the μ-η^2,η^2-peroxo structure with the same constraints (Table 6.1). This unusual disagreement may suggest computational difficulty associated with covalently bonded dicopper sites.

TABLE 6.1. Measured and Calculated ^{18}O EIEs on Reversible Dioxygen Binding to Hemocyanin

Model			^{18}O EIE_{exp}[a]	^{18}O EIE_{calc}[b]
oxyHc	**o**	**p**	1.0184(23)	**o**: $1.0200^{c,d}$
				p: $1.0284^{d,e}$

[a] Experimental data are from Ref. 28.
[b] Gaussian03 calculations[39] performed as outlined in the text and detailed in Ref. 6.
[c] The o = hypothetical bis-μ-oxo structure.
[d] Identical results for restricted and unrestricted singlet states.
[e] The p = experimental side-on peroxo structure.

6.4.2. Experimental Studies

Though the computations on oxy Hc appear to overestimate values the ^{18}O EIE_{exp}, the experimental value, may be questioned because the efficiency of O_2 release was not evaluated.[28] Thus, the disagreement may be attributed to error in the ^{18}O EIE caused by solutions that are not completely equilibrated and/or the incomplete recovery of O_2 from the metal–dioxygen complex.

Additional experimental and computational results obtained for the reversible formation of mononuclear CuO_2 complexes are collected in Table 6.2. Note that these comparisons represent only a fraction of the reductive O_2 activation reactions of model complexes and proteins containing other redox metals. The ^{18}O EIEs have been used to characterize end-on (η^1) and side-on (η^2) bonding modes, as well as superoxide (O_2^{-I}) and peroxide (O_2^{-II}) structures. In general, the extent of internal electron redistribution, from metal to oxygen appears to be manifest in the magnitude of the ^{18}O EIE due to the strength of bonding interactions.

Calculated temperature profiles of the ^{18}O EIEs reveal a maximum normal ^{18}O EIE for each reversible $Cu-O_2$ coordination reaction (Fig. 6.1).[31,32] As depicted, the ^{18}O EIEs are inverse (<1) at the lowest temperature and gradually increase through a normal (>1) maximum before decreasing to unity as the temperature approaches

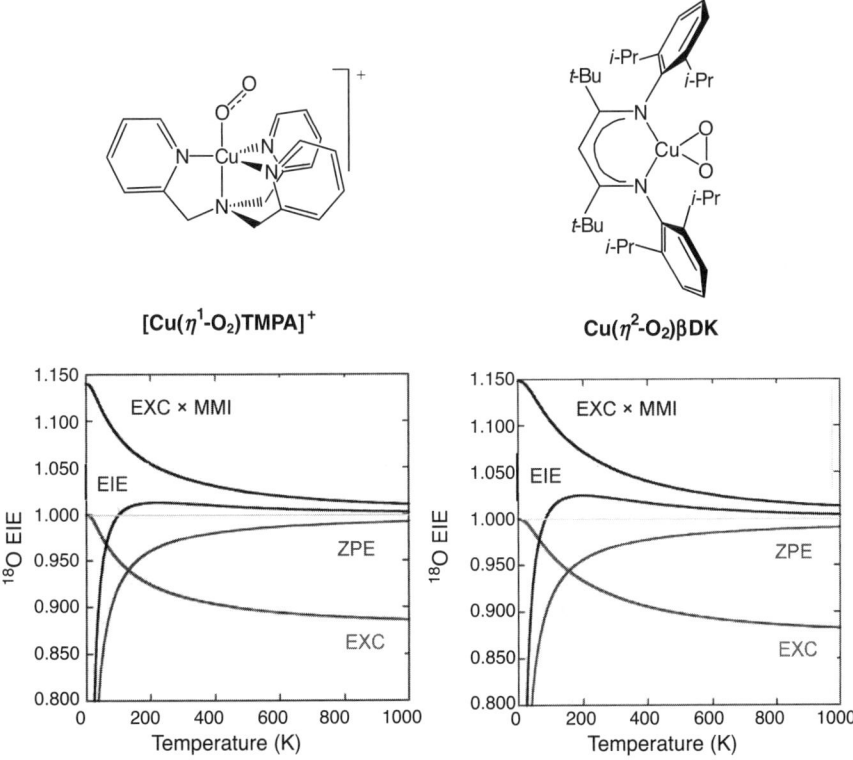

FIGURE 6.1. Temperature profiles of ^{18}O EIE_{calc} and partition functions for $Cu-O_2$ adducts.

TABLE 6.2. Equilibrium Isotope Effects for the Reversible Formation of CuO$_2$ Structures.[a]

Reduced Form	Oxygenated Form	^{18}O EIE$_{exp}$	^{18}O EIE$_{calc}$[b]
[Cu^I(TMG$_3$Tren)]$^{+}$[c]		1.0148 ± 0.0012[d]	1.0129[e]
[Cu^I(TMPA)]$^{+}$[f]		<1.0165[g]	1.0128[h]
Cu^I(βDK)(thf)[i]		n.d.[j,k]	1.0239[l]
Cu^I(t-Bu-Tp)(dmf)[m]		n.d.[j,n]	1.0251[o]

[a] See Ref. 40.
[b] Calculated in vacuum unless noted.
[c] TMG$_3$Tren = 1,1,1-tris{2-[n^2- (1,1,3,3-tetramethylguanidine)]ethyl}amine.
[d] Measured in DMF (dimethylformamide, solvent) at 261 K.
[e] The PCM calculation with acetone at 261 K (see Ref. 40).
[f] TMPA = tris (2-pyridylmethyl)amine.
[g] Estimated from the observed ^{18}O KIE in dimethylsulfoxide (solvent) (DMSO) at 295 K (see Ref. 41).
[h] The maximum ^{18}O EIE at 222 K.
[i] β-Diketiminate = βDK and thf = tetrahydrofuran (ligand).
[j] Not determined = n.d.
[k] The O$_2$ binding is kinetically irreversible. The ^{18}O KIE was determined to be 1.0024 ± 0.0019 in dimethylformamide (solvent) (DMF) at 228 K.
[l] The maximum ^{18}O EIE at 173 K.
[m] Hydrotris(3-tert-butyl-pyrazolyl)borate = t-Bu-Tp and dmf = dimethylformamide (ligand).
[n] No reaction between O$_2$ and the analogous 3-admantyl, 5-isopropyl trispyrazolyl borate complex: Cu^I(Ad,i-Pr-Tp)(dmf) could be observed in DMF at 228 K (see Ref. 42).
[o] The maximum ^{18}O EIE at 197 K.

infinity. The transition from inverse to normal arises from a changeover, where the dominant enthalpic effect (ZPE) at low temperature gives way to the entropic effect (EXC × MMI) at high-temperature. At the high-temperature limit, all vibrational modes are populated causing the ^{18}O EIE to vanish.

The temperature profiles associated with the ^{18}O EIEs and the constituent isotopic partition functions are shown below for two structurally distinct copper–oxygen complexes.[41] The size of the ^{18}O EIE, together with the temperature at which it reaches a maximum, differentiates the end-on from the side-on bound O_2 structures. The end-on structures exhibit smaller ^{18}O EIEs that reach a maximum at higher temperatures due to the population of isotopically sensitive low-frequency vibrational modes.[5,9]

The explanation[5–7] for the inverse ZPE and large normal EXC × MMI contributions is different from that proposed earlier,[21,29,43] where the focus was primarily upon reduction of the O–O force constant.[44] Though the decrease in O–O force constant, and attendant decrease in zero-point energy level splitting, is believed to be the origin of most heavy atom isotope effects (>1), this is an oversimplification and misconception for reactions, when O_2 binds to a heavier metal fragment, as illustrated below (Fig. 6.2).

Upon coordination to a reduced copper center, the rotations and translations present in free O_2 are transformed into low-frequency vibrations. These new Cu–O_2 bond(s) contribute to an increased molecular force constant offsetting the weakened O–O bond. The end result is a net increase in the zero-point energy level splitting associated with the CuO$_2$ relative to O_2. Such behavior gives rise to an inverse ZPE term calculated from the net isotope shift of vibrational modes within the product relative to the reactant via Eq. 6.5. This effect is consistent with the favorable enthalpy of O_2 binding. Loss of the mass-dependent rotational and translation modes and population of new low-frequency vibrational modes together contribute to a large normal entropic isotope effect expressed as EXC × MMI. It is this product of Eqs. 6.6 and 6.7 that make the overall ^{18}O EIE >1 at temperatures above ∼100 K. In the low-temperature limit, smaller entropic terms and inverse enthalpic terms can result in ^{18}O EIEs <1.

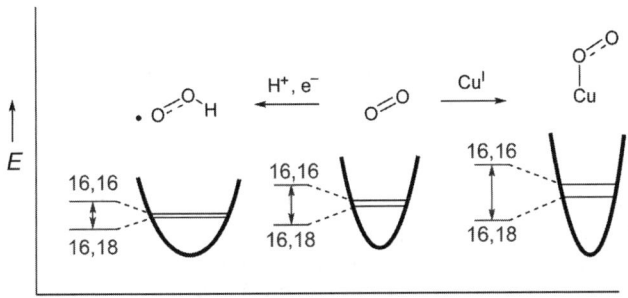

Nuclear coordinate

FIGURE 6.2. Predicted trends in zero-point-energy level splitting upon reducing O_2.

6.5. OXYGEN KINETIC ISOTOPE EFFECTS

Oxygen kinetic isotope effects (^{18}O KIEs) have provided a new and unprecedented approach to studying the reactivity of O_2 and $O_2^{\bullet-}$ with synthetic copper complexes and copper proteins. By virtue of being a competitive measurement, the ^{18}O KIE reports on all microscopic steps that contribute to the bimolecular rate constant, designated $k_{cat}/K_M(O_2)$ for steady-state enzymatic reactions and k_{O2} for the analogues reactions in solution. The competition between heavy and light isotopologues ($^{16}O^{16}O$ vs $^{16}O^{18}O$) present at natural abundance results in an ^{18}O KIE, which reflects all steps beginning with encounter and leading up to, and including, the first kinetically irreversible step.

Though experimental ^{18}O KIEs have been determined for reactions between O_2 and Cu(I), as well as $O_2^{\bullet-}$ and Cu(II), computational methods have not advanced to the stage where transition state structures and vibrational frequencies can be predicted reliably for such reactions. Preliminary analysis using published B3LYP calculations[45] suggest an ^{18}O KIE on O_2 coordination to $Cu^I(\beta DK)(MeCN)$ close to 1.005. In spite of the poorer performance of B3LYP as compared to mPWPW91,[6] the computed value is in reasonable agreement with the experimental ^{18}O KIE of 1.0024 ± 0.0019 determined in DMF at 228 K.[46] In another case, where O_2 reacts with an imidazole-coordinated Cu(I) center in the reduced form of copper zinc superoxide dismutase at ambient temperature, the ^{18}O KIE $= 1.0044 \pm 0.0016$.[47] These reactions have been proposed to occur by inner-sphere ET.

In studies of other copper enzymes,[48] ^{18}O KIEs reportedly range from 1.0075 to 1.0185 for O_2 reduction by oxidases[49–53] and from 1.0173 to 1.0256 for reactions of monooxygenases[54,55] involving cleavage of the O−O bond. The ^{18}O KIEs for both enzyme classes have been proposed to derive from multistep pathways involving a CuO_2 intermediate, formed reversibly prior to rate-limiting ET or PCET. Therefore, the "observed ^{18}O KIE" is the product of a pre-equilibrium ^{18}O EIE and an ^{18}O KIE upon a single, irreversible, ET or PCET step.

In a related study of O_2 coordination to a Pd(0) complex, Landis et al.[56] used a method where the transition state was located from the minimum energy crossing points of calculated singlet and triplet surfaces. The ^{18}O KIE was predicted to be 1.026 for the proposed rate-limiting reorganization of a preformed $Pd^I(\eta^1\text{-}O_2^{-I})$ intermediate to the stable $Pd^{II}(\eta^2\text{-}O_2^{-II})$ product.[57] Although the contribution from the reaction coordinate frequency does not appear to have been considered in this study, the DFT calculations revealed the transition states for the sequential and concerted pathways to be very close in energy,[57] consistent with the prediction of similar ^{18}O KIEs$_{calc}$.

The transition state theory formalism in Eq. 6.9 has been used to calculate ^{18}O KIEs from the isotope effect on the reaction coordinate frequency (v_{RC}), which is the mode that converts a translation into a metal–O_2 vibration, together with the isotope effect on the pseudo-equilibrium constant for forming the transition state (K_{TS}). In the expression for the ^{18}O KIE$_{calc}$ (Eq. 6.10) the imaginary vibration due to the reaction coordinate has been removed from K_{TS}, which is analyzed in the same manner as the ^{18}O EIE in Eq. 6.5.

$$^{18}O\ KIE = \left(^{16,16}v_{RC}/^{16,18}v_{RC}\right)\left(^{16,16}K_{TS}/^{16,18}K_{TS}\right) \tag{6.9}$$

$$^{16,16}K_{TS}/^{16,18}K_{TS} = ZPE \times EXC \times MMI \tag{6.10}$$

6.5.1. Inner- and Outer-Sphere Electron Transfer

Reactions with O_2 may occur by inner- or outer-sphere ET when the redox partner is a Cu(I) complex. The same is true of reactions with $O_2^{\bullet-}$, where the inner-sphere reaction is likely favored by the formation of a bond to O_2 in the transition state. Many ET reactions involve covalently bound intermediates, even though the $Cu(\eta^1-O_2)$ interaction may not be very strong.[1,40] In contrast, outer-sphere ET inter conversion of O_2 and $O_2^{\bullet-}$ in the absence of a covalent transition state, is likely to be driven by the reaction thermodynamics.

Oxygen-18 KIEs upon outer-sphere ET to O_2 were first interpreted in the context of bond rehybridization and the effect of isotopic substitution upon the O–O stretching frequency.[4] Since the early studies, it has been shown that calculation of the ^{18}O KIE requires that the O–O stretch be treated quantum mechanically; that is, because the $^{16,16}v$ of O_2 ($1556\ cm^{-1}$) and $O_2^{\bullet-}$ ($1064\ cm^{-1}$) exceed the available thermal energy ($>4k_BT$) making the O–O bond essentially frozen at ambient temperature. Classical motions that contribute to the reaction coordinate are defined by low-frequency vibrational and librational modes of the surrounding solvent or protein.

A quantum mechanical formalism has been used to describe the ^{18}O KIE on electron self-exchange, where O_2 reacts with $O_2^{\bullet-}$. The ^{18}O KIE is calculated from isotope-dependent Franck–Condon overlap factors.[58–60] Although simplified by use of an averaged force constant, for the reactant and product states this quantum mechanical treatment reproduces experimental results surprisingly well; the ^{18}O $KIE_{calc} = 1.0270$ at 298 K, corresponds to a 0.1-Å change in O–O bond distance. The calculation is in excellent agreement with the ^{18}O $KIE_{exp} = 1.0279 \pm 0.0006$ measured for the reduction of O_2 by an anionic flavin adenine dinucleotide (FAD) cofactor in glucose oxidase (GO).[58] In this theoretical treatment,[3,4,58] the population of vibrationally excited states is neglected, as is the very small difference in Gibbs free energies of reaction for $^{16,16}O_2$ and $^{16,18}O_2$ isotopologues. Nevertheless, the calculations reproduce the ^{18}O KIEs for outer-sphere ET to O_2.[61] It is also observed, as predicted, that the ^{18}O KIE is insensitive to small changes in reaction driving force ($\sim140\ mV$). Calculations of Franck–Condon overlap factors predict that much larger changes in ΔG°, on the order of 500 mV, are required to cause the ^{18}O KIE to decrease substantially from the parabolic maximum.[3,47]

The theoretical prediction of ^{18}O KIEs is more challenging for inner-sphere ET involving the production or consumption on O_2. One source of ambiguity is the multiconfigurational character of the transition state. There is also the potential for multiple reactive geometries. Modeling the formation of the covalent intermediate(s) is a further complicating factor. Equilibria involving such species cannot be described using the conventional electrostatic "work" terms.[58,61] Yet, as described below, one way to address whether such covalent (e.g., CuO_2) intermediates are formed in ET reactions involves varying the redox potential of the electron donor (Cu^I) and analyzing the impact on the ^{18}O KIE.

6.5.2. Evidence for Dioxygen-Bound Intermediates

Under some circumstances, the comparison of the ^{18}O KIE to the ^{18}O EIE can signal the presence of dioxygen-bound intermediates. When such an intermediate is involved in the reaction mechanism, the observed ^{18}O KIE is the product of a pre-equilibrium ^{18}O EIE corresponding to its formation and the microscopic KIE for its disappearance in the rate-determining step. The isotope effect on the pre-equilibrium is readily calculated from the vibrational frequencies of the ground-state structures, as in Tables 6.1 and 6.2, whereas the ^{18}O KIE must consider the transition state structure via Eq. 6.9.

When O_2 is released from a copper-bound oxygen, superoxide, or peroxide species, the microscopic step is expected to exhibit an inverse (<1) ^{18}O KIE. This reasoning derives from the empirical relationship ^{18}O EIE $> {}^{18}O$ KIE, observed in studies of O_2 binding to transition metal complexes.[6,7,62] Similar behavior appears to influence the ^{18}O KIEs determined for dicopper enzymes like tyrosinase,[14] which approach the magnitude of the ^{18}O EIE observed for oxyHc.[29] On similar grounds, the dissociation of $O_2^{\bullet-}$ from a CuO_2 complex is predicted to exhibit a normal ^{18}O KIE in contrast to the inverse KIE and EIE that characterize $O_2^{\bullet-}$ assocation.[41]

A CuO_2 intermediate has been inferred on the basis of ^{18}O KIEs measured for the reaction of potassium superoxide (KO_2) with Cu(II) tris(2-pyridylmethyl $[Cu^{II}(TMPA)]^{2+}$ and 2-pyridylethyl) and $[Cu^{II}(TEPA)]^{2+}$ amine complexes; both possessing noncoordinating trifluoromethanesulfonate (triflate, OTf) counterions.[41] These relatively high-potential Cu(II) complexes convert $O_2^{\bullet-}$ to O_2 in DMSO at a rate much greater than that predicted by the Marcus theory of outer-sphere ET. The ^{18}O KIEs were determined under kinetically irreversible conditions, where O_2 was collected from the reaction mixture as it formed. The results in Figure 6.3 indicate that the ^{18}O content of the unreacted $O_2^{\bullet-}$ decreases as the reaction progresses, consistent with an inverse ^{18}O KIE. The resulting narrow range of ^{18}O KIEs, from 0.984 to 0.989,

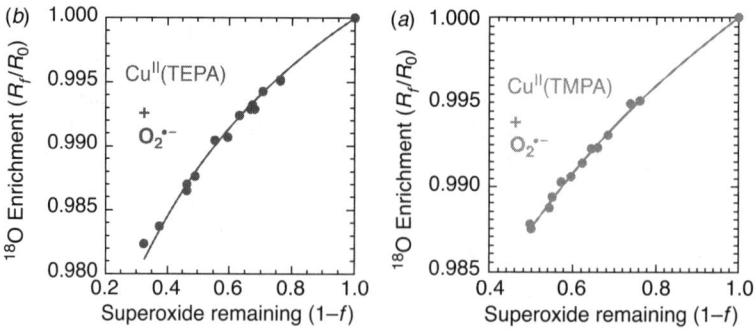

FIGURE 6.3. Isotope fractionation upon reacting (a) $[Cu^{II}(TMPA)]^{2+}$ and (b) $[Cu^{II}(TEPA)]^{2+}$ with KO_2 at 295 K in DMSO.

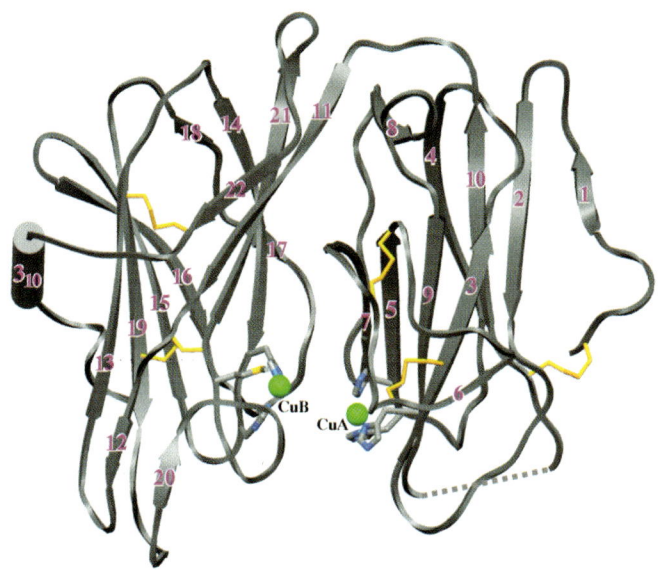

Figure 1.2 A representation of the catalytic core of PHM. (*See text for full caption.*)

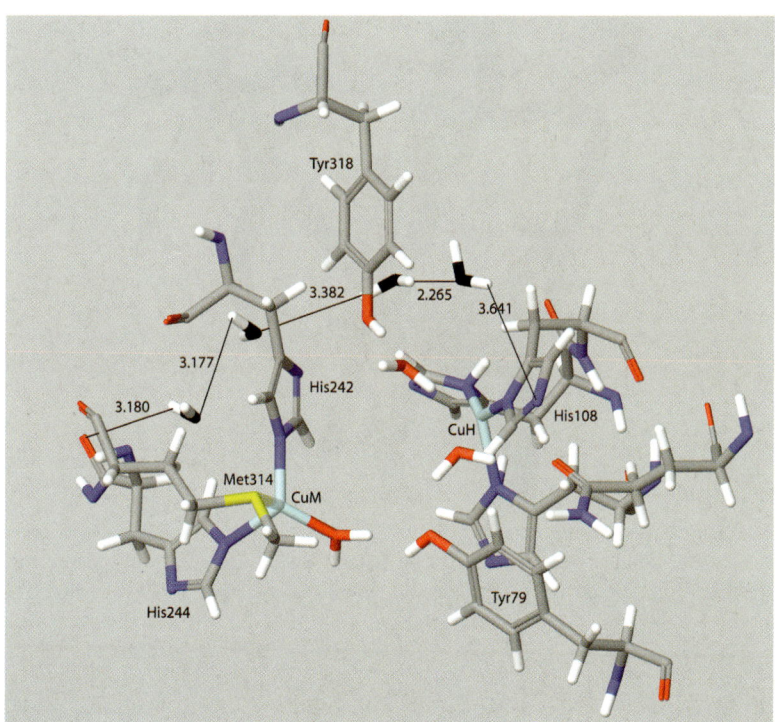

Figure 1.6 Active-site structure of reduced PHM recreated from PDB: 3PHM.[55] (*See text for full caption.*)

Active Sites

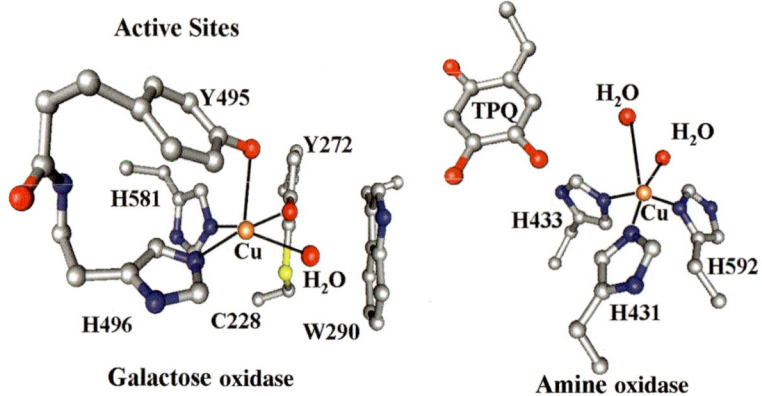

Galactose oxidase

Amine oxidase

Figure 3.1 Active-site structures of GO from *Hypomyces rosellus* (PDB ID: 1GOG) and AO from *Arthrobacter globiformis* (PDB ID: 1AV4).

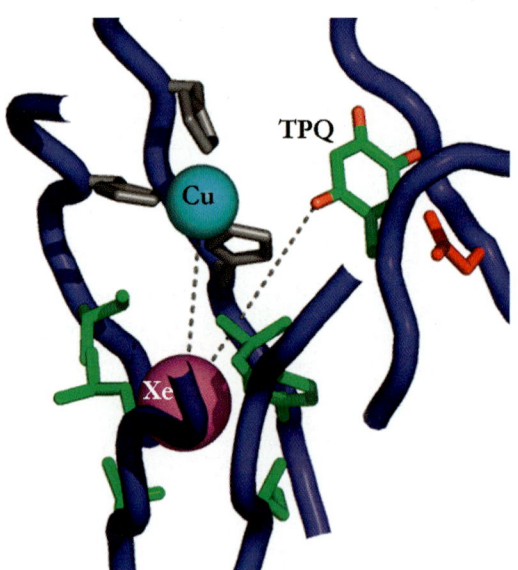

Figure 3.12 Xenon-binding site in PSAO. The pink sphere represents the consensus Xe binding site found in AGAO, PSAO, PPLO, and BSAO and the cyan sphere represents the copper site. The hydrophobic residues are colored as green.

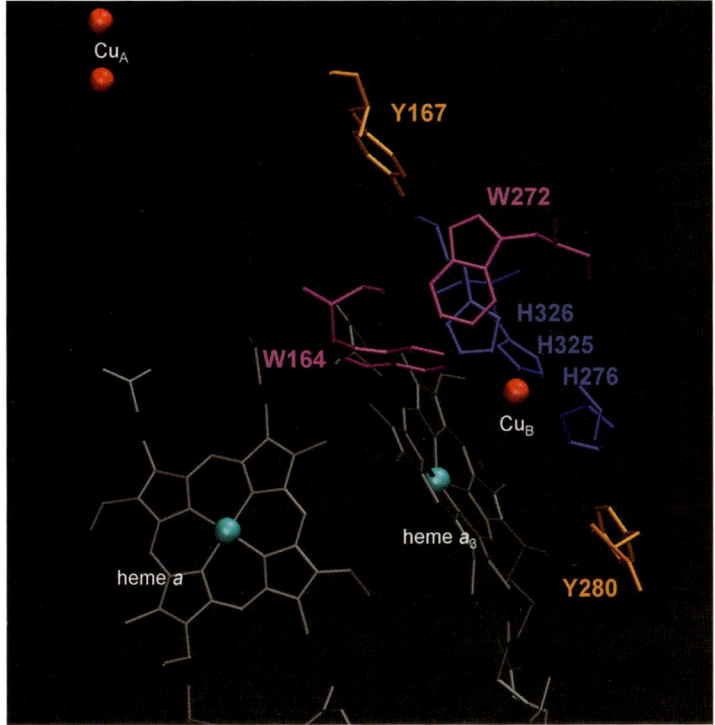

Figure 4.1 The active site of cytochrome *c* oxidase (C$_c$O) highlighting the conserved aromatic residues in Type A and B oxidases (with the exception of W164, which occurs only in Type A oxidases) and all four redox-active metal centers.

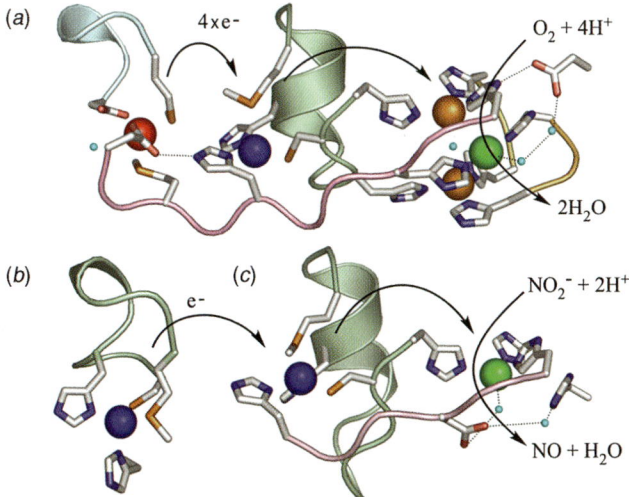

Figure 5.1 The copper-binding sites of CueO (*a*), psuedoazurin (*b*), and nitrite reductase (*c*). The T1, T2, T3, and the substrate Cu ions are colored blue, green, orange, and red, respectively.

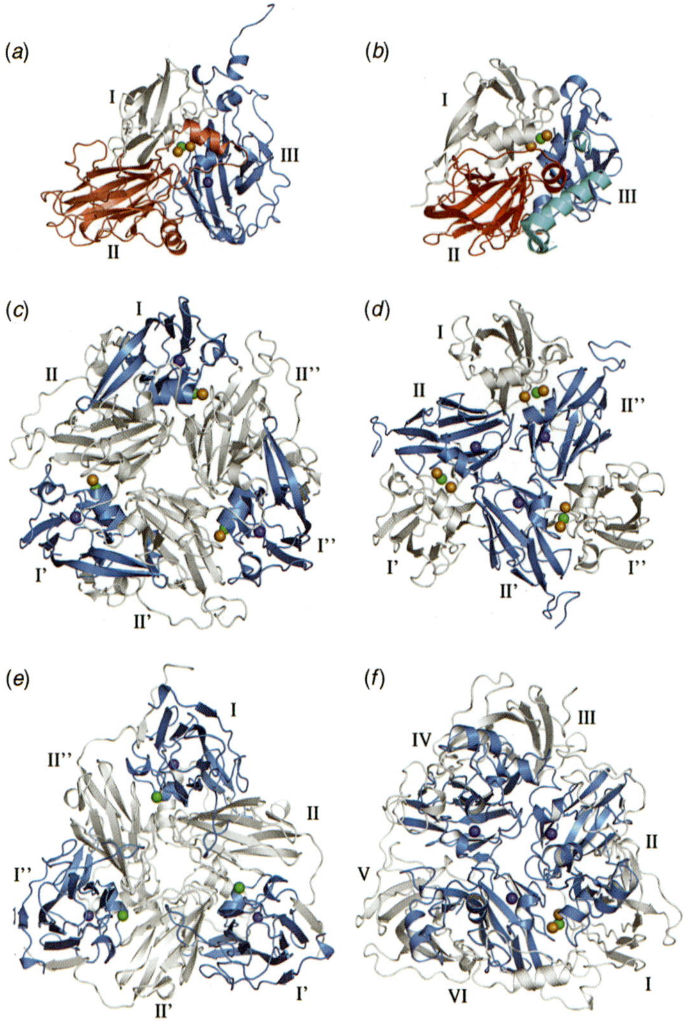

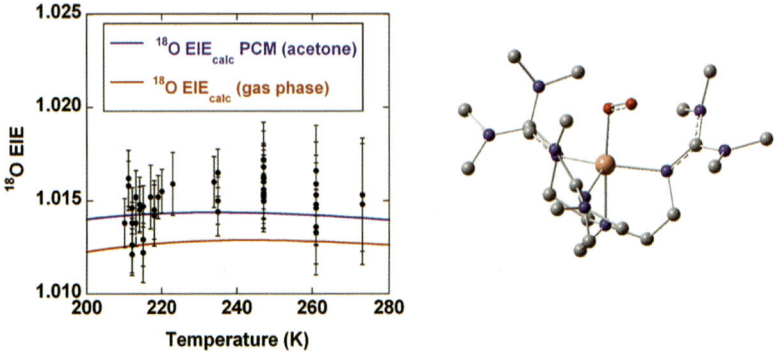

Figure 5.3 Crystal structures of two-, three-, and six-domain MCOs and NIR: (*a*) AOase, (*b*) CueO, (*c*) BCO, (*d*) SLAC, (*e*) AxNIR, and (*f*) hCp. (*See text for full caption.*).

Figure 6.5 The ^{18}O EIEs on reversible formation of $[Cu(\eta^1\text{-}O_2)(TMG_3Tren)]^+$. The polarized continuum model (PCM) was used to calculate the energy minimized structure.

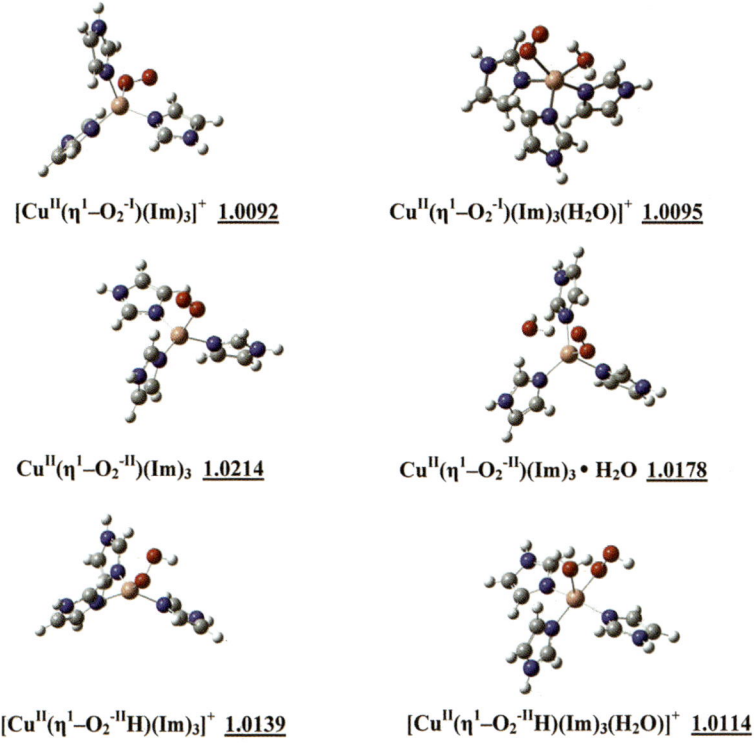

$[Cu^{II}(\eta^1-O_2^{-I})(Im)_3]^+$ __1.0092__

$Cu^{II}(\eta^1-O_2^{-I})(Im)_3(H_2O)]^+$ __1.0095__

$Cu^{II}(\eta^1-O_2^{-II})(Im)_3$ __1.0214__

$Cu^{II}(\eta^1-O_2^{-II})(Im)_3 \cdot H_2O$ __1.0178__

$[Cu^{II}(\eta^1-O_2^{-II}H)(Im)_3]^+$ __1.0139__

$[Cu^{II}(\eta^1-O_2^{-II}H)(Im)_3(H_2O)]^+$ __1.0114__

Figure 6.8 Proposed CuO_2 intermediates (Im = imidazole).

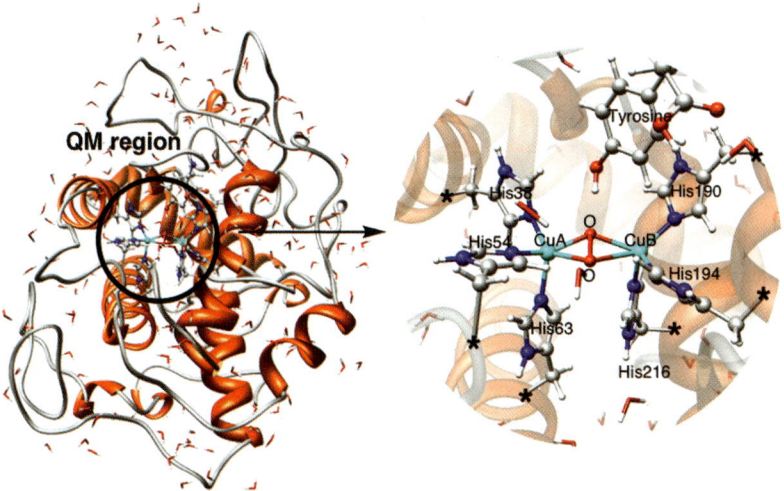

Figure 7.6 The QM/MM optimized structure of the oxy form of tyrosinase, in which the asterisks indicates the QM/MM border.

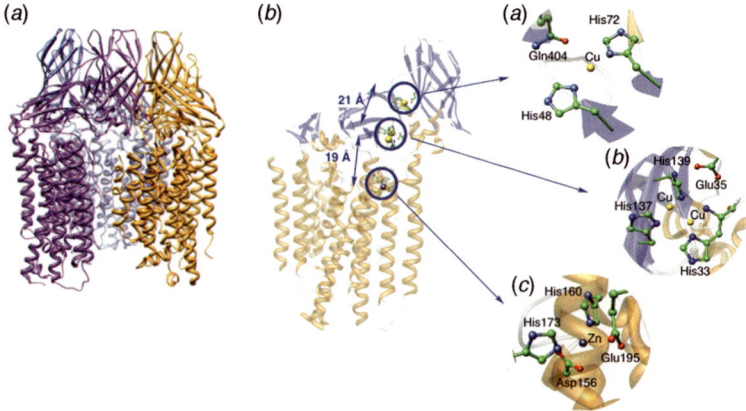

Figure 7.10 (*a*) X-ray crystal structure of the pMMO trimer. (*See text for full caption.*)

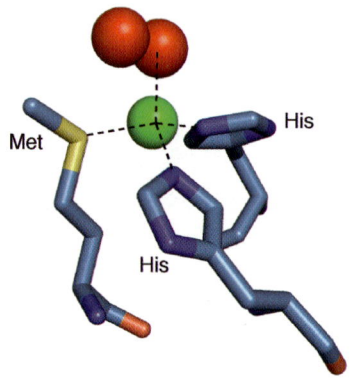

Figure 8.2 The active site structure of peptidylglycine α-hydroxylating monooxygenase.

$[(F_8)Fe^{III}-(O_2^{2-})-Cu^{II}(TMPA)]^+$ (**10** $Fe^{III}-(O_2^{2-})-Cu^{II}$)

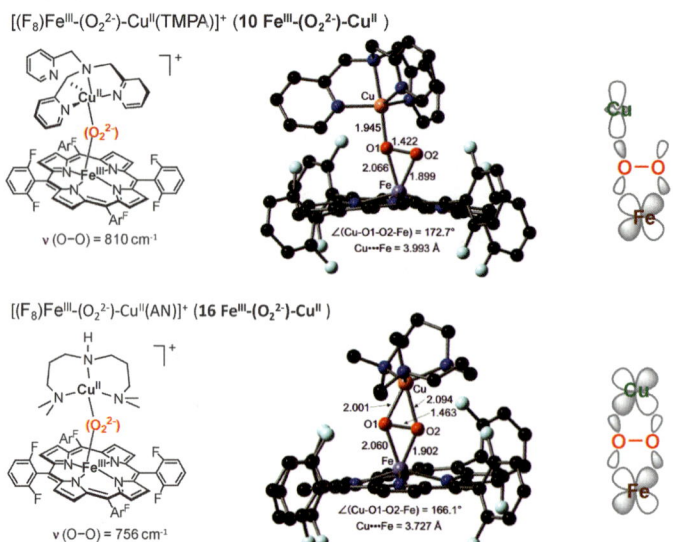

$[(F_8)Fe^{III}-(O_2^{2-})-Cu^{II}(AN)]^+$ (**16** $Fe^{III}-(O_2^{2-})-Cu^{II}$)

Figure 9.5 Unrestricted DFT (BS = broken symmetry, $S_T = 2$) optimized structures of (center) and schematic representation of the major orbital contributions in **10** $Fe^{III}-(O_2^{2-})-Cu^{II}$ and **16** $Fe^{III}-(O_2^{2-})-Cu^{II}$ (center). BS = broken.

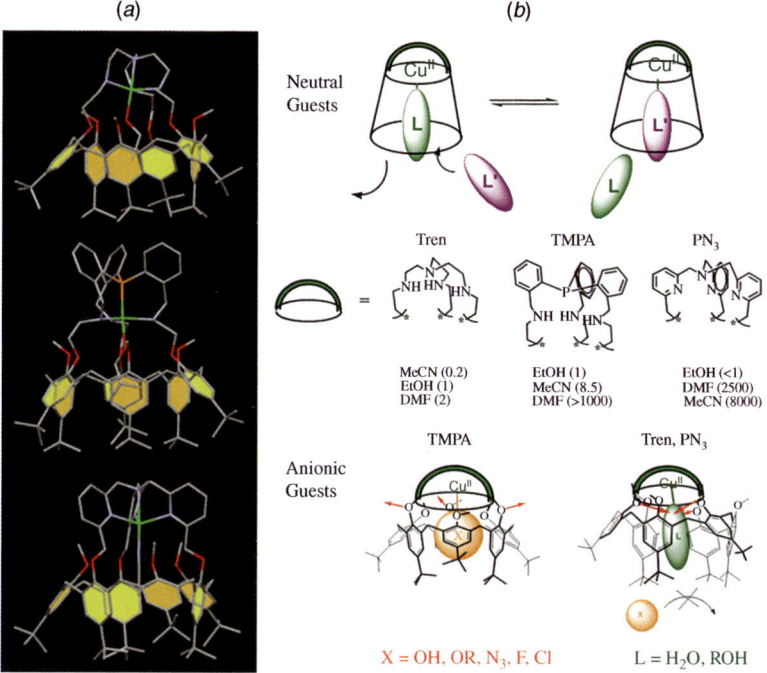

Figure 10.8 Supramolecular control of neutral and anionic guest binding. (*See text for full caption.*)

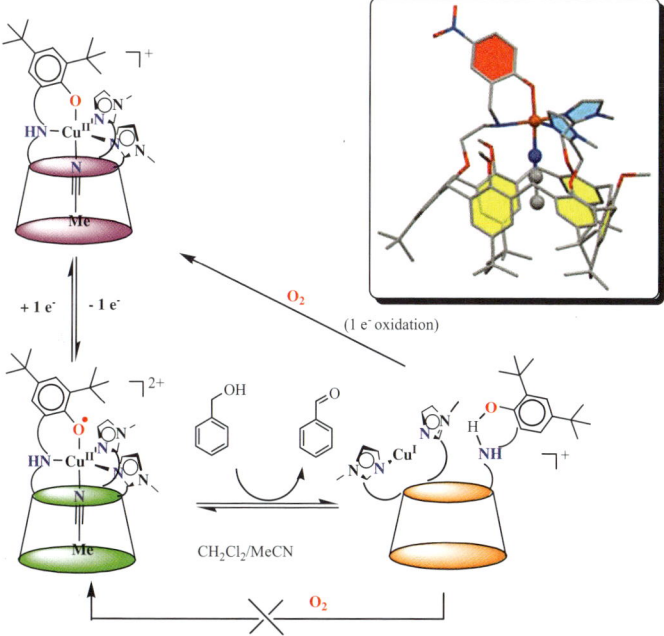

Figure 10.19 Redox bahavior and reactivity of Cu complexes based on second generation ligand, model of the active site of GO. *Inset*: The XRD structure of the *p*-NO₂ArOCu(II) complex.

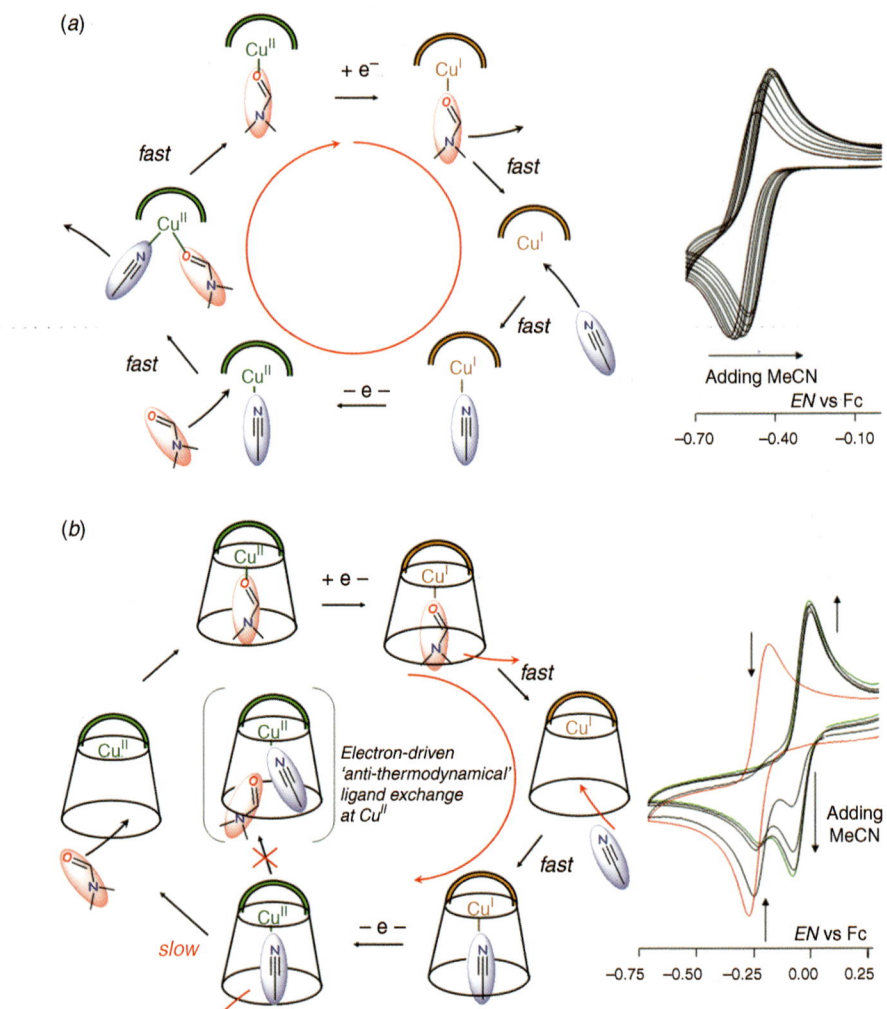

Figure 10.11 Electron exchange coupled to ligand exchange and associated CVs at a Pt electrode in CH$_2$Cl$_2$: Comparative behavior of a calix-based complex (*b*) and a "classical" complex (i.e., deprived of cavity, *a*). The cavity enforces a dissociative process, which slows down ligand exchange at the Cu(II) state, thus leading to a redox-driven ligand exchange (DMF for MeCN).

were determined for the Cu(II) complexes shown below and two other analogues containing a coordinated N_3^-.

Thus, the ^{18}O KIEs are quite insensitive to changes in reaction driving force. The $\Delta G°$ for $O_2^{•-}$ oxidation varies from -6 to -19 kcal/mol, as estimated from the Cu(II)/(I) redox potentials, which span ~ 0.7 V, and the $O_2/O_2^{•-}$ redox potential in DMSO.[63] The pre-equilibrium formation of the precursor complex is calculated to account for an additional electrostatic attraction in this solvent.[41]

In the Cu(II) + superoxide reactions examined,[41] the ^{18}O KIEs are invariant to reaction driving force, as well as changes in coordination geometry. Similar results are obtained for Cu(II) complexes with azide (N_3^-) or solvent (DMSO or DMF) occupying the fifth coordination site. These results suggest that the major changes in bonding occur in the pre-equilibrium step. A dominant contribution from the associated ^{18}O EIE explains how the ^{18}O KIEs on $O_2^{•-}$ oxidation can be uninfluenced by large changes in reaction driving force. Furthermore, the inverse ^{18}O KIE on $CuO_2(TMPA)$ formation is 0.984, which is consistent with the ^{18}O EIE calculated for O_2 coordination to the Cu(I) complex (see Table 6.2) and correction for the ^{18}O EIE on conversion of $O_2^{•-}$ to O_2. Electrostatic effects, together with the thermodynamically downhill nature of the reaction, results in unexpected reversibility of $O_2^{•-}$ binding and a transition state that resembles the structure of the putative CuO_2 intermediate rather than the product O_2 (Fig. 6.4).

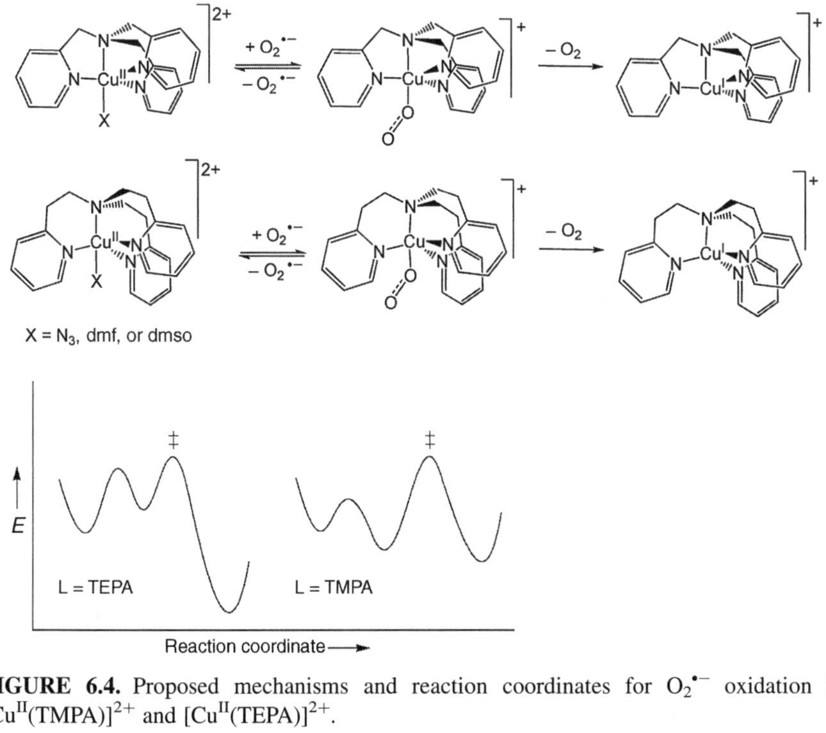

X = N₃, dmf, or dmso

L = TEPA

L = TMPA

Reaction coordinate ⟶

FIGURE 6.4. Proposed mechanisms and reaction coordinates for $O_2^{•-}$ oxidation by $[Cu^{II}(TMPA)]^{2+}$ and $[Cu^{II}(TEPA)]^{2+}$.

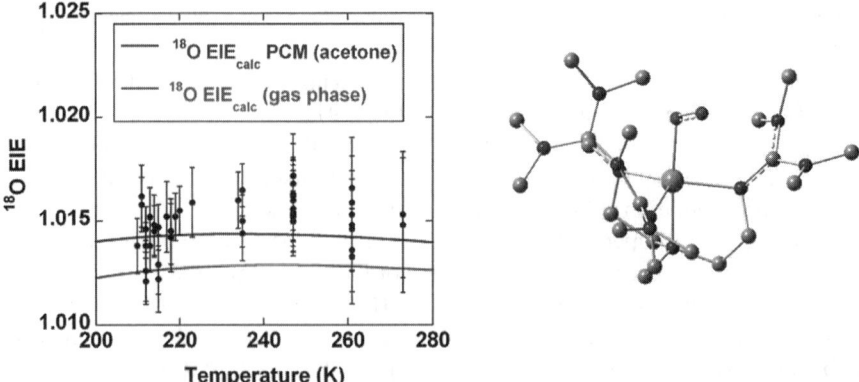

FIGURE 6.5. The ^{18}O EIEs on reversible formation of $[Cu(\eta^1\text{-}O_2)(TMG_3Tren)]^+$. The polarized continuum model (PCM) was used to calculate the energy minimized structure. (See color insert.)

The observation of $[CuO_2(TMPA)]^+$ as an intermediate by cryogenic stopped-flow spectrophotometry supports the proposed inner-sphere ET mechanism.[41] Upon reacting $[Cu^{II}(TMPA)]^{2+}$ with $O_2^{\bullet-}$ in a 4 : 1 DMF/THF mixture at 193 K, an optical spectrum appears that is virtually identical to the spectrum generated from the reaction of $Cu^I(TMPA)$ with O_2.[64] This spectrum resembles the crystallographically characterized end-on superoxide species, $[Cu(\eta^1\text{-}O_2)(TMG_3Tren)]^+$ where $TMG_3Tren = 1,1,1\text{-tris}\{2\text{-}[N^2\text{-}(1,1,3,3\text{-tetramethylguanidino})]ethyl\}amine$,[65] and is considerably different from spectra associated with η^2-superoxide and peroxide species.[42]

The $[Cu(\eta^1\text{-}O_2)(TMG_3Tren)]^+$ complex has been the subject of an independent isotopic study that characterized the reversible O_2 binding and formation of a paramagnetic species that adopts a triplet ground state. The latter was a first of a kind observation for a CuO_2 adduct.[40,66] The ^{18}O EIEs determined for O_2 binding between 213 and 273 K average 1.0148 ± 0.0012. These experimental data are in good agreement with the ^{18}O EIE_{calc} derived from vibrational frequencies of the energy minimized $[Cu(\eta^1\text{-}O_2)(TMG_3Tren)]^+$ structure with a triplet spin state (Fig. 6.5).[40] Good agreement is observed between the computed and experimental $O-O$ stretching frequencies for $[Cu(\eta^1\text{-}O_2)(TMG_3Tren)]^+$ and for the other CuO_2 adducts in Table 6.2.[40]

6.6. MECHANISMS OF COPPER ENZYMES

Since its inception, the competitive ^{18}O isotope fractionation technique has been applied to an increasing number of dioxygen-utilizing enzymes, catalysts that process reactive oxygen species, and various entities capable of oxidizing H_2O.[15,16,67,68] The first measurements actually date from 1959 when Feldman et al.[14] reported oxygen isotope fractionation by enzymes now known to have dicopper active sites (e.g., tyrosinase, catecholase, and cytochrome oxidase) capable of binding O_2. The ^{18}O KIEs were determined to fall within the range of 1.010–1.016. The ^{18}O $EIE_{exp} = 1.0184$ measured several years later for the structurally related oxyhemocyanin (cf. Fig. 6.1) reflects the proposed upper limit.[29] The observation that the magnitude of the ^{18}O KIE is limited by the associated ^{18}O EIE is consistent with a single-step O_2 coordination reaction, where the transition state adopts a structure intermediate of the reactant and product.[5–7]

In the following sections, ^{18}O KIEs on enzymatic reactions involving O_2 and $O_2^{\bullet-}$ are interpreted using ^{18}O EIEs as boundary conditions. Two disparate classes of copper proteins are described. Further discussions of ^{18}O KIEs pertaining to other copper oxidases and monooxygenases are described elsewhere.[3,4,21,54,55] Here it is noted simply that the copper monooxygenase reactions, which use exogenous reductants to break the $O-O$ bond, might not be interpretable using the assumption that the ^{18}O KIE is limited by the ^{18}O EIE, because of the isotopically sensitive nature of the reaction coordinate.[38,67]

6.6.1. Copper Amine Oxidases

The copper amine oxidases (CAOs) catalyze the dioxygen-dependent deamination of primary amines to aldehydes and the production of H_2O_2. The enzymes utilize a trihydroxyphenylalanine quinone (TPQ) cofactor derived from oxidation of a tyrosine residue by the nearby active site Cu(II).[69] Interestingly, the Cu(II) is proposed to be directly involved in the oxidation of tyrosine using O_2.[69–71]

The TPQ cofactor is reduced by primary amines to the two-electron ($2e^-$) reduced aminoquinol, which is subsequently reoxidized by O_2. The oxidative and reductive phases of catalysis are kinetically independent of one another. They are referred to as "half-reactions", where the cofactor is either oxidized or reduced. In the oxidative half-reaction, O_2 oxidizes the $2e^-$ reduced cofactor, TPQ_{red}, forming H_2O_2 and TPQ_{ox} as products. The mechanism of this reaction has been the subject of ongoing debate. There are two proposals that differ with respect to the role of the active-site copper ion. Dooely and co-workers[72,73] and Roth and co-workers[49] have proposed mechanisms wherein the $2e^-$ reduced protein undergoes rapid ET or PCET leading to the formation of a Cu(I) and TPQ semiquinone state: $Cu^{II}/TPQ_{red} \rightarrow Cu^{I}/TPQ_{sq}$ (Fig. 6.6). The O_2 reacts rapidly with the Cu(I) reversibly forming a CuO_2 intermediate that is subsequently reduced by a rate-limiting ET or PCET step. Alternatively, Klinman et al.[71] has maintained the proposal of an outer-sphere mechanism involving direct ET from the TPQ_{red} to O_2.

An important objective has been to characterize the reduced form of CAO that reacts with O_2. Complications arise because of the very rapid internal redox chemistry, which interconverts the Cu^{II}/TPQ_{red} and Cu^{I}/TPQ_{sq}. Several crystallographic studies have been performed on the fully oxidized CAOs, while fewer have examined the reduced and/or partially reduced forms of the proteins.[74] In the former, two water ligands are bound cis to one another at the active site Cu(II), which coordinates three histidines and adopts a distorted square-pyramidal geometry.

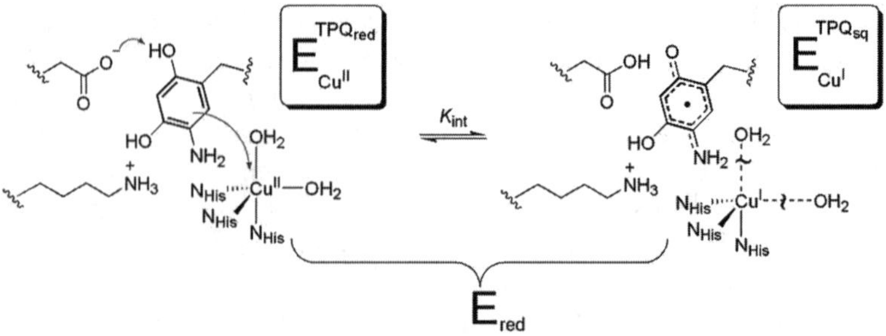

FIGURE 6.6. Rapid internal PCET in CAOs, where E = enzyme.

Reduction by dithionite under anaerobic conditions results in the release of water and adoption of a tri-coordinate geometry.[75] Reduction of the enzyme can also be achieved anaerobically using the primary amine substrate. This procedure is especially informative for CAOs that exist as equilibrium mixtures of Cu^{I}/TPQ_{sq} and Cu^{II}/TPQ_{red}.[48,73]

The Cu^{I}/TPQ_{sq} state is favored thermodynamically relative to the Cu^{II}/TPQ_{red} in pea seedling amine oxidase (PSAO),[49] and in lentil seedling amine oxidase (LSAO)[76,77] at physiological pH. The equilibrium constant is closer to unity for Arthrobacter globiformus (AGAO), a bacterial enzyme in which the proton-coupled redox potentials are apparently well matched.[78] For reasons not understood, the Cu^{II}/TPQ_{red} is dominant in bovine serum amine oxidase (BSAO) and to some extent in Hansenula polymorpha-1 (HPAO-1).[50,51,79] A second-sphere aspartate to asparagine mutant of HPAO-1 has been prepared and found to have both a greater percentage of Cu^{I}/TPQ_{sq} and greater amine oxidase activity.[50] In addition to second sphere interactions, the TPQ can rotate about its $C_{\alpha}–C_{\beta}$ bond in a manner that changes hydrogen-bonding patterns within the active site, affecting protonation states which tune the redox characteristics of the Cu and TPQ cofactors.

Upon removal of the metal cofactor and replacement of the active site, Cu(II) with Co(II), the TPQ_{sq} becomes difficult to detect.[79–81] Yet it is unclear what happens if the protein is reconstituted with Co(III) instead of Co(II) under anaerobic rather than aerobic conditions. Experimental methods for generating the reduced apoprotein differ and inconsistencies between laboratories appear to reflect the structural integrity of the reconstituted protein, as well as the use of anaerobic techniques. In accord with a recently proposed mechanism,[49,82] the kinetically competent cobalt-reconstituted amine oxidase may exist in the Co^{II}/TPQ_{sq} state formed upon oxidation of Co(II) to Co(III) in the enzyme active site.

The Co(II) reconstituted CAO may be oxidized by O_2 or by the H_2O_2 produced during enzyme turnover. Though reconstitution of HPAO-1 and AGAO is believed to give an active enzyme, the precise role of the cobalt ion during the oxidative half-reaction has yet to be explored. In one notable stopped-flow spectrophotometric study,[83] a rate constant for the single turnover reaction of reduced Co(II) reconstituted CAO with $O_2(k_{O_2})$ is on the order of $10^2–10^3$ $M^{-1}s^{-1}$ (i.e., at least 10^3 times smaller than the reaction of the native Cu(II) containing enzyme.[84] A commensurate decrease in the steady-state $k_{cat}/K_M(O_2)$ has also been observed upon cobalt reconstitution.[83,84] Interestingly, the diminution in $k_{cat}/K_M(O_2)$ observed in AGAO is larger than in HPAO-1 where $K_M(O_2)$ is proposed to approximate $K_d(O_2)$. This interpretation requires O_2 binding with significant affinity to a non-metal site on the protein.[71]

In each of the mechanisms proposed for CAO, the active-site metal has a different influence on O_2 reactivity. In PSAO and AGAO, the Cu^{II}/TPQ_{red} gives rise to the Cu^ITPQ_{sq} that reacts directly with O_2 to reversibly form a CuO_2 intermediate followed by reduction in the rate-limiting step. In the alternative proposal for HPAO, O_2 reacts directly with TPQ_{red} by outer-sphere ET.[85] Thermodynamic analysis has indicated that the outer-sphere mechanism requires an unfavorable deprotonation of TPQ_{red} prior to ET.[49]

The ^{18}O KIEs of ~ 1.010 reported for BSAO and HPAO are nearly three times smaller than values of ~ 1.028 expected for enzyme catalyzed outer-sphere ET.[58,61] It is easier to reconcile the small ^{18}O KIE with a mechanism involving CuO_2 intermediate formation. The extent to which this reaction may be reversible in the above-mentioned CAOs is not known. Yet, in PSAO and AGAO, there is compelling evidence that the $k_{cat}/K_M(O_2)$ and accompanying ^{18}O KIEs of 1.0136^{49} and 1.0175^{86} arise from rate-limiting reduction of a CuO_2 intermediate by ET or PCET (Fig. 6.7).

One reason for favoring rate-limiting ET (as shown below) over PCET is the very small solvent KIE, $^{D_2O}k_{cat}/K_M(O_2) = 1.2 \pm 0.1$, spanning the pH range from 6 to 9.5.[49] The ^{18}O KIEs and $k_{cat}/K_M(O_2)$ are both pH independent while the pre-equilibrium PCET designated in Figure 6.7 favors the Cu^I/TPQ_{sq} as the pH is lowered.[49,87] These results implicate rate-limiting reduction of the CuO_2 intermediate because changing the relative concentrations of Cu^{II}/TPQ_{red} and Cu^I/TPQ_{sq} has no effect on the rate of O_2 consumption.

In addition to analysis of steady-state intermediates, pH and temperature studies, DFT calculations have provided boundary conditions for the expected ^{18}O KIEs; 1.0092 is the lower limit and 1.0240 is the upper limit, in agreement with the experimental results (Fig. 6.8). Calculations were performed for the η^1-superoxide

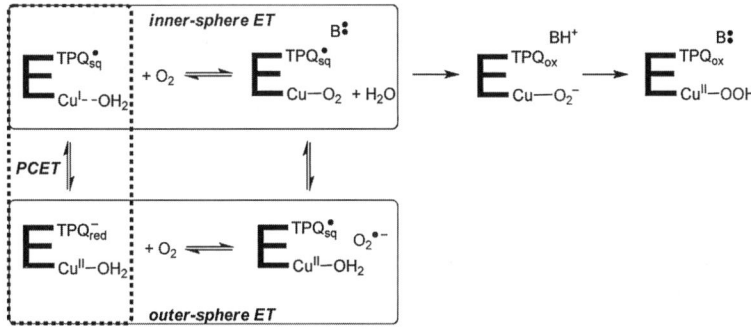

FIGURE 6.7. Proposed mechanism of CuO_2 formation and its rate-limiting reduction in COAs.

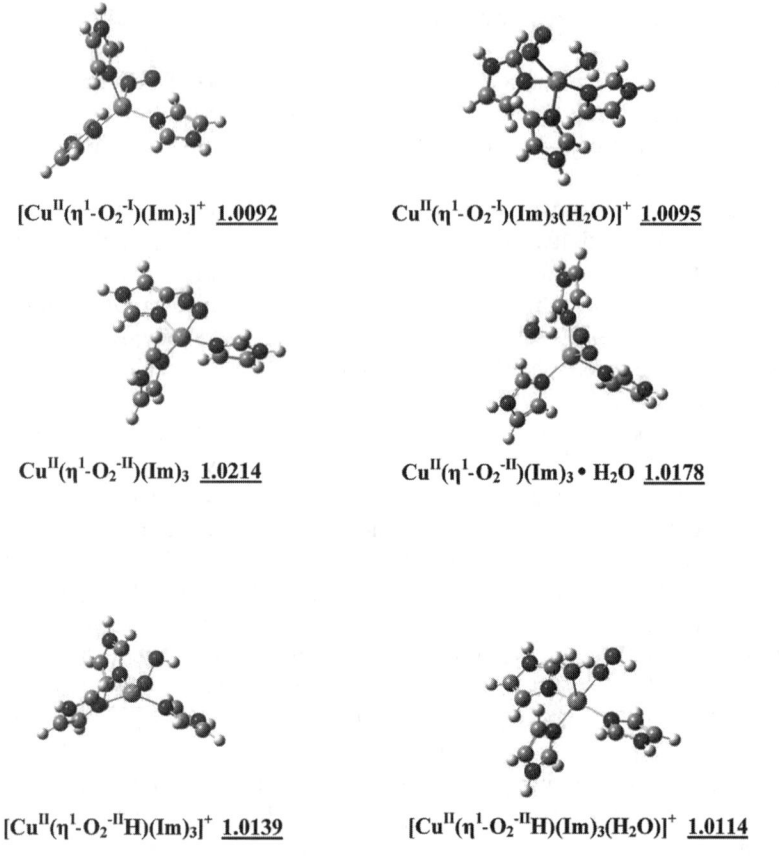

$[Cu^{II}(\eta^1\text{-}O_2^{-I})(Im)_3]^+$ __1.0092__ $Cu^{II}(\eta^1\text{-}O_2^{-I})(Im)_3(H_2O)]^+$ __1.0095__

$Cu^{II}(\eta^1\text{-}O_2^{-II})(Im)_3$ __1.0214__ $Cu^{II}(\eta^1\text{-}O_2^{-II})(Im)_3 \cdot H_2O$ __1.0178__

$[Cu^{II}(\eta^1\text{-}O_2^{-II}H)(Im)_3]^+$ __1.0139__ $[Cu^{II}(\eta^1\text{-}O_2^{-II}H)(Im)_3(H_2O)]^+$ __1.0114__

FIGURE 6.8. Proposed CuO_2 intermediates (Im = imidazole). (See Ref. 88.) (see color insert.)

species with a triplet ground state[40] and η^1-peroxide and hydroperoxide species with doublet ground states. Calculations on hypothetical η^2-peroxide complexes gave somewhat larger ^{18}O EIEs of 1.027–1.031.[5,49] Such species would seem unlikely to be involved in oxidase catalysis due to the tight binding of O_2.[2] The primary CuO_2 intermediate generated from Cu^I/TPQ_{sq} and O_2 is not expected to accumulate due to the unfavorable energetics of CuO_2 formation at ambient temperatures,[1,40] as well as the anticipated rapid reduction of the intermediate CuO_2 by the TPQ_{sq}.

6.6.2. Copper Zinc Superoxide Dismutase

Superoxide dismutases (SODs) found in nearly all forms of life, catalyze the disproportionation: $2O_2^{\bullet-} + 2H^+ \rightarrow O_2 + H_2O_2$ using a redox active copper, nickel, iron, or manganese ion.[89] The CuZnSOD and MnSOD are found in mammals where their main purpose appears to be deffending the cell against oxidative damage by the perhydroxyl radical ($HO_2^\bullet$). The enzymes may also catalyze the formation of H_2O_2 for cell signaling. The metal ion coordination geometry is believed to tune redox potentials and pK_a values in order to facilitate disproportonation at diffusion controlled rates.[90] The ET mechanisms, together with the factors that result in dramatic rate acceleration of the enzymatic reaction, relative to spontaneous disproportionation in aqueous solution, have not been understood. These fundamental aspects of enzyme activity and stability underlie the involvement of CuZnSOD in various neurodegenerative diseases.[91]

Catalysis by CuZnSOD involves cyclical oxidative and reductive "half-reactions" defined analogously to the amine oxidases. In the reductive half-reaction, the Cu(II) enzyme oxidizes $O_2^{\bullet-}$ to O_2, while in the oxidative half-reaction the Cu(I) enzyme reduces $O_2^{\bullet-}$ to H_2O_2.[92] Because $O_2^{\bullet-}$ is the substrate for both the oxidative and reductive reactions, it has been a challenge to resolve rate constants, intermediates, and the timing of electron- and proton-transfer steps. Isotopic fractionation studies have provided evidence of rate-limiting proton transfer (PT) in the oxidative half-reaction and rate-limiting inner-sphere ET in the reductive half-reaction (Fig. 6.9).

Studies of oxygen isotope fractionation using bovine CuZnSOD were undertaken to address the ET and PT mechanisms during SOD catalysis.[47] Experiments were conducted at pH 10 in either borate or carbonate buffer, where $O_2^{\bullet-}$ disproportionation is relatively pH independent; that is, the rate constant is only slightly diminished from that at physiological pH. Competitive experiments with ferricytochrome c demonstrated that all of the O_2 produced came from the CuZnSOD reaction rather than the spontaneous disproportionation, which can occur in the presence of trace metals. The SOD reaction proceeds rapidly to 100% conversion allowing the ratio of fractionation factors, $R(O_2)/R(H_2O_2)$ to be determined from the $^{18}O/^{16}O$ in the products of the oxidative and reductive half-reactions. The ratio of KIEs on the oxidative and reductive reactions was obtained directly and is referred throughout as β (Eq. 6.11). The β value, determined to be 1.0104 ± 0.0012, is due to the ratio of two inverse ^{18}O KIEs.

$$\beta = \frac{\text{KIE}(Cu^I + O_2^{\bullet-})}{\text{KIE}(Cu^{II} + O_2^{\bullet-})} = \frac{R(O_2)}{R(H_2O_2)} \quad (6.11)$$

The individual contributions to β were resolved by examining the reaction of $Cu^I ZnSOD$ with O_2.[47] The transition state of this reaction is related to that of the

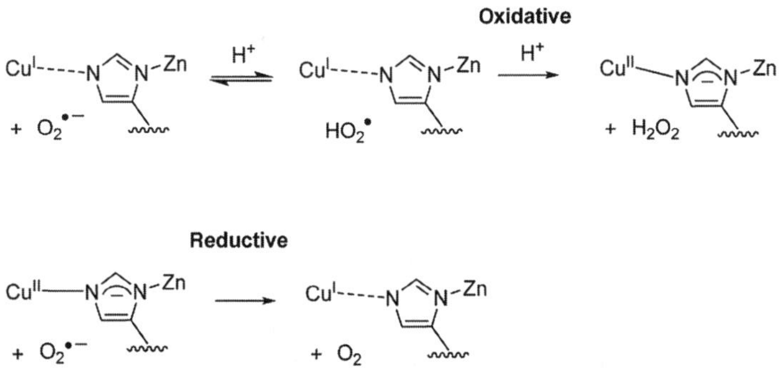

FIGURE 6.9. Oxidative and reductive half-reactions of CuZnSOD.

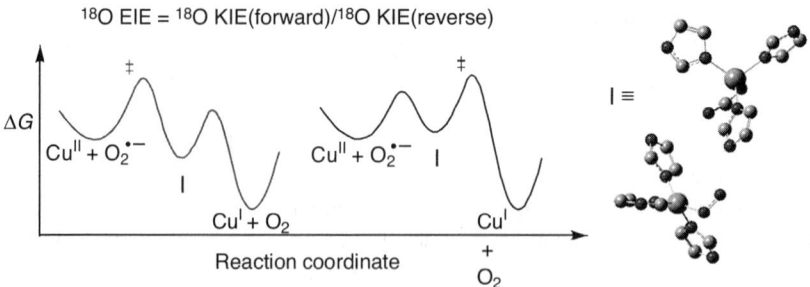

FIGURE 6.10. Proposed mechanism of the CuZnSOD involving a CuO$_2$ complex.

reductive half-reaction between Cu^{II}ZnSOD and O$_2^{\bullet-}$ by the principle of microscopic reversibility (Fig. 6.10).[47] The stoichiometric oxidation of the prereduced enzyme at pH 10 indicated an ^{18}O KIE $= 1.0044 \pm 0.0016$. On the basis of previous observations for outer-sphere ET to O$_2^{\bullet-}$ the observed ^{18}O KIE would be expected to be normal and much larger in magnitude.[6,7,58,61] (see Refs. 3 and 40).

The observed ^{18}O KIE for the Cu^{I}ZnSOD reduction of O$_2$ is more consistent with inner-sphere ET involving a CuO$_2$ intermediate.[93] Since microscopic reversibility stipulates that the transition state is the same in the forward and reverse directions, the ^{18}O KIE on the reaction, where O$_2^{\bullet-}$ is oxidized by Cu^{II}(SOD) can be calculated.[3,47] Taking the observed ^{18}O KIE for O$_2$ reduction (1.0044) and dividing it by the ^{18}O EIE for O$_2 + e^- \rightarrow$ O$_2^{\bullet-}$ (1.033) gives the ^{18}O KIE $= 0.972$ for O$_2^{\bullet-}$ oxidation by the enzyme. This value is less inverse than the ^{18}O KIEs of 0.984–0.989 determined for O$_2^{\bullet-}$ oxidation by synthetic Cu(II) model complexes.[41] Though the mechanisms are believed to be similar, the more inverse magnitude of the ^{18}O KIE for the enzymatic reaction suggests a transition state with greater Cu^I–O$_2^0$ character. In spite of the very rapid rates exhibited by the enzyme, the free energy of the reaction (ΔG°), dictated by the aqueous redox potentials of the copper ion and O$_2$, is estimated to be less favorable than that of the cationic Cu(II) tris-2-pyridyl amine complexes in DMSO, pointing to electrostatic barrier-lowering effects.

The analysis above allows β to be calculated the ^{18}O KIE $= 0.982$ on the oxidative half-reaction (i.e., the reduction of O$_2^{\bullet-}$ by Cu^{I}ZnSOD).[47] The significantly inverse ^{18}O KIE is inconsistent with reduction to a peroxide species, where the O–O bond would be weakened relative to O$_2^{\bullet-}$ resulting in a normal ^{18}O KIE on the formation of a Cu^{II}–O$_2^{-II}$ species or HO$_2^-$. The inverse ^{18}O KIE is more consistent with rate-determining proton transfer to form HO$_2^\bullet$, where the ^{18}O EIE$_{calc} = 0.978$.[47] Due to its positive redox potential, HO$_2^\bullet$ is a viable intermediate and should easily reoxidize Cu^{I}ZnSOD as proposed in Figure 6.9.

Considering the estimated rate constant for the O$_2^{\bullet-}$ reduction, as well as the pK_a of HO$_2^\bullet$, the proposed protonation step is likely to be facilitated by preassociation of O$_2^{\bullet-}$ with a positively charged residue in the enzyme active site. The conserved arginine previously associated with electrostatic catalysis is the likely candidate.[94,95] It is also worth noting that the reverse reaction of H$_2$O$_2$ with Cu^{II}ZnSOD results in the accumulation of $HO_2^{\bullet}$; this reactive oxygen species would be expected to damage the enzyme active site, possibly explaining the oxidative protein modification,[96] which has been implicated in loss of the active-site metal(s) and the formation of misfolded protein aggregates associated with neurodegenerative states.[97]

6.7. CONCLUSIONS

Competitive ^{18}O kinetic and equilibrium isotope effects have been used to characterize reactions of Cu(I) and (II) species with natural abundance molecular oxygen, and superoxide. The oxygen isotope effects can be assimilated with other experimental results including vibrational spectroscopy, and crystallographic analysis. The use of oxygen isotope effects as diagnostics of structure and mechanism additionally provide calibration for electronic structure calculations at the density functional level of theory. Most importantly, the technique offers a unique approach to comparing reactivity of small molecules to that which occurs in enzymes.

Competitive isotope fractionation measurements have been performed on numerous reactions involving synthetic copper complexes and copper-containing proteins. Evidence for CuO_2 intermediates has been obtained in many of these cases due to the magnitudes of the ^{18}O KIEs. The magnitudes of these KIEs may be related to ^{18}O EIEs which are readily calculated for CuO_2 intermediates that precede rate-limiting transformations.

It is generally found that the structure of CuO_2 intermediates can be associated with the magnitude of the isotope effects. The reversible formation of an end-on bound $Cu^{II}(\eta^1-O_2^{-1})$ intermediate is generally characterized by a smaller isotope effect (<1.015) than the $2e^-$ reduced peroxide or side-on bound analogue (>1.020). Increase in extent of ET, in the absence of proton transfer, increases the size of the isotope effect; whereas comparable ^{18}O EIEs are observed for the formation of η^1-superoxide and η^1-hydroperoxide species.

Oxygen isotope fractionation provides a means to understand the mechanistic involvement of CuO_2 species during catalysis. One point that has become clear in the progress of these studies is that ^{18}O KIEs that are significantly smaller ($\sim 2\times$) than the relevant ^{18}O EIEs signal irreversible O_2 coordination. In contrast, ^{18}O KIEs that exceed the related equilibrium values computed for CuO_2 structures reveal rate limitation by a downstream step, such as ET or PCET. For these reasons, experimental measurements combined with DFT calculations of minimum energy structures and the accompanying vibrational frequencies are indispensable to mechanistic analyses.

Good agreement between experimental and computational ^{18}O isotope effects on the reversible formation of ground-state (e.g., CuO_2) structures has been obtained. Efforts are now focused on calculating transition states for O_2 activation. Progress in this area will provide new means for calibrating calculated electronic and vibrational structures and improving applications of DFT above and beyond its current level.

ACKNOWLEDGMENTS

Support from the National Science Foundation CAREER award CHE-0449900, Department of Energy grant DE-FG02-09ER16094, Petroleum Research Fund grant 50046-ND3, an Alfred P. Sloan Fellowship and a Camille and Henry Dreyfus Foundation Teacher-Scholar Award is gratefully acknowledged.

REFERENCES

1. Karlin, K. D.; Kaderli, S.; Zueberbuehler, A. D. Kinetics and thermodynamics of copper (I)/dioxygen interaction. *Acc. Chem. Res.* **1997**, *30*, 139–147.

2. Cramer, C. J.; Tolman, W. B. Mononuclear Cu-O_2 complexes: geometries, spectroscopic properties, electronic structures, and reactivity. *Acc. Chem. Res.* **2007**, *40*, 601–608.

3. Roth, J. P. Advances in studying bioinorganic reaction mechanisms: isotopic probes of activated oxygen intermediates in metalloenzymes. *Curr. Opin. Chem. Biol.* **2007**, *11*, 142–150.

4. Roth, J. P.; Klinman, J. P. Oxygen-18 isotope effects as a probe of enzymatic activation of molecular oxygen. *Isotope Effects in Chemistry and Biology* Kohen, A.; Limbach H-H., Eds., CRC Press, Boca Raton, FL, 2006, pp 645–669.

5. Roth, J. P. Heavy atom isotope effects as probes of small molecule activation. *Physical Inorganic Chemistry: Methods* Bakac, A., Ed., John Wiley & Sons, Inc. New York, 2010, pp 425–457.

6. Ashley, D. C.; Brinkley, D. W.; Roth, J. P. Oxygen isotope effects as structural and mechanistic probes in inorganic oxidation chemistry. *Inorg. Chem.* **2010**, special forum on "Oxygen Reduction and Activation", *49*, 2661–3675.

7. Roth, J. P. Oxygen isotope effects as probes of electron transfer mechanisms and structures of activated O_2. *Acc. Chem. Res.* **2009**, *42*, 399–408.

8. Szilagyi, R. K.; Solomon, E. I. Electronic structure and its relation to function in copper proteins. *Curr. Op. Chem. Biol.* **2002**, *6*, 250–258.

9. Whittaker, J. W.; Whittaker, M. M. Radical copper oxidases, one electron at a time. *Pure App. Chem.* **1998**, *70*, 903–910.

10. Gamez, P.; Aubel, P. G.; Driessen, W. L.; Reedijk, J. Homogeneous bio-inspired copper-catalyzed oxidation reactions. *Chem. Soc. Rev.* **2001**, *30*, 376–385.

11. Que, L., Jr; Tolman, W. B. Biologically inspired oxidation catalysis. *Nature (London)* **2008**, *455*, 333–340.

12. Bigeleisen, J.; Hom, R. C.; Ishida, T. Isotope chemistry and molecular structure. Carbon and oxygen isotope chemistry. *J. Chem. Phys.* **1976**, *64*, 3303–3310.

13. Richet, P.; Bottinga, Y.; Javoy, M. A review of hydrogen, carbon, nitrogen, oxygen, sulfur, and chlorine stable isotope fractionation among gaseous molecules. *Ann. Rev. Earth Plan. Sci.* **1977**, *5*, 65–110.

14. Feldman, D. E.; Yost, H. T. Jr.; Benson, B. E. Oxygen isotope fractionation in reactions catalyzed by enzymes. *Science* **1959**, *129*, 146–147.

15. Dole, M. R., DeForest, P.; Muchow, G. R.; Comte, C. Isotopic composition of oxygen in the catalytic decomposition of hydrogen peroxide. *J. Chem. Phys.* **1952**, *20*, 961–968.

16. Guy, R. D.; Fogel, M. F.; Berry, J. A.; Hoering, T.C. Isotope fractionation during oxygen production and consumption by plants. *Prog. Photosynth. Res.* **1987**, *3*, 597–600.

17. Kreckl, W.; Kexel, H.; Melzer, E.; Schmidt, H. L. Oxygen isotope effects on the ribulosebisphosphate oxygenase reaction. *J. Biol. Chem.* **1989**, *264*, 10982–10986.

18. Cahill, A. E.; Taube, H. The use of heavy oxygen in the study of reactions of hydrogen peroxide. *J. Am. Chem. Soc.* **1952**, *74*, 2312–2318.

19. Knoller K.; Vogt, C.; Richnow, H-H.; Weise, S. M. Sulfur and oxygen isotope fractionation during benzene, toluene, ethyl benzene, and xylene degradation by sulfate-reducing bacteria. *Env. Sci Tech.* **2006**, *40*, 3879–3885.

20. Sturchio, N. C.; Boehlke, J. K.; Beloso, A. D., Jr.; Streger, S. H.; Heraty, L. J.; Hatzinger, P. B. Oxygen and chlorine isotopic fractionation during perchlorate biodegradation: Laboratory results and implications for forensics and natural attenuation studies. *Env. Sci Tech.* **2007**, *41*, 2796–2802.

21. Klinman, J. P.; Berry, J. A.; Tian, G. New probes of oxygen binding and activation: Application to dopamine β-monooxygenase in *Bioinorganic Chemistry of Copper* Karlin, K. D.; Z. Tyeklar, Eds., Chapman & Hall: New York, **1993**: pp 151–63.

22. McKinney, C. R.; McCrea, J. M.; Epstein, S.; Allen, H. A.; Urey, H. C. Improvements in mass spectrometers for the measurement of small differences in isotope abundance ratios. *Rev. Sci. Instrum.* **1950**, *21*, 724–730.

23. Smirnov, V. V.; Brinkley, D. W.; Lanci, M. P.; Karlin, K. D.; Roth, J. P. Probing metal-mediated O2 activation in chemical and biological systems. *J. Mol. Cat. A.* **2006**, *251*, 100–107.

24. Bigeleisen, J.; Goeppert-Mayer, M. Calculation of equilibrium constants for isotopic exchange reactions. *J. Chem. Phys.* **1947**, *15*, 261–267.

25. Smirnov, V. V.; Lanci, M. P.; Roth, J. P. Computational modeling of oxygen isotope effects on metal mediated O2 activation at varying temperatures. *J. Phys. Chem. A* **2009**, *113*, 1934–1945.

26. Slaughter, L. M.; Wolczanski, P. T.; Klinckman, T. R.; Cundari, T. R. Inter- and intramolecular experimental and calculated equilibrium isotope effects for (silox)$_2$(^{t}Bu$_3$SiND)TiR + RH (silox = ^{t}Bu$_3$SiO): Inferred kinetic isotope effects for RH/D addition to transient (silox)$_2$Ti:NSi^{t}Bu$_3$. *J. Am. Chem. Soc.* **2000**, *122*, 7953–7975.

27. Huskey, P. W. Origins and interpretations of heavy atom isotope effects. *Enzyme Mechanism from Isotope Effects* Cook, P. F., Ed., CRC Press, Boca Raton, FL, 1991, pp 37–72.

28. Tian, G.; Klinman, J. P. Discrimination between ^{16}O and ^{18}O in oxygen binding to the reversible oxygen carriers hemoglobin, myoglobin, hemerythrin, and hemocyanin: A new probe for oxygen binding and reductive activation by proteins. *J. Am. Chem. Soc.* **1993**, *115*, 8891–8897.

29. Wolfsberg, M. Hook, W. A.; Paneth, P.; Rebelo, L. P. N. *Isotope Effects in the Chemical, Geological and Biosciences* Springer, New York, 2010, pp 71–75.

30. Ling, J.; Nestor, L.P.; Czernuszewicz, R. S.; Spiro, T. G.; Fraczkiewicz, R.; Sharma, K. D.; Loehr, T. M.; Sanders-Loehr, J. Common oxygen binding site in hemocyanins from arthropods and mollusks. Evidence from Raman spectroscopy and normal coordinate analysis. *J. Am. Chem. Soc.* **1994**, *116*, 7682–7691.

31. Henson, M. J.; Mahadevan, V.; Stack, T. D. P.; Solomon, E. I. A new Cu(II) side-on peroxo model clarifies the assignment of the oxyhemocyanin Raman spectrum. *Inorg. Chem.* **2001**, *40*, 5068–5069.

32. Cramer, C. J.; Wloch, M.; Piecuch, P.; Puzzarini, C.; Gagliardi, L. Theoretical models on the Cu_2O_2 torture track: Mechanistic implications for oxytyrosinase and small-molecule analogues. *J. Phys. Chem. A* **2006**, *110*, 1991–2004.

33. Op't Holt, B.T.; Dustman, J.; Solomon, E. I. Structure/function relationships in binuclear copper(II) dioxygen-binding proteins. *Chemtracts* **2006**, *19*, 435–444.

34. Lanci, M. P.; Roth, J. P. Oxygen isotope effects upon reversible O_2-binding reactions: Characterizing mononuclear superoxide and peroxide structures. *J. Am. Chem. Soc.* **2006**, *128*, 16006–16007.

35. Adamo, C.; Barone, V. Exchange functionals with improved long-range behavior and adiabatic connection methods without adjustable parameters: the mPW and mPW1PW models. *J. Chem. Phys.* **1998**, *108*, 664–675.

36. Stevens, W. J.; Krauss, M.; Basch, H.; Jasien, P. G. Relativistic compact effective potentials and efficient, shared-exponent basis sets for the third-, fourth-, and fifth-row atoms. *Can. J. Chem.* **1992**, *70*, 612–630.

37. Hehre, W. J.; Radom, L.; Schleyer, P. V. R.; Pople, J. A. *Ab Initio Molecular Orbital Theory*; John Wiley & Sons, Inc., New York, 1986.

38. Halfen, J. A.; Mahapatra, S.; Wilkinson, E. C.; Kaderli, S.; Young, V. G., Jr.; Que, L., Jr.; Zuberbuehler, A. D.; Tolman, W. B. Reversible cleavage and formation of the dioxygen O-O bond within a dicopper complex. *Science* **1996**, *271*, 1397–400.

39. Molecular geometries were optimized using mPWPW91 in Gaussian03: Frisch, M. J.; et al. Gaussian 03W, Revision C.02; Gaussian, Inc.: Pittsburgh, PA, 2003.

40. Lanci, M. P.; Smirnov, V. V.; Cramer, C. J.; Gauchenova, E. V.; Sundermeyer, J.; Roth, J. P. Isotopic probing of molecular oxygen activation at copper(I) sites. *J. Am. Chem. Soc.* **2007**, *129*, 14697–14709.

41. Smirnov, V. V.; Roth, J. P. Evidence for Cu-O_2 intermediates in superoxide oxidations by biomimetic copper(II) complexes. *J. Am. Chem. Soc.* **2006**, *128*, 3683–3695.

42. Sarangi, R.; Aboelella, N.; Fujisawa, K.; Tolman, W. B.; Hedman, B.; Hodgson, K. O.; Solomon, E. I. X-ray absorption edge spectroscopy and computational studies on $LCuO_2$ species: Superoxide-Cu^{II} versus peroxide-Cu^{III} bonding. *J. Am. Chem. Soc.* **2006**, *128*, 8286–8296.

43. Burger, R. M.; Tian, G.; Drlica, K. Oxygen isotope effect on activated bleomycin stability. *J. Am. Chem. Soc.* **1995**, *117*, 1167–1168.

44. Wolfsberg, M. Theoretical evaluation of experimentally observed isotope effects. *Acc. Chem. Res.* **1972**, *5*, 225–233.

45. Aboelella, N. W.; Kryatov, S. V.; Gherman, B. F.; Brennessel, W. W.; Young, V. G., Jr.; Sarangi, R.; Rybak-Akimova, E. V.; Hodgson, K. O.; Hedman, B.; Solomon, E. I.; Cramer, C. J.; Tolman, W. B. Dioxygen activation at a single copper site: Structure, bonding, and mechanism of formation of 1:1 $Cu-O_2$ adducts. *J. Am. Chem. Soc* **2004**, *126*, 16896–16911.

46. Lanci, M. P. Ph.D. Dissertation Johns Hopkins University, 2008.

47. Smirnov, V. V.; Roth, J. P. Mechanisms of electron transfer in catalysis by copper zinc superoxide dismutase. *J. Am. Chem. Soc.* **2006**, *128*, 16424–16425.

48. Klinman, J. P. Mechanisms whereby mononuclear copper proteins functionalize organic substrates. *Chem. Rev.* **1996**, *96*, 2541–2561.

49. Mukherjee, A.; Smirnov, V. V.; Lanci, M. P.; Brown, D. E.; Shepard, E. M.; Dooley, D. M.; Roth, J. P. Inner-sphere mechanism for molecular oxygen reduction catalyzed by copper amine oxidases. *J. Am. Chem. Soc.* **2008**, *130*, 9459–9473.

50. Welford, R. W. D.; Lam, A.; Mirica, L. M.; Klinman, J. P. Partial conversion of Hansenula polymorpha amine oxidase into a "plant" amine oxidase: Implications for copper chemistry and mechanism. *Biochemistry* **2007**, *46*, 10817–10827.

51. Su, Q.; Klinman, J. P. Probing the mechanism of proton coupled electron transfer to dioxygen: The oxidative half-reaction of bovine serum amine oxidase. *Biochemistry* **1998**, *37*, 12513–12525.

52. Humphreys, K. J.; Mirica, L. M.; Wang, Y.; Klinman, J. P. Galactose oxidase as a model for reactivity at a copper superoxide center. *J. Am. Chem. Soc.* **2009**, *131*, 4657–4663.

53. Roth, J. P.; Dooley, D. M. et al., in preparation.

54. Tian, G.; Berry, J. A.; Klinman, J. P. Oxygen-18 kinetic isotope effects in the dopamine β-monooxygenase reaction: Evidence for a new chemical mechanism in non-heme, metallomonooxygenase. *Biochemistry* **1994**, *33*, 226–234.

55. Klinman, J. P. The copper-enzyme family of dopamine β-monooxygenase and peptidylglycine α-hydroxylating monooxygenase: Resolving the chemical pathway for substrate hydroxylation. *J. Biol. Chem.* **2006**, *281*, 3013–3016.

56. Landis, C. R.; Morales, C. M.; Stahl, S. S. Insights into the spin-forbidden reaction between L_2Pd^0 and molecular oxygen. *J. Am. Chem. Soc.* **2004**, *126*, 16302–16303.

57. Popp, B. V.; Wendlandt, J. E.; Landis, C. R.; Stahl, S. S. Reaction of molecular oxygen with an NHC-coordinated Pd^0 complex: computational insights and experimental implications. *Angew. Chem., Int. Ed.* **2007**, *46*, 601–604.

58. Roth, J. P.; Wincek, R.; Nodet, G.; Edmondson, D. E.; McIntire, W. S.; Klinman, J. P. Oxygen isotope effects on electron transfer to O_2 probed using chemically modified flavins bound to glucose oxidase. *J. Am. Chem. Soc.* **2004**, *126*, 15120–15131.

59. Buhks, E.; Bixon, M.; Jortner, J.; Navon, G. Quantum effects on the rates of electron-transfer reactions. *J. Phys. Chem.* **1981**, *85*, 3759–3762.

60. Buhks, E.; Bixon, M.; Jortner, J. Deuterium isotope effects on outer-sphere electron-transfer reactions. *J. Phys. Chem.* **1981**, *85*, 3763–3766.

61. Roth, J. P.; Klinman, J. P. Catalysis of electron transfer during activation of O2 by the flavoprotein glucose oxidase. *Proc. Natl. Acad. Sci. USA* **2003**, *100*, 62–67.

62. Lanci, M. P.; Brinkley, D. W.; Stone, K. L.; Smirnov, V. V.; Roth, J. P. Structures of transition states in metal-mediated O_2-activation reactions. *Angew. Chem., Int. Ed.* **2005**, *44*, 7273–7276.

63. Sawyer, D. T.; Sobkowiak, A.; Roberts, J. L. Jr., *Electrochemistry for Chemists*, John Wiley & Sons, Inc., New York: 1995, pp 370–375.

64. Zhang C. X.; Kaderli S.; Costas M.; Kim E-I.; Neuhold Y-M.; Karlin K. D.; Zuberbuhler, A. D. Copper(I)-dioxygen reactivity of [(L)Cu(I)](+) (L = tris(2-pyridylmethyl)amine): kinetic/thermodynamic and spectroscopic studies concerning the formation of Cu-O_2 and Cu_2-O_2 adducts as a function of solvent medium and 4-pyridyl ligand substituent variations. *Inorg. Chem.* **2003**, *42*, 1807–1824.

65. Schatz, M.; Raab, V.; Foxon, S. P.; Brehm, G.; Schneider, S.; Reiher, M.; Holthausen, M. C.; Sundermeyer, J.; Schindler, S. Dioxygen complexes: Combined spectroscopic and theoretical evidence for a persistent end-on copper superoxo complex. *Angew. Chem., Int. Ed.* **2004**, *43*, 4360–4363.

66. Pantazis, D. A.; McGrady, J. E. On the nature of the bonding in 1:1 adducts of O_2. *Inorg. Chem.* **2003**, *42*, 7734–7736.

67. Roth, J. P.; Cramer, C. J. Direct examination of H_2O_2 activation by a heme peroxidase. *J. Am. Chem. Soc.* **2008**, *130*, 7802–7803.

68. Guy, R. D.; Fogel, M. L.; Berry, J. A. Photosynthetic fractionation of the stable isotopes of oxygen and carbon. *Plant Phys.* **1993**, *101*, 37–47.

69. Klinman, J. P. How many ways to craft a cofactor? *Proc. Natl. Acad. Sci. USA* **2001**, *98*, 14766–14768.

70. Schwartz, B.; Dove, J. E.; Klinman, J. P. Kinetic analysis of oxygen utilization during cofactor biogenesis in a copper-containing amine oxidase from yeast. *Biochemistry* **2000**, *39*, 3699–3707.

71. Klinman J. P. The multi-functional topa-quinone copper amine oxidases. *Biochim Biophys Acta* **2003**, *1647*, 131–137.

72. Dooley D. M., McGuirl, M. A.; Brown, D. E.; Turowski, P. N.; McIntire, W. S.; Knowles, P. F. A Cu(I)-semiquinone state in substrate-reduced amine oxidase. *Nature (London)* **1991**, *349*, 262–264.

73. Turowski, P. N.; McGuirl, M. A.; Dooley, D. M. Intramolecular electron transfer rate between active-site copper and topa quinone in pea seedling amine oxidase. *J. Biol. Chem.* **1993**, *268*, 17680–17682.

74. Wilmot, C. M.; Hajdu, J.; McPherson, M. J.; Knowles, P. F.; Phillips, S. E. V. Visualization of dioxygen bound to copper during enzyme catalysis. *Science* **1999**, *286*, 1724–1728.

75. Dooley, D. M.; Scott, R. A.; Knowles, P. F.; Colangelo, C. M.; McGuirl, M. A.; Brown, D. E. Structures of the Cu(I) and Cu(II) forms of amine oxidases from X-ray absorption spectroscopy. *J. Am. Chem. Soc.* **1998**, *120*, 2599–2605.

76. Medda, R.; Padiglia, A.; Bellelli, A.; Sarti, P.; Santanche, S.; Agro, A. F.; Floris, G. Intermediates in the catalytic cycle of lentil (Lens esculenta) seedling copper-containing amine oxidase. *Biochem. J.* **1998**, *332*, 431–437.

77. Padiglia, A.; Medda, R.; Pedersen, J. Z.; Agro, A. F.; Lorrai, A.; Murgia, B.; Floris, G. Effect of metal substitution in copper amine oxidase from lentil seedlings. *J. Biol. Inorg. Chem.* **1999**, *4*, 608–613.

78. Shepard, E. M.; Dooley, D. M. Intramolecular electron transfer rate between active-site copper and TPQ in Arthrobacter globiformis amine oxidase. *J. Biol. Inorg. Chem.* **2006**, *11*, 1039–1048.

79. Padiglia, A.; Medda, R.; Bellelli, A.; Agostinelli, E.; Morpurgo, L.; Mondovi, B.; Agro, A. F.; Floris, G. The reductive and oxidative half-reactions and the role of copper ions in plant and mammalian copper amine oxidases. *Eur. J. Inorg. Chem.* **2001**, *1*, 35–42.

80. Smith, M. A.; Pirrat, P.; Pearson, A. R.; Kurtis, C. R. P.; Trinh, C. H.; Gaule, T. G.; Knowles, P. F.; Phillips, S. E. V.; McPherson, M. J. Exploring the roles of the metal ions in Escherichia coli copper amine oxidase. *Biochemistry* **2010**, *49*, 1268–1280.

81. Mills, S. A.; Goto, Y.; Su, Q.; Plastino, J.; Klinman, J. P. Mechanistic comparison of the cobalt-substituted and wild-type copper amine oxidase from Hansenula polymorpha. *Biochemistry* **2002**, *41*, 10577–10584.

82. Shepard, Eric M.; Okonski, Kristina M.; Dooley, David M. Kinetics and spectroscopic evidence that the Cu(I)-semiquinone intermediate reduces molecular oxygen in the oxidative half-reaction of Arthrobacter globiformis amine oxidase. *Biochemistry* **2008**, *47*, 13907–13920.

83. Kishishita, S.; Okajima, T.; Kim, M.; Yamaguchi, H.; Hirota, S.; Suzuki, S.; Kuroda, S.; Tanizawa, K.; Mure, M. Role of copper ion in bacterial copper amine oxidase: Spectroscopic and crystallographic studies of metal-substituted enzymes. *J. Am. Chem. Soc.* **2003**, *125*, 1041–1105.

84. Takahashi, K.; Klinman, J. P. Relationship of stopped flow to steady state parameters in the dimeric copper amine oxidase from Hansenula polymorpha and the role of zinc in inhibiting activity at alternate copper-containing subunits. *Biochemistry* **2006**, *45*, 4683–4694.

85. Mills, S. A.; Klinman, J. P. Evidence against reduction of Cu^{2+} to Cu^+ during dioxygen activation in a copper amine oxidase from yeast. *J. Am. Chem. Soc.* **2000**, *122*, 9897–9904.

86. Mukherjee, A. M. Ph.D. Dissertation Johns Hopkins University (2009)

87. See Chapter 3 by Dooley and co-workers in this text.

88. Structures were considered with methyl imidazole and imidazole coordinated though the δ position in the DFT calculations of Ref 49.

89. Cabelli, D. E.; Riley, D.; Rodriguez, J. A.; Valentine, J. S.; Zhu, H. Models of superoxide dismutases. In *Biomimetic Oxidations Catalyzed by Transition Metal Complexes* Meunier, B.Ed. Imperial College Press, London: 2000, pp 461–508.

90. Miller, A-F. Redox tuning over almost 1 V in a structurally conserved active site: Lessons from Fe-containing superoxide dismutase. *Acc. Chem. Res.* **2008**, *41*, 501–510.

91. Valentine, J. S.; Doucette, P. A.; Potter, S. Z. Copper-zinc superoxide dismutase and amyotrophic lateral sclerosis. *Ann. Rev. Biochem.* **2005**, *74*, 563–593.

92. Rabani, J.; Klug, D.; Fridovich, I. Decay of the HO_2 and O_2^- radicals catalyzed by superoxide dismutase. A pulse radiolytic investigation. *Israel J. Chem.* **1972**, *10*, 1095–1106.

93. Fee, J. A.; Bull, C. Steady-state kinetic studies of superoxide dismutases. Saturative behavior of the copper- and zinc-containing protein. *J. Biol. Chem.* **1986**, *261*, 13000–13005.

94. Polticelli, F.; Battistoni, A.; O'Neill, P.; Rotilio, G.; Desideri, A. Identification of the residues responsible for the alkaline inhibition of Cu,Zn superoxide dismutase: a site-directed mutagenesis approach. *Protein Sci.* **1996**, *5*, 248–253.

95. Malinowski, D. P.; Fridovich, I. Chemical modification of arginine at the active site of the bovine erythrocyte superoxide dismutase. *Biochemistry* **1979**, *18*, 5908–5916.

96. Uchida, K.; Kawakishi, S. Identification of oxidized histidine generated at the active site of Cu,Zn-superoxide dismutase exposed to H_2O_2. *J. Biol. Chem.* **1994**, *269*, 2405–2410.

97. Valentine, J. S. Do oxidatively modified proteins cause ALS? *Free Rad. Biol. Med.* **2002**, *33*, 1314–1320.

7

THEORETICAL ASPECTS OF DIOXYGEN ACTIVATION IN DICOPPER ENZYMES

KAZUNARI YOSHIZAWA

Institute for Materials Chemistry and Engineering and International Research Center for Molecular Systems, Kyushu University, Fukuoka 819-0395, Japan

7.1. INTRODUCTION

The ground-state electron configuration of the dioxygen (O_2) molecule with 12 valence electrons is based on Figure 7.1. The triplet state of dioxygen ($^3\sum_g^-$) is represented by configuration $1\sigma_g^2 2\sigma_u^{*2} 3\sigma_g^2 1\pi_u^4 2\pi_g^{*2}$; it contains two unpaired electrons with parallel electron spins in the $2\pi_g^*$ orbitals. The bond order of O_2 is $1/2(8-4) = 2$, where 8 is the number of electrons in the bonding orbitals ($1\sigma_g$, $3\sigma_g$,

Copper-Oxygen Chemistry, First Edition. Edited by Kenneth D. Karlin and Shinobu Itoh.
© 2011 John Wiley & Sons, Inc. Published 2011 by John Wiley & Sons, Inc.

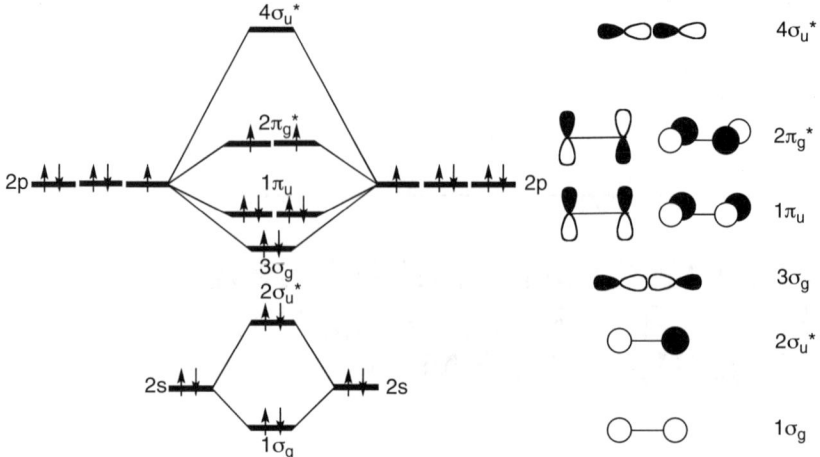

FIGURE 7.1. Fragment molecular orbital analysis of the dioxygen molecule, in which ∗ indicates antibonding orbitals.

and $1\pi_g$) and 4 is that in the antibonding orbitals ($2\sigma_u^*$, $2\pi_g^*$, and $4\sigma_u^*$), in which the asterisk ∗ indicates antibonding orbitals with respect to the O−O bond. If the $2\pi_g^*$ orbitals are fully occupied by four electrons, its bond order is 1, and if the $4\sigma_u^*$ orbital is further occupied by two electrons, its bond order is 0, i.e., the chemical bond of O_2 is cleaved completely. The two ($2e^-$)- and four ($4e^-$)-electron reductions of O_2, which lead to peroxide and dioxide, respectively, are important in their great oxidizing power to organic substrates because their ground states are singlet.

The reactions of the O_2 molecule with organic substrates do not take place under ambient conditions because typical organic molecules have singlet ground states and their reactions with O_2 are spin-forbidden.

$$\frac{1}{2}\,{}^3O_2 + {}^1X \rightarrow {}^1XO \qquad (7.1)$$

The cleavage of the dioxygen O−O bond occurring on transition metal active sites in biological systems is one of the most important steps in the catalytic processes for O_2 activation. Metalloenzymes having one or two copper ions in their active sites utilize the oxidizing power of O_2 for respiration and for conversion of substrates. For example, hemocyanins are well-known oxygen carrier proteins found in mollusks and arthropods, and tyrosinase oxidizes tyrosine to form dopaquinone.[1-5] These enzymes have dicopper active sites that can bind O_2 to form a peroxo species.

A dicopper enzyme, hemocyanin, shows interesting spectroscopic features in its oxidized form. For example, oxyhemocyanin exhibits an extremely intense absorption band at 350 nm, and a very low O−O stretching frequency of $\sim$750 cm.[6-15] These spectroscopic features have been proposed to be characteristic of a dimetal complex with a μ-η^2:η^2-O_2 binding mode. From X-ray structural analyses, Kitajima et al.[16-20] established this type of binding mode in dicopper model complexes.

Since oxytyrosinase exhibits very similar spectroscopic features, the active site of oxytyrosinase is proposed to have a dicopper core structure similar to that of oxyhemocyanin.[7–15] The intense absorption band of these compounds indicates that the peroxide with a μ-η^2:η^2-O_2 mode is a very strong σ-donor ligand to two Cu(II) atoms, while the low O—O stretching frequency is due to the weak O—O bond. According to Solomon and co-workers,[7–15] the low O—O stretching frequency shows the peroxide to act as a π-acceptor ligand, while the π back-bonding shifts electron density into the $4\sigma_u^*$ orbital (of the peroxide) that is strongly antibonding with respect to the O—O bond. The bound O_2 is thus activated to be cleaved.

Tolman, Que, and their co-workers[21,22] demonstrated through a dicopper model complex similar to the active site of hemocyanin that there is an interesting equilibrium between $[Cu_2(\mu$-η^2:η^2-$O_2)]^{2+}$ and $[Cu_2(\mu$-$O)_2]^{2+}$ modes in solutions, as shown in Scheme 7.1. This equilibrium is shifted toward a μ-η^2:η^2-O_2 core structure in dichloromethane, while it is reversely shifted toward a $(\mu$-$O)_2$ core structure in tetrahydrofuran (THF). *Ab initio* calculations[23] showed that dicopper model complexes with μ-η^2:η^2-O_2 and $(\mu$-$O)_2$ cores are very close in energy, but the dioxo forms are better solvated than the peroxo forms by 5.6–8.4 kcal/mol. However, detailed analyses of the reaction coordinate for O_2 cleavage on actual dicopper enzymes and model systems still seem to be lacking, concerning the reason why such a nuclear distortion induces significant electron transfer (ET).

From extended X-ray absorption fine structure (EXAFS) and Mössbauer spectroscopic analyses, intermediate Q of methane monooxygenase (MMO), which can directly activate methane, is likely to have an $Fe_2(\mu$-$O)_2$ "diamond-core" structure.[24] An O_2 evolving tetranuclear manganese cluster in photosystem II also has been proposed to consist of two $Mn_2(\mu$-$O)_2$ diamond-core components.[25–28] Possible processes for the O_2 generation in this system have been investigated in terms of orbital interactions by Hoffmann and co-workers.[29] Thus, such dimetal diamond-core structures seem to be important entities for the active sites in multinuclear metalloenzymes that use various transition metals.

The catalytic reactions mediated by copper-based enzymes are initiated by important two-step reactions involving O_2. The first of these is an initial binding step to a dicopper active site to form a peroxo species; the second is a cleavage step of the dioxygen O—O bond. Theoretical analyses of this binding mode of O_2 to the dicopper active site are important for the better understanding of the chemistry of hemocyanin and tyrosinase. The μ-η^2:η^2-O_2 binding mode mentioned above is one example for the metabolism of dicopper enzymes; there is still another possible structure. It is the μ-η^1:η^1-O_2 mode. This binding mode was also found in dicopper model complexes by Karlin and co-workers.[30] Moreover, such a binding mode has

SCHEME 7.1.

been crystallographically established[31–33] and theoretically analyzed[34–36] for diiron peroxo model complexes of MMO.[37–40] However, the naturally occurring peroxo intermediates in dicopper enzyme systems in general adopt the μ-η^2:η^2-O_2 mode.

Dioxygen O−O bond cleavage performed at the active centers of metalloenzymes is an important, but unknown, step in the catalytic cycles of oxygen-activating enzymes. A purpose of this part of the chapter is to understand and analyze, from a qualitative point of view and by consideration of orbital interactions, how ET occurs from the dicopper active site to O_2 and how O_2 cleavage proceeds from a dicopper peroxo species. Accordingly, we examined a reaction pathway from a dicopper model peroxo complex with the μ-η^2:η^2-O_2 mode to a corresponding dioxo complex with the $(\mu$-$O)_2$ mode.[41]

7.2. DICOPPER MODELS OF DIOXYGEN ACTIVATION

To investigate the dioxygen O−O bond cleavage process in the dicopper active centers of hemocyanin and tyrosinase, we considered two theoretical model complexes, as indicated below in structures **I** and **II**. Structure **I** models a dicopper peroxo complex with a μ-η^2:η^2-O_2 core structure. The terminal ligands of these are modeled by NH_3 because the coordination sphere of hemocyanin is composed of histidine. In these models, five-coordinations around each copper atom is assumed on the basis of an earlier theoretical model.[42] We set the Cu−O (bridge) distances as 1.85 Å. The Cu−N(axial) distances are taken as 2.26 Å and those of Cu−N(equatorial) as 1.95 Å. These Cu−N bond distances have been well characterized by X-ray structural analyses using model complexes. The reason the apical Cu−N bonds are longer than the equatorial bonds has been discussed in terms of orbital interactions. The N(axial)−Cu−N(equatorial) and N(equatorial)−Cu−N(equatorial) angles were optimized to be 98.0° and 94.8°, respectively, using the extended Hückel method.[44,45] The O−O distances in the peroxo and dioxo forms are assumed to be 1.4 and 2.4 Å, respectively. The resultant Cu−Cu distances for structures **I** and **II** become 3.43 and 2.78 Å, respectively. These models belong to the C_{2h} point group, with the C_2 axis being along the O−O bond. Since the two copper atoms have a 1 + charge and the ligands are all neutral, the total charge of structures **I** and **II** is 2 + .

We would like to look at the general electronic features of structures **I** and **II** to analyze the change in the electronic structure along the reaction coordinate from **I** to **II**. Molecular orbital levels are calculated and then filled with electrons. The extended Hückel method is not reliable for bonding energies, but it is a good model for general orbital energy trends, orbital interactions, and major charge shifts. All calculations were carried out by using YAeHMOP[43–45] with standard parameters, which were collected by Alvarez,[46] for copper, oxygen, nitrogen, carbon, and hydrogen.

7.3. REACTION PATHWAY FOR DIOXYGEN CLEAVAGE

To understand the essential orbital interactions for the bonding in a peroxo model complex structure **I**, this complex is partitioned into $Cu_2(NH_3)_6$ and O_2 fragments. Although fragment molecular orbital (FMO) analyses for this binding mode have been discussed by Kitajima et al.,[42] let us look at the metal–ligand interactions in detail. In Figure 7.2, the molecular orbitals of $[Cu(NH_3)_3]_2(O_2)^{2+}$ are constructed by

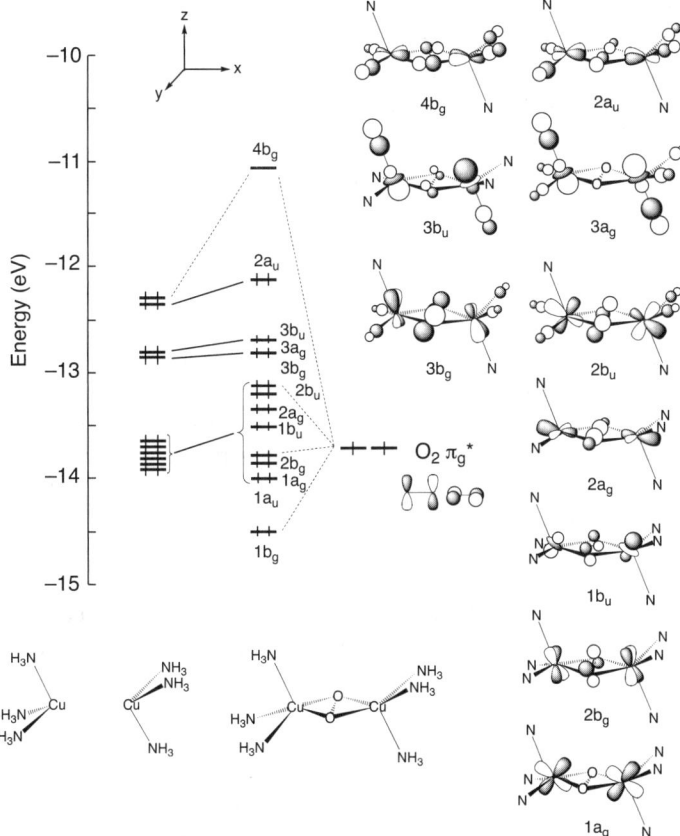

FIGURE 7.2. Orbital interaction diagram for a dicopper peroxo model with a $\mu\text{-}\eta^2{:}\eta^2\text{-}O_2$ core structure (**I**).

allowing two $[Cu(NH_3)_2]^{2+}$ fragments and O_2 to interact.[41] A total of 10 d-block orbitals are shown in this illustration. The degenerate $1\pi_u$ set of O_2 (not shown in this illustration) contributes to the b_u orbitals, as can be seen from the d-block orbitals indicated in Figure 7.2.

Some of the low-lying six d-block orbitals interact with the dioxygen $2\pi_g^*$ orbitals, which will be occupied by electrons. Thus, while the $2\pi_g^*$ orbitals of O_2 begin in the neutral diatomic state with two electrons, two more electrons are effectively transferred from the d-block of the dicopper fragment to O_2, to form O_2^{2-}. The net charge of one oxygen atom in this peroxo model is calculated as -0.49. In this peroxo model complex, there is still one O_2 orbital that is not occupied (i.e., the antibonding "$4\sigma_u^*$") orbital). The bond order, which is a measure of the net bonding, of the peroxide is thus effectively 1. To cleave the dioxygen O—O bond completely, it is necessary to fill the antibonding "$4\sigma_u^*$" orbital with two electrons.

In the next step on the catalytic cycle of the dicopper enzyme, the reductive cleavage of the peroxide O—O bond is supposed to take place through a still unknown reaction mechanism. It has been demonstrated that O—O bond cleavage can occur in a model complex similar to oxyhemocyanin, as mentioned above.[21,22] The reaction coordinate for the O_2 cleavage can be modeled by the distortion from **I** to **II** in our theoretical models. An effective way to push the two oxygen atoms apart is to decrease the Cu—O—Cu angles, θ, as a parameter, while keeping the Cu—O bond distances to be 1.85 Å, as illustrated in **III**.

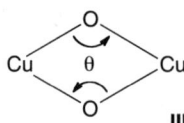

III

In Figure 7.3, we show a Walsh-diagram analysis along an O—O bond cleavage pathway from **I** to **II** as a function of θ defined in **III**.[41] The related O—O distance is also given in this illustration. This distortion corresponds to a reaction coordinate we expect to describe O—O bond cleavage in the dicopper model complex. Dioxygen cleavage occurs with a decrease in θ in this diagram proceeding from a peroxo (Fig. 7.3a) to a dioxo species (Fig. 7.3b).

The $2\pi_g^*$ orbital set, which is distributed into some b_g orbitals, as indicated in Figure 7.2, is not so dramatically changed by the distortion. For example, the $3b_g$ orbital (at -13.2 eV in Fig. 7.2), which is of antibonding character between dioxygen $\pi_g^*(\sigma)$ and dicopper d_{xy}, is not remarkably changed along the reaction coordinate. These b_g orbitals are all occupied in this diagram except for a high-lying one, so that the $2\pi_g^*$ set is almost fully occupied along the reaction coordinate in contrast to the $4\sigma_u^*$ orbital. This orbital is not occupied in the peroxo complex, but is occupied in the dioxo one, as described below.

When we look at the Walsh diagram from right to left, an a_u orbital remarkably decreases across the dicopper d-block orbitals, although, as a consequence of the symmetry, a crossing with some of these orbitals is avoided. This a_u orbital is the antibonding $4\sigma_u^*$ orbital of O_2. The $4\sigma_u^*$ of the peroxo form located at -6 eV

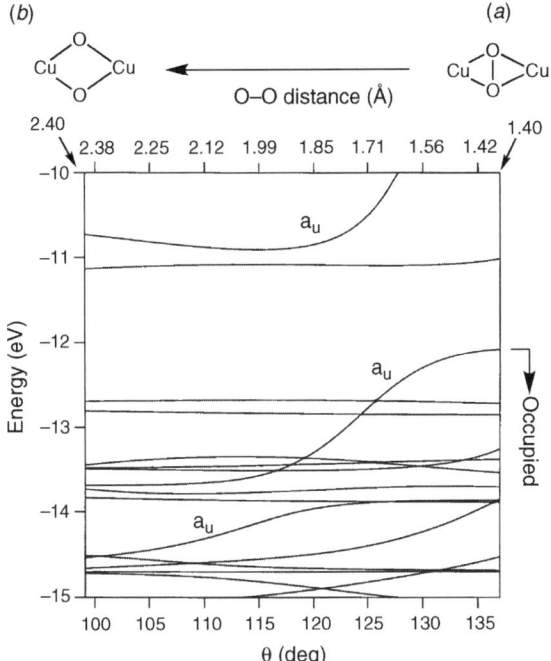

FIGURE 7.3. Walsh diagram along an O–O bond cleavage pathway from **I** to **II** as a function of O–O distance.

(outside the energy window of Fig. 7.3) is indicated in Scheme 7.2 left. As the O atoms are pushed apart, the $4\sigma_u^*$ orbital goes down in energy, due to a decrease in the strong antibonding interactions.

To break the O–O bond of the peroxo complex completely, the a_u ($4\sigma_u^*$) orbital should be occupied by two more electrons, as mentioned above. As the O–O distance is increased, the antibonding interactions between the oxygen atoms are significantly decreased. Thus, the $4\sigma_u^*$ orbital lies at -14.5 eV when the O–O distance becomes 2.4 Å in the dioxide model, **II**. As a consequence, while in the peroxo model, **I**, the π_g^* orbitals are occupied by $4e^-$ to create O_2^{2-}, two more electrons are effectively transferred from the low-lying d-block orbitals to the $4\sigma_u^*$ orbital to create $2O^{2-}$ in **II**.

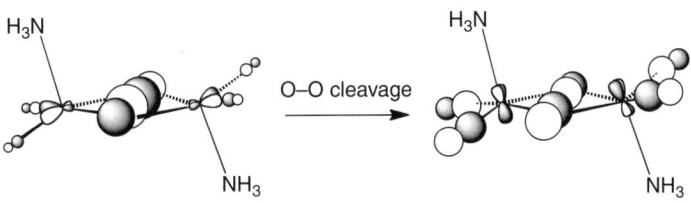

SCHEME 7.2.

In fact, the net charge of one oxygen in the peroxo complex is -0.49 while that in the corresponding dioxo complex is -1.33.

It is essential to evaluate the activation energy for the O−O bond cleavage for a better understanding of the mechanism of O_2 activation by the dicopper enzymes. Clamer, et al.[23] did not refer to the activation energy in their recent *ab initio* work. It was found from the qualitative calculations that the reaction should proceed with no cost of activation energy along the O−O bond cleavage pathway. Although the extended Hückel method is not good at calculating absolute energy difference between states, the activation energy is likely to be small even if it exists, because this reaction is symmetry allowed judging from the profile of Figure 7.2.

Next, let us look at an FMO analysis for the dioxo complex **II**, shown in Figure 7.4.[41] The orbital interactions between the two oxygen atoms are very weak in this complex, and consequently, the bonding and antibonding levels are decreased in energy. Therefore the $4\sigma_u{}^*$ orbital significantly goes down to cross the copper

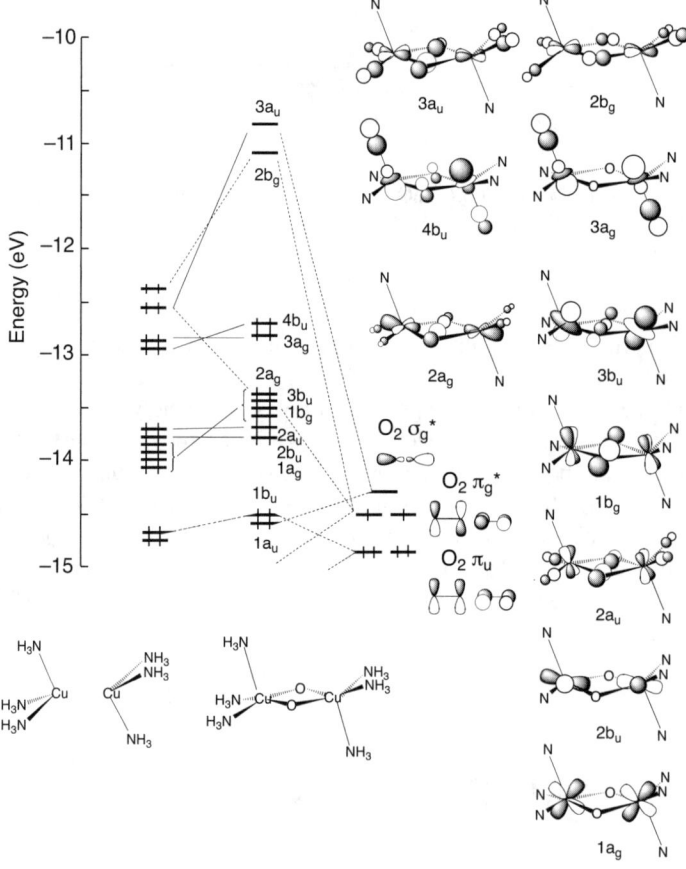

FIGURE 7.4. Orbital interaction diagram for a dicopper dioxo model with a $\mu\text{-}\eta^2{:}\eta^2\text{-}O_2$ core structure (**II**).

d-block orbitals when the O$-$O distance becomes $\sim$1.7 about 1.8 Å (see Fig. 7.3). In this way, two electrons are effectively transferred from the copper d-block orbitals to the $4\sigma_u^*$ orbital reducing O_2^{2-} to $2O^{2-}$. Thus, almost all of the antibonding orbitals of O_2 are filled so that the dioxygen O$-$O bond is reductively cleaved forming the dioxo species.

For a closer inspection of the bonding and antibonding interactions between the two oxygen atoms, MOOP (molecular orbital overlap population) analyses were carried out. Figure 7.5 shows such analyses for the dicopper model complexes with O$-$O distances of 1.4, 1.8, and 2.4 Å. The MOOP plots for the O$-$O bond show the contributions of individual molecular orbitals to a Mulliken overlap population. In these, bonding contributions are plotted to the right and antibonding are to the left. The dotted lines mark the highest occupied molecular orbital (HOMO) levels; the dashed lines going up are the integrations of the bonding and antibonding contributions.

It is helpful to look at the MOOP plots together with the Walsh diagram shown in Figure 7.3 and the FMO analyses in Figures 7.2 and 7.3. At an O$-$O distance of 1.4 Å, the HOMO is $2a_u$ (at -12.1 eV), which is slightly antibonding between the oxygen atoms (see Fig. 7.2). However, $3a_u$ (at -6 eV), which is actually the antibonding $4\sigma_u^*$ of the O_2 molecule (see Scheme 7.2 left), is unoccupied so that the net integration at the HOMO level is positive (see Fig. 7.5a). This means that the peroxo species with an O$-$O distance of 1.4 Å still has a net bonding character, although its degenerate π_g^* orbital set is fully occupied. This result is consistent with a qualitative picture that the peroxo species has a bond order of 1. As the O$-$O distance is increased, orbital levels related to the oxygen atoms go down, and in particular the antibonding $4\sigma_u^*$ orbital crosses with the HOMO level at an O$-$O distance of $\sim$1.7 Å. Consequently, the

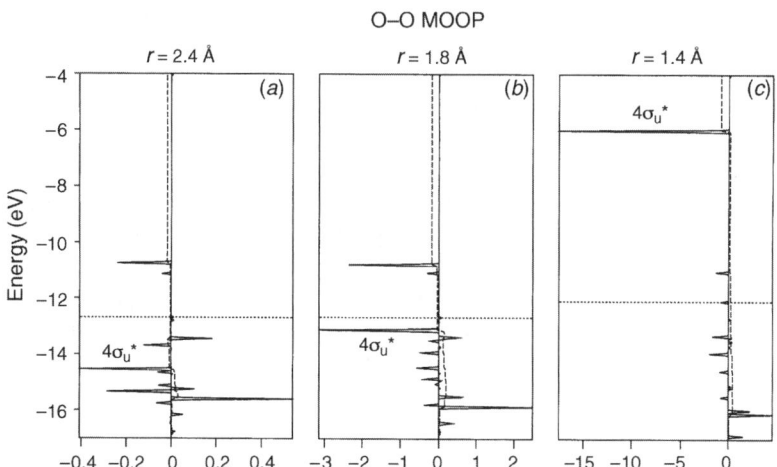

FIGURE 7.5. Molecular orbital overlap population curves for the O$-$O bond in dicopper model complexes with O$-$O distances (r) of (a) 1.4, (b) 1.8, and (c) 2.4 Å. Bonding contributions are plotted to the right and antibonding ones to the left. The dotted lines mark the HOMO levels; the dashed lines going up are the integrations of the bonding and antibonding contributions.

antibonding contribution gets more and more dominant below the HOMO level. At an O$-$O distance of 1.8 Å, the net integration at the HOMO level is significantly reduced to be minus (Fig. 7.5b) in contrast to that at an O$-$O distance of 1.4 Å being plus.

Let us look at the MOOP plot for the dioxo species with an O$-$O distance of 2.4 Å. The HOMO is 4b$_u$ (at -12.7 eV; see Fig. 7.4), in which the contribution from the oxygen atoms is not significant (see Fig. 7.5a). Most of the orbital levels related to the oxygen atoms go down, due to the long O$-$O distance, and lie below the HOMO level. As a consequence, the net integration at the HOMO level is minus, as shown in Figure 7.5a. Therefore the dioxo species with an O$-$O distance of 2.4 Å has no net O$-$O bonding interactions.

7.4. STRUCTURE OF TYROSINASE

Matoba et al.[47] recently succeeded in the crystallization of some forms of tyrosinase from *streptomyces castaneoglobisporus* with a caddie protein (ORF378; ORF = open reading frame) and carried out X-ray crystal structural analyses at high resolutions of up to 1.2 Å. Tyrosinase is an essential enzyme for all organisms.[6-15,48-53] It catalyzes the conversion of tyrosine to dopaquinone (Scheme 7.3), which is a precursor of the melanine pigment. Since melanin is a key pigment of some phenomena (e.g., suntan), skin disorder, and bruising of fruits, tyrosinase has attracted much attention from cosmetology, medicine, and agriculture with respect to the control of the melanin pigment synthesis.

Tyrosinase, catechol oxidase, and hemocyanin are classified into the type 3 copper protein family, as mentioned earlier. These three enzymes have very similar dicopper active sites, but their enzymatic functions are different. Tyrosinase initiates the synthesis of melamine by catalyzing the hydroxylation of monophenols to *o*-diphenols (cresolase activity) and the subsequent 2e$^-$ oxidation to *o*-quinones (catecholase activity) with O$_2$.[48] On the other hand, catechol oxidase has the catecholase activity, but lacks the cresolase activity. Hemocyanin plays the role of an oxygen carrier in crustaceans like hemoglobin in mammals, but is considered not to have any catalytic activity. These different enzymatic functions are believed to derive from accessibility to the active site. Therefore the X-ray crystal structural analyses of tyrosinase,[47] hemocyanin,[54-56] and catechol oxidase[57-58] are essential to compare the correlation between their functions and surrounding environments of the dicopper active site. In this sense, the recent X-ray structural analysis of tyrosinase is of great use to elucidate the biological mechanism for the conversion of tyrosine to dopaquinone. From

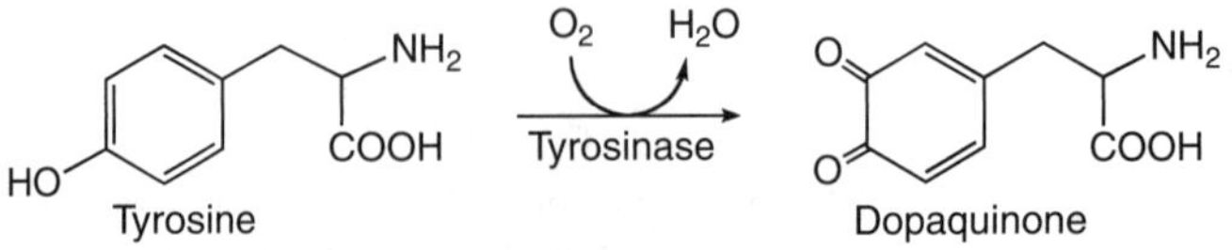

SCHEME 7.3.

spectroscopic measurements, the presence of oxy-tyrosinase, in which O_2 is bound to the dicopper site, was demonstrated during the catalytic cycle of tyrosinase.[59,60] Moreover, some sequence analysis data implied that six histidine residues should exist in the active site of tyrosinase as ligands of the two copperions.[58,61,62] Measured reaction rates of tyrosinase from kinetic analysis indicate that the oxidation of o-diphenols is much more rapid than the o-hydroxylation of mono-phenols ($k_{oxidation} = 10^7 \text{ s}^{-1}, k_{hydroxylation} = 10^3 \text{ s}^{-1}$).[63] This information is useful for mechanistic consideration based on quantum chemical calculations.

Dicopper model complexes that mimic the active site of the type 3 copper protein family have played an essential role to better understand the structure–function relationships of the dicopper enzymes.[7,42,64] Prior to protein X-ray structural anal-yses, spectroscopic features of the model complexes were found to be in good agreement with those of oxy-hemocyanin, catechol oxidase, and tyrosinase, suggest-ing that O_2 is bound to the dicopper site as peroxide in a μ-η^2:η^2 side-on bridging mode in these enzymes. The (μ-η^2:η^2-peroxo)dicopper(II) complexes supported by a variety of tridentate and bidentate nitrogen ligands were reported to provide important information about the effects of the ligand on the structure and physicochemical properties of the dicopper peroxo complexes.[65–69] Many reactivity studies on the dicopper peroxo complexes and substrate phenol derivatives demonstrated that C−C coupling dimer products may be formed rather than the oxygenation product catechols.[70–76] These observations tell us that the reactions of phenols and the dicopper peroxo complexes involve phenolic O−H bond activation to generate stable phenoxyl radicals as intermediates.

An X-ray crystal structural analysis of the oxy form of tyrosinase was successfully made together with ORF378, a tyrosine residue that extends to the substrate-binding pocket and shields the catalytically active dicopper site. After dissociation of this caddie protein, the active site is accessible to substrates. It is therefore reasonable to remove the ORF378 moiety in quantum mechanical/molecular mechanical (QM/MM) calculations. In view of the structure, one histidine residue (His54) that coordinates to CuA comes from a protein loop, while the other five histidine residues (His38, His63, His190, His194, and His216) come from surrounding α-helix struc-tures. Matoba et al.[47] propose that the His54 residue should act as a catalytic base because of the flexibility of the loop structure. It is interesting to turn our attention to this statement on the basis of detailed QM/MM calculations (Fig. 7.6).[77]

7.5. MECHANISMS OF TYROSINASE BY DFT CALCULATIONS

Recently, computational quantum chemistry in terms of density functional theory (DFT) calculations has been of great use in the mechanistic analysis of metalloen-zymes. With the use of DFT calculations Siegbahn and Wirstram[78] showed that the bis (μ-oxo)Cu^{III}Cu^{III} state should not be involved in the catalytic mechanism of tyros-inase because the bis(μ-oxo)Cu^{III}Cu^{III} state lies above the μ-η^2:η^2 peroxo Cu^{II}Cu^{II} state in energy. Moreover, Siegbahn[79] showed that a O_2 species attacks the phenolate ring, which is then followed by O−O bond cleavage (see Fig. 7.7). Density functional

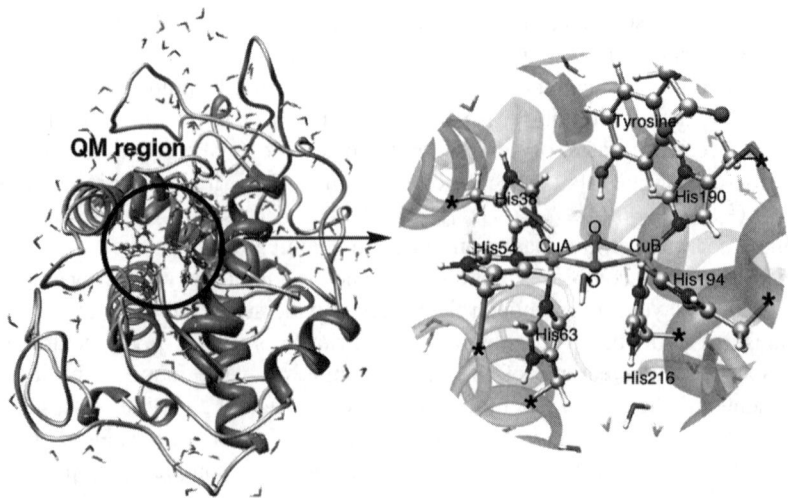

FIGURE 7.6. The QM/MM optimized structure of the oxy form of tyrosinase, in which the asterisks indicates the QM/MM border. (see color insert.)

theory calculations gave a reasonable barrier for the O_2 attack of only 12.3 kcal/mol, provided one of the copper ligands is able to move substantially away from its direct copper coordination. This result can be achieved with six histidine ligands even if these ligands are held in their positions by the enzyme, but can also be achieved if one of the copper atoms only has two histidine ligands and the third ligand is water. In the latter study,[79] a calculation model with five imidazole ligands for histidine and one water ligand was set up on the basis of limited structural information about tyrosinase at that time, when no X-ray structural analysis was available. Now, we have better structural information for reasonably developing a quantum chemical study for the catalysis of tyrosinase.

On the basis of the X-ray structural analysis,[47] it is possible to perform QM/MM calculations, which can reasonably take account of important effects of the surrounding amino-acid residues, hydrogen bonding, and protein environment. Under these situations, we studied the mechanism for the biological conversion of tyrosine to dopaquinone on the basis of whole enzyme calculations to throw light on a recent mechanistic proposal for this enzyme from a different point of view. A model of tyrosinase–substrate complex shown in Figure 7.6 was prepared by using the crystal structure of the oxy form of tyrosinase determined by X-ray diffraction at a resolution of 1.80 Å.[47,80,81] An initial geometry was obtained from the crystal structure by removal of ORF378 and neighboring water molecules. Hydrogen atoms were added based on the geometry specified in the residue database and first relaxed by using steepest descent minimization with the peptide held fixed, and then tyrosine was added in the neighborhood of the dicopper site. The ball-and-stick representation in Figure 7.6 indicates atoms in the QM region and the ribbon representation indicates atoms in the MM region. Geometry optimizations were done by using the ONIOM–QM/MM method.[82,83]

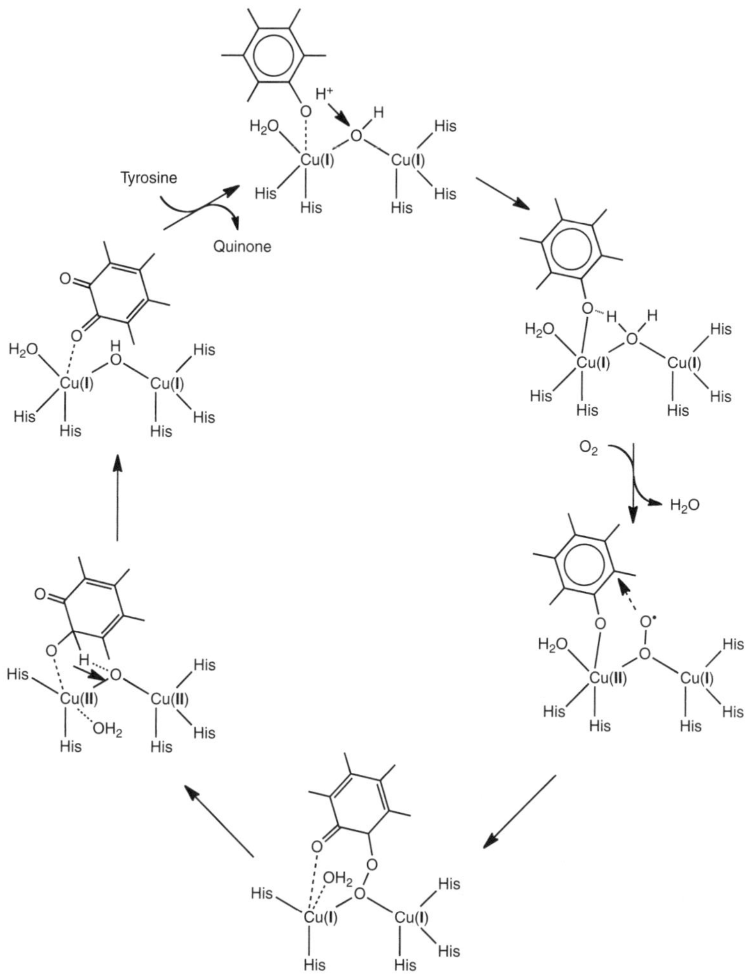

FIGURE 7.7. A catalytic mechanism of the conversion of tyrosine to dopaquinone by tyrosinase proposed by Siegbahn.[79]

The mechanism we propose for the catalytic function of tyrosinase is summarized in Figure 7.8.[77] When substrate tyrosine comes into contact with the μ-η^2:η^2 peroxo $Cu^{II}Cu^{II}$ site, one of the Cu—O bonds is cleaved while the O—O bond remains unchanged. Then the O—H bond of tyrosine is cleaved by the resultant Cu—O species. At the same time, the resulting phenoxo ligand is bound to a copper ion. The O—O bond is subsequently cleaved while phenoxyl radical is released from the copper site. After the O—O bond cleavage, the bridging oxo species attacks an ortho position of the benzene ring. In the final stages of the reaction, dopaquinone is formed with the involvement of a histidine residue (His54) that comes from a flexible-loop structure. This histidine residue plays an important role in the proton transfer processes.

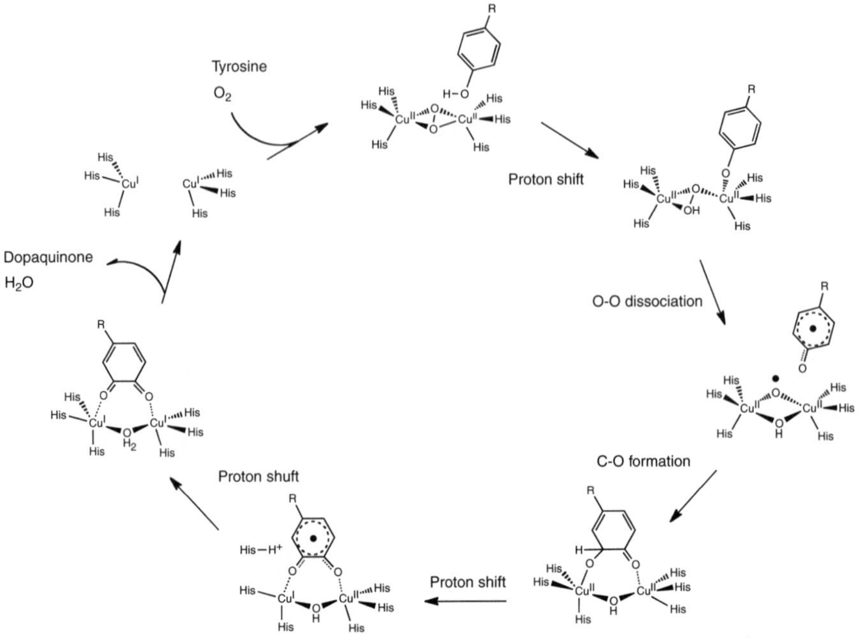

FIGURE 7.8. A catalytic mechanism of the conversion of tyrosine to dopaquinone by tyrosinase by Inoue et al.[77]

Thus, the overall reaction consists of five elementary reactions; a computed energy diagram is shown in Figure 7.9.[77] The energy profile is quite reasonable as a biochemical process that occurs under physiological conditions because the energy barriers for the transition states are <15 kcal/mol. The rate-determining step is found to be the O$-$O bond cleavage step with an activation energy of 14.9 kcal/mol. We characterized this rate-determining step in the catalytic cycle of tyrosinase by calculating the reaction rate with transition state theory.[77] In a previous DFT study,[79] the O$-$O dissociation step was calculated to be endothermic by 23 kcal/mol, and therefore this mechanism was ruled out. In contrast, in our mechanism the peroxo species **1** is protonated as a result of the heterolysis of the phenolic O$-$H bond of substrate; therefore the O$-$O(H) bond in the hydroperoxo species **2** is significantly activated. Since the O$-$O(H) bond is weak in comparison with the peroxo O$-$O bond, our computational result is reasonable. Calculated Mulliken atomic spin densities of the bridging oxo ligand and the tyrosine moiety in intermediate **3** are 0.76 and -0.99, respectively. Thus, the O$-$O bond dissociation results in the formation of a copper–oxygen radical and a phenoxyl radical. These radical species are coupled to form a C$-$O bond at an ortho position of the benzene ring in the following reaction step. This process is similar to that taking place in the cytochrome P450 chemistry, the so-called oxygen-rebound mechanism.[85] Note that the stable phenoxyl radical plays an essential role in the proposed reaction pathway and that this radical coupling is well regulated in the protein environment. Involvement of phenoxyl radical is also

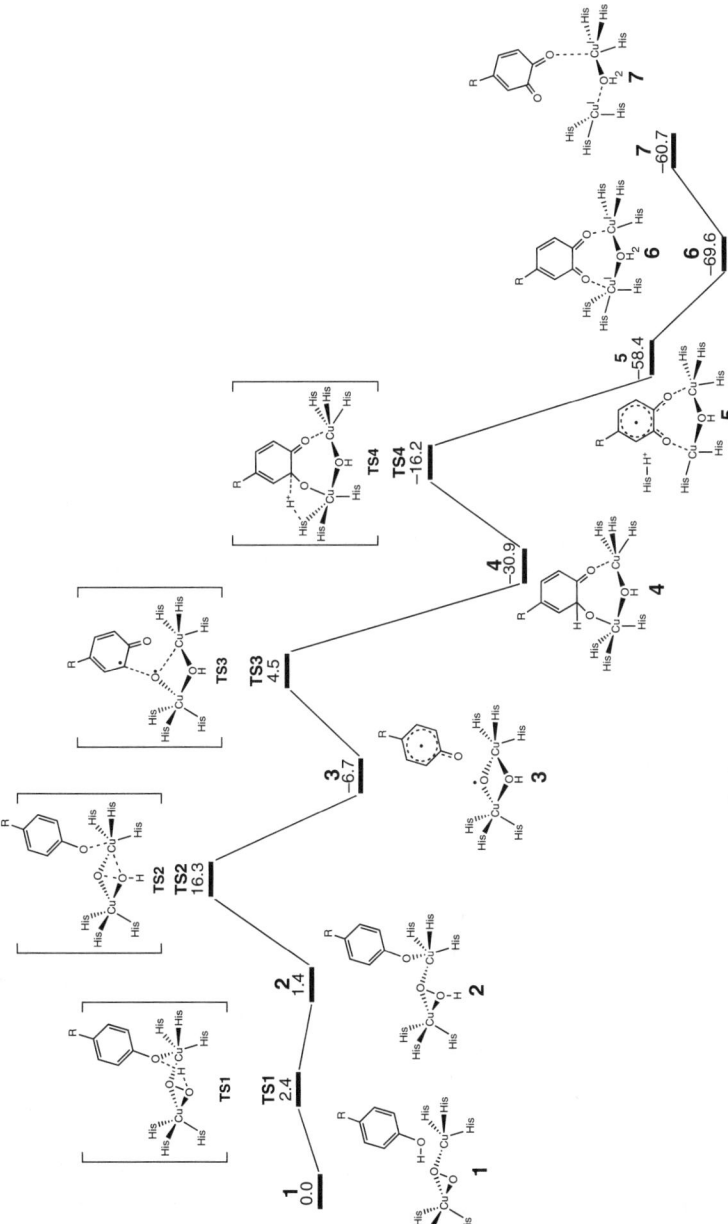

FIGURE 7.9. An energy diagram for the conversion of tyrosine to dopaquinone. Energies are in kilocalories per mol (kcal/mol).

TABLE 7.1. Calculated Reaction Rates at 298 K as a Function of Activation Energy (ΔE)

ΔE (kcal/mol)	k (s^{-1})
10	$2.89 \times 10^{+5}$
11	$5.34 \times 10^{+4}$
12	$9.87 \times 10^{+3}$
13	$1.82 \times 10^{+3}$
14	$3.37 \times 10^{+2}$
15	$6.23 \times 10^{+1}$
16	$1.15 \times 10^{+1}$
17	2.13×10^{0}
18	3.93×10^{-1}
19	7.26×10^{-2}
20	1.34×10^{-2}

suggested in the chemistry of dicopper model complexes because C–C coupling dimer products are observed in extensive reactivity studies.[69–76]

Table 7.1 summarizes calculated reaction rates at 298 K as a function of activation energy. Rodríguez-López et al.[63] reported the reaction rate for the o-hydroxylation of monophenols by tyrosinase to be 10^3 s^{-1}. In view of the data given in Table 7.1, we estimate that this value should correspond to the activation energy of 13–14 kcal/mol, which is in excellent agreement with a calculated value of 14.9 kcal/mol for **TS2**. Thus, the reaction pathway we propose is reasonable from the point of view of reaction rate.

7.6. DICOPPER SITE OF PARTICULATE METHANE MONOOXYGENASE

Methanotrophic bacteria that use methane as their sole source of carbon and energy convert methane to methanol, formaldehyde, formic acid, and finally carbon dioxide efficiently using molecular O_2.[37–40] In the first stages of the metabolic pathway of methanotrophic bacteria, MMO plays an essential role in the oxidation of methane to methanol under physiological conditions.[86,87] Since methane is the most inert hydrocarbon, the C–H dissociation energy being 104 kcal/mol, one cannot efficiently convert methane to methanol at ambient temperature in a direct process. The biological hydroxylation of methane by MMO is of great interest in the field of applied chemistry as well. The biological methane hydroxylation is performed by methanotrophs (e.g., *Methylococcus capsulatus* (Bath) and *Methylosinus trichosporium* OB3b). These methanotrophic bacteria possess two different forms of MMO; one is cytoplasmic (soluble) MMO (sMMO) that has a dinuclear iron active site and the other is membrane-bound particulate MMO (pMMO) that has mono- and multi-copper active sites.[88,89] Note that the representation of the two forms varies, depending on surrounding conditions (e.g., copper concentration).[90–92] The sMMO play an essential role in methane hydroxylation under conditions of copper limitation.

On the other hand, pMMO is significantly activated by sufficient addition of copper to the growth medium; however, the biochemistry, structure, and mechanism of the predominant pMMO enzyme have remained unclear despite considerable research efforts.

In 2005, Lieberman and Rosenzweig[93] reported an X-ray crystal structure analysis of pMMO at 2.8-Å resolution for the first time. The monomeric structure consisting of three subunits, pmoA ($\sim$24 kDa), pmoB ($\sim$47 kDa), and pmoC ($\sim$22 kDa), forms a trimeric arrangement (see Fig. 7.10). Each protomer houses three metal centers in the crystal structure; the first metal center is occupied by a zinc species; the second metal center is occupied by a mononuclear copper species; the third one involves a dinuclear copper species. The monocopper and dicopper active sites are located in the soluble regions of the pmoB subunit, and the monocopper site is found 21 Å apart from the dicopper site. The zinc site, which lies in the membrane between pmoA and pmoC, is 19-Å apart from the dicopper center and 32-Å apart from the monocopper center. According to EXFAS data, although copper ion at the monocopper site is observed in purified pMMO from *Methylococcus capsulatus* (Bath), pMMOs derived from other methanotrophs do not store the monocopper site.[94] The zinc ion arises from crystallization buffer and is not detected in X-ray absorption spectroscopy (XAS) measurements.[95] Therefore this site probably is substituted by another metal ion *in vivo* (e.g., Cu or Fe). Hakemian et al.[96] suggested that the zinc site is occupied by copper in purified pMMO from *Methylosinus trichosporium* OB3b.

There always has been a great deal of controversy with regard to the structure of the pMMO active site, and to my best knowledge there are four kinds of active sites proposed so far; (1) mixed-valent trinuclear Cu$^{(II)}$ species,[97,98] (2) mononuclear Cu$^{(II)}$ species,[99–103] (3) mixed-valent dinuclear Cu^{II}Cu^{III} species,[104] and (4) mixed-metal

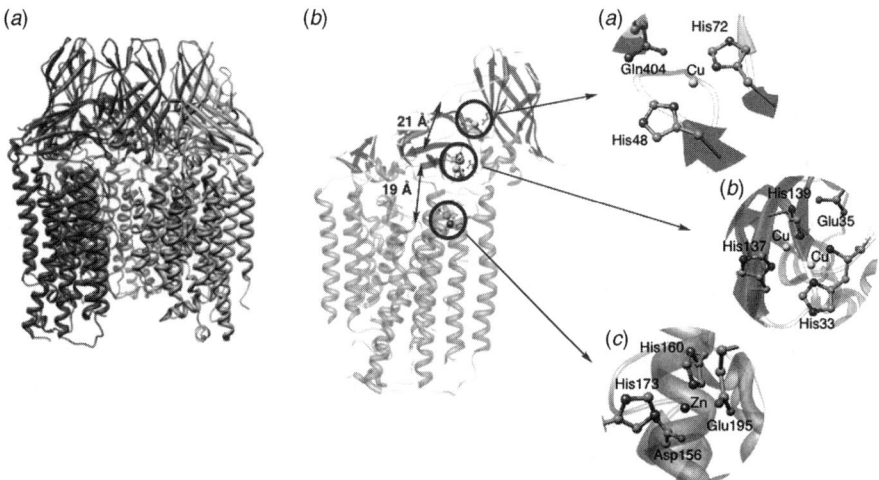

FIGURE 7.10. (*a*) X-ray crystal structure of the pMMO trimer. Three protomers are shown in purple, orange, and dark blue. (*b*) A protomer of pMMO, composed of three subunits pmoA, pmoB, and pmoC, has (*a*) monocopper, (*b*) dicopper, and (*c*) zinc sites. (See color insert.)

CuFe species.[105,106] Balasubramanian et al.[107] recently reported that the pMMO activity is dependent on copper, not iron, and that the copper active site is located in the soluble domains of the pmoB subunit, in which there are monocopper and dicopper active sites. Disruption of each copper center in the soluble fragments of pmoB by mutagenesis indicates that the active site is a dicopper center.

7.7. A METHANE HYDROXYLATION MECHANISM OF pMMO

According to pervious computational results of model-complex studies,[108–111] mono-, di-, and tricopper species possibly contribute to O_2 activation and selective methane oxidation. Chan and co-workers[112–116] reported that a trinuclear $Cu^{II}Cu^{II}Cu^{II}$ active site was detectable from X-ray adsorption edge and electron spin resonance (ESR) spectroscopic experiments. They suggested that the tricopper cluster is located in a hydrophilic cavity that consists of metal-binding residues, including His38, Met42, Met45, Asp47, Trp48, Asp49, and Glu100 from pmoA and Glu154 from pmoC. Chen and Chan[111] proposed that a bis(μ_3-oxo)trinuclear $Cu^{II}Cu^{II}Cu^{III}$ model was the best cluster for the hydroxylation of methane in a concerted oxo-insertion way. Although this proposal is very interesting and fascinating, the trinuclear copper species is not found in the crystal structure of pMMO reported so far.

Thus, one can confine the thinking to the dicopper site on the basis of the X-ray crystal structure.[93] In a previous study, we performed DFT calculations for the conversion of methane to methanol at simple mixed-valent $Cu^{II}Cu^{III}$ model complexes that have two or three ammonia and one hydroxo ligands.[108] Furthermore, we proposed a nonradical mechanism for methane hydroxylation using a more realistic monocopper model having His48, His72, and Glu75 and a dicopper model having His33, His137, His139, and Glu35 on the basis of the observed monocopper and dicopper sites, respectively.[109] As a result, the dicopper species is more likely to play a role in the activation of substrate methane than the monocopper species. A computed energy diagram for methane hydroxylation by the mixed-valent bis(μ-oxo)$Cu^{II}Cu^{III}$ species is shown in Figure 7.11, in which the doublet potential energy surface plays an essential role, whereas the quartet potential energy surface in the dicopper species is high lying in energy. The activation energy for the C$-$H dissociation step at the dinuclear copper site was reasonable for a biological process taking place under physiological conditions; however, the activation barrier for the next rebound step that leads to the formation of the C$-$O bond for product methanol was higher in energy than the first C$-$H dissociation barrier. The electronic and structural properties of the dinuclear copper species of bis(μ-oxo)$Cu^{II}Cu^{III}$ and $Cu^{III}Cu^{III}$ are discussed with respect to the C–H bond activation of methane. The bis(μ-oxo)$Cu^{II}Cu^{III}$ species is highly reactive and considered to be an active species for the conversion of methane to methanol by pMMO, whereas the bis(μ-oxo)$Cu^{III}Cu^{III}$ species is unable to react with methane as it is.[110]

The mechanism is similar to the $Cu^{III}$$-$oxo species.[109] The dicopper–oxo species [**Oxo(d)**] optimized by using a small model agrees well with the one optimized in the protein environment about the coordination bonds around the two copper atoms.

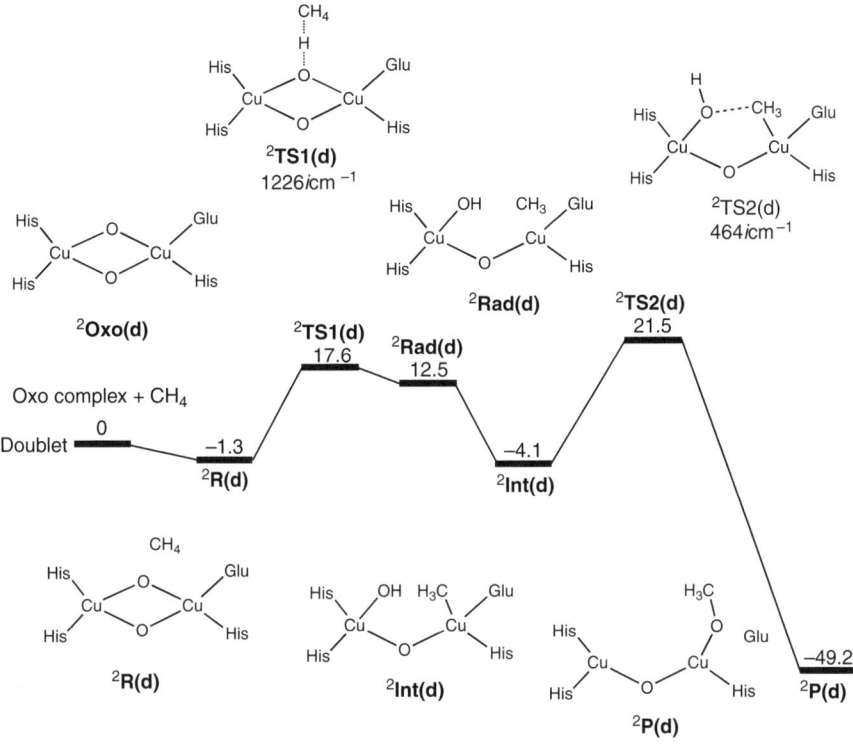

FIGURE 7.11. Energy diagram for the conversion of methane to methanol by the bis(μ-oxo) $Cu^{II}Cu^{III}$ species of pMMO. Units in kilocalories per mol (kcal/mol).

Methane is weakly bound to the right-side copper center of the dicopper species [**R(d)**], and after that hydrogen-atom abstraction from the methane takes place via **TS1(d)**. The activation energy in this process is computed to be 17.6 kcal/mol relative to the dissociation limit on the doublet potential energy surface. The resultant methyl radical [**Rad(d)**] is also trapped at the right-side copper center to form a nonradical intermediate [**Int(d)**]. Since the methyl radical is very unstable compared to other alkyl radicals, the formation of this nonradical species is reasonable from the viewpoint of energy.[116] The final step is a recombination between the OH ligand located at the left-side copper center and the CH_3 ligand located at the right-side copper center via **TS2(d)**, which is computed to be 21.5 kcal/mol relative to the dissociation limit. Therefore the rate-determining step of the reaction pathway is the recombination step in this mechanism. The formation of the product [**P(d)**] is computed to be 49.2-kcal/mol exothermic relative to the dicopper-oxo species + CH_4. Unfortunately, this mechanism is not consistent with the kinetic isotope effect (KIE) measurements,[117] which tells us that the rate-determining step is the initial step relevant to the C−H bond activation of methane. Therefore, this mechanism should be reconsidered; we should take the effects of the neighboring amino acid residues, hydrogen bonding, and the protein environment into account in detail using QM/MM calculations to improve the mechanism for methane hydroxylation.[118]

7.8. CONCLUDING REMARKS

The electronic structures of a dicopper peroxo complex with a μ-η^2:η^2-O_2 mode and a corresponding dioxo complex with a $(\mu$-$O)_2$ diamond core have been theoretically analyzed in terms of orbital interactions. Dioxygen O−O bond cleavage is a very important step in the catalytic cycles of many kinds of metalloenzymes for O_2 activation. The bond cleavage on a dicopper model complex was discussed by distorting a peroxo model complex to a corresponding dioxo complex along a supposed reaction coordinate describing dioxygen O−O bond cleavage. The FMO, Walsh diagram, and MOOP analyses have shed new light on the mechanism of ET from the d block of the metal active center to O_2 and how the O_2 activation occurs on a dicopper enzyme model. Qualitative calculations have shown that the activation energy for O−O bond cleavage on a dicopper model complex is likely to be small even if it exists because this reaction is symmetry allowed judging from the Walsh diagram along an O−O bond cleavage pathway from the peroxide with the μ-η^2:η^2-O_2 mode to the dioxide with the bis(μ-oxo)Cu^{III}Cu^{III} mode. We therefore think that O−O bond cleavage should proceed with no cost to activation energy on the dicopper active site of actual enzyme systems. The qualitative orbital analysis is useful for a better understanding of O_2 activation by transition metal active centers.

A possible mechanism for the biological conversion of tyrosine to dopaquinone was discussed by using QM/MM calculations, which can reasonably take environmental effects into account. The QM/MM study supports the recent mechanistic proposal on the enzymatic function of tyrosinase. The $(\mu$-η^2:η^2-peroxo)dicopper(II) species plays a central role in the catalysis of tyrosinase. A stable phenoxyl radical is involved in the reaction pathway. A series of tyrosinase reactions involve five steps, including proton transfer from the phenolic O−H bond to the dioxygen moiety, O−O bond dissociation of the hydroperoxo species, C−O bond formation at an ortho position of the benzene ring, proton migrations mediated by His54, and quinone formation. The energy profile of the calculated reaction pathway is reasonable as elementary processes that occur under physiological conditions. Detailed analyses of the energy profile demonstrate that the O−O bond dissociation is the rate-determining step. It was demonstrated that the His54 residue, which is flexible because of a loop structure in the protein, would play a role as a general base in the proton migrations in the final stages of the reaction. The activation energy for the O−O bond dissociation computed to be 14.9 kcal/mol is in good agreement with a measured kinetic constant $(k = 10^3 \, s^{-1})$.[63]

Finally the mechanism of methane hydroxylation at the mononuclear and dinuclear copper sites of pMMO was considered. Dioxygen is incorporated into the dicopper site of pMMO to form a $(\mu$-η^2:η^2-peroxo)dicopper species, which is then transformed into a bis(μ-oxo)Cu^{II}Cu^{III} species. The reactivity of the bis(μ-oxo) Cu^{II}Cu^{III} species is sufficient for the conversion of methane to methanol. The reactivity of the mixed-valent bis(μ-oxo)$Cu^{II}Cu^{III}$ species is higher than that of the bis (μ-oxo)$Cu^{III}Cu^{III}$ species in general because in the former species the amplitude of the σ^* Singly occupied molecular orbital (SOMO) localized on the bridging oxo moieties plays an essential role in the homolytic $C-H$ cleavage. The methyl radical is trapped to the mononuclear and dinuclear copper sites to form intermediates that involve OH and CH_3 ligands. Recombination of the resultant OH and CH_3 ligands takes place in a nonradical manner at a metal active center to form a final methanol complex.

ACKNOWLEDGMENTS

K.Y. thanks Yoshihito Shiota, Takehiro Ohta, and Toshinori Inoue for their great efforts. This work was supported by Grants-in-Aid for Scientific Research (Nos. 18GS0207 and 22245028) from the Japan Society for the Promotion of Science, the Kyushu University Global COE Project, the Nanotechnology Support Project, the MEXT Project of Integrated Research on Chemical Synthesis, and CREST of the Japan Science and Technology Cooperation.

REFERENCES

1. Linder, M. C.; Goode, C. A. *Biochemistry of Copper*; Plenum: New York, 1991.

2. Karlin, K. D.; Tyeklar, Z., Eds.; *Bioinorganic Chemistry of Copper*; Chapman & Hall: New York, 1993.

3. Reedijk, J., Ed.; *Bioinorganic Catalysis*; Marcel Dekker: New York, **1993**.

4. Kitajima, N.; Moro-oka, Y. Copper–Dioxygen Complexes. Inorganic and Bioinorganic Perspectives. *Chem. Rev.* **1994**, *97*, 737–757.

5. Kaim, W.; Schwederski, B. *Bioinorganic Chemistry: Inorganic Elements in the Chemistry of Life*; John Wiley & Sons, Inc. New York, 1994.

6. Solomon, E. I.; Sundaram, U. M.; Machonkin, T. E. Multicopper Oxidases and Oxygenases. *Chem. Rev.* **1996**, *96*, 2563–2606.

7. Solomon, E. I.; Tuczek, F.; Root, D. E.; Brown, C. A. Spectroscopy of Binuclear Dioxygen Complexes. *Chem. Rev.* **1994**, *94*, 827–856.

8. Solomon, E. I.; Lowery, M. D. Electronic Structure Contributions to Function in Bioinorganic Chemistry. *Science*, **1993**, *259*, 1575–1581.

9. Tuczek, F.; Solomon, E. I. Charge-Transfer States of Bridged Transition Metal Dimers: Mono- vs Binuclear Copper Azide Systems with Relevance to Oxy-Hemocyanin. *Inorg. Chem.* **1993**, *32*, 2850–2862.

10. Solomon, E. I.; Baldwin, M. J.; Lowery, M. D. Electronic Structures of Active Sites in Copper Proteins: Contributions to Reactivity. *Chem. Rev.* **1992**, *92*, 521–542.

11. Tuczek, F.; Solomon, E. I. Single-Crystal Polarized Absorption Spectroscopic Study of the Electronic Structure of μ-1,2-Peroxo Binuclear Cobalt Complexes. *Inorg. Chem.* **1992**, *31*, 944–953.

12. Ross, P. K.; Solomon, E. I. An Electronic Structural Comparison of Copper–Peroxide Complexes of Relevance to Hemocyanin and Tyrosinase Active Sites. *J. Am. Chem. Soc.* **1991**, *113*, 3246–3259.

13. Woolery, G. L.; Powers, L.; Winkler, M.; Solomon, E. I.; Lerch, K.; Spiro, T. G. Extended X-ray Absorption Fine Structure Study of the Coupled Binuclear Copper Active Site of Tyrosinase from *Neurospora Crassa. Biochim. Biophys. Acta* **1984**, *788*, 155–161.

14. Himmelwright, R. S.; Eickman, N. C.; Lubein, C. D.; Lerch, K.; Solomon, E. I. Chemical and Spectroscopic Studies of the Binuclear Copper Active Site of *Neurospora* Tyrosinase: Comparison to Hemocyanins. *J. Am. Chem. Soc.* **1980**, *102*, 7339–7344.

15. Eickman, N. C.; Solomon, E. I.; Larrabee, J. A.; Spiro, T. G.; Lerch, K. Ultraviolet Resonance Raman Study of Oxytyrosinase. Comparison with Oxyhemocyanins. *J. Am. Chem. Soc.* **1978**, *100*, 6529–6531.

16. Kitajima, N.; Koda, T.; Hashimoto, S.; Kitagawa, T.; Moro-oka, Y. An Accurate Synthetic Model of Oxyhaemocyanin. *J. Chem. Soc., Chem. Commun.* **1988**, 151–152.

17. Kitajima, N.; Fujisawa, K.; Moro-oka, Y.; Toriumi, K. μ-η_2:η_2-Peroxo Binuclear Copper Complex, $[Cu(HB(3,5-iPr_2pz)_3)]_2(O_2)$. *J. Am. Chem. Soc.* **1989**, *111*, 8975–8976.

18. Kitajima, N.; Koda, T.; Iwata, Y.; Moro-oka, Y. Reaction Aspects of a μ-Peroxo Binuclear Copper(II) Complex. *J. Am. Chem. Soc.* **1990**, *112*, 8833–8839.

19. Kitajima, N.; Koda, T.; Hashimoto, S.; Kitagawa, T.; Moro-oka, Y. Synthesis and Characterization of the Dinuclear Copper(II) Complexes $[Cu(HB(3,5-Me_2pz)_3)]_2X$ ($X = O^{2-}$, $(OH)_2^{2-}$, CO_2^{2-}, O_2^{2-}). *J. Am. Chem. Soc.* **1991**, *113*, 5664–5671.

20. Kitajima, N.; Fujisawa, K.; Moro-oka, Y. Formation and Characterization of a Mononuclear (Acylperoxo)copper(II) Complex. *Inorg. Chem.* **1990**, *29*, 357–358.

21. Mahapatra, S.; Halfen, J. A.; Wilkinson, E. C.; Pan, G.; Cramer, C. J.; Que, Jr., L.; Tolman, W. B. A New Intermediate in Copper Dioxygen Chemistry: Breaking the O–O Bond To Form a $\{Cu_2(\mu\text{-}O)_2\}^{2+}$ Core. *J. Am. Chem. Soc.* **1995**, *117*, 8865–8866.

22. Halfen, J. A.; Mahapatra, S.; Wilkinson, E. C.; Kaderli, S.; Que, Jr., L.; Tolman, W. B. Reversible Cleavage and Formation of the Dioxygen O—O Bond Within a Dicopper Complex. *Science* **1996**, *271*, 1397–1400.

23. Cramer, C. J.; Smith, B. A.; Tolman, W. B. *Ab Initio* Characterization of the Isomerism between the μ-η_2:η_2-Peroxo- and Bis(μ-oxo)dicopper Cores. *J. Am. Chem. Soc.* **1996**, *118*, 11283–11287.

24. Shu, L.; Nesheim, J. C.; Kauffmann, K.; Münck, E.; Lipscomb, J. D.; Que, Jr., L. An $Fe_2^{IV}O_2$ Diamond Core Structure for the Key Intermediate Q of Methane Monooxygenase. *Science* **1997**, *275*, 515–518.

25. Pecoraro, V. L.; Baldwin, M. J.; Gelasco, A. Interaction of Manganese with Dioxygen and Its Reduced Derivatives. *Chem. Rev.* **1994**, *94*, 807–826.

26. Sauer, K.; Yachandra, V. K.; Britt, R. D.; Klein, M. P. The Photosynthetic Water Oxidation Complex Studied by EPR and X-ray Absorption Spectroscopy, *Manganese Redox Enzymes*; V. L. Pecoraro, Ed.; VCH; New York, 1992; pp 141–175.

27. Yachandra, V. K.; DeRose, V. J.; Latimer, M. J.; Mukerji, I.; Sauer, K.; Klein, M. P. Where Plants Make Oxygen: A Structural Model for the Photosynthetic Oxygen-Evolving Manganese Cluster. *Science* **1993**, *260*, 675–679.

28. Wieghardt, K. The Active Sites in Manganese-Containing Metalloproteins and Inorganic Model Complexes. *Angew. Chem. Int. Ed. Engl.* **1989**, *28*, 1153–1172.

29. Proserpio, D. M.; Hoffmann, R.; Dismukes, G. C. Molecular Mechanism of Photosynthetic Oxygen Evolution. A Theoretical Approach. *J. Am. Chem. Soc.* **1992**, *114*, 4374–4382.

30. Jacobson, R. R.; Tyeklar, Z.; Farooq, A.; Karlin, K. D.; Liu, S.; Zubieta, J. A Copper–Oxygen (Cu_2-O_2) Complex. Crystal Structure and Characterization of a Reversible Dioxygen Binding System. *J. Am. Chem. Soc.* **1988**, *110*, 3690–3692.

31. Ookubo, T.; Sugimoto, H.; Nagayama, T.; Masuda, H.; Sato, T.; Tanaka, K.; Maeda, Y.; Okawa, H.; Hayashi, Y.; Uehara, A.; Suzuki, M. *cis*-μ-1,2-Peroxo Diiron Complex: Structure and Reversible Oxygenation. *J. Am. Chem. Soc.* **1996**, *118*, 701–702.

32. Dong, Y.; Shiping, Y.; YoungJr., V. G.; Que, Jr., L. Crystal Structure Analysis of a Synthetic Non-Heme Diiron–O_2 Adduct: Insight into the Mechanism of Oxygen Activation. *Angew. Chem. Int. Ed. Engl.* **1996**, *35*, 618–620.

33. Kim, K.; Lippard, S. J. Structure and Mössbauer Spectrum of a (μ-1,2-Peroxo)bis (μ-carboxylato)diiron(III) Model for the Peroxo Intermediate in the Methane Monooxygenase Hydroxylase Reaction Cycle. *J. Am. Chem. Soc.* **1996**, *118*, 4914–4915.

34. Yoshizawa, K.; Hoffmann, R. Dioxygen Binding to Dinuclear Iron Centers on Methane Monooxygenase Models. *Inorg. Chem.* **1996**, *35*, 2409–2410.

35. Yoshizawa, K.; Yamabe, T.; Hoffmann, R. Possible Intermediates for the Conversion of Methane to Methanol on Dinuclear Iron Centers of Methane Monooxygenase Models. *New J. Chem.* **1997**, *21*, 151–161.

36. Yoshizawa, K.; Ohta, T.; Yamabe, T.; Hoffmann. R. Dioxygen Cleavage and Methane Activation on Diiron Enzyme Models: A Theoretical Study. *J. Am. Chem. Soc.* **1997**, *119*, 12311–12321.

37. Que, Jr., L.; Ho, R. Y. N. Dioxygen Activation by Enzymes with Mononuclear Non-Heme Iron Active Sites. *Chem. Rev.* **1996**, *96*, 2607–2624.

38. Wallar, B. J.; Lipscomb, J. D. Dioxygen Activation by Enzymes Containing Binuclear Non-Heme Iron Clusters. *Chem. Rev.* **1996**, *96*, 2625–2658.

39. Feig, A. L.; Lippard, S. J. Reactions of Non-Heme Iron(II) Centers with Dioxygen in Biology and Chemistry. *Chem. Rev.* **1994**, *94*, 759–805.

40. Lippard, S. J. Oxo-Bridged Polyiron Centers in Biology and Chemistry. *Angew. Chem. Int. Ed. Engl.* **1988**, *27*, 344–361.

41. Yoshizawa, K.; Ohta, T.; Yamabe, T. A Theoretical Study of Dioxygen Activation on the Dicopper Enzyme Models. *Bull. Chem. Soc. Jpn.* **1997**, *70*, 1911–1917.

42. Kitajima, N.; Fujisawa, K.; Fujimoto, C.; Moro-oka, Y.; Hashimoto, S.; Kitagawa, T.; Toriumi, K.; Tatsumi, K.; Nakamura, A. A New Model for Dioxygen Binding in Hemocyanin. Synthesis, Characterization, and Molecular Structure of the μ-η_2:η_2 Peroxo Dinuclear Copper(II) Complexes, [Cu(HB(3,5-R_2pz)$_3$)]$_2$(O_2) (R = isopropyl and Ph). *J. Am. Chem. Soc.* **1992**, *114*, 1277–1291.

43. Extended Hückel calculations were carried out using "YAeHMOP": Landrum, G. Cornell University, Ithaca, New York, 1995.

44. See Hoffmann, R. An Extended Hückel Theory. I. Hydrocarbons. *J. Chem. Phys.* **1963**, *39*, 1397–1412.

45. Hoffmann, R.; Lipscomb, W. N. Theory of Polyhedral Molecules. I. Physical Factorizations of the Secular Equation. *J. Chem. Phys.* **1962**, *36*, 2179–2189.

46. Alvarez, S., Table of parameters for the extended Hückel Method, Universitat de Barcelona, Barcelona, 1993.

47. Matoba, Y.; Kumagai, T.; Yamamoto, A.; Yoshitsu, H.; Sugiyama, M. Crystallographic Evidence That the Dinuclear Copper Center of Tyrosinase Is Flexible during Catalysis. *J. Biol. Chem.* **2006**, *281*, 8981–8990.

48. Garcia-Borron, J. C.; Solano, F. Molecular Anatomy of Tyrosinase and Its Related Proteins: Beyond the Histidine-Bound Metal Catalytic Center. *Pigm. Cell Res.* **2002**, *15*, 162–173.

49. Land, E. J.; Ramsden, C. A.; Riley, P. A. Tyrosinase Autoactivation and the Chemistry of *ortho*-Quinone Amines. *Acc. Chem. Res.* **2003**, *36*, 300–308.

50. Halaouli, S.; Asther, M.; Sigoillot, J. C.; Hamdi, M.; Lomascolo, A. Fungal Tyrosinases: New Prospects in Molecular Characteristics, Bioengineering and Biotechnological Applications. *J. App. Microbiol.* **2006**, *100*, 219–232.

51. Marusek, C. M.; Trobaugh, N. M.; Flurkey, W. H.; Inlow, J. K. Comparative Analysis of Polyphenol Oxidase from Plant and Fungal Species. *J. Inorg. Biochem.* **2006**, *100*, 108–123.

52. Wang, N.; Hebert, D. N. Tyrosinase Maturation through the Mammalian Secretory Pathway: Bringing Color to Life. *Pigm. Cell. Res.* **2006**, *19*, 3–18.

53. Claus, H.; Decker, H. Bacterial Tyrosinases. *Syst. Appl. Microbiol.* **2006**, *29*, 3–14.

54. Gaykema, W. P. J.; Hol, W. G. J.; Vereijken, J. M.; Soeter, N. M.; Bak, H. J.; Beintema, J. J. 3.2 Å Structure of the Copper-Containing, Oxygen-Carrying Protein *Panulirus Interruptus* Haemocyanin. *Nature (London)* **1984**, *309*, 23–29.

55. Magnus, K. A.; Hazes, B.; Ton-That, H.; Bonaventura, C.; Bonaventura, J.; Hol, W. G. J. Crystallographic Analysis of Oxygenated and Deoxygenated States of Arthropod Hemocyanin Shows Unusual Differences. *Proteins* **1994**, *19*, 302–309.

56. Cuff, M. E.; Miller, K. I.; van Holde, K. E.; Hendrickson, W. A. Crystal Structure of a Functional Unit from *Octopus* Hemocyanin. *J. Mol. Biol.* **1998**, *278*, 855–870.

57. Klabunde, T.; Eicken, C.; Sacchettini, J. C.; Krebs, B. Crystal Structure of a Plant Catechol Oxidase Containing a Dicopper Center. *Nat. Struct. Biol.* **1998**, *5*, 1084–1090.

58. Gerdemann, C.; Eicken, C.; Krebs, B. The Crystal Structure of Catechol Oxidase: New Insight into the Function of Type-3 Copper Proteins. *Acc. Chem. Res.* **2002**, *35*, 183–191.

59. Jolley, R. L., Jr., Evans, L. H.; Mason, H. S. Reversible Oxygenation of Tyrosinase. *Biochem. Biophys. Res.* **1972**, *46*, 878–884.

60. Jolley, R. L., Jr., ; Evans, L. H.; Makino, N.; Mason, H. S. *Oxytyrosinase. J. Biol. Chem.* **1974**, *249*, 335–345.

61. Jackman, M. P.; Hajnal, A.; Lerch, K. Albino Mutants of *Streptomyces Glaucescens* Tyrosinase. *Biochem. J.* **1991**, *274*, 707–713.

62. van Gelder, C. W. G.; Flurkey, W. H.; Wichers, H. J. Sequence and Structural Features of Plant and Fungal Tyrosinases. *Phytochemistry* **1997**, *45*, 1309–1323.

63. Rodríguez-López, J. N.; Tudela, J.; Varón, R.; García-Carmona, F.; García-Cánovas, F. Analysis of a Kinetic Model for Melanin Biosynthesis Pathway. *J. Biol. Chem.* **1992**, *267*, 3801–3810.

64. Blackburn, N. J.; Strange, R. W.; Farooq, A.; Hake, M. S.; Karlin, K. D. X-ray Absorption Studies of Three-Coordinate Dicopper(I) Complexes and Their Dioxygen Adducts. *J. Am. Chem. Soc.* **1988**, *110*, 4263–4272.

65. Mirica, L. M.; Ottenwaelder, X.; Stack, T. D. P. Structure and Spectroscopy of Copper–Dioxygen Complexes. *Chem. Rev.* **2004**, *104*, 1013–1046.

66. Lewis, E. A.; Tolman, W. B. Reactivity of Dioxygen–Copper Systems. *Chem. Rev.* **2004**, *104*, 1047–1076.

67. Hatcher, L. Q.; Karlin, K. D. Ligand Influences in Copper–Dioxygen Complex-Formation and Substrate Oxidations. *Adv. Inorg. Chem.* **2006**, *58*, 131–184.

68. Itoh, S.; Fukuzumi, S. Monooxygenase Activity of Type 3 Copper Proteins. *Acc. Chem. Res.* **2007**, *40*, 592–600.

69. Kitajima, N.; Koda, T.; Iwata, Y.; Moro-oka, Y. Reaction Aspects of a μ-Peroxo Binuclear Copper(II) Complex. *J. Am. Chem. Soc.* **1990**, *112*, 8833–8839.

70. Paul, P. P.; Tyeklár, Z.; Jacobson, R. R.; Karlin, K. D. Reactivity Patterns and Comparisons in Three Classes of Synthetic Copper–Dioxygen {Cu$_2$–O$_2$} Complexes: Implication for Structure and Biological Relevance. *J. Am. Chem. Soc.* **1991**, *113*, 5322–5332.

71. Mahapatra, S.; Halfen, J. A.; Wilkinson, E. C., Jr., Tolman, W. B. Modeling Copper-Dioxygen Reactivity in Proteins: Aliphatic C−H Bond Activation by a New Dicopper (II)–Peroxo Complex. *J. Am. Chem. Soc.* **1994**, *116*, 9785–9786.

72. Obias, H. V.; Lin, Y.; Murphy, N. N.; Pidcock, E.; Solomon, E. I.; Ralle, M.; Blackburn, N. J.; Neuhold, Y.-M.; Zuberbühler, A. D.; Karlin, K. D. Peroxo-, Oxo-, and Hydroxo-Bridged Dicopper Complexes: Observation of Exogenous Hydrocarbon Substrate Oxidation. *J. Am. Chem. Soc.* **1998**, *120*, 12960–12961.

73. Halfen, J. A.; Young, V. G., Jr.; Tolman. W. B. An Unusual Ligand Oxidation by a (μ-η$_2$:η$_2$-Peroxo)dicopper Compound: 1° > 3° C−H Bond Selectivity and a Novel Bis (μ-alkylperoxo)dicopper Intermediate. *Inorg. Chem.* **1998**, *37*, 2102–2103.

74. Mahadevan, V.; DuBois, J. L.; Hedman, B.; Hodgson, K. O.; Stack, T. D. P. Exogenous Substrate Reactivity with a [Cu(III)$_2$O$_2$]$^{2+}$ Core: Structural Implications. *J. Am. Chem. Soc.* **1999**, *121*, 5583–5584.

75. Mahadevan, V.; Henson, M. J.; Solomon, E. I.; Stack, T. D. P. Differential Reactivity between Interconvertible Side-On Peroxo and Bis-μ-oxodicopper Isomers Using Peralkylated Diamine Ligands. *J. Am. Chem. Soc.* **2000**, *122*, 10249–10250.

76. Osako, T.; Ohkubo, K.; Taki, M.; Tachi, Y.; Fukuzumi, S.; Itoh, S. Oxidation Mechanism of Phenols by Dicopper–Dioxygen (Cu$_2$/O$_2$) Complexes. *J. Am. Chem. Soc.* **2003**, *125*, 11027–11033.

77. Inoue, T.; Shiota, Y.; Yoshizawa, K. Quantum Chemical Approach to the Mechanism for the Biological Conversion of Tyrosine to Dopaquinone. *J. Am. Chem. Soc.* **2008** *130*, 16890–16897.

78. Siegbahn, P. E. M.; Wirstam, M. Is the Bis-μ-Oxo Cu$_2$(III,III) State an Intermediate in Tyrosinase? *J. Am. Chem. Soc.* **2001**, *123*, 11819–11820.

79. Siegbahn, P. E. M. The Catalytic Cycle of Tyrosinase: Peroxide Attack on the Phenolate Ring Followed by O−O Bond Cleavage. *J. Biol. Inorg. Chem.* **2003**, *8*, 567–576.

80. Decker, H.; Schweikardt, T.; Tuczek, F. The First Crystal Structure of Tyrosinase: All Questions Answered? *Angew. Chem., Int. Ed.* **2006**, *45*, 4546–4550.

81. Decker, H.; Schweikardt, T.; Nillius, D.; Salzbrunn, U.; Jaenicke, E.; Tuczek, F. Similar Enzyme Activation and Catalysis in Hemocyanins and Tyrosinases. *Gene* **2007**, *398*, 183–191.

82. Maseras, F.; Morokuma, K. IMOMM: A New Integrated *Ab Initio* + Molecular Mechanics Geometry Optimization Scheme of Equilibrium Structures and Transition States. *J. Comput. Chem.* **1995**, *16*, 1170–1179.

83. Svensson, M.; Humbel, S.; Froese, R. D. J.; Matsubara, T.; Sieber, S.; Morokuma, K. ONIOM: A Multilayered Integrated MO + MM Method for Geometry Optimizations

and Single Point Energy Predictions. A Test for Diels–Alder Reactions and Pt(P(t-Bu)₃)₂
+ H₂ Oxidative Addition. *J. Phys. Chem.* **1996**, *100*, 19357–19363.

84. Frost, A. A.; Pearson, R. G. *Kinetics and Mechanism*; John Wiley & Sons, Inc.: New York, 1961.

85. Groves, J. T.; Han, Y.-Z. Models and Mechanism of Cytochrome P450 Action In Cytochrome P450: Structure, Mechanism, and Biochemistry, 2nd ed.; Ortiz de Montellano, P. R. Ed.; Plenume: New York, 1995; pp 3–48.

86. Dalton, H. The Leeuwenhoek Lecture 2000 The natural and unnatural history of methane-oxidizing bacteria. *Philos. Trans. R. Soc. London, Ser. B* **2005**. *360*, 1207–1222.

87. Hanson, R. S; Hanson, T. E. Methanotrophic Bacteria. *Microbiol. Rev.* **1996**, *60*, 439–471.

88. Chan, S. I.; Chen, K. H.-C.; Yu, S. S.-F.; Chen, C.-L.; Kuo, S. S.-J. Toward Delineating the Structure and Function of the Particulate Methane Monooxygenase from Methanotrophic Bacteria. *Biochemistry* **2004**, *43*, 4421–4430.

89. Lieberman, R. L.; Shrestha, D. B.; Doan, P. E.; Hoffman, B. M.; Stemmler, T. L.; Rosenzweig, A. C. Bioinorganic Chemistry Special Feature: Purified particulate methane monooxygenase from *Methylococcus capsulatus* (Bath) is a dimer with both mononuclear copper and a copper-containing cluster. *Proc. Natl. Acad. Sci. USA* **2003**, *100*, 3820–3825.

90. Murrell, J. C.; McDonald, I. R.; Gilbert, B. Regulation of expression of methane monooxygenases by copper ions. *Trends Microbiol.* **2000**, *8*, 221–225.

91. Prior, S. D.; Dalton, H. The Effect of Copper Ions on Membrane Content and Methane Monooxygenase Activity in Methanol-grown Cells of Methylococcus capsulatus (Bath). *J. Gen. Microbiol.* **1985**, *131*, 155–163.

92. Stanley, S. H.; Prior, S. D.; Leak, D. J.; Dalton, H. Copper stress underlies the fundamental change in intracellular location of methane mono-oxygenase in methane-oxidizing organisms: Studies in batch and continuous cultures. *Biotechnol. Lett.* **1983**, *5*, 487–492.

93. Lieberman, R.; Rosenzweig, A. C. Crystal structure of a membrane-bound metalloenzyme that catalyses the biological oxidation of methane. *Nature (London)* **2005**, *434*, 177–182.

94. Hakemian, A. S.; Rosenzweig, A. C. The Biochemistry of Methane Oxidation. *Ann. Rev. Biochem.* **2007**, *76*, 223–241.

95. Lieberman, R. L.; Kondapalli, K. C.; Shrestha, D. B.; Hakemian, A. S.; Smith, S. M.; Telser, J.; Kuzelka, J.; Gupta, R.; Borovik, A. S.; Lippard, S. J.; Hoffman, B. M.; Rosenzweig, A. C.; Stemmler, T. L. Characterization of the Particulate Methane Monooxygenase Metal Centers in Multiple Redox States by X-ray Absorption Spectroscopy. *Inorg. Chem.* **2006**, *45*, 8372–8381.

96. Hakemian, A. S.; Kondapalli, K. C.; Telser, J.; Hoffman, B. M.; Stemmler, T. L.; Rosenzweig, A. C. The Metal Centers of Particulate Methane Monooxygenase from Methylosinus trichosporium OB3b. *Biochemistry* **2008**, *47*, 6793–6801.

97. Nguyen, H.-H. T.; Shiemke, A. K.; Jacobs, S. J.; Hales, B. J.; Lidstrom, M. E.; Chan, S. I. The nature of the copper ions in the membranes containing the particulate methane monooxygenase from Methylococcus capsulatus (Bath). *J. Biol. Chem.* **1994**, *269*, 14995–15005.

98. Nguyen, H.-H. T.; Nakagawa, K. H.; Hedman, B.; Elliott, S. J.; Lidstrom, M. E.; Hodgson, K. O.; Chan, S. I. X-ray Absorption and EPR Studies on the Copper Ions Associated with

the Particulate Methane Monooxygenase from Methylococcus capsulatus (Bath). Cu(I) Ions and Their Implications. *J. Am. Chem. Soc.* **1996**, *118*, 12766–12776.

99. Yuan, H.; Collins, M. L. P.; Antholine, W. E. Low-Frequency EPR of the Copper in Particulate Methane Monooxygenase from Methylomicrobium albus BG8. *J. Am. Chem. Soc.* **1997**, *119*, 5073–5074.

100. Takeguchi, M.; Miyakawa, K.; Okura, I. Purification and properties of particulate methane monooxygenase from Methylosinus trichosporium OB3b. *J. Mol. Catal. A* **1998**, *132*, 145–153.

101. Lemos, S. S.; Yuan, H.; Perille-Collins, M. L. P. Review of multifrequency EPR of copper in particulate methane monooxygenase. *Curr. Top. Biophys.* **2002**, *26*, 43–48.

102. Choi, D.-W.; Kunz, R. C.; Boyd, E. S.; Semrau, J. D.; Antholin, W. E.; Han, J.-I.; Zahn, J. A.; Boyd, J. M.; de la Mora, A. M.; DiSpirito, A. A. The Membrane-Associated Methane Monooxygenase (pMMO) and pMMO–NADH:Quinone Oxidoreductase Complex from Methylococcus capsulatus Bath. *J. Bacteriol.* **2003**, *185*, 5755–5764.

103. Basu, P.; Katterle, B.; Andersson, K. K.; Dalton, H. The membrane-associated form of methane mono-oxygenase from Methylococcus capsulatus (Bath) is a copper/iron protein. *Biochem. J.* **2003**, *369*, 417–427.

104. Elliott, S. J.; Zhu, M.; Tso, L.; Nguyen, H.-H. T.; Yip, J. H.-K.; Chan, S. I. Regio- and Stereoselectivity of Particulate Methane Monooxygenase from Methylococcus capsulatus (Bath). *J. Am. Chem. Soc.* **1997**, *119*, 9949–9955.

105. Zahn, J. A.; DiSpirito, A. A. Membrane-associated methane monooxygenase from Methylococcus capsulatus (Bath). *J. Bacteriol.* **1996**, *178*, 1018–1029.

106. Kim, H. J.; Graham, D. W.; DiSpirito, A. A.; Alterman, M. A.; Galeva, N.; Larive, C. K.; Asunskis, D.; Sherwood, P. M. A. Methanobactin, a Copper-Acquisition Compound from Methane-Oxidizing Bacteria. *Science* **2004**, *305*, 1612–1615.

107. Balasubramanian, R.; Smith, S. M.; Rawat, S.; Yatsunyk, L. A.; Stemmler, T. L.; Rosenzweig, A. C. Oxidation of methane by a biological dicopper centre. *Nature (London)* **2010**, *465*, 115–119.

108. Yoshizawa, K.; Suzuki, A.; Shiota, Y.; Yamabe, T. Conversion of Methane to Methanol on Diiron and Dicopper Enzyme Models of Methane Monooxygenase: A Theoretical Study on a Concerted Reaction Pathway. *Bull. Chem. Soc. Jpn.* **2000**, *73*, 815–827.

109. Yoshizawa, K.; Shiota, Y. Conversion of Methane to Methanol at the Mononuclear and Dinuclear Copper Sites of Particulate Methane Monooxygenase (pMMO): A DFT and QM/MM Study. *J. Am. Chem. Soc.* **2006**, *128*, 9873–9881.

110. Shiota, Y.; Yoshizawa, K. Comparison of the Reactivity of Bis(μ-oxo)Cu^{II}Cu^{III} and Cu^{III}Cu^{III} Species to Methane. *Inorg. Chem.* **2009**, *48*, 838–845.

111. Chen, P. P.-Y.; Chan, S. I. Theoretical modeling of the hydroxylation of methane as mediated by the particulate methane monooxygenase. *J. Inorg. Biochem.* **2006**, *100*, 801–809.

112. Chen K. H.-C.; Chen C.-L.; Tseng, C.-F.; Yu, S. S.-F.; Ke, S.-C.; Lee, J.-F.; Nguyen, H.-H. T.; Elliott, S. J.; Alben, J. O.; Chan, S. I. The Copper Clusters in the Particulate Methane Monooxygenase (pMMO) from Methylococcus casulatus (Bath). *J. Chin. Chem. Soc.* **2004**, *51*, 1081–1098.

113. Hung, S. C.; Chen C.-L.; Chen K. H.-C.; Yu, S. S.-F.; Chan, S. I. The Catalytic Cop per Clusters of the Particulate Methane Monooxygenase from Methanotrophic Bacteria:

Electron Para magnetic Resonance Spectral Simulations. *J. Chin. Chem. Soc.* **2004**, *51*, 1229–1244.

114. Chan, S. I.; Wang, V. C.-C.; Lai, J. C.-H.; Yu, S. S.-F.; Chen, P. P.-Y.; Chen K. H.-C.; Chen C.-L.; Chan, M. K. J. Redox Potentiometry Studies of Particulate Methane Monooxygenase: Support for a Trinuclear Copper Cluster Active Site. *Angew. Chem., Int. Ed.* **2007**, *46*, 1992–1994.

115. Chan, S. I.; Yu, S. S.-F. Controlled Oxidation of Hydrocarbons by the Membrane-Bound Methane Monooxygenase: The Case for a Tricopper Cluster. *Acc. Chem. Res.* **2008**, *41*, 969–979.

116. Yoshizawa, K. Non-Radical Mechanism for Methane Hydroxylation by Iron-Oxo Complexes. *Acc. Chem. Res.* **2006**, *39*, 375–382.

117. Wilkinson, B.; Zhu, M.; Priestley, N. D.; Nguyen, H.-H. T.; Morimoto, H.; Williams, P. G.; Chan, S. I.; Floss, H. G. A Concerted Mechanism for Ethane Hydroxylation by the Particulate Methane Monooxygenase from Methylococcus capsulatus (Bath). *J. Am. Chem. Soc.* **1996**, *118*, 921–922.

118. Hori, K.; Shiota, Y.; Yoshizawa, K. submitted.

8

CHEMICAL REACTIVITY OF COPPER ACTIVE-OXYGEN COMPLEXES

SHINOBU ITOH

Department of Material and Life Science, Division of Advanced Science and Biotechnology, Graduate School of Engineering, Osaka University, 2-1 Yamada-oka, Suita, Osaka 565–0871, Japan

Copper-Oxygen Chemistry, First Edition. Edited by Kenneth D. Karlin and Shinobu Itoh.
© 2011 John Wiley & Sons, Inc. Published 2011 by John Wiley & Sons, Inc.

8.1. INTRODUCTION

Copper complexes of active-oxygen species, such as superoxide ($O_2^{\bullet-}$), hydroperoxide (HOO^-), alkyl- or acylperoxide (ROO^-), and oxide (O^{2-}) have been invoked as key reactive intermediates in a wide variety of redox reactions involved not only in biological systems, but also in numerous catalytic oxidation processes.[1–16] Representative examples of such complexes are shown in Figure 8.1, most of which are well characterized by X-ray crystallographic analysis and/or several spectroscopic techniques.[7,8,17]

The $Cu-O_2$ adduct initially generated in the reaction of Cu(I) complex (LCu^I) and molecular oxygen (O_2) is a mononuclear Cu(II)–superoxo complex.[9] So far, both end-on and side-on superoxo Cu(II) complexes designated as Cu_1S^E and Cu_1S^S, respectively (Fig. 8.1), have been detected directly by spectroscopic methods ultraviolet–visible (UV–vis), resonance Raman (RR), etc.) in some ligand sysetms,[18–34] and one of each complex has been isolated by using sterically demanding supporting ligands and is structurally characterized by X-ray crystallographic analysis.[22,35] The binding mode of O_2 (end-on vs. side-on) to the copper ion, as well as the spin state (singlet vs. triplet) of Cu_1S^E, is controlled by the supporting ligands (L), but the details about the ligand roles are not yet understood clearly.[36] An X-ray crystallographic analysis of peptidylglycine α-amidating monooxygenase (PAM, see Chapter 1) has revealed the existence of a Cu_1S^E type oxygen adduct in the enzyme active site.[37] In a unique case with monoanionic didentate ligands (e.g., β-diketiminate derivatives), mononuclear copper(III)–peroxo complexes with a side-on binding mode Cu_1P^S has been obtained in the reaction of a Cu(I) precursor and O_2, where two electrons are transferred from the Cu(I) ion to molecular oxygen.[38–41] On the other hand, the reaction of Cu(II) complexes and superoxide ($O_2^{\bullet-}$) has also been investigated in connection with the chemical functions of copper–zinc superoxide dismutase (SOD),[42–48] where mononuclear copper(II)–superoxo adducts have in some cases also been characterized by spectroscopic methods.[43,48]

One-electron reduction and subsequent protonation of copper(II)–superoxo complexes or hydrogen atom ($H^{\bullet}$) abstraction from a certain substrate by the copper(II)–superoxo complexes will result in formation of copper(II)–hydroperoxo complexes Cu_1P^H.[49–58] Alternatively, Cu_1P^H can be prepared by the reaction of Cu(II) complexes and hydrogen peroxide (H_2O_2) in the presence of base, the so-called shunt pathway.[59–75] Copper(II)–alkylperoxo complexes Cu_1P^R and acylperoxo complexes Cu_1P^{Ac} have been prepared in a similar manner using alkyl hydroperoxides

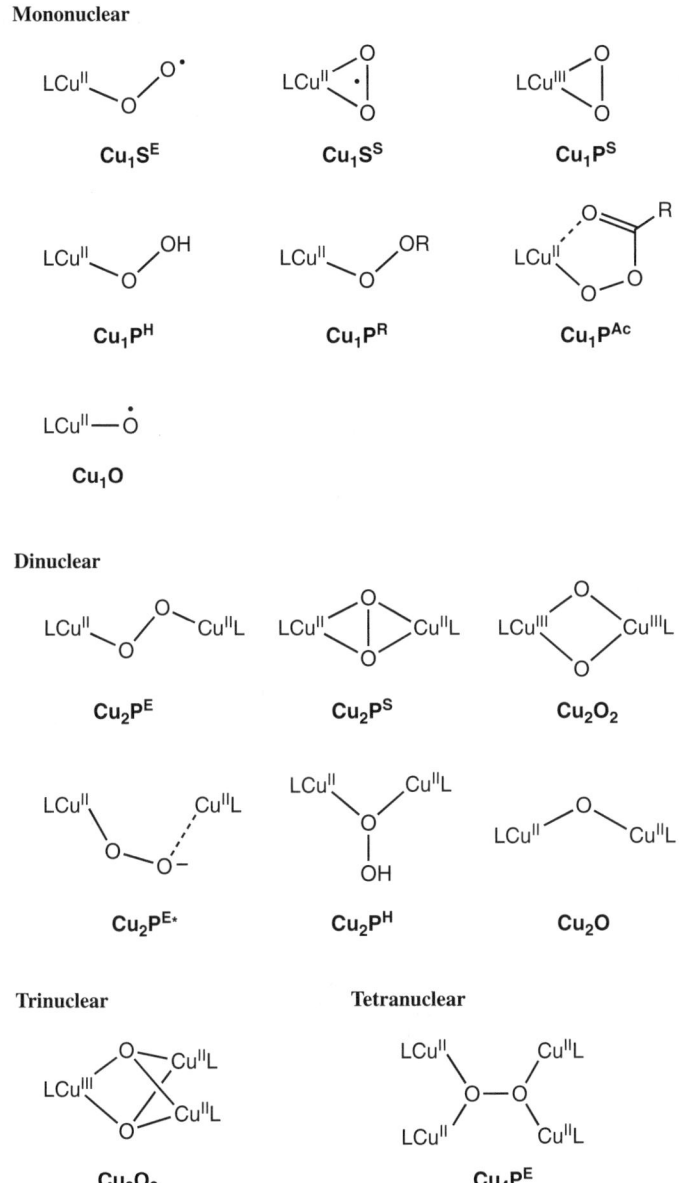

FIGURE 8.1. Active-oxygen complexes of copper (L denotes supporting ligand).

and peracids instead of H_2O_2, respectively.[76–81] These copper(II)–peroxo complexes **Cu₁Pᴴ**, **Cu₁Pᴿ**, and **Cu₁Pᴬᶜ** have been postulated as the key intermediates involved in several copper-catalyzed oxidation reactions. The O–O bond homolysis of these copper(II)–peroxo complexes may produce the more reactive oxidant

$LCu^{II}-O^{•}$ ($\mathbf{Cu_1O}$). However, such an oxyl radical species has never been isolated and has only been detected in a gas-phase reaction by mass spectrometry (MS).[82]

In many cases, mononuclear copper(II)–superoxo complexes, $\mathbf{Cu_1S^E}$ and $\mathbf{Cu_1S^S}$, further react *in situ* with another Cu(I) precursor complex to give end-on and side-on μ-peroxo dicopper(II) complexes, $\mathbf{Cu_2P^E}$ and $\mathbf{Cu_2P^S}$.[1,7,8,10,83] These complexes also have been isolated in some ligand systems and their structures have been determined by X-ray crystallographic analysis.[29,84–93] The crystal structures of the side-on peroxo dicopper(II) species $\mathbf{Cu_2P^S}$ involved in the active sites of hemocyanin and tyrosinase (type-3 copper proteins) were solved later.[94–96]

From the side-on peroxo dicopper(II) complexes $\mathbf{Cu_2P^S}$, homolytic cleavage of the $O-O$ bond takes place in certain ligand systems to provide bis(μ-oxo)dicopper(III) complexes $\mathbf{Cu_2O_2}$,[7,8,97–101] some of which have also been crystallized and structurally characterized by X-ray crystallographic analysis.[38,102–109] Since the intrinsic thermal stability of $\mathbf{Cu_2P^S}$ and $\mathbf{Cu_2O_2}$ complexes are relatively similar, the equilibrium position between them is largely affected by the structure of the supporting ligands, solvents, and the counteranions used for the preparation of these complexes.[7,97,102]

When bidentate alkylamine ligands carrying smaller *N*-alkyl substituents are employed, a $\mathbf{Cu_2O_2}$ intermediate generated may further react with an extra Cu(I) precursor complex to give a mixed-valent trinuclear Cu(II,II,III) complex, $\mathbf{Cu_3O_2}$.[110,111] In addition, formation of tetranuclear copper(II)–peroxo complexes, $\mathbf{Cu_4P^E}$, have been reported in certain ligand systems.[112,113]

So far, a large number of review articles have been published to provide very detailed information about the structures and physicochemical properties, as well as the reactivity of the copper active-oxygen complexes shown in Figure 8.1. Especially, two *Chemical Review* articles (Stack and co-workers[7] and Tolman and Lewis[8]) cover almost all of the information about the structures, spectroscopic characteristics, ligand effects, formation mechanism, and reactivity of the copper–dioxygen complexes reported before 2004.[7,8] Thus, this chapter mainly focuses on recent advances on the *chemical reactivity* of the series of *well-characterized* mononuclear copper active-oxygen complexes (Fig. 8.1). Reactivity of *well-defined* dinuclear copper active oxygen complexes reported after 2004 are also included. Dioxygen chemistry of the copper–heme hetero dinuclear system is summarized in Chapter 9, and biological, as well as theoretical, studies of the copper enzymes themselves are summarized in Chapters 1–7.

8.2. REACTIVITY OF MONONUCLEAR COPPER ACTIVE-OXYGEN SYSTEM

8.2.1. Copper(II) End-on Superoxo Complexes ($\mathbf{Cu_1S^E}$)

Mononuclear copper(II)–superoxo complexes are the species formed initially in the reaction of Cu(I) complexes and O_2. These complexes have attracted much recent attention as a possible key reactive intermediate of dopamine β-monooxygenase

(DβM, EC 1.14.17.1) and peptidylglycine α-hydroxylating monooxygenase (PHM, EC 1.14.17.3) (details of these enzymes are presented in Chapter 1.[14,37,114–117] As mentioned in the Introduction, mononuclear copper(II)–superoxo complexes are in general unstable and transiently formed during the course of the reactions of Cu(I) complexes and O_2 or in the reactions of Cu(II) complexes and superoxide $O_2^{\cdot-}$. The superoxo complexes generated *in situ* readily react with another Cu(I) precursor to generate μ-peroxo dicopper(II) complexes or rapidly decompose to unidentified products. Thus, almost nothing had been known about the intrinsic *chemical reactivity* of the mononuclear copper(II)–superoxo complexes until recently. Nonetheless, the spectroscopic features of **Cu_1S^E** (end-on superoxo complex), as well as the kinetics and thermodynamics for the reversible O_2 binding process have been well examined by using low-temperature stopped-flow techniques, and these data (UV–vis, K_{eq}, k_{on}, and k_{off}) are summarized in the literature.[7,19,21,27,28,118–120]

During the passed few years, significant progress has been made in the chemistry of **Cu_1S^E** complexes. Relatively stable mononuclear end-on superoxo–copper(II) complexes, **Cu_1S^E**, were prepared, and their structure, spectroscopic characteristics, as well as reactivity, have been explored (**1–3^X** in Chart 8.1).[30,32–35] These superoxo complexes were prepared by the reaction of O_2 and the Cu(I) precursor complexes supported by TMG$_3$tren [tris(tetramethylguanidino)tren], NMe$_2$TPA [tris(4-dimethylaminopyridin-2-ylmetyl)amine], and L3^X [1-(*p*-substituted-2-phenethyl)-5-(2-pyridin-2-ylethyl)-1,5-diazacyclooctane] ligands, respectively, at low temperature.

Complex **1** exhibits an intense absorption band at 447, 680, and 780 nm ($\varepsilon_{447} =$ 3400 $M^{-1} cm^{-1}$) and an isotope sensitive RR band at $\nu(^{16}O-^{16}O) = 1117 cm^{-1}$ [$\nu(^{18}O-^{18}O) = 1059 cm^{-1}$].[30] The shape of the UV–vis spectrum of **1** resembles those of the copper(II)–superoxo complexes detected during the course of the reactions of the Cu(I) complexes supported by the tripodal tetradentate TPA [tris(pyridin-2-ylmethyl) amine] ligand[19] and its derivatives,[28,118] as well as the tripodal N-alkylated tren ligands [tris(2-aminoethyl)amine] (Me$_6$tren,[23,27] Bz$_3$tren,[121] and Me$_3$Bz$_3$tren[29]; Bz $=$ benzyl).[9] The O$-$O bond stretching vibration and the isotope shift of compound **1** are also similar to those of the previously reported Me$_6$tren and Me$_3$Bz$_3$tren complexes.[27,29] Compound **1** has a triplet ground state ($S = 1$, paramagnetic), showing proton nuclear magnetic resonance (^{1}H NMR signals in an expanded region from -9 to 300 ppm, and exhibits reversibility in deoxygen-binding.[32] Finally, the end-on (η^1) superoxide binding mode and the trigonal bipyramidal (tbp) core structure have been confirmed by X-ray crystallographic analysis on [Cu(TMG$_3$tren)(O_2)]SbF$_6$ (**1**• SbF$_6$).[35]

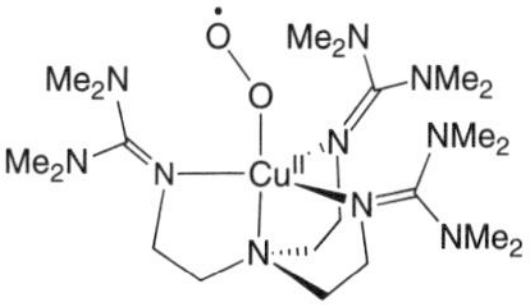

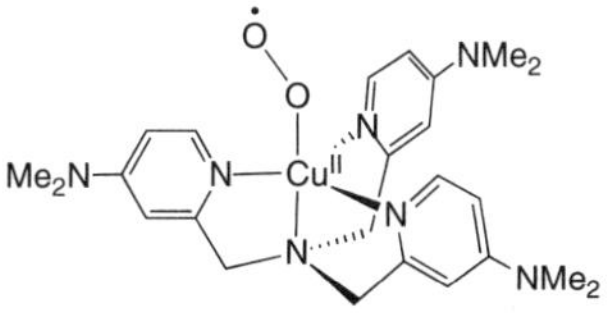

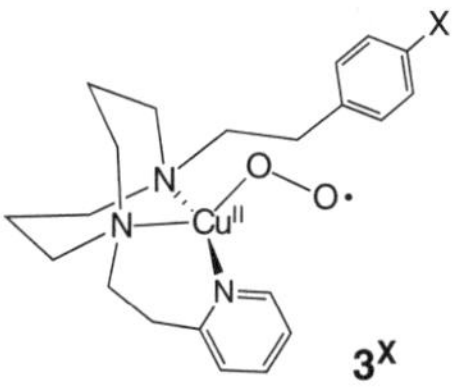

(X = OMe, Me, H, Cl, No$_2$)

CHART 8.1 Mononuclear copper(II) end-on superoxo complexes (**Cu_1S^E**).

The TPA ligand with a 4-dimethylamino group on the pyridine donor groups (NMe$_2$TPA) prevents the further reaction of generated **Cu$_1$S^E** with another Cu(I) precursor, thus stabilizing the mononuclear end-on superoxo–copper(II) complex **2**; $\lambda_{max} = 418$ nm ($\varepsilon = 4300\,M^{-1}\,cm^{-1}$), 615 (1100), and 767 (840); $\nu(^{16}O-^{16}O) = 1121\,cm^{-1}$, $\nu(^{18}O-^{18}O) = 1058$, $\nu(Cu-^{16}O) = 472$, $\nu(Cu-^{18}O) = 452$.[33] Probably, the electron-donating NMe$_2$ groups enhance the reactivity of the Cu(I) complex toward O$_2$ and makes the dioxygen-binding process practically irreversible.

In contrast to the tripodal tetradentate TPA and tren ligand systems that provide the trigonal bipyramidal Cu(II) end-on superoxide complexes **1** and **2**, tridentate ligand L3^X, consisting of a relatively rigid 1,5-diazacyclooctane framework and a pyridylethyl arm, affords a mononuclear Cu(II) end-on superoxide complex with a distorted tetrahedral gemometry.[34] The ligand has a unique structural feature stabilizing the copper complexes with lower coordination numbers. For example, the Cu(I) complex of L3^X exhibits a three coordinate T-shape structure without any external coligand even though the complex is made in the strongly coordinative solvent MeCN, and the Cu(II) complexes, such as [Cu^{II}(L3^X)(Cl)]$^+$ and [Cu^{II}(L3^X)(OAc)]$^+$, show a four-coordinate distorted tetrahedral geometry (usually, the Cl$^-$ ligand tends to act as a bridging ligand to afford dimeric Cu(II) complexes and AcO$^-$ acts as a didentate ligand exhibiting an η^2-binding mode).[34] In general, flexible tridentate alkylamine ligands afford four-coordinate tetrahedral Cu(I) and five-coordinate square pyramidal or trigonal bipyramidal Cu(II) complexes, including external coligand(s).

The reaction of [Cu^I(L3^H)]$^+$ and O$_2$ at low temperature affords a mononuclear copper(II)–superoxo complex **3^X** exhibiting characteristic absorption bands at 397 nm ($\varepsilon = 4200\,M^{-1}\,cm^{-1}$), 570 (850), and 705 (1150). The final spectrum is negligibly affected by changing the solvent (acetone-MeCN, CH$_2$Cl$_2$, and propionitrile), suggesting that no solvent coordination to the metal center takes place as in the case of the Cu(I) precursor. Moreover, similar spectra were obtained in the reactions with a series of **1^X** (X = OMe, Me, H, Cl, and NO$_2$) and O$_2$. The final spectrum of **3^X** is rather similar to that of the above-mentioned Cu(II) end-on (η^1) superoxo complexes **Cu$_1$S^E**, but is significantly different from that of the Cu(II) side-on (η^2) superoxo complex **Cu$_1$S^S** (see section 8.2.2).[22] The complex generated using $^{16}O_2$ shows isotope-sensitive RR bands at 1033 and 457 cm^{-1}, which shift to 968 and 442 cm^{-1}, respectively, upon $^{18}O_2$ substitution. The peak positions, as well as the associated isotope shifts, are similar to those of the reported copper(II)–superoxo complexes, and can be assigned to the O$-$O and Cu$-$O stretching vibrations, respectively. Exactly the same Raman bands are obtained in other solvent systems, also indicating no solvent coordination to the metal center. The fine-structure electron spin resonance (ESR) spectrum of **3^H** in the parallel microwave excitation mode at 3 K, clearly indicates that the superoxo complex is spin–triplet in the ground state.[34] The estimated distance between the two unpaired electron spins is 2.73 Å, which is nearly identical to the distance between the copper and the distal oxygen atom of the end-on superoxide ligand in the active site of peptidylglycine α-hydroxylating monooxygenase (PHM) (2.78 Å, Fig. 8.2).[37] Such a good agreement between the estimated spin–spin distance and the Cu$-$O$_{distal}$ distance in the enzymatic system further supports the end-on binding mode of O$_2^{\cdot -}$.

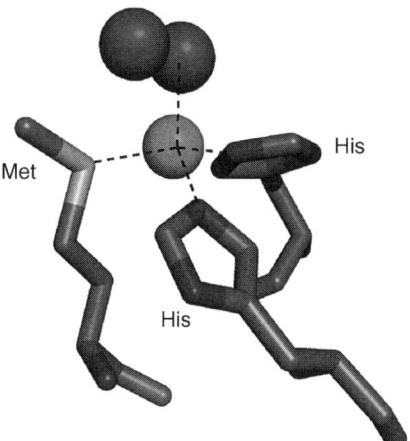

FIGURE 8.2. The active site structure of peptidylglycine α-hydroxylating monooxygenase. (see Ref. 37.) (See color insert.)

Reactivity of $\mathbf{Cu_1S^E}$ has also been investigated to obtain possible insights into the reactive intermediate of DβM and PHM. Both compounds **1** and **2** abstract hydrogen atoms from phenolic substrates generating the corresponding phenoxyl radical intermediates, which further undergo several types of reactions to give quinones, ortho-hydroperoxylated products, and C–C coupling dimers.[33,51] Notably, in the reaction of compound **1** and phenols, aliphatic hydroxylation takes place on one of the methyl groups of the tetramethylguanidino group of TMG$_3$tren ligand to give compound **4** in an ~80% yield (Scheme 8.1).[51] The same reaction (methyl hydroxylation of the supporting ligand) takes place from superoxo complex **1** in the presence of other hydrogen-atom donors, such as TEMPO-H (2,2,6,6-tetramethylpyperidine-1-hydroxide) or in the reaction of Cu(II) complex supported by TMG$_3$tren and H$_2$O$_2$. On the other hand, however, thermal decomposition of **1** without any hydrogen-atom donor, such as phenols and TEMPO-H, does not cause the ligand hydroxylation.[51] These results suggest that Cu(II) hydroperoxo complex $[Cu^{II}(TMG_3tren)(OOH)]^+$ ($\mathbf{Cu_1P^H}$ in Fig. 8.1) generated *in situ* is involved as a key intermediate (Scheme 8.1). Moreover, the same reaction (ligand hydroxylation) occurs when $[Cu^I(TMG_3tren)]^+$ is treated with iodosylbenzene PhIO.[51] The result further implicates that the reaction involves the $Cu^{II}–O^{\bullet}$ or $Cu^{III}=O$ type intermediate ($\mathbf{Cu_1O}$ in Fig. 8.1) generated by the O–O bond homolysis of $[Cu^{II}(TMG_3tren)(OOH)]^+$ ($\mathbf{Cu_1P^H}$) as the real reactive species for the methyl hydroxylation reaction. Recent density functional theory (DFT) calculation studies on this reaction suggest that the hydrogen atom of the ligand methyl group is abstracted by hydroxyl radical (HO$\bullet$) species that is formed by O–O homolysis of $\mathbf{Cu_1P^H}$ and then the generated $\mathbf{Cu_1O}$ just rebinds to the carbon center radical to give the hydroxylated product.[122] This mechanism looks similar to so-called Mechanism III or IV of DβH and PHM described in the Klinman review article,[114] where the C–H bond activation of the substrate occurs after the O–O bond cleavage of a $Cu^{II}–OOH$ ($\mathbf{Cu_1P^H}$) intermediate. However, those mechanisms

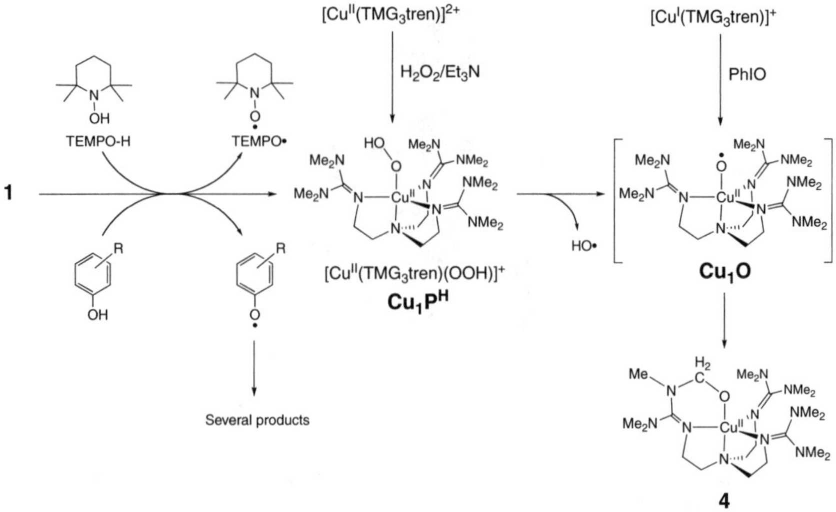

SCHEME 8.1. Cu/O$_2$ chemistry of the TMG$_3$tren ligand system.

(Mechanism III or IV) have recently been demonstrated to be inconsistent with the experimental results of the enzymatic reactions.[114]

In contrast to the reactivity of superoxo–copper(II) complex **1**, which itself does not exhibit ligand hydroxylation reactivity, but needs a hydrogen-atom donor to effect the aliphatic ligand hydroxylation (Scheme 8.1), decomposition of copper(II)–superoxo complex **3^X** gives aliphatic ligand hydroxylation product **5^X** in ~30% yield even in the absence of any hydrogen-atom donor (Scheme 8.2).[34] The reaction obeys first-order kinetics and shows a kinetic deuterium isotope effect of 4.1 at $-60°C$, indicating a rate-limiting intramolecular C−H bond cleavage to occur.[34] In addition, Hammett analysis on the ligand hydroxylation reaction (plot of log k_{obs} vs. σ^+) using the series of para-substituted ligands L3^X (X = OMe, Me, H, Cl, and NO$_2$) provides the Hammett ρ value as -0.5 ($R = 0.99$), suggesting a homolytic C−H bond cleavage. Furthermore, the corresponding copper(II)–hydroperoxo

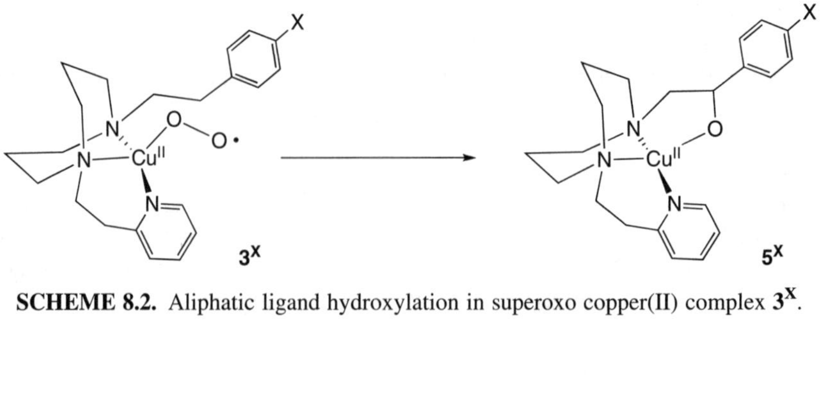

SCHEME 8.2. Aliphatic ligand hydroxylation in superoxo copper(II) complex **3^X**.

complex $L3^H Cu^{II}-OOH$ generated *in situ* by the reaction of 3^H and TEMPO-H does not cause any ligand hydroxylation reaction under the same experimental conditions. All these results suggest that the benzylic ligand hydroxylation reaction involves rate-limiting hydrogen-atom abstraction by the weakly electrophilic copper(II)–superoxo species 3^X itself. This mechanism is similar to the so-called Mechanism I (recently most accepted mechanism) described in Klinman's review article,[114] where copper(II)–superoxo species Cu_1S^E directly contributes to the substrate hydroxylation.

An oxidative ligand modification is also observed in the oxygenation reaction of Cu(I) complexes supported by the calix[6]arene derivatives encapsulated by the tren or TPA ligand.[123,124] In these reactions, hydrogen-atom abstraction from the methylene linker connecting the metal ligand moiety and calix[6]arene by a generated copper(II)–superoxo species occurs, although the copper(II)–superoxo intermediate could not be detected during the course of the reaction (evolution of superoxide O_2^- occurs when coordinative solvent, such as MeCN, is used); (for details, see Chapter 10 by Reinaud et al).[123,124] Intramolecular hydrogen-atom abstraction from the ligand is also seen in the reaction of Cu(I) complexes supported by the TPA derivatives, $L = H_2BPPA$ and BNPA (Chart 8.2), and O_2.[49,50] In these cases as well, the copper(II)–superoxo intermediate could not be directly detected, but could be trapped by DMPO (5,5-dimethyl-1-pyrroline N-oxide).[49,50] The product generated at low temperature is a $1:1$ mixture of the copper(II)–hydroperoxide complex $[Cu^{II}(L)OOH]^+ Cu_1P^H$ and $[Cu^{II}(L^-)]^+$. A plausible reaction pathway is one that involves hydrogen-atom abstraction by the initially generated superoxo intermediate Cu_1S^E from the ligand N–H group in another Cu(I) precursor and subsequent electron transfer (ET) from the Cu(I) center to the nitrogen-centered radical generated (Scheme 8.3). Then, a solvent proton may be abstracted by the resulting amide anion. In this case, however, no ligand oxygenation occurs from the generated hydroperoxo complex Cu_1P^H (Scheme 8.3).

In summary, mononuclear Cu(II) end-on superoxo complexes (Cu_1S^E) do exhibit an ability for hydrogen-atom abstraction reactions from C–H, O–H, and N–H groups of the substrate (mostly supporting ligands), suggesting a possible contribution of Cu_1S^E in the enzymatic reactions of DβM and PHM, as has been suggested by DFT calculations.[122,125–127] However, the mechanism of the oxygen-rebound process that follows is controversial and requires further investigations in model

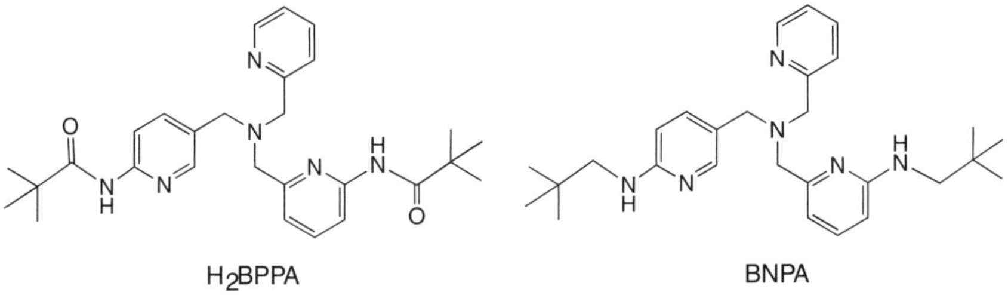

CHART 8.2 TPA ligands having amide and amine substituents.

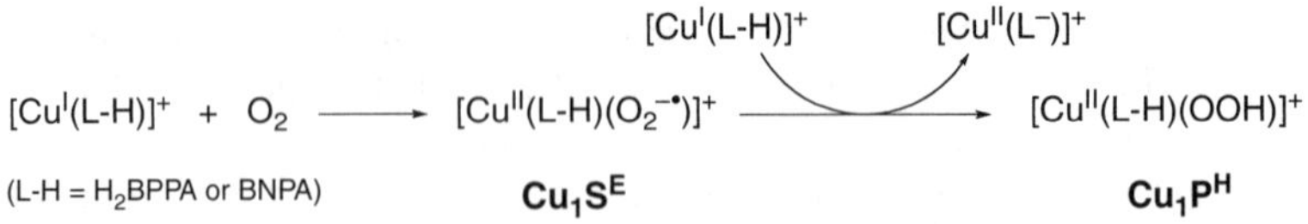

$[Cu^I(L\text{-}H)]^+$ $[Cu^{II}(L^-)]^+$

$[Cu^I(L\text{-}H)]^+ + O_2 \longrightarrow [Cu^{II}(L\text{-}H)(O_2^{-\bullet})]^+ \longrightarrow [Cu^{II}(L\text{-}H)(OOH)]^+$

(L-H = H$_2$BPPA or BNPA)   **Cu$_1$S^E**   **Cu$_1$P^H**

SCHEME 8.3. Cu/O$_2$ chemistry of the H$_2$BPPA and BNPA ligand systems.

systems. An end-on superoxo Cu(II) complex $[Cu^{II}(DPH_2^-)(O_2^-)(NEt_3)]$ supported by *N,N*-bis(2-hydrozy-3,5-di-*tert*-butylphenyl)amine (DPH$_3$), exhibiting charge transfer (CT) bands at 524 and 650 nm and an infrared (IR) stretch $\nu(^{16}O-^{16}O)$ at 964 cm^{-1} $[\nu(^{18}O-^{18}O) = 909$ cm$^{-1}]$, was also reported.[24] However, reactivity of the superoxo complex toward external substrates has yet to be examined in detail.

8.2.2. Copper(II) Side-On Superoxo Complexes (Cu$_1$S^S)

The first mononuclear Cu(II) active-oxygen complex isolated and structurally characterized is the side-on superoxo complex **Cu$_1$S^S** with a distorted square pyramidal geometry (**6** in Chart 8.3).[22] The bond length (1.22 Å), determined at room temperature, may underestimate the true O–O distance because of vibrational motion within the **CuO$_2$** core.[128] Compound **6** was prepared by the oxygenation reaction of the Cu(I) complex supported by a sterically demanding monoanionic hydrotris(pyrazoly)borate ligand HB(3-*t*-Bu-5-*i*-Pr-pz)$_3^-$ (pz = 3,5, -diphenylpyrazoli; *t*-Bu = *tert*-butyl; *i*-Pr = isopropyl) at a low temperature ($-50°$C), where dimerization of the Cu(II) complexes is prohibited *in the solid state* by the bulky *t*-Bu substituent at the 3-position of the pyrazole ring [when the substituent R^1 is less hindered, e.g., an *i*-Pr group, the main product is the $(\mu\text{-}\eta^2\text{:}\eta^2\text{-peroxo})$dicopper(II) complex **Cu$_2$P^S**].[87] However, it was later found that the *tert*-butyl group of the ligand is not sufficient to protect the dimerization reaction *in solution*, thus providing a mixture of the monomeric superoxo complex and the side-on peroxo dicopper(II) complex.[25] Thus, they reinvestigated the spectroscopic features of the mononuclear copper(II)–superoxo complex (**7**) having a more bulky ligand HB(3-Ad-5-*i*-Pr-pz)$_3^-$ bearing peripheral adamantyl substituents (Chart 8.3).[25]

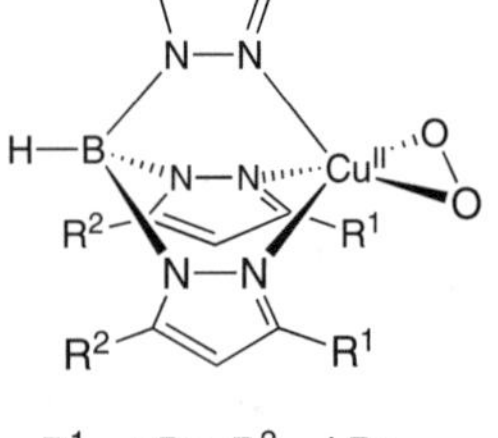

6 = R^1 = *t*-Bu, R^2 = *i*-Pr
7 = R^1 = adamantyl, R^2 = *i*-Pr

CHART 8.3 Mononuclear copper(II) side-on superoxo complexes (**Cu$_1$S^S**).

Compound **7** is a diamagnetic species ($S = 0$, singlet ground state) and exhibits an RR band at $1043\,cm^{-1}$ due to the O−O bond stretching vibration of the side-on superoxo ligand that shifts to $984\,cm^{-1}$ upon ^{18}O-substitution. On the other hand, it shows very weak d–d bands at 383, 452, 699, and 980 nm with extinction coefficients $< 300\,M^{-1}\,cm^{-1}$, whereas its ligand-to-metal change transfer (LMCT) band shifts to a significantly higher energy ($\lambda < 300\,nm$) due to the strong covalency between the Cu d_{xy} and the superoxide $\pi_\sigma{}^*$ orbitals. Detailed analysis of the electronic structure of the side-on superoxo complex $[Cu(\eta^2\text{-}O_2)(HB(3\text{-Ad-}5\text{-}i\text{-Pr-pz})_3)]$ have been described.[25]

With regard to chemical reactivity of **Cu₁Sˢ**, no experimental data is available. The possibility of **Cu₁Sˢ** as a reactive intermediate for the enzymatic C−H activation in PHM was examined by DFT calculations.[129]

8.2.3. Copper(III) Side-On Peroxo Complexes (Cu₁Pˢ)

Monoanionic β-diketiminate ligands with bulky aromatic N-substituents and related anilido imine ligands provide a mononuclear copper–dioxygen adduct with a different electronic structure, a diamagnetic mononuclear copper(III) side-on peroxo complex (**Cu₁Pˢ**) (**8–12**, Chart 8.4). The monoanionic didentate ligand may stabilize the higher oxidation state of Cu(III) to produce **Cu₁Pˢ** instead of **Cu₁Sˢ**, thus two electrons are formally transferred from Cu(I) to O_2. In this case, the ligand substituents play important roles in determining the structure of the resulting copper–dioxygen complex.[38–41] Namely, β-diketiminate ligands without the α-alkyl substituents on the carbon framework or those with smaller ortho-(2,6)-substituents (Me or Et instead of i-Pr) on the aromatic groups attached to the nitrogen atoms gave bis (μ-oxo)dicopper(III) complexes (**Cu₂O₂**).[130] The mechanism of formation of **Cu₁Pˢ** supported by the β-diketiminate ligands has also been investigated in detail, by kinetic analysis and DFT calculations.[39]

The electronic structure of **Cu₁Pˢ** has been confirmed by the observed O−O bond length of 1.392 Å in the crystals of **9** and **11**, as well as using Cu K- and L-edge X-ray absorption spectroscopy (XAS) in combination with valence bond configuration-interaction (VBCI) simulations and spin-unrestricted broken symmetry DFT calculations.[39,40,131] Cramer and co-workers[128,132] reported a linear correlation between the

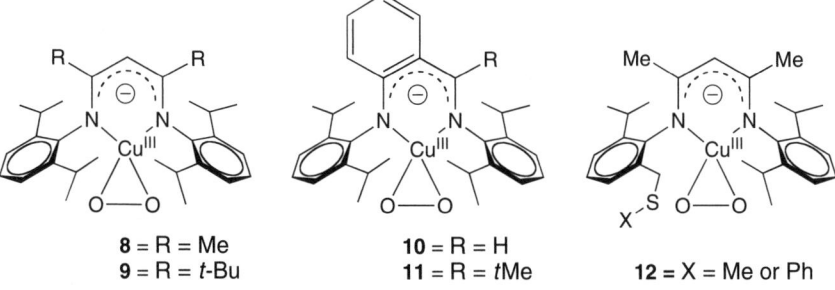

8 = R = Me
9 = R = t-Bu

10 = R = H
11 = R = tMe

12 = X = Me or Ph

CHART 8.4 Mononuclear copper(III) side-on peroxo complexes (**Cu₁Pˢ**).

O—O bond length (Å) and the O—O bond stretching vibration energy (cm^{-1}) in the series of 1:1 metal-dioxygen complexes with the side-on (η^2) binding mode. From the DFT calculations, they concluded that peroxide and superoxide formulations are two limits of a continuum defined by the degree of mixing between metal and ligand orbitals, and that, in principle, transition metal complexes could sample all O—O bond orders along this continuum. However, McGrady and Pantazis[133] argued that the peroxide and superoxide are fundamentally distinct states, differing both in their symmetry and their multiplicity.

Compounds **8** and **9** decompose upon warming to give a complicated mixture of products. External substrates (e.g., phenols, phenolates, thioanisole, cyclohexane, ferrocene, and HBF_4) do not react with **8** and **9** at temperatures as high as $-60°C$, whereas phosphines (e.g., $PMePh_2$) displaces bound O_2 to give the corresponding copper(I)–phosphine adduct.[134] On the other hand, addition of $[Cu^I(MeCN)_4]^+$ into a solution of **8** at $-80°C$ in tetrahydrofuran (THF) gives a Cu(II)-o-iminosemiquinonato derivative **13**, which can be isolated in the presence of an added coligand (pz in Scheme 8.4).[134] Although mechanistic details for the formation of **13** have yet to be clarified, the authors suggested that $[Cu^I(MeCN)_4]^+$ leads to formation of a species in which a bound O_2 is activated for attack at the aryl ring of the β-diketiminate ligand, perhaps as a (μ-η^2:η^2-peroxo)dicopper(II) (**Cu_2P^S**) or bis(μ-oxo)dicopper(III) (**Cu_2O_2**) unit. After electrophilic attack at the aryl ring, an NIH shift of an i-pr group and subsequent oxidation would generate product complex **13**.[134] Complex **8** has also been applied for the preparation of an asymmetric bis(μ-oxo)dicopper(III) and a hetero-bimetallic bis(μ-oxo) complex ($Cu^{III}(\mu$-O)$_2Ni^{III}$).[135] Overall, no significant chemical reactivity of **Cu_1P^S** toward external substrates has yet to be reported.

The activation of O_2 by dopamine β-monooxygenase (DβM) and PHM occurs at a copper site ligated by two histidine imidazoles and a methionine thioether, which is unusual because such thioether ligation is not present in other dioxygen-activating copper proteins. To assess the possible role of the thioether ligand in O_2 activation by DβM and PHM, β-diketiminate ligands carrying a thioether functional group at one of the ortho positions of the aryl substituents have been developed and their dioxygen-reactivity has been examined. Oxygenation reaction of the supported Cu(I) complexes also provides a mononuclear side-on peroxo–copper(III) complex **12** (Scheme 8.5). Although there is no coordinative interaction between the sulfur atom of the thioether group and the Cu(III) ion, compound **12** exhibits a somewhat unique behavior. Namely, in contrast to the case of **8–11** lacking the thioether group, purging the solution of **12** with argon results in conversion to bis(μ-oxo)dicopper(III) (**Cu_2O_2**) species **14** (Scheme 8.5).[41] The thioether group may promote release of O_2 from the

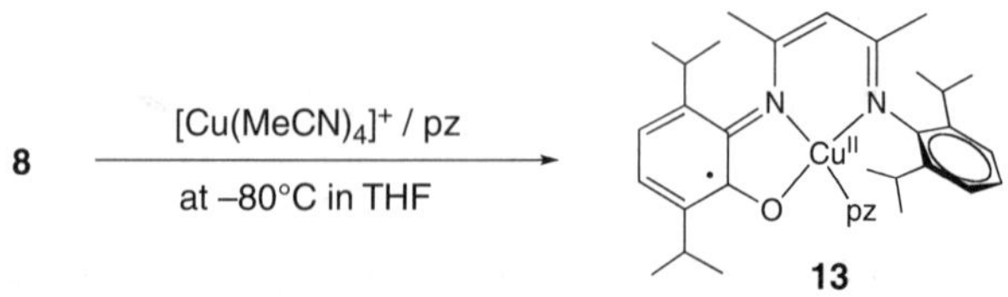

SCHEME 8.4. Decomposition product of mononuclear copper(III) side-on peroxo complex **8**.

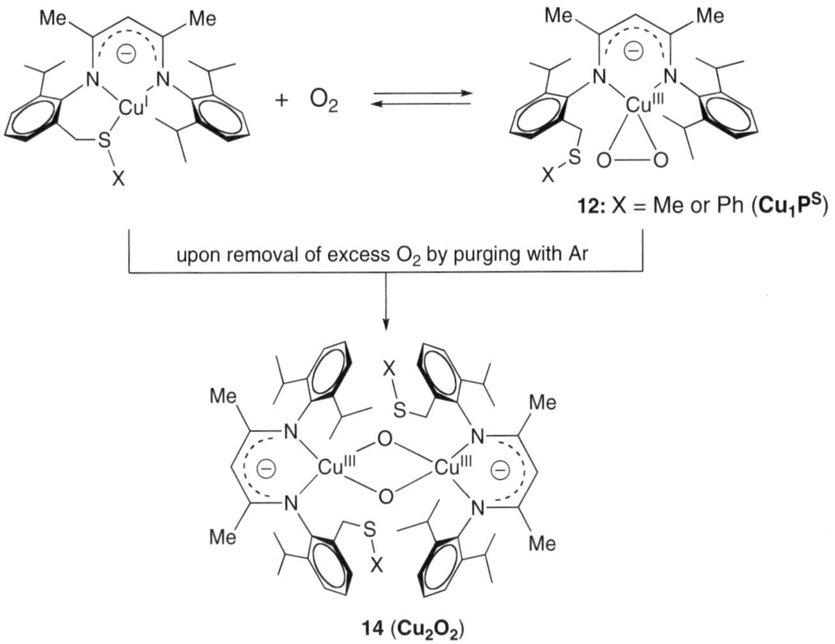

12: X = Me or Ph (**Cu₁Pˢ**)

upon removal of excess O_2 by purging with Ar

14 (Cu₂O₂)

SCHEME 8.5. Cu/O_2 chemistry of the β-diketiminate ligand system carrying a thioether functional group.

1:1 Cu/O_2 adduct **12**, which facilitates trapping of the resulting Cu(I) complex to yield the **Cu₂O₂** species **14** (Scheme 8.5).[41]

Effects of thioether groups on copper(I)–dioxygen reactivity and/or spectroscopic features of generated dicopper(II)–peroxo species have also been investigated in other sulfur-containing ligand systems, even though no mononuclear copper–dioxygen adduct was detected in those systems.[136–141]

8.2.4. Copper(II) Hydroperoxo Complexes (Cu₁Pᴴ)

One-electron reduction and subsequent protonation of a copper(II)–superoxo species will result in formation of copper(II)–hydroperoxo complex **Cu₁Pᴴ**. Such chemistry can be seen in some dinuclear Cu(I) systems, although the resulting oxygen inter-mediates are not mononuclear copper species (Scheme 8.6).[52,53]

Thus, oxygenation of the dinuclear Cu(I) complex supported by a phenolate ligand having bis[2-(pyridin2-yl)ethyl]amine tridentate metal binding moiety at its ortho positions gave dicopper(II)–peroxo complex **15** of the type **Cu₂Pᴱ***[see Chart 8.1, $\lambda_{max} = 505$ nm $(\varepsilon = 6300\,M^{-1}\,cm^{-1})$, 610 nm $(2400\,M^{-1}\,cm^{-1})$; $\nu_{Cu-O(16)} = 488\,cm^{-1}$, $\nu_{Cu-O(18)} = 464\,cm^{-1}$, $\nu_{O(16)-O(16)} = 803\,cm^{-1}$, $\nu_{O(16)-O(16)} = 750\,cm^{-1}$] (Scheme 8.6).[142] Protonation of **15** with HBF₄ gave dinuclear copper(II)–hydroperoxo complex **16** of type **Cu₂Pᴴ**[ESR silent; $\lambda_{max} = 395$ nm $(\varepsilon = 8000\,M^{-1}\,cm^{-1})$ (Scheme 8.6). The same species can be generated directly from the dinuclear Cu⁽ᴵ⁾–CO complex supported by the neutral phenol ligand, in which the

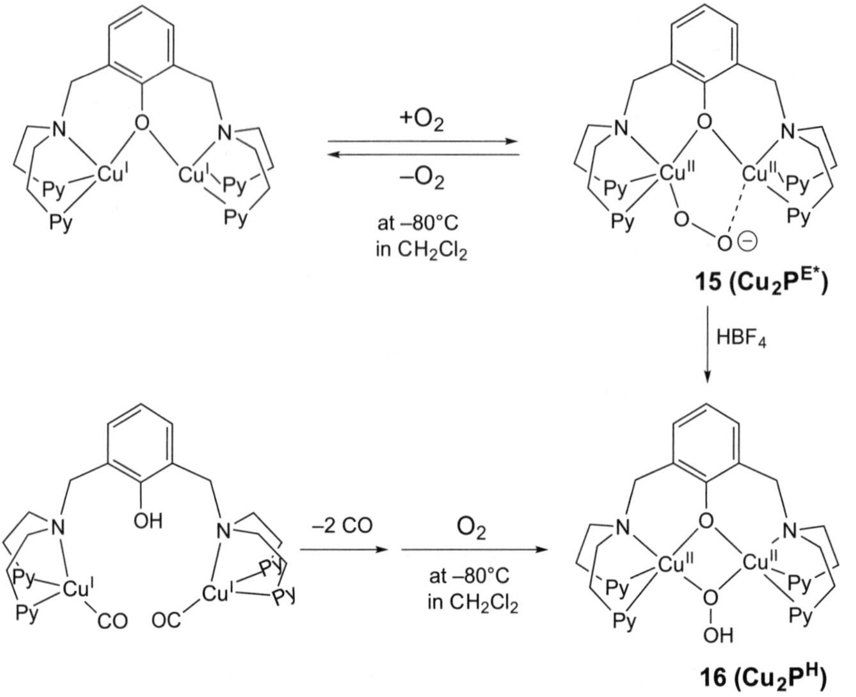

SCHEME 8.6. Generation of dicopper(II) peroxo complex **15** of the type Cu_2P^{E*} and dinuclear copper(II) hydroperoxo complex **16** of type Cu_2P^{H}.

phenolic proton is used for the protonation of peroxide adduct (Scheme 8.6). The similar reaction takes place with the related dinucleating ligands with phenol linker,[54–58] and the spectral data of resulting Cu_2P^{E*} and Cu_2P^{H} species are summarized in the literature.[7]

In a mononuclear copper sysetm, Cu_1P^{H} is also generated in the reaction of some Cu(I) complexes and O_2, where the initially formed Cu_1S^{E} intermediates abstract a hydrogen atom from external substrates (Schemes 8.1 and 8.3). In most cases, however, such an ET and subsequent protonation ($e^- + H^+$) or hydrogen-atom ($H^{\cdot}$) transfer from external substrates to Cu_1S^{E} is difficult to control. Thus, the reaction of Cu(II) complexes and hydrogen peroxide (H_2O_2) is generally employed for the preparation of Cu_1P^{H} (Scheme 8.7).[59–75] In this case, an amine base (B:) is added to abstract a proton from H_2O_2. Copper(II)–hydroperoxo complexes Cu_1P^{H} generally exhibit an RR band at $\sim850\,cm^{-1}$ $[\Delta\nu\,(^{16}OH_2 - {}^{18}OH_2) = \sim50\,cm^{-1}]$ due to the O−O

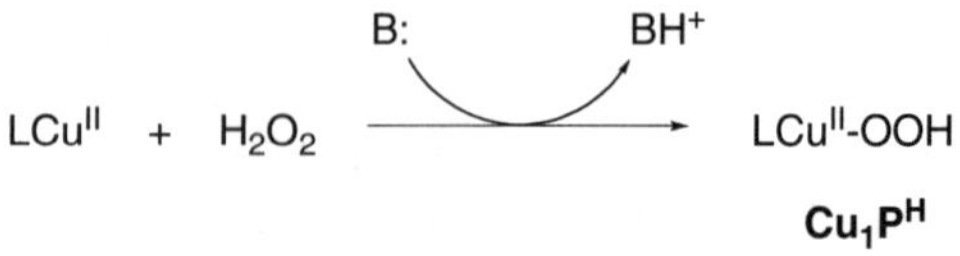

SCHEME 8.7. Generation of mononuclear copper(II) hydroperoxo complex Cu_1P^{H} from the reaction of copper(II) complex and H_2O_2.

bond stretching vibration of the end-on hydroperoxo ligand and an LMCT band at 350–400 nm ($\varepsilon \sim 10^3\,M^{-1}\,cm^{-1}$).[9]

Reactivity of $\mathbf{Cu_1P^H}$ has long attracted much interests since these have been suggested as one of the key reactive intermediates involved in DβM and PHM (Mechanism III described in Klinman's review article[114]), as well as in the oxidative damage of DNA catalyzed by Cu complexes.[2,143–145] However, little had been known about the intrinsic reactivity of $\mathbf{Cu_1P^H}$ until recently.

Masuda and co-workers [67] examined the effects of hydrogen-bonding interaction on the stability of $\mathbf{Cu_1P^H}$ using well-designed tripodal tetradentate ligands. They found that the hydrogen-bonding interaction with the proximal oxygen of the hydroperoxo group results in stabilization of the complex (Scheme 8.8). This result enabled them to determine the first crystal structure of $\mathbf{Cu_1P^H}$.[59] They also demonstrated that the hydrogen-bonding interaction with the distal oxygen, on the other hand, reduces the stability of $\mathbf{Cu_1P^H}$ (Scheme 8.8).[66] Kodera et al.[62] demonstrated that a sulfur ligation to Cu(II) (**17** in Scheme 8.8) also stabilizes $\mathbf{Cu_1P^H}$, providing some insights into the effect of methionine ligand in DβM and PHM. Unfortunately, however, the effects of these hydrogen-bonding interactions and the sulfur ligation on the reactivity of $\mathbf{Cu_1P^H}$ toward external substrates have yet to be examined.

With regard to the reactivity of $\mathbf{Cu_1P^H}$ toward external substrates, Masuda and co-workers[65] demonstrated that Cu(II) complex **18**, supported by a bis(pyridin-2-ylmethyl)amine tridentate ligand with bulky *tert*-butyl substituent, can act as an efficient turnover catalyst in the oxygenation of sulfides to sulfoxides by H_2O_2. Involvement of $\mathbf{Cu_1P^H}$ type species has been invoked, since formation of copper(II)–hydroperoxo complex **19** was detected in the reaction of **18** and H_2O_2 (Scheme 8.9).[65] However, direct reaction between **19** and sulfide substrates has not been examined.

Karlin and co-workers[72–74] have developed a series of TPA ligand derivatives, in which *p-tert*-butylphenyl (TPA$^{t\text{-BP}}$) and *N,N*-dialkylamino (TPA^{NR2}, R = Me or CH_2Ph) groups are attached at the 6-position of one of the pyridine donor groups (Chart 8.5). These substituents are used as internal substrates to examine the reactivity of generated active oxygen species in the reaction of Cu(II) complexes and H_2O_2. In the case of the TPA$^{t\text{-BP}}$, reaction of the Cu(II) precursor complex and H_2O_2 (5 equiv) in the presence of Et$_3$N in acetone at -80°C provides a green solution exhibiting an absorption band at 380 nm ($\varepsilon = 1500\,M^{-1}\,cm^{-1}$) and a typical Cu(II) axial ESR

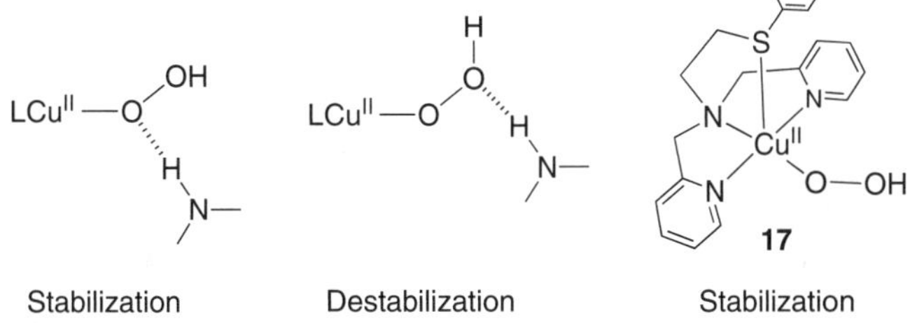

| Stabilization | Destabilization | Stabilization |

SCHEME 8.8. Mononuclear copper(II) hydroperoxo complexes ($\mathbf{Cu_1P^H}$).

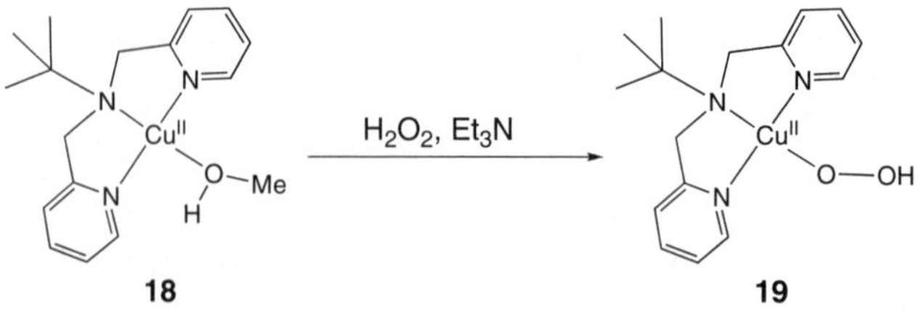

SCHEME 8.9. Generation of mononuclear copper(II) hydroperoxo complex **19**.

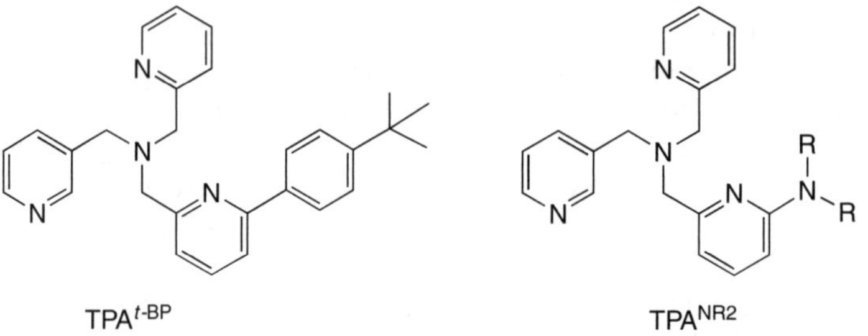

CHART 8.5 TPA ligands with a bound substrate moiety.

spectrum, $g_{\parallel} = 2.245$, $g_{\perp} = 2.042$, and $A_{\parallel} = 180$ G. These spectral features together with the electronspray ionization–mass spectroscopy (ESI–MS) analysis including ^{18}O isotope labeling experiments suggested the formation of an $\mathbf{Cu_1P^H}$ type intermediate, although no RR data are available.[72] Upon warming of the green solution, aromatic ligand hydroxylation takes place at the ortho-position of the t-BP group in $\sim$20%. Since such an aromatic ligand hydroxylation reaction does not occur in the bis(μ-oxo)dicopper(III) complex ($\mathbf{Cu_2O_2}$) supported by the same ligand, the authors concluded that $\mathrm{Cu^{II}OOH}$ ($\mathbf{Cu_1P^H}$) or a high-valent copper–oxo species ($\mathbf{Cu_1O}$) generated by the O–O cleavage of $\mathbf{Cu_1P^H}$ is the reactive species for the aryl hydroxylation reaction (Scheme 8.10).[72]

The reaction of the Cu(II) complex and H_2O_2 was also examined using TPA$^{\mathrm{NR2}}$.[73,74] Formation of an $\mathbf{Cu_1P^H}$ type intermediate was also implicated by the observed UV–vis ($\lambda_{\mathrm{max}} = \sim$380 nm, $\varepsilon = 1400$ M^{-1} cm^{-1}), ESI–MS analysis including ^{18}O labeling, and an axial Cu(II) ESR spectrum, even though no Raman data were available. In this case, an oxidative N-dealkylation reaction takes place at the

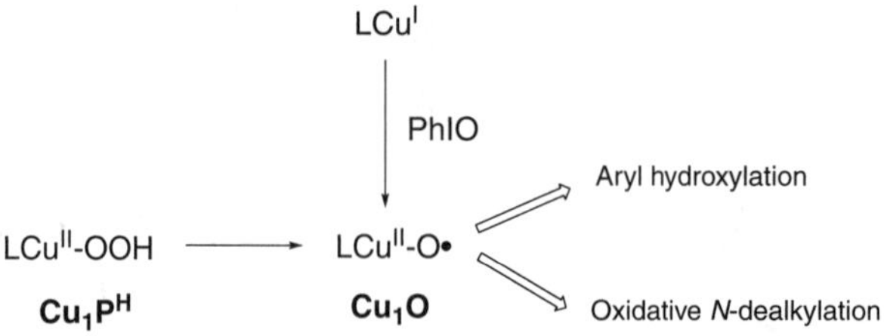

SCHEME 8.10. Reactivity of mononuclear copper(II) hydroperoxo complex $\mathbf{Cu_1P^H}$ supported by TPA$^{t\text{-BP}}$ and TPA$^{\mathrm{NR2}}$.

dialkylamine substituent in ~50%.[73,74] In this case as well, the product distribution pattern is different from the one that arose from the thermal decomposition of the bis (μ-oxo)dicopper(III) complex (**Cu$_2$O$_2$**) supported by the same ligand.[74] Notably, the oxidative N-dealkylation reaction also proceeds when the Cu(I) precursor complex of the same ligand is treated with PhIO, and a set of mass peaks corresponding to the molecular weight of LCuO (**Cu$_1$O**) has been detected by ESI–MS analysis.[74] From these results, together with a kinetic deuterium isotope effect of ~2, obtained using a perdeuterated ligand, the authors suggested a possible contribution of a high-valent copper-oxo species (**Cu$_1$O**) for the oxidation N-dealkylation reaction (Scheme 8.10).[73,74]

As presented above, most of the reactions between Cu(II) complexes and H$_2$O$_2$ in the presence of base, initially afford Cu(II) hydroperoxo complexes (**Cu$_1$P^H**), as indicated in Scheme 8.7. In some instances, however, reduction of Cu(II) to Cu(I) by H$_2$O$_2$ occurs when the supporting ligand can stabilize the lower oxidation state of Cu(I).[64,146]

8.2.5. Copper(II) Alkylperoxo Complexes (Cu$_1$P^R)

Similar to the case of copper(II)–hydroperoxo complex **Cu$_1$P^H**, copper(II) alkylperoxo complexes **20** and **21** (**Cu$_1$P^R**, see Chart 8.6) can be generated by the reaction of

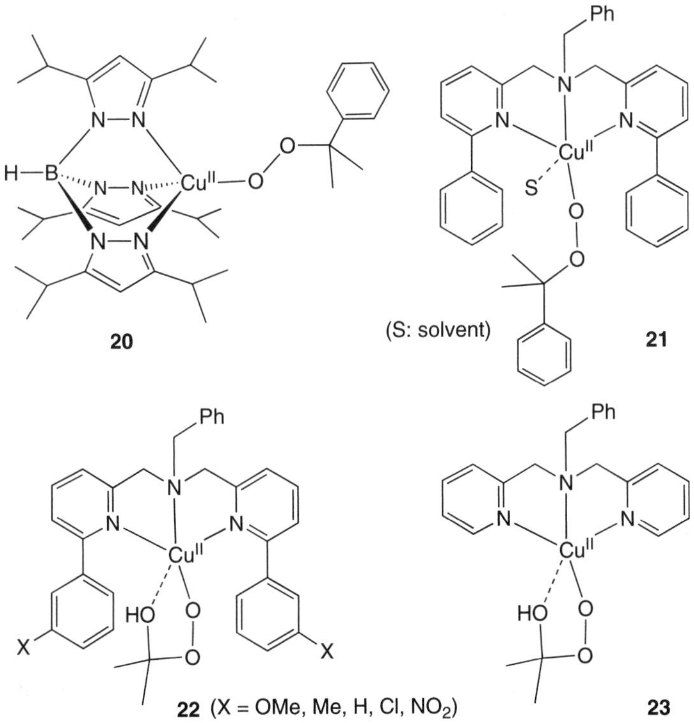

CHART 8.6 Mononuclear copper(II) alkylperoxo complexes (**Cu$_1$P^R**).

Cu(II) precursor complexes and alkylhydroperoxide (ROOH) in the presence of base.[78,81] The crystal structure of **20** clearly shows its distorted tetrahedral geometry with an O−O bond length of 1.460 Å.[78] Compound **20** exhibits a typical $S = 1/2$ axial ESR spectrum with a small $A_{\parallel}$ value (56 G), indicating a similar distorted tetrahedral geometry in solution.[80] Compound **20** shows a relatively intense LMCT band at 600 nm ($\varepsilon = 5410\,M^{-1}\,cm^{-1}$) together with a weak d−d band at 800 nm ($\varepsilon = 580\,M^{-1}\,cm^{-1}$).[80] The RR spectrum of **20** is rather complicated exhibiting several isotope sensitive Raman bands at 550–850 cm^{-1} {$\nu = 843$ cm^{-1} [$\Delta\nu(^{16}O-^{18}O) = 26$ cm^{-1}], 809 (28), 756 (4), 645 (16), 551 (8), and 536 (7)} due to mixed O−O, C−O, C−C, and Cu−O stretching vibrations, as well as C−C−C and O−C−C deformation modes of the alkylperoxo moiety.[80]

Alkylperoxo complex **21** also exhibits multiple Raman bands as in the case of **20**; $\nu = 885$ cm^{-1} [$\Delta\nu(^{16}O-^{18}O) = 30$ cm^{-1}], 841 (33), 608 (11), 529 (5), and 485 (11).[81] However, the UV–vis spectrum of complex **21** is different from that of **20**, exhibiting a characteristic absorption band at 465 nm ($\varepsilon = 1100\,M^{-1}\,cm^{-1}$) attributable to the LMCT transition from the alkylperoxo ligand to Cu(II) and a weak d−d band at 725 nm ($\varepsilon = 320\,M^{-1}\,cm^{-1}$).[81] The difference in UV–vis between **20** and **21** could be attributed to the different geometry of the copper centers; tetrahedral versus tetragonal for **20** and **21**, respectively. In fact, the ESR spectrum of **21** [$g_1 = 2.250$, $g_2 = 2.065$, $g_3 = 2.030$, $A_1 = 160$, $A_2 = 7$, $A_3 = 5$ G)] is typical of Cu(II) complexes with a tetragonal geometry.[81]

Notably, the reaction of Cu(II) complexes supported by bis(pyridin-2-ylmethyl) amine tridentate ligands and H$_2$O$_2$ in acetone containing a base gives unique 2-hydroxy-2-hydroperoxypropane (HHPP) adducts **22** and **23** (Chart 8.6).[147,148] The HHPP adducts show a LMCT band $\sim$ 400 nm together with a broad d−d band > 600 nm [**22**: 420 nm, $\varepsilon = 1350\,M^{-1}\,cm^{-1}$ (LMCT); 630 nm, $\varepsilon = 200\,M^{-1}\,cm^{-1}$ (d−d), **23**: 375 nm, $\varepsilon = 2500\,M^{-1}\,cm^{-1}$ (LMCT); 650 nm, $\varepsilon = 220\,M^{-1}\,cm^{-1}$ (d−d)] and the multiple RR bands due to the mixed O−O/C−O/C−C and Cu−O vibrations [**22**: $\nu = 855$ cm^{-1} [$\Delta\nu(^{16}O - ^{18}O) = 30$ cm^{-1}], 823 (20), 792 (7), and 545 (20), **23**: $\nu = 879$ cm^{-1} [$\Delta\nu(^{16}O-^{18}O) = 29$ cm^{-1}], 864 (31), and 847 (28)]. The spin quantization of the axial Cu^{II} ESR spectra of both **22** and **23** unambiguously demonstrated the mononuclearity of the complexes, and the ESI–MS analysis confirmed the incorporation of the acetone molecule into the Cu(II) complex **22**. Thus, it becomes apparent that the reactivity of the Cu(II) complexes toward H$_2$O$_2$ is largely affected by the supporting ligands; hydroperoxo complex formation (Scheme 8.7) versus HHPP adduct formation.

The reactivity of **Cu$_1$P^R** supported by hydrotris(pyrazoly)borate ligand HB(3,5-i-Prpz)$_3^-$ has been mainly examined by using the $tert$-butyl-hydroperoxo adduct **20'** (LCu−OOt-Bu).[78] While compound **20'** is fairly stable below −20°C, it gradually decomposes at room temperature to give t-BuOH and t-BuOOC$_5$H$_{11}$.[78] The formation of t-BuOOC$_5$H$_{11}$ may indicate that t-BuOO$^{\bullet}$ is an initial intermediate in the decomposition reaction of **20'**. The radical generation suggests the Cu−O bond cleavage in **20'**. In this case, the formation of t-BuOH can be explained either by the redox decomposition of t-BuOOC$_5$H$_{11}$ or by the homolysis of the O−O bond in **20'** to afford t-BuO$^{\bullet}$, which subsequently abstracts H$^{\bullet}$ from the solvent. In the presence of external substrate, such as 2,4,6-tri-*tert*-butylphenol, tetramethylene sulfoxide, and PPh$_3$, the decomposition rate of **20'** is accelerated, and the corresponding oxidation products, 2,4,6-tri-*tert*-butylphenoxyl radical, tetrahydrothiophene 1,1-dioxide, and O=PPh$_3$ are produced, respectively.[78] On the other hand, the addition of cyclohexene did not affect the consumption rate of **20'**, but provided the products expected by classical radical-type reactions initiated by *t*-BuOO$^•$ or *t*-Bu$^•$. Thus, the role of **20'** in the above reactions is the generation of *t*-BuOO$^•$ and/or *t*-BuO$^•$, and the products are determined by the reactivity of the peroxyl and alkoxyl radicals generated.

Reactivity of another alkylperoxo–copper(II) complex (**21**) having a tetragonal geometry has also been examined in detail (Scheme 8.11).[81] Complex **21** gradually decomposes obeying first-order kinetics even at −40°C to give bis(μ-hydroxo) dicopper(II) complex **24** in a 64% isolated yield. Notably, acetophenone [PhC(O)Me] was produced in 92% yield from the final reaction mixture. This result clearly demonstrates that O−O bond homolysis of the peroxo moiety of **21** occurs, since it is well known that the cumyloxyl radical quickly undergoes β-scission to give acetophenone.[149,150] In fact, putative intermediate LCu(II)−O$^•$ (**Cu$_1$O, 25**), generated by the O−O bond homolysis of **21**, could be trapped by the reaction with 5,5-dimethyl-1-pyrroline-*N*-oxide (DMPO), a well-known radical trap reagent, to give the 1 : 1 adduct **26**.[49,50,151,152] Addition of AcrH$_2$ (10-methyl-9,10-dihydroacridine) into the acetonitrile solution of **21** at −40°C causes a formation of AcrH$^+$ (*N*-methylacridinium ion) as an oxidation product in a 49% yield based on **21**. In addition, a significantly large kinetic deuterium isotope effect of $k_2{}^H/k_2{}^D = 19.2$ was obtained at −40°C when AcrD$_2$ (AcrD$_2$: 9,9-dideuterated derivative) was used in place of AcrH$_2$. This finding clearly indicates that a hydrogen-transfer process is involved in the rate-determining

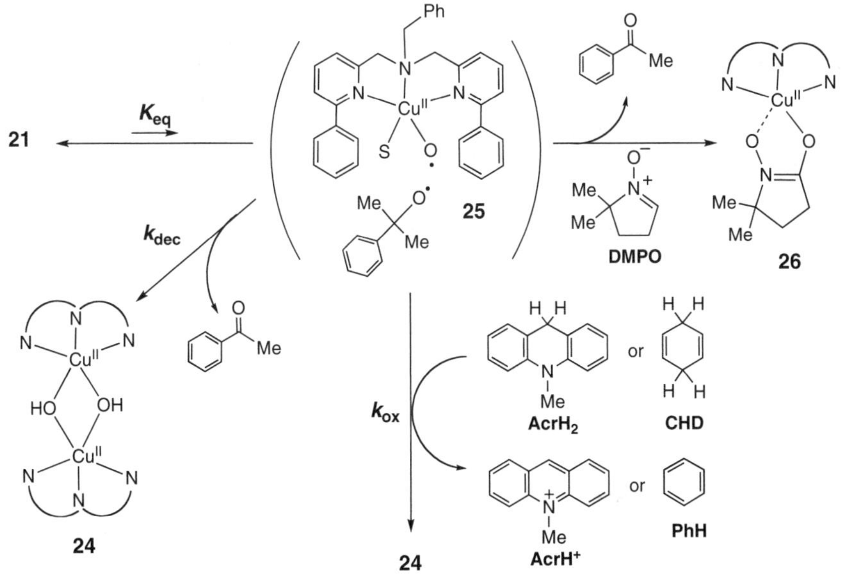

SCHEME 8.11. Reactivity of mononuclear copper(II) alkylperoxo complex **21**.

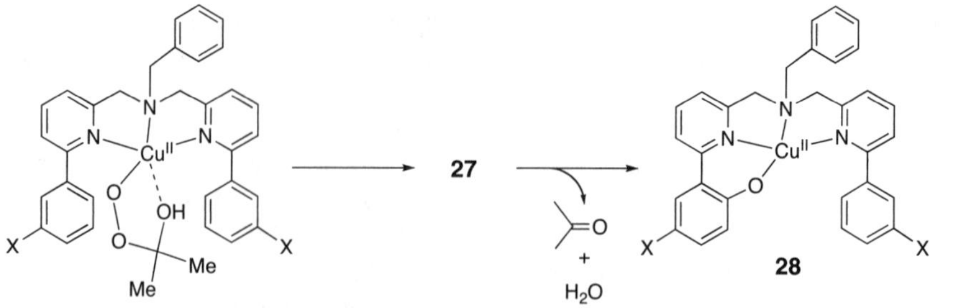

22 (X = OMe, Me, H, Cl, and NO$_2$)

SCHEME 8.12. Aromatic ligand hydroxylation in HHPP-adduct **22**.

step of the C−H bond activation of AcrH$_2$ by **21**. Similarly, oxidation of 1,4-cyclohexadiene (CHD) proceeds smoothly to give benzene. These results suggest a possible contribution of **Cu$_1$O** (**25**) to the C−H bond activation process, although the timing of the O−O bond homolysis of **21**, "prior to" or "concerted with" the substrate C−H bond activation, are not clear at present.

In the case of the HHPP-adduct (**22**) (Chart 8.6), ligand hydroxylation takes place at the ortho-position of one of the phenyl substituents attached to the pyridine donor groups to give **28** (Scheme 8.12).[147] During the course of the reaction, a still unidentified intermediate, **27**, is observed. This finding is in sharp contrast to the case of simple alkylperoxo complex **21**, with which no such aromatic ligand hydroxylation is observed (Scheme 8.11). Detailed spectroscopic and kinetic analyses including Hammett analysis (X = OMe, Me, H, Cl, and NO$_2$; plot of log k_{obs} vs. σ^+; $\rho = -2.2$) and kinetic deuterium isotope effect (KIE = 1.0) obtained with perdeuterated ligand L^H-d_{10} (replacing all protons of the 6-phenyl groups), as well as an isotope labeling experiment using H$_2$^{18}O$_2$, suggested that the reaction involves an electrophilic aromatic substitution mechanism. Theoretical studies at the DFT level implicated the existence of conjugate acid–base catalysis in the O−O bond cleavage and C−O bond-formation steps, as illustrated in Scheme 8.13. This finding may be a reason for the different reactivity of **21** versus **22**.[148]

The HHPP adduct **23**, which does not have phenyl substituents on the pyridine donor groups of the ligand, also decomposes at low temperature to give a

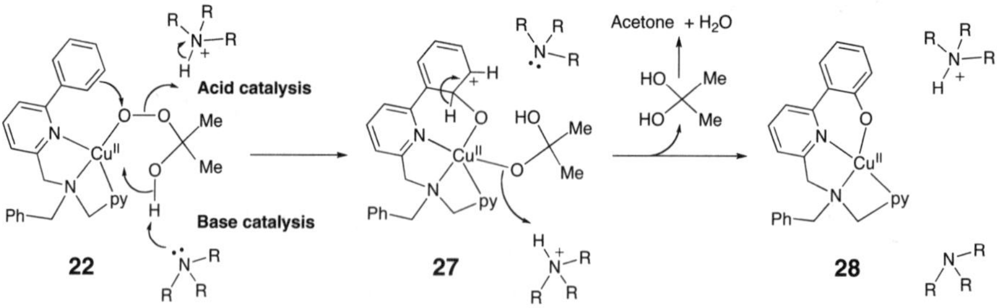

SCHEME 8.13. Aromatic ligand hydroxylation mechanism in HHPP-adduct **22**.

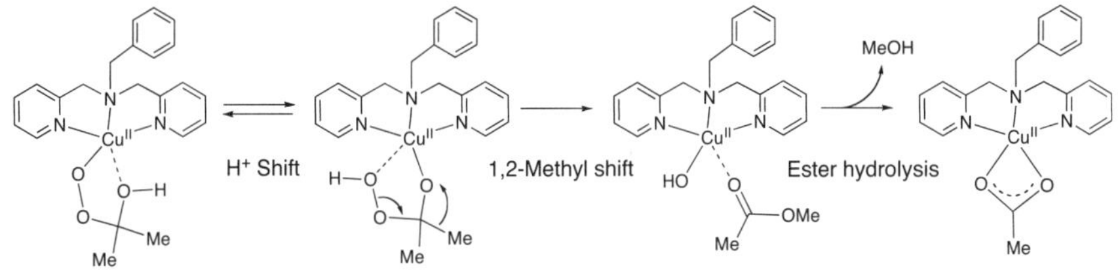

SCHEME 8.14. Baeyer-Villiger type 1,2-methyl shift in HHPP-adduct **23**.

copper(II)–acetate complex (**29**). Here one of the oxygen atoms of the acetate coligand originates from the added H_2O_2.[148] In this case, a mechanism involving a Baeyer–Villiger type 1,2-methyl shift from the HHPP adduct and subsequent ester hydrolysis, has been proposed on the basis of DFT calculations. In this case as well, conjugate acid–base catalysis is implicated to be involved in the 1,2-methyl shift process (Scheme 8.14).[148]

8.2.6. Copper(II) Acylperoxo Complexes (Cu_1P^{Ac})

So far, little attention has been focused on the chemistry of copper(II)–acylperoxo complex Cu_1P^{Ac} (Chart 8.7). Complex **30** has been prepared by the reaction of the peroxo complex **15** (see Scheme 8.6, $Cu_2P^{E^*}$) and *m*-chlorobenzoyl chloride or by the reaction of a hydroxide bridged dicopper(II) complex of the same ligand with *m*-chloroperbenzoic acid (*m*-CPBA) at low temperature.[76] The crystal structure of **30** clearly shows the existence of the μ-1,1-bound acylperoxo group bridging two Cu(II) ions with a distorted square pyramidal geometry.[76] Acylperoxo complex **30** exhibits a reactivity toward PPh_3 to give $O=PPh_3$ quantitatively.[76]

The first *mononuclear* Cu(II) acylperoxo complex **31** has been prepared by the reaction of the di(μ-hydroxo)dicopper(II) complex of HB(3,5-*i*-Prpz)$_3^-$ ligand

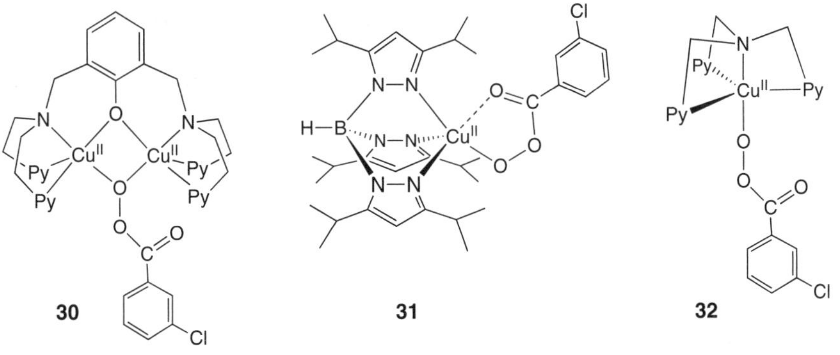

(Py = 2-pyridyl)

CHART 8.7 Copper(II) acylperoxo complexes.

and m-CPBA.[77] The ESR spectrum indicated a square pyramidal mononuclear Cu(II) structure.[77] Complex **31** also reacted with PPh$_3$ to give O=PPh$_3$, but its reactivity is lower than that of m-CPBA itself.[77] Another mononuclear acylperoxo complex having trigonal bipyramidal structure **32** can be generated by the low-temperature reaction of the peroxo-dicopper(II) species [(Cu^{II}(TPA))$_2$(μ-1,2-O$_2$)]$^{2+}$ (type **Cu$_2$P^E**, Fig. 8.1) with m-CPBA, where H$_2$O$_2$ is a byproduct.[79] Similar reactions between the Cu(II) complexes supported by 1,2-dimethylimidazole (Me$_2$ lm) ligands and by the bis[2-(pyridin-2-yl)ethyl][(2-hydroxyphenyl)methyl]amine (CPY2-O$^-$) ligand with m-CPBA have been examined and PPh$_3$ reactivity of the generated acylperoxo complexes have been compared.[79]

8.2.7. Copper Oxo Species (Cu$_1$O)

Mononuclear copper oxo species (**Cu$_1$O**: [Cu^{II}–O$^{-\bullet}$ or Cu^{III}=O^{2-}]$^+$) have long been postulated as a possible reactive intermediate in several reaction systems, such as those introduced in this chapter (e.g., see Schemes 8.1, 8.10, and 8.11), as well as in other synthetic oxidation reactions.[153–160] Theoretical studies on the enzymatic reactions catalyzed by PHM, DβH, and particulate methane monooxygenase (pMMO) have also suggested the possible contribution of **Cu$_1$O** as a key reactive intermediate for the aliphatic C–H bond activation process.[125,126,161] Details of the theoretical studies are presented in Chapter 7. However, such a mononuclear copper–oxo species has never been isolated and only has been detected in a gas-phase reaction.[82]

Tolman and co-workers[162] studied the oxygenation reaction of copper(I)–α-ketocarboxylate complexes **33** and **34** to find oxidative decarboxylation of the α-ketocarboxylate ligand and arene hydroxylation of the supporting ligands (Scheme 8.15). The electron-donating substituent X on the phenyl group enhances the reaction, suggesting that the active oxygen species generated had an electrophilic character.

Theoretical studies on this reaction have suggested that the reaction proceeds in a manner largely analogous to those of similar Fe^{II}–α-ketocarboxylate systems;[163–166] that is, by initial attack of a coordinated O$_2$ molecule on a ketocarboxylate ligand with concomitant decarboxylation. Subsequently, a **Cu$_1$O** type species, best described as Cu^{II}–O$^{\bullet-}$ with triplet ground states, is generated by O–O bond cleavage, and this oxygenates the phenyl substituent of the supporting ligand (Scheme 8.16).[167]

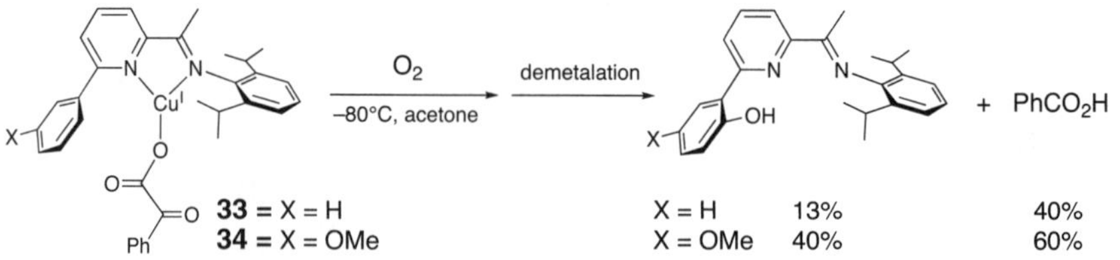

SCHEME 8.15. Aromatic ligand hydroxylation in copper(I)–α-ketocarboxylate complexes **34** and **35**.

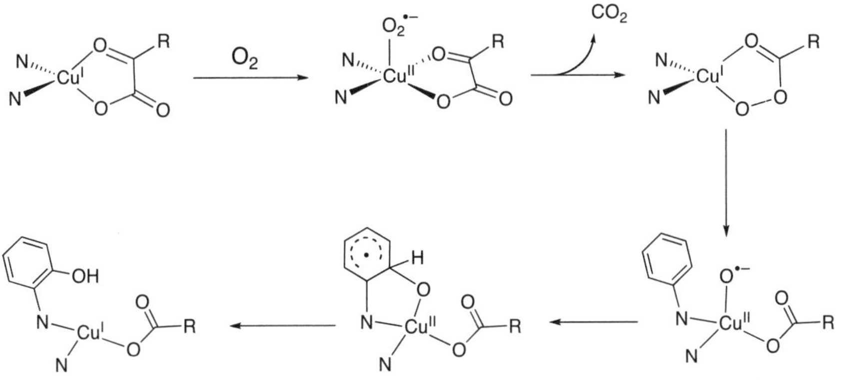

SCHEME 8.16. Aromatic ligand hydroxylation mechanism in copper(I)$-\alpha$-ketocarboxylate complex.

Tolman and co-workers [168] also investigated the reactions of a series of Cu(I) complexes supported by the β-diketiminate and imino-pyridine N_2 didentate ligand derivatives (cf. Chart 8.4 and Scheme 8.15) with pyridine- and trialkylamine-N-oxides, as well as with PhIO to find an aromatic ligand hydroxylation and formation of a bis(μ-oxo)dicopper(III) complex ($\mathbf{Cu_2O_2}$). These results also suggest the formation of $\mathbf{Cu_1O}$ type reactive intermediate during the course of the reaction.

8.3. REACTIVITY OF DINUCLEAR COPPER ACTIVE-OXYGEN SYSTEM

A huge amount of information about the structure, physicochemical properties, reversibility in dioxygen-binding, ligand effects, and reactivity of the dinuclear Cu(II) peroxo ($\mathbf{Cu_2P^E}$ and $\mathbf{Cu_2P^S}$) and bis(μ-oxo)dicopper(III) ($\mathbf{Cu_2O_2}$) complexes has been accumulated during the past three decades. This information has already been well-documented in a large number of review articles.[1,7,8,10–12,16,17,83,97–101,119,169–180] Thus, in this chapter, recent advances in the *reactivity studies* of the dinuclear copper active oxygen complexes are summarized together with the recent topics of related chemistry.

8.3.1. (*trans*-μ-1,2-Peroxo)dicopper(II) Complex ($\mathbf{Cu_2P^E}$)

In 1988, Karlin and co-workers[84] reported the first end-on μ-peroxo dicopper(II) complex $\mathbf{Cu_2P^E}$ supported by the TPA ligand that is the first copper active oxygen complex structurally characterized by X-ray crystallographic analysis (compound **35** in Chart 8.8). The Cu(II) ions exhibit a trigonal bipyramidal geometry, and the O$-$O bond length is 1.43 Å, which is typical for a peroxo ligand.[84] The two cupric ions are antiferromagnetically coupled through the end-on peroxo bridge, making the complex EPR-silent.[84] Complex **35** exhibits RR bands at 832 cm^{-1} (788 cm^{-1} with $^{18}O_2$) and 561 cm^{-1} (535 cm^{-1} with $^{18}O_2$), which have been assigned as the O$-$O and Cu$-$O stretching vibrational modes, respectively.[181] Peroxo-to-copper(II) CT bands appear

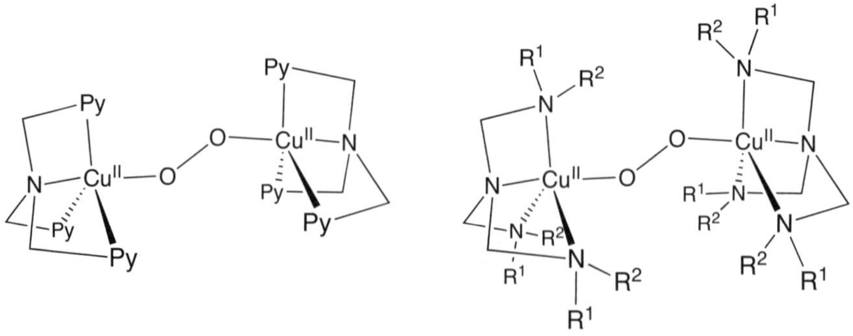

35 = Py = 2-pyridyl

36 = R^1 = H, R^2 = –CH$_2$Ph (Bn = Benzyl)

37 = R^1 = R^2 = –CH$_3$

CHART 8.8 (trans-μ-1,2-Peroxo)dicopper(II) Complex (**Cu$_2$P^E**).

at 525 nm ($\varepsilon = 11{,}500 \,\mathrm{M}^{-1}\,\mathrm{cm}^{-1}$) and ~590 nm ($7600 \,\mathrm{M}^{-1}\,\mathrm{cm}^{-1}$, shoulder), together with a d–d band at 1,035 nm ($160 \,\mathrm{M}^{-1}\,\mathrm{cm}^{-1}$) of the trigonal bipyramidal copper(II).[84,181] Since then, several **Cu$_2$P^E**-type complexes having similar structural and spectroscopic features have been reported, and the kinetic and thermodynamic parameters for the formation process, as well as ligand effects, were explored in detail.[7,8,139,141,182–185] The second crystal structure of **Cu$_2$P^E** supported by the Bn$_3$tren ligand (compound **36**) was reported by Suzuki and co-worker[86] in 2004.[29] Notably, Schindler and co-workers[86] found that the peroxo complex **37**, which is supported by a similar tren ligand (Me$_6$tren), as well as Karlin's peroxo complex (**35**) are significantly stabilized when BPh$_4^-$ is employed as the counteranion; the solid sample of these peroxo complexes are stable at room temperature for a year and no decomposition is observed even at 70°C. In these cases, the large counteranions BPh$_4^-$ shield the Cu$_2$O$_2$ core in the solid state to protect undesirable intermolecular collisions that leads to decomposition of the complex.

Karlin and co-workers[186] systematically investigated the chemical reactivity of **Cu$_2$P^E**, as well as that of the asymmetric dicopper(II)–peroxo complex **Cu$_2$P$^{E^*}$** (compound **15** in Scheme 8.6) to find their intrinsic *nucleophilic* reactivity. They found that **15** reacts with stoichiometric amount of HBF$_4$ to give the μ-1,1-OOH complex **16** (**Cu$_2$P^H**) (Scheme 8.6) and with acyl chloride to give acylperoxo complex **30** (Chart 8.7).[76] Addition of excess H$^+$ to either **Cu$_2$P^E** or **Cu$_2$P$^{E^*}$** releases the peroxide ligand yielding H$_2$O$_2$ and the oxidized dicopper(II) complexes. Both complexes react with CO$_2$ forming peroxycarbonate complex [LCu$^{II}_2$(CO$_4$)]$^{2+}$, which thermally decomposes to the carbonate complex [LCu$^{II}_2$(CO$_3$)]$^{2+}$ with liberation of O$_2$. The reaction of the end-on peroxo complexes with PPh$_3$ results in formation of PPh$_3$ adducts of the Cu(I) species and the release of O$_2$. Furthermore, the reaction with 2,4-di-*tert*-butylphenol causes acid–base chemistry where the phenol derivative serves as a proton donor.[186] All these results are consistent with the *nucleophilic character* of the end-on peroxo dicopper(II) complexes with no interesting oxidation chemistry observed. Nonetheless, recent reactivity studies have

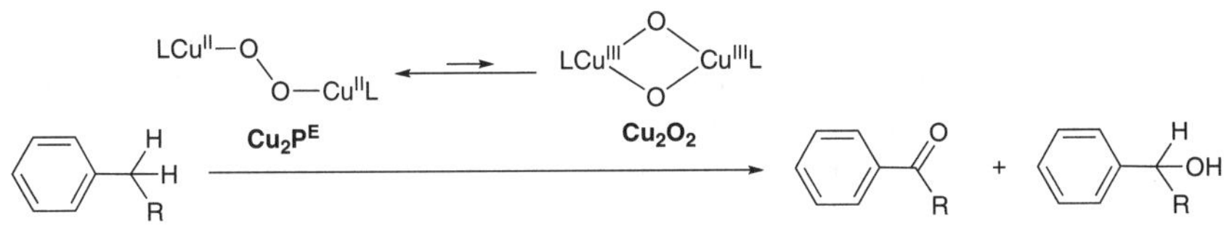

SCHEME 8.17. Oxygenation of toluene by dicopper(II) peroxo complex Cu_2P^E.

shown that Cu_2P^E can oxidize toluene to give benzaldehyde as a major product together with benzyl alcohol as a minor product.[86,187] Mechanistic studies, however, suggested that a bis(μ-oxo)dicopper(III) (Cu_2O_2) generated from Cu_2P^E *in situ* is more likely to be the real active species for the C—H activation (Scheme 8.17).[187]

8.3.2. Reactivity of (μ-η^2:η^2-Peroxo)dicopper(II) (Cu_2P^S) and Bis(μ-oxo)dicopper(III) (Cu_2O_2) Complexes

The (μ-η^2:η^2-peroxo)dicopper(II) complexes (Cu_2P^S), which were first reported by Kitajima et al. in 1989,[87] is the most extensively studied copper–dioxygen complex due to its strong relevance to the peroxo intermediate involved in the oxy-forms of hemocyanin, tyrosinase, and catechol oxidase (type-3 copper proteins).[176] The Cu_2P^S complexes exhibit characteristic CT bands due to peroxide π_σ^* and π_v^* orbitals to the copper $d_{x^2-y^2}$ orbital ($\lambda_{max} = 340$–380 nm; $\varepsilon = 18{,}000 \sim 25{,}000$ M^{-1} cm^{-1} and $\lambda_{max} = 510$–550 nm; $\varepsilon = 1000$ M^{-1} cm^{-1}, respectively) and an isotope sensitive RR band ~ 730–760 cm^{-1} [$\Delta(^{16}O-^{18}O) = \sim 40$ cm^{-1}] due to the O–O bond stretching vibration of the side-on peroxide ligand.[7] So far, six crystal structures of Cu_2P^S have been reported, the Cu_2O_2 core structures of which (O–O and Cu–Cu distances, distortion of the copper center from square pyramidal geometry, and bending of Cu_2O_2 core plane) are somewhat different depending on the supporting ligands.[87,89,91–93,188] Notable features of Cu_2P^S recently reported are the high thermal stability and high reversibility of the dioxygen-binding of Kodera's complex supported by a hexapyridine ligand[92] and the butterfly shaped Cu_2O_2 core bridged by a carboxylate coligand with alkylamine didentate supporting ligands, as shown in Scheme 8.18.[93] In the later case,

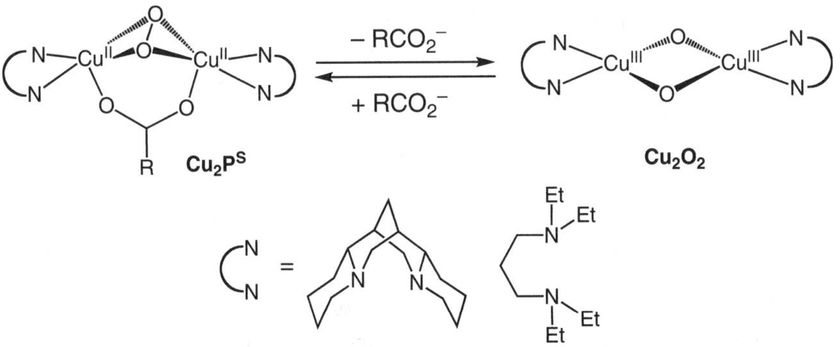

SCHEME 8.18. Reversible conversion between Cu_2P^S and Cu_2O_2 with external carboxylate ligand.

bis(μ-oxo)dicopper(III) (**Cu₂O₂**) became a major component when the carboxylate bridging ligand is absent.[189,190] Ligand effects on the structure and spectroscopic features of **Cu₂Pˢ** complexes are detailed in the literature.[7,191–197]

As frequently mentioned in the preceding sections, **Cu₂Pˢ** is converted to **Cu₂O₂** or exists in equilibrium with **Cu₂O₂** in certain ligand systems. The **Cu₂O₂** exhibits two intense LMCT bands at $\sim$300 and $\sim$400 nm and an isotope sensitive RR band at $\sim$600 cm^{-1} [$\Delta(^{16}O-^{18}O) = \sim$25 cm^{-1}], due to the **Cu₂O₂** core vibration,[97,101,102] and crystal structures of some **Cu₂O₂** complexes, have also been reported.[38,102–108] The structure and spectroscopic features of **Cu₂O₂**, as well as the kinetics and the thermodynamics of its formation process and ligand effects, have been detailed.[7,8,184,198–209] Most of the **Cu₂O₂** compounds are thermally unstable and thus decompose to bis(μ-hydroxo)dicopper(II) complexes at room temperature. Recently, it has been reported that the stability of **Cu₂O₂** is significantly improved by using a sterically and electronically constrained didentate ligand, bis(guanidine) N^1,N^3-bis[bis(2,2,6,6-tetramethyl-piperidin-1-yl)methylene]propane-1,3-diamine (**46** in Chart 8.11).[210]

The **Cu₂Pˢ** and **Cu₂O₂** complexes supported by several kinds of ligands have been shown to exhibit a variety of reactivity toward *external substrates*, as well as *intramolecular oxidative ligand modification*, as summarized in Scheme 8.19.[8,10,11,17,109,178,184,200,203,204,211–229] Details about the reactivity of these complexes are available from recent review articles.[8,17] Briefly stated, both complexes have *electrophilic* character in nature which is in sharp contrast to the *nucleophilic* character of the end-on peroxo species **Cu₂Pᴱ**.[186]

8.3.2.1. Aromatic Ligand Hydroxylation.

Among the series of reactions listed in Scheme 8.19, the aromatic ligand hydroxylation in the *m*-xylyl-based dinucleating ligand sysetm [Eq. 8.2 in Scheme 8.19], which was first reported by Karlin and co-workers[230–232] in the early 1980s (Scheme 8.20), has been studied in detail, since the reaction is strongly related to the chemical function of tyrosinase (phenol monooxygenase activity).[176]

Mechanistic studies on this reaction led to the following conclusions; (1) a migration of the aromatic methyl group occurs when R^1 = Me (route b, NIH shift[233]), (2) no kinetic deuterium isotope effect is observed (R^1 = H vs. D), (3) electronic donating para-substituents accelerates the reaction rate (R^2 = *t*-Bu > H > F > NO₂), and (4) a side-on peroxo intermediate **38** (**Cu₂Pˢ**) is detected in the case of R^2 = NO₂.[234–236] These results suggest that the aromatic ligand hydroxylation reaction involves an electrophilic attack on the ligand arene ring by the (μ-η^2:η^2-peroxo)dicopper(II) intermediate, **38**, as indicated in Scheme 8.20. Similar aromatic ligand hydroxylation reactions have been observed in a variety of *m*-xylyl-based dinucleating ligand systems, the details of which are introduced in the literature.[8,11,176,237,238]

Notably, a bis(μ-oxo)dicopper(III) complex (**39**, **Cu₂O₂**) supported by 2-(diethyl-aminomethyl)-6-phenylpyridine didentate ligands also induce aromatic ligand hydroxylation (Scheme 8.21).[239] The reactivity patterns (ligand substituent and kinetic deuterium isotope effects) are similar to those of Karlin's system (Scheme 8.20),

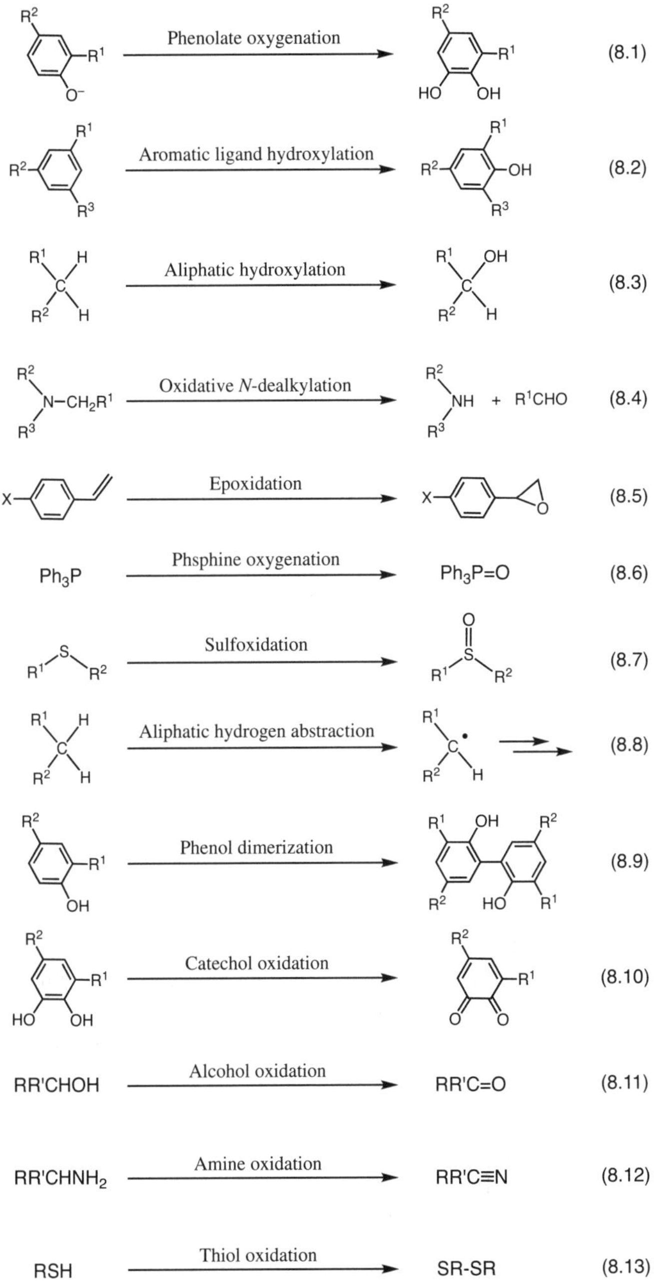
Phenolate oxygenation (8.1)
Aromatic ligand hydroxylation (8.2)
Aliphatic hydroxylation (8.3)
Oxidative N-dealkylation (8.4)
Epoxidation (8.5)
Ph3P Phsphine oxygenation Ph3P=O (8.6)
Sulfoxidation (8.7)
Aliphatic hydrogen abstraction (8.8)
Phenol dimerization (8.9)
Catechol oxidation (8.10)
RR'CHOH Alcohol oxidation RR'C=O (8.11)
RR'CHNH2 Amine oxidation RR'C≡N (8.12)
RSH Thiol oxidation SR-SR (8.13)

SCHEME 8.19. Reactivity of $\mathbf{Cu_2P^S}$ and $\mathbf{Cu_2O_2}$.

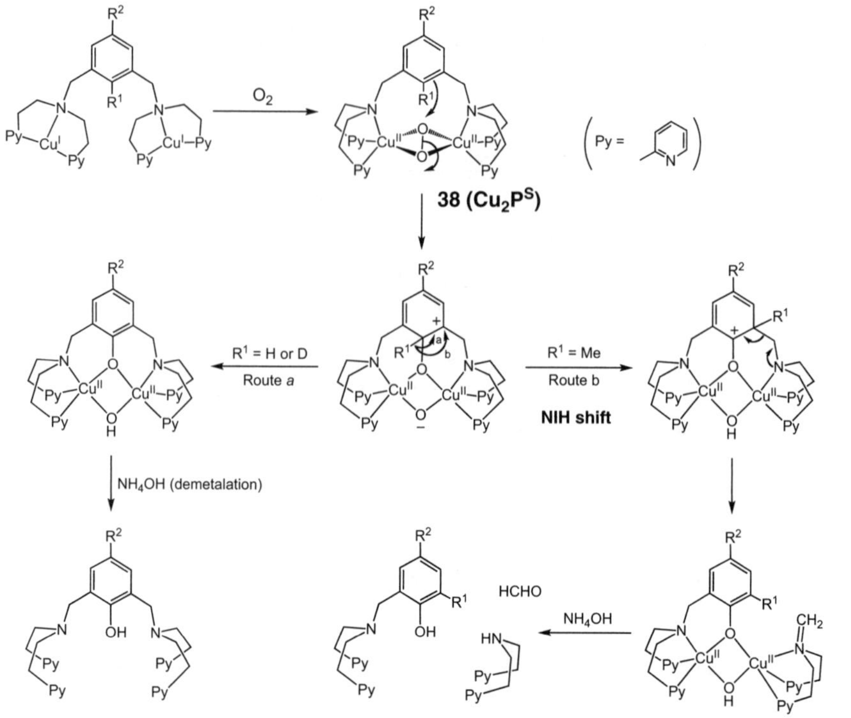

SCHEME 8.20. Aromatic ligand hydroxylation in $(\mu-\eta^2:\eta^2$-peroxo)dicopper(II) complex **38**.

suggesting that the reaction involves a similar electrophilic aromatic substitution mechanism by a bis(μ-oxo)dicopper(III) intermediate (**39, Cu$_2$O$_2$**).

8.3.2.2. Epoxidation. Suzuki and co-workers[218] found a similar aromatic ligand hydroxylation reaction using a *m*-xylyl-based dinucleating ligand with bis(6-methyl-pyridin-2-ylmethyl)amine metal-binding moieties (Scheme 8.22). In this case, addition of olefins causes external substrate oxygenation to give the corresponding

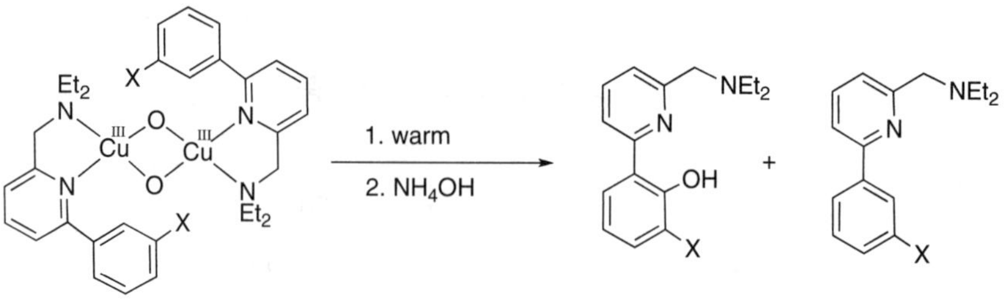

SCHEME 8.21. Aromatic ligand hydroxylation in bis(μ-oxo)dicopper(III) complex **39**.

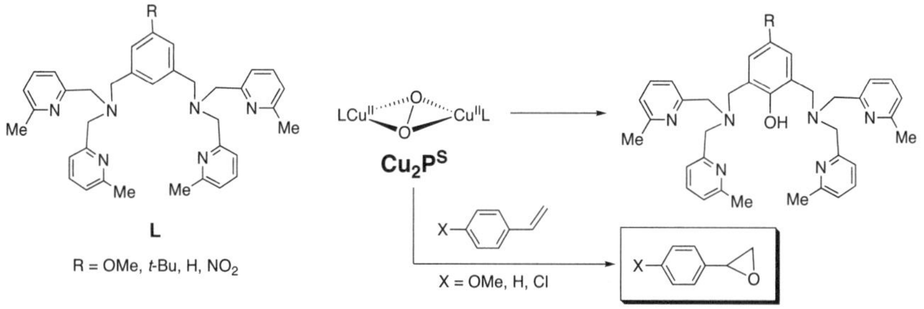

SCHEME 8.22. Epoxidation by $(\mu-\eta^2:\eta^2$-peroxo)dicopper(II) complex.

epoxides, in which incorporation of oxygen-atom from $\mathbf{Cu_2P^S}$ has been confirmed by using $^{18}O_2$. The rate constant for the epoxidation increases as the electron-donating power of the para-substituent of styrene (X = OMe, H, and Cl) increases. A Hammett plot gives $\rho = -1.9$ at $-60°C$, which is almost the same as that for the aromatic ligand hydroxylation ($\rho = -1.9$) and as those for other aromatic ligand hydroxylation in the m-xylyl-based dinucleating ligand systems.[218] Thus, both reactions may involve rate-limiting electrophilic attack of the peroxo ligand to the sp^2 carbon of the substrate.

8.3.2.3. Dimerization of Phenols.

The reaction of $\mathbf{Cu_2P^S}$ with phenol derivatives has also been investigated in many model systems, because the reaction may be more relevant to the phenolase reaction of tyrosinase (type-3 copper monooxygenase catalyzing the monooxygenation reaction of phenols), as compared to the aromatic ligand hydroxylation reactions (Schemes 8.20–8.22). In most cases, however, the reaction of $\mathbf{Cu_2P^S}$ and *neutral phenols* affords C–C coupling dimer products (Eq. 8.9, Scheme 8.19). This finding clearly indicates that the reaction involves phenolic O–H bond activation generating phenoxyl radical intermediates, which spontaneously collapse with each other to give the C–C coupling dimer products. Detailed mechanistic studies including kinetic deuterium isotope effects and Marcus analysis by Itoh and co-workers[214] suggested that these reactions proceed via a PCET (proton coupled electron transfer) mechanism as shown in Scheme 8.23. In this case as well, the contribution of a bis(μ-oxo)dicopper(III) ($\mathbf{Cu_2O_2}$) existing in solution as a minor component has been suggested as the actual active speceis.

8.3.2.4. Oxygenation of Phenolates.

When the *phenolate* derivatives (deprotonated form of the phenols) are employed instead of *neutral phenols*, monooxygenation reaction takes place to give the corresponding catechols without substrate C–C coupling and dimer formation (Eq. 8.1, Scheme 8.19).[215,216,240,241] Kinetic studies on the reaction of $\mathbf{Cu_2P^S}$ and phenolate derivatives containing various para-substituents have been carried out using compounds **40** and **41** (Chart 8.9) to demonstrate that; (1) the reaction obeys first-order kinetics with respect to the peroxo species $\mathbf{Cu_2P^S}$, and a plot of the observed first-order rate constant k_{obs} against the substrate concentration afforded a Michaelis–Menten type saturation curve, suggesting complex formation between $\mathbf{Cu_2P^S}$ and the substrate prior to the oxygen atom transfer

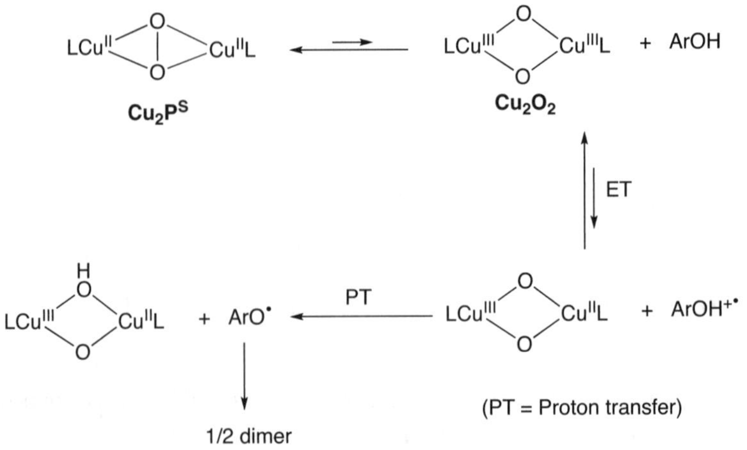

SCHEME 8.23. Dimerization of phenols by **Cu₂Pˢ** and **Cu₂O₂**.

reaction; (2) 'Hammett analysis (plot of log k vs Hammett constant σ^+) gives a relatively small ρ value, about -2; and (3) there is little kinetic deuterium isotope effect (k_H/k_D), when a perdeuterated substrate is employed. Furthermore, the origin of the oxygen atom incorporated into the product has been confirmed to be derived from the peroxo ligand in **Cu₂Pˢ** following an isotope labeling experiment using $^{18}O_2$. All these results strongly suggest that the phenolate oxygenation reaction involves an electrophilic aromatic substitution mechanism, as in the case of the above aromatic ligand hydroxylation (Scheme 8.24). Itoh and co-workers demonstrated that the reactivity of the series of phenol derivatives in enzyme (protein) systems (tyrosinase and hemocyanin) is the same as that in the model reactions, demonstrating that the enzymatic reaction also involves the same electrophilic aromatic substitution mechanism.[178,242–246]

As described above, the oxidation of neutral phenols by **Cu₂Pˢ** and/or **Cu₂O₂** proceeds via a PCET mechanism (Scheme 8.23), whereas the oxygenation of

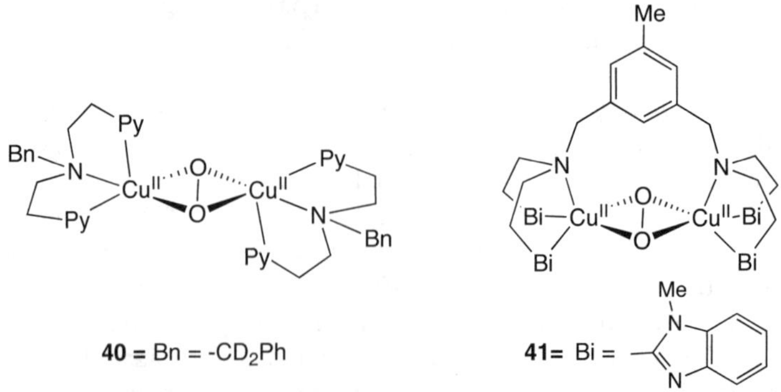

CHART 8.9 ($\mu-\eta^2{:}\eta^2$-Peroxo)dicopper(II) complexes **40** and **41**.

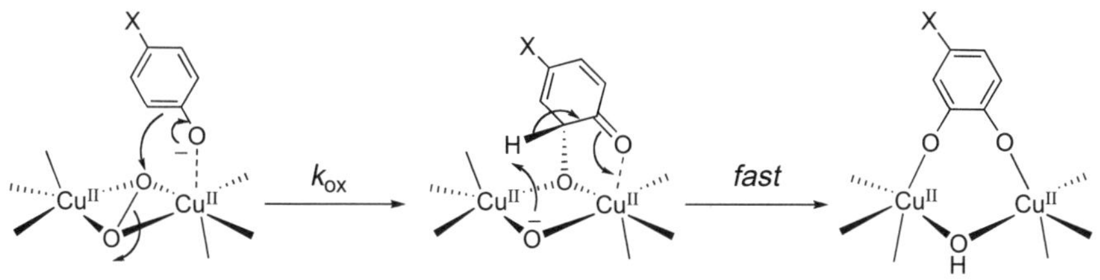

Cu$_2$P^S   (X = *t*-Bu, Me, Br, Cl, F, COMe, COOMe)

SCHEME 8.24. Oxygenation of phenolates by $(\mu-\eta^2:\eta^2$-peroxo)dicopper(II) complex **Cu$_2$P^S**.

phenolates to catechols by **Cu$_2$P^S** involves an electrophilic aromatic substitution mechanism (ionic mechanism). These results may indicate that the 1e$^-$ reduction potentials (E^0_{red}) of both **Cu$_2$P^S** and **Cu$_2$O$_2$** are lower than the 1e$^-$ oxidation potentials (E^0_{ox}) of phenol(ate)s, even though accurate E^0_{red} values have yet to be determined. In the oxidation of neutral phenols, the initial ET from the phenol substrates to **Cu$_2$P^S** and/or **Cu$_2$O$_2$** complexes may be energetically uphill, whereas the subsequent proton transfer to generate the phenoxyl radical may be highly downhill in rendering the overall reaction to completion. In the case of phenolate substrates, the ET from the phenolate to the **Cu$_2$P^S** complexes may still be uphill to prohibit direct formation of the phenoxyl radical from phenolate, whereas the anionic reaction (electrophilic aromatic substitution reaction) proceeds much faster, leading to phenolate oxygenation to catechols, as illustrated in Scheme 8.24.

In the oxygenation reaction of phenolates to catechols by the peroxo complexes **40** and **41 (Cu$_2$P^S)**, no intermediate has been detected even at to $-78 \sim -94°$C.[216,240,241] Itoh et al. also reported that no catechol formation is observed, when the phenolate is directly treated with a bis(μ-oxo)dicopper(III) complex **42 (Cu$_2$O$_2$)** supported by a pyridylethylamine didentate ligand (Chart 8.10) under the same reaction conditions.[241] Thus, the O$-$O bond cleavage and C$-$O bond-formation processes in the oxygenation reaction of phenolates to catechols is considered to proceed via a concerted manner, as illustrated in Scheme 8.24.

In 2005, Stack and co-workers[222] demonstrated that a bis(μ-oxo)dicopper(III) intermediate **Cu$_2$O$_2$** could be detected when a $[(\mu-\eta^2:\eta^2$-peroxo)dicopper(II) **43** (**Cu$_2$P^S**)] supported by the ethylenediamine derivative (Chart 8.10) was treated with phenolate substrates in 2-methyltetrahydrofurane at $-120°$C. The **Cu$_2$O$_2$** complex

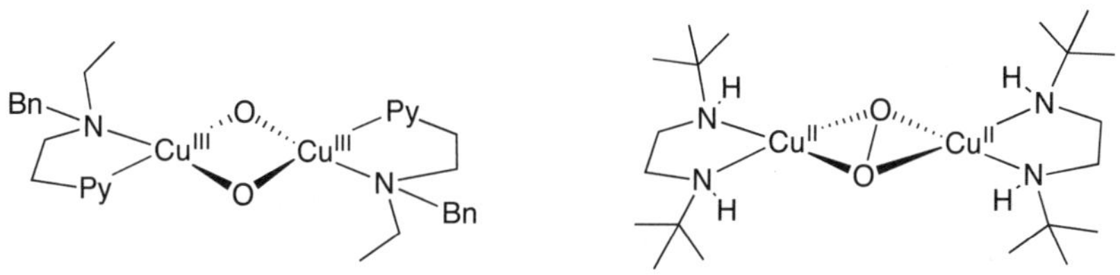

42 = Bn = $-$CD$_2$Ph   **43**

CHART 8.10 Bis(μ-oxo)dicopper(III) complex **42** and $(\mu-\eta^2:\eta^2$-peroxo)dicopper(II) **43**.

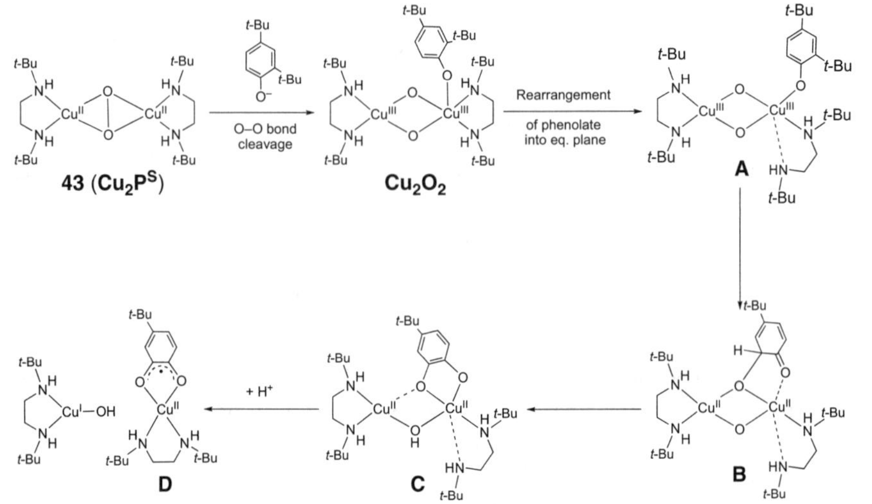

SCHEME 8.25. Oxygenation mechanism of phenolates by $(\mu-\eta^2:\eta^2\text{-peroxo})$dicopper(II) complex **43**.

generated has been well characterized by UV–vis, RR, and X-ray absorption [X-ray absorption spectrum, (XAS) and extended X-ray absorption fine structure (EXAFS)] spectroscopic techniques, and Hammett analysis using a series of para-substituted phenolates that give a ρ value of -2.2. This value is nearly identical to that found for the tyrosinase model reactions.[222] In this case, both the catechol and o-quinone products are obtained in a 1: 1 ratio.[222] On the basis of detailed spectroscopic and computational studies, a reaction mechanism, shown in Scheme 8.25 has been derived, where intermediates **A**, **C**, and **D** could be detected by UV–vis spectroscopy.[247]

Axial binding of an exogenous phenolate to **43**, followed by a rapid rearrangement of the side-on peroxy dicopper(II) core to a bis(μ-oxo)dicopper(III) core and an equatorial positioning of coordinated phenolate yields intermediate **A**. Resonance Raman and UV–vis data clearly identify **A** as a phenolate ligated to a single copper center of the bis(μ-oxo)dicopper(III) core, which is further supported by DFT calculations. The RR profile experiments indicate that the plane of the phenolate ring is oriented between 45 and 90° relative to the plane of the Cu_2O_2 core. This orientation also aligns the phenolate-bonded highest-occupied molecular orbital (HOMO) and the Cu_2O_2 based lowest-occupied molecular orbital, (LUMO) in a manner that facilitates electrophilic attack by the proximal O atom on the ortho-carbon of the phenolate. The subsequent ortho-hydroxylation of the ring is accomplished via an electrophilic aromatic substitution mechanism, as has been well established by the previous studies using the side-on peroxo–dicopper(II) complexes and in the tyrosinase and hemocyanin systems.[216,241–245] This mechanism is supported by the experimentally observed inverse 2° KIE of the *ortho*-C–H bond, indicating a hybridization change at the carbon from sp^2 to sp^3. The phenolate-bonded Cu is reduced through the Copper–phenolate bond and the other Cu is reduced

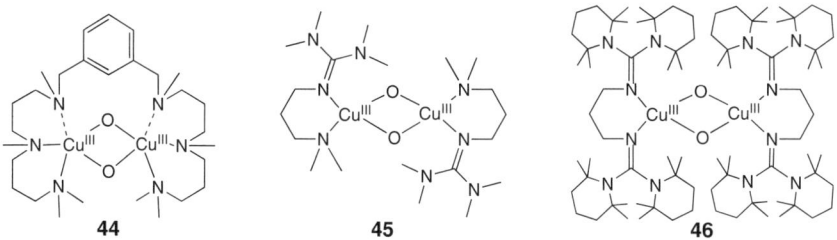

CHART 8.11 Bis(μ-oxo)dicopper(III) complexes **44**, **45**, and **46**.

through the partially formed *ortho*-C–O proximal bond giving intermediate **B**, and subsequent proton transfer and aromatization affords **C**, which exhibits no intense absorption feature in the visible region. Addition of 1 equiv of proton to **C** produces **D**, a copper(II)–semiquinone, and a copper(I)–diamine–OH$_2$ monomer.[247]

With compound **43**, a bis(μ-oxo)dicopper(III) (**Cu$_2$O$_2$**) intermediate is only observed after phenolate binding at very low temperature (Scheme 8.25). Recently, similar phenolate oxygenation reactions have been observed using bis(μ-oxo)dicopper(III) complexes **44** and **45** (Chart 8.11), where the importance of phenolate binding to the **Cu$_2$O$_2$** core for the oxygenation reaction has been demonstrated by kinetic and spectroscopic analyses.[223,224] In fact, **Cu$_2$O$_2$** complex **46** does not induce phenolate oxygenation, probably due to steric hindrance with the four bulky guanidine substituents, which prevent substrate binding. These results together with Tolman's aromatic ligand hydroxylation reaction (Scheme 8.21) may suggest that a **Cu$_2$O$_2$** species is the actual active species in the tyrosinase reaction. However, such a **Cu$_2$O$_2$** species has never been observed in the enzymatic systems.[242–245,248] Thus, this issue requires further studies.

8.3.2.5. Aliphatic C–H Bond Activation (Oxidative N-Dealkylation and Aliphatic Hydroxylation).

As described in Section 8.3.2.4 aromatic ligand hydroxylation reactions had been examined most extensively as model reactions for copper monooxygenases before the early 1990s. In the mid-1990s, Tolman and co-workers[249] first reported on *aliphatic ligand* C–H *bond activation* (Eq. 8.8 in Scheme 8.19) by **47** (**Cu$_2$P^S**) supported by 1,4,7-triisopropyl-1,4,7-triazacyclononane (R$_3$TACN). Such an aliphatic C–H bond activation eventually leads to an oxidative *N*-dealkylation of the aliphatic ligand substituent (Eq. 8.4 in Schemes 8.19 and 8.26).[250] Soon after that, Itoh et al.[251] observed *aliphatic ligand hydroxylation* by **49** (**Cu$_2$P^S**) supported by *N,N*-bis[2-(pyridin-2-yl)ethyl]-2-phenylethylamine (Py2Phe) (Eq. 8.3 in Scheme 8.19 and 8.27). Detailed mechanistic studies on these reactions demonstrated that the real active species are **48** and **50** (**Cu$_2$O$_2$**) generated *in situ* via O–O bond homolysis of **47** and **49** (**Cu$_2$P^S**), respectively (Schemes 8.26 and 8.27).[102,103,250,252,253] Kinetic data for the aliphatic *N*-dealkylation and hydroxylation reactions of a variety of ligand systems are well summarized in the review article by Tolman and Lewis.[8]

More recently, aliphatic C–H bond activation mechanisms by **Cu$_2$P^S** and/or **Cu$_2$O$_2$** have been investigated in detail for a number of ligand sysetms.[213,217,218,221]

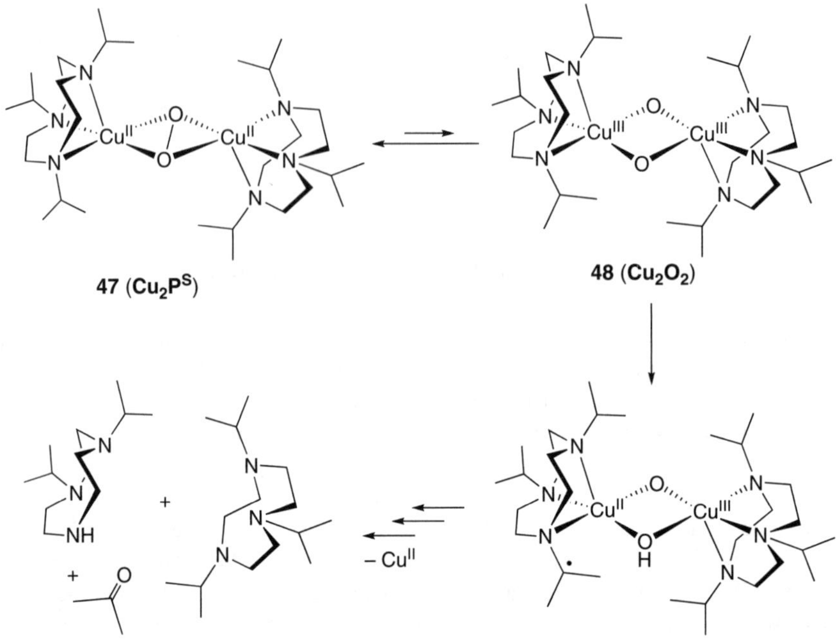

SCHEME 8.26. Oxidative N-dealkylation in $(\mu-\eta^2:\eta^2$-peroxo)dicopper(II) complex **47** via bis(μ-oxo)dicopper(III) intermediate **48**.

Karlin and co-workers demonstrated that a series of peroxo complexes $\mathbf{Cu_2P^S}$ (**51^R**, Chart 8.12) can oxidize external substrates, such as tetrahydrofuran (THF) and N,N-dimethylaniline (DMA). Based on the results of kinetic analysis on the above reactions together with product analysis on the oxidation of (N-cyclopropyl-N-

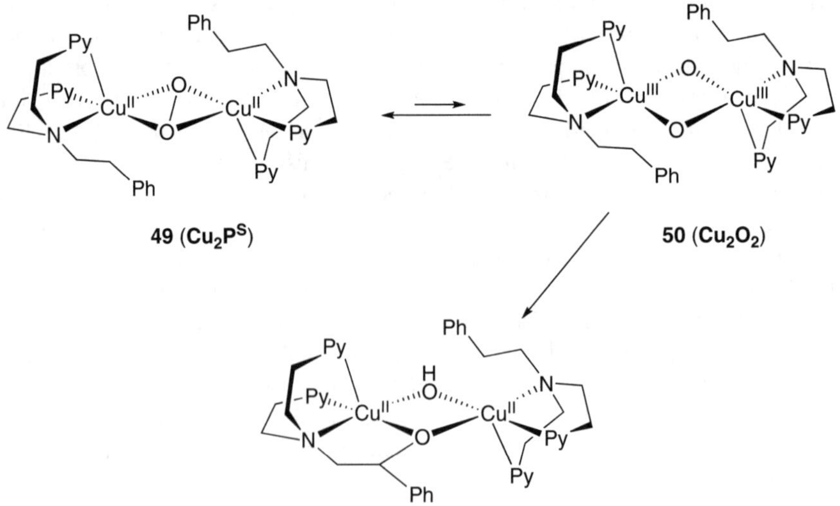

SCHEME 8.27. Aliphatic ligand hydroxylation in $(\mu-\eta^2:\eta^2$-peroxo)dicopper(II) complex **49** via bis(μ-oxo)dicopper(III) intermediate **50**.

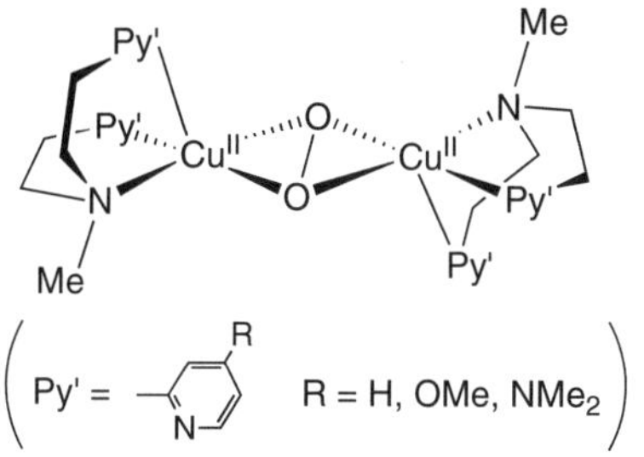

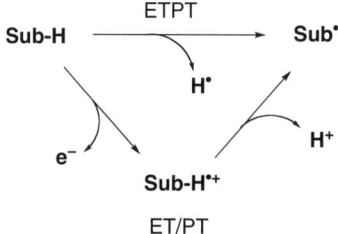

CHART 8.12 $(\mu-\eta^2:\eta^2\text{-Peroxo})$dicopper(II) complexes **51**[R].

methylaniline (CMA) and (p-methoxyphenyl)-2,2-dimethylpropanol (MDP), mechanistic probes for PCET processes, they suggested that at low thermodynamic driving force for substrate oxidation, a *consecutive* electron-transfer proton-transfer (ET/PT) mechanism is operable, but once ET becomes prohibitively uphill, the *concerted* electron transfer proton-transfer (ETPT) pathway occurs (Scheme 8.28).[213,217]

Suzuki, Fukuzumi, and co-workers[221] also reported that a $(\mu-\eta^2:\eta^2\text{-peroxo})$ dicopper(II) complex **Cu₂Pˢ**, supported by the *m*-xylyl dinucleating ligand shown in Scheme 8.22, can activate the C-H bonds of a variety of external aliphatic substrates having different bond-dissociation energies (BDE = ca. 75–92 kcal/mol), such as 10-methyl-9,10-dihydroacridine (AcrH₂), 1,4-cyclohexadiene (1,4-CHD), 9,10-dihydroanthracene (9,10-DHA), fluorene, tetralin, toluene, and THF. A linear correlation between the logarithm of the oxidation rate constants and the BDEs of the C−H bond of the substrates together with relatively large deuterium kinetic isotope effects k_2^H/k_2^D (13 for 9,10-DHA, 29 for toluene, and ~34 for THF at −70°C and ~9 for AcrH₂ at −94°C), clearly indicated that hydrogen-atom abstraction from the substrates is the rate-determining step. It also has been shown that the reactivity of the **Cu₂Pˢ** species is similar to that of cumylperoxyl radical.[218,221] In the reactions of the bis$(\mu\text{-oxo})$dicopper(III) complex **42** (Chart 8.10) with 1,4-CHD and AcrH₂, on the other hand, Itoh and co-workers[229] suggested that the $(\mu\text{-oxo})(\mu\text{-oxyl}$

SCHEME 8.28. Consecutive electron-transfer proton-transfer (ET/PT) pathway and concerted electron-transfer proton-transfer (ETPT) pathway.

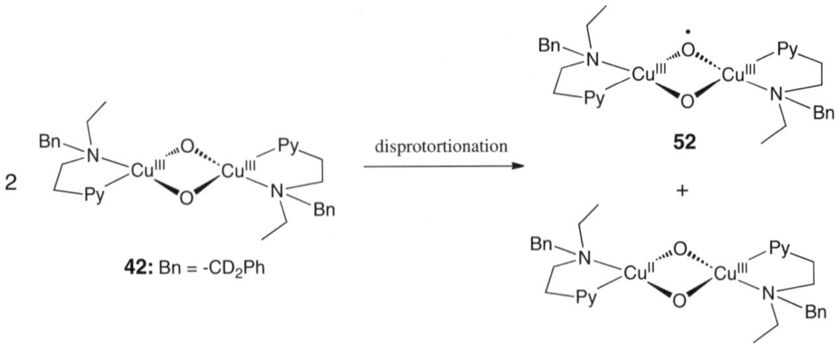

SCHEME 8.29. Disproportionation of bis(μ-oxo)dicopper(III) complex **42** generating (μ-oxo)(μ-oxyl radical)dicopper(III) species **52**.

radical)dicopper(III) species (**52**) that is generated *in situ* via disproportionation of two molecules of bis(μ-oxo)dicopper(III) complex **42** (Scheme 8.29) is involved as an actual reactive intermediate.

Hydroxylation of cyclohexane and adamantine with H_2O_2 catalyzed by β-diketiminatocopper(II) complexes also has been reported, where Cu_2O_2 was suggested as a key reactive intermediate.[203]

8.3.2.6. Other Reactions (Oxygenation of PPh₃, Sulfoxidation, Dehydrogenation of Catechol, Alcohol, Amine, and Thiol).

Oxygen atom transfer reactions from Cu_2P^S and/or Cu_2O_2 to phosphine derivatives (particularly triphenyl phosphine PPh₃) giving phosphine oxides (Eq. 8.6 in Scheme 8.19) is frequently employed as a probe to examine their oxygenation ability.[228] However, the mechanism of the phosphine oxygenation has yet to be addressed in detail.

Oxygenation of sulfides to sulfoxides by bis(μ-oxo)dicopper(III) complex **42** (Cu_2O_2) (Eq. 8.7 in Scheme 8.19) has been investigated in detail to show that the reaction consists of two distinct steps, where the first quick process is association of the substrate to Cu_2O_2 and the second slow process is intramolecular oxygen-atom transfer from the copper-oxo species to the substrate in the associated complex. The rate constant of the second slow process is rather insensitive to the oxidation potential of the substrates, suggesting that the oxo-transfer reaction proceeds via a direct oxygen-atom transfer mechanism.[225] Casella and co-workers[220] recently demonstrated a catalytic sulfoxidation reaction with a dinuclear copper complex supported by a *m*-xylyl dinucleating ligand carrying bis(benzimidazol-2-ylmethyl)amine metal-binding moieties in the presence of hydroxylamine as an external reductant under aerobic conditions. In this case, (μ-η^2:η^2-peroxo)dicopper(II) **41** (Cu_2P^S, see Chart 8.9) has been suggested to be involved as the key reactive intermediate.

Dehydrogenation of catechols, alcohols, amines, and thiols by Cu_2P^S and/or Cu_2O_2 complexes (Eqs. 8.10–8.13 in Scheme 8.19) have also been examined briefly in some ligand systems, although mechanistic details of each reaction have not been examined.[109,184,212,226,227]

8.3.3. Dinuclear Copper(II)–Hydroperoxo Complexes (Cu$_2$P^H)

Karlin and et al.[52,53] demonstrated that (μ-η^1:η^1-hydroperoxo)dicopper(II) complexes (**16, Cu$_2$P^H**, see Scheme 8.6) reacts with PPh$_3$ or sulfides quantitatively to given the monooxygenated product [O=PPh$_3$ or RS(O)R] and a (μ-hydroxo)dicopper(II) complex, while the dicopper(II)–peroxo complex (**15, Cu$_2$P$^{E^*}$**, also see Scheme 8.6) reacts with triphenylphosphine (PPh,) liberating O$_2$ quantitatively and providing the PPh$_3$ adduct of dicopper(I) complex. Similar reactivity of **Cu$_2$P^H** also has been observed in the related phenoxo-bridged dinucleating ligand systems.[54–56]

The reaction of the dicopper(II) complex supported by the *m*-xylyl-based dinucleating ligand with H$_2$O$_2$ resulted in aromatic ligand hydroxylation, as seen in the reaction of the dicopper(I) complex of the same ligand with O$_2$ (Scheme 8.20).[254] In this case, a (μ-1,1-hydroperoxo)dicopper(II) species (**Cu$_2$P^H**) is implicated as a key reactive intermediate that performs electrophilic aromatic substitution reaction.[254] In a related ligand system with *N*-methylbenzimidazolyl instead of pyridyl donors, double arene hydroxylation occurs upon reaction of its dicopper(II) complex with H$_2$O$_2$ (Scheme 8.30).[255] On the basis of the results of a mechanistic study, attack of a (μ-1,1-hydroperoxide)dicopper intermediate (**53**, type **Cu$_2$P^H**) at the 5-position has been proposed, with functionalization of the 2-position subsequently occurring.[255]

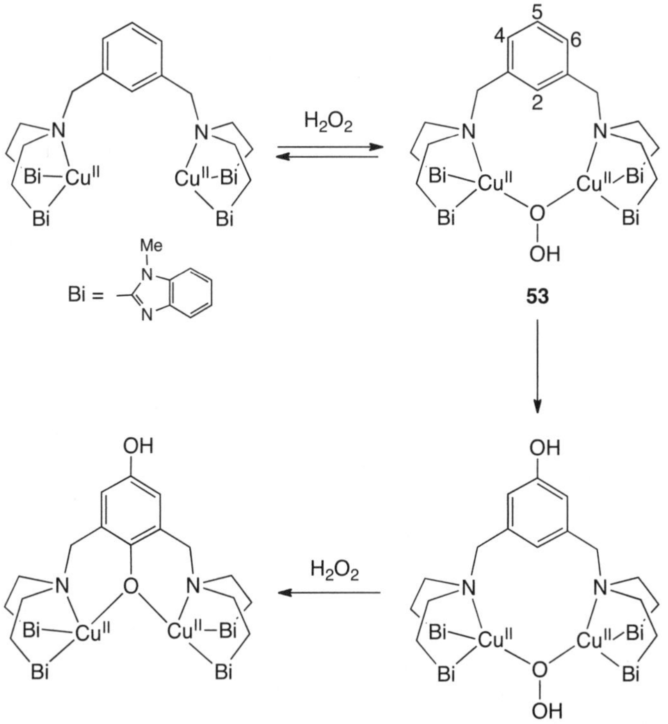

SCHEME 8.30. Dinuclear copper(II)-hydroperoxo complex **53** (**Cu$_2$P^H**) that induces double aromatic ligand hydroxylation.

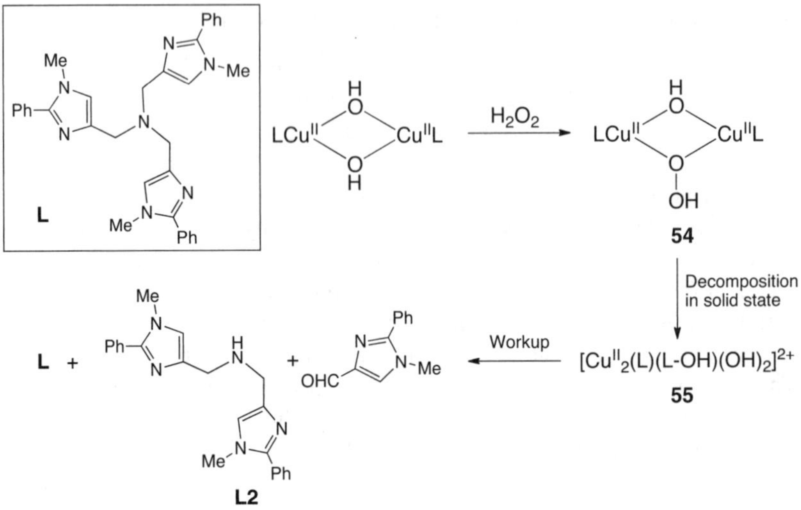

SCHEME 8.31. (μ-1,1-Hydroperoxo)(μ-hydroxo)dicopper(II) complex **54** (**Cu$_2$P^H**) that induces oxidative *N*-dealkylation.

Suzuki and co-workers[68] isolated and structurally characterized a (μ-1,1-hydroperoxo)(μ-hydroxo)dicopper(II) complex (**54**, type **Cu$_2$P^H**) supported by a tetradentate tripodal ligand possessing three sterically bulky imidazolyl groups [tris (1-methyl-2-phenylimidazol-4-ylmethyl)amine] labeled L (Scheme 8.31). Complex **54** is generated by the reaction of the bis(μ-hydroxo)dicopper(II) complex with H$_2$O$_2$ in acetonitrile at $-40°$C. The crystal structure of **54** revealed that one pendant arm of the ligand is free from coordination, producing a hydrophobic cavity around the Cu$_2$(μ-1,1-OOH)(μ-OH) core. Thus, the compound is relatively stable, and H/D and ^{16}O/^{18}O exchange reactions are supressed. Decomposition of a solid sample of **54** at $60°$C induces aliphatic ligand hydroxylation at one of the methylene groups of the pendant arms to give complex **55**, from which an *N*-dealkylated ligand, bis(1-methyl-2-phenyl-4-imidazolylmethyl)amine (L2), is obtained upon workup.[68] This aliphatic hydroxylation performed in **54** is in marked contrast to the arene hydroxylation reported for the above mentioned (μ-1,1-hydroperoxo)dicopper(II) complexes with a xylyl linker.

Karlin and co-workers[57,58] recently reported another type of reactivity of **Cu$_2$P^H** supported by a new dinucleating ligand PD'OH (complex **56^R** in Scheme 8.32). The (μ-η^1:η^1-hydroperoxo)dicopper(II) complexes (**56^R**) (R = $-$H, $-$Me or $-$Ph); λ_{max} = 407 nm; ν_{O-O} = 870 cm^{-1} is generated by reacting a precursor dicopper(I) complex [Cu$^I_2$(PD'OH)(CH$_3$CN)$_2$](ClO$_4$)$_2$ with O$_2$ in nitrile solvents at $-80°$C. Upon warming, hydroperoxo complex **56^R** decomposes to induce hydroxylation of the copper-bound organocyanides, providing the corresponding aldehyde and CN$^-$. The thermal decomposition of **56^R** also leads to a trinuclear Cu(II) product **58**, which apparently derives from both oxidative *N*-dealkylation and then oxidative dehydrogenation of the linker moiety of PD'OH ligand (the chloride bound to Cu ion presumably derives from the CH$_2$Cl$_2$ solvent). With an excess of PPh$_3$ added to **56^R**, a binuclear Cu(I) complex (**57**) with a cross-linked PD'OH ligand is also produced. The newly formed

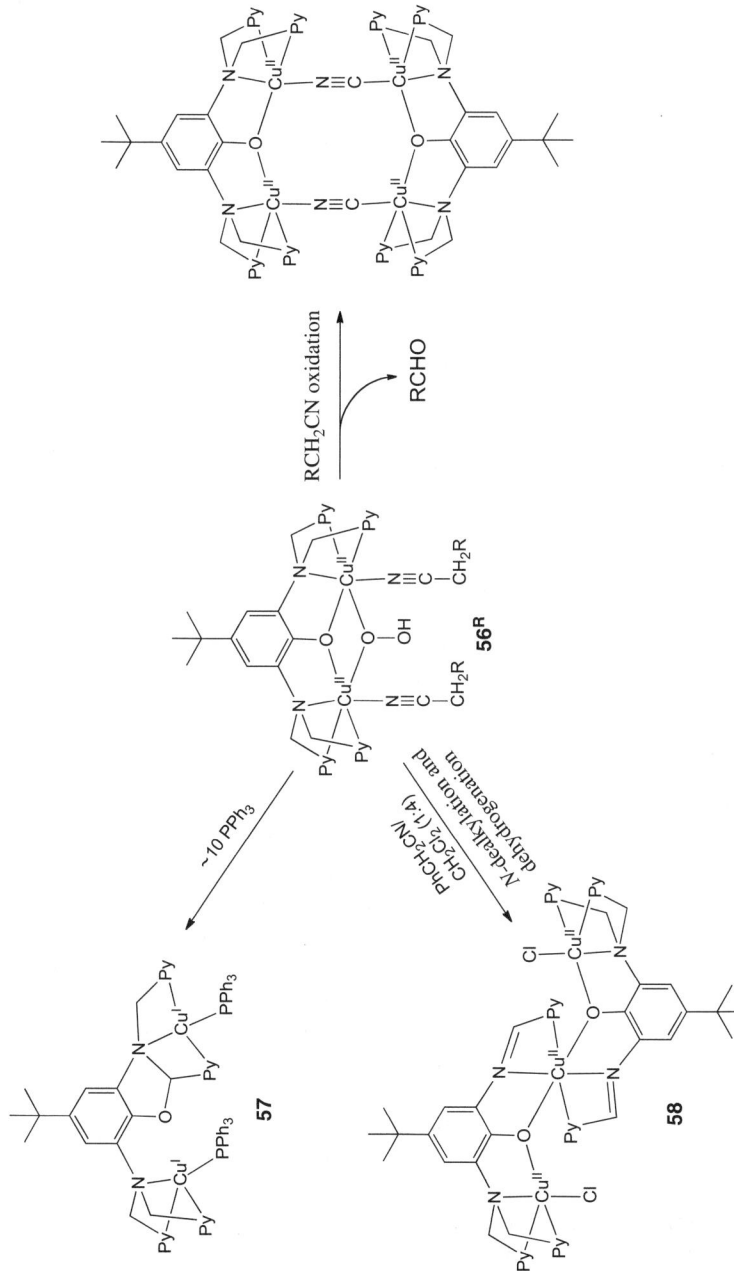

SCHEME 8.32. (μ-1,1-Hydroperoxo)(μ-phenoxo)dicopper(II) complex 56^R (Cu_2P^H) that induces oxidative N-dealkylation and nitrile oxygenation.

C–O bond in **57** and an apparent KIE ($k_H/k_D = 2.9 \pm 0.2$) in the benzylcyanide ($R = -C_6H_5$) oxidation reaction suggest that a common ligand-based radical is formed during the thermal decay of **56^R**. Although further studies are needed, O–O bond homolysis may initiate the radical formation on the ligand.[58]

8.3.4. Dinuclear Copper(II) Oxo Complexes (Cu$_2$O)

Zeolites containing transition metal ions often show promising activity as heterogeneous catalysts for selective oxidation reactions. In Cu–ZSM-5, calcination with O_2 results in the formation of an active oxygen species exhibiting a characteristic absorption band at 440 nm. This speceis has been found to be a key intermediate in both the direct decomposition of NO and N_2O and the selective oxidation of methane into methanol.[256–259] So far, various types of mononuclear and dinuclear copper active oxygen species have been proposed mainly based on the results of DFT calculations, but no consensus has been obtained. Very recently, Schoonheydt and co-workers[260–262] identified the active site in Cu–ZSM-5 as a bent $[\text{Cu–O–Cu}]^{2+}$ core (**Cu$_2$O**) by means of an RR techniques. Such a speceis also could be involved in particulate methane monooxygenase (pMMO).

8.4. SUMMARY

Since the early 1980s, copper–dioxygen chemistry has been growing steadily and is constituting one of the most important research area not only in bioinorganic chemistry, but also in coordination chemistry. During this period, there have been several breakthroughs in copper–dioxygen chemistry, such as the first crystal structure determination of **Cu$_2$P^E** by Karlin and co-workers[84] the correct prediction of the peroxo intermediates **Cu$_2$P^S** of tyrosinase and hemocyanin using a simple model compound by Kitajima et al.,[87] discovery of **Cu$_2$O$_2$** by Tolman and co-workers[102] and the crystal structure determination of **Cu$_1$S^E** by Schindler and co-workers[35]. The isolation and X-ray structure determination of other mononuclear (**Cu$_1$S^S**, **Cu$_1$P^S**, **Cu$_1$P^H**, **Cu$_1$P^R**, and **Cu$_1$P^{Ac}**), dinuclear (**Cu$_2$P^H**), and multinuclear (**Cu$_3$O$_2$** and **Cu$_4$P^E**) copper-active oxygen complexes are also noteworthy.[7,8] Stimulated by those structures, many researchers have explored the detailed electronic structures, spectroscopic characteristics, and reactivity of the copper-active oxygen complexes, providing significant and important insights into the reactive intermediates involved not only in biological systems, but also in numerous copper-catalyzed oxidation reactions. Synthetic chemists have also been trying to develop various types of supporting ligands to control the structure and reactivity of the generated copper–active oxygen species. Thanks to those efforts in various research fields, the chemistry of *dinuclear* copper-active oxygen species has now matured. However, there still remain a lot to be done in *mononuclear* and *multinuclear* copper-active oxygen chemistry. After the 30 year anniversary of copper–dioxygen chemistry, we should also expand our research direction to more applied fields, such as developing more effective catalysts for synthetic organic chemistry, more efficient devices for a fuel cell, and more useful reagents for pharmaceutical applications. Hopefully, information presented in this chapter will help in the development of such future works in copper–dioxygen chemistry.

REFERENCES

1. Kitajima, N.; Moro-oka, Y. Copper–dioxygen complexes. Inorganic and bioinorganic perspectives. *Chem. Rev.* **1994**, *94*, 737–757.

2. Klinman, J. P. Mechanisms whereby mononuclear copper proteins functionalize organic substrates. *Chem. Rev.* **1996**, *96*, 2541–2562.

3. Solomon, E. I.; Sundaram, U. M.; Machonkin, T. E. Multicopper oxidases and oxygenases. *Chem. Rev.* **1996**, *96*, 2563–2606.

4. McGuirl, M. A.; Dooley, D. M. Copper-containing oxidases. *Curr. Opin. Chem. Biol.* **1999**, *3*, 138–144.

5. Prigge, S. T.; Mains, R. E.; Eipper, B. A.; Amzel, L. M. New insights into copper monooxygenases and peptide amidation: Structure, mechanism and function. *Cell. Mol. Life Sci.* **2000**, *57*, 1236–1259.

6. Solomon, E. I.; Chen, P.; Metz, M.; Lee, S. K.; Palmer, A. E. Oxygen binding, activation, and reduction to water by copper proteins. *Angew. Chem. Int. Ed.* **2001**, *40*, 4570–4590.

7. Mirica, L. M.; Ottenwaelder, X.; Stack, T. D. P. Structure and spectroscopy of copper-dioxygen complexes. *Chem. Rev.* **2004**, *104*, 1013–1045.

8. Lewis, E. A.; Tolman, W. B. Reactivity of dioxygen-copper systems. *Chem. Rev.* **2004**, *104*, 1047–1076.

9. Itoh, S. Mononuclear copper active-oxygen complexes. *Curr. Opin. Chem. Biol.* **2006**, *10*, 115–122.

10. Hatcher, L. Q.; Karlin, K. D. In *Advances in Inorganic Chemistry*; van Eldik, R., Reedijk, J. Eds.; Academic Press: Amsterdam, The Netherlands, 2006; Vol. *58;* pp 131–184.

11. Battaini, G.; Granata, A.; Monzani, E.; Gullotti, M.; Casella, L.In *Advances in Inorganic Chemistry*; van Eldik, R., Reedijk, J. Eds.; Academic Press: Amsterdam, The Netherlands, 2006; Vol. *58;* pp 185–233.

12. Koval, I. A.; Gamez, P.; Belle, C.; Selmeczi, K.; Reedijk, J. Synthetic models of the active site of catechol oxidase: mechanistic studies. *Chem. Soc. Rev.* **2006**, *35*, 814–840.

13. Balasubramanian, R.; Rosenzweig, A. C. Structural and mechanistic insights into methane oxidation by particulate methane monooxygenase. *Acc. Chem. Res.* **2007**, *40*, 573–580.

14. Bollinger, J.; Martin, J.; Krebs, C. Enzymatic C-H activation by metal-superoxo intermediates. *Curr. Opin. Chem. Biol.* **2007**, *11*, 151–158.

15. Punniyamurthy, T.; Rout, L. Recent advances in copper-catalyzed oxidation of organic compounds. *Coord. Chem. Rev.* **2008**, *252*, 134–154.

16. Himes, R. A.; Karlin, K. D. Copper–dioxygen complex mediated C-H bond oxygenation: relevance for particulate methane monooxygenase (pMMO). *Curr. Opin. Chem. Biol.* **2009**, *13*, 119–131.

17. Hatcher, L. Q.; Karlin, K. D. Oxidant types in copper-dioxygen chemistry: the ligand coordination defines the Cu_n-O_2 structure and subsequent reactivity. *J. Biol. Inorg. Chem.* **2004**, *9*, 669–683.

18. Thompson, J. S. Copper–dioxygen chemistry. Synthesis, spectroscopy, and properties of a copper(II) superoxide complex. *J. Am. Chem. Soc.* **1984**, *106*, 4057–4059.

19. Karlin, K. D.; Wei, N.; Jung, B.; Kaderli, S.; Zuberbühler, A. D. Kinetic, thermodynamic, and spectral characterization of the primary copper–oxygen (Cu-O$_2$) adduct in a reversibly formed and structurally characterized peroxo-dicopper(II) complex. *J. Am. Chem. Soc.* **1991**, *113*, 5868–5870.

20. Mahroof-Tahir, M.; Karlin, K. D. A dinuclear mixed-valence Cu(I)/Cu(II) complex and its reversible reactions with dioxygen: Generation of a superoxodicopper(II) species. *J. Am. Chem. Soc.* **1992**, *114*, 7599–7601.

21. Karlin, K. D.; Wei, N.; Jung, B.; Kaderli, S.; Niklaus, P.; Zuberbühler, A. D. Kinetics and thermodynamics of formation of copper-dioxygen adducts: oxygenation of mononuclear copper(I) complexes containing tripodal tetradentate ligands. *J. Am. Chem. Soc.* **1993**, *115*, 9506–9514.

22. Fujisawa, K.; Tanaka, M.; Morooka, Y.; Kitajima, N. A Monomeric side-on superoxocopper(Ii) complex: Cu(O$_2$)(HB(3-tBu-5-iPrpz)$_3$). *J. Am. Chem. Soc.* **1994**, *116*, 12079–12080.

23. Becker, M.; Heinemann, F. W.; Schindler, S. Reversible binding of dioxygen by a copper (I) complex with tris(2-dimethylaminoethyl)amine (Me$_6$tren) as a ligand. *Chem.-Eur. J.* **1999**, *5*, 3124–3129.

24. Chaudhuri, P.; Hess, M.; Weyhermuller, T.; Wieghardt, K. Aerobic oxidation of primary alcohols by a new mononuclear Cu^{II}-radical catalyst. *Angew. Chem. Int. Ed. Engl.* **1999**, *38*, 1095–1098.

25. Chen, P.; Root, D. E.; Campochiaro, C.; Fujisawa, K.; Solomon, E. I. Spectroscopic and electronic structure studies of the diamagnetic side-on Cu^{II}-superoxo complex Cu(O$_2$) [HB(3-R-5-iPrpz)$_3$]: Antiferromagnetic coupling versus covalent delocalization. *J. Am. Chem. Soc.* **2003**, *125*, 466–474.

26. Jazdzewski, B. A.; Reynolds, A. M.; Holland, P. L.; YoungJr., V. G.; Kaderli, S.; Zuberbuhler, A. D.; Tolman, W. B. Copper(I)–phenolate complexes as models of the reduced active site of galactose oxidase: synthesis, characterization, and O$_2$ reactivity. *J. Biol. Inorg. Chem.* **2003**, *8*, 381–393.

27. Weitzer, M.; Schindler, S.; Brehm, G.; Schneider, S.; Hormann, E.; Jung, B.; Kaderli, S.; Zuberbühler, A. D. Reversible binding of dioxygen by the copper(I) complex with tris(2-dimethylaminoethyl)amine (Me$_6$tren) ligand. *Inorg. Chem.* **2003**, *42*, 1800–1806.

28. Zhang, C. X.; Kaderli, S.; Costas, M.; Kim, E.; Neuhold, Y. M.; Karlin, K. D.; Zuberbühler, A. D. Copper(I)–dioxygen reactivity of (L)Cu(I)$^+$ (L = tris(2-pyridylmethyl)amine): Kinetic/thermodynamic and spectroscopic studies concerning the formation of Cu$-$O$_2$ and Cu$_2$$-O_2$ adducts as a function of solvent medium and 4-pyridyl ligand substituent variations. *Inorg. Chem.* **2003**, *42*, 1807–1824.

29. Komiyama, K.; Furutachi, H.; Nagatomo, S.; Hashimoto, A.; Hayashi, H.; Fujinami, S.; Suzuki, M.; Kitagawa, T. Dioxygen reactivity of copper(I) complexes with tetradentate tripodal ligands having aliphatic nitrogen donors: Synthesis, structures, and properties of peroxo and superoxo complexes. *Bull. Chem. Soc. Jpn.* **2004**, *77*, 59–72.

30. Schatz, M.; Raab, V.; Foxon, S. P.; Brehm, G.; Schneider, S.; Reiher, M.; Holthausen, M. C.; Sundermeyer, J.; Schindler, S. Combined spectroscopic and theoretical evidence for a persistent end-on copper superoxo complex. *Angew. Chem. Int. Ed.* **2004**, *43*, 4360–4363.

31. Smirnov, V. V.; Roth, J. P. Evidence for Cu−O$_2$ intermediates in superoxide oxidations by biomimetic copper(II) complexes. *J. Am. Chem. Soc.* **2006**, *128*, 3683–3695.

32. Lanci, M. P.; Smirnov, V. V.; Cramer, C. J.; Gauchenova, E. V.; Sundermeyer, J.; Roth, J. P. Isotopic probing of molecular oxygen activation at copper(I) sites. *J. Am. Chem. Soc.* **2007**, *129*, 14697–14709.

33. Maiti, D.; Fry, H. C.; Woertink, J. S.; Vance, M. A.; Solomon, E. I.; Karlin, K. D. A 1:1 Copper-dioxygen adduct is an end-on bound superoxo copper(II) complex which undergoes oxygenation reactions with phenols. *J. Am. Chem. Soc.* **2007**, *129*, 264–265.

34. Kunishita, A.; Kubo, M.; Sugimoto, H.; Ogura, T.; Sato, K.; Takui, T.; Itoh, S. Mononuclear copper(II)–superoxo complexes that mimic the structure and reactivity of the active centers of PHM and DβM. *J. Am. Chem. Soc.* **2009**, *131*, 2788–2789.

35. Würtele, C.; Gaoutchenova, E.; Harms, K.; Holthausen, M. C.; Sundermeyer, J.; Schindler, S. Crystallographic characterization of a synthetic 1:1 end-on copper dioxygen adduct complex. *Angew. Chem. Int. Ed.* **2006**, *45*, 3867–3869.

36. de la Lande, A.; Gérard, H.; Moliner, V.; Izzet, G.; Reinaud, O.; Parisel, O. Theoretical modelling of tripodal CuN$_3$ and CuN$_4$ cuprous complexes interacting with O$_2$, CO or CH$_3$CN. *J. Biol. Inorg. Chem.* **2006**, *11*, 593–608.

37. Prigge, S. T.; Eipper, B. A.; Mains, R. E.; Amzel, L. M. Dioxygen binds end-on to mononuclear copper in a precatalytic enzyme complex. *Science* **2004**, *304*, 864–867.

38. Aboelella, N. W.; Lewis, E. A.; Reynolds, A. M.; Brennessel, W. W.; Cramer, C. J.; Tolman, W. B. Snapshots of dioxygen activation by copper: The structure of a 1:1 Cu/O$_2$ adduct and its use in syntheses of asymmetric bis(μ-oxo) complexes. *J. Am. Chem. Soc.* **2002**, *124*, 10660–10661.

39. Aboelella, N. W.; Kryatov, S. V.; Gherman, B. F.; Brennessel, W. W.; Young, V. G.; Sarangi, R.; Rybak-Akimova, E. V.; Hodgson, K. O.; Hedman, B.; Solomon, E. I.; Cramer, C. J.; Tolman, W. B. Dioxygen activation at a single copper site: Structure, bonding, and mechanism of formation of 1:1 Cu-O$_2$ adducts. *J. Am. Chem. Soc.* **2004**, *126*, 16896–16911.

40. Reynolds, A. M.; Gherman, B. F.; Cramer, C. J.; Tolman, W. B. Characterization of a 1:1 Cu-O$_2$ adduct supported by an anilido imine ligand. *Inorg. Chem.* **2005**, *44*, 6989–6997.

41. Aboelella, N. W.; Gherman, B. F.; Hill, L. M. R.; York, J. T.; Holm, N.; Young, V. G.; Cramer, C. J.; Tolman, W. B. Effects of thioether substituents on the O$_2$ reactivity of β-diketiminate-Cu(I) complexes: Probing the role of the methionine ligand in copper monooxygenases. *J. Am. Chem. Soc.* **2006**, *128*, 3445–3458.

42. Valentine, J. S.; Curtis, A. B. Convenient preparation of solutions of superoxide anion and the reaction of superoxide anion with a copper(II) complex. *J. Am. Chem. Soc.* **1975**, *97*, 224–226.

43. Nappa, M.; Valentine, J. S.; Miksztal, A.; Schugar, H. J.; Isied, S. S. Reactions of superoxide in aprotic solvents. A superoxo complex of copper(II) rac-5,7,7,12,14, 14-hexamethyl-1,4,8,11-tetraazacyclotetradecane. *J. Am. Chem. Soc.* **1979**, *101*, 7744–7746.

44. Weinstein, J.; Bielski, B. H. J. Reaction of superoxide radicals with copper(II)–histidine complexes. *J. Am. Chem. Soc.* **1980**, *102*, 4916–4919.

45. O'Young, C.-L.; Lippard, S. J. Reactions of superoxide anion with copper(II) salicylate complexes. *J. Am. Chem. Soc.* **1980**, *102*, 4920–4924.

46. Bailey, C. L.; Bereman, R. D.; Rillema, D. P. Redox and spectral properties of cobalt(II) and copper(II) tetraazaannulene complexes: {H$_2$[Me$_4$(RBzo)$_2$[14]tetraeneN$_4$]} (R = H, CO$_2$CH$_3$). Evidence for superoxide ligation and reduction. *Inorg. Chem.* **1986**, *25*, 3149–3153.

47. Cabelli, D. E.; Bielski, B. H. J.; Holcman, J. Interaction between copper(II)–arginine complexes and HO$_2$/O$_2$- radicals, a pulse radiolysis study. *J. Am. Chem. Soc.* **1987**, *109*, 3665–3669.

48. Nishida, Y.; Unoura, K.; Watanabe, I.; Yokomizo, T.; Kato, Y. Colored species formation between mononuclear copper(II) complex and superoxide anion. *Inorg. Chim. Acta* **1991**, *181*, 141–143.

49. Fujii, T.; Yamaguchi, S.; Funahashi, Y.; Ozawa, T.; Tosha, T.; Kitagawa, T.; Masuda, H. Mononuclear copper(II)–hydroperoxo complex derived from reaction of copper(I) complex with dioxygen as a model of DβM and PHM. *Chem. Commun.* **2006**, 4428–4430.

50. Fujii, T.; Yamaguchi, S.; Hirota, S.; Masuda, H. H-atom abstraction reaction for organic substrates via mononuclear copper(II)–superoxo species as a model for DβM and PHM. *Dalton Trans.* **2008**, 164–170.

51. Maiti, D.; Lee, D.-H.; Gaoutchenova, K.; Würtele, C.; Holthausen, M. C.; Sarjeant, A. A. N.; Sundermeyer, J.; Schindler, S.; Karlin, K. D. Reactions of a copper(II) superoxo complex lead to C−H and O−H substrate oxygenation: Modeling copper-monooxygenase C−H hydroxylation. *Angew. Chem. Int. Ed.* **2008**, *47*, 82–85.

52. Karlin, K. D.; Cruse, R. W.; Gultneh, Y. Dioxygen copper reactivity - A hydroperoxo dicopper(II) complex. *J. Chem. Soc., Chem. Commun.* **1987**, 599–600.

53. Karlin, K. D.; Ghosh, P.; Cruse, R. W.; Farooq, A.; Gultneh, Y.; Jacobson, R. R.; Blackburn, N. J.; Strange, R. W.; Zubieta, J. Dioxygen—copper reactivity— Generation, characterization, and reactivity of a hydroperoxo—dicopper(II) complex. *J. Am. Chem. Soc.* **1988**, *110*, 6769–6780.

54. Sorrell, T. N.; Vankai, V. A. Synthesis and dioxygen reactivity of dinuclear copper-phenolate and copper-phenol complexes with pyrazole and pyridine donors. *Inorg. Chem.* **1990**, *29*, 1687–1692.

55. Mahroof-Tahir, M.; Murthy, N. N.; Karlin, K. D.; Blackburn, N. J.; Shaikh, S. N.; Zubieta, J. New thermally stable hydroperoxo- and peroxo-copper complexes. *Inorg. Chem.* **1992**, *31*, 3001–3003.

56. Murthy, N. N.; Mahroof-Tahir, M.; Karlin, K. D. Dicopper(I) complexes of unsymmetrical binucleating ligands and their dioxygen reactivities. *Inorg. Chem.* **2001**, *40*, 628–635.

57. Li, L.; Sarjeant, A. A. N.; Vance, M. A.; Zakharov, L. N.; Rheingold, A. L.; Solomon, E. I.; Karlin, K. D. Exogenous nitrile substrate hydroxylation by a new dicopper-hydroperoxide complex. *J. Am. Chem. Soc.* **2005**, *127*, 15360–15361.

58. Li, L.; Sarjeant, A. A. N.; Karlin, K. D. Reactivity study of a hydroperoxodicopper(II) complex: Hydroxylation, dehydrogenation, and ligand cross-link reactions. *Inorg. Chem.* **2006**, *45*, 7160–7172.

59. Wada, A.; Harata, M.; Hasegawa, K.; Jitsukawa, K.; Masuda, H.; Mukai, M.; Kitagawa, T.; Einaga, H. Structural and spectroscopic characterization of a mononuclear hydroperoxo-copper(II) complex with tripodal pyridylamine ligands. *Angew. Chem. Int. Ed. Engl.* **1998**, *37*, 798–799.

60. Ohta, T.; Tachiyama, T.; Yoshizawa, K.; Yamabe, T.; Uchida, T.; Kitagawa, T. Synthesis, structure, and H_2O_2-dependent catalytic functions of disulfide-bridged dicopper(I) and related thioether-copper(I) and thioether-copper(II) complexes. *Inorg. Chem.* **2000**, *39*, 4358–4369.

61. Ohtsu, H.; Itoh, S.; Nagatomo, S.; Kitagawa, T.; Ogo, S.; Watanabe, Y.; Fukuzumi, S. Characterization of imidazolate-bridged Cu(II)–Zn(II) heterodinuclear and Cu(II)–Cu(II) homodinuclear hydroperoxo complexes as reaction intermediate models of Cu, Zn-SOD. *Chem. Commun.* **2000**, 1051–1052.

62. Kodera, M.; Kita, T.; Miura, I.; Nakayama, N.; Kawata, T.; Kano, K.; Hirota, S. Hydroperoxo-copper(II) complex stabilized by N_3S-type ligand having a phenyl thioether. *J. Am. Chem. Soc.* **2001**, *123*, 7715–7716.

63. Ohtsu, H.; Itoh, S.; Nagatomo, S.; Kitagawa, T.; Ogo, S.; Watanabe, Y.; Fukuzumi, S. Characterization of imidazolate-bridged dinuclear and mononuclear hydroperoxo complexes. *Inorg. Chem.* **2001**, *40*, 3200–3207.

64. Osako, T.; Nagatomo, S.; Tachi, Y.; Kitagawa, T.; Itoh, S. Low-temperature stopped-flow studies on the reactions of copper(II) complexes and H_2O_2: The first detection of a mononuclear copper(II)–peroxo intermediate. *Angew. Chem. Int. Ed.* **2002**, *41*, 4325–4328.

65. Fujii, T.; Naito, A.; Yamaguchi, S.; Wada, A.; Funahashi, Y.; Jitsukawa, K.; Nagatomo, S.; Kitagawa, T.; Masuda, H. Construction of a square-planar hydroperoxo-copper(II) complex inducing a higher catalytic reactivity. *Chem. Commun.* **2003**, 2700–2701.

66. Yamaguchi, S.; Nagatomo, S.; Kitagawa, T.; Funahashi, Y.; Ozawa, T.; Jitsukawa, K.; Masuda, H. Copper hydroperoxo species activated by hydrogen-bonding interaction with its distal oxygen. *Inorg. Chem.* **2003**, *42*, 6968–6970.

67. Yamaguchi, S.; Wada, A.; Nagatomo, S.; Kitagawa, T.; Jitsukawa, K.; Masuda, H. Thermal stability of mononuclear hydroperoxocopper(II) species. Effects of hydrogen bonding and hydrophobic field. *Chem. Lett.* **2004**, *33*, 1556–1557.

68. Itoh, K.; Hayashi, H.; Furutachi, H.; Matsumoto, T.; Nagatomo, S.; Tosha, T.; Terada, S.; Fujinami, S.; Suzuki, M.; Kitagawa, T. Synthesis and reactivity of a (μ-1, 1-hydroperoxo) (μ-hydroxo)dicopper(II) complex: Ligand hydroxylation by a bridging hydroperoxo ligand. *J. Am. Chem. Soc.* **2005**, *127*, 5212–5223.

69. Osako, T.; Nagatomo, S.; Kitagawa, T.; Cramer, C. J.; Itoh, S. Kinetics and DFT studies on the reaction of copper(II) complexes and H_2O_2. *J. Biol. Inorg. Chem.* **2005**, *10*, 581–590.

70. Yamaguchi, S.; Kumagai, A.; Nagatomo, S.; Kitagawa, T.; Funahashi, Y.; Ozawa, T.; Jitsukawa, K.; Masuda, H. Synthesis, characterization, and thermal stability of new mononuclear hydrogenperoxocopper(II) complexes with N_3O-type tripodal ligands bearing hydrogen-bonding interaction sites. *Bull. Chem. Soc. Jpn.* **2005**, *78*, 116–124.

71. Cheruzel, L. E.; Cecil, M. R.; Edison, S. E.; Mashuta, M. S.; Baldwin, M. J.; Buchanan, R. M. Structural and spectroscopic characterization of copper(II) complexes of a new bisamide functionalized imidazole tripod and evidence for the formation of a mononuclear end-on Cu-OOH species. *Inorg. Chem.* **2006**, *45*, 3191–3202.

72. Maiti, D.; Lucas, H. R.; Sarjeant, A. A. N.; Karlin, K. D. Aryl hydroxylation from a mononuclear copper-hydroperoxo species. *J. Am. Chem. Soc.* **2007**, *129*, 6998–6999.

73. Maiti, D.; Sarjeant, A. A. N.; Karlin, K. D. Copper(II)–hydroperoxo complex induced oxidative *N*-dealkylation chemistry. *J. Am. Chem. Soc.* **2007**, *129*, 6720–6721.

74. Maiti, D.; Sarjeant, A. A. N.; Karlin, K. D. Copper-hydroperoxo-mediated *N*-debenzylation chemistry mimicking aspects of copper monooxygenases. *Inorg. Chem.* **2008**, *47*, 8736–8747.

75. Kamachi, T.; Lee, Y. M.; Nishimi, T.; Cho, J.; Yoshizawa, K.; Nam, W. Combined experimental and theoretical approach to understand the reactivity of a mononuclear Cu(II)–hydroperoxo complex in oxygenation reactions. *J. Phys. Chem. A* **2008**, *112*, 13102–13108.

76. Ghosh, P.; Tyeklar, Z.; Karlin, K. D.; Jacobson, R. R.; Zubieta, J. Dioxygen-copper reactivity: X-ray structure and characterization of an (acylperoxo)dicopper complex. *J. Am. Chem. Soc.* **1987**, *109*, 6889–6891.

77. Kitajima, N.; Fujisawa, K.; Moro-oka, Y. Formation and characterization of a mononuclear (acylperoxo)copper(II) complex. *Inorg. Chem.* **1990**, *29*, 357–358.

78. Kitajima, N.; Katayama, T.; Fujisawa, K.; Iwata, Y.; Morooka, Y. Synthesis, molecular structure, and reactivity of (alkylperoxo)copper(II) complex. *J. Am. Chem. Soc.* **1993**, *115*, 7872–7873.

79. Sanyal, I.; Ghosh, P.; Karlin, K. D. Mononuclear copper(II)–acylperoxo complexes. *Inorg. Chem.* **1995**, *34*, 3050–3056.

80. Chen, P.; Fujisawa, K.; Solomon, E. I. Spectroscopic and theoretical studies of mononuclear copper(II) alkyl- and hydroperoxo complexes: Electronic structure contributions to reactivity. *J. Am. Chem. Soc.* **2000**, *122*, 10177–10193.

81. Kunishita, A.; Ishimaru, H.; Nakashima, S.; Ogura, T.; Itoh, S. Reactivity of mononuclear alkylperoxo copper(II) complex. O–O bond cleavage and C–H bond activation. *J. Am. Chem. Soc.* **2008**, *130*, 4244–4245.

82. Schröder, D.; Holthausen, M. C.; Schwarz, H. Radical-like activation of alkanes by the ligated copper oxide cation (phenanthroline)CuO$^+$. *J. Phys. Chem. B* **2004**, *108*, 14407–14416.

83. Itoh, S.; Fukuzumi, S. Dioxygen activation by copper complexes. Mechanistic insights into copper monooxygenases and copper oxidases. *Bull. Chem. Soc. Jpn.* **2002**, *75*, 2081–2095.

84. Jacobson, R. R.; Tyeklar, Z.; Farooq, A.; Karlin, K. D.; Liu, S.; Zubieta, J. A Cu$_2$–O$_2$ complex. Crystal structure and characterization of a reversible dioxygen binding system. *J. Am. Chem. Soc.* **1988**, *110*, 3690–3692.

85. Tyeklar, Z.; Jacobson, R. R.; Wei, N.; Murthy, N. N.; Zubieta, J.; Karlin, K. D. Reversible-reaction of O$_2$ (and CO) with a copper(I) complex - X-ray structures of relevant mononuclear Cu(I) precursor adducts and the *trans-(μ-1,2-peroxo)*dicopper(II) product. *J. Am. Chem. Soc.* **1993**, *115*, 2677–2689.

86. Wurtele, C.; Sander, O.; Lutz, V.; Waitz, T.; Tuczek, F.; Schindler, S. Aliphatic C–H bond oxidation of toluene using copper peroxo complexes that are stable at room temperature. *J. Am. Chem. Soc.* **2009**, *131*, 7544–7545.

87. Kitajima, N.; Fujisawa, K.; Morooka, Y.; Toriumi, K. μ-η^2:η^2-Peroxo binuclear copper complex, [Cu(HB(3,5-iPr$_2$pz)$_3$)]$_2$(O$_2$). *J. Am. Chem. Soc.* **1989**, *111*, 8975–8976.

88. Kitajima, N.; Fujisawa, K.; Fujimoto, C.; Morooka, Y.; Hashimoto, S.; Kitagawa, T.; Toriumi, K.; Tatsumi, K.; Nakamura, A. A new model for dioxygen binding in hemocyanin. Synthesis, characterization, and molecular structure of the μ-η^2:η^2-peroxo dinuclear copper(II) complexes, [Cu(HB(3,5-R2pz)$_3$)]$_2$(O$_2$) (R = isopropyl and Ph). *J. Am. Chem. Soc.* **1992**, *114*, 1277–1291.

89. Kodera, M.; Katayama, K.; Tachi, Y.; Kano, K.; Hirota, S.; Fujinami, S.; Suzuki, M. Crystal structure and reversible O_2-binding of a room temperature stable μ-η^2:η^2-peroxodicopper(II) complex of a sterically hindered hexapyridine dinucleating ligand. *J. Am. Chem. Soc.* **1999**, *121*, 11006–11007.

90. Hu, Z.; Williams, R. D.; Tran, D.; Spiro, T. G.; Gorun, S. M. Re-engineering enzyme-model active sites: reversible binding of dioxygen at ambient conditions by a bioinspired copper complex. *J. Am. Chem. Soc.* **2000**, *122*, 3556–3557.

91. Hu, Z.; George, G. N.; Gorun, S. M. Fluorine encapsulation and stabilization of biologically relevant low-valence copper-oxo cores. *Inorg. Chem.* **2001**, *40*, 4812–4813.

92. Kodera, M.; Kajita, Y.; Tachi, Y.; Katayama, K.; Kano, K.; Hirota, S.; Fujinami, S.; Suzuki, M. Synthesis, structure, and greatly improved reversible O_2 binding in a structurally modulated μ-η^2:η^2-peroxodicopper(II) complex with room-temperature stability. *Angew. Chem. Int. Ed.* **2004**, *43*, 334–337.

93. Funahashi, Y.; Nishikawa, T.; Wasada-Tsutsui, Y.; Kajita, Y.; Yamaguchi, S.; Arii, H.; Ozawa, T.; Jitsukawa, K.; Tosha, T.; Hirota, S.; Kitagawa, T.; Masuda, H. Formation of a bridged butterfly-type μ-η^2:η^2-peroxo dicopper core structure with a carboxylate. *J. Am. Chem. Soc.* **2008**, *130*, 16444–16445.

94. Cuff, M. E.; Miller, K. I.; van Holde, K. E.; Hendrickson, W. A. Crystal structure of a functional unit from octopus hemocyanin. *J. Mol. Biol.* **1998**, *278*, 855–870.

95. Magnus, K. A.; Hazes, B.; Ton-That, H.; Bonaventura, C.; Bonaventura, J.; Hol, W. G. J. Crystallographic analysis of oxygenated and deoxygenated states of arthropod hemocyanin shows unusual differences. *Proteins: Structure, Function, Geneti.* **1994**, *19*, 302–309.

96. Matoba, Y.; Kumagai, T.; Yamamoto, A.; Yoshitsu, H.; Sugiyama, M. Crystallographic evidence that the dinuclear copper center of tyrosinase is flexible during catalysis. *J. Biol. Chem.* **2006**, *281*, 8981–8990.

97. Tolman, W. B. Making and breaking the dioxygen O–O bond: New insights from studies of synthetic copper complexes. *Acc. Chem. Res.* **1997**, *30*, 227–237.

98. P. L. Holland, W. B. Tolman, Dioxygen activation by copper sites: relative stability and reactivity of (μ-η^2:η^2-peroxo)- *d* and bis(*m*-oxo)dicopper cores. *Coord. Chem. Rev.* **1999**, *190–192*, 855–869.

99. Blackman, A. G.; Tolman, W. B.In *Metal-Oxo and Metal-Peroxo Species in Catalytic Oxidations*; Springer-Verlag Berlin: Berlin, **2000**; Vol. *97;* pp 179–211.

100. Stack, T. D. P. Complexity with simplicity: a steric continuum of chelating diamines with copper(I) and dioxygen. *Dalton Trans.* **2003**, 1881–1889.

101. Que, Jr., L.; Tolman, W. B. Bis(μ-oxo)dimetal "diamond" cores in copper and iron complexes relevant to biocatalysis. *Angew. Chem. Int. Ed.* **2002**, *41*, 1114–1137.

102. Halfen, J. A.; Mahapatra, S.; Wilkinson, E. C.; Kaderli, S.; Young, V. G.; Que, Jr., L.; Zuberbühler, A. D.; Tolman, W. B. Reversible cleavage and formation of the dioxygen O–O band within a dicopper complex. *Science* **1996**, *271*, 1397–1400.

103. Mahapatra, S.; Halfen, J. A.; Wilkinson, E. C.; Pan, G. F.; Wang, X. D.; Young, V. G.; Cramer, C. J.; Que, Jr., L. Tolman, W. B. Structural, spectroscopic, and theoretical characterization of bis(μ-oxo)dicopper complexes, novel intermediates in copper-mediated dioxygen activation. *J. Am. Chem. Soc.* **1996**, *118*, 11555–11574.

104. Mahadevan, V.; Hou, Z. G.; Cole, A. P.; Root, D. E.; Lal, T. K.; Solomon, E. I.; Stack, T. D. P. Irreversible reduction of dioxygen by simple peralkylated diamine-copper(I)

complexes: Characterization and thermal stability of a $Cu_2(\mu\text{-}O)_2{}^{2+}$ core. *J. Am. Chem. Soc.* **1997**, *119*, 11996–11997.

105. Mahapatra, S.; Young, V. G.; Kaderli, S.; Zuberbühler, A. D.; Tolman, W. B. Tuning the structure and reactivity of the $Cu_2(\mu\text{-}O)_2{}^{2+}$ core: Characterization of a new bis(μ-oxo) dicopper complex stabilized by a sterically hindered dinucleating bis(triazacyclononane) ligand. *Angew. Chem. Int. Ed. Engl.* **1997**, *36*, 130–133.

106. Hayashi, H.; Fujinami, S.; Nagatomo, S.; Ogo, S.; Suzuki, M.; Uehara, A.; Watanabe, Y.; Kitagawa, T. A Bis(μ-oxo)dicopper(III) complex with aromatic nitrogen donors: structural characterization and reversible conversion between copper(I) and bis (μ-oxo)dicopper(III) species. *J. Am. Chem. Soc.* **2000**, *122*, 2124–2125.

107. Straub, B. F.; Rominger, F.; Hofmann, P. A neutral dicopper(III) bis(μ-oxo) complex from a copper(I) ethylene iminophosphanamide and O_2. *Chem. Commun.* **2000**, 1611–1612.

108. Mizuno, M.; Hayashi, H.; Fujinami, S.; Furutachi, H.; Nagatomo, S.; Otake, S.; Uozumi, K.; Suzuki, M.; Kitagawa, T. Ligand effect on reversible conversion between copper(I) and bis(μ-oxo)dicopper(III) complex with a sterically hindered tetradentate tripodal ligand and monooxygenase activity of bis(μ-oxo)dicopper(III) complex. *Inorg. Chem.* **2003**, *42*, 8534–8544.

109. Cole, A. P.; Mahadevan, V.; Mirica, L. M.; Ottenwaelder, X.; Stack, T. D. P. Bis(μ-oxo) dicopper(III) complexes of a homologous series of simple peralkylated 1,2-diamines: Steric modulation of structure, stability, and reactivity. *Inorg. Chem.* **2005**, *44*, 7345–7364.

110. Cole, A. P.; Root, D. E.; Mukherjee, P.; Solomon, E. I.; Stack, T. D. P. Trinuclear intermediate in the copper-mediated reduction of O_2: Four electrons from three coppers. *Science* **1996**, *273*, 1848–1850.

111. Taki, M.; Teramae, S.; Nagatomo, S.; Tachi, Y.; Kitagawa, T.; Itoh, S.; Fukuzumi, S. Fine-tuning of copper(I)–dioxygen reactivity by 2-(2-pyridyl)ethylamine bidentate ligands. *J. Am. Chem. Soc.* **2002**, *124*, 6367–6377.

112. Reim, J.; Krebs, B. A thermally stable peroxocopper(II) complex with unusual μ_4 coordination of the peroxo ligand. *Angew. Chem. Int. Ed. Engl.* **1994**, *33*, 1969–1971.

113. Meyer, F.; Pritzkow, H. μ_4-Peroxo versus bis(μ_2-hydroxo) cores in structurally analogous tetracopper(II) Complexes. *Angew. Chem. Int. Ed.* **2000**, *39*, 2112–2115.

114. Klinman, J. P. The copper-enzyme family of dopamine β-monooxygenase and peptidylglycine α-hydroxylating monooxygenase: Resolving the chemical pathway for substrate hydroxylation. *J. Biol. Chem.* **2006**, *281*, 3013–3016.

115. Evans, J. P.; Ahn, K.; Klinman, J. P. Evidence that dioxygen and substrate activation are tightly coupled in dopamine β-monooxygenase: Implications for the reactive oxygen species. *J. Biol. Chem.* **2003**, *278*, 49691–49698.

116. Francisco, W. A.; Wille, G.; Smith, A. J.; Merkler, D. J.; Klinman, J. P. Investigation of the pathway for inter-copper electron transfer in peptidyglycine α-amidating monooxygenase. *J. Am. Chem. Soc.* **2004**, *126*, 13168–13169.

117. Bauman, A. T.; Yukl, E. T.; Alkevich, K.; McCormack, A. L.; Blackburn, N. J. The hydrogen peroxide reactivity of peptidylglycine monooxygenase supports a Cu (II)–superoxo catalytic intermediate. *J. Biol. Chem.* **2006**, *281*, 4190–4198.

118. Lee, D.-H.; Wei, N.; Murthy, N. N.; Tyeklar, Z.; Karlin, K. D.; Kaderli, S.; Jung, B.; Zuberbühler, A. D. Reversible O_2 binding to a dinuclear copper(I) complex with

linked tris(2-pyridylmethyl)amine units: kinetic–thermodynamic comparisons with mononuclear analogs. *J. Am. Chem. Soc.* **1995**, *117*, 12498–12513.

119. Karlin, K. D.; Kaderli, S.; Zuberbühler, A. D. Kinetics and thermodynamics of copper(I)/dioxygen interaction. *Acc. Chem. Res.* **1997**, *30*, 139–147.

120. Weitzer, M.; Schatz, M.; Hampel, F.; Heinemann, F. W.; Schindler, S. Low temperature stopped-flow studies in inorganic chemistry. *J. Chem. Soc., Dalton Trans.* **2002**, 686–694.

121. Schatz, M.; Becker, M.; Walter, O.; Liehr, G. n.; Schindler, S. Reactivity towards dioxygen of a copper(I) complex of tris(2-benzylaminoethyl)amine. *Inorg. Chim. Acta* **2001**, *324*, 173–179.

122. Poater, A.; Cavallo, L. Probing the mechanism of O₂ activation by a copper(I) biomimetic complex of a C−H hydroxylating copper monooxygenase. *Inorg. Chem.* **2009**, *48*, 4062–4066.

123. Izzet, G.; Zeitouny, J.; Akdas-Killig, H.; Frapart, Y.; Ménage, S.; Douziech, B.; Jabin, I.; Le Mest, Y.; Reinaud, O. Dioxygen activation at a mononuclear Cu(I) center embedded in the calix[6]arene-tren core. *J. Am. Chem. Soc.* **2008**, *130*, 9514–9523.

124. Thiabaud, G.; Guillemot, G.; Schmitz-Afonso, I.; Colasson, B.; Reinaud, O. Solid-state chemistry at an isolated copper(I) center with O₂. *Angew. Chem. Int. Ed.* **2009**, *48*, 7383–7386.

125. Kamachi, T.; Kihara, N.; Shiota, Y.; Yoshizawa, K. Computational exploration of the catalytic mechanism of dopamine β-monooxygenase: Modeling of Its mononuclear copper active sites. *Inorg. Chem.* **2005**, *44*, 4226–4236.

126. Crespo, A.; Marti, M. A.; Roitberg, A. E.; Amzel, L. M.; Estrin, D. A. The catalytic mechanism of peptidylglycine α-hydroxylating monooxygenase investigated by computer simulation. *J. Am. Chem. Soc.* **2006**, *128*, 12817–12828.

127. de la Lande, A.; Parisel, O.; Gérard, H.; Moliner, V.; Reinaud, O. Theoretical exploration of the oxidative properties of a [(tren^{Me1})CuO₂]⁺ adduct relevant to copper monooxygenase enzymes: Insights into competitive dehydrogenation versus hydroxylation reaction pathways. *Chem. Eur. J.* **2008**, *14*, 6465–6473.

128. Cramer, C. J.; Tolman, W. B.; Theopold, K. H.; Rheingold, A. L. Variable character of O−O and M−O bonding in side-on η2 1:1 metal complexes of O₂. *Proc. Natl. Acad. Sci. USA* **2003**, *100*, 3635–3640.

129. Chen, P.; Solomon, E. I. Oxygen activation by the noncoupled binuclear copper site in peptidylglycine α-hydroxylating monooxygenase. Reaction mechanism and role of the noncoupled nature of the active site. *J. Am. Chem. Soc.* **2004**, *126*, 4991–5000.

130. Spencer, D. J. E.; Reynolds, A. M.; Holland, P. L.; Jazdzewski, B. A.; Duboc-Toia, C.; Le Pape, L.; Yokota, S.; Tachi, Y.; Itoh, S.; Tolman, W. B. Copper chemistry of β-diketiminate ligands: Monomer/dimer equilibria and a new class of bis(μ-oxo) dicopper compounds. *Inorg. Chem.* **2002**, *41*, 6307–6321.

131. Sarangi, R.; Aboelella, N.; Fujisawa, K.; Tolman, W. B.; Hedman, B.; Hodgson, K. O.; Solomon, E. I. X-ray absorption edge spectroscopy and computational studies on LCuO₂ species: Superoxide-Cu^{II} versus peroxide-Cu^{III} bonding. *J. Am. Chem. Soc.* **2006**, *128*, 8286–8296.

132. Gherman, B. F.; Cramer, C. J. Modeling the peroxide/superoxide continuum in 1:1 side-on adducts of O₂ with Cu. *Inorg. Chem.* **2004**, *43*, 7281–7283.

133. Pantazis, D. A.; McGrady, J. E. On the nature of the bonding in 1: 1 adducts of O₂. *Inorg. Chem.* **2003**, *42*, 7734–7736.

134. Reynolds, A. M.; Lewis, E. A.; Aboelella, N. W.; Tolman, W. B. Reactivity of a 1 : 1 copper-oxygen complex: isolation of a Cu(II)-o-iminosemiquinonato species. *Chem. Commun.* **2005**, 2014–2016.

135. Aboelella, N. W.; York, J. T.; Reynolds, A. M.; Fujita, K.; Kinsinger, C. R.; Cramer, C. J.; Riordan, C. G.; Tolman, W. B. Mixed metal bis(μ-oxo) complexes with $[CuM(\mu\text{-}O)_2]^{n+}$ (M = Ni(III) or Pd(II)) cores. *Chem. Commun.* **2004**, 1716–1717.

136. Casella, L.; Gullotti, M.; Bartosek, M.; Pallanza, G.; Laurenti, E. Model monooxygenase reactivity by binuclear two-coordinate copper(I) complexes extends to new ligand systems containing nitrogen and sulphur donors. *J. Chem. Soc., Chem. Commun.* **1991**, 1235–1237.

137. Alzuet, G.; Casella, L.; Villa, M. L.; Carugo, O.; Gullotti, M. Copper monooxygenase models. Aromatic hydroxylation by a dinuclear copper(I) complex containing methionine sulfur ligands. *J. Chem. Soc., Dalton Trans.* **1997**, 4789–4794.

138. Champloy, F.; Benali-Cherif, N.; Bruno, P.; Blain, I.; Pierrot, M.; Reglier, M.; Michalowicz, A. Studies of copper complexes displaying N_3S coordination as models for Cu-B center of dopamine β-hydroxylase and peptidylglycine α-hydroxylating monooxygenase. *Inorg. Chem.* **1998**, *37*, 3910–3918.

139. Hatcher, L. Q.; Lee, D. H.; Vance, M. A.; Milligan, A. E.; Sarangi, R.; Hodgson, K. O.; Hedman, B.; Solomon, E. I.; Karlin, K. D. Dioxygen reactivity of a copper(I) complex with a N3S thioether chelate; Peroxo-dicopper(II) formation including sulfur-ligation. *Inorg. Chem.* **2006**, *45*, 10055–10057.

140. Lee, Y.; Lee, D.-H.; Narducci Sarjeant, A. A.; Zakharov, L. N.; Rheingold, A. L.; Karlin, K. D. Thioether sulfur oxygenation from O_2 or H_2O_2 reactivity of copper complexes with tridentate N_2S thioether ligands. *Inorg. Chem.* **2006**, *45*, 10098–10107.

141. Lee, D.-H.; Hatcher, L. Q.; Vance, M. A.; Sarangi, R.; Milligan, A. E.; Narducci Sarjeant, A. A.; Incarvito, C. D.; Rheingold, A. L.; Hodgson, K. O.; Hedman, B.; Solomon, E. I.; Karlin, K. D. Copper(I) complex O_2-reactivity with a N_3S thioether ligand: a copper-dioxygen adduct including sulfur ligation, ligand oxygenation, and comparisons with all nitrogen ligand analogues. *Inorg. Chem.* **2007**, *46*, 6056–6068.

142. Pate, J. E.; Cruse, R. W.; Karlin, K. D.; Solomon, E. I. Vibrational, electronic, and resonance Raman spectral studies of $[Cu_2(YXL\text{-}O^-)O_2]^+$, a copper(II) peroxide model complex of oxyhemocyanin. *J. Am. Chem. Soc.* **1987**, *109*, 2624–2630.

143. Li, L.; Murthy, N. N.; Telser, J.; Zakharov, L. N.; Yap, G. P. A.; Rheingold, A. L.; Karlin, K. D.; Rokita, S. E. Targeted guanine oxidation by a dinuclear copper(II) complex at single stranded/double stranded DNA junctions. *Inorg. Chem.* **2006**, *45*, 7144–7159.

144. Thyagarajan, S.; Murthy, N. N.; Sarjeant, A. A. N.; Karlin, K. D.; Rokita, S. E. Selective DNA strand scission with binuclear copper complexes: Implications for an active Cu_2O_2 species. *J. Am. Chem. Soc.* **2006**, *128*, 7003–7008.

145. Zhu, Q.; Lian, Y. X.; Thyagarajan, S.; Rokita, S. E.; Karlin, K. D.; Blough, N. V. Hydrogen peroxide and dioxygen activation by dinuclear copper complexes in aqueous solution: Hydroxyl radical production initiated by internal electron transfer. *J. Am. Chem. Soc.* **2008**, *130*, 6304–6305.

146. Ghattas, W.; Giorgi, M.; Mekmouche, Y.; Tanaka, T.; Rockenbauer, A.; Réglier, M.; Hitomi, Y.; Simaan, A. J. Identification of a copper(I) intermediate in the conversion of 1-aminocyclopropane carboxylic acid (ACC) into ethylene by Cu(II)-ACC complexes and hydrogen peroxide. *Inorg. Chem.* **2008**, *47*, 4627–4638.

147. Kunishita, A.; Teraoka, J.; Scanlon, J. D.; Matsumoto, T.; Suzuki, M.; Cramer, C. J.; Itoh, S. Aromatic hydroxylation reactivity of a mononuclear Cu(II)–alkylperoxo complex. *J. Am. Chem. Soc.* **2007**, *129*, 7248–7249.

148. Kunishita, A.; Scanlon, J. D.; Ishimaru, H.; Honda, K.; Ogura, T.; Suzuki, M.; Cramer, C. J.; Itoh, S. Reactions of copper(II)−H_2O_2 adducts supported by tridentate bis(2-pyridylmethyl)amine ligands: Sensitivity to solvent and variations in ligand substitution. *Inorg. Chem.* **2008**, *47*, 8222–8232.

149. Avila, D. V.; Brown, C. E.; Ingold, K. U.; Lusztyk, J. Solvent effects on the competitive β-scission and hydrogen atom abstraction reactions of the cumyloxyl radical. Resolution of a long-standing problem. *J. Am. Chem. Soc.* **1993**, *115*, 466–470.

150. Adachi, S.; Nagano, S.; Ishimori, K.; Watanabe, Y.; Morishima, I.; Egawa, T.; Kitagawa, T.; Makino, R. Roles of proximal ligand in heme proteins: replacement of proximal histidine of human myoglobin with cysteine and tyrosine by site-directed mutagenesis as models for P-450, chloroperoxidase, and catalase. *Biochemistry* **1993**, *32*, 241–252.

151. Vicic, D. A.; Jones, W. D. Evidence for the existence of a late-metal terminal sulfido complex. *J. Am. Chem. Soc.* **1999**, *121*, 4070–4071.

152. Hetterscheid, D. G. H.; Bens, M.; Bruin, B. d. Ir^{I}(ethene): Metal or carbon radical? Part II: Oxygenation via iridium or direct oxygenation at ethene? *Dalton Trans.* **2005**, 979–984.

153. Capdevielle, P.; Maumy, M. Selective ortho-hydroxylation of phenols: I - Towards a simple chemical model of tyrosinase. *Tetrahedron Lett.* **1982**, *23*, 1573–1576.

154. Réglier, M.; Amadei, E.; Tadayoni, R.; Waegell, B. Pyridine nucleus hydroxylation with copper oxygenase models. *J. Chem. Soc., Chem. Commun.* **1989**, 447–450.

155. Capdevielle, P.; Sparfel, D.; Barannelafont, J.; Cuong, N. K.; Maumy, M. Copper(I) and copper(II) mediated two-electron oxidations of benzylic alcohols and diaryl acetic acids by trimethylamine N-oxide. *J. Chem. Soc., Chem. Commun.* **1990**, 565–566.

156. Reinaud, O.; Capdevielle, P.; Maumy, M. Copper(II) mediated aromatic hydroxylation by trimethylamine N-oxide. *J. Chem. Soc., Chem. Commun.* **1990**, 566–568.

157. Reinaud, O.; Capdevielle, P.; Maumy, M. 2-(N-amide)-4-nitrophenol: a new ligand for the copper-mediated hydroxylation of aromatics by trimethylamine N-oxide. *J. Mol. Catal.* **1991**, *68*, L13–L15.

158. Rousselet, G.; Capdevielle, P.; Maumy, M. Copper-induced synthesis of iminiums: trimethylamine oxidation or amine N-oxide conversion. *Tetrahedron Lett.* **1995**, *36*, 4999–5002.

159. Buijs, W.; Comba, P.; Corneli, D.; Pritzkow, H. Structural and mechanistic studies of the copper(II)–assisted ortho-hydroxylation of benzoates by trimethylamine N-oxide. *J. Organomet. Chem.* **2002**, *641*, 71–80.

160. Comba, P.; Knoppe, S.; Martin, B.; Rajaraman, G.; Rolli, C.; Shapiro, B.; Stork, T. Copper (II)–mediated aromatic ortho-hydroxylation: A hybrid DFT and *Ab initio* exploration. *Chem. Eur. J.* **2008**, *14*, 344–357.

161. Yoshizawa, K.; Kihara, N.; Kamachi, T.; Shiota, Y. Catalytic mechanism of dopamine β-monooxygenase mediated by Cu(III)-oxo. *Inorg. Chem.* **2006**, *45*, 3034–3041.

162. Hong, S.; Huber, S. M.; Gagliardi, L.; Cramer, C. C.; Tolman, W. B. Copper(I)-α-ketocarboxylate complexes: Characterization and O_2 reactions that yield copper-oxygen intermediates capable of hydroxylating arenes. *J. Am. Chem. Soc.* **2007**, *129*, 14190–14192.

163. Costas, M.; Mehn, M. P.; Jensen, M. P.; Que, Jr., L. Dioxygen activation at mononuclear nonheme iron active sites: Enzymes, models, and intermediates. *Chem. Rev.* **2004**, *104*, 939–986.

164. Abu-Omar, M. M.; Loaiza, A.; Hontzeas, N. Reaction mechanisms of mononuclear non-heme iron oxygenases. *Chem. Rev.* **2005**, *105*, 2227–2252.

165. Vaillancourt, F. d. r. H.; Yeh, E.; Vosburg, D. A.; Garneau-Tsodikova, S.; Walsh, C. T. Nature's inventory of halogenation catalysts: Oxidative strategies predominate. *Chem. Rev.* **2006**, *106*, 3364–3378.

166. Purpero, V.; Moran, G. The diverse and pervasive chemistries of the α-keto acid dependent enzymes. *J. Biol. Inorg. Chem.* **2007**, *12*, 587–601.

167. Huber, S., M.; Ertem, M. Z.; Aquilante, F.; Gagliardi, L.; Tolman, W. B.; Cramer, C. J. Generating Cu^{II}-oxyl/Cu^{III}-oxo species from Cu^{I}-α-ketocarboxylate complexes and O_2: In silico studies on ligand effects and C-H-activation reactivity. *Chem. Eur. J.* **2009**, *15*, 4886–4895.

168. Hong, S. J.; Gupta, A. K.; Tolman, W. B. Intermediates in reactions of copper(I) complexes with *N*-oxides: From the formation of stable adducts to oxo transfer. *Inorg. Chem.* **2009**, *48*, 6323–6325.

169. Karlin, K. D.; Gultneh, Y. *Prog. Inorg. Chem.* **1987**, *35*, 219–327.

170. Sorrell, T. N. Synthetic models for binuclear copper proteins. *Tetrahedron* **1989**, *45*, 3–68.

171. Tyeklár, Z.; Karlin, K. D. Copper dioxygen chemistry—A bioinorganic challenge. *Acc. Chem. Res.* **1989**, *22*, 241–248.

172. Solomon, E. I.; Tuczek, F.; Root, D. E.; Brown, C. A. Spectroscopy of binuclear dioxygen complexes. *Chem. Rev.* **1994**, *94*, 827–856.

173. Karlin, K. D.; Zuberbühler, A. D. *Bioinorganic Catalysis*; Reedijk, J., Bouwman, E., Eds.; Marcel Dekker, Inc.: New York, 1999; pp 469–534.

174. Liang, H.-C.; Dahan, M.; Karlin, K. D. Dioxygen-activating bioinorganic model complexes. *Curr. Opin. Chem. Biol.* **1999**, *3*, 168–175.

175. Zhang, C. X.; Liang, H. C.; Humphreys, K. J.; Karlin, K. D. *Advances in Catalytic Activation of Dioxygen by Metal Complexes*; Simandi, L. I., Ed.; Kluwer Academic Publishers: Dordrecht, The Netherlands, **2003**; pp 79–121.

176. Itoh, S.In *Comprehensive Coordination Chemistry II*; Que, Jr., J. L., Tolman, W. B. Eds.; Elsevier: Amsterdam, 2004 The Netherlands; pp 369–393.

177. Itoh, S.; Tachi, Y. Structure and O_2-reactivity of copper(I) complexes supported by pyridylalkylamine ligands. *Dalton Trans.* **2006**, 4531–4538.

178. Itoh, S.; Fukuzumi, S. Monooxygenase activity of type 3 copper proteins. *Acc. Chem. Res.* **2007**, *40*, 592–600.

179. Kodera, M.; Kano, K. Reversible O_2-binding and activation with dicopper and diiron complexes stabilized by various hexapyridine ligands. Stability, modulation, and flexibility of the dinuclear structure as key aspects for the dimetal/O_2 chemistry. *Bull. Chem. Soc. Jpn.* **2007**, *80*, 662–676.

180. Suzuki, M. Ligand effects on dioxygen activation by copper and nickel complexes: Reactivity and intermediates. *Acc. Chem. Res.* **2007**, *40*, 609–617.

181. Baldwin, M. J.; Ross, P. K.; Pate, J. E.; Tyeklar, Z.; Karlin, K. D.; Solomon, E. I. Spectroscopic and theoretical studies of an end-on peroxide-bridged coupled binuclear

copper(II) model complex of relevance to the active sites in hemocyanin and tyrosinase. *J. Am. Chem. Soc.* **1991**, *113*, 8671–8679.

182. Yamaguchi, S.; Wada, A.; Funahashi, Y.; Nagatomo, S.; Kitagawa, T.; Jitsukawa, K.; Masuda, H. Thermal stability and absorption spectroscopic behavior of (μ-peroxo) dicopper complexes regulated with intramolecular hydrogen bonding interactions. *Eur. J. Inorg. Chem.* **2003**, 4378–4386.

183. Koval, I. A.; Belle, C.; Selmeczi, K.; Philouze, C.; Saint-Aman, E.; Schuitema, A. M.; Gamez, P.; Pierre, J. L.; Reedijk, J. Catecholase activity of a μ-hydroxodicopper(II) macrocyclic complex: structures, intermediates and reaction mechanism. *J. Biol. Inorg. Chem.* **2005**, *10*, 739–750.

184. Maiti, D.; Woertink, J. S.; Sarjeant, A. A. N.; Solomon, E. I.; Karlin, K. D. Copper dioxygen adducts: Formation of bis(μ-oxo)dicopper(III) versus (μ-1,2)peroxodicopper (II) complexes with small changes in one pyridyl-ligand substituent. *Inorg. Chem.* **2008**, *47*, 3787–3800.

185. Lee, Y.; Park, G. Y.; Lucas, H. R.; Vajda, P. L.; Kamaraj, K.; Vance, M. A.; Milligan, A. E.; Woertink, J. S.; Siegler, M. A.; Sarjeant, A. A. N.; Zakharov, L. N.; Rheingold, A. L.; Solomon, E. I.; Karlin, K. D. Copper(I)/O_2 chemistry with imidazole containing tripodal tetradentate ligands leading to μ-1, 2-peroxo-dicopper(II) species. *Inorg. Chem.* **2009**, *48*, 11297–11309.

186. Paul, P. P.; Tyeklár, Z.; Jacobson, R. R.; Karlin, K. D. Reactivity patterns and comparisons in three classes of synthetic copper–dioxygen {Cu_2-O_2} complexes: implication for structure and biological relevance. *J. Am. Chem. Soc.* **1991**, *113*, 5322–5332.

187. Lucas, H. R.; Li, L.; Sarjeant, A. A. N.; Vance, M. A.; Solomon, E. I.; Karlin, K. D. Toluene and ethylbenzene aliphatic C−H bond oxidations initiated by a dicopper(II)-μ-1,2-peroxo complex. *J. Am. Chem. Soc.* **2009**, *131*, 3230–3245.

188. Lam, B. M. T.; Halfen, J. A.; Young, V. G.; Hagadorn, J. R.; Holland, P. L.; Lledos, A.; Cucurull-Sanchez, L.; Novoa, J. J.; Alvarez, S.; Tolman, W. B. Ligand macrocycle structural effects on copper-dioxygen reactivity. *Inorg. Chem.* **2000**, *39*, 4059–4072.

189. Ottenwaelder, X.; Rudd, D. J.; Corbett, M. C.; Hodgson, K. O.; Hedman, B.; Stack, T. D. P. Reversible O−O bond cleavage in copper-dioxygen isomers: Impact of anion basicity. *J. Am. Chem. Soc.* **2006**, *128*, 9268–9269.

190. Funahashi, Y.; Nakaya, K.; Hirota, S.; Yamauchi, O. Tetrahedral distortion in copper(II) complexes of (−)-sparteine and its effect on the oxygen adduct formation. *Chem. Lett.* **2000**, *29*, 1172–1173.

191. Osako, T.; Tachi, Y.; Doe, M.; Shiro, M.; Ohkubo, K.; Fukuzumi, S.; Itoh, S. Quantitative evaluation of d-π interaction in copper(I) complexes and control of copper(I)–dioxygen reactivity. *Chem. Eur. J.* **2004**, *10*, 237–246.

192. Liang, H. C.; Henson, M. J.; Hatcher, L. Q.; Vance, M. A.; Zhang, C. X.; Lahti, D.; Kaderli, S.; Sommer, R. D.; Rheingold, A. L.; Zuberbuhler, A. D.; Solomon, E. I.; Karlin, K. D. Solvent effects on the conversion of dicopper(II) μ-η^2:η^2-peroxo to bis-μ-oxo dicopper(III) complexes: Direct probing of the solvent interaction. *Inorg. Chem.* **2004**, *43*, 4115–4117.

193. Osako, T.; Terada, S.; Tosha, T.; Nagatomo, S.; Furutachi, H.; Fujinami, S.; Kitagawa, T.; Suzuki, M.; Itoh, S. Structure and dioxygen-reactivity of copper(I) complexes supported by bis(6-methylpyridin-2-ylmethyl)amine tridentate ligands. *Dalton Trans.* **2005**, 3514–3521.

194. Sprakel, V. S. I.; Feiters, M. C.; Klaucke, W. M.; Klopstra, M.; Brinksma, J.; Feringa, B. L.; Karlin, K. D.; Nolte, R. J. M. Oxygen binding and activation by the complexes of PY2- and TPA-appended diphenylglycoluril receptors with copper and other metals. *Dalton Trans.* **2005**, 3522–3534.

195. Kunishita, A.; Osako, T.; Tachi, Y.; Teraoka, J.; Itoh, S. Syntheses, structures, and O_2-reactivities of copper(I) complexes with bis(2-pyridylmethyl)amine and bis(2-quinolylmethyl)amine tridentate ligands. *Bull. Chem. Soc. Jpn.* **2006**, *79*, 1729–1741.

196. Hatcher, L. Q.; Vance, M. A.; Sarjeant, A. A. N.; Solomon, E. I.; Karlin, K. D. Copper-dioxygen adducts and the side-on peroxo dicopper(II)/bis(μ-oxo) dicopper(III) equilibrium: Significant ligand electronic effects. *Inorg. Chem.* **2006**, *45*, 3004–3013.

197. Park, G. Y.; Lee, Y.; Lee, D.-H.; Woertink, J. S.; Sarjeant, A. A. N.; Solomon, E. I.; Karlin, K. D. Thioether S-ligation in a side-on μ-η^2:η^2-peroxodicopper(II) complex. *Chem. Commun.* **2010**, *46*, 91–93.

198. Osako, T.; Ueno, Y.; Tachi, Y.; Itoh, S. Structures and redox reactivities of copper complexes of (2-pyridyl)alkylamine ligands. Effects of the alkyl linker chain length. *Inorg. Chem.* **2003**, *42*, 8087–8097.

199. Arii, H.; Saito, Y.; Nagatomo, S.; Kitagawa, T.; Funahashi, Y.; Jitsukawa, K.; Masuda, H. C−H activation by $Cu(III)_2O_2$ intermediate with secondary amino ligand. *Chem. Lett.* **2003**, *32*, 156–157.

200. Herres, S.; Heuwing, A. J.; Flrke, U.; Schneider, J.; Henkel, G. Hydroxylation of a methyl group: synthesis of $[Cu_2(btmmO)_2I]^+$ and of $[Cu_2(btmmO)_2]^{2+}$ containing the novel ligand {bis(trimethylmethoxy)guanidino}propane (btmmO) by copper-assisted oxygen activation. *Inorg. Chim. Acta* **2005**, *358*, 1089–1095.

201. Herres-Pawlis, S.; Florke, U.; Henkel, G. Tuning of copper(I)–dioxygen reactivity by bis (guanidine) ligands. *Eur. J. Inorg. Chem.* **2005**, 3815–3824.

202. Jensen, M. P.; Que, Jr., E. L.; Shan, X. P.; Rybak-Akimova, E.; Que, Jr., L. Spectroscopic and kinetic studies of the reaction of $[Cu^I(6\text{-PhTPA})]^+$ with O_2. *Dalton Trans.* **2006**, 3523–3527.

203. Shimokawa, C.; Teraoka, J.; Tachi, Y.; Itoh, S. A functional model for pMMO (particulate methane monooxygenase): Hydroxylation of alkanes with H_2O_2 catalyzed by β-diketiminatocopper(II) complexes. *J. Inorg. Biochem.* **2006**, *100*, 1118–1127.

204. Kajita, Y.; Arii, H.; Saito, T.; Saito, Y.; Nagatomo, S.; Kitagawa, T.; Funahashi, Y.; Ozawa, T.; Masuda, H. Syntheses, characterization, and dioxygen reactivities of Cu(I) complexes with cis,cis-1,3,5-triaminocyclohexane derivatives: A $Cu(III)_2O_2$ intermediate exhibiting higher C−H activation. *Inorg. Chem.* **2007**, *46*, 3322–3335.

205. Company, A.; Gómez, L.; Mas-Ballesté, R.; Korendovych, I. V.; Ribas, X.; Poater, A.; Parella, T.; Fontrodona, X.; Benet-Buchholz, J.; Solá, M.; Que, Jr., L.; Rybak-Akimova, E. V.; Costas, M. Fast O_2 binding at dicopper complexes containing Schiff-base dinucleating ligands. *Inorg. Chem.* **2007**, *46*, 4997–5012.

206. Petrovic, D.; Hill, L. M. R.; Jones, P. G.; Tolman, W. B.; Tamm, M. Synthesis and reactivity of copper(I) complexes with an ethylene-bridged bis(imidazolin-2-imine) ligand. *Dalton Trans.* **2008**, 887–894.

207. Astner, J.; Weitzer, M.; Foxon, S. P.; Schindler, S.; Heinemann, F. W.; Mukherjee, J.; Gupta, R.; Mahadevan, V.; Mukherjee, R. Syntheses, characterization, and reactivity of copper complexes with tridentate N-donor ligands. *Inorg. Chim. Acta* **2008**, *361*, 279–292.

208. Mandal, S.; De, A.; Mukherjee, R. Formation of $\{Cu^{III}_2(\mu\text{-}O)_2\}^{2+}$ core due to dioxygen reactivity of a copper(I) complex supported by a new hybrid tridentate ligand: Reaction with exogenous substrates. *Chem. Biodivers.* **2008**, *5*, 1594–1608.

209. Hong, S.; Hill, L. M. R.; Gupta, A. K.; Naab, B. D.; Gilroy, J. B.; Hicks, R. G.; Cramer, C. J.; Tolman, W. B. Effects of electron-deficient β-diketiminate and formazan supporting ligands on copper(I)–mediated dioxygen activation. *Inorg. Chem.* **2009**, *48*, 4514–4523.

210. Herres-Pawlis, S.; Binder, S.; Eich, A.; Haase, R.; Schulz, B.; Wellenreuther, G.; Henkel, G.; Rubhausen, M.; Meyer-Klaucke, W. Stabilisation of a highly reactive bis(μ-oxo) dicopper(III) species at room temperature by electronic and steric constraint of an unconventional nitrogen donor ligand. *Chem.-Eur. J.* **2009**, *15*, 8678–8682.

211. De, A.; Mandal, S.; Mukherjee, R. Modeling tyrosinase activity. Effect of ligand topology on aromatic ring hydroxylation: An overview. *J. Inorg. Biochem.* **2008**, *102*, 1170–1189.

212. Zhang, C. X.; Liang, H. C.; Kim, E. I.; Shearer, J.; Helton, M. E.; Kim, E.; Kaderli, S.; Incarvito, C. D.; Zuberbühler, A. D.; Rheingold, A. L.; Karlin, K. D. Tuning copper-dioxygen reactivity and exogenous substrate oxidations via alterations in ligand electronics. *J. Am. Chem. Soc.* **2003**, *125*, 634–635.

213. Shearer, J.; Zhang, C. X.; Hatcher, L. Q.; Karlin, K. D. Distinguishing rate-limiting electron versus H-atom transfers in $Cu_2(O_2)$-mediated oxidative *N*-dealkylations: Application of inter- versus intramolecular kinetic isotope effects. *J. Am. Chem. Soc.* **2003**, *125*, 12670–12671.

214. Osako, T.; Ohkubo, K.; Taki, M.; Tachi, Y.; Fukuzumi, S.; Itoh, S. Oxidation mechanism of phenols by dicopper–dioxygen (Cu_2/O_2) complexes. *J. Am. Chem. Soc.* **2003**, *125*, 11027–11033.

215. Battaini, G.; De Carolis, M.; Monzani, E.; Tuczek, F.; Casella, L. The phenol ortho-oxygenation by mononuclear copper(I) complexes requires a dinuclear $\mu\text{-}\eta^2{:}\eta^2$-peroxodicopper(II) complex rather than mononuclear CuO_2 species. Chem. Commun. **2003**, 726–727.

216. Palavicini, S.; Granata, A.; Monzani, E.; Casella, L. Hydroxylation of phenolic compounds by a peroxodicopper(II) complex: Further insight into the mechanism of tyrosinase. *J. Am. Chem. Soc.* **2005**, *127*, 18031–18036.

217. Shearer, J.; Zhang, C. X.; Zakharov, L. N.; Rheingold, A. L.; Karlin, K. D. Substrate oxidation by copper–dioxygen adducts: Mechanistic considerations. *J. Am. Chem. Soc.* **2005**, *127*, 5469–5483.

218. Matsumoto, T.; Furutachi, H.; Kobino, M.; Tomii, M.; Nagatomo, S.; Tosha, T.; Osako, T.; Fujinami, S.; Itoh, S.; Kitagawa, T.; Suzuki, M. Intramolecular arene hydroxylation versus intermolecular olefin epoxidation by ($\mu\text{-}\eta^2{:}\eta^2$-peroxo)dicopper(II) complex supported by dinucleating ligand. *J. Am. Chem. Soc.* **2006**, *128*, 3874–3875.

219. Mirica, L. M.; Rudd, D. J.; Vance, M. A.; Solomon, E. I.; Hodgson, K. O.; Hedman, B.; Stack, T. D. P. $\mu\text{-}\eta^2{:}\eta^2$-Peroxodicopper(II) complex with a secondary diamine ligand: A functional model of tyrosinase. *J. Am. Chem. Soc.* **2006**, *128*, 2654–2665.

220. Gamba, I.; Palavicini, S.; Monzani, E.; Casella, L. Catalytic sulfoxidation by dinuclear copper complexes. *Chem. Eur. J.* **2009**, *15*, 12932–12936.

221. Matsumoto, T.; Ohkubo, K.; Honda, K.; Yazawa, A.; Furutachi, H.; Fujinami, S.; Fukuzumi, S.; Suzuki, M. Aliphatic C-H bond activation initiated by a ($\mu\text{-}\eta^2{:}\eta^2$-peroxo)dicopper(II) complex in comparison with cumylperoxyl radical. *J. Am. Chem. Soc.* **2009**, *131*, 9258–9267.

222. Mirica, L. M.; Vance, M.; Rudd, D. J.; Hedman, B.; Hodgson, K. O.; Solomon, E. I.; Stack, T. D. P. Tyrosinase reactivity in a model complex: An alternative hydroxylation mechanism. *Science* **2005**, *308*, 1890–1892.

223. Company, A.; Palavicini, S.; Garcia-Bosch, I.; Mas-Balleste, R.; Que, Jr., L.; Rybak-Akimova, E. V.; Casella, L.; Ribas, X.; Costas, M. Tyrosinase-like reactivity in a $Cu^{III}_2(\mu\text{-}O)_2$ species. *Chem.-Eur. J.* **2008**, *14*, 3535–3538.

224. Herres-Pawlis, S.; Verma, P.; Haase, R.; Kang, P.; Lyons, C. T.; Wasinger, E. C.; Florke, U.; Henkel, G.; Stack, T. D. P. Phenolate hydroxylation in a bis(μ-oxo) dicopper(III) complex: Lessons from the guanidine/amine series. *J. Am. Chem. Soc.* **2009**, *131*, 1154–1169.

225. Taki, M.; Itoh, S.; Fukuzumi, S. Oxo-transfer reaction from a bis(μ-oxo)dicopper(III) complex to sulfides. *J. Am. Chem. Soc.* **2002**, *124*, 998–1002.

226. Mahadevan, V.; DuBois, J. L.; Hedman, B.; Hodgson, K. O.; Stack, T. D. P. Exogenous substrate reactivity with a $Cu(III)_2O_2^{2+}$ core: Structural implications. *J. Am. Chem. Soc.* **1999**, *121*, 5583–5584.

227. Berreau, L. M.; Mahapatra, S.; Halfen, J. A.; Houser, R. P.; Young, V. G.; Tolman, W. B. Reactivity of peroxo- and bis(μ-oxo)dicopper complexes with catechols. *Angew. Chem. Int. Ed.* **1999**, *38*, 207–210.

228. Mahadevan, V.; Henson, M. J.; Solomon, E. I.; Stack, T. D. P. Differential reactivity between interconvertible side-on peroxo and bis-μ-oxodicopper isomers using peralkylated diamine ligands. *J. Am. Chem. Soc.* **2000**, *122*, 10249–10250.

229. Taki, M.; Itoh, S.; Fukuzumi, S. C-H bond activation of external substrates with a bis (μ-oxo)dicopper(III) complex. *J. Am. Chem. Soc.* **2001**, *123*, 6203–6204.

230. Karlin, K. D.; Dahlstrom, P. L.; Cozzette, S. N.; Scensny, P. M.; Zubieta, J. Activation of O_2 by a binuclear copper(I) compound. Hydroxylation of a new xylyl-binucleating ligand to produce a phenoxy-bridged binuclear copper(II) complex; X-ray crystal structure of $[Cu_2\{OC_6H_3[CH_2N(CH_2CH_2py)_2]_2\text{-}2,6\}(OMe)]$ (py = 2-pyridyl). *J. Chem. Soc., Chem. Commun.* **1981**, 881–882.

231. Karlin, K. D.; Gultneh, Y.; Hutchinson, J. P.; Zubieta, J. Three-coordinate binuclear copper(I) complex: Model compound for the copper sites in deoxyhemocyanin and deoxytyrosinase. *J. Am. Chem. Soc.* **1982**, *104*, 5240–5242.

232. Karlin, K. D.; Hayes, J. C.; Gultneh, Y.; Cruse, R. W.; McKown, J. W.; Hutchinson, J. P.; Zubieta, J. Copper-mediated hydroxylation of an arene: Model system for the action of copper monooxygenases. Structures of a binuclear Cu(I) complex and its oxygenated product. *J. Am. Chem. Soc.* **1984**, *106*, 2121–2128.

233. Sheldon, R. A.; Kochi, J. K. *Metal-catalyzed oxidations of organic compounds*; Academic Press: New York, 1981.

234. Nasir, M. S.; Cohen, B. I.; Karlin, K. D. Mechanism of aromatic hydroxylation in a copper monooxygenase model system. 1,2-Methyl migrations and the NIH shift in copper chemistry. *J. Am. Chem. Soc.* **1992**, *114*, 2482–2494.

235. Karlin, K. D.; Nasir, M. S.; Cohen, B. I.; Cruse, R. W.; Kaderli, S.; Zuberbuhler, A. D. Reversible dioxygen binding and aromatic hydroxylation in O_2-reactions with substituted xylyl dinuclear Copper(I) complexes: Syntheses and low-temperature kinetic/ thermodynamic and spectroscopic investigations of a copper monooxygenase model system. *J. Am. Chem. Soc.* **1994**, *116*, 1324–1336.

236. Pidcock, E.; Obias, H. V.; Zhang, C. X.; Karlin, K. D.; Solomon, E. I. Investigation of the reactive oxygen intermediate in an arene hydroxylation reaction performed by xylyl-bridged binuclear copper complexes. *J. Am. Chem. Soc.* **1998**, *120*, 7841–7847.

237. Matsumoto, T.; Furutachi, H.; Nagatomo, S.; Tosha, T.; Fujinami, S.; Kitagawa, T.; Suzuki, M. Synthesis and reactivity of (μ-η^2:η^2-peroxo)dicopper(II) complexes with dinucleating ligands: Hydroxylation of xylyl linker with a NIH shift. *J. Organomet. Chem.* **2007**, *692*, 111–121.

238. Sander, O.; Henss, A.; Näther, C.; Würtele, C.; Holthausen, M. C.; Schindler, S.; Tuczek, F. Aromatic hydroxylation in a copper bis(imine) complex mediated by a μ-η^2:η^2 peroxo dicopper core: A mechanistic scenario. *Chem.-Eur. J.* **2008**, *14*, 9714–9729.

239. Holland, P. L.; Rodgers, K. R.; Tolman, W. B. Is the bis(μ-oxo)dicopper core capable of hydroxylating an arene? *Angew. Chem. Int. Ed. Engl.* **1999**, *38*, 1139–1142.

240. Santagostini, L.; Gullotti, M.; Monzani, E.; Casella, L.; Dillinger, R.; Tuczek, F. Reversible dioxygen binding and phenol oxygenation in a tyrosinase model system. *Chem. Eur. J.* **2000**, *6*, 519–522.

241. Itoh, S.; Kumei, H.; Taki, M.; Nagatomo, S.; Kitagawa, T.; Fukuzumi, S. Oxygenation of phenols to catechols by a (μ-η^2:η^2-peroxo)dicopper(II) complex: Mechanistic insight into the phenolase activity of tyrosinase. *J. Am. Chem. Soc.* **2001**, *123*, 6708–6709.

242. Yamazaki, S.; Itoh, S. Kinetic evaluation of phenolase activity of tyrosinase using simplified catalytic reaction system. *J. Am. Chem. Soc.* **2003**, *125*, 13034–13035.

243. Yamazaki, S.; Morioka, C.; Itoh, S. Kinetic evaluation of catalase and peroxygenase activities of tyrosinase. *Biochemistry* **2004**, *43*, 11546–11553.

244. Morioka, C.; Tachi, Y.; Suzuki, S.; Itoh, S. Significant enhancement of monooxygenase activity of oxygen carrier protein hemocyanin by urea. *J. Am. Chem. Soc.* **2006**, *128*, 6788–6789.

245. Suzuki, K.; Shimokawa, C.; Morioka, C.; Itoh, S. Monooxygenase activity of octopus vulgaris hemocyanin. *Biochemistry* **2008**, *47*, 7108–7115.

246. Fujieda, N.; Yakiyama, A.; Itoh, S. Catalytic oxygenation of phenols by arthropod hemocyanin, an oxygen carrier protein, from Portunus trituberculatus. *Dalton Trans.*, *39*, 3083–3092.

247. Op't Holt, B. T.; Vance, M. A.; Mirica, L. M.; Heppner, D. E.; Stack, T. D. P.; Solomon, E. I. Reaction coordinate of a functional model of tyrosinase: Spectroscopic and computational characterization. *J. Am. Chem. Soc.* **2009**, *131*, 6421–6438.

248. Spada, A.; Palavicini, S.; Monzani, E.; Bubacco, L.; Casella, L. Trapping tyrosinase key active intermediate under turnover. *Dalton Trans.* **2009**, 6468–6471.

249. Mahapatra, S.; Halfen, J. A.; Wilkinson, E. C.; Que, Jr., L.; Tolman, W. B. Modeling copper-dioxygen reactivity in proteins: Aliphatic C-H bond activation by a new dicopper (II)–peroxo complex. *J. Am. Chem. Soc.* **1994**, *116*, 9785–9786.

250. Mahapatra, S.; Halfen, J. A.; Tolman, W. B. Mechanistic study of the oxidative N-dealkylation reactions of bis(μ-oxo)dicopper complexes. *J. Am. Chem. Soc.* **1996**, *118*, 11575–11586.

251. Itoh, S.; Kondo, T.; Komatsu, M.; Ohshiro, Y.; Li, C. M.; Kanehisa, N.; Kai, Y.; Fukuzumi, S. Functional model of dopamine β-hydroxylase. Quantitative ligand hydroxylation at the benzylic position of a copper complex by dioxygen. *J. Am. Chem. Soc.* **1995**, *117*, 4714–4715.

252. Itoh, S.; Nakao, H.; Berreau, L. M.; Kondo, T.; Komatsu, M.; Fukuzumi, S. Mechanistic studies of aliphatic ligand hydroxylation of a copper complex by dioxygen: A model reaction for copper monooxygenases. *J. Am. Chem. Soc.* **1998**, *120*, 2890–2899.

253. Itoh, S.; Taki, M.; Nakao, H.; Holland, P. L.; Tolman, W. B.; Que, Jr., L.; Fukuzumi, S. Aliphatic hydroxylation by a bis(μ-oxo)dicopper(III) complex. *Angew. Chem. Int. Ed.* **2000**, *39*, 398–400.

254. Cruse, R. W.; Kaderli, S.; Meyer, C. J.; Zuberbuhler, A. D.; Karlin, K. D. Copper-mediated hydroxylation of an arene - Kinetics and mechanism of the reaction of a dicopper(Ii) meta-xylyl-containing complex with H_2O_2 to yield a phenoxodicopper(Ii) complex. *J. Am. Chem. Soc.* **1988**, *110*, 5020–5024.

255. Battaini, G.; Monzani, E.; Perotti, A.; Para, C.; Casella, L.; Santagostini, L.; Gullotti, M.; Dillinger, R.; Nather, C.; Tuczek, F. A double arene hydroxylation mediated by dicopper (II)–hydroperoxide species. *J. Am. Chem. Soc.* **2003**, *125*, 4185–4198.

256. Groothaert, M. H.; Lievens, K.; Leeman, H.; Weckhuysen, B. M.; Schoonheydt, R. A. An operando optical fiber UV–vis spectroscopic study of the catalytic decomposition of NO and N_2O over Cu-ZSM-5. *J. Catal.* **2003**, *220*, 500–512.

257. Groothaert, M. H.; van Bokhoven, J. A.; Battiston, A. A.; Weckhuysen, B. M.; Schoonheydt, R. A. Bis(μ-oxo)dicopper in Cu-ZSM-5 and its role in the decomposition of NO: A combined in situ XAFS, UV-Vis-Near-IR, and kinetic study. *J. Am. Chem. Soc.* **2003**, *125*, 7629–7640.

258. Groothaert, M. H.; Smeets, P. J.; Sels, B. F.; Jacobs, P. A.; Schoonheydt, R. A. Selective oxidation of methane by the bis(μ-oxo)dicopper core stabilized on ZSM-5 and mordenite zeolites. *J. Am. Chem. Soc.* **2005**, *127*, 1394–1395.

259. Smeets, P. J.; Groothaert, M. H.; Schoonheydt, R. A. Cu based zeolites: A UV-vis study of the active site in the selective methane oxidation at low temperatures. *Catal. Today* **2005**, *110*, 303–309.

260. Woertink, J. S.; Smeets, P. J.; Groothaert, M. H.; Vance, M. A.; Sels, B. F.; Schoonheydt, R. A.; Solomon, E. I. A $[Cu_2O]^{2+}$ core in Cu-ZSM-5, the active site in the oxidation of methane to methanol. *Proc. Nat. Acad. Sci.* **2009**, *106*, 18908–18913.

261. Smeets, P. J.; Woertink, J. S.; Sels, B. F.; Solomon, E. I.; Schoonheydt, R. A. Transition-Metal Ions in Zeolites: Coordination and Activation of Oxygen. *Inorg. Chem.* **2010**, *49*, 3573–3583.

262. Smeets, P. J.; Hadt, R. G.; Woertink, J. S.; Vanelderen, P.; Schoonheydt, R. A. Sels, B. F.; Solomon, E. L. Oxygen Precursor to the Reactive Intermediate in Methanol Synthesis by Cu-ZSM-5. *J. Am. Chem. Soc.* **2010**, *132*, 14736–14738.

9

CYTOCHROME *c* OXIDASE AND MODELS

ZAKARIA HALIME AND KENNETH D. KARLIN

Department of Chemistry, Johns Hopkins University, Baltimore, MD 21218, USA

Copper-Oxygen Chemistry, First Edition. Edited by Kenneth D. Karlin and Shinobu Itoh.
© 2011 John Wiley & Sons, Inc. Published 2011 by John Wiley & Sons, Inc.

9.1. INTRODUCTION

Cytochrome c oxidase (CcO) is one of the most important and most extensively investigated enzymes. Its physiological importance is due to the crucial role it plays as the terminal oxidant in aerobic respiration catalyzing a four-electron and four-proton reduction of molecular oxygen to water.

$$O_2 + 4e^-_{\text{(from cyt}-c\text{ reduced)}} + 8H^+_{\text{in}} \rightarrow 2H_2O + 4H^+_{\text{out}} + 4cyt - c_{\text{oxidized}}$$

Cytochrome c oxidase uses four redox-active metal centers, a mixed-valence binuclear copper center (Cu_A), a low-spin heme a, and the heme a_3/Cu_B binuclear center (Figure 9.1). Dioxygen (O_2) reduction by the enzyme is coupled with proton pumping across the mitochondrial inner membrane. The electrons used for the O_2-reduction are transferred to the O_2-reduction site from cytochrome c molecules located in the intermembrane space, while the protons come from the matrix space.

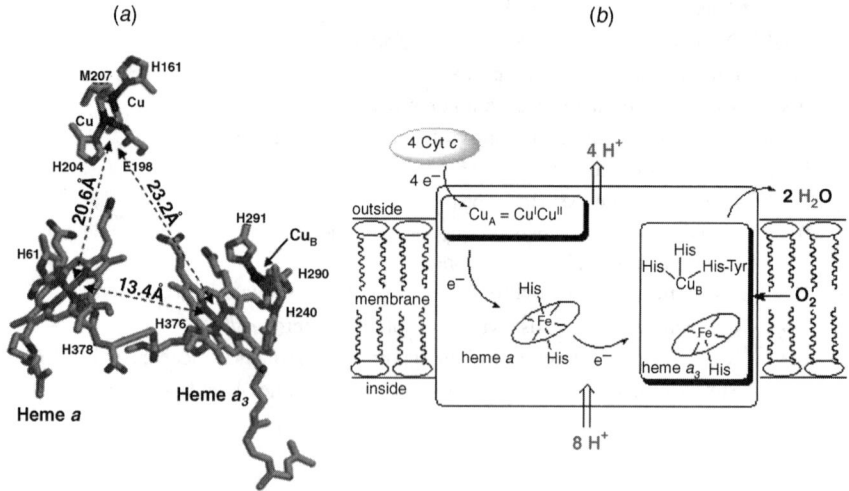

FIGURE 9.1. (a) X-ray structure of the redox active metal centers of CcO from bovine heart. (b) Diagram depicting the important metal centers and the general pathway for electrons transfer (ET) and the proton translocation.

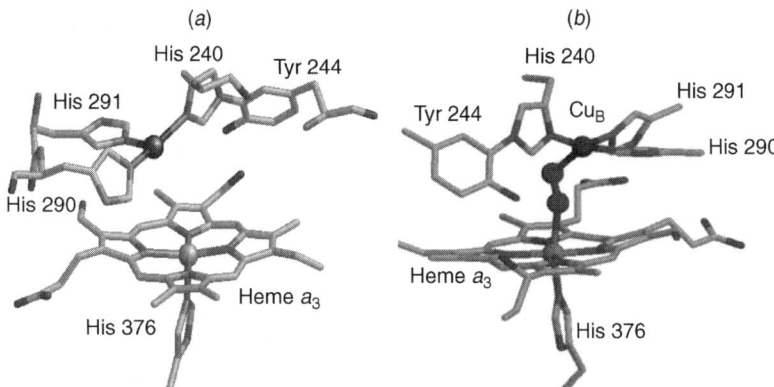

FIGURE 9.2. Diagrams representing the X-ray structures of bovine heart CcO (*a*) fully reduced bimetallic heme a_3/Cu$_B$ center (Fe^{II}··· Cu^I = 5.19 Å) and (*b*) an oxidized peroxo-bridged form. (Adapted from figures in Ref. 4; structures are from Yoshikawa et al., Ref. 3.)

The reduction of each O_2 molecule by this enzyme is coupled to the translocation of four additional protons from the inner side of the membrane. Thus, both dioxygen reduction and proton pumping contribute to the formation of an electrochemical potential gradient across the mitochondrial inner membrane that is utilized to drive adenosine triphosphate (ATP) synthetase activity and thus generate ATP that is stored and used in subsequent metabolic functions.[1–6]

The three-dimensional (3D) structure of the enzyme is now available not only for beef heart,[7] but also for three other CcOs, namely, the aa_3 -type oxidases from *Paracoccus deniftrificans*[2] and from *Rhodobacter sphaeroides*,[8] as weel as from the ba_3-type oxidase from the thermophilic bacterium *Thermus thermophilus*.[9] These structures have been followed by improved higher resolution or further refined structures, as well as fully reduced (also, see Fig. 9.2*a*), carbon monoxide-bound to heme, netrogen oxide-bound to Cu$_B$, peroxo-bridged (also, see Fig. 9.2*b*), and various oxidized derivatives.[3,10–15] Surprisingly, the structures from bacterial and mammalian systems have essentially identical metal centers and 3D structures of the subunits containing those metal centers. The structure of the redox active metal centers of bovine CcO are shown in Figure 9.1. The Cu$_A$, which is the direct electron acceptor from cytochrome c,[16–18] has two copper atoms bridged with two thiolate ligands with Cu^I··· Cu^I distance of 2.58 Å in this fully reduced structure. This center acts as a 1e$^-$ redox center, and the oxidized form is a fully delocalized mixed-valence Cu^I··· Cu^{II} ↔ Cu^{II}··· Cu^I. The electrons received by Cu$_A$ are sequentially transferred to heme a and to the heme a_3/Cu$_B$ binuclear center, where the O_2 binding and reduction occur. Hemes a and a_3 are connected through their axial histidine ligands (His-Phe-His linker), having a heme–heme interplanar angle of 104° and an Fe··· Fe distance of 13.4 Å. Heme a is a low-spin center with two axial histidine ligands, whereas heme a_3 is high spin bound by one histidine, found in close proximity to Cu$_B$. The latter has a tridentate chelation with three histidine ligands; one of these is linked to a tyrosine via a covalent bond between the N$_{\varepsilon 2}$ of His240 and the $C_{\varepsilon 2}$ of Tyr244 (see Section 9.4). The origin and function of this modified tyrosine have provoked considerable interest; it has been proposed to provide an additional site for electron–proton transfer during the dioxygen-reduction catalytic cycle[19–21] (see Section 9.4 and Chapter 4 in this text). The Fe$\cdots$Cu distances in the known X-ray structures vary in the range of 4.9–5.3 Å, depending on the redox states and the particular protein derivative.

The dioxygen-reduction mechanism of C*c*O, which occurs at the heme a_3/Cu$_B$ binuclear center (Figure 9.2), has been subject to a variety of kinetic and spectroscopic investigations.[1–4,21,22] Although not all details concerning the reaction mechanism and nature of intermediates are fully understood, there is a general agreement on certain aspects (Scheme 9.1). Dioxygen reacts with a fully reduced Fe$^{II}\cdots$Cu^{I} binuclear center to form the first detectable transient, the Fe^{III}–(O$_2^{\bullet-}$)$\cdots$Cu^{I} "Oxy" complex **A**. This complex possesses Ultraviolet–visible (UV-vis) and resonance Raman (*rR*) spectroscopic properties very similar to those known for oxy-hemoglobin or myoglobin. A density functional theory (DFT) calculated structure of **A** supports the Fe^{III}(O$_2^{\bullet-}$)$\cdots$Cu^{I} formulation with the Cu ion positioned very close to the

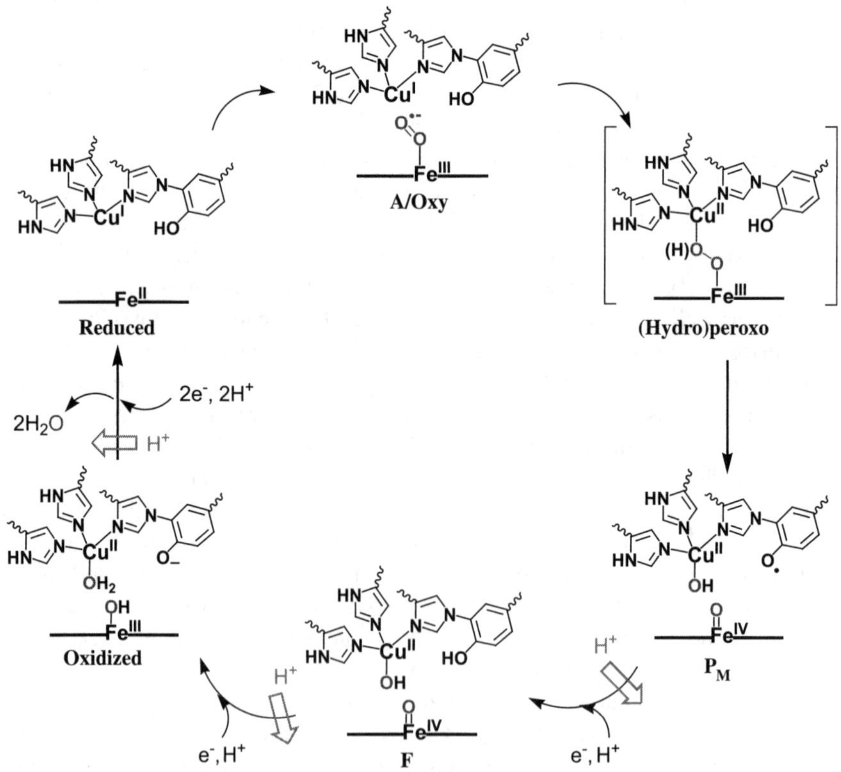

SCHEME 9.1. The proposed dioxygen-reduction mechanism at the heme a_3/Cu$_B$ bimetallic center of C*c*O.

superoxo O atom, $Cu \cdots O = 2.12$ Å.[23] Kinetic studies also suggest that there is an initial interaction between Cu_B and O_2 (not shown in Scheme 9.1) prior to formation of oxy.[4] The key $O-O$ bond cleavage follows in the next step to yield a ferryl oxo $(Fe^{VI}=O)$ species P_M. This transformation seems optimized to prevent leakage of deleterious intermediates, such as a hydroxyl radical or hydrogen peroxide (H_2O_2). Because four electrons are required to break an $O-O$ bond and reduce to the oxidations state level of water, and only three electrons are readily available from the binuclear site (two from iron, $Fe^{II} \rightarrow Fe^{IV}$, and one from the copper $Cu^I \rightarrow Cu^{II}$), it is suggested that a close-by tyrosine (Tyr244) or nearby tryptophan (Trp236)[24] residue efficiently provides the electron needed. A peroxo-bridged $Fe^{III}-(O_2^{2-})-Cu^{II}$ transient has been discussed as possibly forming from **A** prior to P_M formation.[19,25–27] In fact, protein X-ray structures where a peroxide bridges the heme-iron and the Cu_B ion have been characterized (Figure 9.2), and rR data supports its peroxide formulation.[3,14,15] Such a species may (or may not) represent a turnover intermediate. Also discussed and/or calculated in the enzyme reaction mechanism is an ET from $Cu^I{}_B$ in **A**, which is accompanied by proton uptake, to give a µ-hydroperoxo transient, $Fe^{III}_{a3}-O-O(H)-Cu^{II}{}_B$; the protonation event in this scenario would trigger ET (from the heme a_3 and the Tyr244) yielding to $O-O$ cleavage and the formation of the transient P_M.[25,28–30]

Various methodologies have been applied to elucidate the structure and function in numerous enzymes, including protein crystallography, spectroscopic methods, site-directed mutagenesis, mechanistic enzymology, and theoretical calculations. An additional significant factor has been biomimetic inorganic chemistry, which largely involves the synthesis and study of small model molecules that approach or achieve one or more significant properties of a protein active site. The synthetic analogues purpose is to mimic some facets of the biological unit in terms of metallic centers composition, oxidation states, coordination geometry, and the nature of the ligands. The purpose of models is not necessarily to duplicate natural properties, but to sharpen or focus certain questions. The goal is to elucidate fundamental aspects of the structure, spectroscopy, magnetic and electronic structure, reactivity, and chemical mechanism. A synergic approach to the study of metalloenzymes can and has yielded crucial information, because synthetic analogues can be used to investigate the effect of systematic variations in the coordination geometry, ligation, environment, and other factors without being encumbered by the protein. These results reflect the intrinsic properties of the coordination unit unmodified by the protein environment.

Since the subsequently reported CcO X-ray structures, the synthetic approach has evolved from just generating bridged and oxidized forms $(Fe^{III}-X-Cu^{II})$ complexes, where X is a bridging ligand of various types, to modeling functional analogues in the reduced state $Fe^{II} \cdots Cu^I$ and studying their reactivity toward O_2. This new direction is developed in an attempt to answer many questions raised by the X-ray structures and the proposed mechanism:

- How does the proximity of a copper complex affect the heme bonding and the reactivity toward O_2 and vice-versa?

- Dose the O_2 bond preferentially to the heme or to the copper?
- Can bridged peroxo $Fe^{III}-(O_2^{2-})-Cu^{II}$ intermediates be generated, and can such a transient be part of the enzyme mechanism?
- What role does the copper ligand play in the dioxygen-reduction chemistry?
- Can one generate intermediates, probe their spectroscopy, and determine their geometric and the electronic structures?
- Can synthetic models reproduce the $4e^-$ reduction of O_2 to water?
- What is the role of the unusual cross-linked histidine-tyrosine and how does it form?

9.2. CYTOCHROME *c* OXIDASE SYNTHETIC MODEL DERIVATIVES OF PICKET FENCE PORPHYRINS AND THE ELECTROCHEMICAL APPROACH

9.2.1. Early C*c*O Synthetic Models

The reactivity of O_2 with a Fe^{II}/Cu^{I} model of the C*c*O bimetallic active site was first described by the Gunter and Murray[31] research groups in the early 1980s. To provide a coordination site for the copper in a proximity of the porphyrin ring four pyridine (Py) arms were incorporated into a tetrakis(*o*-aminophenyl)porphyrin (TAPP). This adaptation of the already existing "picket fence" concept allowed a synthesis of an Fe^{II}/Cu^{I} complex with a Fe$\cdots$Cu distance close to that observed in the C*c*O active site (Figure 9.2). In the presence or absence of an axial ligand (Me$_2$SO or 1-methylimidazole) at ambient temperature, the binuclear complex [(TAPP)Fe^{II}Cu^{I}(Py$_4$)]$^+$ (**1 Fe^{II}/Cu^{I}**) and the copper-free complex [(TAPP)Fe^{II} (Py$_4$)] (**1 Fe^{II}**) reacted with O_2 to yield oxidized compounds (Scheme 9.2). The reactions were monitored by ultraviolet–visible (UV–vis) and electron paramagnetic resonance (EPR) spectroscopies and magnetic moment measurements. The final products were determined to be a terminal hydroxy [(TAPP)Fe^{III}—OH (Py$_4$)] (**1 Fe^{III}—OH**) and [(TAPP)Fe^{III}—OH/Cu^{II}(Py$_4$)]$^{2+}$ (**1 Fe^{III}—OH/Cu^{II}**) complexe (Scheme 9.2). No electronic–magnetic interaction between iron and copper was observed. Although the researchers did not observe any significant role for the copper ion complex in the reaction of **1 Fe^{II}/Cu^{I}** with O_2 compared to the **1 Fe^{II}** with O_2, this was an important pioneering work.

9.2.2. Models for the Electrochemical Approach

Major advances in understanding heme/Cu/O_2 chemistry have been made in the past three decades through the development of two largely complementary approaches. One approach, described in Section 9.3, involves the isolation and characterization of the O_2 intermediates obtained from stoichiometric reactions of reduced heme/Cu models with O_2 at low temperature. The second approach developed by Collman,

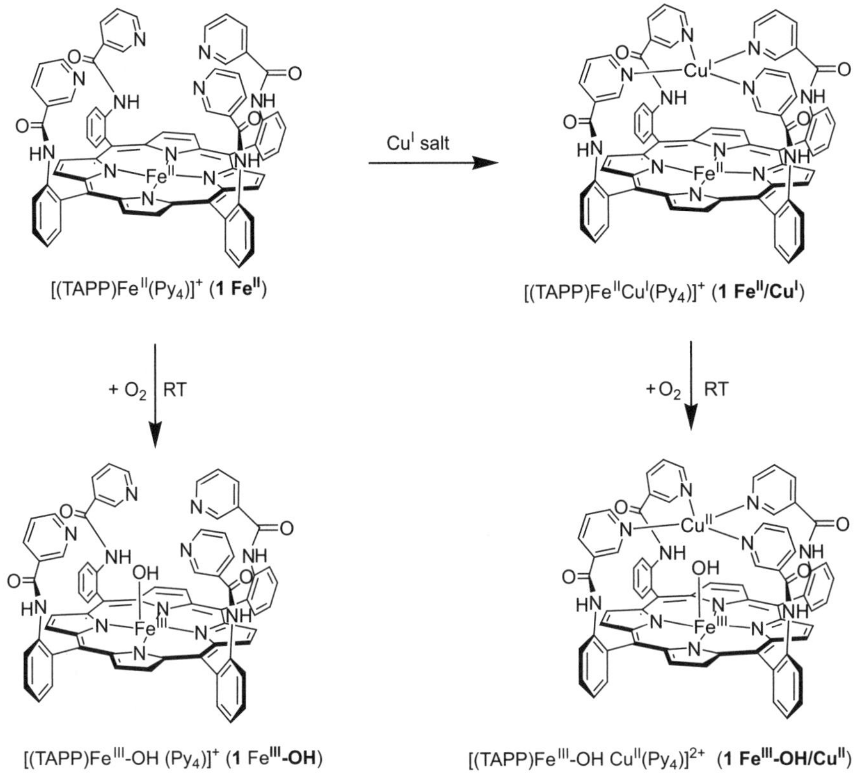

$[(TAPP)Fe^{II}(Py_4)]^+$ (**1 Fe^{II}**)

$[(TAPP)Fe^{II}Cu^{I}(Py_4)]^+$ (**1 Fe^{II}/Cu^{I}**)

$[(TAPP)Fe^{III}-OH (Py_4)]^+$ (**1 Fe^{III}-OH**)

$[(TAPP)Fe^{III}-OH Cu^{II}(Py_4)]^{2+}$ (**1 Fe^{III}-OH/Cu^{II}**)

SCHEME 9.2

Boitrel, and their co-workers (see later) has been to study electrochemical dioxygen-reduction catalyzed by heme/Cu analogues. This latter approach was almost exclusively associated with heme/Cu models using the picket-fence porphyrin concept (Chart 9.1). To adequately reproduce the immediate coordination environment of both metals and the Fe··· Cu distance and to be able to retain the structure integrity under the harsh conditions of the electrochemical catalytic turnover, several binucleating ligands comprised of a porphyrin moiety and covalently attached chelate with at least three coordinating nitrogen atoms were developed. Most of these ligands were also designed with an appended heterocyclic moiety (e.g., imidazole or pyridine) that can serve as a proximal axial ligand for the iron mimicking the heme axial base (His376) in the C*c*O active site.

9.2.3. Dioxygen-Reduction Catalyzed by Simple Iron Porphyrins

The study of dioxygen-reduction by simple iron porphyrins using the biomimetic electrochemical technique has been an essential part of the understanding process for the structure–activity relationship of the heme/Cu site of C*c*O. It has been shown since the early 1980s that simple iron porphyrins,[32,33] when adsorbed on an electrode, can

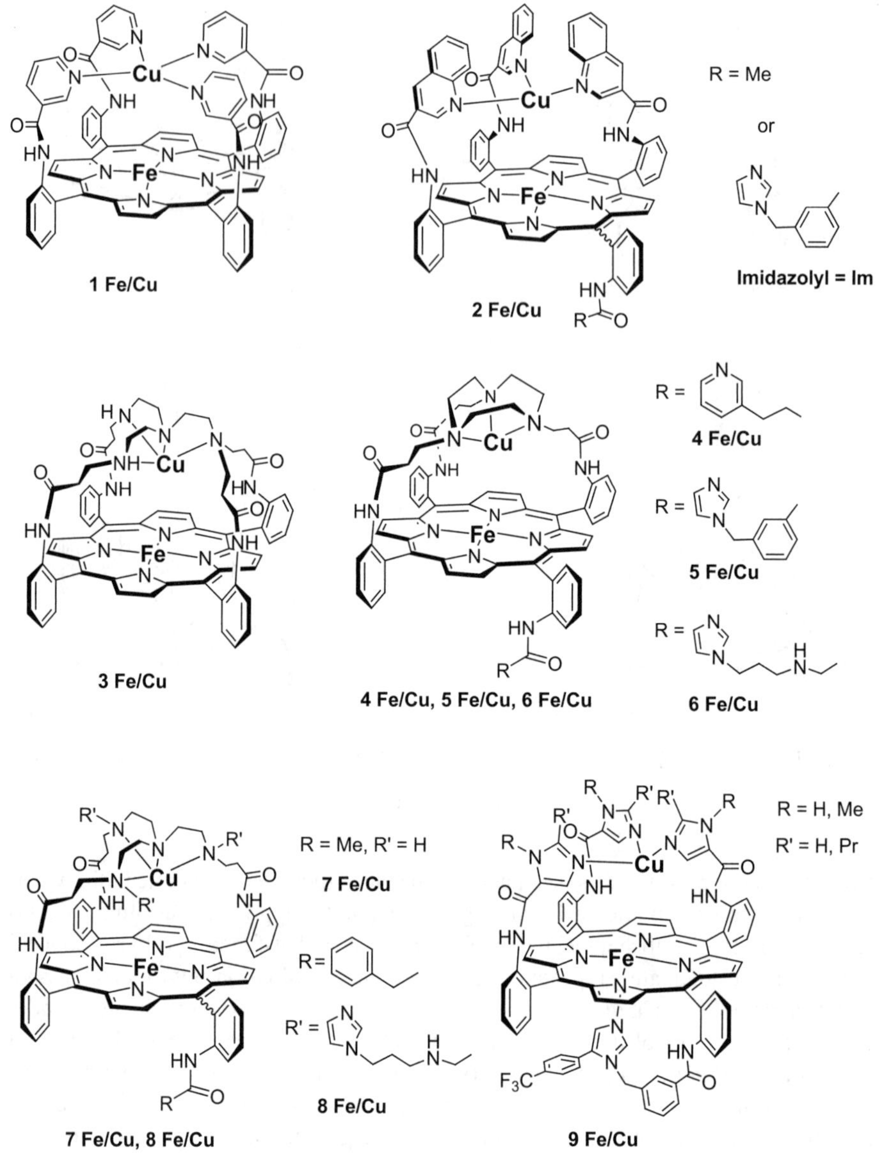

CHART 9.1

catalyze the reduction of O_2 with high selectivity for the $4e^-$ reduction producing H_2O rather than the $2e^-$ reduction yielding H_2O_2. However, the potential at which this reduction occurs ($<0\,V$ vs NHE) seems to be different from the physiological potentials observed for the enzyme ($>240\,mV$ vs NHE), furthermore, this $4e^-$ reduction appears to proceed via intermediates that are inaccessible or nonexistent in enzymatic heme catalysis.

9.2.4. Dioxygen-Reduction Catalyzed by Heme/Cu Models

Surprisingly, later studies by Boitrel's research group,[34] employing ligands such as **2 Fe/Cu** (Chart 9.1), indicated that when Cu is incorporated the heme/Cu models obtained appear to have lower selectivity for the $4e^-$ reduction and they produce mostly H_2O_2 as the $2e^-$ reduction product. Among the hypotheses made to explain such unexpected results is that the Cu does not bind to the O_2 molecule or the Cu ion is no longer coordinated or the Cu in these particular complexes facilitates O–O bond homolysis either in iron-bound peroxo species or in free hydrogen peroxide (H_2O_2), which is generated as a byproduct of dioxygen-reduction.

Collman's group developed a series of capped porphyrin complexes, in which an imidazole or pyridine axial base was covalently bound to one side of a porphyrin, and a 1,4,7-triazacyclononane (TACN) or a N,N',N''-tribenzyltris(aminoethyl) amine (TBTren) copper ligand was attached to the opposite side (Fig. 9.3).[35-37] All of these $Fe^{(II)}/Cu^{(I)}$ complexes bound O_2 irreversibly to form a bridging peroxo intermediate, $Fe^{III}-(O_2^{2-})-Cu^{II}$, with diamagnetic proton nuclear magnetic resonance (^1H NMR) spectra supporting the formulation. However, depending on the ligand environments of the metal centers, they exhibited different chemical reactivity. For example, when the peroxo complexes $[(\alpha_3TACN\beta Im_{alk})Fe^{III}-(O_2^{2-})-Cu^{II}]^+$ (**4 Fe^{III}–(O$_2^{2-}$)–Cu^{II}**) and $[(\alpha_3Tren_{Ph}\beta Im_{alk})Fe^{III}-(O_2^{2-})-Cu^{II}]^+$ (**8 Fe^{III}–(O$_2^{2-}$)–Cu^{II}**) were titrated with Cp_2Co, **4 Fe^{III}–(O$_2^{2-}$)–Cu^{II}** required only 2 equiv of Cp_2Co to regenerate **4 Fe^{II}/Cu^I**, whereas **8 Fe^{III}–(O$_2^{2-}$)–Cu^{II}** needed 4 equiv to form **8 Fe^{II}/Cu^I**.[36] Differences in behavior between **4 Fe^{III}–(O$_2^{2-}$)–Cu^{II}** and **8 Fe^{III}–(O$_2^{2-}$)–Cu^{II}** were also observed in their electrocatalytic reactivities, carried out through the use of rotating ring-disk voltammetry (Scheme 9.3). Complex **8 Fe^{III}–(O$_2^{2-}$)–Cu^{II}** catalyzed the $4e^-$ reduction of O_2 to H_2O, whereas **4 Fe^{III}–(O$_2^{2-}$)–Cu^{II}** predominantly catalyzed the $2e^-$ reduction of O_2 to H_2O_2. Interestingly, a cobalt analogue of **4 Fe^{III}–(O$_2^{2-}$)–Cu^{II}**, $[(\alpha_3TACN\beta Im_{alk})Co^{II}Cu^I]^+$ (**4 Co^{II}/Cu^I**), catalyzed the $4e^-$ reduction of O_2 with very little H_2O_2 leakage.[35] Such observed differences could be attributed to their redox behavior. Cyclic voltammograms of **4 Co^{II}/Cu^I** and **8 Fe^{III}–(O$_2^{2-}$)–Cu^{II}** showed that Co(III/II) or Fe(III/II) redox potentials are higher (more positive, i.e., easier to reduce) than Cu(II/I) potentials. In contrast, a more positive Cu(II/I) potential [if to Fe(III/II)] was observed for **4 Fe^{III}–(O$_2^{2-}$)–Cu^{II}**, allowing Cu to be reduced before Fe in the electrocatalytic cycle. The researchers suggested[38] that prior O_2 binding to Cu(I) [rather than Fe(II)] could lead to a Cu–O_2 moiety that resides nearer the outside of the porphyrin macrocycle cavity, resulting in a $2e^-$ reduction pathway in which only Cu is involved. On the other hand, in complexes **4 Co^{II}/Cu^I** or **8 Fe^{III}–(O$_2^{2-}$)–Cu^{II}**, prior reduction of the Co or Fe center would direct the dioxygen-binding site inside the bimetallic cavity (because the other side of the porphyrin is blocked by an axial ligand), and it could further react with Cu as the copper center is (electrochemically) reduced. Studies of this series of complexes suggested that both the copper and the heme-axial base were important in the $4e^-$ reduction of O_2 and that the catalytic properties of these complexes could be fine-tuned through small structural ligand modifications.

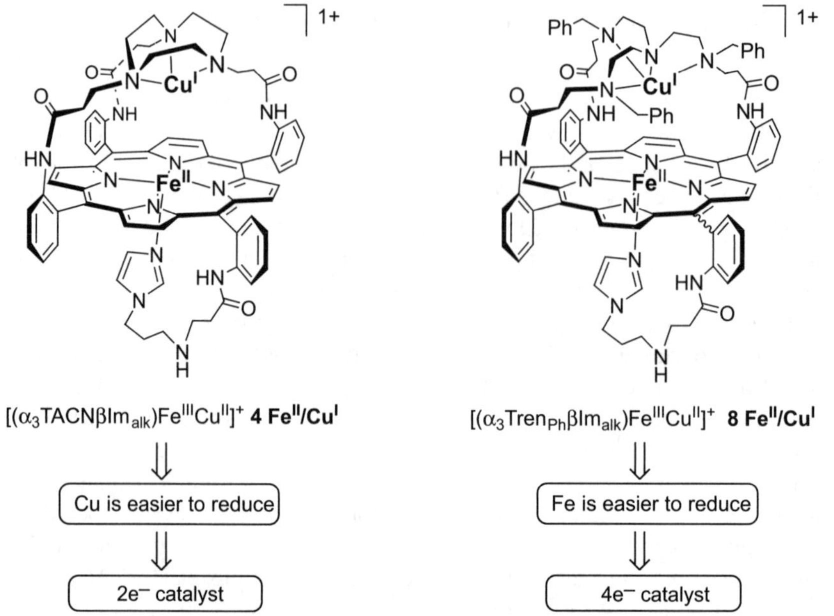

[(α₃TACNβIm_alk)Fe^{III}Cu^{II}]^+ **4 Fe^{II}/Cu^I**   [(α₃Tren_Ph βIm_alk)Fe^{III}Cu^{II}]^+ **8 Fe^{II}/Cu^I**

| Cu is easier to reduce | | Fe is easier to reduce |

| 2e⁻ catalyst | | 4e⁻ catalyst |

SCHEME 9.3

9.2.5. Biomimetic Electrocatalytic Studies and the Role of Cu_B

Somewhat more recent investigations from the Collman research group[39,40] used models where the Cu was bound by three imidazole ligands instead of a TACN or TBTren moiety (**9 Fe/Cu**, Chart 9.1). Since imidazole donors were employed, and there were only three N-ligands for copper, and a fourth imidazole is used as an axial base for the iron as in the enzymes, the authors suggested that these newer complexes were more biomimetic. Indeed, these complexes were very efficient catalysts for the O_2 4e⁻ reduction at physiologically relevant pH and positive potential (>50 mV vs NHE) with respectable selectivity toward the 4e⁻ pathway for at least 10,000 turnovers.[41] These models were employed to study the role of the Cu on the dioxygen-reduction catalysis, and the authors compared the electrochemical behavior of the Fe/Cu and the Fe version of the same biomimetic analogue under conditions of both slow and rapid electron flux.[41,42] This study showed that under fast electron transport the dioxygen-reduction mechanism and turnover were largely independent of the presence of the Cu. However, at low more biologically relevant rates of electron transport, only the Fe/Cu version was active in O_2 reduction, whereas, the Fe version shows very low activity and seemed to undergo rapid degradation. The lack of reduction activity at the regime of slow electron transport in the absence of Cu was explained by the existence of two different reduction mechanisms for the Fe/Cu and Fe version (Fig. 9.3).[5] The hypothesized mechanisms suggested that the lack of activity and the lower stability of the Fe form may have arisen from oxidation of the organic ligand during the O_2 reduction. Also, this degradation occurred during the last

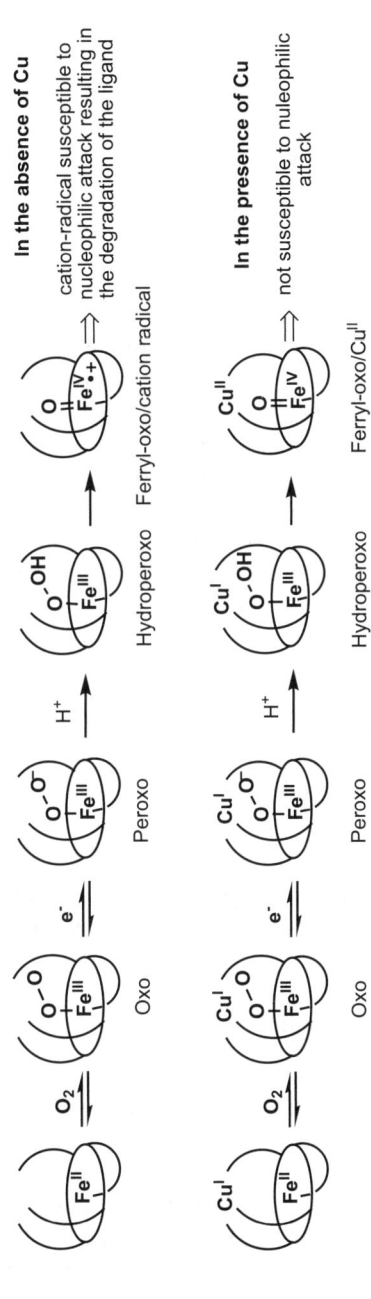

FIGURE 9.3. Proposed mechanism for the electrochemical reaction of **9 Fe^{II}** and **9 Fe^{II}/Cu^I** in the presence of O$_2$ in a regime of slow electron flux (see text).

step when hydroperoxo species formed in the absence of an electron source capable of fast electron delivery [i.e., from a nearby Cu(I)]; then, an unstable reactive ferryl/ cation radical formed and was followed by nucleophilic attack of the solvent and ligand degradation (Fig. 9.3). Is was concluded from these results that in CcO the Cu$_B$ may not participate directly in the O$_2$ bonding and cleavage, but it serves mainly as an electron storage site, which allows decoupling of the dioxygen-reduction rate from the rate of electron delivery from cytochrome c. This study was claimed to show that the Cu is essential to achieve the 4e$^-$ reduction of O$_2$ in the slow electron transport regime. But, of course, in the absence of any strong spectroscopic evidence or more detailed kinetic studies it is difficult to confirm or disprove the hypothesized mechanism(s).

The adaptation of picket-fence porphyrins to the synthesis of biomimetic models for the heme a_3/Cu$_B$ bimetallic center of CcO offered the opportunity to develop a series Fe/Cu models compatible with electrocatalytic studies. These sophisticated models were designed not only to reproduce the immediate coordination environment of both metals and the Cu$\cdots$Fe distance, but also to retain their structure integrity under the electrocatalytic turnover. The study of these complexes using the electrochemical approach has given new insight into the electrochemical behavior of Fe/Cu. However, the complicated synthesis required to prepare such models has limited the variations brought to the Cu coordination site (e.g., geometry, electronic properties, denticity of the Cu ligand), which explains the lack of more complete systematic studies concerning the influence of the copper coordination environment on the dioxygen-reduction chemistry. Reproducing the 4e$^-$ reduction of O$_2$ at physiological pH, potential, and temperature as with the enzyme have been one of the main goals of the electrochemical studies. Consequently, the use of the spectroscopic techniques to detect and/or characterize the reaction intermediates under electrocatalytic conditions remains one of the principal limitation of this method.

9.3. SYNTHETIC MODELS OF CYTOCHROME c OXIDASE BASED ON HEME/O$_2$/Cu ASSEMBLIES AND THE INTERMEDIATES DETECTION APPROACH

9.3.1. Discovery of the Heme/O$_2$/Cu Assemblies

In the early 1990s, Karlin and co-workers[43,44] reported a new approach to mimic the dioxygen-reduction chemistry occurring at the CcO active site. The authors showed that the reaction of O$_2$ with an equimolar mixture of [(TMPA)Cu^I(RCN)]$^+$ and (F$_8$)Fe^{II}, where TMPA = tris(2-pyridylmethyl)amine and F$_8$ = tetrakis(2,6-difluorophenyl)porphyrin], led exclusively to the heteronuclear oxo-bridged heme–copper complex [(F$_8$)Fe^{III}–(O^{2-})–Cu^{II}(TMPA)]$^+$ (**10 Fe^{III}–O–Cu^{II}**), in preference to the homonuclear μ-peroxo heme- or copper-only products. Further investigation of this reaction yielded to the detection of a heterobinuclear heme–peroxo–Cu complex as a low–temperature stable O$_2$ intermediate [(F$_8$)Fe^{III}–(O$_2^{2-}$)–Cu^{II}(TMPA)]$^+$ (**10 Fe^{III}–(O$_2^{2-}$)–Cu^{II}**),[45]

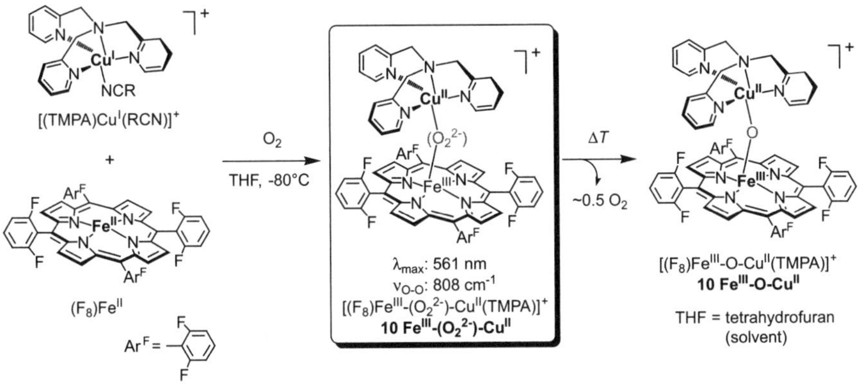

SCHEME 9.4

which thermally transforms to μ-oxo complex $\mathbf{10\,Fe^{III}-O-Cu^{II}}$, while releasing ~0.5 equiv of O_2 (Scheme 9.4).

The formulation of $\mathbf{10\,Fe^{III}-(O_2^{2-})-Cu^{II}}$ was supported by various spectroscopic methods. Dioxygen uptake measurements (through spectrophotometric titrations) and Matrix assisted laser desorption ionisation–time of flight–mass spectrometry (MALDI–TOF–MS) revealed that $\mathbf{10\,Fe^{III}-(O_2^{2-})-Cu^{II}}$ is an O_2 adduct with a stoichiometry of (TMPA)Cu:(F_8)Fe:$O_2 = 1{:}1{:}1$. Resonance Raman spectroscopy further supported the peroxo formulation, exhibiting a $\nu(O-O)$ stretching vibration at $808\,cm^{-1}$ ($\Delta^{18}O_2 = -46\,cm^{-1}$, $\Delta^{16/18}O_2 = -23\,cm^{-1}$). Solution magnetic moment measurements and ^1H NMR investigations indicated that $\mathbf{10\,Fe^{III}-(O_2^{2-})-Cu^{II}}$ has an $S = 2$ spin state ($\mu_{eff} = 5.1\,\mu_B$, $-40°C$), arising from the antiferromagnetic coupling between the high-spin Fe(III) and the Cu(II) centers. Mössbauer spectroscopy revealed that $\Delta E_Q = 1.14$ and $\delta = 0.57$ mm/s (4.2 K, zero field); thus, the high-spin ferric ion possesses an electron-rich peroxide ligand. Formation of $\mathbf{10\,Fe^{III}-(O_2^{2-})-Cu^{II}}$ was studied by stopped-flow UV–vis spectroscopy in the temperature range of -94 to $-75°C$ in acetone as solvent. It revealed that a heme–superoxide intermediate $(S)(F_8)Fe^{III}-(O_2^{2-})$ ($\lambda_{max} = 537$ nm, $S =$ solvent) was generated within the mixing time (~1 ms) prior to reaction with the copper complex and formation of $[(F_8)Fe^{III}-(O_2^{2-})-Cu^{II}(TMPA)]^+$ $(\mathbf{10\,Fe^{III}-(O_2^{2-})-Cu^{II}})$ ($\lambda_{max} = 556$ nm). This transformation of $(S)(F_8)Fe^{III}-(O_2^-)$ to $\mathbf{10\,Fe^{III}-(O_2^{2-})-Cu^{II}}$ could be described by a first-order rate constant with $\Delta H^{\ddagger} = 45 \pm 1$ kJ/mol, $\Delta S^{\ddagger} = -19 \pm 6$ J/mol K, and $k = 0.007\,s^{-1}$ (at $-90°C$).

9.3.2. Intermediates Detection Approach

The nature of this new generation of heme/O_2/Cu assemblies, where the Fe⋯Cu distance and integrity of the complex itself is dictated only by the interactions between the iron, oxygen ligand $[O_2, (O^{2-}),(OH^-)\cdots]$ and the copper, makes these complexes incompatible with the relatively harsh conditions of electrochemical studies described earlier in Section 9.2. On the other hand, heme/O_2/Cu assemblies strategy present a fast and relatively easy way to generate heme/oxygen/

copper adducts, and because the copper ligand is not covalently attached to the heme, this strategy theoretically, offers unlimited possibilities for heme–Cu combinations depending only on the choice of the copper ligand. Thus, an opportunity became available to develop systematic studies to examine the influence of the copper coordination environment on heme–copper dioxygen-reduction chemistry. These advantages, plus the fact that the spectroscopic characteristics and the reactivity of individual heme oxygen adducts and individual copper complexes are already quite well known, make the heme/O_2/Cu assemblies very promising candidates for an intermediate detection approach to study the mechanism of O_2 reduction in CcO active site using synthetic models. Complementary to the electrochemical studies, this approach is based on the generation of heme/oxygen/Cu adducts at low temperature and the study of their structure, spectroscopic properties, magnetic and electronic structure, and reactivity in order to understand the structure and nature of the intermediates present in the enzyme. We should also examine the plausibility of the proposed mechanisms for O_2 reduction at the CcO active site.

9.3.3. Monotethered Heme–Cu Complex Models

Another system associated with the intermediate detection approach is a group of binuclear complexes studied independently by the Karlin[46,47] and Naruta[48–50] research groups, in which a Cu–TMPA moiety is tethered to the heme via one covalently attached linker (**11 Fe^{III}–(O$_2^{2-}$)–Cu^{II}** to **15 Fe^{III}–(O$_2^{2-}$)–Cu^{II}**, Chart 9.2). These complexes present a compromise between the rigid complexes that are derivatives of picket-fence porphyrins and the heme/O_2/Cu assemblies. Karlin and co-workers have developed a binuclear a complex where the Cu–TMPA is covalently attached to the porphyrin through position 5 (**11 Fe^{II}/Cu^{I}**) or position 6 (**12 Fe^{II}/Cu^{I}**) of one pyridyl arm.[46,47] Both complexes react with O_2 and form peroxo intermediates with v(O–O) stretching vibration at 787 cm^{-1} ($\Delta^{18}O_2 = -43$ cm^{-1}) for **11 Fe^{III}–(O$_2^{2-}$)–Cu^{II}** and 809 cm^{-1} ($\Delta^{18}O_2 = -53$ cm^{-1}) for **12 Fe^{III}–(O$_2^{2-}$)–Cu^{II}**. The ^{1}H and ^{2}H NMR, as well as the EPR spectra of **12 Fe^{III}–(O$_2^{2-}$)–Cu^{II}**, indicated that there are high-spin ($S_{total} = 2$) heme–copper dioxygen adducts, where a high-spin Fe ($S = {}^5/_2$) and Cu ($S = {}^1/_2$) are antiferromagnetically coupled. Stopped-flow kinetic studies on the [(^{6}L)Fe^{II}Cu^{I}]$^+$ (**12 Fe^{II}/Cu^{I}**) reaction with O_2 revealed that a new short-lived intermediate ($\lambda_{max} = 538$ nm, thought to be a superoxo species [(^{6}L) Fe^{III}(O$_2^-$)$\cdots$Cu^{I}]$^+$ and/or a bis adduct [(^{6}L)Fe^{III}(O$_2^-$)$\cdots$Cu^{II}(O$_2^-$)]$^+$), plus the peroxo complex **12 Fe^{III}–(O$_2^{2-}$)–Cu^{II}** ($\lambda_{max} = 561$ nm), are independently produced within the mixing time ($\sim$1 ms) of the experiment. The ratio of these two species was dependent on temperature and O_2 concentration. The 538 nm intermediate was favored at low temperatures, and it converted to the peroxo complex **12 Fe^{III}–(O$_2^{2-}$)–Cu^{II}** in a first-order reaction (-94 to $-60°$C), where $\Delta H^{\ddagger} = 37.4 \pm 0.4$ kJ/mol and $\Delta S^{\ddagger} = -28.7 \pm 2.3$ J/mol K. However, ^{1}H NMR spectroscopy indicated that **11 Fe^{III}–(O$_2^{2-}$)–Cu^{II}** is a diamagnetic compound, and further spectrophotometric titration experiments revealed that the stoichiometry of the reaction was ^{5}LFe/Cu:$O_2 = 2:1$. The nature of **11 Fe^{III}–(O$_2^{2-}$)–Cu^{II}** is not understood, but the contrasting features between **11 Fe^{III}–(O$_2^{2-}$)–Cu^{II}** and

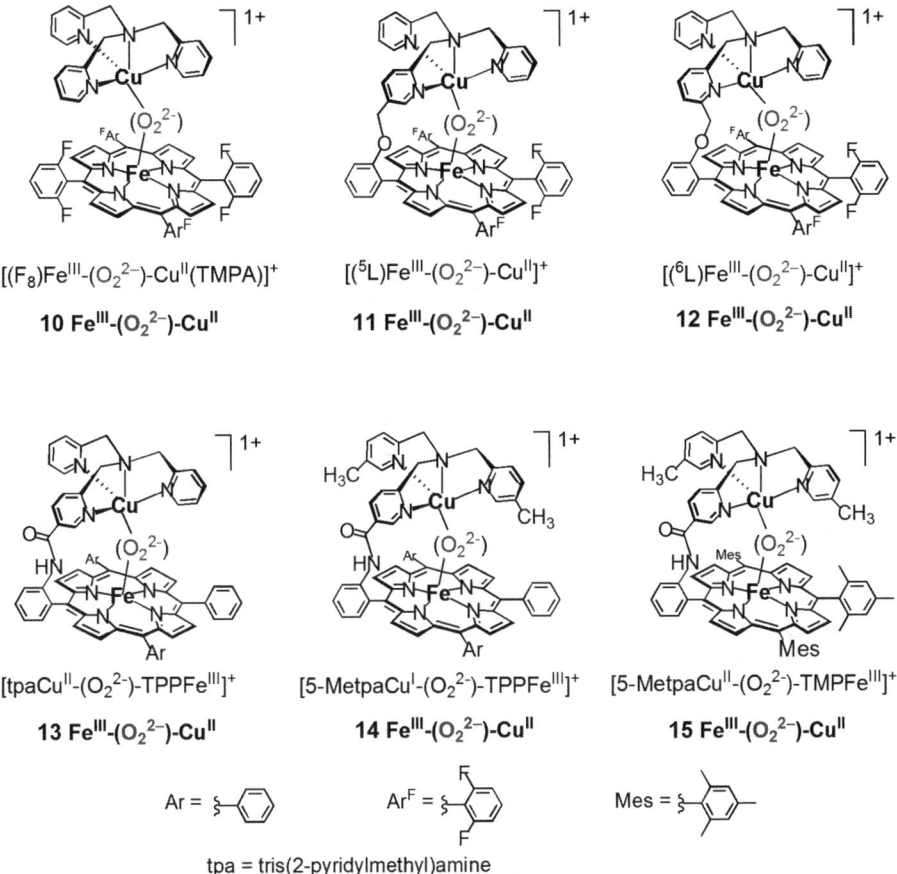

[(F$_8$)Fe^{III}-(O$_2^{2-}$)-Cu^{II}(TMPA)]$^+$

10 Fe^{III}-(O$_2^{2-}$)-Cu^{II}

[(^{5}L)Fe^{III}-(O$_2^{2-}$)-Cu^{II}]$^+$

11 Fe^{III}-(O$_2^{2-}$)-Cu^{II}

[(^{6}L)Fe^{III}-(O$_2^{2-}$)-Cu^{II}]$^+$

12 Fe^{III}-(O$_2^{2-}$)-Cu^{II}

[tpaCu^{II}-(O$_2^{2-}$)-TPPFe^{III}]$^+$

13 Fe^{III}-(O$_2^{2-}$)-Cu^{II}

[5-MetpaCu^I-(O$_2^{2-}$)-TPPFe^{III}]$^+$

14 Fe^{III}-(O$_2^{2-}$)-Cu^{II}

[5-MetpaCu^{II}-(O$_2^{2-}$)-TMPFe^{III}]$^+$

15 Fe^{III}-(O$_2^{2-}$)-Cu^{II}

Ar =

Ar^F =

Mes =

tpa = tris(2-pyridylmethyl)amine

5-metpa = 5-(2'bis((5''-methyl-2''-pyridylmethyl)aminomethyl) pyridine

CHART 9.2

12 Fe^{III}−(O$_2^{2-}$)−Cu^{II} are believed to originate from the differences in their ligand architecture.[51]

Using an approach very similar to that of Karlin and co-workers,[46,47] independently, Naruta and co-workers[48–50] described the synthesis and characterization of three binuclear complexes, where a Cu−TMPA was attached from the 5-position of the pyridine to the heme through an amide linkage. All three reduced heme–copper complexes reacted with O$_2$ and formed μ-peroxo complexes (**13 Fe^{III}−(O$_2^{2-}$)−Cu^{II}**, **14 Fe^{III}−(O$_2^{2-}$)−Cu^{II}**, **15 Fe^{III}−(O$_2^{2-}$)−Cu^{II}**, Chart 9.2) with spectroscopic features similar to those displayed by the μ-peroxo analogues described by Karlin and co-workers.[46,47] The characterization of the X-ray structure of **15 Fe^{III}−(O$_2^{2-}$)−Cu^{II}** represented a remarkable achievement in heme/Cu/O$_2$ chemistry (Fig. 9.4).[50] Resonance Raman spectroscopy of this peroxo complex showed a v(O−O) stretching vibration at 790 cm^{-1} (Δ^{18}O$_2$ = − 44 cm^{-1}) and Mössbauer spectroscopy revealed the existence of a high-spin Fe(III) ion [$\Delta E_Q = 1.17$ mm s^{-1} ($\delta = 0.56$ mm s^{-1})]. However, magnetic data suggested an $S = 2$ spin state with strong antiferomagnetic

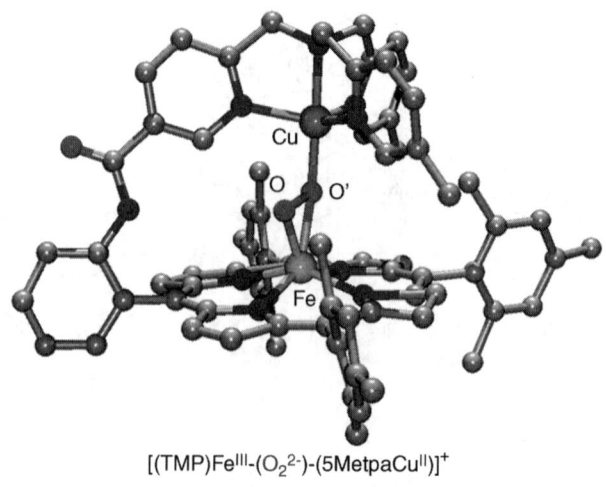

$[(TMP)Fe^{III}-(O_2^{2-})-(5MetpaCu^{II})]^+$

$\nu(O-O) = 790\ cm^{-1}$

FIGURE 9.4. X-ray crystal structure of $[(TMP)Fe^{III}-(O_2^{2-})-(5MetpaCu^{II})]^+$ (**15 Fe^{III}–(O$_2^{2-}$)–Cu^{II}**). TMP = 10,15,10-tris (2,4,6-trimethylphenyl).

coupling between high-spin Fe(III) and Cu(II) atoms through the bridged peroxide ligand. The X-ray structure showed that both oxygen atoms of the peroxide ligand were ligated to the Fe (Fe–O = 1.89 Å, Fe–O' = 2.03 Å), whereas only one oxygen atom was bound to Cu (Cu–O = 1.92 Å, Cu$\cdots$O = 2.66 Å), thus displaying an asymmetric μ-η^2:η^1 coordination mode with a Fe$\cdots$Cu distance of 3.92 Å and O–O bond length of 1.46 Å.

9.3.4. Heme/Cu Models Containing Tridentate Copper Ligands

Heme–peroxo–copper complexes also form with tridentate copper ligands (Chart 9.3).[52–54] These were generated in the same manner as their analogues with a tetradentate copper ligand, by the oxygenation of reduced component mononuclear

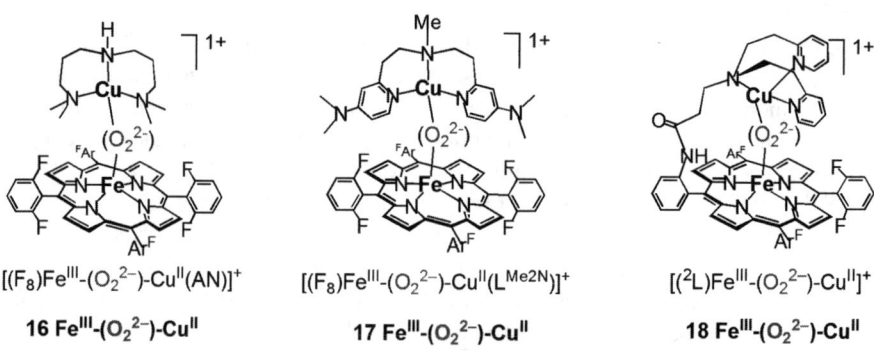

$[(F_8)Fe^{III}-(O_2^{2-})-Cu^{II}(AN)]^+$

16 Fe^{III}-(O$_2^{2-}$)-Cu^{II}

$[(F_8)Fe^{III}-(O_2^{2-})-Cu^{II}(L^{Me2N})]^+$

17 Fe^{III}-(O$_2^{2-}$)-Cu^{II}

$[(^2L)Fe^{III}-(O_2^{2-})-Cu^{II}]^+$

18 Fe^{III}-(O$_2^{2-}$)-Cu^{II}

AN = Bis(3-dimethylaminopropyl)amine

CHART 9.3

TABLE 9.1. The O–O Bond Stretching Values [v(O–O)] of Dioxygen Adducts of a Heme–Copper Model Complexe

Complex	v(O–O) ($\Delta^{18}O_2$)(cm^{-1})	Reference
Tetradentate Copper Ligand		
[(F$_8$)Fe^{III}–(O$_2^{2-}$)–Cu^{II}(TMPA)]$^+$ (**10 Fe^{III}–(O$_2^{2-}$)–Cu^{II}**)	810 (−46)	45
[(^{5}L)Fe^{III}–(O$_2^{2-}$)–Cu^{II}]$^+$ (**11 Fe^{III}–(O$_2^{2-}$)–Cu^{II}**)	809 (−53)	46
[(^{6}L)Fe^{III}–(O$_2^{2-}$)–Cu^{II}]$^+$ (**12 Fe^{III}–(O$_2^{2-}$)–Cu^{II}**)	787 (−43)	47
[tpaCu^{II}–(O$_2^{2-}$)–TPPFe^{III}]$^+$ (**13 Fe^{III}–(O$_2^{2-}$)–Cu^{II}**)	803 (−44)	48
[5-MetpaCu^{II}–(O$_2^{2-}$)–TPPFe^{III}]$^+$ (**14 Fe^{III}–(O$_2^{2-}$)–Cu^{II}**)	793 (−42)	49
[5-MetpaCu^{II}–(O$_2^{2-}$)–TMPFe^{III}]$^+$ (**15 Fe^{III}–(O$_2^{2-}$)–Cu^{II}**)	790 (−44)	50
Tridentate Copper Ligand		
[(F$_8$)Fe^{III}–(O$_2^{2-}$)–Cu^{II}(AN)]$^+$ (**16 Fe^{III}–(O$_2^{2-}$)–Cu^{II}**)	756 (−43)	52
[(F$_8$)Fe^{III}–(O$_2^{2-}$)–Cu^{II}(L^{Me2N})]$^+$ (**17 Fe^{III}–(O$_2^{2-}$)–Cu^{II}**)	752 (−42), 767 (−40)	53
[(^{2}L)Fe^{III}–(O$_2^{2-}$)–Cu^{II}]$^+$ (**18 Fe^{III}–(O$_2^{2-}$)–Cu^{II}**)	747 (−46)	54

heme and copper complexes (**16 Fe^{III}–(O$_2^{2-}$)–Cu^{II}** and **17 Fe^{III}–(O$_2^{2-}$)–Cu^{II}**) or binuclear Fe^{II}–Cu^{I} complexes (**18 Fe^{III}–(O$_2^{2-}$)–Cu^{II}**). The peroxo complexes are also $S = 2$ spin systems, with high-spin Fe$^{(III)}$ coupled to Cu$^{(II)}$. The peroxo formulation was confirmed by NMR and rR spectroscopy.

Stopped-flow kinetic studies of the chemistry of **17 Fe^{III}–(O$_2^{2-}$)–Cu^{II}** and **18 Fe^{III}–(O$_2^{2-}$)–Cu^{II}** revealed that, as with the analogue with tetradentate copper ligand, an initial heme–superoxo (Fe^{III}–(O$_2^-$)$\cdots$Cu^{I}) transient similar to the "Oxy" **A** state described in the enzyme proposed mechanism, precedes the formation of the bridged heme–peroxo–Cu complex.[55,56] Another interesting find is that the values of O–O stretching in all peroxo complexes with tridentate ligands are considerably reduced [v(O–O) < 760 cm^{-1}] from those observed with tetradentate Cu chelates [v(O–O) > 790 cm^{-1}] (Table 9.1). This finding indicated a weakened O–O peroxo bond, undoubtedly due to differing heme–peroxo–copper core structures.

9.3.5. Tetra versus Tridentate Copper Ligands

In Cu/O$_2$ chemistry, it is well established that even subtle differences in ligand–denticity can dramatically change the nature of the copper–dioxygen adduct and its reactivity toward substrates.[57,58] Tetradentate ligands (e.g., TMPA) induce formation of

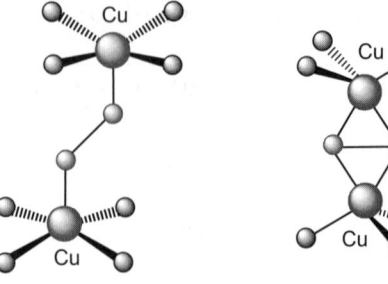

Tetradentate Cu chelate   Tridentate Cu chelate

$v(\text{O-O}) = 805\text{–}830\ \text{cm}^{-1}$ $v(\text{O-O}) = 716\text{–}760\ \text{cm}^{-1}$

$\text{Cu}\cdots\text{Cu} = \sim 4.4\ \text{Å}$ $\text{Cu}\cdots\text{Cu} = \sim 3.5\text{–}3.6\ \text{Å}$

CHART 9.4

end-on ($\mu\text{-}\eta^{1}{:}\eta^{1}$) peroxo dicopper(II) structures (Chart 9.4), whereas tridentate ligands generate side-on ($\mu\text{-}\eta^{2}{:}\eta^{2}$) peroxo dicopper(II) species.[59] The former possess relatively average $v(\text{O}-\text{O})$ values, typically $>800\ \text{cm}^{-1}$ (Chart 9.4). However, the latter side-on complexes have reduced ($< 760\ \text{cm}^{-1}$) $v(\text{O}-\text{O})$ values (Chart 9.4), ascribed to back-bonding from copper to the peroxo antibonding σ^{*} orbital that considerably weakens the O–O bond.[60–62]

Recent studies by Karlin and co-workers[63] had for their goal to compare the molecular and electronic structure of high-spin heme–peroxo–Cu complexes containing N_4 (**10 $\mathbf{Fe^{III}}$–$\mathbf{(O_2^{2-})}$–$\mathbf{Cu^{II}}$**) tetradentate or N_3 (**16 $\mathbf{Fe^{III}}$–$\mathbf{(O_2^{2-})}$–$\mathbf{Cu^{II}}$**) tridentate copper ligands. In addition to the difference in $v(\text{O}-\text{O})$ values (**10 $\mathbf{Fe^{III}}$–$\mathbf{(O_2^{2-})}$–$\mathbf{Cu^{II}}$** : $810\ \text{cm}^{-1}$, **16 $\mathbf{Fe^{III}}$–$\mathbf{(O_2^{2-})}$–$\mathbf{Cu^{II}}$** : $756\ \text{cm}^{-1}$) between the two complexes, extended X-ray absorption fine structure (EXAFS) spectroscopic investigations showed that the Fe$\cdots$Cu distance was greater in **10 $\mathbf{Fe^{III}}$–$\mathbf{(O_2^{2-})}$–$\mathbf{Cu^{II}}$** (Fe$\cdots$Cu $\sim3.72\ \text{Å}$) compared to that in **16 $\mathbf{Fe^{III}}$–$\mathbf{(O_2^{2-})}$–$\mathbf{Cu^{II}}$** (Fe$\cdots$Cu $\sim3.63\ \text{Å}$), possibly reflecting a difference in the structure of the peroxo bridge between the two complexes. More advanced investigations coupling rR and EXAFS spectroscopic data to DFT calculations allowed the authors to propose different structures for **10 $\mathbf{Fe^{III}}$–$\mathbf{(O_2^{2-})}$–$\mathbf{Cu^{II}}$** and **16 $\mathbf{Fe^{III}}$–$\mathbf{(O_2^{2-})}$–$\mathbf{Cu^{II}}$** (Fig. 9.5),[63] and establish $\mu\text{-}(O_2^{2-})$ side-on binding to the Fe$^{(III)}$ and end-on to Cu$^{(II)}$ ($\mu\text{-}\eta^{2}{:}\eta^{1}$) binding for the complex **10 $\mathbf{Fe^{III}}$–$\mathbf{(O_2^{2-})}$–$\mathbf{Cu^{II}}$**, but side-on/side-on ($\mu\text{-}\eta^{2}{:}\eta^{2}$) μ-peroxo coordination for the complex **16 $\mathbf{Fe^{III}}$–$\mathbf{(O_2^{2-})}$–$\mathbf{Cu^{II}}$** (Fig. 9.5). Separate reactivity studies of high-spin heme–peroxo–Cu tridentate versus a tetradentate copper ligand demonstrated that the two types of complexes also react differently in the presence of certain substrates (CO, PPh$_3$, or axial base ligand).[52,63]

9.3.6. Low-Spin Heme–Peroxo–Copper Assembly

Another major difference between the high-spin heme–peroxo–Cu containing tri- or tetradentate copper ligand was that in the presence of an axial base, such as 1,5-dicyclohexylimidazole (DCHIm) or 4-(dimethylamino)-pyridine (DMAP),

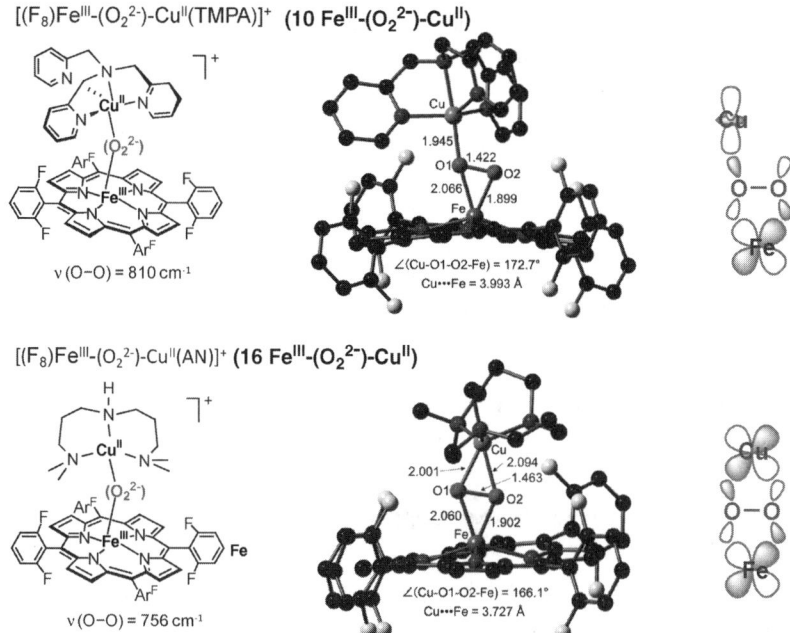

FIGURE 9.5. Unrestricted DFT (BS = broken symmetry, $S_T = 2$) optimized structures of (center) and schematic representation of the major orbital contributions in **10 Fe^{III}–(O$_2^{2-}$)–Cu^{II}** and **16 Fe^{III}–(O$_2^{2-}$)–Cu^{II}** (center). (Adapted from Ref. 63.) (See color insert.)

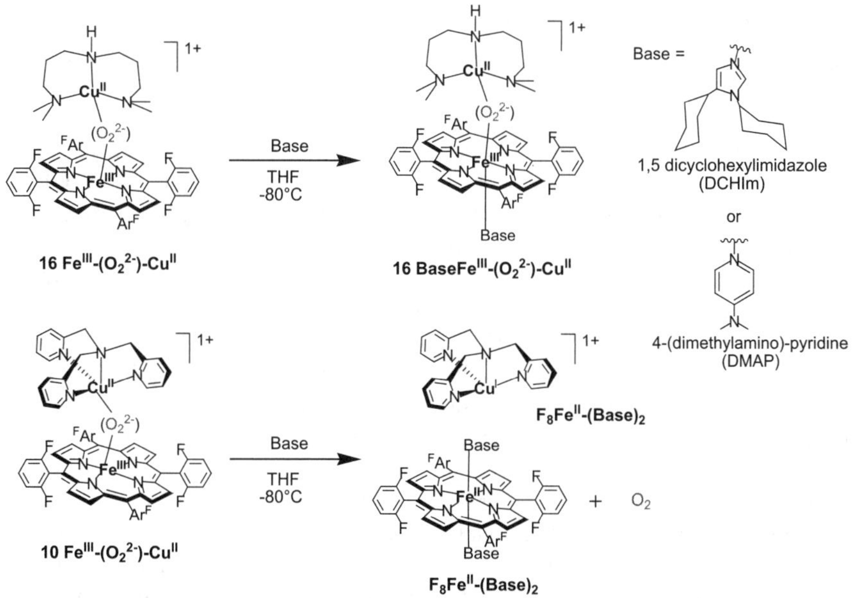

SCHEME 9.5

10 Fe^{III}−(O$_2^{2-}$)−Cu^{II} released O$_2$ and formed of a mixture of [Cu^{I}(TMPA)]$^+$ and a six-coordinate low-spin compound (F$_8$)Fe^{II}−(Base)$_2$ (Scheme 9.5), while **16 Fe^{III}−(O$_2^{2-}$)−Cu^{II}** is transformed to a new low-temperature stable species. This compound has been formulated as a low-spin species [base(F$_8$)Fe^{III}−(O$_2^{2-}$)−Cu^{II}(AN)]$^+$ (**16 baseFe^{III}−(O$_2^{2-}$)−Cu^{II}**), with the complex's diamagnetism confirmed by its ^{2}H and ^{1}H NMR spectra (pyrrole signal ∼8.3 ppm). This behavior thus suggested the presence of antiferromagnetic coupling between the low-spin ($S = {}^1/_2$) six-coordinate Fe(III) and the d^9 Cu(II) (also $S = {}^1/_2$) centers, presumably through a peroxo bridge. The same diamagnetic behavior was observed with low-spin heme–peroxo–copper complexes containing appended axial base, as described by the Naruta[64] and Collman research groups.[35–37] The peroxidic nature of **16 baseFe^{III}−(O$_2^{2-}$)−Cu^{II}**, that is, that the complex possesses a dioxygen-derived moiety with a still intact O−O bond, was also suggested by the reactivity of this low-spin complex with hydrochloric acid. This reaction leds to the expected release of H$_2$O$_2$ in good yield.[63]

9.3.7. Reactivity of Heme–Peroxo–Copper with Phenols as Hydrogen Atom (H$^+$ + e$^-$) Donor

In the proposed mechanism of O$_2$ reduction in the C*c*O active site, it has been postulated that after the formation of a peroxo transient (Scheme 9.1), an active-site tyrosine, found crystallographically to be covalently tethered to a copper-bound histidine, acts as a net hydrogen-atom (H$^+$ + e$^-$) donor. This finding induces the crucial O−O bond cleavage event and the formation of a ferryl oxo (Fe^{IV}=O) intermediate (**P$_M$**) (see Section 9.4). In order to examine this hypothesis, Collman et al.[65] studied the reactivity of a low-spin heme–superoxo–Cu complex [i.e., a LSFe^{III}(O$_2^-$)···Cu^{I} moiety] derivative of a picket-fence porphyrin with an appended axial base, with a phenol as a hydrogen-atom donor. The study revealed that hydrogen-atom abstraction from the phenol does occur leading to the formation of an EPR detectable phenoxyl radical (Scheme 9.6). The authors suggested that this hydrogen-atom abstraction was accompanied by the O−O bond cleavage and the formation of a ferryl oxo intermediate and Cu^{II}−OH, as proposed for the enzyme mechanism. More interestingly, using the heme/O$_2$/Cu assembly approach, Karlin and co-workers[66] recently reported that in the absence of an axial base, high-spin heme–peroxo–Cu complexes, such as **10 Fe^{III}−(O$_2^{2-}$)− Cu^{II}** or **16 Fe^{III}−(O$_2^{2-}$)−Cu^{II}**, do not react with phenols. However, when an exogenous axial base was added to **16 Fe^{III}−(O$_2^{2-}$)−Cu^{II}** to form the low-spin heme–peroxo–copper complex [**16 baseFe^{III}−(O$_2^{2-}$)−Cu^{II}**), the latter reacted with phenols performing an hydrogen-atom abstraction (Scheme 9.6). Evidence for this reactivity comes from the distinctively strong EPR signal ($g \sim 2.0$) of the phenoxyl radical (when 2,4,6-tri-*tert*-butylphenol or 2,6-di-*tert*-buyl-4-methoxyphenol were used as substrate) or from gas chromatography–mass spectrometry (GC–MS) detection of the phenol–phenol coupling product (when 2,4-*t*-Bu$_2$-phenol was used as substrate). To confirm the formation of a ferryl oxo intermediate as a product of the O−O bond cleavage, taking advantage of the adjustable nature of

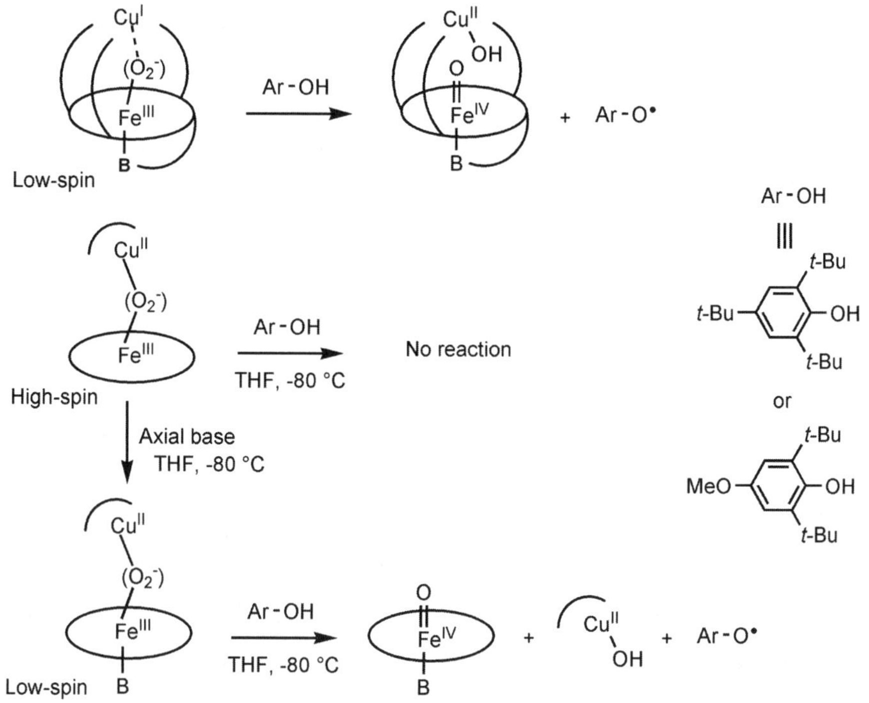

SCHEME 9.6

the heme/oxygen/Cu assemblies, the authors compared the spectroscopic features (UV–vis : Q band $= 540$) and reactivity of this intermediate with an authentic ferryl oxo complex generated separately by the reaction of $[(F_8)Fe^{III}]_2(O_2^{2-})$ with an axial base (e.g., DCHIm) in a noncoordinating solvent.[63,66]

9.4. HIS-TYR CROSS-LINK AT THE CYTOCHROME *c* OXIDASE ACTIVE SITE HEME a_3/Cu$_B$ CENTER

9.4.1. Characterization of His-Tyr Cross-Link At the Heme a_3/Cu$_B$

High-resolution X-ray crystallographic structural characterization of C*c*O has revealed that at the site for the heme a_3/Cu$_B$ binuclear center, where O_2 binding, O—O cleavage, and coupling to membrane proton translocation occurs, is found the unique post-translationally modified covalent cross-link between $C_{\varepsilon 2}$ of Tyr244 (in bovine heart enzyme nomenclature) and $N_{\varepsilon 2}$ of the His240 imidazole group[3]; the latter is one of the endogenous ligands of Cu$_B$. Further evidence for the His-Tyr cross-link have been provided by biochemical studies of C*c*O subunits using electrospray ionization mass spectroscometry (ESI–MS), when a decrease of 2 mass units from the theoretical value (assuming a linear normal peptide) is then explained by His-Tyr N−C covalent bond formation,[67] as two hydrogen atoms are removed. Spectroscopic investigations were also carried out to identify the vibrational features of the cross-linked His-Tyr group. Combining Fourier transform infrared (FTIR) spectroscopy with electrochemical methods or CO photolysis,[68,69] spectroscopic signatures of bi-ring structure (1448, 1550 cm^{-1}) in the covalently cross-linked His-Tyr, have been identified through Tyr isotopic labeling and study of the mutant in which the His-Tyr moiety is absent or modified.[69]

9.4.2. Role of the His-Tyr Cross-Link in the Dioxygen-Reduction Mechanism by CcO

Since its discovery, there have been several suggestions for the role of the His-Tyr cross-link in heme–copper oxidases.[11,19,70–73] Initially, the generation of the cross-link was proposed to have a structural function or modulate one or more of the properties of either residue.[3] Later, the His-Tyr moiety, as mentioned in the Introduction (Section 9.1), was proposed to provide an electron and a proton during the dioxygen-reduction cycle.[19–21]

In the catalytic mechanism of dioxygen-reduction, the cleavage of the O−O bond results in the formation of the intermediate designated as $\mathbf{P_M}$, where the heme a_3 iron is in a ferryl oxo (Fe^{IV}=O) state (Scheme 9.7). Three of the four electrons needed to reach this state are clearly provided by the heme a_3 and Cu$_B$ (where Fe(II) converts to Fe(IV) and Cu$_B$ goes from Cu(I) to Cu(II)). The one electron remaining has been proposed to be provided by the cross-linked His–Tyr moiety. Consequently, the formation of the intermediate $\mathbf{P_M}$ is proposed to be accompanied by the formation of a His-Tyr radical. However, EPR signals that might be expected have not been observed. If a phenoxyl radical was present, the absence of its EPR signal could be attributed to spin coupling between the $S = \frac{1}{2}$ Cu$_B^{2+}$ and the $S = \frac{1}{2}$ tyrosyl radical. One suggestion[20,74] made is that spin coupling might serve to delocalize the oxidizing equivalent so that Cu$_B$ assumes a partial 3+ character, leaving a tyrosinate anion.

A recent report indicates a radical EPR signal can be detected (at 80 K, g_{iso} ~ 2.0055) from a H$_2$O$_2$ treated (pH 6.0) fully oxidized CcO. Although the yield of a

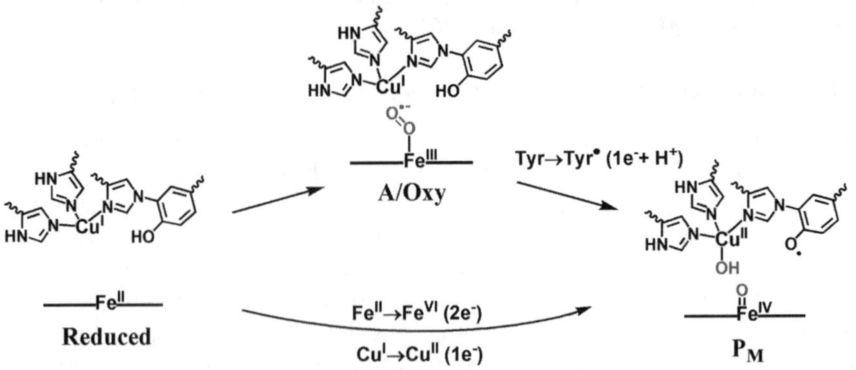

SCHEME 9.7

radical signal is low (20%), it was assigned as a His–Tyr radical, following perturbation of the EPR signal resulting from selective deuteration of the tyrosine.[75] The researchers suggested that the lower yield of radical may represent a partial uncoupling of the Cu_B/His-Tyr moiety.[19,75] Subsequent work showed that this signal may also arise from a tryptophan cation radical[76–79] or another unmodified tyrosyl radical.[80–82] Budiman et al.[81] proposed a radical migration pathway from the cross-linked tyrosine radical to Tyr129 (bovine heart numbering).

Infrared (IR) spectroscopic studies on the *R. sphaeroides* CcO dioxygen reaction demonstrated the presence of two different intermediates (P_M and P_R), where the heme a_3 is in the ferryl oxo state and the Cu_B is in the Cu^{II}–OH state.[82] Optical spectroscopic (UV–vis and IR) differences for these two intermediates have been recently elucidated and were determined to arise from a different number of redox states of the binuclear center, that is, proposed as $[Fe_{a3}^{IV}=O\ Cu_B^{II}-OH-(TyrO^\bullet)]$ for P_M with a neutral tyrosyl radical ($1479\ cm^{-1}$) and $[Fe_{a3}^{IV}=O\ Cu_B^{II}-OH-(TyrO^-)]$ for P_R with a tyrosinate moiety (1455 and $1515\ cm^{-1}$).[77–84]

These results suggest that the cross-linked tyrosine may play a role in O_2 reduction to water. To define its functional significance, several research groups have synthesized simple organic cofactor mimics, as well as copper-containing chelates or heme/Cu models that bear covalently linked imidazole–phenol structures. The major goal in all cases is to produce synthetic mimics of the cross-linked His-Tyr structure and to elucidate their physiochemical properties. It is hoped that such investigations would pave the way for an in-depth understanding of the function and benefit of this novel cofactor in the O_2 reduction event and perhaps its coupling to proton translocation.

9.4.3. Simple Organic Cofactor Models Containing a Mimic of the His-Tyr Cross-Link

Several synthetic methodologies have been considered to construct the N-substituted *o*-hydroxyphenyl–imidazole as a model for the cross-linked histidine-tyrosine. Early syntheses using nucleophilic aromatic substitution,[85–88] Ullmann-type coupling[89,90] or Buchwald coupling[91] have shown major limitations concerning the introduction of an appropriate biomimetic substituent (i.e., 4-alkylimidazole or 4-alkylphenol) because of the high temperature or a specific type of substitution (electron withdrawing) needed to perform the imidazole–phenol coupling.

More efficient methods were established later for generating *N*-arylimidazoles, via coupling of arylboronic acid and aryllead reagents with imidazoles (Reaction 1, Scheme 9.8); these Cu(II) promoted reactions occur under relatively mild reaction

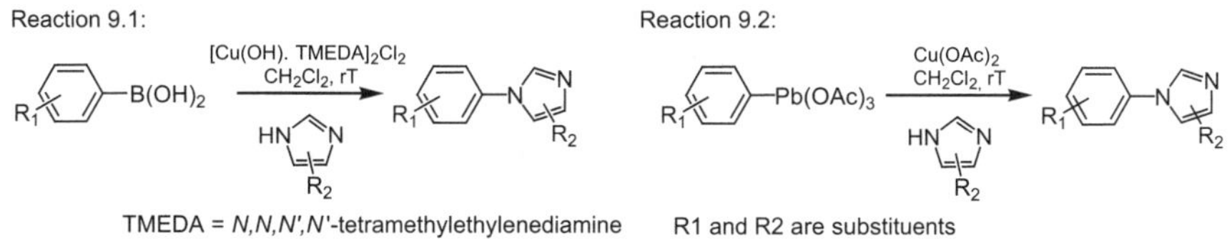

Reaction 9.1:   Reaction 9.2:

TMEDA = *N,N,N′,N′*-tetramethylethylenediamine   R1 and R2 are substituents

SCHEME 9.8

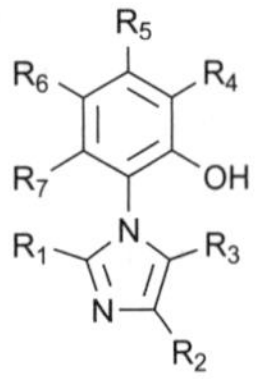

19 = R_1-R_7 = H

20 = R_1 = CO_2Me; R_2-R_7 = H

21 = R_2 = CO_2Me; R_1, R_2-R_7 = H

22 = R_3 = CO_2Me; R_1, R_2, R_4-R_7 = H

23 = R_6 = Me; R_1-R_5, R_7 = H

24 = R_2 = $CH_2C(NHAc)(CO_2Me)$;

    R_1, R_3-R_7 = H

25 = R_2, R_6 = Me; R_1, R_3-R_5, R_7 = H

26 = R_1, R_6 = Me; R_2-R_5, R_7 = H

27 = R_1-R_3 = D; R_6=CD_3; R_4,R_5, R_7 = H

28 = R_1-R_3 = H; R_6=CD_3; R_4,R_5, R_7 = D

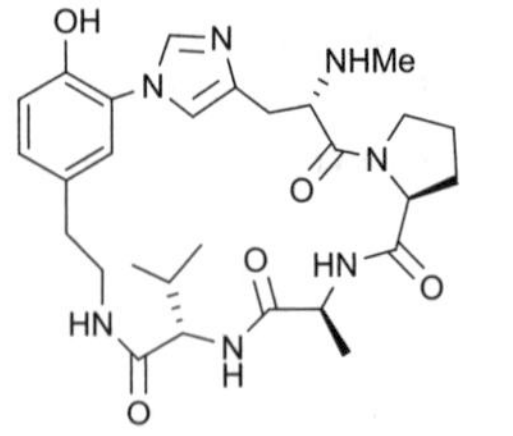

29

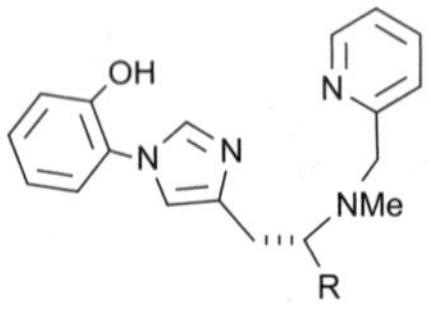

30 = R = CH_2OMe

31 = R = CO_2Me

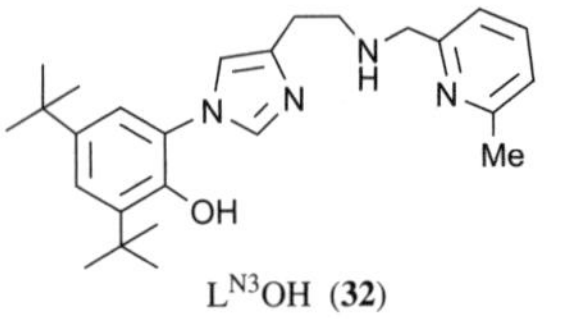

$L^{N3}OH$ (32)

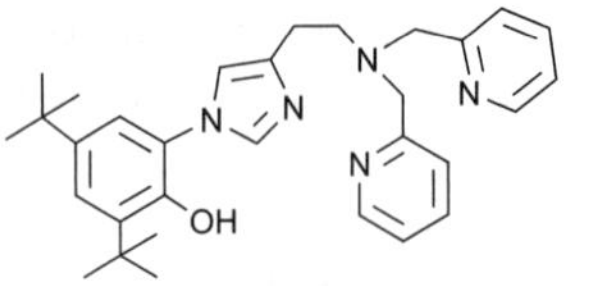

$L^{N4}OH$ (33)

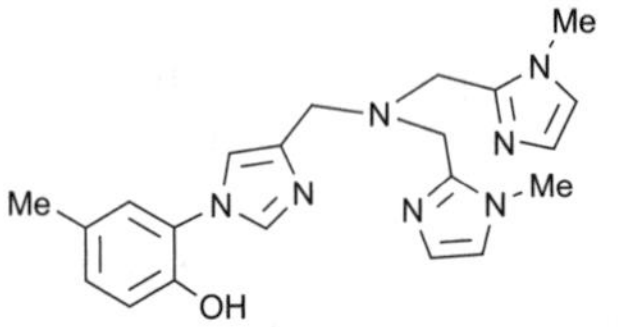

BIAIP (34)

CHART 9.5

conditions.[92–94] However, this procedure has been reported to afford mixtures of regioisomers when the coupling is performed on 4-substituted imidazoles.[91,93] The regioselectivity problem was resolved after Elliott et al.[94] reported the coupling of imidazoles with arylleadtriacetate in the presence of catalytic $Cu(OAc)_2$ (Reaction 9.2, Scheme 9.8), resulting in the exclusive formation of 1,4-disubstituted imidazoles. Arylleadtriacetate is usually prepared either by direct plumbation of aryl halides or through the intermediacy of aryltin compounds.[95] A variety of simple organic models containing cross-linked imidazole–phenol groups have been synthesized[96–99] (**19–28**, Chart 9.5), as well as more sophisticated peptides and ligands for the copper (**29–34**, Chart 9.5)[100–103] using either the arylboronic acid or the arylleadtriacetate method.

One of the principal goals in developing models of the cross-linked His-Tyr is to establish its properties and compare those with corresponding properties of individual imidazole and phenol components. A number of physical methods have been employed, including electrochemical and FTIR, UV–resonance Raman UV–rR, and EPR spectroscopies, to probe the effect of cross-link on the propriety of the imidazole–phenol models with an objective to extend the information obtained to understand the mechanistic role of the cross-linked His-Tyr cofactor in enzyme catalysis. For the organic His-Tyr models, that is, in the absence of copper ion, measurement of imidazole and phenol pK_a values and identification of spectroscopic features for the histidine-tyrosyl radical are of interest.[68,69,96–99]

Significantly reduced values for the phenol and imidazole pK_a in the cross-linked models (**19–24**, Chart 9.5) were observed as determined from spectrophotometric and titrimetric methods. The increased acidity, apparently due to the cross-link, at the imidazole and phenol was attributed to an increased π-delocalization through the two aromatic rings. The UV–rR studies on the cross-linked imidazole–phenol models **19**, **23**, and **26–28** (Chart 9.5) and their isotopically substituted (^{18}O) analogues have demonstrated that some imidazole vibrations are resonance enhanced upon excitation of phenol π–π* transitions, whereas they are absent for corresponding equimolar mixtures of imidazole and *p*-cresol.[98] This result indicates delocalization of π-electrons between the imidazole and phenol.

The EPR investigation of radical species generated by UV photolysis within several cross-link models (**23–25**, **30**, **31**, **34**, Chart 9.5) revealed that the spin distribution on the phenoxyl radical is only modestly perturbed by the imidazole cross-link. However, time-resolved optical absorption studies have shown that the imidazole cross-link causes a substantial red shift of the radical's electronic spectrum.[97] Those authors suggested that this property may have important implications for ET pathways in proteins.[104–106] As it may pertain to heme–copper oxidases (e.g., C*c*O), other researchers[96–98] suggest that the His–Tyr covalent bond may enable enhanced migration of an electron between the two rings.

9.4.4. Copper Complex Models Possessing Cross-Linked Imidazole-Phenol Ligands

Karlin and co-workers[100] described the first example of Cu(I) and Cu(II) possessing an cross-linked imidazole–phenol with either a tridentate **32Cu** or tetradentate **33Cu** pyridylamine-containing chelate (Chart 9.6). The redox properties and O$_2$ reactivity for these complexes are of general interest, because such information may eventually provide insights into the O$_2$ reduction process occurring at the enzyme Cu$_B$ site. Electrochemical comparison between protected (anisole analogues) and the non-protected phenol version of these complexes or closely related imidazole complex

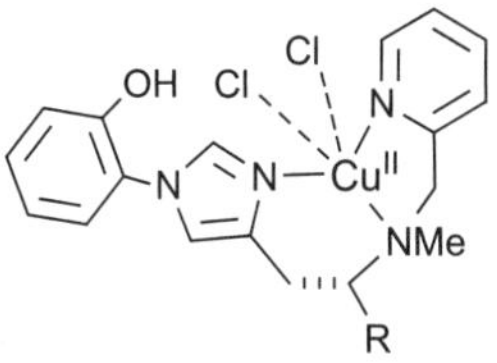

30Cu = R = CH$_2$OMe
31Cu = R = CO$_2$Me

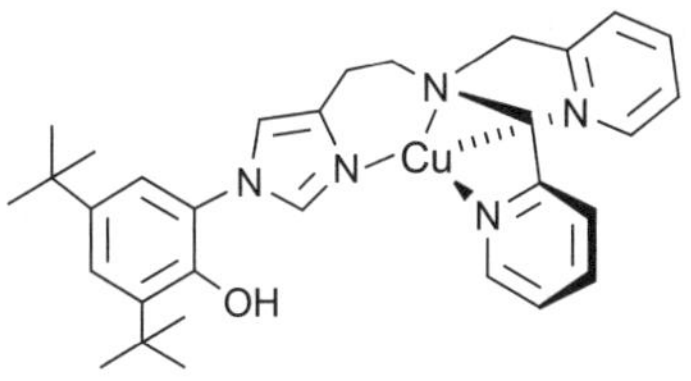

[Cu$^{(I \text{ or } II)}$L^{N4}OH] (**33Cu**)

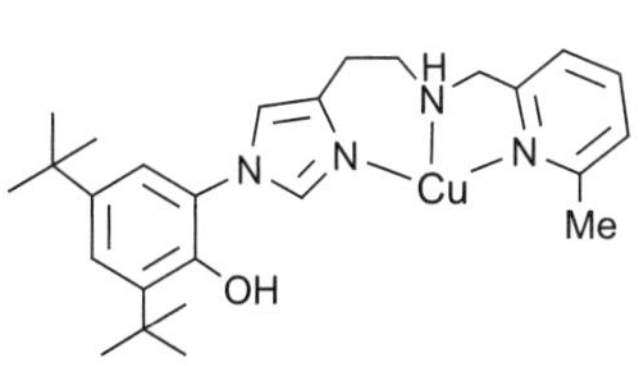

Cu$^{(I \text{ or } II)}$L^{N3}OH (**32Cu**)

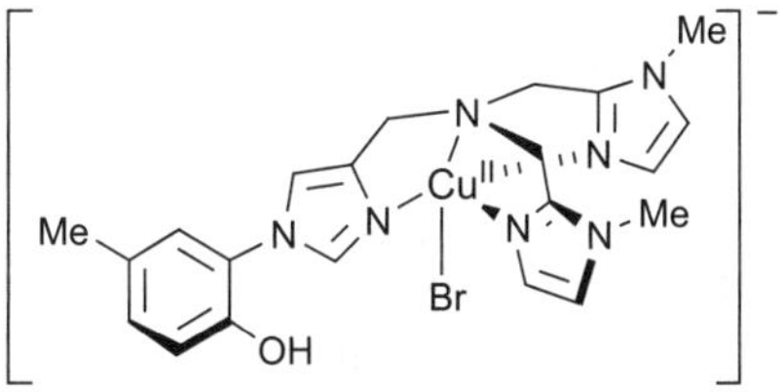

[Cu^{II}BIAIPBr]$^+$ (**34Cu**)

CHART 9.6

(without a phenol moiety) $[(L)Cu^I]^+$, $L = [2$-$(1H$-imidazol-4-yl)-ethyl]-bis-(2-pyr-idylmethyl)amine have shown that the Cu(II)/Cu(I) redox potential is unaffected by the presence of the imidazole–phenol cross-link. Dioxygen reactivity studies of the Cu(I) complex versions were carried out to see if phenoxyl radicals may form. Both tetra and tridentate complexes react with O_2 at low temperature to form spectro-scopically (UV–vis) detected peroxo–dicopper(II) $Cu–O_2–Cu$ intermediates, apparently preorganized for subsequent deprotonation of the cross-linked imidazole–phenol moiety and subsequent phenolate dimer product formation. It was suggested that the phenol moiety acted as a proton donor toward the basic peroxide intermediate species formed.[100] No evidence for formation of a histidine-tyrosyl radical was observed.

In their investigation to elucidate a spectroscopic basis for the putative Tyr^{224} radical and explore the phisycochemical properties of the Tyr-His-Cu_B moiety, the Naruta[101] (**34Cu**, Chart 9.6) and Einarsdòttir[101,103] (**30Cu**, **31Cu**, Chart 9.6) research groups separately developed Cu(II) complexes containing cross-linked imidazole–phenol. To investigate the presence of a phenoxyl radical, the authors combined time-resolved absorbance or EPR spectroscopy with UV light irradiation, where the latter causes photoelectron emission leading to generation of phenoxyl radical. Ultraviolet photolysis of the copper free ligands or the corresponding Zn complexes, generate the expected narrow signal ($g = 2.00$), attributed to the phenoxyl radical. However, EPR spectra recorded before and after UV photolysis of the Cu complexes showed only a typical signal characteristic of oxidized copper, suggesting that the copper is not spin coupled to the phenoxyl radical. An EPR signal from the phenoxyl radical was not observed in the copper complex. This absence of a phenoxyl radical EPR signal in the copper complex versions was attributed either to spin relaxation of the two unpaired electrons or to masking of the narrow phenoxyl radical signal by the strong copper contribution.

In a more recent study combining ^{13}C labeling of the phenol ring (C1′) with photoinduced difference FTIR spectroscopy, Einarsdòttir and co-workers[103] were able to determine the $C–O$ stretching frequency at the radical state in tyrosine ($1517\,cm^{-1}$, $\Delta^{13}C = 16\,cm^{-1}$) and the copper free ligand **30** ($1485\,cm^{-1}$, $\Delta^{13}C = 22\,cm^{-1}$). Based on the presence of similar bands in the ^{12}C-minus-^{13}C-isotope-edited spectra ($1483\,cm^{-1}$, $\Delta^{13}C = 11\,cm^{-1}$) of the copper complex **30Cu**, the authors suggested that, despite the absence of an EPR signal, the phenoxyl radical does form in both the ligand and copper complex.

9.4.5. Heme–Copper Models Possessing Cross-Linked Imidazole-Phenol

In the last few years, new generations of heme a_3/Cu_B models bearing a mimic of the His-Tyr cross-link have appeared. The Naruta[107,108] research group reported O_2 reactivity of heme-based binuclear models (**35 Fe/Cu**, **36 Fe/Cu**, Chart 9.7) incor-porating a cross-linked imidazole–phenol moiety as a cytochrome *c* oxidases Cu_B site mimic.[106,107] Complex **35 Fe^{II}/Cu^I** was reported to react with O_2 to form a low-temperature stable heme–peroxo–Cu complex, however, when an intramolecular axial base was incorporated (**36 Fe/Cu**), the heme–peroxo–Cu intermediate became a

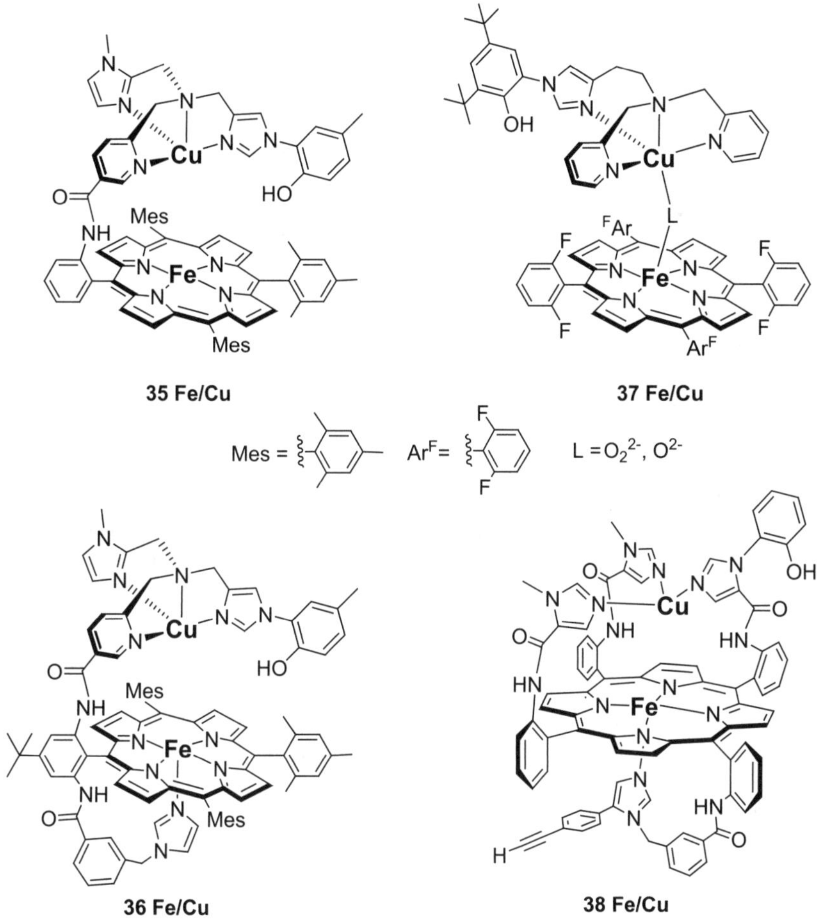

CHART 9.7

short-lived transient and it evolved rapidly to form a low-temperature stable heme–superoxo/Cu(I) complex ($Fe^{III}-(O_2^-)\cdots Cu^I$), an intermediate very similar to that observed for the **Oxy** state in the mechanism of O_2 reduction by CcO. Based on this results, the authors suggested that the copper center not only plays the role of a redox center, but also stabilizes the heme–superoxide intermediate by a possible precoordination of the dioxygen ($Cu^{II}-O_2^-$) before the O_2 is transferred to the heme ($Fe^{III}-O_2^-$) passing by a peroxide intermediate ($[Fe^{III}-(O_2^{2-})-Cu^{II}]$); the phenolic hydroxyl group was proposed to facilitate the conversion of the peroxide into superoxide via hydrogen-bonding stabilization.

Karlin and co-workers[109] also described the reactivity of the heme/Cu assembly in which the imidazole–phenol moiety was employed as part of the copper ligand (**37 Fe/Cu**, Chart 9.7). Spectroscopic investigations including UV–vis, ^{1}H and ^{2}H NMR, EPR, and rR spectroscopies along with spectrophotometric titration revealed that low-temperature oxygenation of a solution with equimolar $Cu^I(L^{N4}OH)$ (**33Cu^I**) and $(F_8)Fe^{II}$ leads to formation of a heme–peroxo–copper species $[(F_8)Fe^{III}-(O_2^{2-})-Cu^{II}(L^{N4}OH)]^+$ (**37 Fe^{III}–(O$_2^{2-}$)–Cu^{II}**), $\nu(O-O) = 813\,cm^{-1}$. Complex **37 Fe^{III}–(O$_2^{2-}$)–Cu^{II}** is an $S = 2$ spin system with strong antiferromagnetic coupling between high-spin Fe(III) and Cu(II) through a bridging peroxide ligand. A very similar complex $[(F_8)Fe^{III}-(O_2^{2-})-Cu^{II}(L^{N4}OMe)]^+$ $[\nu(O-O) = 815\,cm^{-1}]$ was generated by utilizing the phenol protected (anisole) version of compound **33Cu^I**, which indicates that the cross-linked phenol moiety in **37 Fe^{III}–(O$_2^{2-}$)–Cu^{II}** does not interact with the peroxo group that bridges the heme and copper. Preliminary experiments which consisted of the warming of a solution of **37 Fe^{III}–(O$_2^{2-}$)–Cu^{II}** to $-60°C$ did not reveal phenoxyl radical or $Fe^{IV}=O$ formation products that might be expected if a biomimetic $O-O$ cleavage reaction in **37 Fe^{III}–(O$_2^{2-}$)–Cu^{II}** occurred. Yet, the fact that a stable heme–peroxo–copper species could be generated even in the presence of an imidazole–phenol group (i.e., as a possible electron–proton donor source) in close proximity provides new opportunities for studies probing key factors that can trigger the reductive $O-O$ cleavage in CcO model compounds.

Using the electrochemical approach, Colman et al.[110] employed a synthetic model of the heme a_3/Cu$_B$ dinuclear center of CcO with an incorporated His-Tyr mimic (**38 Fe/Cu**, Chart 9.7). As discussed for the copper in Section 9.2.5, under slow rates of ET, the presence of a redox-active phenol to mimic tyrosine was reported to improve the selectivity ($>99\%$) of the model toward $4e^-$ reduction of O_2 to water reducing the formation of physiologically harmful species (H_2O_2, $HO_2^{\cdot}$, $HO^{\cdot}, \cdots$) as products of the partial reduction of O_2. Despite the lack of spectroscopic evidence for the formation of a phenoxyl radical or ferry oxo intermediate, a hydrogen-atom ($e^- + H^+$) donor role, as in the enzyme proposed mechanism, was attributed to the imidazole–phenol cross-link in this model.

9.5. SUMMARY

Site analogue chemistry continues to contribute significantly to our understanding of protein structure, physical properties, and reactivity by facilitating correlation of composition and structure to spectroscopic and magnetic observations. The development of functional models to reproduce some aspects of the protein active-site reactivity has helped us gain precious insights into the mechanism in which these enzymes proceed.

The electrochemical approach to study O_2 reduction by CcO has shown that synthetic heme/Cu analogues are capable of reproducing the enzymatic $4e^-$ reduction of O_2 to H_2O under physiologically relevant conditions of pH, electrochemical potential, and electron flux. The study of the O_2 reactivity of heme only, and heme/Cu synthetic models under different rates of electron flux, have shown that the distal Cu is obligatory for $4e^-$ dioxygen reduction under biologically relevant turnover-determining electron flux. On the basis of these results, it was speculated that Cu$_B$ is essential to achieve the $4e^-$ reduction of O_2 by CcO under slow electron diffusion from cytochrome c in the natural system.

The generation and characterization of heme/Cu/O_2 adducts represents a major tool in the inorganic modeling chemist's hands to understand the mechanism of O_2 reduction by the enzyme. This approach led to the determination of exciting intermediates, such as a heme–superoxide species (with copper present) and heme–peroxo–copper compounds. The latter is directly relevant to an enzyme transient or may be a precursor to such (i.e., by protonation giving a heme hydroperoxo $Fe^{III}-OOH\cdots Cu^{II}$ moiety). The X-ray structure of the heme–peroxo–copper complex reported by Naruta and co-workers[50] confirmed the nature of such species, with an example of an unsymmetrically bridged peroxo ligand bound to the heme in a side-on η^2 fashion, whereas it is η^1-ligated to copper.

The discovery of heme–peroxo–copper assemblies by Karlin[43,46] opened new perspectives for the intermediates detection approach by offering a new and relatively easy way to synthesize heme/Cu/O_2 species. The convenience of heme/copper assemblies promoted the development of systematic studies for the influence of copper coordination environment on the structure and the reactivity of the O_2 adducts. One of these studies, directed toward the understanding of the nature of high-spin heme–peroxo–copper complexes, has revealed that the denticity (tri- vs tetradentate) of the Cu ligand can have a major influence on the reactivity, the molecular structure, and the electronic structure (i.e., bonding) in such species. Combining spectroscopic methods (rR and EXAFS) with DFT calculations, a $\eta^2:\eta^1$-peroxo [side-on to the Fe(III) ion and end-on to Cu(II)] was established for a complex with a tetradentate copper ligand, whereas side-on/ side-on $(\mu-\eta^2:\eta^2)$ μ-peroxo coordination was attributed to a complex with a tridentate copper ligand.

Low-spin heme–peroxo–copper complexes present promising potential given the new reactivity demonstrated by such a species. For a better understanding of differences of the spin state (high vs low-spin), considerable future efforts are needed to determine the geometric and electronic structure of low-spin system in comparison to the structure of high-spin systems. Such studies could present a crucial contribution to determine the importance of the spin state if a bridged peroxo intermediate is identified as an intermediate in the CcO mechanism to reduce O_2 to H_2O.

The discovery of the histidine–tyrosine cross-link in heme–copper oxidases has created a new direction for active-site modeling. Generation and studies of new models incorporating a mimic imidazole–phenol moiety have begun, but much remains for the future in order to clarify the implication of such an unusual cross-link in the mechanism of O_2 reduction. While it is generally accepted that a tyrosine radical forms at some point during the reaction, it remains an open question as to when its proton and electron enter the catalytic cycle and by what means. Specifically, does the active-site oxidant direct hydrogen-atom abstraction from the tyrosine, or does the reaction occur stepwise, and do the transfers occur before or after the $O-O$ bond is cleaved? Stepwise mechanisms would indicate the importance of the Tyr-His cross-link to form an adequate superexchange pathway for ET and would significantly enhance the functional role of copper in mediating the ET. Another very interesting question that should be addressed is how, in the first place, does the His-Tyr cross-link form? Does the oxidative coupling (formally) of a His and Tyr residue require O_2 or H_2O_2? Does it require only Cu_B or both heme and Cu metal centers? Is the proenzyme first reaction turnover (i.e., the post-translational modification step) evolved to form the His-Tyr moiety?

Another exciting future challenge will be to examine the proton and electron transport as essential parts of understanding of the $O-O$ bond cleavage process in CcO. Also here, a number of questions need to be answered, such as what is the order in which the electrons and the protons enter the catalytic cycle? What influence does this order have on the formation of species, such as superoxo, peroxo, or hydroperoxo entities? Also, is this order related to proton pumping by CcO during the dioxygen-reduction process?

REFERENCES

1. Ferguson-Miller, S.; Babcock, G. T. Heme/Copper Terminal Oxidases *Chem. Rev.* **1996**, *96*, 2889–2907.

2. Iwata, S.; Ostermeier, C.; Ludwig, B.; Michel, H. Structure at 2.8 Å resolution of cytochrome c oxidase from *Paracoccus denitrificans*. *Nature (London)* **1995**, *376*, 660–669.

3. Yoshikawa, S.; Shinzawa-Itoh, K.; Nakashima, R.; Yaono, R.; Yamashita, E.; Inoue, N.; Yao, M.; Jei-Fei, M.; Libeu, C. P.; Mizushima, T.; Yamaguchi, H.; Tomizaki, T.; Tsukihara, T. Redox-Coupled Crystal Structure Changes in Bovine Heart Cytochrome c Oxidase. *Science* **1998**, *280*, 1723–1729.

4. Kim, E.; Chufán, E. E.; Kamaraj, K.; Karlin, K. D. Synthetic Models for Heme-Copper Oxidases *Chem. Rev.* **2004**, *104*, 1077–1133.

5. Collman, J. P.; Boulatov, R.; Sunderland, C. J.; Fu, L. Functional analogues of cytochrome c oxidase, myoglobin, and hemoglobin *Chem. Rev.* **2004**, *104*, 561–588.

6. Chufán, E. E.; Puiu, S. C.; Karlin, K. D. Heme–Copper/Dioxygen Adduct Formation, Properties, and Reactivity *Acc. Chem. Res.* **2007**, *40*, 563–572.

7. Tsukihara, T.; Aoyama, H.; Yamashita, E.; Tomizaki, T.; Yamaguchi, H.; Shinzawa-Itoh, K.; Nakashima, R.; Yaono, R.; Yoshikawa, S. Structures of Metal Sites of Oxidized Bovine Heart Cytochrome c Oxidase at 2.8 Å *Science* **1995**, *269*, 1069–1074.

8. Svensson-Ek, M.; Abramson, J.; Larsson, G.; Tornroth, S.; Brzezinski, P.; Iwata, S. The X-ray crystal structures of wild-type and EQ(I-286) mutant cytochrome c oxidases from Rhodobacter sphaeroides *J. Mol. Biol.* **2002**, *321*, 329–339.

9. Soulimane, T.; Buse, G.; Bourenkov, G. P.; Bartunik, H. D.; Huber, R.; Than, M. E. Structure and mechanism of the aberrant ba_3-cytochrome c oxidase from *Thermus thermophilus EMBO J.* **2000**, *19*, 1766–1776.

10. Tsukihara, T.; Aoyama, H.; Yamashita, E.; Tomizaki, T.; Yamaguchi, H.; Shinzawa-Itoh, K.; Nakashima, R.; Yaono, R.; Yoshikawa, S. The Whole Structure of the 13-Subunit Oxidized Cytochrome c Oxidase at 2.8 Å *Science* **1996**, *272*, 1136–1144.

11. Ostermeier, C.; Harrenga, A.; Ermler, U.; Michel, H. Structure at 2.7Å resolution of the *Paracoccus denitrificans* two-subunit cytochrome c oxidase complexed with an antibody F_V fragment *Proc. Natl. Acad. Sci. USA* **1997**, *94*, 10547–10553.

12. Harrenga, A.; Michel, H. The Cytochrome *c* Oxidase from *Paracoccus denitrificans* Does Not Change the Metal Center Ligation upon Reduction *J. Biol. Chem.* **1999**, *274*, 33296–33299.

13. Ohta, K.; Muramoto, K.; Shinzawa-Itoh, K.; Yamashita, E.; Yoshikawa, S.; Tsukihara, T. X-ray structure of the NO-bound Cu_B in bovine cytochrome *c* oxidase *Acta Crystallogr. Sec. F* **2010**, *66*, 251–253.

14. Aoyama, H.; Muramoto, K.; Shinzawa-Itoh, K.; Hirata, K.; Yamashita, E.; Tsukihara, T.; Ogura, T.; Yoshikawa, S. A peroxide bridge between Fe and Cu ions in the O_2 reduction site of fully oxidized cytochrome *c* oxidase could suppress the proton pump. *Proc. Natl. Acad. Sci. USA* **2009**, *106*, 2165–2169.

15. Koepke, J.; Olkhova, E.; Angerer, H.; Müller, H.; Peng, G.; Michel, H. High resolution crystal structure of Paracoccus denitrificans cytochrome *c* oxidase: New insights into the active site and the proton transfer pathways *Biochim. Biophys. Acta (BBA)—Bioenerget.* **2009** *1787*, 635–645.

16. Hill, B. C. The Reaction of the Electrostatic Cytochrome-*c*-Cytochrome Oxidase Complex with Oxygen *J. Biol. Chem.* **1991**, *266*, 2219–2226.

17. Pan, L. P.; Hibdon, S.; Liu, R. Q.; Durham, B.; Millett, F. Intracomplex Electron-Transfer between Ruthenium-Cytochrome-*c* Derivatives and Cytochrome-*c*-Oxidase *Biochemistry* **1993**, *32*, 8492–8498.

18. Malatesta, F.; Nicoletti, F.; Zickermann, V.; Ludwig, B.; Brunori, M. Electron entry in a Cu-A mutant of cytochrome c oxidase from Paracoccus denitrificans Conclusive evidence on the initial electron entry metal center *Febs Lett.* **1998**, *434*, 322–324.

19. Proshlyakov, D. A.; Pressler, M. A.; Babcock, G. T. Dioxygen activation and bond cleavage by mixed-valence cytochrome *c* oxidase *Proc. Natl. Acad. Sci. USA* **1998**, *95*, 8020–8025.

20. Proshlyakov, D. A.; Pressler, M. A.; DeMaso, C.; Leykam, J. F.; DeWitt, D. L.; Babcock, G. T. Oxygen Activation and Reduction in Respiration: Involvement of Redox-Active Tyrosine 244 *Science* **2000**, *290*, 1588–1591.

21. Babcock, G. T. How oxygen is activated and reduced in respiration *Proc. Nat. Acad. Sci. USA* **1999**, *96*, 12971–12973.

22. Michel, H.; Behr, J.; Harrenga, A.; Kannt, A. Cytochrome *c* Oxidase: Structure and Spectroscopy *Annu. Rev. Biophys. Biomol. Struct.* **1998**, *27*, 329–356.

23. Blomberg, L. M.; Blomberg, M. R. A.; Siegbahn, P. E. M. A theoretical study on the binding of O_2, NO and CO to heme proteins *J. Inorg. Biochem.* **2005**, *99*, 949–958.

24. de Vries, S. The role of the conserved tryptophan272 of the Paracoccus denitrificans cytochrome *c* oxidase in proton pumping *Biochim. Biophys. Acta (BBA)—Bioenerget.* **2008** *1777*, 925–928.

25. Blomberg, M. R. A.; Siegbahn, P. E. M.; Wikström, M. A metal-bridging mechanism for $O-O$ bond cleavage in cytochrome c oxidase *Inorg, Chem.* **2003**, *42*, 5231–5243.

26. Ogura, T.; Kitagawa, T. Resonance Raman characterization of the P intermediate in the reaction of bovine cytochrome *c* oxidase *Biochim. Biophys. Acta* **2004**, *1655*, 290–297.

27. Blomberg, M. R. A.; Siegbahn, P. E. M. Different types of biological proton transfer reactions studied by quantum chemical methods *Biochim. Biophys. Acta* **2006** *1757*, 969–980.

28. Himo, F.; Siegbahn, P. E. M. Quantum Chemical Studies of Radical-Containing Enzymes *Chem. Rev.* **2003**, *103*, 2421–2456.

29. Huynh, M. H. V.; Meyer, T. J. Proton-Coupled Electron Transfer *Chem. Rev.* **2007**, *107*, 5004–5064.

30. Yoshioka, Y.; Satoh, H.; Mitani, M. Theoretical study on electronic structures of FeOO, FeOOH, FeO(H₂O), and FeO in hemes: As intermediate models of dioxygen reduction in cytochrome *c* oxidase *J. Inorg. Biochem.* **2007**, *101*, 1410–1427.

31. Gunter, M. J.; Mander, L. N.; Murray, K. S. Oxygenation Reactions of Cytochrome Oxidase Models. Evidence for Hematin Formation in Mononuclear and Heterbinuclear Complexes *J. Chem. Soc. Chem. Commun.* **1981**, 799–801.

32. Shigehara, K.; Anson, F. C. Electrocatalytic Activity of 3 Iron Porphyrins in the Reductions of Dioxygen and Hydrogen-Peroxide at Graphite-Electrodes *J. Phys. Chem.* **1982**, *86*, 2776–2783.

33. Shi, C. N.; Anson, F. C. Catalytic Pathways for the Electroreduction of O-2 by Iron Tetrakis(4-N-Methylpyridyl)Porphyrin or Iron Tetraphenylporphyrin Adsorbed on Edge Plane Pyrolytic-Graphite Electrodes *Inorg. Chem.* **1990**, *29*, 4298–4305.

34. Ricard, D.; Didier, A.; L'Her, M.; Boitrel, B. Application of 3-quinolinoyl picket porphyrins to the electroreduction of dioxygen to water: mimicking the active site of cytochrome *c* oxidase *ChemBioChem* **2001**, *2*, 144–148.

35. Collman, J. P.; Fu, L.; Herrmann, P. C.; Zhang, X. A Functional Model Related to Cytochrome *c* Oxidase and Its Electrocatalytic Four-Electron Reduction of O₂ *Science* **1997**, *275*, 949–951.

36. Collman, J. P.; Fu, L.; Herrmann, P. C.; Wang, Z.; Rapta, M.; Bröring, M.; Schwenninger, R.; Boitrel, B. A Functional Model of Cytochrome *c* Oxidase: Thermodynamic Implications *Angew. Chem. Int. Ed. Engl.* **1998**, *37*, 3397–3400.

37. Collman, J. P.; Schwenninger, R.; Rapta, M.; Bröring, M.; Fu, L. New 1,4,7-triazacylcononane-based functional analogues of the Fe/Cu active site of cytochrome c oxidase: struture, spectroscopy and electrocatalytic reduction of oxygen *Chem. Commun.* **1999**, 137–138.

38. Collman, J. P. Functional Analogs of Heme Protein Active Sites *Inorg. Chem.* **1997**, *36*, 5145–5155.

39. Collman, J. P.; Sunderland, C. J.; Boulatov, R. Biomimetic Studies of Terminal Oxidases: Trisimidazole Picket Metalloporphyrins *Inorg. Chem.* **2002**, *41*, 2282–2291.

40. Collman, J. P.; Sunderland, C. J.; Berg, K. E.; Vance, M. A.; Solomon, E. I. Spectroscopic Evidence for a Heme-Superoxide/Cu(I) Intermediate in a Functional Model of Cytochrome *c* Oxidase *J. Am Chem. Soc.* **2003**, *125*, 6648–6649.

41. Boulatov, R.; Collman, J. P.; Shiryaeva, I. M.; Sunderland, C. J. Functional Analogues of the Dioxygen Reduction Site in Cytochrome Oxidase: Mechanistic Aspects and Possible Effects of Cu_B *J. Am. Chem. Soc.* **2002**, *124*, 11923–11935.

42. Collman, J. P.; Boulatov, R. Electrocatalytic O₂ reduction by synthetic analogues of the heme/Cu site of cytochrome oxidase incorporated in a lipid film *Angew. Chem. Int. Ed. Engl.* **2002**, *41*, 3487–3489.

43. Nanthakumar, A.; Fox., S.; Murthy, N. N.; Karlin, K. D.; Ravi, N.; Huynh, B. H.; Orosz, R. D.; Day, E. P.; Hagen, K. S.; Blackburn, N. J. Oxo- and Hydroxo-Bridged (Porphyrin)iron (III)-Copper(II) Species as Cytochrome *c* Oxidase Models: Acid-Base Interconversions

and X-ray Structure of the Fe(III)-(O^{2-})-Cu(II) Complex *J. Am. Chem. Soc.* **1993**, *115*, 8513–8514.

44. Karlin, K. D.; Nanthakumar, A.; Fox, S.; Murthy, N. N.; Ravi, N.; Huynh, B. H.; Orosz, R. D.; Day, E. P. X-ray Structure and Physical Properties of the Oxo-Bridged Complex [(F8-TPP)Fe-O-Cu(TMPA)]$^+$, F$_8$-TPP = Tetrakis(2,6-difluorophenyl)porphyrinate (2-), TMPA = Tris(2-pyridylmethyl)amine: Modeling the Cyhtochrome *c* Oxidase Fe-Cu Heterodinuclear Active Site *J. Am. Chem. Soc.* **1994**, *116*, 4753–4763.

45. Ghiladi, R. A.; Hatwell, K. R.; Karlin, K. D.; Huang, H.-w.; Moënne-Loccoz, P.; Krebs, C.; Huynh, B. H.; Marzilli, L. A.; Cotter, R. J.; Kaderli, S.; Zuberbühler, A. D. Dioxygen Reactivity of Mononuclear Heme and Copper Components Yielding A High-Spin Heme-Peroxo-Cu Complex *J. Am. Chem. Soc.* **2001**, *123*, 6183–6184.

46. Ghiladi, R. A.; Ju, T. D.; Lee, D.-H.; Moënne-Loccoz, P.; Kaderli, S.; Neuhold, Y.-M.; Zuberbühler, A. D.; Woods, A. S.; Cotter, R. J.; Karlin, K. D. Formation and Characterization of a High-Spin Heme-Copper Dioxygen (Peroxo) Complex *J. Am. Chem. Soc.* **1999**, *121*, 9885–9886.

47. Ju, T. D.; Ghiladi, R. A.; Lee, D.-H.; van Strijdonck, G. P. F.; Woods, A. S.; Cotter, R. J.; Young, J. V. G.; Karlin, K. D. Dioxygen Reactivity of Fully Reduced [LFeII···CuI]$^+$ Complexes Utilizing Tethered Tetraarylpoprhyrinates: Active Site Models for Heme-Copper Oxidases *Inorg. Chem.* **1999**, *38*, 2244–2245.

48. Sasaki, T.; Nakamura, N.; Naruta, Y. Formation and Spectroscopic Characterization of the PeroxoFe^{III}–Cu^{II} Complex. A Modeling Reaction of the Heme-Cu Site in Cytochrome *c* Oxidase *Chem. Lett.* **1998**, 351–352.

49. Naruta, Y.; Sasaki, T.; Tani, F.; Tachi, Y.; Kawato, N.; Nakamura, N. Heme-Cu complexes as oxygen-activating functional models for the active site of cytochrome *c* oxidase. *J. Inorg. Biochem.* **2001**, *83*, 239–246.

50. Chishiro, T.; Shimazaki, Y.; Tani, F.; Tachi, Y.; Naruta, Y.; Karasawa, S.; Hayami, S.; Maeda, Y. Isolation and Crystal Structure of a Peroxo-Bridged Heme-Copper Complex. *Angew. Chem. Int. Ed.* **2003**, *42*, 2788–2791.

51. Obias, H. V.; van Strijdonck, G. P. F.; Lee, D.-H.; Ralle, M.; Blackburn, N. J.; Karlin, K. D. Heterobinucleating Ligand Induced Structural and Chemical Variations in [(L) Fe^{III}–O–Cu^{II}]$^+$ m-Oxo Complexes *J. Am. Chem. Soc.* **1998**, *120*, 9696–9697.

52. Chufan, E. E.; Mondal, B.; Gandhi, T.; Kim, E.; Rubie, N. D.; Moenne-Loccoz, P.; Karlin, K. D. Reactivity Studies on Fe^{III}–(O$_2{}^{2-}$)–Cu^{II} Compounds: Influence of the Ligand Architecture and Copper Ligand Denticity *Inorg. Chem.* **2007**, *46*, 6382–6394.

53. Kim, E.; Helton, M. E.; Wasser, I. M.; Karlin, K. D.; Lu, S.; Huang, H.-w.; Moënne-Loccoz, P.; Incarvito, C. D.; Rheingold, A. L.; Honecker, M.; Kaderli, S.; Zuberbühler, A. D. Superoxo, m-peroxo and m-oxo complexes from heme/O$_2$ and heme-copper/O$_2$ reactivity studies: Copper-ligand influences in cytochrome *c* oxidase models. *Proc. Natl. Acad. Sci. USA* **2003**, *100*, 3623–3628.

54. Kim, E.; Shearer, J.; Lu, S.; Moënne-Loccoz, P.; Helton, M. E.; Kaderli, S.; Zuberbühler, A. D.; Karlin, K. D. Heme/Cu/O$_2$ Reactivity: Change in Fe^{III}–(O$_2{}^{2-}$)–Cu^{II} Unit Peroxo Binding Geometry Effected by Tridentate Copper Chelation *J. Am. Chem. Soc.* **2004**, *126*, 12716–12717.

55. Kopf, M.-A.; Neuhold, Y.-M.; Zuberbühler, A. D.; Karlin, K. D. Oxo and Hydroxo-Bridged Heme-Copper Assemblies formed from Acid–Base or Metal-Dioxygen Chemistry *Inorg. Chem.* **1999**, *38*, 3093–3102.

56. Kim, E.; Helton, M. E.; Lu, S.; Moënne-Loccoz, P.; Incarvito, C. D.; Rheingold, A. L.; Kaderli, S.; Zuberbühler, A. D.; Karlin, K. D. Tridentate copper ligand influences on heme-peroxo-copper formation and properties: Reduced, superoxo, and µ-peroxo iron/copper complexes *Inorg, Chem.* **2005**, *44*, 7014–7029.

57. Hatcher, L. Q.; Karlin, K. D. Ligand influences in copper-dioxygen complex-formation and substrate oxidations *Adv. Inorg. Chem.* **2006**, *58*, 131–184.

58. Hatcher, L. Q.; Karlin, K. D. Oxidant types in copper-dioxygen chemistry: the ligand coordination defines the Cu_n-O_2 structure and reactivity *J. Biol. Inorg. Chem.* **2004**, *9*, 669–683.

59. Mirica, L. M.; Ottenwaelder, X.; Stack, T. D. P. Structure and Spectroscopy of Copper-Dioxygen Complexes *Chem. Rev.* **2004**, *104*, 1013–1045.

60. Ross, P. K.; Solomon, E. I. An Electronic Structural Comparison of Copper-Peroxide Complexes of Relevance to Hemocynin and Tyroinsase Active Sites *J. Am. Chem. Soc.* **1991**, *113*, 3246–3259.

61. Ross, P. K.; Solomon, E. I. Electronic-Structure of Peroxide-Bridged Copper Dimers of Relevance to Oxyhemocyanin *J. Am. Chem. Soc.* **1990**, *112*, 5871–5872.

62. Solomon, E. I.; Tuczek, F.; Root, D. E.; Brown, C. A. Spectroscopy of Binuclear Dioxygen Complexes *Chem. Rev.* **1994**, *94*, 827–856.

63. Halime, Z.; Kieber-Emmons, M. T.; Qayyum, M. F.; Mondal, B.; Gandhi, T.; Puiu, S. C.; Chufán, E. E.; Sarjeant, A. A. N.; Hodgson, K. O.; Hedman, B.; Solomon, E. I.; Karlin, K. D. Heme–Copper–Dioxygen Complexes: Toward Understanding Ligand-Environmental Effects on the Coordination Geometry, Electronic Structure, and Reactivity *Inorg. Chem.* **2010**, *49*, 3629–3645.

64. Liu, J.-G., Naruta, Y., Tani, F. A functional model of the cytochrome c oxidase active site: Unique conversion of a heme-µ-peroxo-Cu^{II} intermediate into heme-superoxo/Cu^{I} *Angew. Chem. Int. Ed.* **2005**, *44*, 1836–1840.

65. Collman, J. P.; Decréau, R. A.; Sunderland, C. J. Single-turnover intermolecular reaction between a Fe^{III}–superoxide–Cu^{I} cytochrome *c* oxidase model and exogeneous Tyr244 mimics. *Chem. Commun.* **2006**, 3894–3896.

66. Ghiladi, R. A.; Kretzer, R. M.; Guzei, I.; Rheingold, A. L.; Neuhold, Y.-M.; Hatwell, K. R.; Zuberbühler, A. D.; Karlin, K. D. Reversible Dioxygen Binding to $(F_8TPP)Fe^{II}$/O_2 Reactivity Studies {F_8TPP = tetrakis(2,6-difluorophenyl)porphyrinate}: Spectroscopic (UV-Visible and NMR) and Kinetic Study of Solvent-Dependent (Fe/O_2 = 1:1 or 2:1) Reversible O_2-Reduction and Ferryl Formation *Inorg. Chem.* **2001**, *40*, 5754–5767.

67. Buse, G.; Soulimane, T.; Dewor, M.; Meyer, H. E.; Blüggel, M. Evidence for a copper-coordinated histidine-tyrosine cross-link in the active site of cytochrome oxidase *Prot. Sci.* **1999**, *8*, 985–990.

68. Hellwig, P.; Pfitzner, U.; Behr, J.; Rost, B.; Pesavento, R. P.; von Donk, W.; Gennis, R. B.; Michel, H.; Ludwig, B.; Maentele, W. Vibrational Modes of Tyrosines in Cytochrome c Oxidase from Paracoccus denitrificans: FTIR and Electrochemical Studies on Tyr-D4-labeled and on Tyr280His and Tyr35Phe Mutant Enzymes *Biochemistry* **2002**, *41*, 9116–9125.

69. Tomson, F.; Bailey, J. A.; Gennis, R. B.; Unkefer, C. J.; Li, Z.; Silks, L. A.; Martinez, R. A.; Donohoe, R. J.; Dyer, R. B.; Woodruff, W. H. Direct Infrared Detection of the Covalently Ring Linked His-Tyr Structure in the Active Site of the Heme–Copper Oxidases *Biochemistry* **2002**, *41*, 14383–14390.

70. Pinakoulaki, E.; Pfitzner, U.; Ludwig, B.; Varotsis, C. The role of the cross-link His-Tyr in the functional properties of the binuclear center in cytochrome c oxidase *J. Biol. Chem.* **2002**, *277*, 13563–13568.

71. Gennis, R. B. Multiple proton-conducting pathways in cytochrome oxidase and a proposed role for the active-site tyrosine *Biochim. Biophys. Acta* **1998**, *1365*, 241–248.

72. Kitagawa, T. Structures of reaction intermediates of bovine cytochrome c oxidase probed by time-resolved vibrational spectroscopy *J. Inorg. Biochem.* **2000**, *82*, 9–18.

73. He, Z. C.; Colbran, S. B.; Craig, D. C. Could redox-switched binding of a redox-active ligand to a copper(II) center drive a conformational proton pump gale? A synthetic model study *Chem. Eur. J.* **2003**, *9*, 116–129.

74. Fabian, M.; Palmer, G. The Interaction of Cyhtochrome Oxidase with Hydrogen Peroxide: The Relationship of Compounds P and F *Biochemistry* **1995**, *34*, 13802–13810.

75. MacMillan, F.; Kannt, A.; Behr, J.; Prisner, T.; Michel, H. Direct Evidence for a Tyrosine Radical in the Reaction of Cytochrome c Oxidase with Hydrogen Peroxide. *Biochemistry* **1999**, *38*, 9179–9184.

76. MacMillan, F.; Budiman, K.; Angerer, H.; Michel, H. The role of tryptophan 272 in the Paracoccus denitrificans cytochrome c oxidase. *FEBS Lett.* **2006**, *580*, 1345.

77. Rigby, S. E. J.; Junemann, S.; Rich, P. R.; Heathcote, P. Reaction of bovine cytochrome c oxidase with hydrogen peroxide produces a tryptophan cation radical and a porphyrin cation radical. *Biochemistry* **2000**, *39*, 5921–5928.

78. Rich, P. R.; Rigby, S. E. J.; Heathcote, P. Radicals associated with the catalytic intermediates of bovine cytochrome c oxidase. *Biochim. Biophys. Acta* **2002**, *1554*, 137–146.

79. Wiertz, F. G. M.; Richter, O.-M. H.; Ludwig, B.; de Vries, S. Kinetic Resolution of a Tryptophan-radical Intermediate in the Reaction Cycle of *Paracoccus denitrificans* Cytochrome c Oxidase *J. Biol. Chem.* **2007**, *282*, 31580–31591.

80. Svistunenko, D. A.; Wilson, M. T.; Cooper, C. E. Tryptophan or tyrosine? On the nature of the amino acid radical formed following hydrogen peroxide treatment of cytochrome c oxidase. *Biochim. Biophys. Acta (BBA)—Bioenerget.* **2004**, *1655*, 372.

81. Budiman, K.; Kannt, A.; Lyubenova, S.; Richter, O. M. H.; Ludwig, B.; Michel, H.; MacMillan, F. Tyrosine 167: The origin of the radical species observed in the reaction of cytochrome c oxidase with hydrogen peroxide in Paracoccus denitrificans. *Biochemistry* **2004**, *43*, 11709–11716.

82. Chen, Y. R.; Gunther, M. R.; Mason, R. P. An electron spin resonance spin-trapping investigation of the free radicals formed by the reaction of mitochondrial cytochrome c oxidase with H_2O_2. *J. Biol. Chem.* **1999**, *274*, 3308–3314.

83. Nyquist, R. M.; Heitbrink, D.; Bolwien, C.; Gennis, R. B.; Heberle, J. Direct observation of protonation reactions during the catalytic cycle of cytochrome c oxidase. *Proc. Natl. Acad. Sci. USA* **2003**, *100*, 8715–8720.

84. Einarsdottir, O.; Szundi, I.; Van Eps, N.; Sucheta, A. P-M and P-R forms of cytochrome c oxidase have different spectral properties. *J. Inorg. Biochem.* **2002**, *91*, 87–93.

85. Cozzi, P.; Carganico, G.; Fusar, D.; Grossoni, M.; Menichincheri, M.; Pinciroli, V.; Tonani, R.; Vaghi, F.; Salvati, P. Imidazol-1-Yl and Pyridin-3-Yl Derivatives of 4-Phenyl-1,4-Dihydropyridines Combining Ca2 + Antagonism and Thromboxane-a(2) Synthase Inhibition. *J. Med. Chem.* **1993**, *36*, 2964–2972.

86. Ohmori, J.; ShimizuSasamata, M.; Okada, M.; Sakamoto, S. Novel AMPA receptor antagonists: Synthesis and structure-activity relationships of 1-hydroxy-7-(1H-imidazol-

1-yl)-6-nitro-2,3(1H, 4H)-quinoxalinedione and related compounds *J. Med. Chem.* **1996**, *39*, 3971–3979.

87. Venuti, M. C.; Stephenson, R. A.; Alvarez, R.; Bruno, J. J.; Strosberg, A. M. Inhibitors of Cyclic-Amp Phosphodiesterase. 3. Synthesis and Biological Evaluation of Pyrido and Imidazolyl Analogs of 1,2,3,5-Tetrahydro-2-Oxoimidazo[2,1-B]Quinazoline *J. Med. Chem.* **1988**, *31*, 2136–2145.

88. Gungor, T.; Fouquet, A.; Teulon, J. M.; Provost, D.; Cazes, M.; Cloarec, A. Cardiotonic Agents - Synthesis and Cardiovascular Properties of Novel 2-Arylbenzimidazoles and Azabenzimidazoles. *J. Med. Chem.* **1992**, *35*, 4455–4463.

89. Jacobs, C.; Frotscher, M.; Dannhardt, G.; Hartmann, R. W. 1-imidazolyl(alkyl)-substituted di- and tetrahydroquinolines and analogues: Syntheses and evaluation of dual inhibitors of thromboxane A(2) synthase and aromatase. *J. Med. Chem.* **2000**, *43*, 1841–1851.

90. Lo, Y. S.; Nolan, J. C.; Maren, T. H.; Welstead, W. J.; Gripshover, D. F.; Shamblee, D. A. Synthesis and Physicochemical Properties of Sulfamate Derivatives as Topical Antiglaucoma Agents. *J. Med. Chem.* **1992**, *35*, 4790–4794.

91. Kiyomori, A.; Marcoux, J. F.; Buchwald, S. L. An efficient copper-catalyzed coupling of aryl halides with imidazoles. *Tetrahedron Lett.* **1999**, *40*, 2657–2660.

92. Lam, P. Y. S.; Clark, C. G.; Saubern, S.; Adams, J.; Winters, M. P.; Chan, D. M. T.; Combs, A. New aryl/heteroaryl C—N bond cross-coupling reactions via arylboronic acid cupric acetate arylation. *Tetrahedron Lett.* **1998**, *39*, 2941–2944.

93. Collman, J. P.; Zhong, M. An Efficient Diamine•Copper Complex-Catalyzed Coupling of Arylboronic Acids with Imidazoles. *Org. Lett.* **2000**, *2*, 1233–1236.

94. Elliott, G. I.; Konopelski, J. P. Complete N-1 Regiocontrol in the formation of N-arylimidazoles. Synthesis of the active site His-Tyr side chain coupled dipeptide of cytochrome *c* oxidase *Org. Lett.* **2000**, *2*, 3055–3057.

95. Kozyrod, R. P.; Morgan, J.; Pinhey, J. T. Reaction of Aryltrialkylstannanes with Lead-Tetraacetate — a Convenient Route to Aryllead Triacetates. *Aust. J. Chem.* **1985**, *38*, 1147–1153.

96. McCauley, K. M.; Vrtis, J. M.; Dupont, J.; Van der Donk, W. A. Insights into the Functional Role of the Tyrosine-Histidine Linkage in Cytochrome *c* Oxidase *J. Am. Chem. Soc.* **2000**, *122*, 2403–2404.

97. Cappuccio, J. A.; Ayala, I.; Elliott, G. I.; Szundi, I.; Lewis, J.; Konopelski, J. P.; Barry, B. A.; Einarsdottir, O. Modeling the Active Site of Cytochrome Oxidase: Synthesis and Characterization of a Cross-Linked Histidine-Phenol *J. Am. Chem. Soc.* **2002**, *124*, 1750–1760.

98. Aki, M.; Ogura, T.; Naruta, Y.; Le, T. H.; Sato, T.; Kitagawa, T. UV Resonance Raman Characterization of Model Compounds of Tyr244 of Bovine Cytochrome *c* Oxidase in Its Neutral, Deprotonated Anionic, and Deprotonated Neutral Radical Forms: Effects of Covalent Binding between Tyrosine and Histidine. *J. Phys. Chem. A* **2002**, *106*, 3436–3444.

99. Collman, J. P.; Wang, Z.; Zhong, M.; Zeng, L. Syntheses and pK(a) determination of 1-(o-hydroxyphenyl)imidazole carboxylic esters. *J. Chem. Soc.-Perkin Trans. 1* **2000**, 1217–1221.

100. Kamaraj, K.; Kim, E.; Galliker, B.; Zakharov, L. N.; Rheingold, A. L.; Zuberbuehler, A. D.; Karlin, K. D. Copper(I) and Copper(II) Complexes Possessing Cross-Linked

Imidazole-Phenol Ligands: Structures and Dioxygen Reactivity. *J. Am Chem. Soc.* **2003**, *125*, 6028–6029.

101. White, K. N.; Sen, I.; Szundi, I.; Landaverry, Y. R.; Bria, L. E.; Konopelski, J. P.; Olmstead, M. M.; Einarsdottir, O. Synthesis and structural characterization of cross-linked histidine-phenol Cu(ii) complexes as cytochrome *c* oxidase active site models. *Chem. Commun.* **2007**, 3252–3254.

102. Nagano, Y.; Liu, J. G.; Naruta, Y.; Ikoma, T.; Tero-Kubota, S.; Kitagawa, T. Characterization of the Phenoxyl Radical in Model Complexes for the Cu$_B$ Site of Cytochrome *c* Oxidase: Steady-State and Transient Absorption Measurements, UV Resonance Raman Spectroscopy, EPR Spectroscopy, and DFT Calculations for M-BIAIP *J. Am. Chem. Soc.* **2006**, *128*, 14560–14570.

103. Mahoney, M. E.; Oliver, A.; Emarsdottir, O.; Konopelski, J. P. Synthesis of a Cyclic Pentapeptide Mimic of the Active Site His-Tyr Cofactor of Cytochrome c Oxidase *J. Org. Chem.* **2009**, *74*, 8212–8218.

104. Ayala, I.; Range, K.; York, D.; Barry, B. A. Spectroscopic properties of tyrosyl radicals in dipeptides *J. Am. Chem. Soc.* **2002**, *124*, 5496–5505.

105. Range, K.; Ayala, I.; York, D.; Barry, B. A. Normal modes of redox-active tyrosine: Conformation dependence and comparison to experiment. *J. Phys. Chem. B* **2006**, *110*, 10970–10981.

106. Range, K.; Ayala, I.; Barry, B. A.; York, D. M. Oft study of the IR and EPR spectra of tyrosinate, tyrosyl radical, and some of their isotopomers *Abstr. Paper Am. Chem Soc.* **2003**, *225*, U769–U769.

107. Liu, J. G.; Naruta, Y.; Tani, F. A functional model of the cytochrome c oxidase active site: Unique conversion of a heme-mu-peroxo-Cu^{II} intermediate into heme-superoxo/Cu^I *Angew. Chem. Int. Ed.* **2005**, *44*, 1836–1840.

108. Liu, J.-G.; Naruta, Y.; Tani, F.; Chishiro, T.; Tachi, Y. Formation and spectroscopic characterization of the dioxygen adduct of a heme-Cu complex possessing a cross-linked tyrosine-histidine mimic: modeling the active site of cytochrome *c* oxidase *Chem. Commun.* **2004**, 120–121.

109. Kim, E.; Kamaraj, K.; Galliker, B.; Rubie, N. D.; Moënne-Loccoz, P.; Kaderli, S.; Zuberbühler, A. D.; Karlin, K. D. Dioxygen reactivity of copper and heme-copper complexes possessing an imidazole-phenol cross-link *Inorg. Chem.* **2005**, *44*, 1238–1247.

110. Collman, J. P.; Devaraj, N. K.; Decreau, R. A.; Yang, Y.; Yan, Y.-L.; Ebina, W.; Eberspacher, T. A.; Chidsey, C. E. D. A Cytochrome c Oxidase Model Catalyzes Oxygen to Water Reduction Under Rate-Limiting Electron Flux. *Science* **2007**, *315*, 1565–1568.

10

SUPRAMOLECULAR COPPER DIOXYGEN CHEMISTRY

JEAN-NOËL REBILLY AND OLIVIA REINAUD

Laboratoire de Chimie et Biochimie Pharmacologiques et Toxicologiques, UMR 8601, Université Paris Descartes, 75006 Paris

Copper-Oxygen Chemistry, First Edition. Edited by Kenneth D. Karlin and Shinobu Itoh.
© 2011 John Wiley & Sons, Inc. Published 2011 by John Wiley & Sons, Inc.

10.1. INTRODUCTION

Copper enzymes are involved in redox processes. In mammals, they catalyze a variety of oxidative transformations, such as $C-H$ oxygenation, dehydrogenation, dioxygenation, or disproportionation (superoxide). Although the Fe content is more than one order of magnitude higher than the Cu content, Cu enzymes have key roles, particularly in the central nervous system. Among these, copper monooxygenases, such as PHM (peptidylglycine α-hydroxylating monooxygenase),[1] DβH (dopamine β-hydroxylase),[2] and TβH (tyramine β-hydroxylase)[3] catalyze a $2e^-$ oxidation process corresponding to the regio- and stereospecific insertion of an oxygen atom into a $C-H$ bond with dioxygen (O_2). They are non-heme systems and for all of them, amino acid residues define the first coordination sphere.

Copper enzyme activity is obviously related to the structure of their active site with a specific coordination sphere and nuclearity. They have been classified in different categories depending on these factors (Fig. 10.1):[4,5] *Type 1 sites* are devoid of a labile site and thus are devoted to "pure" electron transfer (ET). *Type 2 sites* ("normal copper", named after their "normal" d–d visible transition) are mononuclear, based on a polyaza core as a first coordination sphere (e.g., N_3, N_2S, N_2O_2) associated to a labile site for O_2 interaction. The substrate oxidation process is performed either by an associated cofactor, which will lead to its dehydrogenation and formation of hydrogen peroxide (H_2O_2) (Fig. 10.1), as in amine oxidases and galactose oxidase (GO) (the role of copper is then mainly to regenerate the reactive state of the cofactor), or directly by the Cu/O_2 adduct, which is the case of the above-mentionned hydroxylases PHM, DβH, and TβH, with concomitant consumption of $2e^-$ provided by ascorbate and release of water. These enzymes involve in fact two copper centers that can be considered "uncoupled", as the second copper supplies electrons to the first copper reactive site located 11 Å away. *Type 3 sites* (Fig. 10.1) are proper dinuclear copper sites, each metal ion being surrounded by three histidine residues.

This kind of dinuclear system can bind and activate dioxygen (O_2) for oxygen-atom insertion into $C-H$ bonds and/or dehydrogenation, as in tyrosinase and catechol oxidase (phenol hydroxylation and/or oxidation of catechol to *o*-quinone). *Type 4 sites* are trinuclear clusters resulting from the association of one type 2 and one type 3 site.[6] A type 1 site is located further away in the protein.[7] This pattern is found in the

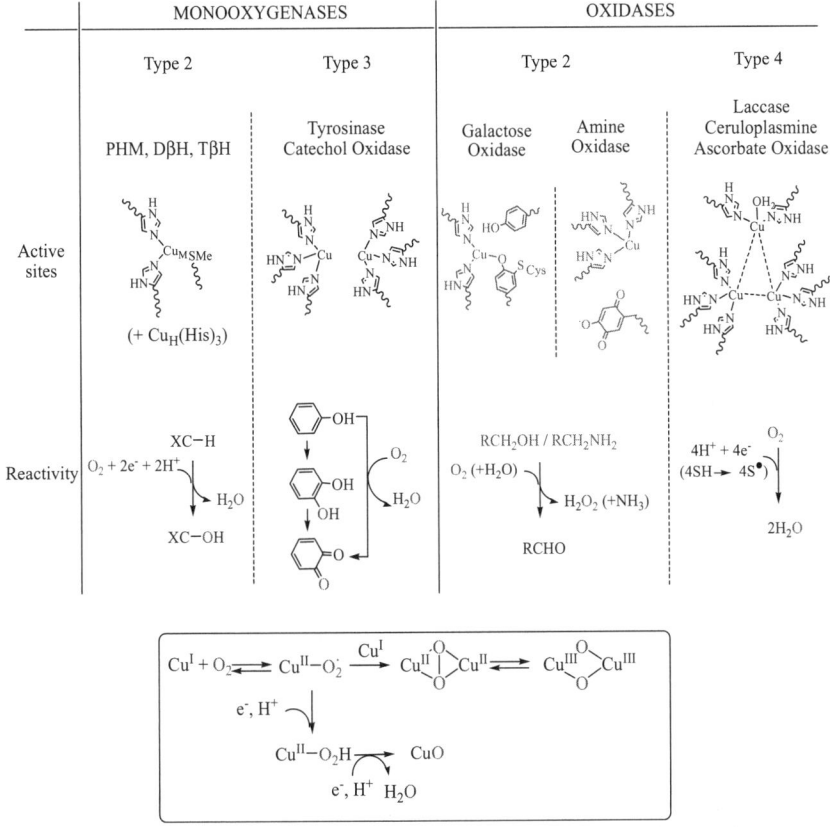

FIGURE 10.1. Environment of copper in the active sites of oxidases and monooxygenases with related transformations performed upon reaction with O₂. *Inset*: Proposed pathways for O₂ activation by Cu.

so-called multicopper oxidases (e.g., laccase, ceruloplasmine, and ascorbate oxidase). As in the case of the Cu_B type 2 site of cytochrome c oxidase (CcO) (which is associated to a heme) of the respiratory chain,[8] the type 4 site is devoted to the 4e⁻ reduction of O₂ into H₂O, whereas the four electrons (and four protons) are pumped from another Cu site, where the substrate interacts through a monoelectronic redox process (Fig. 10.1).

As these metalloenzymes carry out exquisitely selective reactions with impressive turnovers using a nonpolluting combustive, such as O₂, biomimetic models involving copper have been developed, based on type 2–4 sites (Fig. 10.2).[9]

The classical approach consists in reproducing the first coordination sphere of the metal site. Many examples have evidenced the ability of such systems to bind O₂, through a wide range of spectroscopic tools, and the possibility to generate real functional models able to activate O₂ and oxidize a substrate, the substrate being

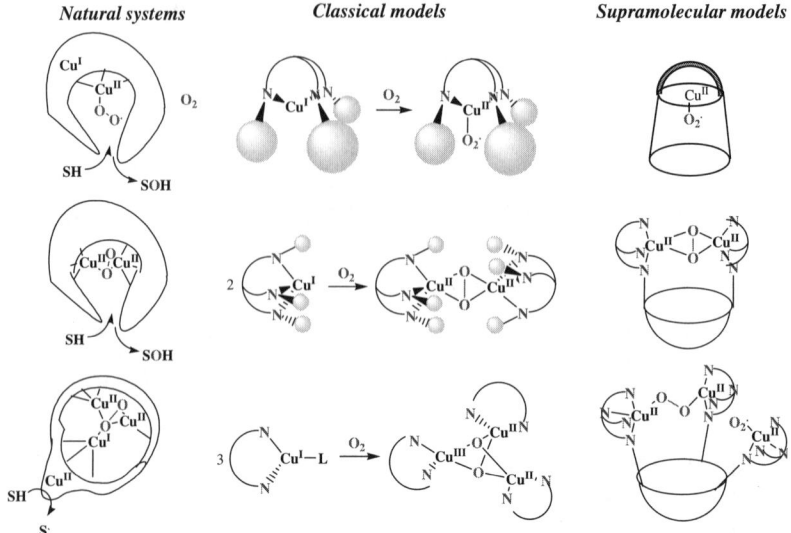

FIGURE 10.2. Models of copper enzymes.

either exogenous or the ligand itself.[9–14] Nevertheless, many shortcomings of these "classical" models remain difficult to address:

- The first is the control of the nuclearity of the complex during the redox process. Indeed, simple ligand systems generate a relatively unconstrained environment around the copper center, which is able to change drastically upon redox switch and O_2 binding. For example, most mononuclear copper systems tend to dimerize upon O_2 reaction in solution (unless the coordination sphere is saturated, which is useless for O_2 activation).[9,15,16]

- The microenvironment provided by the enzymatic pocket defines not only the first coordination sphere, but also a second sphere that has a major influence on the metal ion properties, and particularly guest binding. Likewise, ligand-exchange processes are under the control of the cavity (substrate binding, product release).

- The well-defined copper active sites in enzymes allow fine tuning of the redox properties of the metal during catalysis by imposing geometrical constrains to the metal at the (+ I) and (+ II) oxidation states, which is difficult to achieve with classical ligands in solution.[17–19] This particularity results in a specific interlocking of the receptor and redox metal properties.

- An important shortcoming lies in the reaction kinetics that is lower for models compared to enzymatic systems. Catalysis with "classical" models suffers from the entropic cost related to the approach of the substrate to the active site, which is discarded in natural systems by its optimal preorganization next to the active site. Consequently, copper models can give rise to reactive intermediates that display relatively poor thermal stability as the complexes undergo self-oxidation.[20,21]

- Finally, regio- and stereoselectivity are also difficult to achieve, as the simple tripods lack regio- or stereodiscriminating tools.

Different supramolecular strategies can be used to address each of these issues and will be detailed in the following sections.

10.2. CONTROL OF THE ACTIVE SITE NUCLEARITY

10.2.1. Formation of Dinuclear Complexes: A Thermodynamic Sink

Activation of O_2 involves its partial reduction to produce kinetically reactive species for oxidation catalysis. In copper enzymes, the first step requires interaction of O_2 at a Cu(I) center that is coordinatively unsaturated, generally tricoordinate. Indeed, Cu(I) ions coordinated in a tetrahedral environment provided by a polyaza ligand readily react with O_2 to initially produce a mononuclear adduct that has been identified as a superoxide Cu(II) complex. This transient species is readily trapped in solution by a second Cu(I) center, unless it is sufficiently sterically hindered to disfavor bimolecular reactions. Indeed, in the absence of steric hindrance, dinuclear species, either peroxo Cu(II) or dioxide Cu(III), are readily formed and can be accumulated at low temperature. Hence, the spontaneous evolution of an unconstrained Cu(I) center is to produce dinuclear complexes with O_2, which corresponds to the bielectronic reduction of O_2. Such complexes are reactive *per se* and mimic the reactive intermediate produced at the active site of catecholase and tyrosinase. Consequently, preorganization is not a major requirement for obtaining models of coupled dicopper enzymes (Fig. 10.2).

An interesting, yet unique case of formation of a trinuclear dioxide $Cu^{II}Cu^{II}Cu^{III}$ cluster has been reported with a N_2Cu^{I} complex, but its decomposition leads to dinuclear species, which prevented highlighting its intrinsic oxidative properties.[22] This behavior raises the question of preorganization of the Cu(I) centers in such a way that trinuclear interaction can be favored over dinuclear. In a recent report, formation of O_2 adducts at a trinuclear complex preorganized by a cyclotriveratrylene scaffold has been described, with one peroxo-bridged dicopper unit next to a superoxide monocopper adduct. The reactivity of such a cluster has not yet been identified.[23]

Many dinuclear copper biomimetic systems have been reported, which are formed from a single organic molecule bearing two polydentate coordinating ligands connected by an organic spacer.[9–15] Some groups have focused their attention on the design of systems where a receptacle (e.g., a molecular basket) is appended to the active copper complex. Their findings will be exposed and discussed in Section 10.5 devoted to reactivity.

10.2.2. Confinement of the Metal in a Mononuclear Environment by a Molecular Host

The classical strategy to prevent dimerization of copper complexes is to introduce bulky substituents next to the donor atoms, which in return locks the access of

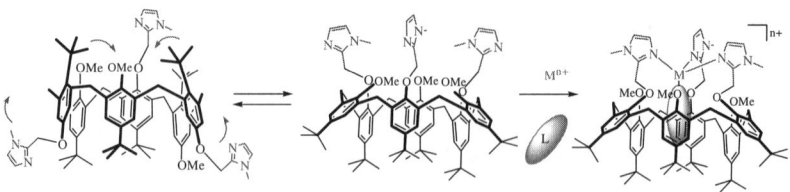

FIGURE 10.3. Conformation freezing of calix-ligands into a cone by metal coordination

substrates to the metal site. The use of a molecular host in the vicinity of the metal ion can greatly reduce the flexibility of the first coordination sphere and limit the coordination modes.

Nuclearity control has been intensely studied in the case of functionalized calix[6] arenes. The basic macrocycles are extremely flexible, as the p-t-Bu-aromatic units can rotate "through the annulus" of the calixarene small rim. One strategy to constrain the calix[6]arene into the cone conformation required to play the role of a molecular host is to link three alternate phenolic units out of six via coordination bonds to a common metal center (Fig. 10.3).

As the coordination properties of the two oxidation states of copper are very different, various environments have been observed. With Cu(I), a calixarene presenting three nitrogenous arms at the small rim gives rise to ill-defined complexes, where the metal ion oscillates between a two- and three-coordination state. The calixarene then remains conformationaly mobile. The condition to constrain it in a cone is to introduce a fourth donor that can play the role of a guest ligand. All three nitrogenous arms then wrap around Cu(I), which adopts a tetrahedral (Td) geometry with its fourth labile site oriented inward, and is then controlled by the well-defined cone-shaped calixarene cavity (Fig 10.4). Good guests are nitriles or CO, depending on the nature of the donor arms: with the stronger σ-donor imidazoles, the good π-acceptor CO efficiently seals the system into a tetrahedron, whereas with the weaker pyridyl σ-donors, a relatively stronger σ-donor, such as a nitrile, is required. This shaping role has been called a "shoetree" effect. Coordination to Cu(II), on the other hand, readily gives rise to five-coordinate species with one guest ligand shaping the calixarene and a water molecule capping the system in the trans position to the inner guest, outside the cavity (Fig. 10.4). The resulting geometry is a distorted square-based pyramid (SBP). The reason for obtaining five-coordinate species stems from the Jahn–Teller distortion, which favors tetragonal geometries [square planar (SP) or square-based pyramid (SBP)], and from the geometrical constraint imposed by the calixarene macrocycle, which discards the SP coordination. Indeed, all three aromatic donors (e.g., imidazole) cannot lie in an equatorial position of a square plane, but they can adapt to the SBP geometry, with one of them sitting on the SBP axis, and the other two on equatorial positions. Two equatorial sites remain open: one in an endo position, which is controlled by the calixarene core and can be occupied by a variety of neutral molecules provided they fit into the cavity (nitriles, alcohols, amides, etc.); the other in an exo position, which can be occupied only by very small ligands (e.g., a water

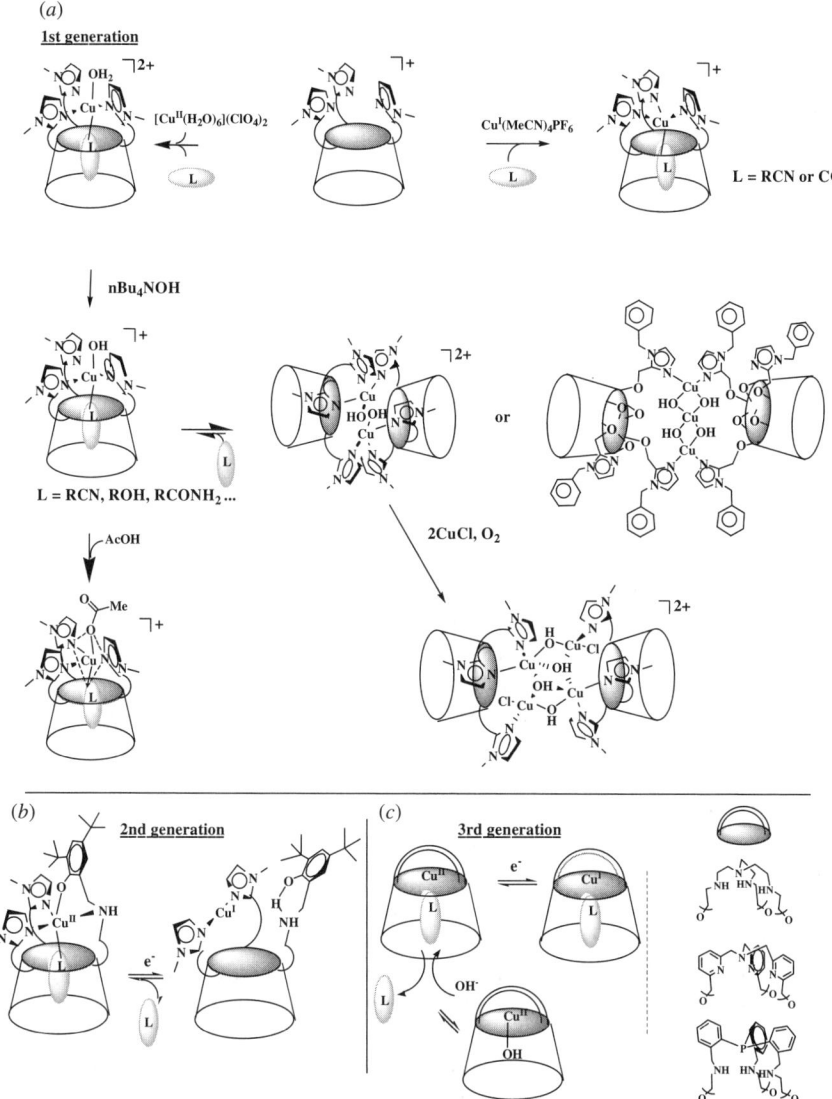

FIGURE 10.4. Coordination modes of three generations of calix[6] ligands. (*a*) first, (*b*) second, and (*c*) third generation.

molecule). This first generation of calix[6]arene ligands has been baptized "funnel complexes" because an organic ligand has to go through the funnel designed by the cone of the calixarene in order to reach the metal center (Fig. 10.4).[24–34]

Hence, in the absence of anions (see Section 10.3.2), these architectures present the advantages of bulky mononuclear copper enzyme models. They confine the metal in a mononuclear environment, and dimerization pathways are prevented by steric hindrance of the ligand. But unlike bulky models, the metal center remains accessible to exogenous molecules through the cavity. They constitute a sophisticated structural model for type 2 sites of enzymes (e.g., PHM and DβH).[1,35]

Attempts to deprotonate the capping water ligand in order to obtain the corresponding hydroxo complex invariably led to the formation[36] of (μ-hydroxo) bridged species. The complexes present a Cu_xO_y core held between two head-to-head calix[6] arenes cones (Fig. 10.4). Interestingly, electron paramagnetic resonance (EPR) studies evidenced the existence of a mononuclear–dinuclear equilibrium in solution, which can be displaced in favor of the mononuclear complex in the presence of organic guest ligands (Fig. 10.4a).[37] In some instances, however, these dinuclear complexes can also evolve toward trinuclear complexes in the presence of excess copper or even to tetranuclear complexes in the presence of chloride anions.[38] This propensity to form polynuclear architectures under basic conditions results from the flexibility of the system, the combined basic and bridging character of hydroxide, and the reluctance of the calixarene structure to bind anionic ligands inside the cavity. This result is due to the π-basic character of the aromatic walls and the crown of phenolic oxygen atoms at the small rim (see Section 10.3.2).

As nuclearity control is condition dependent with this first generation of calix ligands, an additional donor has been appended to the small rim for enhanced stability. This binding arm replaces the external aquo ligand in the metal coordination sphere, thus capping the system in order to enforce endo binding of exogenous ligands. This class of compounds has been described as second generation".[39–41] With a phenate capping group, the nuclearity is well controlled at the Cu(II) oxidation state,[39] whereas one labile site remains accessible to guest molecules. However, this strong donor environment is not favorable to Cu(I) and the control of the geometry is lost upon reduction to the Cu(I) state, due to possible decoordination of one or two hard donor sites of the calix ligand (Fig. 10.4.b).

In order to better control the geometry and the nuclearity of the complexes at both oxidation states and in the presence of anions, a third class of calix ligands was developed by covalent bridging of three alternate phenolic groups by a polydentate chelating ligand (Fig. 10.4c). In other words, the three binding arms of the first generation ligands are now covalently linked together by a trivalent bridging atom, either a nitrogen or a phosphorus.[42–45] Mononuclearity is thus fully controlled at both redox states and in the presence of anions. With the tris[(2-pyridyl)methyl] amine (tmpa) cap, the copper center is constrained in a trigonal environment, either four or five-coordinate, whereas the more flexible tris(2-aminethyl)amine (tren) cap allows distortion toward tetragonal for the Cu(II) state. Interestingly, the PN_3 cap imposes coordination of a phosphorous donor, even to Cu(II), thus providing a very unusual environment associating a soft donor to a hard acceptor. In all cases, a labile site is oriented toward the center of the calix cavity thus allowing full control of the interaction between a guest and metal ion at both oxidation states, through the funnel.

Several platforms (e.g., cyclodextrins,[46] cyclotriveratrylenes,[47] and resorcinarenes[48] have been used to construct a biomimetic coordination core, but none of them allows the cavity control of the labile site. Recently, however, a novel supramolecular copper complex has been reported, based on a molecular basket, derived from a C_{3v} tris(norbornadiene) platform. It is functionalized by three pyridyl groups able to bind a metal center. Nuclear magnetic resonance (NMR) studies of the Cu(I) complex evidenced helical folding of the platform upon coordination of the three pyridyl groups to Cu(I) in a mononuclear environment forcing the metal toward the inner cavity. Ligand-exchange involving coordination to Cu(I) and encapsulation of acetonitrile inside the cavity was also observed, but no Cu(II) complexes has been reported sofar.[49]

10.2.3. Confinement into a Porous Solid Matrix

The use of porous materials (e.g., molecular sieves or zeolites), or even metal-organic frameworks (MOFs), allows us to control, to a certain extend, the nuclearity and the environment of the metal site. Indeed, when confined inside a limited space provided by the pores of the material, the metal complexes are isolated from each other, as they are by the protein backbone of enzymes. However, the control of the geometry and the coordination sphere is difficult to ensure. Therefore, nonbiomimetic strong and rigid ligands allowing maintainance of a stable donor environment to the metal center have most often been used. Their reactivity will be discussed in Section 10.5. The ligand can also be grafted onto the walls of the cavity. Recent approaches consist in functionalizing pores in order to both append an active site and to tune the hydrophobic character of the local environment.[50] The Molecular Stencil Patterning technique, which allows coverage of the internal surface with two different functions, has been used to pattern pores with trimethylsilyl and bromopropyl groups. The latter is an anchor to graft a polydentate ligand [e.g., a bis(pyridine ethylenediamine) (N_4) with a surface coverage of 12.5%]. Metalation with copper leads, as evidenced by EPR, to a mixture of a pseudo-octahedral Cu(N_4O_2) complexes with an elongated Cu$-$O distance and a weaker binding to surface silanol, and Cu(N_2O_3) due to partial protonation of the N_4 donor core (Fig. 10.5).

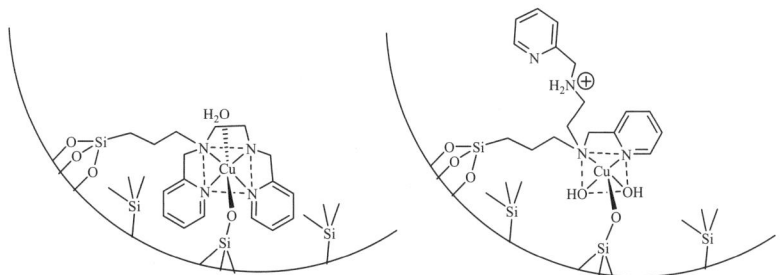

FIGURE 10.5. Binding modes of a grafted copper ligand into the pores of a porous material via molecular stencil patterning. (see Ref. 50.)

10.3. INTERLOCKING METAL BINDING AND CAVITY EFFECT

With the calix[6]ligands, the metal site is deeply embedded at the bottom of the cavity. The cavity itself then participates to the metal-donor environment, which interlocks electronic effects and binding properties of the "funnel-like" systems.

10.3.1. Supramolecular Control of the Guest Exchange Process

In order to reach the metal center, the guest must go through the funnel. With the classical calix systems, three *t*-Bu (*tert*-butyl) substituents constitute a gate that must be open (Fig. 10.6). In order to evaluate the impact of this gate on the properties of the metal ion, two different complexes have been studied for the ligand-exchange process, one presenting the *t*-Bu door, one without. Comparative thermodynamic and kinetic data of guest exchange were obtained by competition experiments with nitrilo guests followed by NMR spectroscopy in CD_2Cl_2.[27] This study evidenced a huge impact on both thermodynamics and kinetics of the exchange process. First of all, it has been shown that in all cases, the process is dissociative. Indeed, the cone shape of the cavity precludes simultaneous binding of two guest ligands in the endo position. With the *t*-Bu door, the order of affinity is EtCN > MeCN ≫ PhCN, while without this *t*-Bu door, it becomes EtCN > PhCN > MeCN, which is associated with a kinetic increase by two orders of magnitude (as measured in the case of MeCN exchange). The thermodynamic impact is explained by the fact that the bulky PhCN bumps into the *t*-Bu sitting in the "in" position at the large rim. The kinetic difference

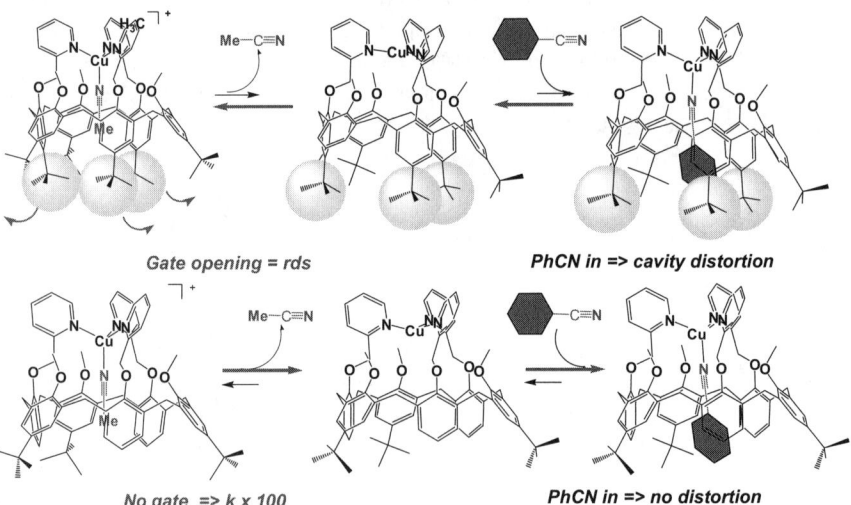

FIGURE 10.6. Gate effect in the calix-tris(pyridine) family: removal of three *t*-Bu groups at the large rim (bottom) opens the cavity access, reverses the binding affinity toward bulky guests and leads to a 100-fold increase of exchange kinetics.

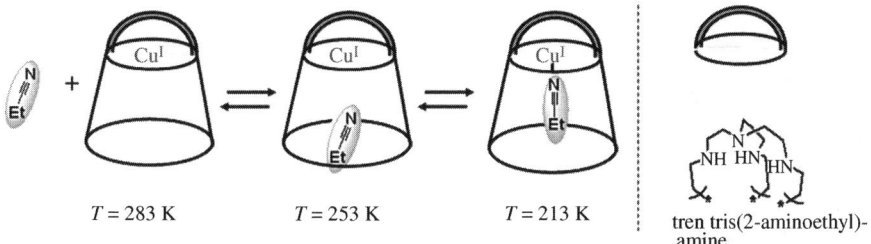

$T = 283$ K $T = 253$ K $T = 213$ K   tren tris(2-aminoethyl)-
                                                                           amine

FIGURE 10.7. Trapping an intermediate along the path of ligand exchange.

stems from the "gate-opening" effect that is necessary when the t-Bu groups close the cavity. This phenomenon presents some reminiscence with the dynamic behavior of access channels to enzyme active sites.

With the more rigid capped calix–tren system (Fig. 10.7), this gate effect has allowed us to trap an intermediate state along the way of the guest exchange process.[50] Due to the strong donor environment provided by the tren cap, the coordination link of nitrile to the Cu(I) center is very weak, labile, and observable only at very low temperature (in CD_2Cl_2). At intermediate temperatures, however (250–240 K), a new species can be observed, corresponding to a nitrilo molecule (EtCN) included in the cavity, but not bound to the metal (Fig. 10.7). Thermodynamic analyses revealed that stabilizing interactions (among which CH-π) due to the guest inclusion inside the calix cavity are actually stronger than the Cu$-$NCEt coordination link.

10.3.2. First and Second Sphere Effects on Guest Binding

Selective binding of neutral guests in calix complexes results from a combination of effects from the first coordination sphere (electronic effects due to the donor cap), second coordination sphere (the oxygen-rich small rim of the calixarene providing hydrogen-bonding sites and dipolar interactions), and cavity (CH–π interaction with the cavity, shape selectivity). Analysis of the competitive binding of neutral guest molecules presenting different functionalities (a nitrile, an alcohol, and an amide) has allowed us probing of the metal-center properties. Hence, the affinity order of a Cu(II) center for MeCN, EtOH, and DMF has been determined through ultraviolet–visible (UV–vis) titration for three different caps of the third generation ligands: tren, PN$_3$, TMPA (see Fig. 10.8), and has evidenced unexpected behaviors: (1) the selectivity order for tren MeCN < EtOH < DMF is similar to that observed for the tris-imidazole system and stands in line with what is expected for a "classical" quite strong Lewis acid (e.g., Cu^{II}) (2) for PN$_3$, the surprising switch of affinity order between EtOH and MeCN and an impressive affinity for DMF was ascribed to an enhancement of the electronic density at Cu(II) by a ligand-to-metal charge transfer (LMCT) due to its axial coordination to the P(Ar)$_3$ group;[52,53] (3) with TMPA, Cu(II) displayed a spectacular preference for the weaker donor MeCN, which suggests a strong effect of the cavity on the substrate shape as the TMPA unit rigidifies the calixarene core much more than the alkyl amino-based caps.[54]

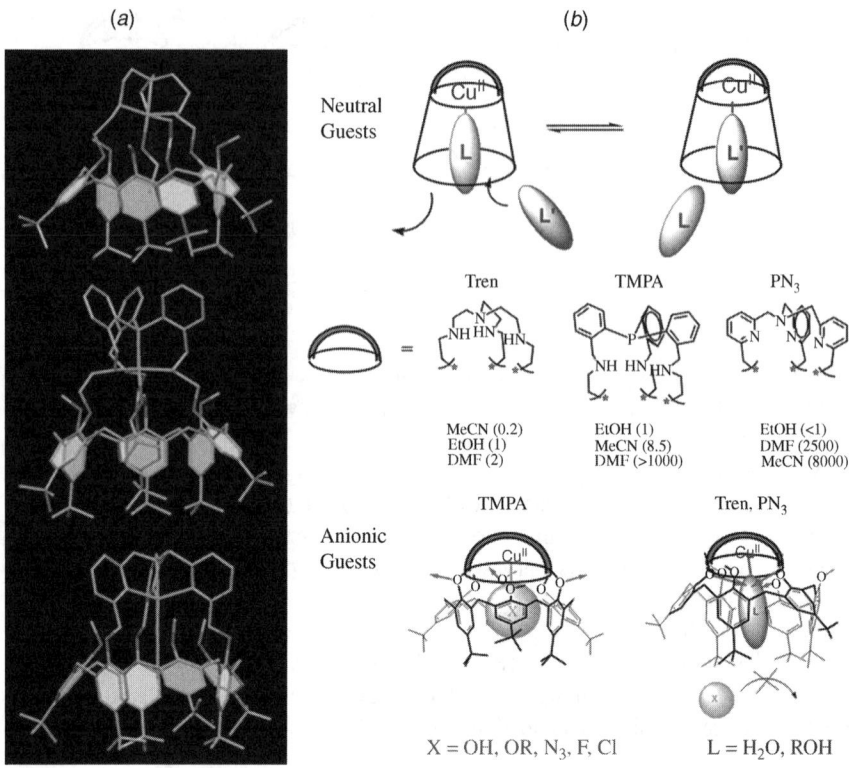

FIGURE 10.8. Supramolecular control of neutral and anionic guest binding. (*a*) The X-ray diffraction (XRD) structures of tren-, PN$_3$-, and TMPA based Cu(II) complexes with EtOH, DMF (dimethylformamide), and MeCN as guests, respectively. (*b*) Guest exchange of neutral and anionic guests (the relative affinities for neutral guests are indicated in brackets). TMPA stands for tris[(2-pyridyl)methyl]amine and Tren stands for tris(2-aminoethyl)amine. (see color insert.)

As observed with first generation calix ligands, dicationic complexes formed with ligand calix–tren and PN$_3$ are reluctant to bind anions inside the cavity (Fig. 10.8). The origin of this phenomenon is proposed to be structural. With the more flexible systems (tris-imidazole, tren, and PN$_3$), the calix adopts a flattened cone conformation with the anisole units connected to the nitrogenous arms pointing their lone pair toward the donor atom of the guest ligand. In contrast, the TMPA core imposes to the same three units to stand in the opposite direction, the corresponding oxygen atom orienting their dipoles toward the outside. As a result, the electron-rich environment provided by the small rim in the former cases prevents coordination of anions, whereas the Cu(II) center embedded in the TMPA capped calixarene is highly sensitive to anions and readily binds hydroxide, alkoxides, azides, and halides.[53] Such contrasted behaviors illustrate the importance of the second coordination sphere generated by the supramolecular environment.

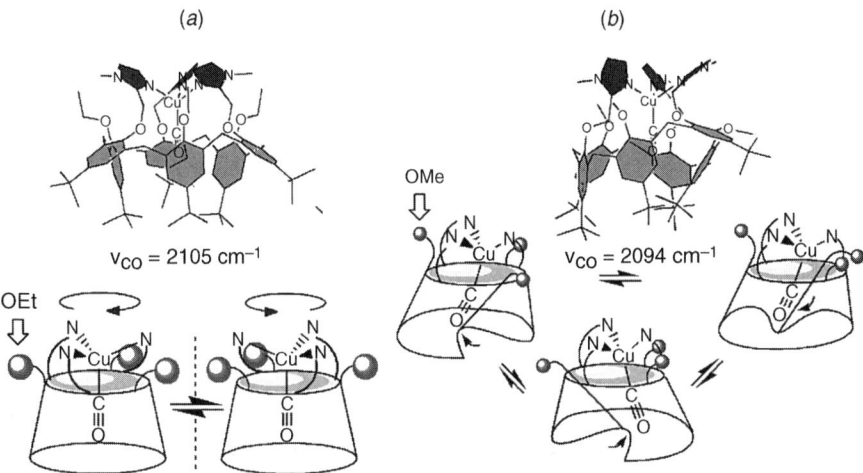

FIGURE 10.9. Coupled conformational and electronic properties in the calix–tris(imidazole) Cu(I) family controlled by the bulkiness of small-rim substituents (*a*) C_3 symmetry complex (OEt substituents) and (*b*) dissymmetrical complex (OMe substituents).

10.3.3. Cavity Effect on the Electronic Properties of the Metal

Carbon monoxide is an analogue of O_2, which does not display redox properties and its coordination to Cu(I) allows us to probe its electron density via vibrational spectroscopy. The metal must be electron rich enough to allow π back-donation to CO. Consequently, calix–tris(pyridine)Cu(I) does not bind CO, whereas the calix–tris(imidazole)Cu(I) does, yielding a Td environment involving the three imidazole arms, as shown by NMR spectroscopy.[28] Low-temperature studies, however, revealed the presence of two different conformations for the calix skeleton: one being of regular C_3 symmetry (favored by the larger OEt substituents at the small rim, see Fig. 10.9*a*), the other being dissymmetrical due to the partial inclusion in the cavity of one *t*-Bu large rim substituent (favored by the smaller OMe substituents at the small rim, see Fig. 10.9*b*). Interestingly, the conformational difference is associated to a ν_{CO} shift, which evidences interdependence between the metal ion electron density and the local environment. Such a behavior is reminiscent of that observed for Cu–CO coordination in C*c*O: The vibrational frequency is split in two, which was ascribed to coordination flexibility at the copper site in the nonpolar environment of the binding pocket.[55]

10.4. SUPRAMOLECULAR CONTROL OF THE REDOX PROCESS

Enzymatic catalysis involving Cu and O_2 is based on the Cu(II)/Cu(I) redox process, which is controlled by constrains imposed by the surrounding protein matrix. The oxidation state of copper highly influences the geometrical preference of the metal. Cu(I) (d^{10}) does not exhibit any crystal field and is thus generally found in a trigonal or tetrahedral environment (often four coordinate, Td) if unconstrained. The Cu(II) (d^9) is subjected to the Jahn–Teller effect and tends to adopt a tetragonal geometry (often five-coordinate, SBP). Thus, important rearrangements arise during the ET process.

10.4.1. Cavity-Controlled Electron Exchange Pathways

In order to evidence trends, three families of first generation calix[6] complexes were studied by Le Mest's group using cyclic voltammetry (CV): tris(pyridine), tris (imidazole), and tris(alkylamine) systems. To describe the ET associated with conformational changes, a square-scheme mechanism has been proposed by Rorabacher (Fig. 10.10). The system operates via the path that involves the most stable of the two possible metastable species in the square scheme. The most common path is pathway A, as Cu(I) readily adapts to SBP and Td geometries in unconstrained environments. Few examples of path B are reported.[56]

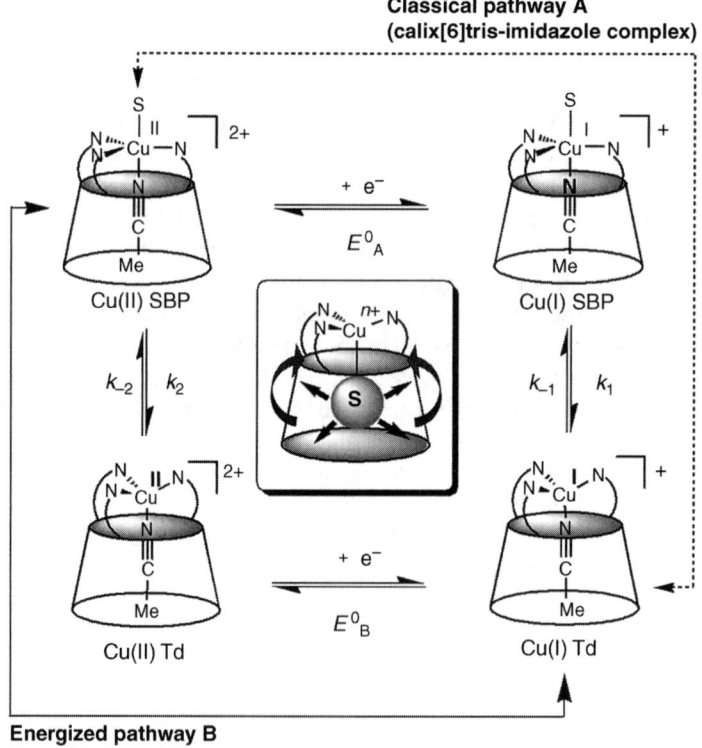

FIGURE 10.10. Redox square scheme summarizing the electrochemical behavior of first-generation complexes, calix–tris(imidazole)Cu(I/II) and calix–tris(pyridine)Cu(I/II).

The presence of a guest molecule (MeCN) inside the cavity is necessary to stabilize a defined conformation in both oxidation states, as well as in the intermediate states. The guest plays the role of a "shoetree" molecule that allows adaptation of the calixarene conformation during the electrochemical process. This concept is reminiscent of the induced fit behavior in biological systems. Hence, completely different behavior is observed depending on the donor character of the binding sites at the small rim. The more donor the N_3 environment, the weaker the Cu–MeCN bond. With pyridine arms (relatively weak donors), anchoring of MeCN is strong. As MeCN also interacts will the aromatic walls of the cavity via CH–π interactions, this generates a supramolecular stress on the metal that favors the Td geometry. The Cu(II) will thus be able to adapt to both SBP and Td geometries, and the ET follows pathway B (Fig. 10.10). It is noteworthy that the Cu(II) intermediate in a Td environment has an enhanced oxidizing power.[33] With the stronger donor imidazole at the small rim, anchoring of MeCN is loose, the supramolecular stress is absent, and Cu(II) prefers the SBP geometry, leading the ET along the classical pathway A (Fig. 10.10).[32] With alkylamino arms, the donor character totally inhibits guest coordination yielding a square planar or SBP arrangement that disfavors the Cu(I) state and makes the complex strongly reducing.[32]

10.4.2. Redox-Driven Ligand Exchange

The cavity has a crucial role in the ligand exchange/redox behavior of the third-generation complexes with respect to their host-free analogues. In the presence of DMF or EtOH guests in a CH_2Cl_2 solution, the behavior of calix–trenCu(II) is electrochemically irreversible, displaying a reduction peak for the guest-bound cupric complex and an oxidation peak for the guest-free cuprous complex [at -0.98 and 0.06 V vs Fc (ferrocene) respectively with DMF].[57] This result is in agreement with the poor affinity of Cu(I) for O-donor ligands. After reduction, the guest ligand is expelled to yield calix–trenCu(I), which has an empty cavity. Upon reoxidation, the guest-free Cu(II) species generated is unstable and is reconverted to the guest-bound cupric complex.

A comparative study of TMPA based complexes with and without calixarene cavity has further evidenced interesting phenomena when two competitive guest ligands are added, one with a good affinity for the Cu(I) state (MeCN), the other for the Cu(II) state (DMF). With the host-free classical complex TMPACu, ligand exchange is fast at both Cu(I) and Cu(II) states and as a consequence, the system is electrochemically reversible. In contrast, the behavior of the supramolecular analogue calix–TMPACu(II) displays an electrochemically irreversible behavior: At the time scale of the voltammetry, ligand exchange is fast at the Cu(I) state, but totally blocked at Cu(II) (Fig. 10.11). This finding is ascribed to the fact that guest exchange at five-coordinate Cu(II) sites is generally associative, which is made impossible by the presence of the cavity that only allows one guest in.[58]

An interesting consequence of the slow exchange process at the Cu(II) state is the possible redox-driven ligand interconversion, which has been observed by CV. At the Cu(II) state, the complex displays a stronger affinity for DMF and $Cu^{II}(DMF)$

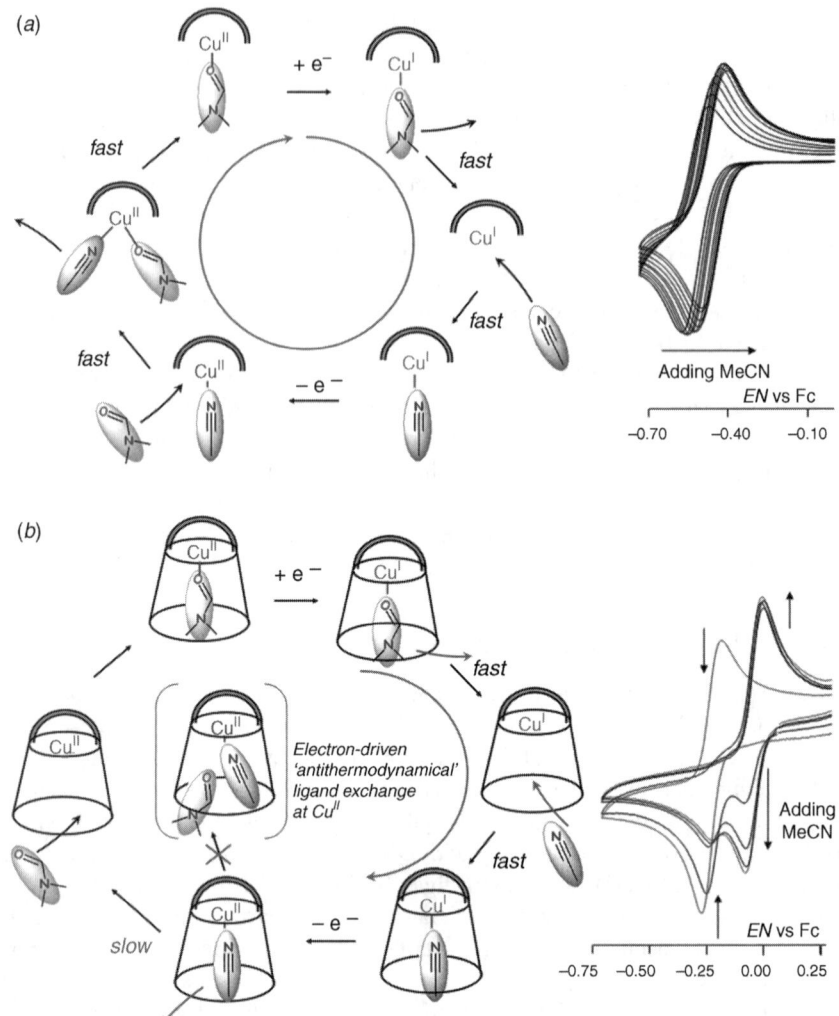

FIGURE 10.11. Electron exchange coupled to ligand exchange and associated CVs at a Pt electrode in CH_2Cl_2: Comparative behavior of a calix-based complex (*b*) and a "classical" complex (i.e., deprived of cavity, *a*). The cavity enforces a dissociative process, which slows down ligand exchange at the Cu(II) state, thus leading to a redox-driven ligand exchange (DMF for MeCN). (see color insert.)

is the only detected species in solution. Upon reduction, DMF is expelled by the Cu(I) complex, due to its poor affinity for the metal center and readily replaced by MeCN. Upon reoxidation, $Cu^{II}(MeCN)$ is formed and detected by CV on the next reduction scan. A good ligand has then been exchanged for a weaker one at the cupric site by redox switch, which can be considered antithermodynamical (Fig. 10.11*b* right).[57]

10.5. DIOXYGEN ACTIVATION AND REACTIVITY

10.5.1. Heterogenous Catalysts

10.5.1.1. Zeolites. This approach, which is based on the incorporation of metal catalysts in porous materials started developing in the mid-1990s with molecular sieves in order to obtain so-called zeozymes or zeolite mimics of enzymes.[59,60] Since then, various examples have been reported using a variety of metal complexes and chemical oxidants with the goal of obtaining efficient catalysts. Dicopper acetate dimers were incorporated in molecular sieves Y, MCM-42, and VPI-5. Incorporated complexes were characterized by elemental analysis, UV–vis and infrared (IR) spectroscopy, and the dimeric nature of the site was confirmed by EPR.[61] Depending on the zeolite, 1 out of 12, or 1 out of 100 unit cells is occupied by a copper dimer.

Other types of copper complexes have been incorporated in zeolites, in particular copper phtalocyanines in zeolites X and Y[62] and M-salen (salen = bis(salicylidene ethylenediamine; M = Co, Cu) into zeolite X (Fig. 10.12a). In this last case, the pore size of 7.4 Å just allows us to accommodate the complexes in the pores while preventing them to diffuse through the narrow windows.[63] The supramolecular environment can have dramatic consequences on the reactivity, as evidenced by the example of the Cu(salen) complex, which is included in zeolite NaX. This result

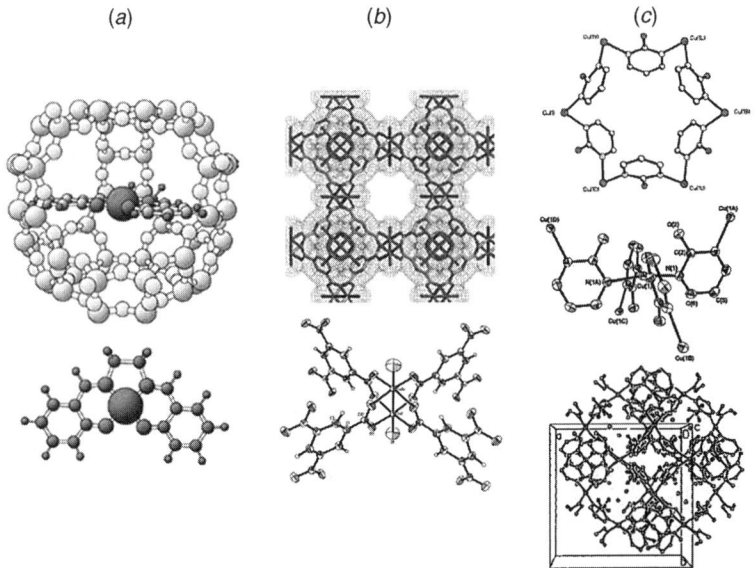

FIGURE 10.12. (*a*) The M-salen entrapped in zeolite NaX. (see Ref. 63). (*b*) Structure of the CuBTC framework showing Cu$_2$ paddlewheel nodes reminiscent of copper acetate. (see Ref. 65.) (*c*) Arrangement of metal centers in [Cu(2-pymo)$_2$] and view of the porous structure. (see Ref. 66.)

shows a lowered activity (almost three times) from the neat complex in cyclooctane aerobic oxidation. The EPR of the complex immobilized in zeolites gives a signal typical of an isolated species, where the axial and in-plane ligand field is weakened. The broadening of the hyperfine structure is characteristic of a micro-strain from the inorganic environment. This result indicates an interaction between the complex and the matrix and possible deformation of the coordination sphere upon encapsulation.[63]

Catalysis with dicopper acetate dimers in molecular sieves Y, MCM-42, and VPI-5 was tested for phenolase (from Tyrosine, phenol, o-, and m-cresol) and catecholase activity, and showed an enhancement of the TON (turnover number) compared to neat copper acetate (10-fold increase with Cu loaded MCM-42, where dimers are separated by 30 unit cells on average). This effect was ascribed to the isolation of the copper dimers from each other, imposed by the inorganic matrix. Those systems also exhibit substrate selectivity and regiospecificity. While phenols are oxidized, aromatics that do not bear a hydroxyl groups are not affected. Most interestingly, monophenols are selectively hydroxylated in the ortho position.[61]

Copper phtalocyanines were incorporated in zeolites X and Y in order to use a bio-inspired catalyst for aerobic radical reactivity. Oxidation reactions were carried out with O_2 as a reagent, in the presence of small quantities of t-Bu-hydroperoxide as an initiator. Incorporation showed an increase in turnover frequencies of several orders of magnitude with respect to the neat complexes.[61]

Benzene hydroxylation was performed using a bimetallic VO_x/Cu catalyst supported on mesoporous SBA-15. It selectively carries out the hydroxylation of benzene as it poorly reacts with phenol under the same conditions, avoiding further oxidation reactions. It also activates O_2 preferentially with respect to H_2O_2, and is much more efficient than mono- (Cu, VO_x) or bimetallic VO_x/M (M = Fe, Ni, Cr, Mn, Zn, Ag) systems of the family, suggesting a synergy between Cu and V might be at work in the mechanism. Again, this approach is probably not biomimetic, but bioinspired, as a pathway involving a hydroxyl radical, where the rate-limiting step would be the O_2 activation and formation of the radical was suggested.[64]

The Cu loaded zeolite Cu–ZSM-5 was characterized by various techniques [extended X-ray absorption fine structure (EXAFS), EPR, transmission electron microscopy (TEM), UV–vis], which evidenced the formation of a mono(μ-oxo)-dicopper core during activation of the system at 723 K under an O_2 flow, a UV band at 440 nm characteristic of this type of adduct being detected under these conditions. Deactivation pathways of intermediates are limited due to isolation of the complex inside the pores and oxidizing power can be used for substrate conversion.[68] Copper-loaded zeolite Cu–ZSM-5 reacts with methane starting from 448 K. Methanol was detected as the only product, but is trapped inside the mesoporous material. Isotope labeling of O_2 and CH_4 reagents resulted in the incorporation of the labeled atoms into the CH_3OH product. The reactivity is highly dependent on the host. Good activity is observed with mordenite and ZSM-5 (similar Si/Al ratios), but low CH_3OH yields are obtained with amorphous silica and zeolite Y (different Si/Al ratios from ZSM-5), evidencing the importance of tuning the pore hydrophilic character.[68] Ethane is converted to ethanol and ethanal in a 1:4 ratio.

10.5.1.2. Metal-Organic Frameworks. As MOFs display metals as part of the host structure in the as-made material, it is thus possible to take advantage of their presence as a potential catalytic site. Framework Cu_3BTC_2 is a three-dimensional (3D) porous material, where metallic nodes are copper dimers bridged by four carboxylate functions, and spacers are benzenetricarboxylates (Fig. 10.12b). The nodes are reminiscent of the copper acetate structure that was incorporated in zeolites (see Section 10.5.1.1).[69] Here [Cu(2-pymo)$_2$], where 2-pymo = 2-hydroxypyrimidinolate, displays accessible copper sites in the material walls, and activity in the oxidation of tetralin (Fig. 10.12c).[70] At 90°C under aerobic conditions, tetralin T is first converted into T−OOH, which is the only product in this first stage. Then, the reaction mixture evolves toward the formation of T−OH and T=O, T=O is the main product.[70] The activity of framework Cu_3BTC_2 was tested for the oxidation of hydroquinone to *p*-benzoquinone under aerobic conditions, and showed a kinetic increase with respect to isostructural frameworks built from Ni and Co.[69]

All in all, few examples have been reported with Cu, and even fewer when oxidation reactions are carried out using molecular O_2 as a reagent.[71]

10.5.2. Biomimetic Catalysts with an Appended Substrate-Binding Site

The hydrophobic pocket of metalloenzymes plays a crucial role in assigning a specific activity to the copper site. Despite having a similar active core, the absence of a significant binding pocket confers hemocyanin, a role in O_2 transport, while it can extend to hydroxylation reactivity when a binding site is located next to the active site as in tyrosinase and catechol oxidase.

10.5.2.1. Calix[4]arenes. Calix[4]arenes have been used by the Gutsche's group[71] in order to generate copper complexes at the large rim of a potential receptor to which (bis-pyrazolylethyl)amino or (bis-pyridylethyl)amino arms have been grafted. Monomeric and dimeric Cu(I) complexes based on calix[4]arenes were studied under O_2 and reactivity was evidenced on the (bis-pyridylethyl)amino compounds. The monofunctionalized calixarene was hydroxylated at the p-benzylic position of the calixarene. The difunctionalized one was only monohydroxylated at the same position (Fig. 10.13).[72] This result was interpreted as an intramolecular oxidation from a transient dicopper peroxo adduct. When reactions were attempted in the presence of potential guests in tetrahydrofuran (THF), no oxidation product could be detected starting from benzene (that forms an endo complex with calix[4]arene in the solid state) or isopropylbenzene. This result was ascribed to a binding competition with solvent molecules in solution, due to relatively poor hosting properties of the calix[4] core in organic solvents.

10.5.2.2. Diphenylglycoluryl-Based Platforms. Molecular clips, which are diphenylglycoluryl-based platforms, have been designed by Nolte and Sijbesma[71] (Fig. 10.14a). They proved to be excellent receptors for di(hydroxy)benzene derivatives (e.g., resorcinol), which displays an association constant of $2600\,M^{-1}$ in CH_2Cl_2. Binding involves hydrogen bonding between phenol groups and the urea

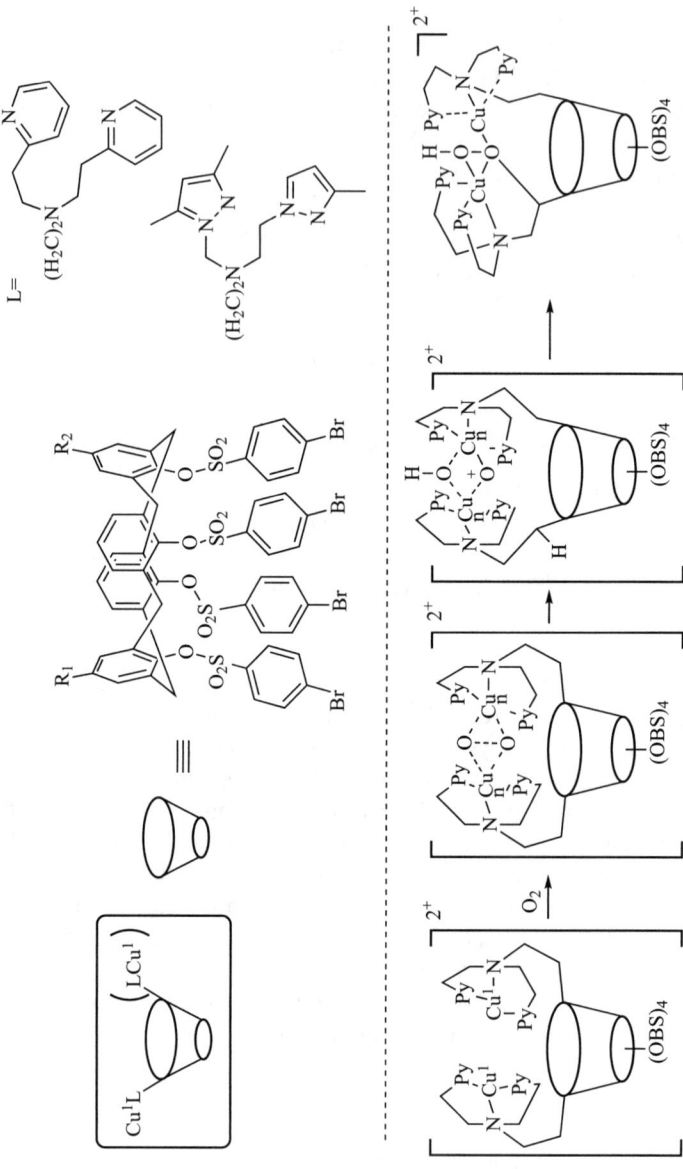

FIGURE 10.13. Calix[4]arene-based molecular baskets and proposed mechanism of activation of O_2 at a dinuclear Cu(I) system appended to a calix[4] arene, leading to hydroxylation of the ligand. (see Ref. 72.)

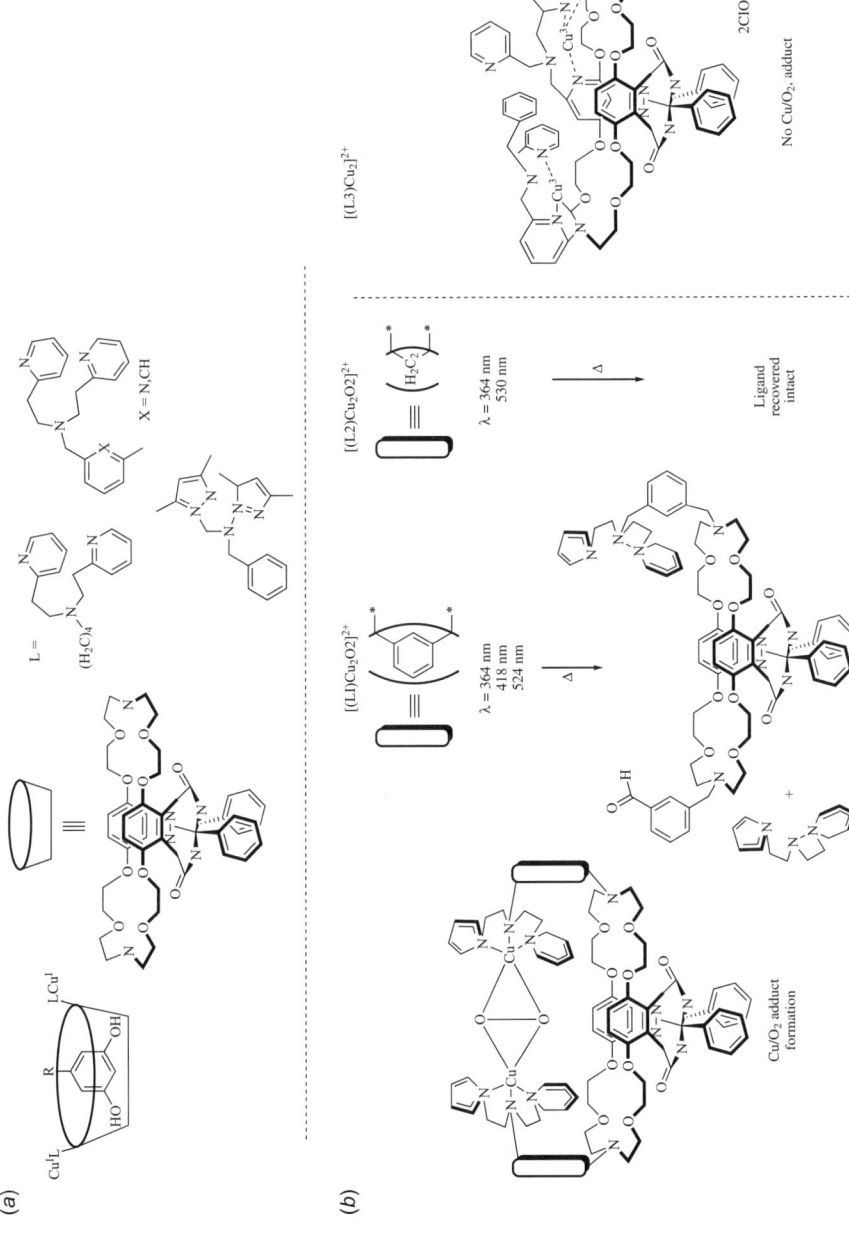

FIGURE 10.14. (*a*) diphenylglycoluryl-based selective receptor of 1,3-dihydroxyaryl compounds (molecular basket) (see Refs. 75 and 76). (*b*) Formation of Cu(I)/O$_2$ adducts with basket-based dicopper(I) complexes and their absorption maxima. (see Refs. 75 and 76.)

functions of the clip, π–π stacking of the guest with the aromatic walls of the receptor, and an optimal filling of the cavity. Electron-withdrawing groups on the aromatic rings enhance the hydrogen bonding strength leading to an association constant of $10^5 \, M^{-1}$ in the case of 1,3-dihydro-5-cyanobenzene.[74] The clips have been functionalized with oxyethylene loops to which two aza arms providing a polydentate ligand at each side have been grafted in order to generate dinuclear Cu complexes above the basket (Fig. 10.14).[74]

With the basket bearing two bis(pyrazolylamine) groups, the dicopper(II) complex was tested for its activity in benzyl alcohol oxidation. It exhibited a massive increase (four orders of magnitude) of the oxidation rate with substrates showing shape complementarity with the receptor (i.e., hydroxy- and dihydroxybenzyl alcohols). Based on UV–vis experiments proving the substrate was bound to copper via its benzyl alcohol function, multipoint recognition was suggested as an explanation for the specific rate increase.[75] This example evidenced the efficiency of preorganizing a substrate in front of a metallic active site with these supramolecular receptors. This approach was further investigated with dinuclear Cu(I) systems, and their O_2 adducts, as a reactive site in oxidation catalysis.

A metastable adduct was observed by UV–vis spectroscopy in CH_2Cl_2 at $-85°C$, upon oxygenation of a $(L1)Cu_2^I$ solution, which displayed patterns reminiscent of hemocyanine and PY2-based dicopper complexes. Thus they were attributed to a dicopper(II) peroxo core with the peroxo bound in a η_2:η_2 fashion (Fig. 10.14).[76] When the temperature was raised to $-60°C$, the adduct decomposed rapidly, yielding a green solution. After work up, PY2 and an aldehyde were identified as products of the oxidative mono-N-dealkylation of the ligand. This finding suggests that the system behaves as a monooxygenase and that the oxidation process is intramolecular, as no bis-N-dealkylation product is observed (Fig. 10.14).

Reactions toward exogenous substrates bearing a dihydroxobenzene moitie were conducted, but no oxidation product could be identified. This result might be ascribed to the formation of polymeric species resulting from radical coupling, as was often observed in the reactions of Cu_2O_2 with phenols. The analogous $(L2) \, Cu_2^I$ was synthesized in order to remove the oxidizable position of the ligand (Fig. 10.14).[78] As in the former case, upon oxygenation of the cuprous complex at low temperature in various solvents, an adduct corresponding to a dicopper(II) η_2:η_2 peroxo was formed intramolecularly. After warming and work up, the ligand was recovered intact, validating the approach.[77] Various types of biomimetic oxidation reactions were then tested on substrates able to bind to the receptor: benzylic hydroxylation, aromatic hydroxylation, and epoxidation. Unfortunately, only unidentified products were obtained, which is attributed to radical coupling reactions leading to oligo- or polymeric products with phenolic substrates. To test the hypothesis of radical coupling reactions, 2,4-di-t-Bu-phenol was used as a substrate. Twenty-eight percent of coupling products were obtained, representing a TON of 1.4 at $-80°C$ and up to 10 after warming up to room temperature under O_2. Radical coupling interferes with the expected hydroxylation activity. Catecholase activity was tested on 3,5-di-t-Bu-catechol (DTBC), which is oxidized to 3,5-di-t-Bu-quinone (DTBQ) by O_2 in the presence of a pair of Cu(II) ions. With 50 equiv of substrate, after 80 min, conversion was twice as high with the basket complex than with the control $Cu(OAc)_2 \cdot H_2O$. The initial rate is also more than five times higher. A biphasic behavior is observed, interpretated as an initial fast oxidation reaction followed by a slow reoxidation of Cu(I). Further experiments evidenced inhibition of the catalyst by the DTBQ product.

In the TMPA appended diphenylglycoluril basket (L3), no $Cu(I)/O_2$ adduct was observed, due to the influence of the vicinal cavity. Indeed, when Cu(I) is bound to TMPA, one of the pyridines of each TMPA is dangling free and dynamic inclusion of one of the coordinated pyridyl occurs (Fig. 10.14). It can probably be attributed to the impossibility for the two Cu sites to adopt the required relative orientation to bind O_2, due to the attachment of the basket platform.[77]

10.5.2.3. A Metallic Anchor. This concept of preorganized substrate-binding site also has been explored by Casella and co-workers[79,80] with, however, a different strategy. They designed ligands bearing three coordination sites, with the aim of generating a dinuclear peroxo Cu(II) species in close proximity of a third Cu center, the role of which is to anchor the substrate through a coordination link.

Casella and co-workers[79–82] reported a series of organic ligands that were used to generate di- and trinuclear copper systems (Fig. 10.15).

Catecholase activity tested under anaerobic conditions with complex $[Cu_3PHI]^{6+}$ and DTBC as a substrate indicates that in the redox process forming DTBQ, one CuA and one CuB are reduced. This result led them to propose that catechol binding occurs between those two sites (Fig. 10.15).[79,82] Aerobic kinetic experiments with chiral catechols displayed enantiodifferentiation between L- and D-DOPA (3,4-dihydroxyphenylanaline) that lies in the difference in the binding constants of the two enantiomers. Anaerobic 1:1 binding isotherms with substrates Tyr, His, Ala shows a preferential binding of D isomers. Thus, a proposed interpretation is the coordination of the catechol group to one A and one B site during the oxidation process, while the amino acid moiety is anchored to the remaining A site. The vicinity of the A site asymmetric carbon may then be responsible for the binding stereoselectivity (Fig 10.15). When chiral groups are located at site B instead of site A, enantioselection

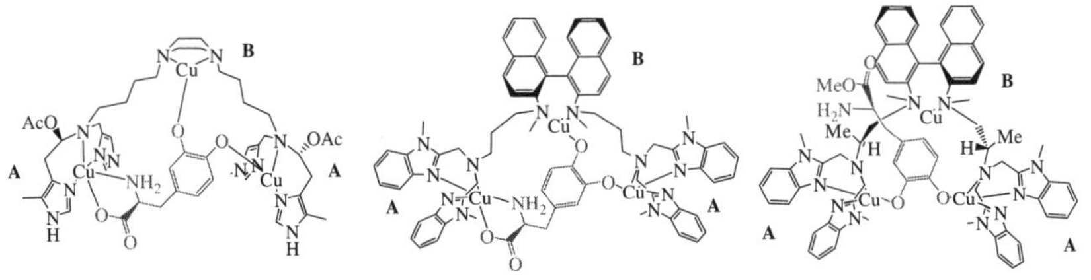

PHI Ligand   (*R*)-DABN-3Bz Ligand   (*R*)-DABN-L-Ala-Bz4 Ligand

FIGURE 10.15. Different trinuclear complexes used in catecholase-type oxidations by Casella and co-workers (see Refs. 77–82) with proposed coordination modes of substrate. PHI = [piperazine-1,4-bis(4-{*N*-[1-acetoxy-3-(1-methyl-1*H*-imidazol-4-yl)]-2-propyl}-*N*-(1-methyl-1*H*-imidazol-2-ylmethyl)aminobutyl)] and DABN stands for the 1,1'-binaphthalenyl-2,2'-diamino group.

drops with the trinuclear compound $[(R)\text{-DABN-3BzCu}_3]^{6+}$, while it remains non-neglectable for the dinuclear $[(R)\text{-DABN-3BzCu}_2]^{4+}$ complex.[78] In $[(R)\text{-DABN-L-}$Ala-Bz$_4$Cu$_3]^{6+}$, chirality is present at both sites A and B and their distance is shortened. The EPR indicates that catechol binding occurs between two A sites (Fig. 10.15). So, chiral recognition must occur between the amino acid and the binaphtyl moities. Noticeable enantioselectivities were observed with DOPA methylester and norepinephrine and are related to a difference in binding constants. Nevertheless, there is no evidence of direct binding of the amino group of the substrate to site B.[81]

10.5.3. Supramolecular Control of Reactivity and Stabilization of Intermediates

Supramolecular systems can provide environments where reactive intermediates are stabilized, preventing undesired deactivation pathways.

This phenomenon was particularly well evidenced with the example of dicopper bis-μ-oxo species. Complex $(Bn)_3TACNCu_2\text{-}(\mu\text{-O})_2{}^{2+}$ displays a half-time of 7 s at $-10°C$ due to self-oxidative decomposition. Dendritic analogues $(L_n)_3TACNCu_2\text{-}(\mu\text{-O})_2{}^{2+}$ ($n = 2,3$, Fig. 10.16) display increased stabilities of 24 s ($n = 2$), and 3075 s ($n = 3$). This effect is entropic, as self-oxidation requires accessibility of the benzylic arms to the Cu_2O_2 core, which is prevented by the dense packing of the benzylic subunits as the dendrimer generation increases.[83]

Reactivity can also be activated–inhibited by the interaction between the metal site and the guest. Reaction of Cu(I) with ligand (L) depicted in Figure 10.17 generates a coordination cage of stoichiometry Cu_4L_4.[84] Copper centers form a regular pyramid, while the C_3 symmetry of the tridentate ligand allows it to bind one copper ion via one nitrogen at each corner of the triangular faces of the pyramid. The diameter of the generated cavity is 7 Å. Depending on the Cu counterions, the cavity is filled either by one of them, or by MeOH when an obvious sterical mismatch exists (triflates, p-toluenesulfonates, etc.).

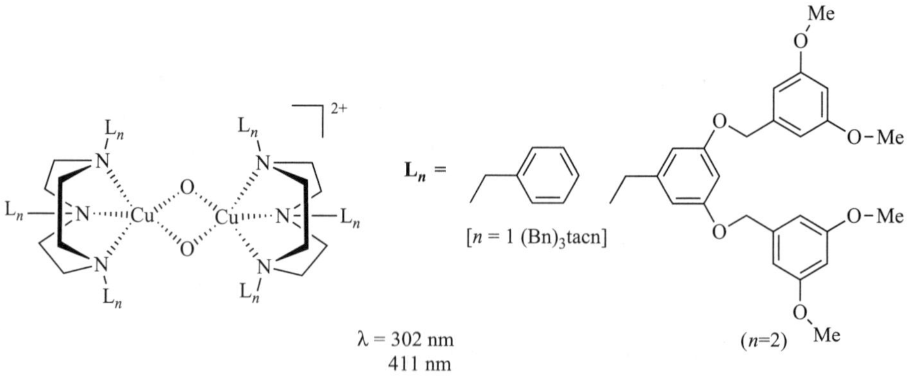

FIGURE 10.16. Formation of Cu(I)/O$_2$ adducts in the [L$_{n3}$TACNCu]$_2$ dendritic family. (see Ref. 83.)

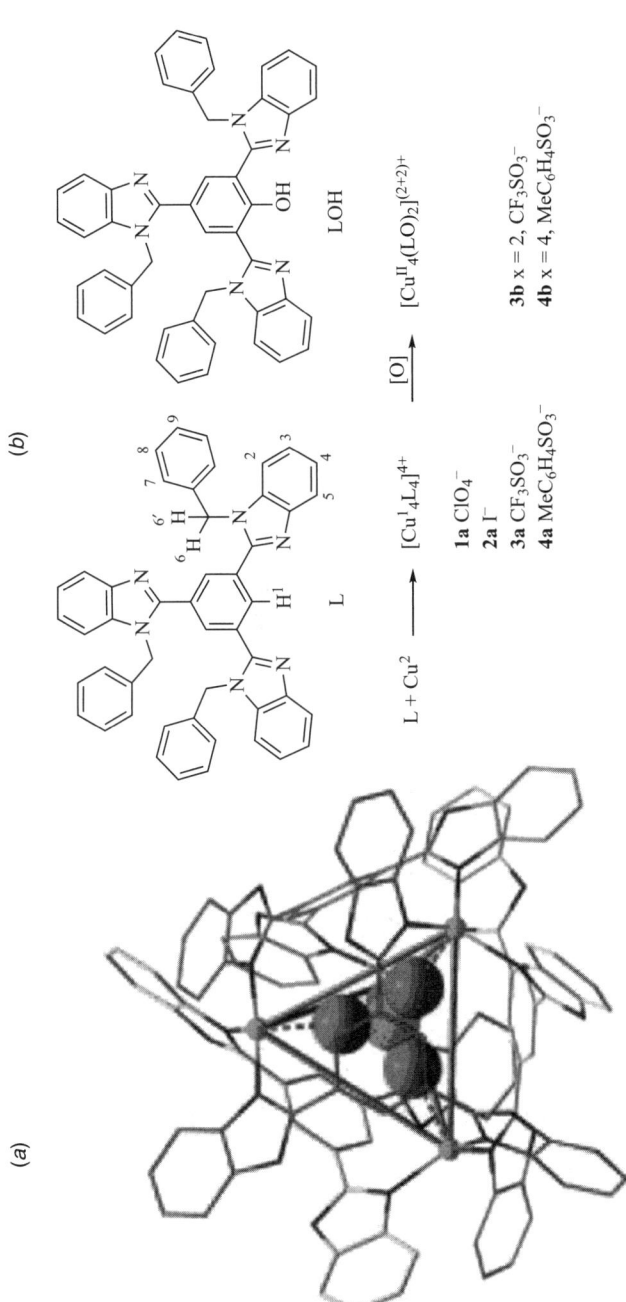

FIGURE 10.17. Structure of the tetranuclear cluster (*a*) formed from reaction of ligand L (*b*) with a Cu(I) perchlorate salt. (see Ref. 84.)

345

Only in the cases where counterions are not included is Cu(I) oxidized to Cu(II) within a few days, with concomitant hydroxylation of the ligand, giving rise to new dinuclear and tetranuclear species. When MeOH is encapsulated, copper centers remain accessible to O_2, thus allowing its activation toward O-insertion into a C—H bond. Spherical anionic guests, such as iodide or perchlorates, fill the cavity and interact with the four Cu centers, thus stabilizing all of them. This finding prevents O_2 interaction with the metal, and leads to redox-inert architectures.

10.5.4. Funnel Complexes: Toward Biomimetic O_2 Catalysis

10.5.4.1. First Generation. Despite their labile binding site, cuprous complexes based on first generation calix ligands are extremely air stable. Reactivity was thus tested by using the peroxidic shunt starting from the cupric complexes and H_2O_2.[30] Addition of 3 equiv of H_2O_2 to calix–tris(pyridine)Cu(II) in CH_2Cl_2/EtOH or MeCN/ benzene resulted in the formation of acetaldehyde and acetic acid from EtOH, and of phenol and benzoquinone from benzene (Fig. 10.18). Interestingly, no catechol or *o*-quinone was detected. Using calix–tris(imidazole)Cu(II) as a catalyst, similar results were observed for benzene, but no acetic acid was formed from EtOH. Lower turnovers were also obtained, all of which can be ascribed to its lower oxidation power [0.2 vs 0.9 V for calix–tris(pyridine)Cu(II) (reference Fc^+/Fc) where ferricinum=Fc^+]. Low-temperature ($-78°C$) UV–vis spectroscopy revealed an intense absorption at 375 nm reminiscent of that observed for a N_4Cu^{II}—OOH

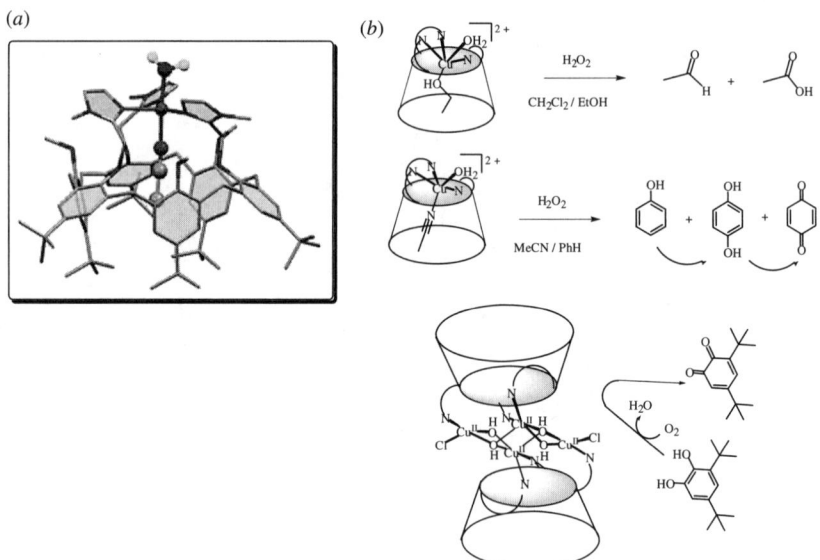

FIGURE 10.18. (*a*) Catalytic reactions performed by calix–tris(pyridine)Cu(II) via the peroxide shunt pathway; (*b*) catecholase activity of the tetranuclear cluster obtained by reaction of O_2 with a 1:1 mixture of complex calix–tris(imidazole)Cu(I)(Cl) and Cu(I)(MeCN)$_4$PF$_6$. *Inset*: XRD structure of the calix–tris(imidazole)Cu(II) complex.

complex and attributed to a LMCT HOO → Cu^{II}. This intermediate could be responsible for the oxidation of the substrate.

A tetranuclear complex displaying two dicopper sites that could model type 3 copper sites can be formed from ligand calix–tris(imidazole). Its catecholase activity was tested on 3,5-di-t-Bu-catechol (dtbc; Fig 10.18b). The Cu(II) complex turned out to be a good oxidant leading to the formation of quinone, but the regeneration of the Cu(II) state with O$_2$ is slow and leads to sluggish catalytic activity. In this example, substrate access to the active site can only take place outside the cavity of the calixarene cones, which already include one of their imidazole arm.

10.5.4.2. Second Generation. The cupric complex of second generation calix–ligand presenting a phenolate group as a capping ligand was tested as a GO model.[39] One-electron oxidation of the Cu(II) phenolate complex at -45°C led to a green species that displays an intense π–π* phenoxyl absorption and concomitant bleaching of its EPR signal. These patterns attest to the formation of a Cu(II)–(phenoxyl radical) complex that is similar to that observed in GO and models. Upon reaction with benzyl alcohol, the Cu(I) complex is formed along with 1 equiv of benzaldehyde. Reoxidation in air regenerates the initial purple Cu(II) phenolate complex and not the Cu(II)–(phenoxyl radical) necessary to close the catalytic cycle (Fig. 10.19).

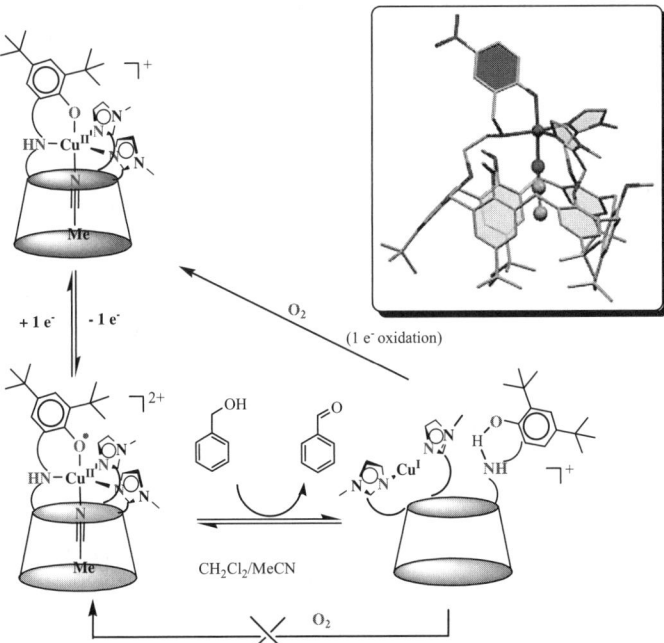

FIGURE 10.19. Redox bahavior and reactivity of Cu complexes based on second generation ligand, model of the active site of GO. *Inset*: The XRD structure of the p-NO$_2$ArOCu^{II} complex. (See color insert.)

Such reactivity with benzyl alcohol is not observed with the Zn analogue despite a similar PhO$^-$/PhO$^\bullet$ redox potential and evidences an inner-sphere mechanism involving coordination of the substrate to the Cu center. This second generation Cu complex mimics GO in both its spectroscopic and oxidative signatures, but also in the supramolecular features related to the substrate-binding site. Nevertheless, reactivity control is lost at the Cu(I) state, and prevents catalytic activity.

10.5.4.3. Third Generation. Optimal control of the metal site at both oxidation states has been obtained with third generation calix[6] complexes. Indeed, at both Cu(I) and Cu(II) states, the metal ion remains coordinated to the polyaza cap closing the calixarene small rim. Hence, the reactivity of the calix–tren, $-$PN$_3$, and $-$TMPA Cu(I) complexes (Fig. 10.4) toward O$_2$, has been explored and all three members displayed very different behaviors. The Calix–trenCu(I), complex in line with the strong electron-donating tren cap, is revealed to be very sensitive to O$_2$ and reacts within seconds even at very low temperature to give Cu(II) complexes. The Calix–PN$_3$Cu(I), complex showed in contrast, to be totally insensitive to O$_2$, whatever the conditions of T or solvent, and remained unchanged after weeks. Obviously, the replacement of a nitrogen atom (in tren) by a soft donor, such as phosphorus (in PN$_3$) in an apical position, strongly affects the properties of the copper ion. Finally, calix–TMPACu(I) displayed remarkable stability versus O$_2$ in solution, like PN$_3$, but reacts in the solid state to produce a new Cu(I) complex in which the ligand has been very cleanly oxygenated and dehydrogenated (CH$_2 \rightarrow$ C=O). Such a process, which corresponds to a formal 4e$^-$ oxidation mediated by an isolated Cu(I) ion, is rare and has been the subject of an in-depth study.

In a noncoordinating solvent, calix–TMPACu(I) readily coordinates CO, but the presence of only 1 equiv of MeCN fully inhibits the process, due to the tremendous affinity of this complex for MeCN as a guest.[85] In CDCl$_3$ or CD$_2$Cl$_2$ and in the strict absence of MeCN (and CO), the cuprous complex, in spite of its empty cavity and apical free coordination site, remains insensitive to O$_2$, even after weeks. Surprisingly, in the solid state, although the complex remains colorless and EPR silent showing the oxidation state remains $+$I, the Cu(I) complex undergoes structural changes within 8 h of air exposure. Electrospray ionization mass spectrometry (ESI–MS) displays peaks at M $+$ 14 ($+$O,-2H) and M $+$ 16 ($+$O). Infrared spectroscopy reveals a new stretch at 1738 cm^{-1} corresponding to an ester function, leading to the hypothesis that a methylene group linked to one calix–phenol unit has been oxidized, which was confirmed by deuteration of these groups (Fig. 10.20). Kinetic data indicated a parabolic dependence over time, characteristic of a diffusion controlled process with fast H abstraction. The partially labeled complex calix–[(H$_2$)$_2$(D$_2$)]TMPA Cu(I) complex was used to determine intramolecular kinetic isotope effect (KIE) values, that were found to be 21 at room temperature (RT), reaching 29 at 277 K. With activation enthalpies (derived from an Arrhenius plot) exceeding the zero-point energy between C$-$H and C$-$D bonds and preexponential factors lower than normal values, it was concluded that a tunneling phenomenon, also observed in natural systems, was probably at work in the oxidation process. No further oxidation of the small rim of the calixarene was observed, which is ascribed to the electron-withdrawing effect of

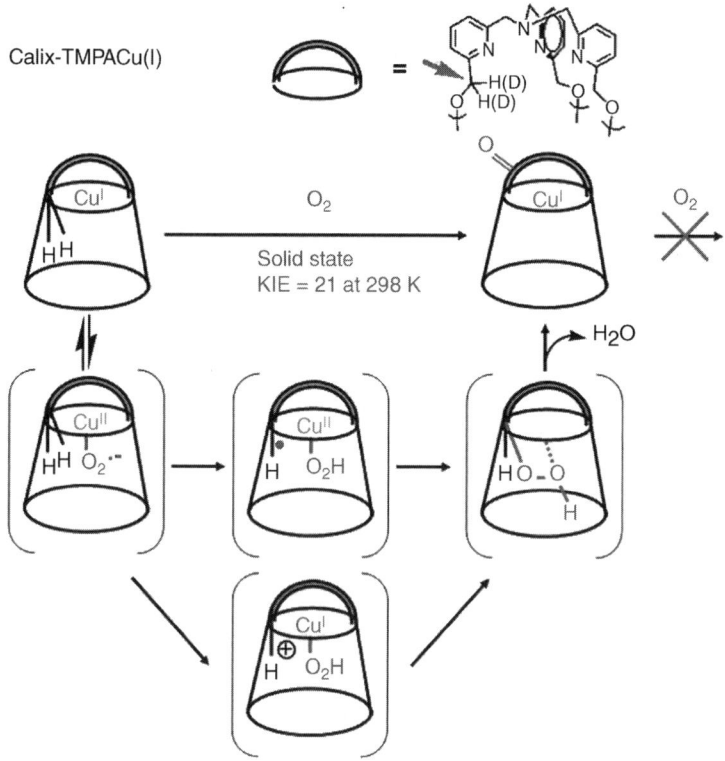

FIGURE 10.20. Oxygen activation in the solid state by a third generation calix complex leading to a $4e^-$ oxidation of the ligand (the arrow on the cap indicates the site of oxygenation, and deuterium labeling), and proposed mechanism (species in brackets have not been detected).

the newly created C=O group that lowers the electron density on the Cu center, thus disfavoring further O_2 activation.

The peroxidic shunt pathway was examined in solution. When H_2O_2 (a few molar equiv) was added to a solution containing the Cu(I) complex, oxidation to Cu(II) was quickly observed and was associated with the formation of a few percent of the alcohol derivative (M + 16), but no keto product. When the cupric aquo complex itself $\{[Cu(Calix[6]TMPA)(H_2O)]^{2+}\}$ was reacted with $10\,M$ equiv of H_2O_2 in the presence of Et_3N (2 equiv), the corresponding hydroxo–Cu(II) complex $\{[Cu(calix[6]$ $TMPA)(OH)]^+\}$ was produced with, again, only a trace of the M + 16 product and no keto derivative. In contrast, however, when the dicationic cupric complex was reacted with KO_2 (3 equiv) and crown ether (18-C-6), both M + 14 and M + 16 products (~ 10 and 5%, respectively) were detected. All this shows that a Cu(II) hydroperoxide proposed to be formed during the peroxidic shunt pathway ($Cu^{II} + H_2O_2 \rightarrow$ Cu^{II}–OOH) leads to poor, and in any case, different reactivity compared to the Cu(I)/O_2 chemistry. This finding suggests a Cu(II) superoxide [Cu(I) + $O_2 \rightarrow$ Cu(II) $O_2^{\bullet}$] to be a reactive intermediate. A possible mechanism (Fig. 10.20) relies on the

$2e^-$ oxidation of the proximal $C-H$ bond by the mononuclear center $[Cu^{II}O_2^-]^+$, either through a radical pathway or through direct hydride abstraction. The resulting putative alkylhydroperoxide Cu(I) intermediate then evolves intramolecularly into a keto product with H_2O release. Kinetics have indicated that the $C-H$ bond cleavage is a relatively fast process, hence suggesting that the $[CuO_2]^+$ intermediate is quite reactive.

Contrarily to the Cu(I) complexes based on the first generation of calix ligands displaying a tris(imidazole) or tris(pyridine) core, or to the TMPA- and PN$_3$-capped ligands of the third generation, the calix–trenCu(I) complex appeared to react very fast in solution with O_2 in various solvents, even at very low temperature, and without accumulation of any detectable intermediate. As a first analysis, such a high reactivity can be ascribed to the donating strength of the tren cap. However, whereas tren^{R}Cu(I) complexes deprived of cavity were described as stable in air in the solid state, solid calix–trenCu(I) turns instantaneously green in contact with air, indicating formation of Cu(II) species, as in solution.[86] Nevertheless, mechanism and products have been found to be condition dependent. When carried out in MeCN at low temperature, the calix ligand remains intact during the oxidation process. The EPR studies in the presence of a radical trap evidenced the release of superoxide. When carried out in CH_2Cl_2 at low temperature, postoxygenation EPR measurements revealed the presence of several Cu(II) complexes resulting from the alteration of the calix ligand. The ESI–MS of the organic extracts displayed a peak at $(M + 14)$ corresponding to a $4e^-$ oxidation $(+O, -2H)$. At room temperature, an additional peak at $(M + 28)$ was detected, corresponding to a further $4e^-$ oxidation (Fig. 10.21). When carried out in a CH_2Cl_2/acetone (1:1 v/v) mixture at 200 K, the most intense mass peak was observed at $(M + 16)$, which corresponds to just one O atom insertion. The $^{18}O_2$ labeling experiments confirmed that the inserted O atom came from O_2. The ESI–MS and IR data are consistent with the oxidation of the methylene group linked to the phenolic oxygen atom, as in the above-described case of calix–TMPA.

In the case of tren, the reaction between the strongly reducing Cu(I) center and O_2, identified as an inner-sphere process, is also proposed to transiently lead to a $[CuO_2]^+$ adduct. According to all previously reported studies on N_4Cu/O_2, its nature can be assigned as a Cu(II) superoxide entity resulting from a $1e^-$ exchange between Cu(I) and O_2. However, such an intermediate could not be observed, even at low T. Such a surprising high instability may be due to the anionic character of the transiently formed superoxide entity and its repulsive interaction with the second coordination sphere provided by the oxygen-rich calixarene small rim. Hence, the transiently formed $O_2^{\bullet-}$ species is either ejected from the calixarene in the presence of a coordinating solvent (MeCN) in a very fast process without any damage to the ligand, or, in the absence of a guest ligand to displace it, attacks a $C-H$ bond and formation of a transient Cu(I) hydroperoxide is proposed, as in the TMPA case (Fig. 10.21). Its intramolecular evolution in CH_2Cl_2 results in the insertion of one O and removal of two H atoms in the calix cap (to give the keto product), whereas it is readily trapped by acetone, thus leading to the alcohol as a major product.

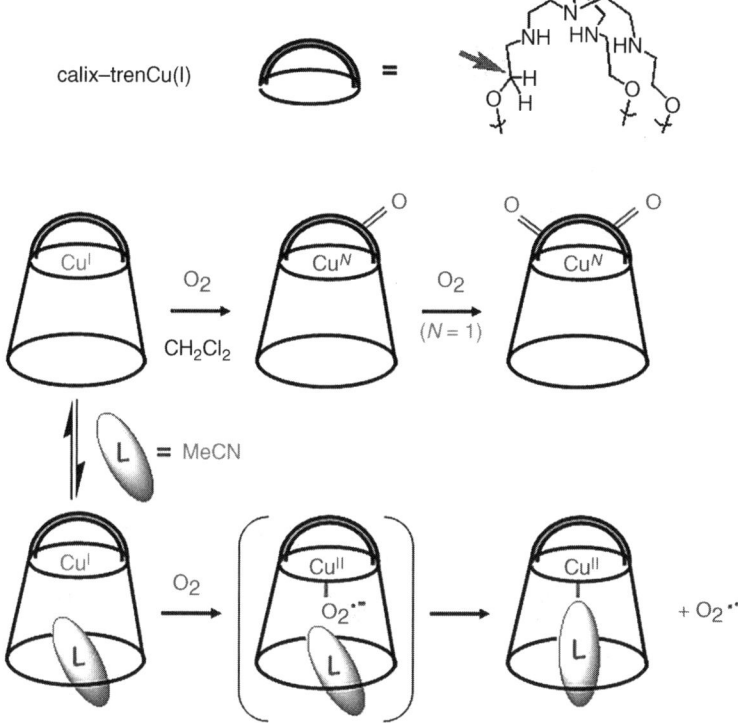

FIGURE 10.21. Oxygen activation in solution by a third generation calix complex, calix–trenCu(I) and deactivation pathway through guest binding thus trapping the putative superoxide transient adduct the arrow on the cap indicates the sites of oxygenation.

All these results unambiguously demonstrate that O_2 interaction at an isolated Cu (I) center can give rise to a species reactive enough to break a $C-H$ bond, at least in substrates activated by a heteroatom (O) or a π system in α-position. It also shows the possible oxygenation of an organic moiety *without* the need of an external electron input. Such observations highly substantiate the hypothesis according to which, in copper monooxygenases, such as DβM and PHM, the $[CuO_2]^+$ adduct first attacks the $C-H$ bond of the substrate, *before* the electron input from the second Cu center. In a more general way, it shows that a single Cu(I) center in interaction with O_2 can mediate an even ET process without the assistance of a redox cofactor, which is a key point for the development of a controlled catalytic process devoted to the oxidation of organic substrates. Indeed, the insertion of two O atoms on the same ligand (in the case of calix–tren) associated with the loss of four hydrogen atoms (formally an $8e^-$ oxidation process associated to O_2 reduction) indicates that the reaction can run catalytically. It indicates that if the oxidation were not directed toward the ligand itself, but toward a guest substrate, it could act catalytically. Future work now must be focusing on the reactivity of such isolated Cu centers toward exogenous guest substrates.

10.6. CONCLUSION AND FUTURE PROSPECTS

The building of supramolecular systems in biomimetic chemistry was predictable as it constitutes a second stage of complexity inspired by natural systems. From nonsupramolecular Cu complexes that proved efficient for modeling the first-coordination sphere of enzymes, the use of secondary noncovalent interactions and cavities allowed us to introduce new degrees of control for the properties of reactive systems.

- **Nuclearity Control Remains an Important Challenge**

Monocopper centers definitely require protection and embedment. This finding has led to the strategy of the "funnel complexes", in which the metal ion is buried at the bottom of a half-open cavity. They are synthetically tunable, but only to a certain degree. Indeed, modifying the cavity shape requires changing the macrocyclic platform to which the biomimetic core is grafted. As a matter of fact, there are not many of these and much organic synthetic effort is required.

For polynuclear sites, the most important requirement is the preorganization of the copper sites with respect to each other. Although this is not an absolute requirement for dinuclear chemistry, it is certainly for trinuclear clusters, which definitely require a platform that orient three Cu centers next to each other (and avoids the dinuclear thermodynamic sink). There is very little work yet published for this topic, which is partly due to the difficult task of identification of the solution species.

Another approach consists of the isolation of a Cu center or a cluster into a porous matrix. The major difficulty stems from the interaction of the matrix with the Cu complex. As a result, most studies have selected complexes, where the metal ion(s) is strongly held in a rigid macrocycle or by hard donors, which are far from being biomimetic. These "bioinspired systems" have nevertheless given promising results in term of oxidation catalysis. Grafting a relatively soft and flexible ligand onto the wall of the porous matrix is an elegant approach, but is still at its infancy.

- **The Microenvironment Is Key for Reactivity**

It is now established that a well-defined second coordination sphere has a decisive role on the metal ion properties and thus reactivity: Hydrogen-bonding sites, dipoles, hydrophobic environments tune the metal ion affinity for exogenous ligands. Embedment into a cavity (a funnel) also has a strong impact on ligand exchange processes, which is naturally related to substrate access and binding–product release and exchange mechanism. Interlocking cavity effects and electronic effects is the basis for controlling the redox properties. Conversely, a redox change within a well-controlled environment may lead to energized states.

However, reproducing the microenvironment found in the active site is extremely difficult. Up to now, first approaches have highlighted the importance of the correlated factors, but we are far from a precise mimic of the proteic chamber.

• Substrate Binding and Positioning Is Key for Efficiency and Selectivity

Several systems have been elaborated on to anchor a substrate next to a metal center. One strategy is to append a basket to the metal complex, the other is to confine the metal ion itself into the basket, which then becomes a funnel. There are, however, severe problems that need to be addressed: competition with solvent, flexibility of the pocket, directional control of the metal reactive site toward the substrate, and so on. With the funnel approach, one step further must be explored as they display remarkable binding properties for guest ligands, which, however, deactivate the metal center. The dream thus remains to direct the reactivity of an embedded copper ion toward a guest substrate, rather than to the ligand itself!

Future design of selective Cu/O_2 catalysts will need to gather not only one, but several supramolecular strategies discussed above in order to address these different issues simultaneously. Host design can also be expected to introduce a further degree of complexity via directional and orientational binding that would allow generating regio- or stereoselective catalysts. Obviously, the elaboration of supramolecular systems as mimics of metalloenzyme active sites is still at its infancy, and much work remains to be done.

ABBREVIATIONS

CcO	Cytochrome c oxidase
CV	Cyclic voltammetry
DABN	1,1'-Binaphthalenyl-2,2'-diamino group
3D	Three dimensional
DβH	Dopamine β-hydroxylase
DMF	Dimethylformamide
DMSO	Dimethyl sulfoxide
DOPA	3,4-Dihydroxyphenylalanine
DTBC	3,5-Di-*tert*-butylcatechol
DTBQ	3,5-Di-*tert*-butylquinone
EPR	Electronic paramagnetic resonance
ESI–MS	Electrospray ionization mass spectrometry
ET	Electron transfer
EXAFS	Extended X-ray absorption spectroscopy
FC	Ferrocene
FC^+	Ferricinium
GO	Galactose oxidase
KIE	Kinetic isotopic effect

LMCT	Ligand-to-metal charge transfer
MOF	Metal organic framework
NMR	Nuclear magnetic resonance
PHI	[piperazine-1,4-bis(4-{N-[1-acetoxy-3-(1-methyl-1H-imidazol-4-yl)]-2-propyl}-N-(1-methyl-1H-imidazol-2-ylmethyl) aminobutyl)]
PHM	Peptidylglycine α-hydroxylating monooxygenase
RT	Room temperature
salen	Bis(salicylidene ethylenediamine)
SBP	Square-based pyramidal
SP	Square planar
TBP	Trigonal bipyramidal
TβH	Tyramine β-hydroxylase
Td	Tetrahedral
TEM	Transmission electron microscopy
THF	Tetrahydrofurane
tmpa	Tris[(2-pyridyl)methyl]amine
TON	Turnover number
tren	Tris(2-aminoethyl)amine
XRD	X-ray diffraction
UV–vis	Ultraviolet visible

REFERENCES

1. Prigge, S. T.; Eipper, B. A.; Mains, R. E.; Amzel, L. M. Dioxygen binds end-on to mononuclear copper in a precatalytic enzyme complex *Science* **2004**, *304*, 864–867.

2. Klinman, J. P. The copper-enzyme family of dopamine beta-monooxygenase and peptidylglycine alpha-hydroxylating monooxygenase: Resolving the chemical pathway for substrate hydroxylation *J. Biol. Chem.* **2006**, *281*, 3013–3016.

3. Hess, C. R.; Wu, Z. N.; Ng, A.; Gray, E. E.; McGuirl, M. A.; Klinman, J. P. Hydroxylase activity of Met471Cys tyramine beta-monooxygenase *J. Am. Chem. Soc.* **2008**, *130*, 11939–11944.

4. Koval, I. A.; Gamez, P.; Belle, C.; Selmeczi, K.; Reedijk, J. Synthetic models of the active site of catechol oxidase: mechanistic studies *Chem. Soc. Rev.* **2006**, *35*, 814–840.

5. Holm, R. H.; Kennepohl, P.; Solomon, E. I. Structural and functional aspects of metal sites in biology *Chem. Rev.* **1996**, *96*, 2239–2314.

6. Solomon, E. I.; Sundaram, U. M.; Machonkin, T. E. Multicopper oxidases and oxygenases *Chem. Rev.* **1996**, *96*, 2563–2605.

7. Lieberman, R. L.; Rosenzweig, A. C. Crystal structure of a membrane-bound metalloenzyme that catalyses the biological oxidation of methane *Nature (London)* **2005**, *434*, 177–182.

8. Ludwig, B.; Bender, E.; Arnold, S.; Huttemann, M.; Lee, I.; Kadenbach, B. Cytochrome c oxidase and the regulation of oxidative phosphorylation *ChemBioChem* **2001**, *2*, 392–403.

9. Lewis, E. A.; Tolman, W. B. Reactivity of dioxygen-copper systems *Chem. Rev.* **2004**, *104*, 1047–1076.

10. Karlin, K. D.; Hayes, J. C.; Gultneh, Y.; Cruse, R. W.; McKown, J. W.; Hutchinson, J. P.; Zubieta, J. Copper-mediated hydroxylation of an arene: model system for the action of copper monooxygenases. Structures of a binuclear Cu(I) complex and its oxygenated product *J. Am. Chem. Soc.* **1984**, *106*, 2121–2128.

11. Becker, M.; Schindler, S.; Karlin, K. D.; Kaden, T. A.; Kaderli, S.; Palanche, T.; Zuberbuhler, A. D. Intramolecular ligand hydroxylation: Mechanistic high-pressure studies on the reaction of a dinuclear copper(I) complex with dioxygen *Inorg. Chem.* **1999**, *38*, 1989–1995.

12. Cruse, R. W.; Kaderli, S.; Karlin, K. D.; Zuberbuhler, A. D. Kinetic and thermodynamic studies on the reaction of O_2 with two dinuclear copper(I) complexes *J. Am. Chem. Soc.* **1988**, *110*, 6882–6883.

13. Karlin, K. D.; Cohen, B. I.; Jacobson, R. R.; Zubieta, J. Dioxygen-copper reactivity: hydroxylation-induced methyl migration in a copper monooxygenase model system *J. Am. Chem. Soc.* **1987**, *109*, 6194–6196.

14. Karlin, K. D.; Nasir, M. S.; Cohen, B. I.; Cruse, R. W.; Kaderli, S.; Zuberbuhler, A. D. Reversible dioxygen binding and aromatic hydroxylation in O_2-reactions with substituted xylyl dinuclear copper(I) complexes: syntheses and low-temperature kinetic/ thermodynamic and spectroscopic investigations of a copper monooxygenase model system *J. Am. Chem. Soc.* **1994**, *116*, 1324–1336.

15. Mirica, L. M.; Ottenwaelder, X.; Stack, T. D. P. Structure and spectroscopy of copper-dioxygen complexes *Chem. Rev.* **2004**, *104*, 1013–1045.

16. Hatcher, L. Q.; Karlin, K. D. Oxidant types in copper-dioxygen chemistry: the ligand coordination defines the Cu_n-O_2 structure and subsequent reactivity *J. Biol. Inorg. Chem.* **2004**, *9*, 669–683.

17. Gray, H. B.; Malmstrom, B. G.; Williams, R. J. P. Copper coordination in blue proteins *J. Biol. Inorg. Chem.* **2000**, *5*, 551–559.

18. DeBeer George, S.; Basumallick, L.; Szilagyi, R. K.; Randall, D. W.; Hill, M. G.; Nersissian, A. M.; Valentine, J. S.; Hedman, B.; Hodgson, K. O.; Solomon, E. I. Spectroscopic investigation of stellacyanin mutants: axial ligand interactions at the blue copper site *J. Am. Chem. Soc.* **2003**, *125*, 11314–11328.

19. Li, H.; Webb, S. P.; Ivanic, J.; Jensen, J. H. Determinants of the relative reduction potentials of type-1 copper sites in proteins *J. Am. Chem. Soc.* **2004**, *126*, 8010–8019.

20. Mahapatra, S.; Halfen, J. A.; Tolman, W. B. Mechanistic study of the oxidative N-dealkylation reactions of bis(μ-oxo)dicopper complexes *J. Am. Chem. Soc.* **1996**, *118*, 11575–11586.

21. Halfen, J. A.; Mahapatra, S.; Wilkinson, E. C.; Kaderli, S.; Young, V. G., Jr.; Que, L., Jr.; Zuberbuhler, A. D.; Tolman, W. B. Reversible cleavage and formation of the dioxygen O-O bond within a dicopper complex *Science* **1996**, *271*, 1397–1400.

22. Cole, A. P.; Root, D. E.; Mukherjee, P.; Solomon, E. I.; Stack, T. D. P. Trinuclear intermediate in the copper-mediated reduction of O_2: four electrons from three coppers *Science* **1996**, *273*, 1848–1850.

23. Maiti, D.; Woertink, J. S.; Ghiladi, R. A.; Solomon, E. I.; Karlin, K. D. Molecular oxygen and sulfur reactivity of a cyclotriveratrylene derived trinuclear copper(I) complex *Inorg. Chem.* **2009**, *48*, 8342–8356.

24. Sénèque, O.; Rondelez, Y.; Le Clainche, L.; Inisan, C.; Rager, M. N.; Giorgi, M.; Reinaud, O. Calix[6]arene-based N₃-donors - A versatile supramolecular system with tunable electronic and steric properties - Study on the formation of tetrahedral dicationic zinc complexes in a biomimetic environment *Eur. J. Inorg. Chem.* **2001**, 2597–2604.

25. Blanchard, S.; Le Clainche, L.; Rager, M. N.; Chansou, B.; Tuchagues, J. P.; Duprat, A. F.; Le Mest, Y.; Reinaud, O. Calixarene-based copper(I) complexes as models for monocopper sites in enzymes *Angew. Chem. Int. Ed. Engl.* **1998**, *37*, 2732–2735.

26. Blanchard, S.; Rager, M. N.; Duprat, A. F.; Reinaud, O. A novel calix[6]arene-based mononuclear copper(I) complex that exhibits chirality at low temperature *New J. Chem.* **1998**, *22*, 1143–1146.

27. Rondelez, Y.; Rager, M. N.; Duprat, A.; Reinaud, O. Calix[6]arene-based cuprous "funnel complexes": a mimic for the substrate access channel to metalloenzyme active sites *J. Am. Chem. Soc.* **2002**, *124*, 1334–1340.

28. Rondelez, Y.; Sénèque, O.; Rager, M. N.; Duprat, A. F.; Reinaud, O. Biomimetic Copper(I)-CO complexes: a structural and dynamic study of a calix[6]arene-based supramolecular system *Chem. Eur. J.* **2000**, *6*, 4218–4226.

29. Le Clainche, L.; Giorgi, M.; Reinaud, O. Synthesis and characterization of a novel calix[4]arene-based two-coordinate copper(I) complex that is unusually resistant to dioxygen *Eur. J. Inorg. Chem.* **2000**, 1931–1933.

30. Le Clainche, L.; Giorgi, M.; Reinaud, O. Novel biomimetic calix[6]arene-based copper(II) complexes *Inorg. Chem.* **2000**, *39*, 3436.

31. Le Clainche, L.; Rondelez, Y.; Sénèque, O.; Blanchard, S.; Campion, M.; Giorgi, M.; Duprat, A. F.; Le Mest, Y.; Reinaud, O. Calix[6]arene-based models for mono-copper enzymes: a promising supramolecular system for oxidation catalysis *C. R. Acad. Sci. Ser. II C.* **2000**, *3*, 811–819.

32. Le Poul, N.; Campion, M.; Douziech, B.; Rondelez, Y.; Le Clainche, L.; Reinaud, O.; Le Mest, Y. Monocopper center embedded in a biomimetic cavity: From supramolecular control of copper coordination to redox regulation *J. Am. Chem. Soc.* **2007**, *129*, 8801–8810.

33. Le Poul, N.; Campion, M.; Izzet, G.; Douziech, N.; Reinaud, O.; Le Mest, Y. Electrochemical behavior of the tris(pyridine)-Cu funnel complexes: An overall induced-fit process involving an entatic state through a supramolecular stress *J. Am. Chem. Soc.* **2005**, *127*, 5280–5281.

34. Sénèque, O.; Giorgi, M.; Reinaud, O. Bio-inspired calix[6]arene-zinc funnel complexes *Supramol. Chem.* **2003**, *15*, 573.

35. Klinman, J. P. Mechanisms whereby mononuclear copper proteins functionalize organic substrates *Chem. Rev.* **1996**, *96*, 2541–2561.

36. Izzet, G.; Frapart, Y. M.; Prange, T.; Provost, K.; Michalowicz, A.; Reinaud, O. X-ray diffraction and EXAFS studies of hydroxo-Cu(II) complexes based on a calix[6]arene-N₃ ligand: evidence for a mononuclear-dinuclear equilibrium controlled by supramolecular features *Inorg. Chem.* **2005**, *44*, 9743–9751.

37. Izzet, G.; Akdas, H.; Hucher, N.; Giorgi, M.; Prange, T.; Reinaud, O. Supramolecular assemblies with calix[6]arenes and copper ions: from dinuclear to trinuclear linear arrangements of hydroxo-Cu(II) complexes *Inorg. Chem.* **2006**, *45*, 1069–1077.

38. Sénèque, O.; Campion, M.; Douziech, N.; Giorgi, M.; Riviere, E.; Journaux, Y.; Le Mest, Y.; Reinaud, O. Supramolecular assembly with calix[6]arene and copper ions - Formation of a novel tetranuclear core exhibiting unusual redox properties and catecholase activity *Eur. J. Inorg. Chem.* **2002**, 2007–2014.

39. Sénèque, O.; Campion, M.; Douziech, B.; Giorgi, M.; Le Mast, Y.; Reinaud, O. Supramolecular control of an organic radical coupled to a metal ion embedded at the entrance of a hydrophobic cavity *Dalton Trans.* **2003**, 4216–4218.

40. Sénèque, O.; Rager, M. N.; Giorgi, M.; Prange, T.; Tomas, A.; Reinaud, O. Biomimetic zinc funnel complexes based on calix[6]N$_3$ArO ligands: an acid-base switch for guest binding *J. Am. Chem. Soc.* **2005**, *127*, 14833–14840.

41. Rondelez, Y.; Li, Y.; Reinaud, O. An efficient route to disymmetrically substituted calix[6] arenes. Synthesis of novel ligands presenting a N$_2$S or N$_3$CO$_2$ binding core *Tet. Lett.* **2004**, *45*, 4669–4672.

42. Jabin, I.; Reinaud, O. First C_{3v} symmetrical calix[6](aza)crown *J. Org. Chem.* **2003**, *68*, 3416–3419.

43. Darbost, U.; Giorgi, M.; Reinaud, O.; Jabin, I. A novel C_{3v} symmetrical calix[6](aza) cryptand with a remarkably high and selective affinity for small ammoniums *J. Org. Chem.* **2004**, *69*, 4879–4884.

44. Zeng, X. S.; Coquiere, D.; Alenda, A.; Garrier, E.; Prange, T.; Li, Y.; Reinaud, O.; Jabin, I. Efficient synthesis of calix[6]tmpa: a new calix[6]azacryptand with unique conformational and host–guest properties *Chem. Eur. J.* **2006**, *12*, 6393–6402.

45. Zeng, X. S.; Hucher, N.; Reinaud, O.; Jabin, I. A novel receptor based on a C-3v-symmetrical PN3-calix 6 cryptand *J. Org. Chem.* **2004**, *69*, 6886–6889.

46. Special issue on cyclodextrins *Chem. Rev.* **1998**, *98*, 1741–2076.

47. Collet, A. Cyclotriveratrylenes and Cryptophanes *Tetrahedron* **1987**, *43*, 5725–5759.

48. Cram, D. J.; Cram, J. M.; Stoddart, J. F. *Container Molecules and Their Guests, Monographs in: Supramolecular Chemistry*; The Royal Society of Chemistry: Cambridge, UK, 1994.

49. Rieth, S.; Yan, Z. Q.; Xia, S. J.; Gardlik, M.; Chow, A.; Fraenkel, G.; Hadad, C. M.; Badjic, J. D. Molecular encapsulation via metal-to-ligand coordination in a Cu(I)-folded molecular basket *J. Org. Chem.* **2008**, *73*, 5100–5109.

50. Abry, S.; Thibon, A.; Albela, B.; Delichere, P.; Banse, F.; Bonneviot, L. Design of grafted copper complex in mesoporous silica in defined coordination, hydrophobicity and confinement states *New J. Chem.* **2009**, *33*, 484–496.

51. Izzet, G.; Rager, M. N.; Reinaud, O. Insights into the binding properties of a cuprous ion embedded in the tren cap of a calix[6]arene and supramolecular trapping of an intermediate *Dalton Trans.* **2007**, 771–780.

52. Izzet, G.; Zeng, X.; Over, D.; Douziech, B.; Zeitouny, J.; Giorgi, M.; Jabin, I.; Le Mest, Y.; Reinaud, O. First insights into the electronic properties of a Cu(II) center embedded in the PN$_3$ cap of a calix[6]arene-based ligand *Inorg. Chem.* **2007**, *46*, 375–377.

53. Over, D.; de la Lande, A.; Zeng, X. S.; Parisel, O.; Reinaud, O. Replacement of a Nitrogen by a Phosphorus Donor in Biomimetic Copper Complexes: a Surprising and Informative Case Study with Calix 6 arene-Based Cryptands *Inorg. Chem.* **2009** *48*, 4317–4330.

54. Izzet, G.; Zeng, X. S.; Akdas, H.; Marrot, J.; Reinaud, O. Drastic effects of the second coordination sphere on neutral vs. anionic guest binding to a biomimetic Cu(II) center embedded in a calix[6]aza-cryptand *Chem. Commun.* **2007**, 810–812.

55. Alben, J. O.; Moh, P. P.; Fiamingo, F. G.; Altschuld, R. A. Cytochrome-oxidase (A3) heme and copper observed by low-temperature Fourier-transform infrared-spectroscopy of the Co complex *Proc. Natl. Acad. Sci. USA* **1981**, *78*, 234–237.

56. Rorabacher, D. B. Electron transfer by copper centers *Chem. Rev.* **2004**, *104*, 651–697.

57. Izzet, G.; Douziech, E.; Prange, T.; Tomas, A.; Jabin, I.; Le Mest, Y.; Reinaud, O. Calix[6] tren and copper(II): a third generation of funnel complexes on the way to redox calix-zymes *Proc. Natl. Acad. Sci. USA* **2005**, *102*, 6831–6836.

58. Le Poul, N.; Douziech, B.; Zeitouny, J.; Thiabaud, G.; Colas, H.; Conan, F.; Cosquer, N.; Jabin, I.; Lagrost, C.; Hapiot, P.; Reinaud, O.; Le Mest, Y. Mimicking the protein access channel to a metal center: effect of a funnel complex on dissociative versus associative copper redox chemistry *J. Am. Chem. Soc.* **2009**.

59. De Vos, D. E.; Dams, M.; Sels, B. F.; Jacobs, P. A. Ordered mesoporous and microporous molecular sieves functionalized with transition metal complexes as catalysts for selective organic transformations *Chem. Rev.* **2002**, *102*, 3615–3640.

60. Robert, R.; Ratnasamy, P. Activation of dioxygen by copper complexes incorporated in molecular sieves *J. Mol. Catal. A: Chem.* **1995**, *100*, 93–102.

61. Chavan, S.; Srinivas, D.; Ratnasamy, P. Structure and catalytic properties of dimeric copper(II) acetato complexes encapsulated in zeolite-Y *J. Catal.* **2000**, *192*, 286–295.

62. Raja, R.; Ratnasamy, P. Oxidation of cyclohexane over copper phthalocyanines encapsulated in zeolites *Catal. Lett.* **1997**, *48*, 1–10.

63. Poltowicz, J.; Pamin, K.; Tabor, E.; Haber, J.; Adamski, A.; Sojka, Z. Metallosalen complexes immobilized in zeolite NaX as catalysts of aerobic oxidation of cyclooctane *Appl. Catal., A* **2006**, *299*, 235–242.

64. Gu, Y. Y.; Zhao, X. H.; Zhang, G. R.; Ding, H. M.; Shan, Y. K. Selective hydroxylation of benzene using dioxygen activated by vanadium-copper oxide catalysts supported on SBA-15 *Appl. Catal., A* **2007**, *328*, 150–155.

65. Chui, S. S.-Y.; Lo, S. M.-F.; Charmant, J. P. n. H.; Orpen, A. G.; Williams, I. D. A chemically functionalizable nanoporous material $[Cu_3(TMA)_2(H_2O)_3]_n$ *Science* **1999**, *283*, 1148–1150.

66. Tabares, L. C.; Navarro, J. A. R.; Salas, J. M. Cooperative guest inclusion by a zeolite analogue coordination polymer. Sorption behavior with gases and amine and group 1 metal salts *J. Am. Chem. Soc.* **2001**, *123*, 383–387.

67. Woertink, J. S.; Smeets, P. J.; Groothaert, M. H.; Vance, M. A.; Sels, B. F.; Schoonheydt, R. A.; Solomon, E. I. A $[Cu_2O]^{2+}$ core in Cu-ZSM-5, the active site in the oxidation of methane to methanol *Proc. Natl. Acad. Sci. USA* **2009**, *106*, 18908–18913.

68. Groothaert, M. H.; Smeets, P. J.; Sels, B. F.; Jacobs, P. A.; Schoonheydt, R. A. Selective oxidation of methane by the bis(μ-oxo)dicopper core stabilized on ZSM-5 and mordenite zeolites *J. Am. Chem. Soc.* **2005**, *127*, 1394–1395.

69. Wu, Y.; Qiu, L. G.; Wang, W.; Li, Z. Q.; Xu, T.; Wu, Z. Y.; Jiang, X. Kinetics of oxidation of hydroquinone to p-benzoquinone catalyzed by microporous metal-organic frameworks $M_3(BTC)_2$ M = copper(II), cobalt(II), or nickel(II); BTC = benzene-1,3,5-tricarboxylate using molecular oxygen *Trans. Met. Chem.* **2009**, *34*, 263–268.

70. Xamena, F.; Casanova, O.; Tailleur, R. G.; Garcia, H.; Corma, A. Metal organic frameworks (MOFs) as catalysts: A combination of Cu^{2+} and Co^{2+} MOFs as an efficient catalyst for tetralin oxidation *J. Catal.* **2008**, *255*, 220–227.

71. Lee, J.; Farha, O. K.; Roberts, J.; Scheidt, K. A.; Nguyen, S. T.; Hupp, J. T. Metal-organic framework materials as catalysts *Chem. Soc. Rev.* **2009**, *38*, 1450–1459.

72. Xie, D. J.; Gutsche, C. D. Synthesis and reactivity of calix[4]arene-based copper complexes *J. Org. Chem.* **1998**, *63*, 9270–9278.

73. Sijbesma, R. P.; Nolte, R. J. M. Binding of dihydroxybenzenes in synthetic molecular clefts *J. Org. Chem.* **1991**, *56*, 3122–3124.

74. Reek, J. N. H.; Priem, A. H.; Engelkamp, H.; Rowan, A. E.; Elemans, J.; Nolte, R. J. M. Binding features of molecular clips. Separation of the effects of hydrogen bonding and pi-pi interactions *J. Am. Chem. Soc.* **1997**, *119*, 9956–9964.

75. Martens, C. F.; Gebbink, R.; Feiters, M. C.; Nolte, R. J. M. Shape-selective oxidation of benzylic alcohols by a receptor functionalized with a dicopper(II) pyrazole complex *J. Am. Chem. Soc.* **1994**, *116*, 5667–5670.

76. Gebbink, J. M. K.; Martens, C. F.; Feiters, M. C.; Karlin, K. D.; Nolte, R. J. M. Novel molecular receptors capable of forming Cu_2O_2 complexes. Effect of preorganization on O_2 binding *Chem. Commun.* **1997**, 389–390.

77. Sprakel, V. S. I.; Feiters, M. C.; Klaucke, W. M.; Klopstra, M.; Brinksma, J.; Feringa, B. L.; Karlin, K. D.; Nolte, R. J. M. Oxygen binding and activation by the complexes of PY2- and TPA-appended diphenylglycoluril receptors with copper and other metals *Dalton Trans.* **2005**, 3522–3534.

78. Sprakel, V. S. I.; Elemans, J.; Feiters, M. C.; Lucchese, B.; Karlin, K. D.; Nolte, R. J. M. Synthesis and characterization of PY2- and TPA-appended diphenylglycoluril receptors and their bis-Cu(I) complexes *Eur. J. Org. Chem.* **2006**, 2281–2295.

79. Gullotti, M.; Santagostini, L.; Pagliarin, R.; Palavicini, S.; Casella, L.; Monzani, E.; Zoppellaro, G. Ligand binding, conformational and spectroscopic properties, and biomimetic monooxygenase activity by the trinuclear copper-PHI complex derived from L-histidine *Eur. J. Inorg. Chem.* **2008**, 2081–2089.

80. Mimmi, M. C.; Gullotti, M.; Santagostini, L.; Saladino, A.; Casella, L.; Monzani, E.; Pagliarin, R. Stereoselective catalytic oxidations of biomimetic copper complexes with a chiral trinucleating ligand derived from 1,1-binaphthalene *J. Mol. Catal. A: Chem.* **2003**, *204*, 381–389.

81. Mutti, F. G.; Zoppellaro, G.; Gullotti, M.; Santagostini, L.; Pagliarin, R.; Andersson, K. K.; Casella, L. Biomimetic modelling of copper enzymes: synthesis, characterization, EPR analysis and enantioselective catalytic oxidations by a new chiral trinuclear copper(II) complex *Eur. J. Inorg. Chem.* **2009**, 554–566.

82. Santagostini, L.; Gullotti, M.; Pagliarin, R.; Monzanic, E.; Casella, L. Enantiodifferentiating catalytic oxidation by a biomimetic trinuclear copper complex containing L-histidine residues *Chem. Commun.* **2003**, 2186–2187.

83. Enomoto, M.; Aida, T. Self-assembly of a copper-ligating dendrimer that provides a new non-heme metalloprotein mimic: "dendrimer effects" on stability of the bis(μ-oxo) dicopper(III) core *J. Am. Chem. Soc.* **1999**, *121*, 874–875.

84. He, Q.-T.; Li, X.-P.; Liu, Y.; Yu, Z.-Q.; Wang, W.; Su, C.-Y. Copper(I) cuboctahedral coordination cages: host–guest dependent redox activity *Angew. Chem. Int. Ed.* **2009**, *48*, 6156–6159.

85. Thiabaud, G.; Guillemot, G.; Schmitz-Afonso, I.; Colasson, B.; Reinaud, O. Solid-state chemistry at an isolated copper(I) center with O_2 *Angew. Chem., Int. Ed.* **2009**, *48*, 7383–7386.

86. Izzet, G.; Zeitouny, J.; Akdas-Killig, H.; Frapart, Y.; Menage, S.; Douziech, B.; Jabin, I.; Le Mest, Y.; Reinaud, O. Dioxygen activation at a mononuclear Cu(I) center embedded in the calix[6]arene-tren core *J. Am. Chem. Soc.* **2008**, *130*, 9514–9523.

11

ORGANIC SYNTHETIC METHODS USING COPPER OXYGEN CHEMISTRY

MARISA C. KOZLOWSKI

Department of Chemistry, Roy and Diana Vagelos Laboratories,
University of Pennsylvania, Philadelphia, PA 19104

Copper-Oxygen Chemistry, First Edition. Edited by Kenneth D. Karlin and Shinobu Itoh.
© 2011 John Wiley & Sons, Inc. Published 2011 by John Wiley & Sons, Inc.

11.1. INTRODUCTION AND ORGANIZATION

Enzymatic oxidizing systems based upon copper represent a powerful set of catalysts able to direct elegant redox chemistry in a highly site-selective and stereoselective manner on simple, as well as highly functionalized molecules, using the simplest and cleanest of oxidants, molecular oxygen, as the terminal oxidant.[1-8] This ability has inspired organic chemists to discover small molecule catalysts that can emulate such processes.[9-15] In this chapter, copper-catalyzed reactions requiring oxygen are studied. The focus will be on preparatively useful constructions and transformations versus model studies of enzymatic systems or copper catalysts that cause decomposition reactions. Most, but not all, of these examples are based upon processes defined in biological catalysts, as described in other chapters of this text.

This chapter is organized into two distinct subtypes of reactions catalyzed by copper using molecular oxygen. In the first, oxidase-type reactions, the oxygen is used as a sink into which electrons flow during a redox process. Consequently, the substrate or substrates become oxidized in some manner, usually involving the loss of protons. The fate of the oxygen in such a transformation is as hydrogen peroxide (H_2O_2) or water. In contrast, oxygenase-type reactions involve direct incorporation of one or both of the oxygen atoms of molecular oxygen into the product itself.

The focus of this chapter is on copper-catalyzed homogenous processes that utilize oxygen as the terminal oxidant. Mixed-metal systems of copper (e.g., the Wacker process) are not included here. Reactions of related oxidants (H_2O_2, t-BuOOH, etc.) are also not considered here, but lead references can be found in some of the many other excellent reviews of the field.[9-15] This overview is not exhaustive, but representative discoveries from all the reaction types are surveyed with a focus on synthetically useful processes.

11.2. OXIDASE-TYPE REACTIONS

11.2.1. Alkynes

11.2.1.1. Glaser–Hay Reaction. The dimerization of terminal alkynes through oxidative homocoupling to give 1,3-diynes is an important C−C bond-formation reaction and has been extensively reviewed.[16] It has been employed for a number of applications, including the construction of linear π-conjugated acetylenic oligomers, polymers, and the synthesis of natural products. The reaction dates back to 1869 when Carl Glaser[17] observed the formation of a precipitate upon the reaction of a terminal alkyne with a Cu(I) salt in the presence of aqueous ammonia; after air oxidation, the oxidative homocoupling product was isolated (Scheme 11.1). A number of improvements on the original method have emerged, including the Eglington procedure, which allows the reaction to be run under homogeneous conditions with an excess of a Cu(II) salt as the oxidant (Scheme 11.1).[18] The Hay procedure permits the use of catalytic amounts of Cu(I) tertiary amine complex and dioxygen (O_2) as the terminal oxidant (Scheme 11.1)[19]; pyridine or *N,N,N',N'*-tetramethylethylenediamine (TMEDA) are particularly effective as the amine additives.

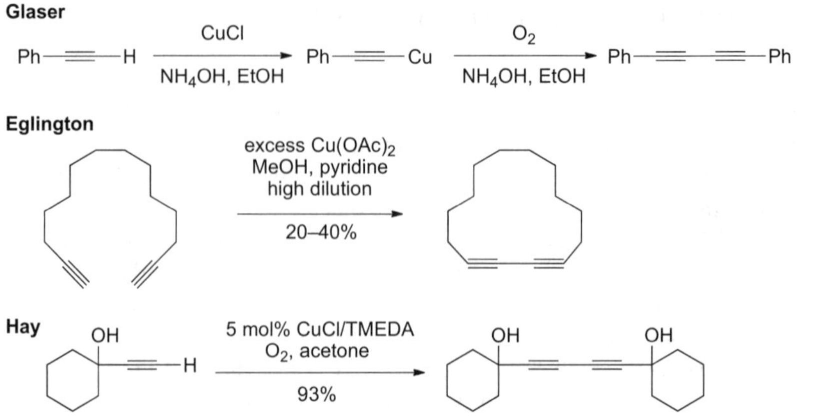

Glaser

$$Ph\text{—}\!\!\equiv\!\!\text{—}H \xrightarrow[\text{NH}_4\text{OH, EtOH}]{\text{CuCl}} Ph\text{—}\!\!\equiv\!\!\text{—}Cu \xrightarrow[\text{NH}_4\text{OH, EtOH}]{\text{O}_2} Ph\text{—}\!\!\equiv\!\!\text{—}\!\!\equiv\!\!\text{—}Ph$$

Eglington

excess Cu(OAc)$_2$
MeOH, pyridine
high dilution

20–40%

Hay

5 mol% CuCl/TMEDA
O$_2$, acetone

93%

SCHEME 11.1. Evolution of the "Glaser–Hay" process.

The reaction is tolerant of many functional groups including unprotected hydroxyls (see Scheme 11.1). The poor transfer ability of alkynes from copper in conjugate addition also allows functional compatibility with enones and carbonyls. Homocoupling predominates and homocoupled products are formed even with mixtures of alkynes. Yields are typically high under mild conditions. The appearance of an *Organic Syntheses* description is a manifestation of the high reliability of the process (Scheme 11.2).[20]

Numerous applications of the Glaser–Hay reaction have been implemented. With natural products, the method proved valuable in approaching the 1,3,5,7-octatetrayne structures, the caryoynencins, which show potent antibacterial activity. Notably, the conjugated polyynes are highly unstable, polymerizing readily, and present a formidable synthetic challenge. The mild nature and versatility of the Hay protocol for alkyne coupling is seen in the high yield with which the requisite tetrayne was generated en route to the caryoynencins (Scheme 11.3).[21]

The Glaser–Hay reaction can also be employed to form polymers and is remarkably tolerant of functionality. For example, terminal alkynes in organotransition metal complexes undergo copper-catalyzed couplings. Bunz[22] employed Hay conditions to prepare molecular rods incorporating cyclopentadienyl π-complexes that have interesting material properties (Scheme 11.4).

There have been surprisingly few mechanistic investigations of alkyne couplings, given their wide applicability in synthesis, and the potential for improved conditions that a more complete mechanistic understanding could provide. The classical oxidative alkyne couplings have been the most extensively studied, yet there are

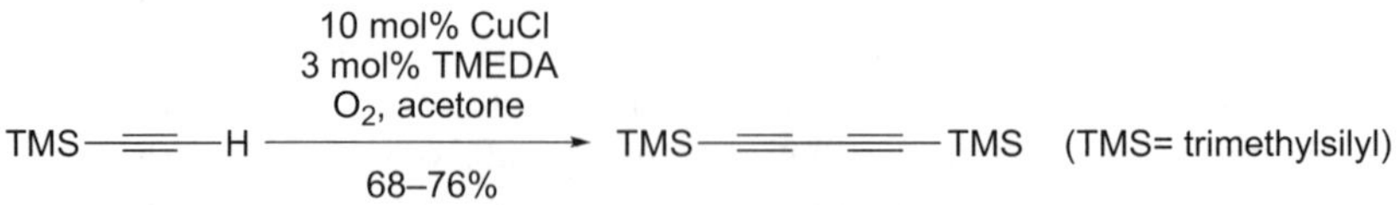

$$\text{TMS}\text{—}\!\!\equiv\!\!\text{—}H \xrightarrow[\text{68–76\%}]{\substack{\text{10 mol\% CuCl} \\ \text{3 mol\% TMEDA} \\ \text{O}_2\text{, acetone}}} \text{TMS}\text{—}\!\!\equiv\!\!\text{—}\!\!\equiv\!\!\text{—}\text{TMS} \quad (\text{TMS= trimethylsilyl})$$

SCHEME 11.2. Preparative scale Hay coupling.

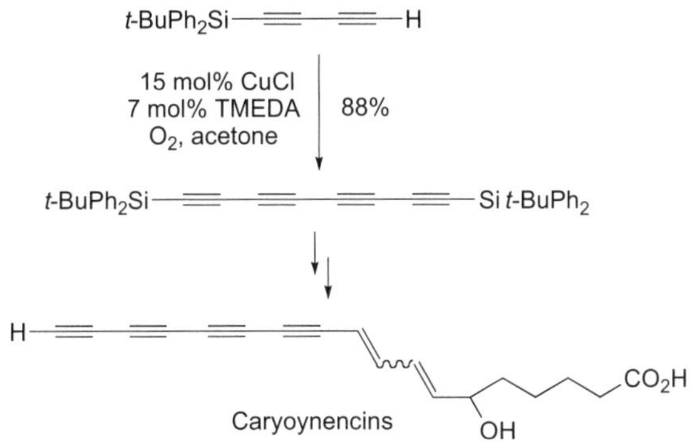

SCHEME 11.3. Glaser–Hay coupling en route to the caryoynencins.

several hypotheses regarding the oxidation state and intermediary copper structures. The bulk of the evidence supports at least two different mechanisms, where deprotonation precedes metal complexation or accompanies it, depending on the reaction conditions.[16] Under acidic conditions (pH 3) coupling is slow, but activity can be recovered with a Cu(I) salt. These observations support alkyne precomplexation to form Cu(I) π-complexes that activate the alkyne toward deprotonation. In contrast, the rate of coupling was faster in basic media and for more acidic terminal alkynes indicating that a direct deprotonation can also occur.

With mixtures of alkynes, homocoupled products are formed, pointing away from a mechanism invoking alkynyl radicals, which would react in a statistical manner. In alkaline solution, only Cu(II) species are observed by electron spin resonance (ESR) and O_2 likely reoxidizes the Cu(I) rapidly to Cu(II). Under acidic conditions, Cu(I) reoxidation is significantly slower and alkyne complexation likely occurs and facilitates deprotonation. With a second-order rate dependence on alkyne concentration, the mechanism in Scheme 11.5 has been proposed for the alkyne dimerization.

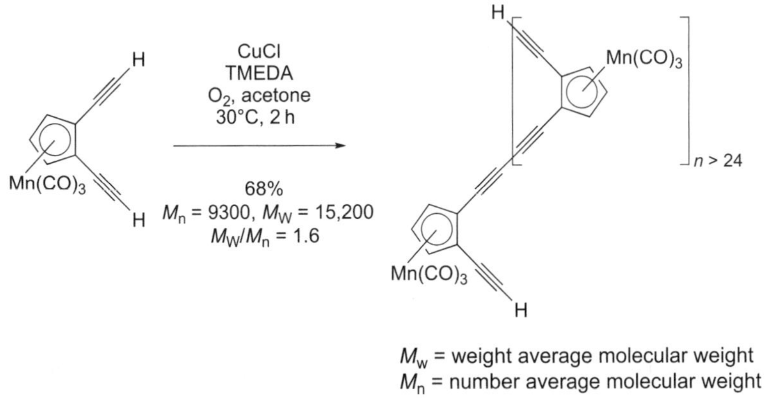

M_w = weight average molecular weight
M_n = number average molecular weight

SCHEME 11.4. Hay coupling of stable organometallic adducts to form polymers.

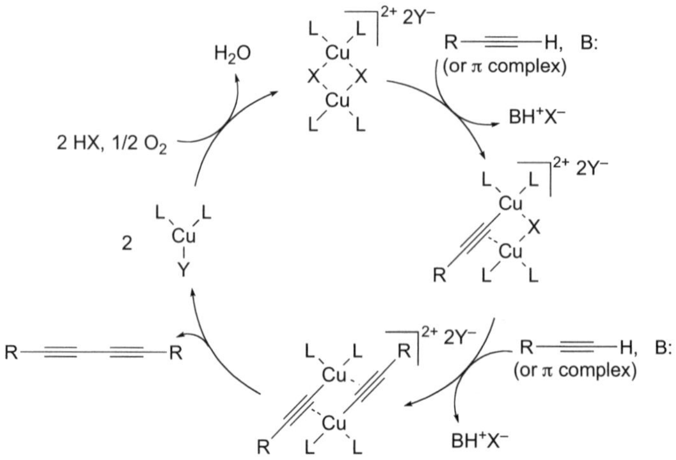

SCHEME 11.5. Mechanism of the Hay version of alkyne dimerization.

Recently, involvement of three copper ions [two Cu(I) and one Cu(II)] per alkyne unit in the rate-limiting stage of the oxidative dimerization has been proposed.[23] Clearly, the current understanding of the mechanism of copper-mediated oxidative acetylenic coupling remains incomplete. Apparently, the precise experimental conditions of base, ligand, and counterion have profound effects. It is not surprising, therefore, that publications continue to appear providing further modifications that improve the outcome.[24]

11.2.1.2. Cross-Coupling. The cross-coupling of two different alkynes is difficult; typically, homocoupling predominates (see Section 11.2.1.1). However, cross-reaction with non-alkynyl nucleophiles has been demonstrated. For example, Stahl and co-wokers[25] accomplished the cross-coupling of alkynes with amides in the presence of a moderately strong base (Na_2CO_3) that presumably assists in deprotonation of the amide (Scheme 11.6). Accordingly, effective nucleophiles in the reactions exhibit a pK_a in the range of 15–23 in dimethyl sulfoxide (DMSO) solution. Interestingly, not all substrates with a pK_a in this range, including pyridone (17.0) and acetanilide (21.5), are effective. The major competing reaction is homodimerzation of the alkyne; this byproduct could be reduced to <5% by utilizing 5 equiv of the nitrogen nucleophile and by slow addition of the alkyne to the reaction mixture over 4 h.

A very interesting cross-coupling variant between alkynes and trifluoromethyl anions has been reported using stoichiometric copper.[26] Again, formation of the alkyne homodimer proved to be highly favorable. In this instance, TMS–CF_3 is used as the source of the trifluoromethyl anion and "$CuCF_3$" is generated *in situ*.

Both of the cross-couplings described in this section may proceed via a bimetallic mechanism similar to that of the Hay alkyne dimerization (see Scheme 11.5). In this case, the redox couple is between two Cu(II) and two Cu(I). Alternately, the lesser stability of the mixed complex due to one less π-bonding interaction may allow the pathway in Scheme 11.7 to function. In this monometallic pathway,

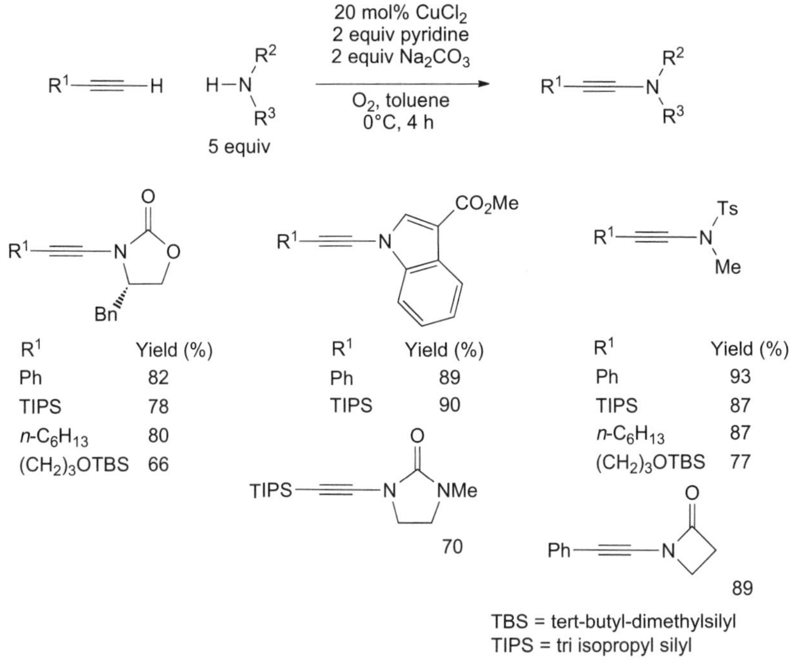

SCHEME 11.6. Oxidative amination of terminal alkynes.

oxidation presumably provides a Cu(III) species that undergoes reductive elimination to a Cu(I) species that is rapidly reoxidized to a Cu(II) intermediate. In both of these mechanisms, the main competing process is formation of a diyne. Any selective process must compete effectively with the facile insertion of copper into the C—H bond of an alkyne.

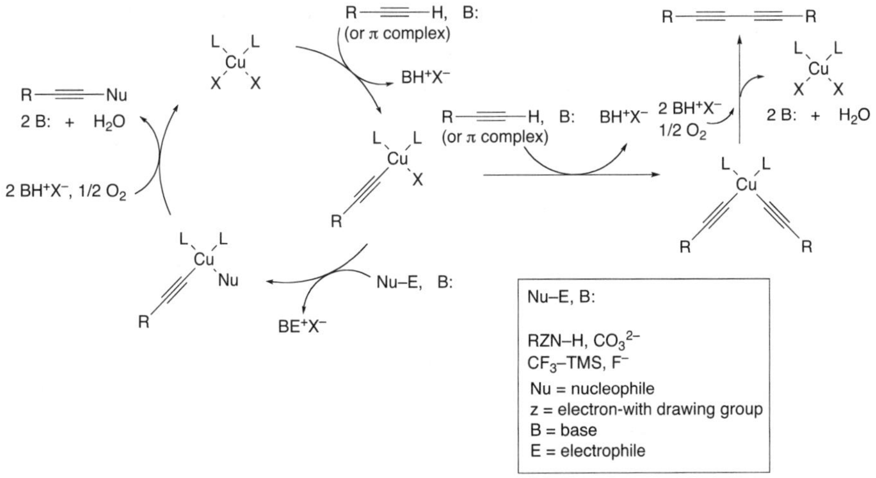

SCHEME 11.7. Possible alkyne-anion coupling mechanism.

Further work to improve the mechanistic understanding of these transformations is needed and would provide a framework for developing improved alkyne cross-coupling methods for a broader range of nucleophiles. Other challenges include lowering the excess of the non-alkyne couple partner and reducing the amount of copper catalyst needed.

11.2.2. Anion Coupling

11.2.2.1. Aryl or Alkenyl Anion Coupling. Lithio organocuprates are readily generated reagents of enormous utility in organic synthesis due to their unique reactivity. Furthermore, these reagents readily undergo oxidation with O_2 resulting in coupling between the ligands on copper and providing a useful avenue for homocoupling of aryl groups.[27,28] With mixed-ligand reagents (i.e., two different aryl groups in the copper reagent), statistical ratios of three products are usually produced. However, Lipshutz et al.[29-31] discovered that the controlled formation of diaryl higher order cuprates can lead to consistently high levels of unsymmetrical ligand coupling. Upon exposure of such reagents to ground-state molecular oxygen, good yields of the unsymmetrical biaryl Ar−Ar' can be realized. Several different types of aryl substrates have been examined, including naphthalenes and heteroaryls, to assess the generality of this method.

An example of this oxidative coupling in total synthesis can be found in the atropodiastereoselective synthesis of calphostin A, one of the perylenequinone natural products (Scheme 11.8).[32,33] Although excellent atropdiastereoselectivity was observed in this biaryl coupling, the absolute configuration of the newly formed axis of chirality was opposite to that required for synthesis of the calphostins. The opposite (S)-enantiomer substrate yielded the (S,S,S_a)-product that was ultimately converted into calphostin A.

In a further advance, conditions have been found where pregenerated anions can be oxidatively coupled using a substoichiometric amount of a copper catalyst. For example, aryl zinc reagents generated from aryl halides and Zn(0) undergo highly efficient homocoupling in the presence of 10 mol% CuBr, air or O_2, and 20–50 mol% of a dinitroarene (Scheme 11.9).[34] The methodology was found to be applicable to a wide range of aryl, heteroaryl, vinyl, and benzyl bromides. Particularly noteworthy is the dimerization in the presence of ketones, a functional group that would not be compatible with arylmagnesium halides or aryl lithiums even at low temperature. In

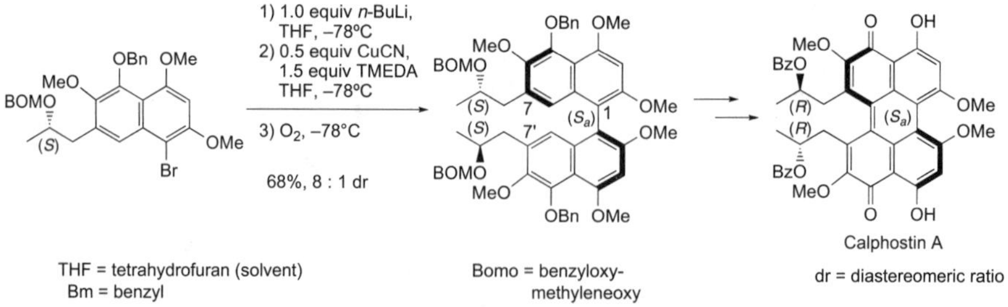

THF = tetrahydrofuran (solvent)
Bm = benzyl

Bomo = benzyloxy-methyleneoxy

dr = diastereomeric ratio

SCHEME 11.8. Oxidative diarylcyanocuprate coupling en route to calphostin A.

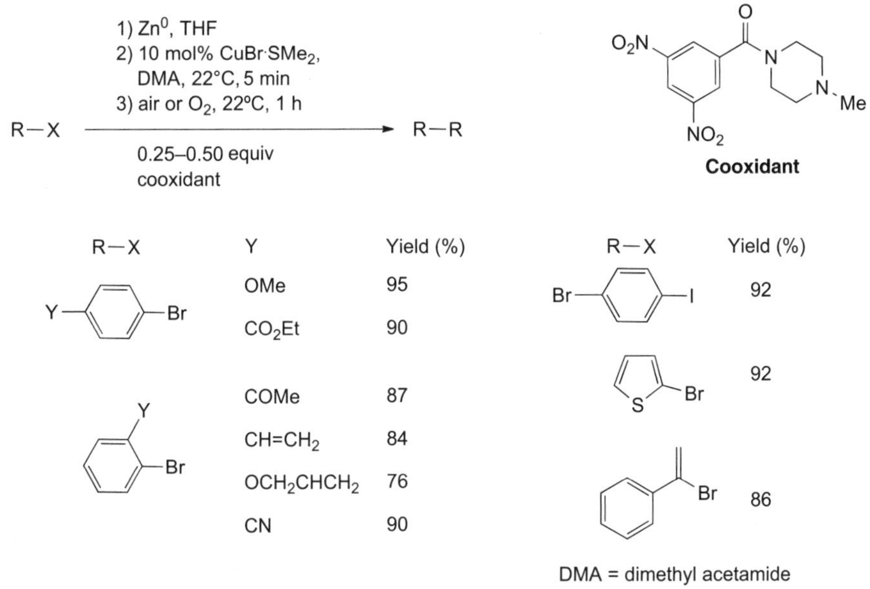

SCHEME 11.9. Copper-catalyzed reaction of aryl and alkenyl zincs with oxygen.

further contrast to the higher order cuprate couplings described above, low temperatures are not required.

This new method was also found to be useful in the synthesis of medium-sized rings in high yields, as illustrated in the total synthesis of buflavine, an *Amaryllidaceae* alkaloid with anti-serotonin properties (Scheme 11.10).[34]

For the transformations in Schemes 11.9 and Scheme 11.10, the use of an inert atmosphere compromised the yield that could be recovered if more of the dintroarene cooxidant was used or if the reaction mixture was placed under an atmosphere of dry air or molecular oxygen. When an oxygen atmosphere was used without the arene oxidant present, then significant quantities of phenolic products were produced. If the reaction mixture was rigorously degassed, substoichiometric quantities of the cooxidant were ineffective. The use of styrene and allyl substrates points away from a

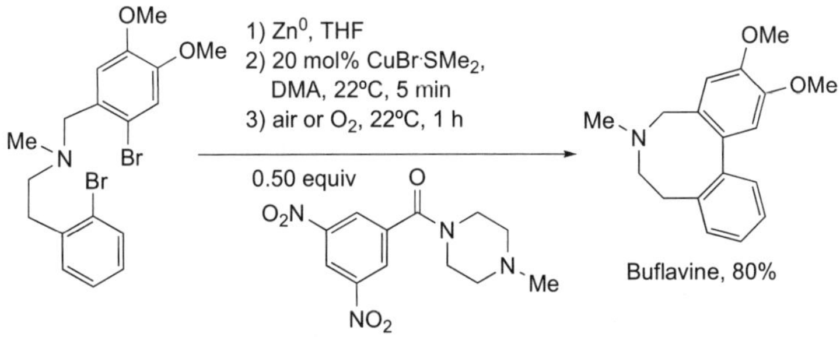

SCHEME 11.10. Copper-catalyzed intramolecular coupling of aryl zincs.

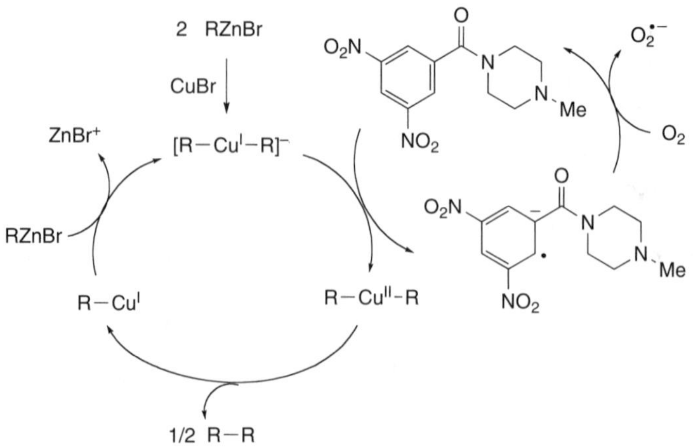

SCHEME 11.11. Aryl zinc coupling with catalytic copper and a cooxidant.

simple radical termination mechanism. These results suggest that the radical anion of the cooxidant is able to catalyze the reduction of molecular oxygen and compete with the formation of undesired products in such reactions (Scheme 11.11). Aryl zinc halides are not oxidized at an appreciable rate under the reaction conditions; this observation is consistent with this mechanism.

In the above cases, the arylzinc reagent needs to be preformed prior to treatment with the copper/cooxidant/O_2 combination. A system combining *in situ* deprotonation and oxidative dimerization has been studied by Daugulis and Do.[35] Hindered amide anions with LiCl and/or $ZnCl_2$ additives gave rise to an aryl anion species that was less prone to oxygenation and more likely to form a copper adduct that would undergo oxidative coupling (Scheme 11.12).

11.2.2.2. Benzyl or Allyl Anion Coupling. The oxidative coupling of zinc reagents using a copper catalyst, oxygen, and dinitroarene cooxidant, as outlined in Schemes 11.9–11.11, also proved applicable to other readily formed organozincs including allyl zinc reagents. Notably, the coupling proceeded to provide predominately a "head-to-tail" coupling rather than either of the two possible symmetrical coupling products (Scheme 11.13).[34]

11.2.2.3. Enolate Coupling. Organozinc reagents can also be readily formed by a Reformatsky reaction from the corresponding α-haloketones. These zinc enolates were highly efficient in homocoupling with the copper catalyst, oxygen, and dinitroarene cooxidant system (Scheme 11.14).[34] The high diastereoselectivity for the racemic over meso products is likely due to equilibration because the isolated ratios were similar to the thermodynamic ratios.

Direct oxidative enolate coupling with the more acidic malonate substrates was demonstrated over 35 years ago using a system very similar to that found in the Glaser–Hay coupling (see Section 11.2.1.1). Namely, a Cu–TMEDA catalyst with

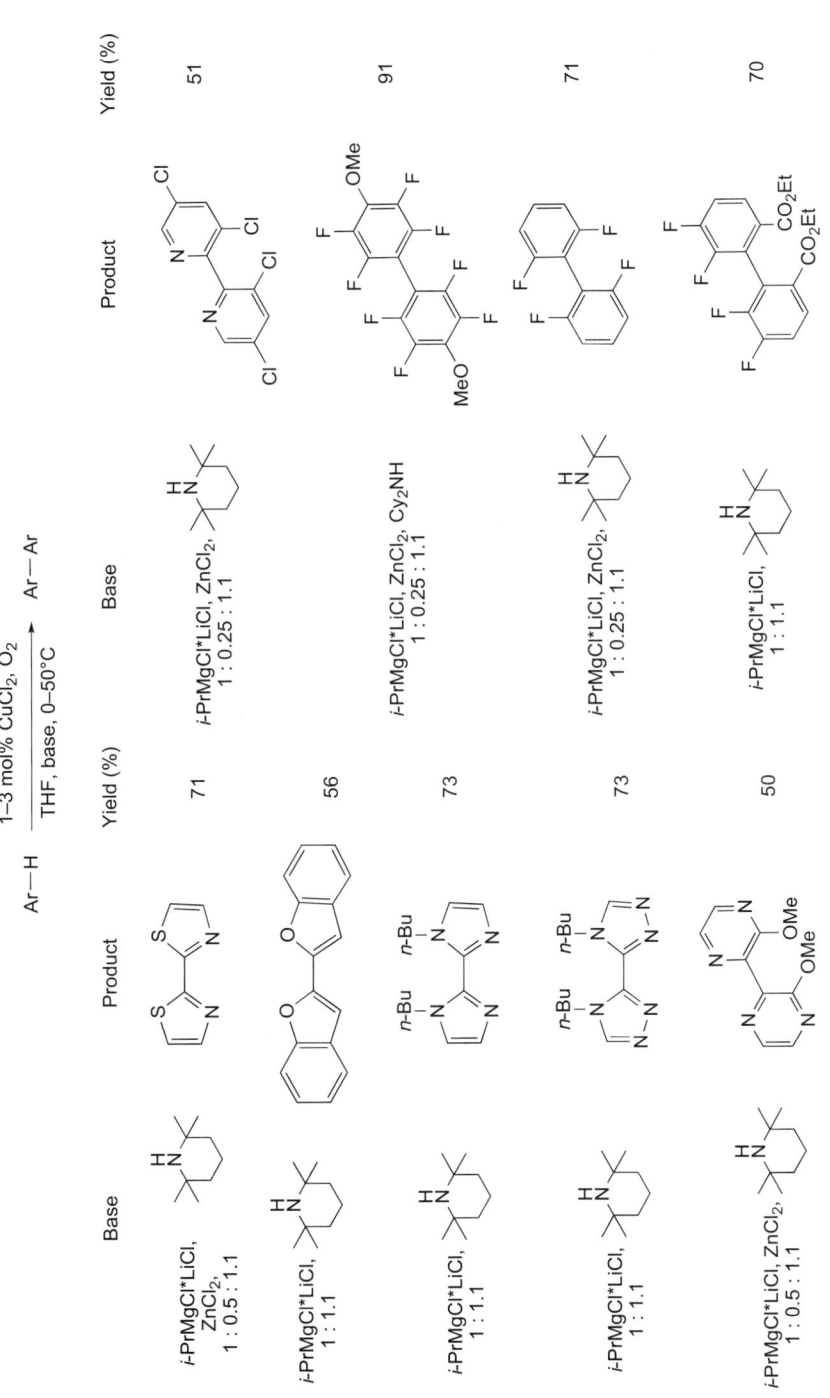

SCHEME 11.12. *In situ* deprotonation, symmetric dimerization with catalytic copper.

Cy = cyclohexyl

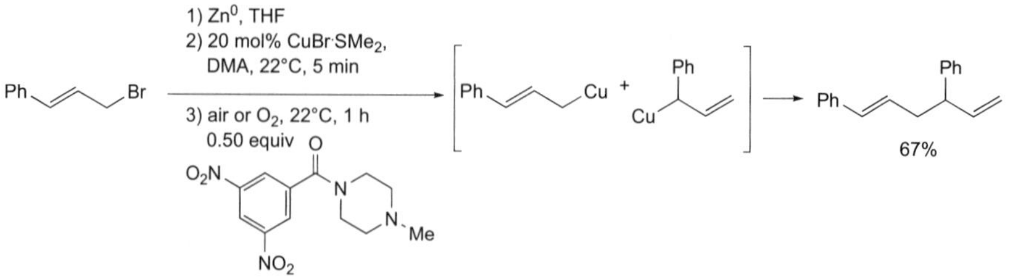

SCHEME 11.13. Copper-catalyzed reaction of benzyl and allyl zincs with oxygen.

oxygen is highly efficient in the oxidative dimerization of α-cyanoesters (Scheme 11.15).[36,37] The transformation is proposed to occur via deprotonation of the acidic substrate, and then single electron transfer (ET) to the copper generating a radical that subsequently dimerizes. Such a mechanism may also operate in the reaction outlined in Scheme 11.14. Use of para substituents on the phenyl prevents polymerization from that position that is reactive due to resonance delocalization of the intermediate radical. Heat causes the product to dissociate back to the radicals, and the uniform 3:2 meso:C_2 product ratio likely reflects the thermodynamic ratio.

In a further report, α-cyanoesters containing a chiral auxiliary were successfully dimerized using a copper TMEDA catalyst with oxygen (Scheme 11.16).[38] However, the meso compound again predominated and diastereoselection imparted by the chiral ester on the C_2 isomer was modest. Furthermore, the corresponding malonates or malonamides underwent decarboxylation in the presence of the copper catalyst.

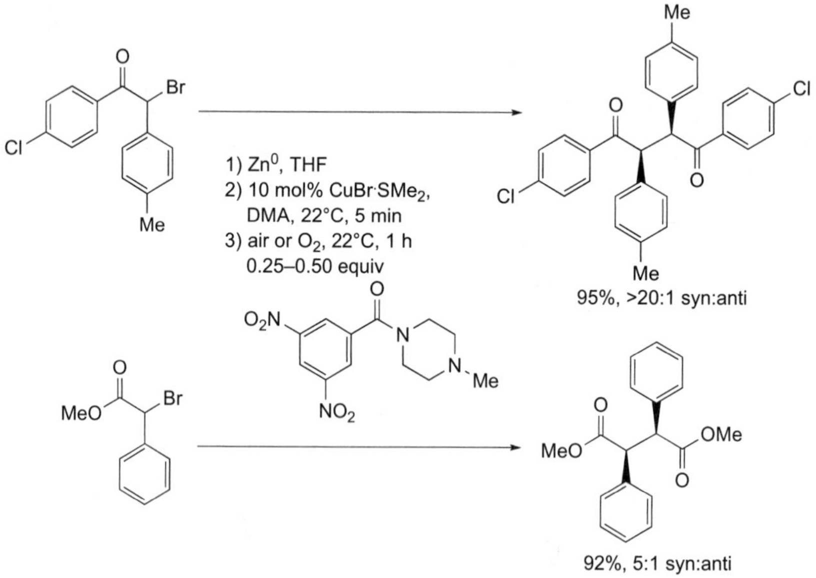

SCHEME 11.14. Copper-catalyzed reaction of zinc enolates with oxygen.

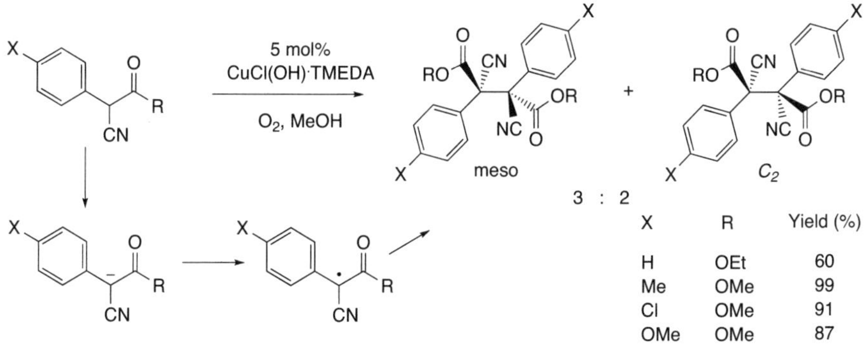

X	R	Yield (%)
H	OEt	60
Me	OMe	99
Cl	OMe	91
OMe	OMe	87
H	Me	79
Me	Me	83
H	Ph	68
Me	NH_2	74

SCHEME 11.15. Oxidative enolate coupling of α-cyanoesters.

The oxidative transformation of aryl anions, allyl anions, and enolates have much greater potential. For example, aryl cyanation has been effected via oxidative decomposition of arylcyanocuprates. While the efficiency of this process is so far low ($\sim$30%), this result highlights the unconventional pathways that are available by combining copper and oxygen.[39]

11.2.3. Bond Formation from Boronic Acids

Independent reports by Chan et al.,[40] Evans et al.,[41] and Lam et al.[42,43] utilizing stoichiometric copper reagents to effect formation of aryl C–N and C–O bonds in 1998 transformed the field of heteroatom arylation reactions. These developments led to new mild methods for C–N, C–O, and C–S bond-forming reactions, which have proven to have broad generality (Scheme 11.17).[44–48] In addition, reports of C–C bond formation from two boronic acids with copper catalysts have grown out of this work.

11.2.3.1. The C–N Bond Formation. A major advance in the field of oxidative coupling with boronic acids was reported by Collman and Zhong.[49] Recognizing that

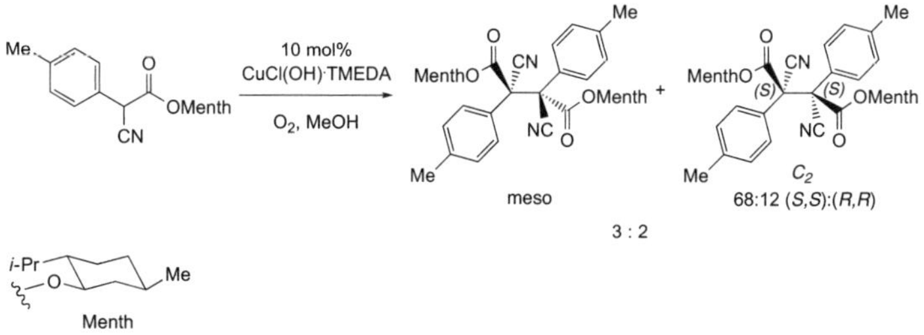

SCHEME 11.16. Diastereoselective oxidative enolate coupling of α-cyanoesters.

$$Ar-B(OH)_2 \quad + \quad HX-R \quad \xrightarrow[\quad O_2 \quad]{\text{cat "Cu"}} \quad Ar-X-R$$

$$X = NR', O, S$$

SCHEME 11.17. Boronic acids in copper-catalyzed oxidative bond formation with oxygen, nitrogen, and sulfur nucleophilic substrates.

the readily available Cu(OH)Cl(TMEDA) catalyst had been successfully employed in aerobic oxidative coupling of 2-naphthols,[50] this same catalyst was employed in this oxidative transformation leading to a copper-catalyzed modified Ullmann reaction (Scheme 11.18). While the copper catalyzed N-arylations with the related aryllead species had been reported earlier, the reactions of the boronic acids proceeded under milder conditions and without the use of toxic lead species.[51]

The authors speculated on a mechanism stemming from Evans' postulate for coupling aryl boronic acids with phenols.[41] The reaction also proceeds under air, but with a lower yield than that with O_2. An elegant mechanism study by Stahl and co-workers[52] revealed an isolable Cu(III) aryl species that would combine with an acidic nitrogen species to generate an N-aryl (Scheme 11.19). More acidic nitrogen species reacted more rapidly, suggesting that the nitrogen nucleophile undergoes deprotonation before or during the rate-limiting step of the reaction. A reactivity pattern for heterocycles (carbazole > imidazole > indole ~ pyrrole > triazole ≫ tetrazole) in N-arylation reactions has emerged based on nucleophilicity, complexing ability of catalyst, and acidity.[48]

Overall, the kinetic data and electronic effects are consistent with at least two different mechanisms for C−N bond formation: (1) a three-centered C−N reductive

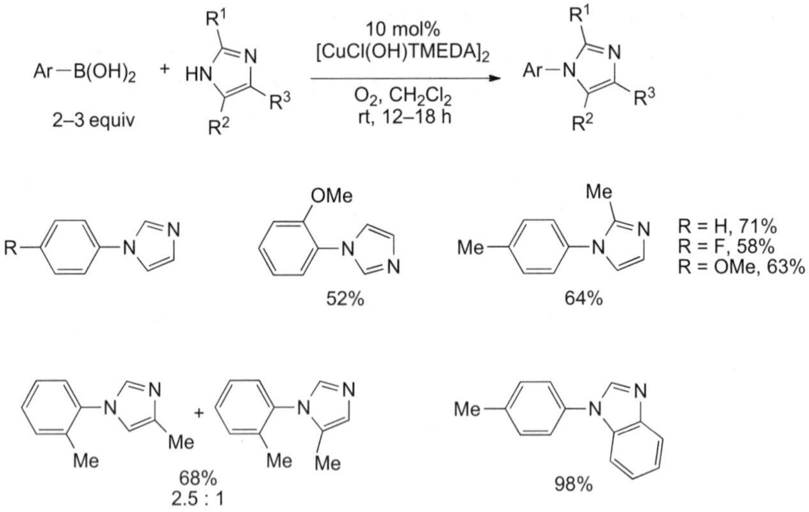

SCHEME 11.18. Copper-catalyzed coupling of arylboronic acids with azaheterocycles.

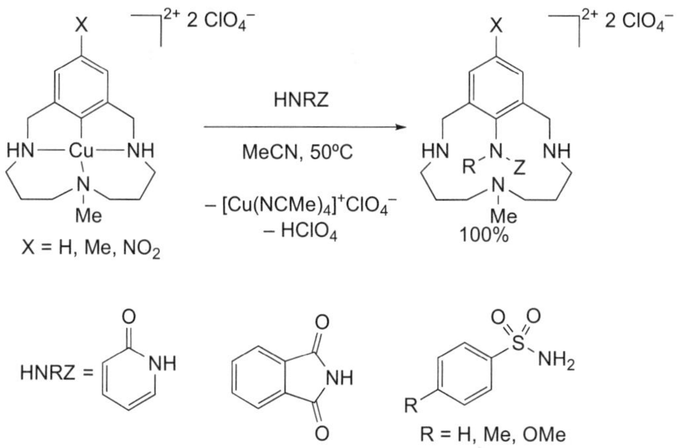

SCHEME 11.19. Isolated Cu(III) intermediates that undergo N-arylation.

elimination from an (unobserved) Cu(III)(aryl)(amidate) intermediate (Scheme 11.20) or (2) bimolecular nucleophilic attack of an amidate at the aryl carbon to displace the aryl–Cu bond.[52] Support for the former is found in another report in which the evidence indicates coordination of the nitrogen nucleophile to the Cu center prior to C−N bond formation in a related copper-catalyzed coupling reaction.[53]

Since the Collman and Zhong[49] discovery of copper-catalyzed N-arylations, numerous nitrogen compounds have been employed with an array of boronic acid

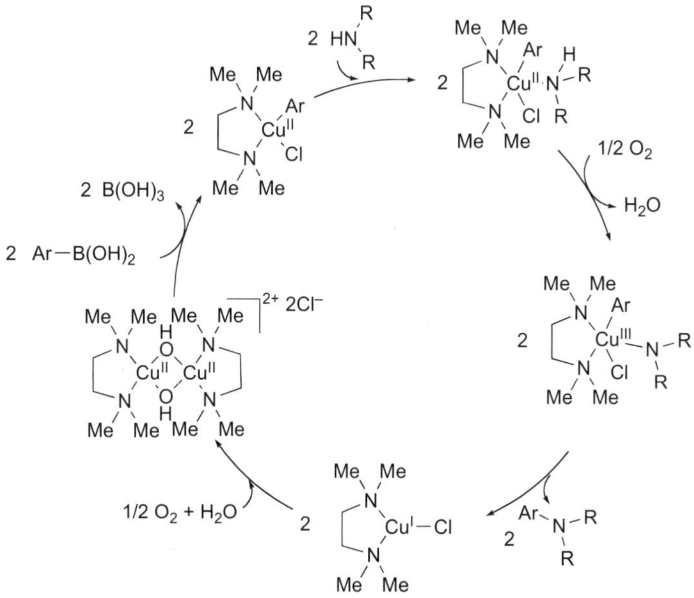

SCHEME 11.20. Proposed mechanism for the copper-catalyzed N-arylation.

derivatives. Permuations include the use of various copper catalysts with vinyl boronic acids,[54] trifluoroborates,[55] pentafluorophenylboronic acid,[56] other borates,[57,58] amines,[55–57,59–63] anilines,[55–57,59–63] amides,[57,59,62,63] sulfonamides,[59,62,63] heterocycles,[57,59,60,63–67] and sulfoximines.[68]

While the more reactive boronic acid precursors permit heterocycle *N*-arylations to proceed at lower temperatures, the formation of the hindered C−N biaryls remains a challenge,[44–48] but recent efforts indicate that very hindered C−N biaryls can indeed be generated under mild conditions.[69] In addition to the examples and citations above, several excellent reviews have appeared describing the oxidative copper-catalyzed C−N bond formation with boronic acids.[44–48] This transformation has been particularly useful in the synthesis of heterocyclic medicinal chemistry agents.[44–48]

11.2.3.2. The C−O Bond Formation. With the realization of C−N coupling using a copper catalyst under aerobic conditions[49] (see Section 11.2.3.1), the next frontier became a simple catalytic procedure for oxidative coupling of alcohols with aryl boronic acids, given that the version with stoichiometric copper proceeded readily using a variety of phenols and *N*-hydroxysuccinimides.[40,41,45,70] A catalytic protocol utilizing a stoichiometric chemical oxidant to regenerate the copper catalyst was hampered by competitive oxidation of the boronic acid and the need to reoptimize the oxidant with each nucleophile.[63] In a breakthrough, Batey and Quach[71] found that trifluoroborates were highly effective using catalytic Cu(OAc)$_2$ and DMAP. Aliphatic, allylic, and internal propargylic alcohols, as well as phenols, underwent cross-coupling without complication to aryl and alkenyl trifluoroborates providing the aryl ethers and vinyl ethers (Scheme 11.21).

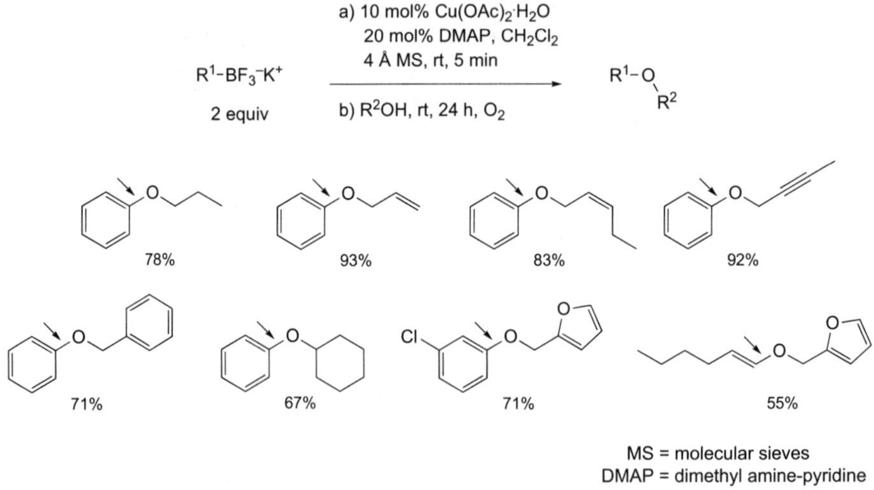

SCHEME 11.21. Copper-catalyzed aryl and vinyl etherification (arrow indicates the bond formed).

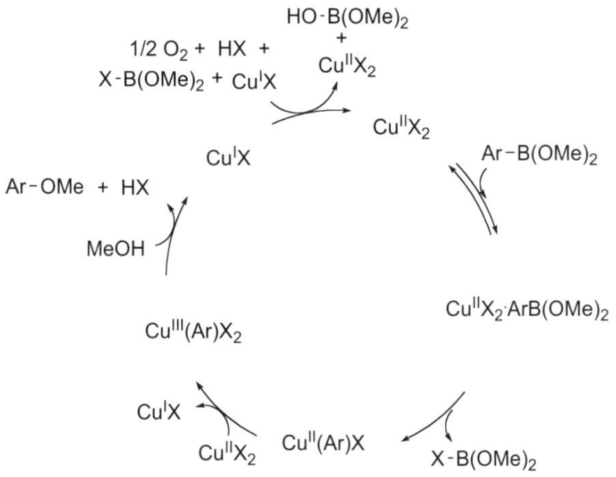

SCHEME 11.22. Proposed mechanism of copper-catalyzed etherification with boronic acids.

Investigation of the mechanism of the etherification reaction by Stahl co-workers[72] revealed that, under rigorously anaerobic conditions, a 2:1 Cu(II):product stoichiometry occurs. Further kinetic and structural experiments supported transmetalation of the aryl group to copper as the turnover limiting step, and confirmed that most of the copper in solution existed as Cu(II) with weak ligands. On this basis the mechanism in Scheme 11.22 was proposed for this transformation.

Other reagents aside from boronic acids have been investigated in etherification and the Ph$_4$BiF reagent has proven to be suitable for the etherification of a wide range of alcohols including tertiary alcohols.[73,74] This mild transformation, which proceeds at ambient temperature, is particularly mild, tolerating chelating substrates and various functionality including ketone, carboxylic ester, amide, tertiary amine, and silyloxy groups. The downsides are that only one aryl group transfers from the Ph$_4$BiF reagent and its preparation requires several steps.

11.2.3.3. The C−S Bond Formation. As was the case for the C−N and C−O bond-forming reactions, initial C−S bond-forming reactions employed stoichiometric copper reagents.[75] Turnover in this transformation was difficult because the Cu(II) species proposed to be involved (see Scheme 11.20) can oxidize thiols. To circumvent this problem, a preactivated N-thiol substrate has been employed.[76] While catalytic copper conditions could be achieved, this change perturbed the reaction so that it was not an oxidative transformation. Thus oxygen is not required.

The use of oxidized sulfur species has been more successful. A mild copper-based protocol for the synthesis of aryl and vinyl sulfones has been reported involving the cross-coupling of aryl and vinyl boronic acids with sodium sulfinate salts (Scheme 11.23).[77] Catalytic amounts of Cu(II) acetate with 1,10-phenanthroline as ligand are employed in the presence of 4-Å molecular sieves. In the case of aryl boronic acids, the major side products in the reactions were symmetric copper-based cross-couplings to biaryls, and oxygenation reactions to form phenols and biaryl ethers.

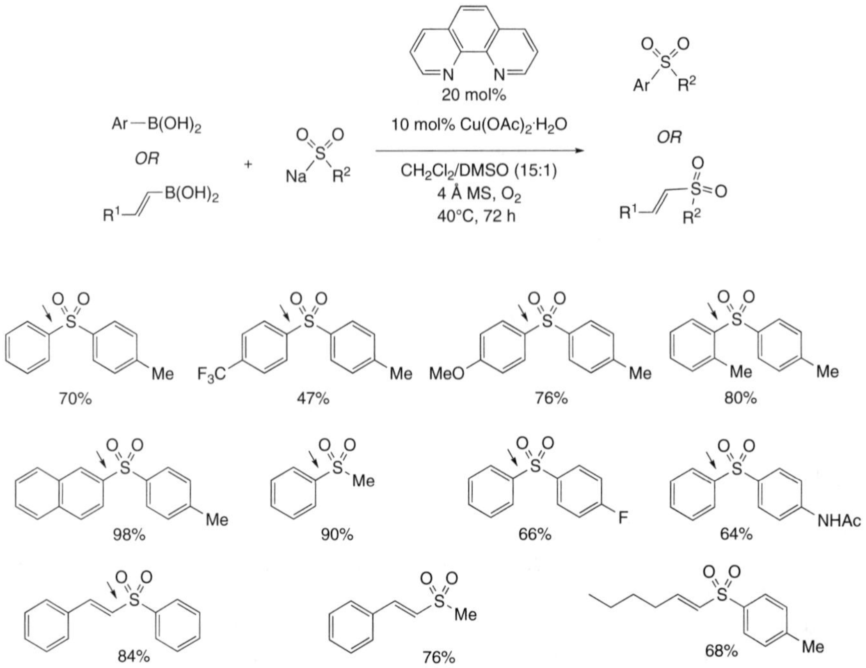

SCHEME 11.23. Copper-catalyzed sulfonylation (arrow indicates the bond formed).

11.2.3.4. The C−C Bond Formation.

The first report of the oxidative dimerization of arylboronic acids came from Demir et al.[78] in 2003 showing that Cu(OAc)$_2$ under oxygen was effective. While most of the examples utilized high-catalyst loading (50 mol%), lower loadings (10 mol%) worked well except with ortho-substituted substrates. This process is not as facile as the related C−N, C−O, and C−S bond-forming reactions because the homodimers are typically not reported as a major byproduct in these other processes.

In 2009, an improved protocol for oxidative dimerization was reported relying on catalysts that would facilitate transmetalation.[79] Ultimately, the Cu(II) bis-μ-hydroxo adduct with a phenathroline ligand proved the most effective, providing the dimeric products under ambient conditions with as little as 2 mol% of the copper catalyst (Scheme 11.24).

In contrast to the mechanisms proposed above for C−N and C−O bond formation (see Sections 11.2.3.1 and 11.2.3.2) involving a Cu(III) to Cu(I) redox event, Kirai and Yamamoto[79] proposed a bimetallic mechanism involving two Cu(II) to Cu(I) redox events (Scheme 11.25). This sequence was proposed in order to accommodate the most ready pathway for aryl boronic acid incorporation via exchange of a hydroxyl ligand.

Another example illustrating the potential for oxidative homodimerization to occur during other couplings is seen in Scheme 11.26.[56] While pentafluorophenyl boronic acid underwent clean aryl amination with anilines (see Section 11.2.3.1),

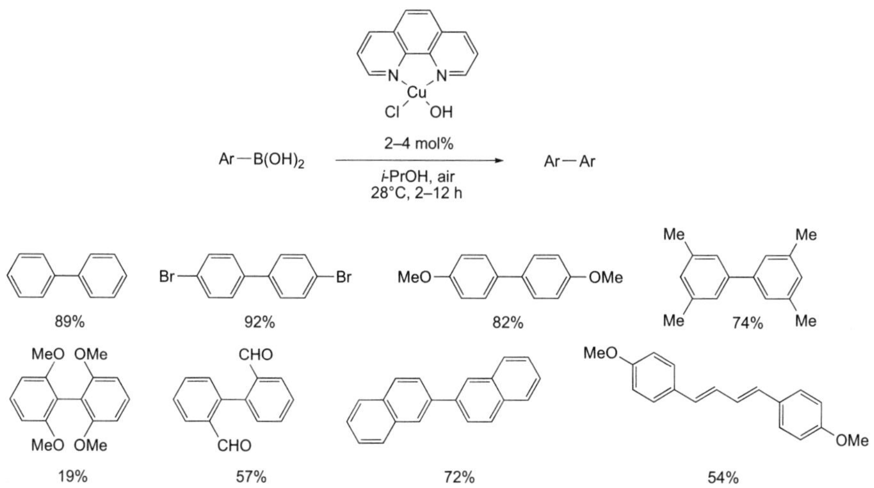

SCHEME 11.24. Oxidative dimerization of arylboronic acids with a copper(II) bis-μ-hydroxo catalyst.

alkylamines gave rise to a quite different product resulting from oxidative dimerization followed by amination of one of the fluorides. This reaction pathway is supported by formation of the final product from isolated homodimer and none from the independently generated pentafluoroaniline compounds.

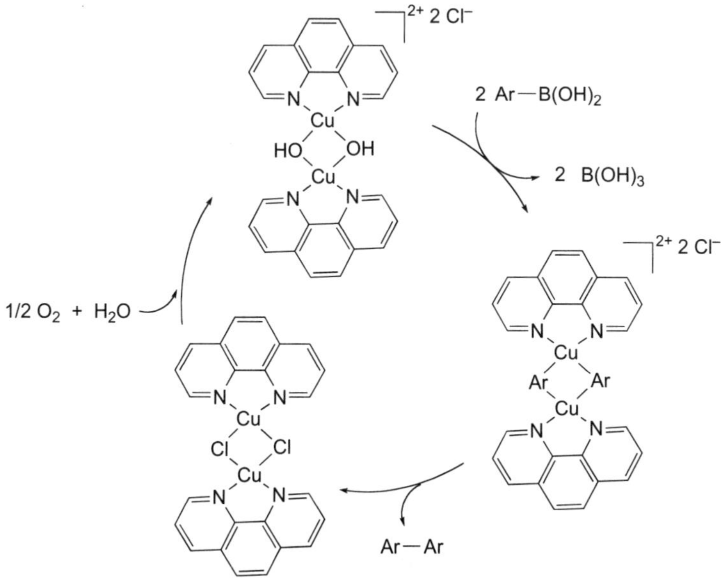

SCHEME 11.25. Potential mechanism for oxidative dimerization of aryl boronic acids.

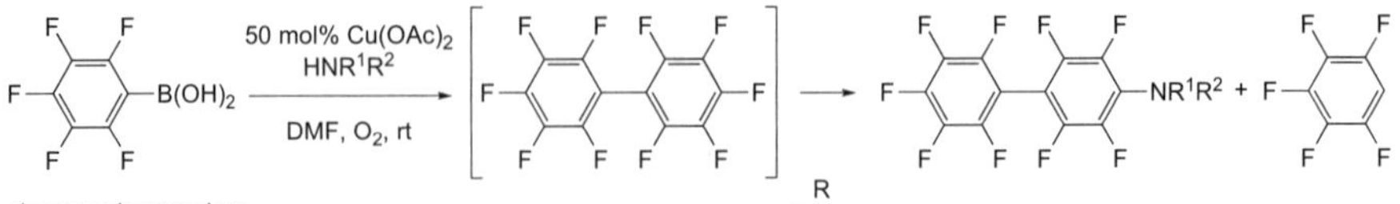

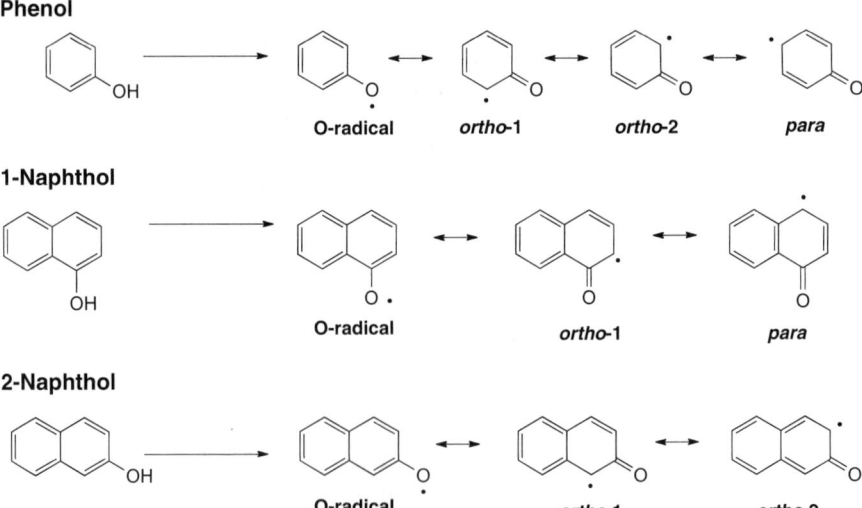

SCHEME 11.26. Tandem oxidative dimerization–amination of an aryl boronic acid.

Another tandem process has been reported by Liebeskind and co-workers,[80] wherein a nonoxidative C−C bond-forming reaction to generate a ketone from a thioester and an aryl boronic acid is accompanied by oxidative coupling of the resultant thiolate with an aryl boronic acid.

11.2.4. Reactions of Phenols–Naphthols

The oxidative coupling of phenols and naphthols has a long history in both biosynthesis studies and in organic synthesis.[81–87] Due to a favorable 1e⁻ phenolic oxidation to stabilized radical species (Scheme 11.27), these transformations can be carried out under mild reaction conditions that tolerate many functional groups. This versatility distinguishes oxidative biaryl coupling from other chiral biaryl synthetic methods such as nucleophilic aromatic substitution, Kumada coupling, Negishi coupling, and Suzuki coupling.[88–90] In addition, these transformations often

Phenol

O-radical   *ortho*-1   *ortho*-2   *para*

1-Naphthol

O-radical   *ortho*-1   *para*

2-Naphthol

O-radical   *ortho*-1   *ortho*-2

SCHEME 11.27. Radicals arising from phenols, 1-napthhols, and 2-naphthols.

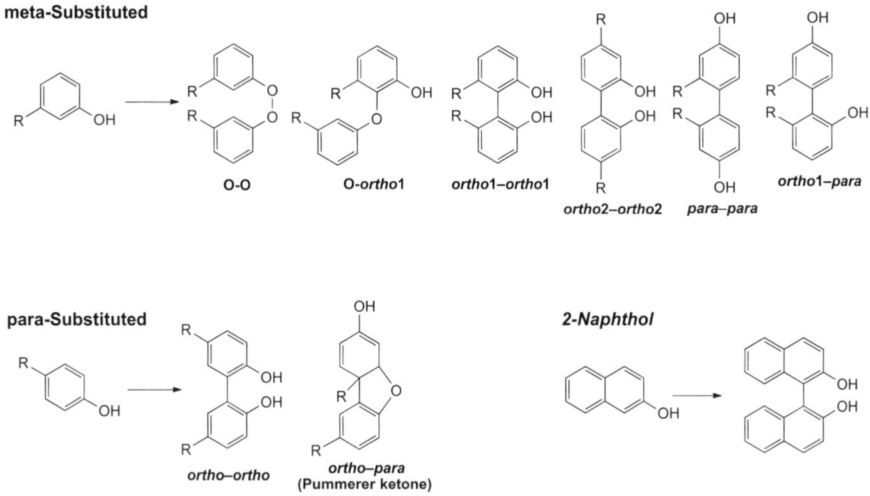

meta-Substituted

O-O   O-ortho1   ortho1–ortho1   ortho2–ortho2   para–para   ortho1–para

para-Substituted   **2-Naphthol**

ortho–ortho   ortho–para (Pummerer ketone)

SCHEME 11.28. Different coupling patterns for phenols.

introduce functionality at nonfunctionalized centers, thereby eliminating the need for complex prefunctionalized starting materials (halides, boronic acids, etc.).

Conversely, the lack of prefunctionalization limits the regioselective control. The substrate typically dictates the available coupling products, though in some cases regioselectivity can be controlled by the catalyst. The mechanism invokes deprotonation and electron abstraction *or* hydrogen radical abstraction. Regardless, meta-coupling is unattainable. For example, the ortho- and para-radicals of phenols have similar stability (Scheme 11.27) leading to several possible product outcomes (Scheme 11.28).[91] Even the para-position of para-substituted phenols, which is ostensibly blocked, can couple to form a Pummerer ketone product (Scheme 11.28).[92,93] The most well-behaved substrates are the 2-naphthols that couple exclusively at the 1-position to yield 1,1'-BINOL (BINOL = 1,1'-bi-2-naphthol) derivatives (Scheme 11.28). Comparison of 1- and 2-naphthol reveals that this selectivity arises from the greater stability of the 2-naphthol ***ortho-1*** in which the radical is benzylic and aromaticity is retained in one ring (Scheme 11.27). Thus, an ongoing challenge in phenol coupling is the selective coupling of multiple sterically and electronically comparable positions.

11.2.4.1. The C–C Bond Formation: Biaryl Coupling

11.2.4.1.1. Racemic and Achiral Couplings. Copper catalysts using oxygen have proven highly effective in phenol coupling. In cases where substitution blocks reaction at two of the three ring sites, highly regioselective couplings can be obtained (Scheme 11.29).[94–96] Even so, overoxidation can occur when the substitution pattern permits the formation of diphenoquinones (Scheme 11.29a). Numerous bioinspired catalysts have been developed that effect phenol coupling.[1–15] Even so, the

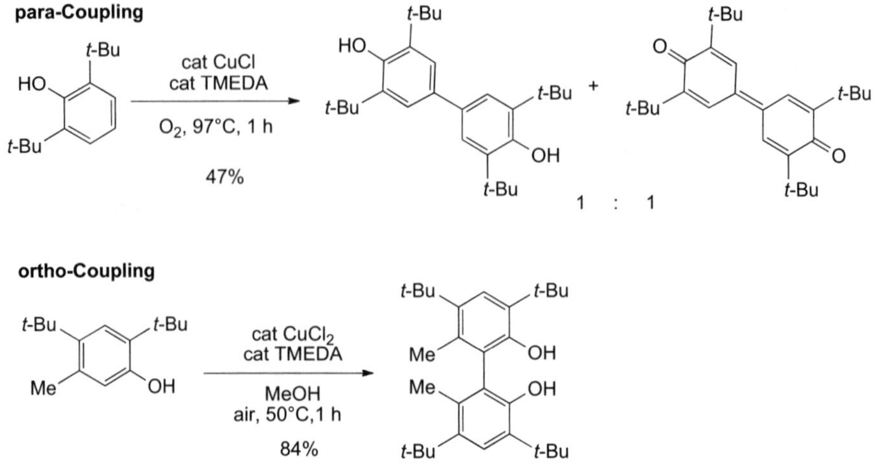

para-Coupling

47%

1 : 1

ortho-Coupling

84%

SCHEME 11.29. Copper-catalyzed phenol couplings.

development of catalysts that are effective with substrates where the substitution pattern permits more than one product (i.e., 2-*tert*-butylphenol) remains a challenge.

Together with the utility of the resultant BINOL products, the suitability of 2-naphthols for oxidative coupling (see Scheme 11.27) has driven much work in this area. For example, the use of stoichiometric Cu reagents in 2-naphthol coupling is well established.[97–102] Nakajima and co-workers[45] reported a breakthrough in using a simple achiral copper catalyst, CuCl(OH)TMEDA, combined with oxygen as the terminal oxidant for the coupling of a broad range of electron-poor and electron-rich 2-naphthols (Scheme 11.30). Air could be used in place of oxygen, but the reactions were slower.

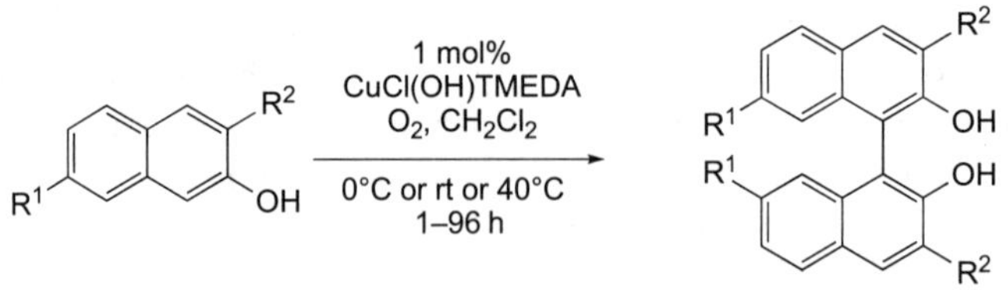

R^1	R^2	Yield (%)
H	H	90
H	Me	92
OMe	H	96
H	CO$_2$Me	99
9-Phenanthrol		79

SCHEME 11.30. Formation of binaphthols and related compounds via oxidative coupling with copper catalysts and oxygen.

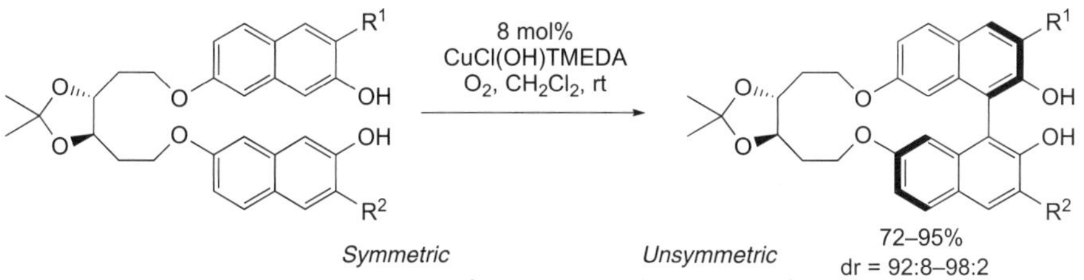

Symmetric		Unsymmetric	
R^1	R^2	R^1	R^2
H	H	H	Ph
Ph	Ph	H	i-Pr
CH$_2$OMOM	CH$_2$OMOM	H	Et$_3$Si
		Ph	i-Pr
		CH$_2$OMOM	i-Pr

72–95%
dr = 92:8–98:2

MOM = methoxy methyl ether

SCHEME 11.31. Diastereoselective intramolecular naphthol coupling

This technology has proved broadly useful as seen by the range of compounds that can be made using this very simple catalyst system. Examples include the ligands phosphonyl BINOL[103] and BICOL (bis carbazole diol),[104] as well as the natural product flavanthrin.[105]

11.2.4.1.2. Diastereoselective Couplings. Diastereoselective couplings of 2-naphthols are one effective means to generate 1,1'-binaphthols that contain a defined chiral axis. For example, Lipshutz et al.[106,107] linked two naphthols with a chiral tether (Scheme 11.31). The subsequent oxidative coupling with CuCl(OH) TMEDA provided the product as a single diastereomer.

Examples of the coupling of chiral naphthols without the use of a tether have also been reported, but the diastereocontrol in formation of the new stereoaxis is not as effective.[108]

11.2.4.1.3. Asymmetric Couplings. The potential for chiral catalysts to generate axial chiral products is high due to the utility of the products as ligands and as precursors in biomimetic synthesis. Indeed, enzymatic systems have been reported that catalyze asymmetric oxdiatve phenolic coupling.[109,110] Nakajima et al.[111,112] made a key breakthrough in this area using chiral copper catalysts under oxygen to generate chiral 1,1'-binaphthols. While the sparteine-derived catalyst displayed only modest selectivity, the prolyldiamines were more effective (Scheme 11.32). However, the reaction was only effective with substrates containing a coordinating group in the 3-position, and a maximum enantiomeric excess (ee) of 78% was reported. Without the additional coordinating group, the interaction between the chiral catalyst and the substrate (Scheme 11.32) after the redox event is too weak to permit an effective stereochemical transfer.

Kozlowski and co-workers[113,114] discovered that diaza-*cis*-decalin as a ligand for the copper catalyst both improved the reactivity and also moved the selectivity to a

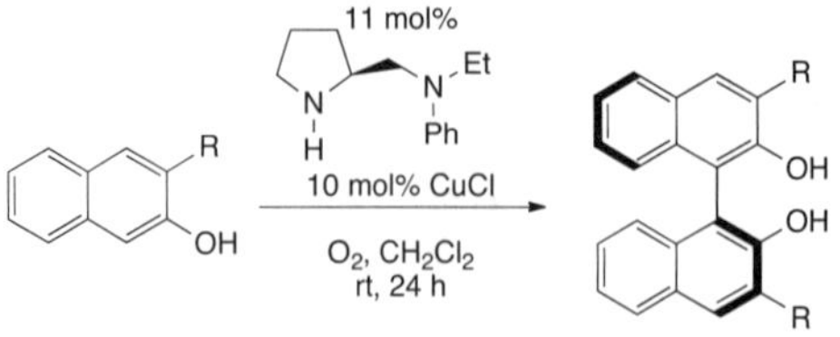

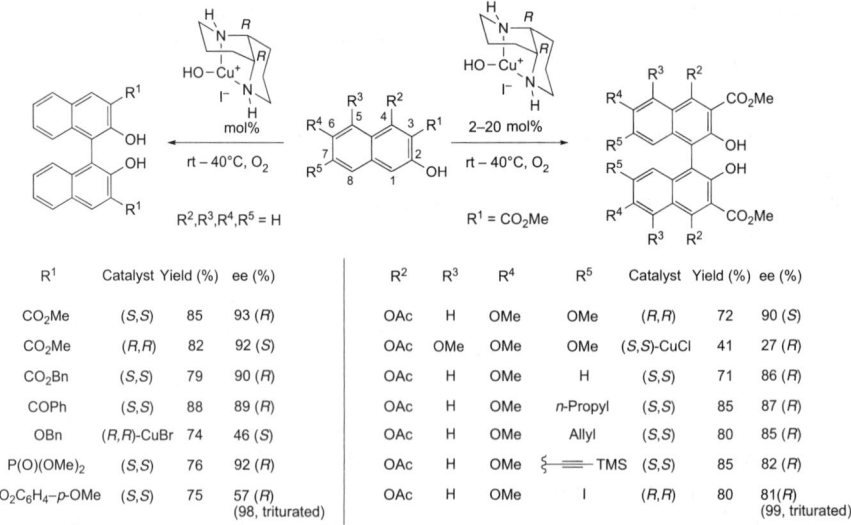

R	Yield (%)	ee (%)
CO$_2$Me	85	78
CO$_2$Et	77	73
CO$_2$Bn	77	76
CO$_2$t-Bu	69	58
H	89	17
i-Pr	58	5
OBn	95	24

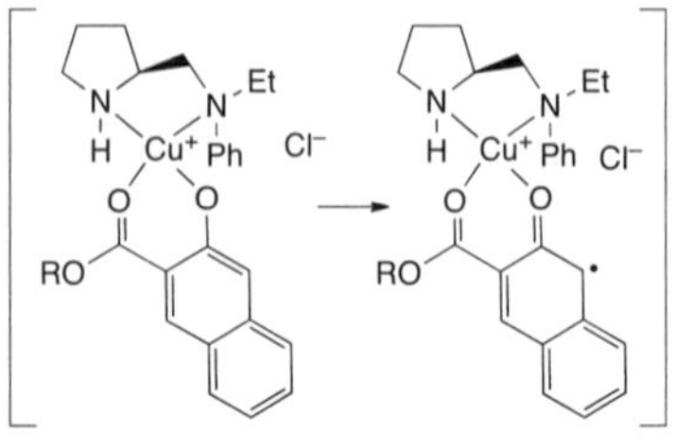

SCHEME 11.32. Scope of first copper-catalyzed enanatioselective naphthol coupling.

usable range for a number of synthetic ventures (Scheme 11.33). Both enantiomers of the chiral diamine were obtained by resolution allowing both enantiomers of the products to be selectively generated. A coordinating group at the 3-position was still required, but considerable variation was tolerated, permitting the synthesis of various sulfonyl and phosphonyl BINOL derivatives. In addition, highly substituted 2-naphthols could be coupled to provide enantioenriched decasubstituted 1,1'-binaphthalenes. Very electron-rich substrates (entries 12–13) did undergo atropisomerization during the reaction, presumably by further oxidation of the product, which eroded the enantioselectivity.

With this method established, several natural products were produced including nigerone (Scheme 11.34),[115,116] In this instance, flavasperone was subjected to the

R^1	Catalyst	Yield (%)	ee (%)
CO$_2$Me	(S,S)	85	93 (R)
CO$_2$Me	(R,R)	82	92 (S)
CO$_2$Bn	(S,S)	79	90 (R)
COPh	(S,S)	88	89 (R)
OBn	(R,R)-CuBr	74	46 (S)
P(O)(OMe)$_2$	(S,S)	76	92 (R)
SO$_2$C$_6$H$_4$–p-OMe	(S,S)	75	57 (R) (98, triturated)

R^2	R^3	R^4	R^5	Catalyst	Yield (%)	ee (%)
OAc	H	OMe	OMe	(R,R)	72	90 (S)
OAc	OMe	OMe	OMe	(S,S)-CuCl	41	27 (R)
OAc	H	OMe	H	(S,S)	71	86 (R)
OAc	H	OMe	n-Propyl	(S,S)	85	87 (R)
OAc	H	OMe	Allyl	(S,S)	80	85 (R)
OAc	H	OMe	⸱≡—TMS	(S,S)	85	82 (R)
OAc	H	OMe	I	(R,R)	80	81(R) (99, triturated)

SCHEME 11.33. Copper–diaza-*cis*-decalin catalyst in asymmetric naphthol couplings.

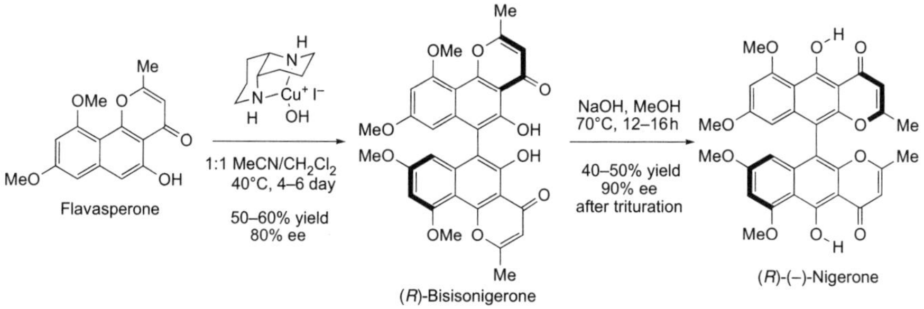

SCHEME 11.34. Asymmetric naphthol coupling in the synthesis of nigerone.

chiral diaza-*cis*-decalin copper catalyst under oxygen to provide an isomer of nigerone, bisisonigerone, which was subsequently isomerized to complete the first synthesis of the natural product (−)-nigerone. The oxidative coupling has proven to be stereoregular for >50 substrates, which was useful in securing the correct absolute configuration of natural nigerone.

The utility of this oxidative coupling method was also seen in the syntheses of several perylenequinones[117] and perylenequinone natural products (Scheme 11.35).[118–123] This method provided access to either enantiomer of the axial chiral bisiodide

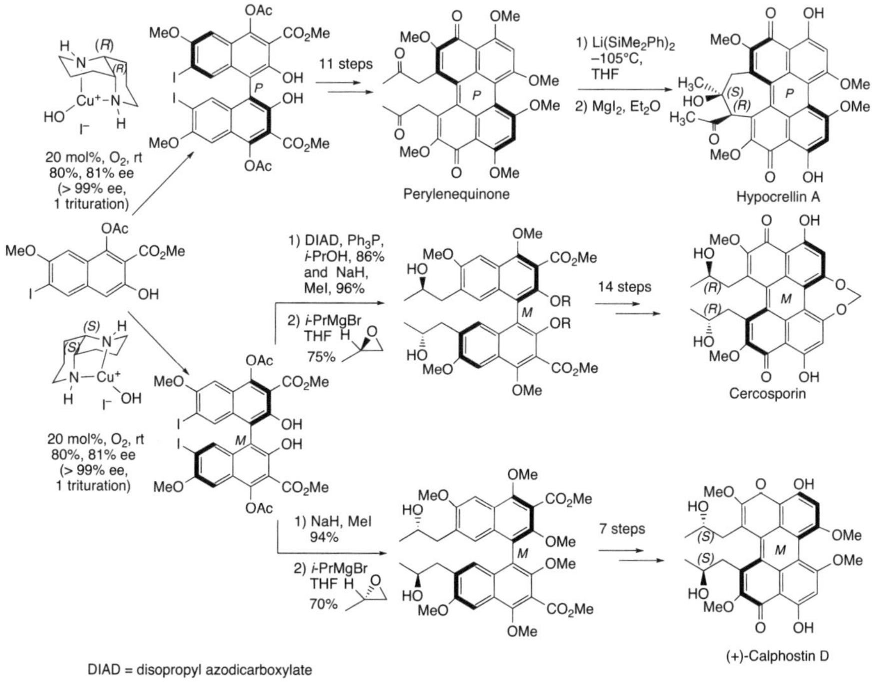

DIAD = disopropyl azodicarboxylate

SCHEME 11.35. Asymmetric naphthol coupling in the synthesis of perylenequinone natural products.

illustrated in Scheme 11.35. From the *P*-enantiomer, a helical chiral perylenequinone was generated that was devoid of additional stereocenters. Such enantiopure compounds had not been generated previously, so this endeavor provided evidence that these compounds are atropisomerically stable at ambient temperatures.[117] A further dynamic stereochemistry aldol transformation upon this perylenequinone provided the correct configuration of the two centrochiral stereocenters and secured the natural product hypocrellin A.[118] From the *M*-enantiomer of the axial chiral bisiodide Scheme 11.35, a biscuprate could be generated and subjected to either enantiomer of propylene oxide providing the diastereomeric series leading to cercosporin, phleichrome, and calphostin D with complete stereocontrol.[118–123]

Mechanism studies of the copper diaza-*cis*-decalin-catalyzed naphthol coupling revealed that the turnover limiting step was an oxidation. However, different initial burst phases were observed when Cu(I) and Cu(II) precatalysts were employed. Detailed kinetic studies provided support for the mechanism outlined in Scheme 11.36, where the copper complex acts as a self-processing catalyst generating a cofactor (NapH^{OX}) from a molecule of substrate.[124,125] Notably, similar behavior is seen with the enzymatic copper catalyst amine oxidase.[126–128] Unfortunately, the C–C coupling event is after the turnover limiting step and few details are known. Roithova et al.[129] found that the corresponding CuCl(OH)TMEDA catalyst oxidatively couples 2-naphthol via a dimeric copper species in the gas phase. Whether the more hindered diaza-*cis*-decalin copper catalyst follows the same pathway in solution is not known at this point.

Additional copper-catalyst systems have been described more recently for 2-naphthol couplings using oxygen. A BINAM (1,1'–bi-2-aminonaphthalene) derived copper catalyst provides slightly higher enantioselectivity than the diaza-*cis*-decalin catalyst with the same chelating substrate, 3-methoxycarbonyl-2-naphthol.[130] Martell and co-workers[131] described a biscopper salan [reduced salen (Salicaldehyde etrylenediamine bisimede)] type of adduct that is the only reported copper catalyst to provide high selectivity in the oxidative coupling of nonchelating substrates, such as 2-naphthol. The natural product rigidanthrin has been synthesized using a very simple

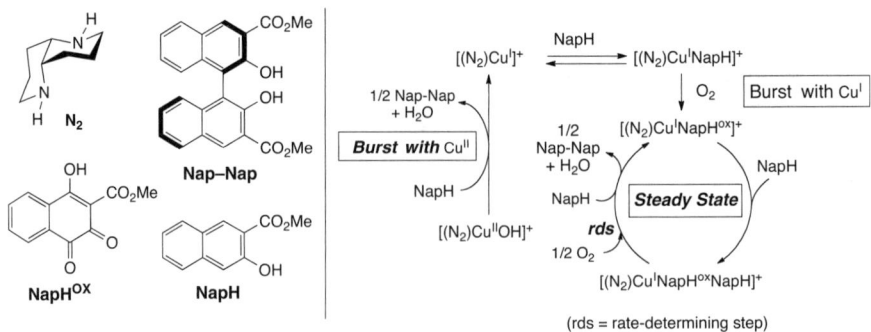

SCHEME 11.36. Mechanism of the copper diaza-*cis*-decalin-catalyzed naphthol coupling.

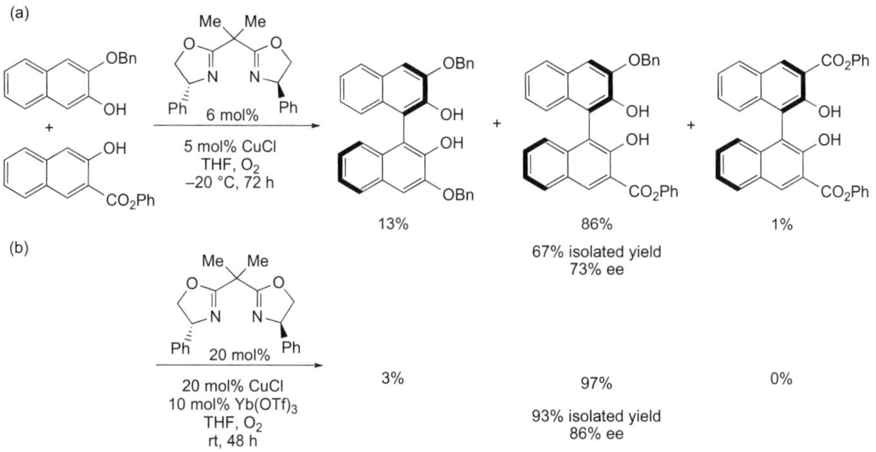

SCHEME 11.37. Enantioselective cross-coupling of naphthols.

copper catalyst derived from a proline methyl ester with a reported high enantioselectivity, which was determined from the optical rotation compared to that of the natural source.[132]

Much has been accomplished in asymmetric phenolic coupling, but many unconquered frontiers remain. For example, control of regioselection in systems other than 2-naphthol is unsolved. In addition, no catalytic asymmetric oxidative phenol couplings have been reported.

11.2.4.1.4. Heterocouplings. Oxidative cross-coupling of two different substrates is a significant challenge. While copper catalysts can partner an electron-rich substrate with an electron-poor substrate, chemoselectivity and enantioselectivity are only moderate (Scheme 11.37a).[113,114,133–135] However, the addition of a Lewis acid, such as Yb(OTf)$_3$, when OTf = tritlate, can improve the outcome significantly (Scheme 11.37b).[136]

Notably, this improved heterocoupling with Lewis acid additive also translated to racemic reactions employing CuCl(OH)TMEDA catalyst and oxygen. A mechanism that accounts for the improvement provided by a simple Lewis acid is outlined in Scheme 11.38. The more electron-rich substrate is oxidized first by the copper catalyst while the Lewis acid selectively coordinates the less electron-rich chelating substrate. Coupling is then followed by a second single-electron oxidation to provide the diketo tautomer of the product. Finally, tautomerization of this form converts the two centrochiral stereocenters into the axial stereochemistry. Many questions remain about this mechanism as the copper catalysts typically do not provide high selectivity with nonchelating substrates.

11.2.4.1.5. Oligomerization and Polymerization. Oligomerization and polymerization of 2-naphthols and 2,2′-binaphthols has been well studied. An example from

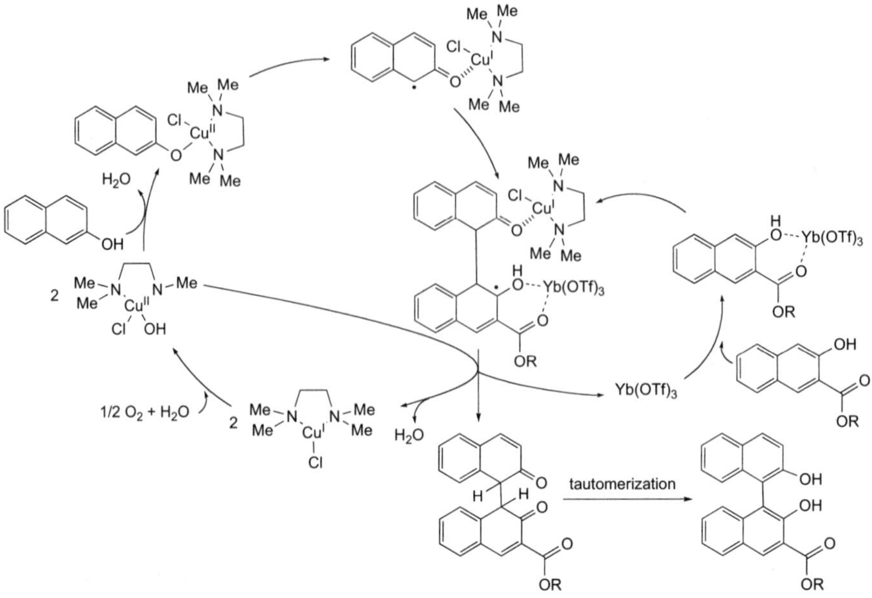

SCHEME 11.38. Mechanism of cross-coupling with Lewis acid.

Okamoto and co-workers[137–140] illustrates how a chiral catalyst can be combined with a chiral substrate leading to polymer products with some level of control over the newly formed axial chiral bonds (Scheme 11.39).

Selective heterocouplings (see Section 11.2.4.1.4) have also been observed in oligomerizaton processes. As with the dimerization, an improvement in the oligomerization selectivity was seen by adding Yb(OTf)$_3$.[141,142]

A combination of catalytic asymmetric naphthol coupling and Glaser–Hay coupling (see Section 11.2.1.1) has been achieved using only the copper diaza-*cis*-decalin complex as catalyst. In doing so, chiral binaphthyl polymers can be prepared directly from achiral naphthol substrates in the presence of oxygen (Scheme 11.40).[143,144] The relative reaction rates of various moieties with the chiral

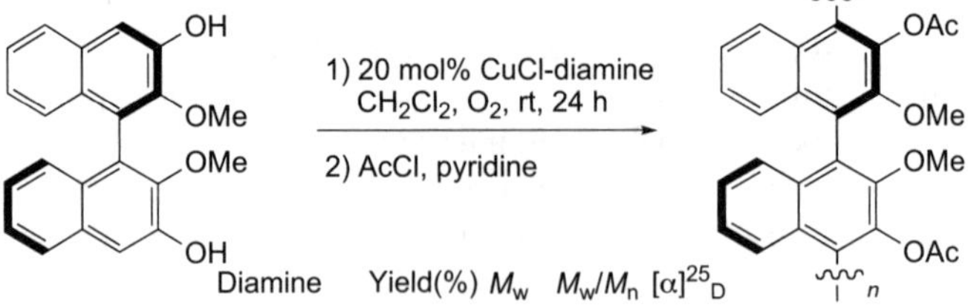

Diamine	Yield(%)	M_w	M_w/M_n	$[\alpha]^{25}_D$
TMEDA	71	5200	1.6	145
(−)-Sparteine	61	3400	2.3	185

SCHEME 11.39. Polymerization of a chiral binaphthol substrate.

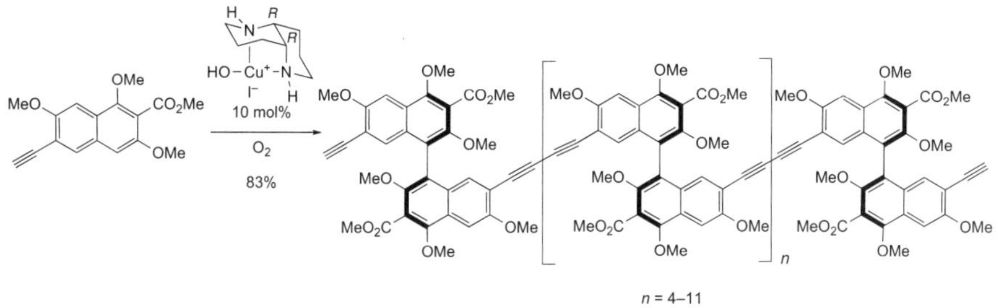

SCHEME 11.40. Tandem asymmetric naphthol coupling and Glaser–Hay coupling.

catalyst follows the order: benzyl cyanides ≫ aryl alkynes > electron-rich 2-naphthols > electron-deficient 2-naphthols > alkyl alkynes. Due to high chemoselectivity, this approach is useful for the organized assembly of multifunctional substrates in a single operation.

11.2.4.2. The C−C Bond Formation: Alkenylphenol Coupling. Scheme 11.41 Alkenylphenols can also undergo the types of oxidations described Section 11.2.4.1.

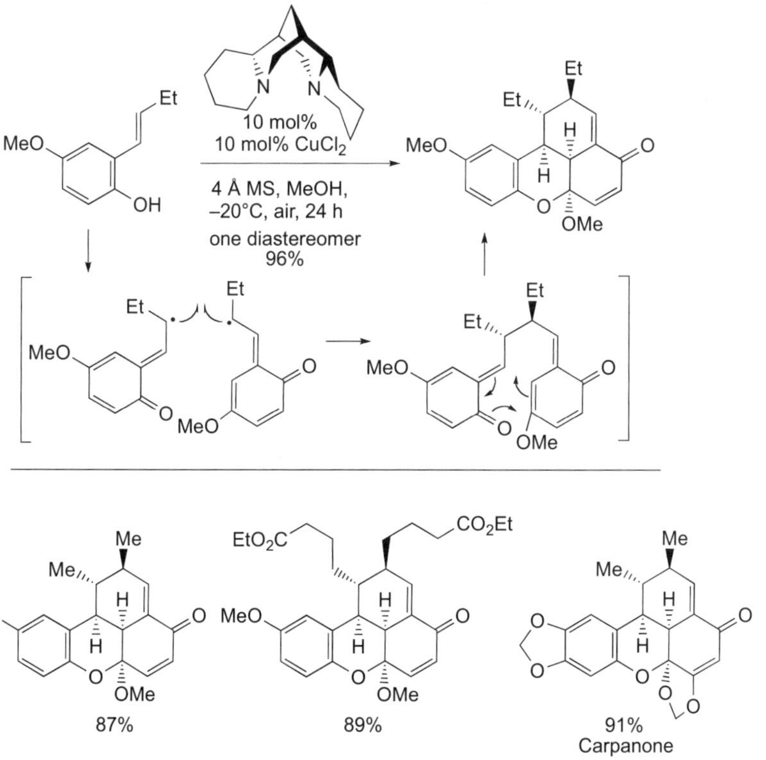

SCHEME 11.41. Oxidative alkenylphenol coupling to carpanone.

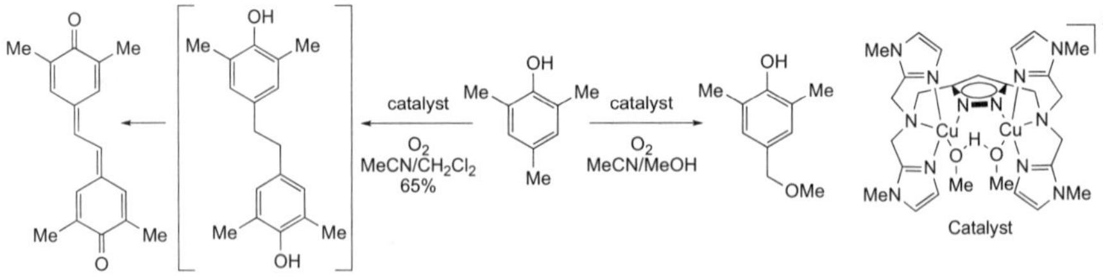

SCHEME 11.42. Oxidative benzylic C–O and C–C coupling of a phenol.

Migration of the radical outside of the ring then allows β,β-phenolic coupling (Scheme 11.41). Subsequent intramolecular hetero-Diels–Alder reaction then constructs a complex tetracyclic structure. With catalytic $CuCl_2$ and a sparteine ligand under air, the coupling is highly efficient providing carpanone and related unnatural congeners as single diastereomers in yields exceeding 85%.[145,146] Unfortunately, <5% ee was observed when employing chiral amine ligands under a variety of reaction conditions, indicating no influence of the chiral catalyst environment on the C–C bond-forming events.

11.2.4.3. The C–C Bond Formation: Benzylic Coupling. A highly preorganized bioinspired dicopper complex with imidazole ligands catalyzes the selective activation of the para-benzylic position of 2,4,6-trimethylphenol under aerobic conditions (Scheme 11.42).[147] The selective oxidation of a single benzylic center is difficult and precedent for C–C coupling of benzylic centers is even rarer. Formation of a benzylic radical would account for generation of either the stilbenequinone (up to 65%) or 4-methoxymethyl-2,6-dimethylphenol depending on the solvent used. Subsequent oxidation of the initial C–C coupling adduct requires the copper catalyst in order to form stilbenequinone.

11.2.4.4. The C–O Bond Formation: Benzylic Coupling. A related, selective benzylic oxidation resulting in C–O bond formation is illustrated in Scheme 11.43.

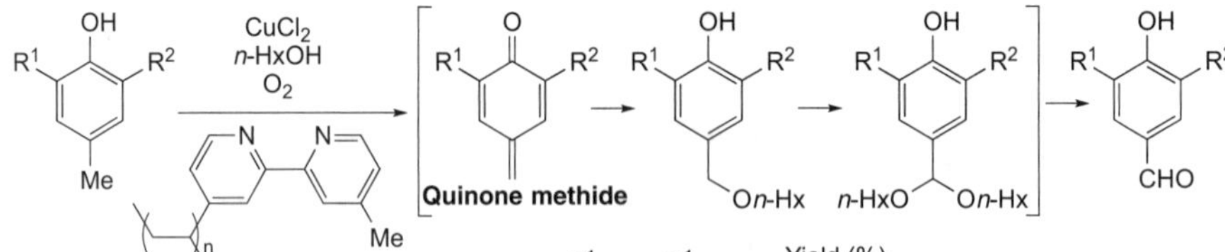

(Hx = hexyl)

R^1	R^1	Yield (%) Aldehyde
Me	Me	92
Me	t-Bu	66
Me	Ph	54
Me	On-Hx	94
OMe	OMe	47

SCHEME 11.43. Oxidative benzylic C–O coupling of a phenol.

With a $CuCl_2$ catalyst supported on a bipyridine containing polymer, 2,6-disubstituted 4-methylphenols could be oxidized to the corresponding 4-hydroxybenzaldehydes in good yields.[148] The polymer catalyst could be recovered and reused without a noticeable loss in activity. In this case, the reaction is proposed to proceed via a *p*-quinone methide intermediate.

11.2.4.5. The C–O Bond Formation: Polymerization. The oxidative coupling of phenols, as shown schematically in Scheme 11.44 for 2,6-dimethylphenol, can be catalyzed by copper–amine complexes, such as Cu(I)Cl–pyridine, under oxygen.[149] The two main products of this reaction are a diphenoquinone, arising from C–C coupling of two phenol moieties, and a C–O coupled product that can react further and will ultimately yield a polymer, the linear poly(phenylene ether) (PPE). The resulting polymer shows very good mechanical properties and chemical stability, even at elevated temperatures, and is therefore an important engineering plastic. With unhindered substrates, the polymer forms in high yield and molecular weight with very little of the C–C coupling ($<5\%$). Even so, the diphenoquinone may degrade the polymer upon further processing. As such, processes that result in low yields of the diphenoquinone are desirable. Mechanistic studies have revealed that the polymer process is complex, resulting from a series of reactions with phenoxonium and quinone ketal intermediates.[150]

Hindered diamine catalysts have also been useful in this regard. For example, a regioregular polymer can be formed with the more hindered substrate, 2,4-dimethylphenol, using 0.1–1 mol% of a CuCl–TMEDA catalyst under oxygen.[151]

11.2.4.6. Halogenation. A simple, low cost and highly selective method for the synthesis of monobromophenols from phenol and electron-rich phenolic compounds, has been developed (Scheme 11.45).[152] Bromide ions are used as halogenating agents, O_2 as a final oxidant, and $Cu(OAc)_2$ as a catalyst. Stahl and co-workers[153] provide compelling evidence for the formation of Br_2 under the reaction conditions, which suggests an electrophilic bromination pathway where oxygen and copper act to convert bromide anions to bromine. A similar method has been reported for oxidative chlorination.[154,155] In this case, it is unlikely that $CuCl_2$ disproportionates chloride to

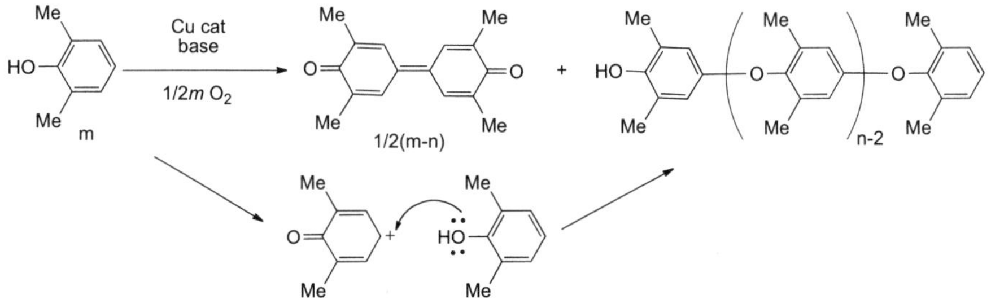

SCHEME 11.44. Oxidative phenol polymerization via C–O coupling.

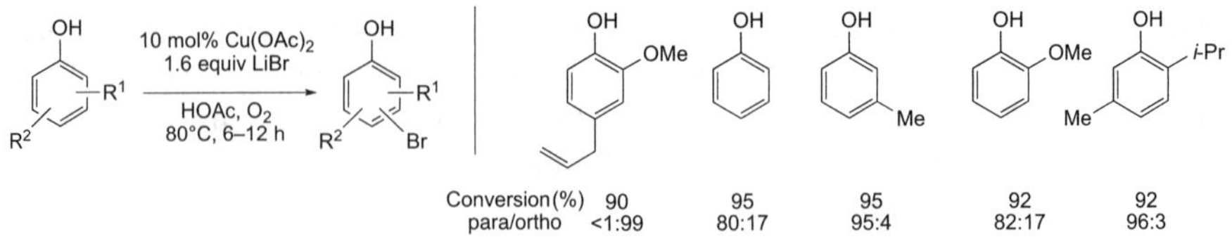

Conversion(%)	90	95	95	92	92
para/ortho	<1:99	80:17	95:4	82:17	96:3

SCHEME 11.45. Oxidative halogenation of phenols.

chlorine and the involvement of a phenolic radical formed by copper-catalyzed oxidation of phenol is likely. For the halogenation of non-phenolic arenes see Section 11.2.6.1.

11.2.4.7. ortho-Quinones from Catechols.

While only tyrosinase catalyzes the hydroxylation of phenols, as well the subsequent oxidation of catechols to quinones, this latter activity is also displayed by catechol oxidase. The low redox potential of catechols to the corresponding quinones has resulted in the development of numerous small molecule biomimetic catalysts.[2,156,157] Most of these studies have centered on 3,5-di-*tert*-butylcatechol because the bulky *tert*-butyl groups prevent alternative reaction pathways (Scheme 11.46) although other catechols can be oxidized by simple copper catalysts, such as copper(II) nitrate.[158]

In a recent advance, chiral copper complexes have been shown to catalyze an enantioselective oxidative kinetic resolution of the chiral catechol 3-(3,4-dihydroxyphenyl) alanine (DOPA) under an oxygen atmosphere (Scheme 11.47).[159] The k_{rel} value of ~7 was measured for the transformation. The difference in reactivity is both a manifestation of the greater affinity of the D-DOPA–OMe in binding the chiral complex and in a faster rate of reaction of the D-DOPA–OMe adduct.

11.2.5. Reactions of Anilines

11.2.5.1. The C–C Bond Formation: Biaryl Coupling.

While the successful oxidative coupling of naphthols and phenols has been accomplished with copper catalysts using oxygen (see Section 11.2.4.1), the same transformation of 2-naphthylmines and anilines does not proceed well. With stoichiometric copper reagents, some couplings have been achieved although carbazole and other byproducts are often observed.[101,160,161] One of the best results to date has resulted in the formation of

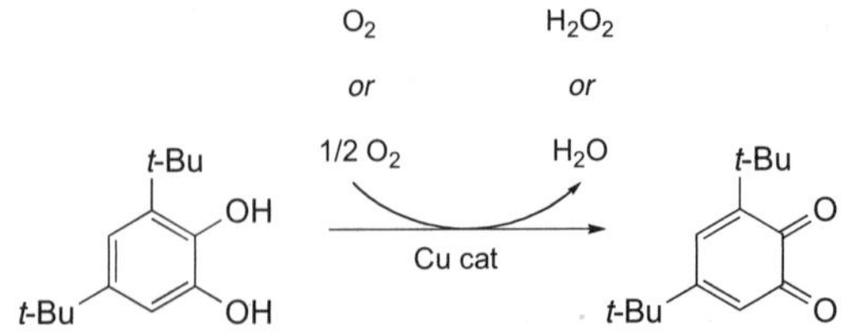

SCHEME 11.46. Catechol oxidation to an *o*-quinone.

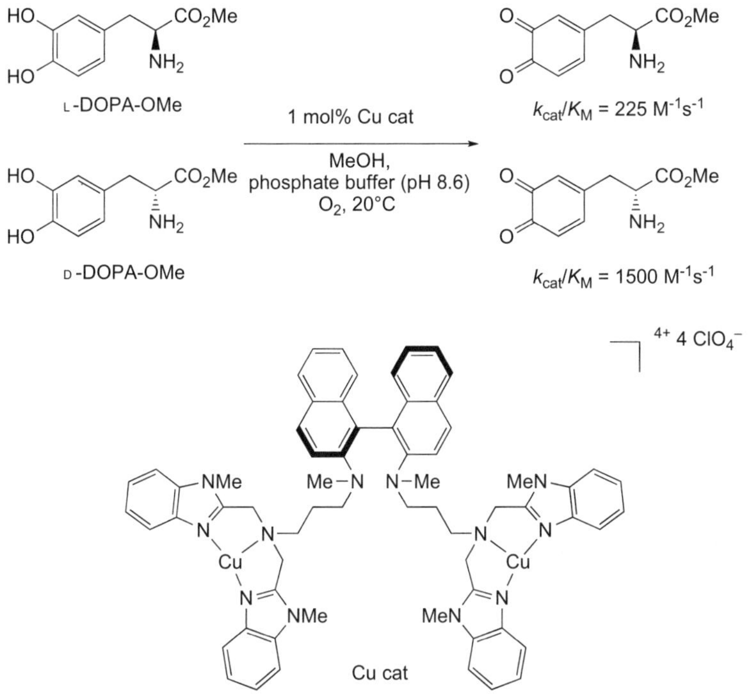

SCHEME 11.47. Oxidative kinetic resolution of DOPA.

(S)-DM-DABN (3,3'-dimethyl-2,2'-diamino-1,1'-binaphthyl) with moderate yield and low selectivity (Scheme 11.48).[162]

11.2.5.2. The C—N Bond Formation: Polymerization. Many catalysts have been used to generate important conducting polymers, the polyanilines, from the corresponding anilines. However, the oxidative polymerization of anilines using copper catalysts with oxygen is relatively underexplored. In one report, the dimer of aniline [N-(4-aminophenyl)aniline] was converted to the conducting emeraldine salt using

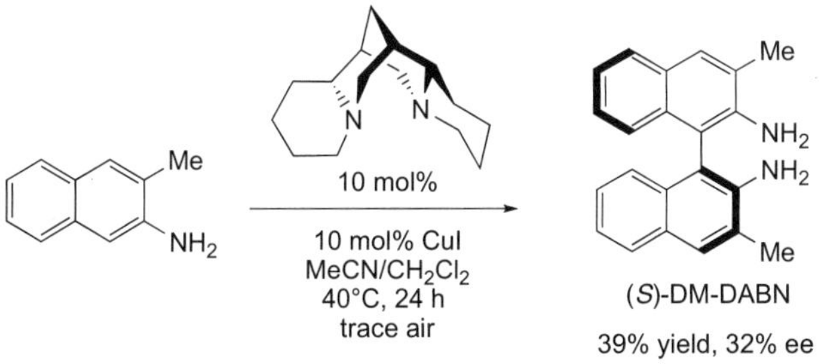

SCHEME 11.48. Asymmetric oxidative coupling of a 2-naphthylamine.

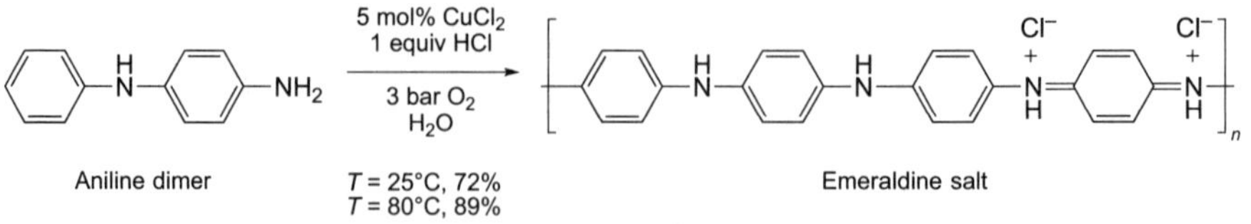

SCHEME 11.49. Oxidative polymerizaton to form polyaniline.

oxygen as the oxidant in aqueous solution under mild conditions (Scheme 11.49).[163] Importantly, the uncatalyzed oxidation by O_2 takes place with a modest yield (13%) and a strong catalytic effect was observed with various copper salts allowing yields of up to 89%.

11.2.5.3. The N—N Bond Formation: Diazo Formation. Instead of the C—C or C—N bond coupling with anilines observed above (see Section 11.2.5.1 and 11.2.5.2), it is also possible to selectively undergo N—N bond coupling with copper catalysts. Using oxygen as the oxidant, symmetric diazobenzenes were formed from primary anilines in much improved yields with copper catalysts when employing *n*-BuMgBr as a base (Scheme 11.50).[164] A likely mechanism is formation of the aniline radical from the deprotonated aniline followed by dimerization to the hydrazine. Further deprotonation and oxidation then yields the diazobenzene. Support for this mechanism can be found in the facile copper oxidations of hydrazines to diazo compounds.[165]

11.2.5.4. Halogenation. Analogous to the halogenation of phenols (see Section 11.2.4.6), the oxidative chlorination and bromination of anilines has been reported.[166] In this case, the intervention of an aniline radical is proposed similar to the phenol radical in the phenolic oxidative halogenation.

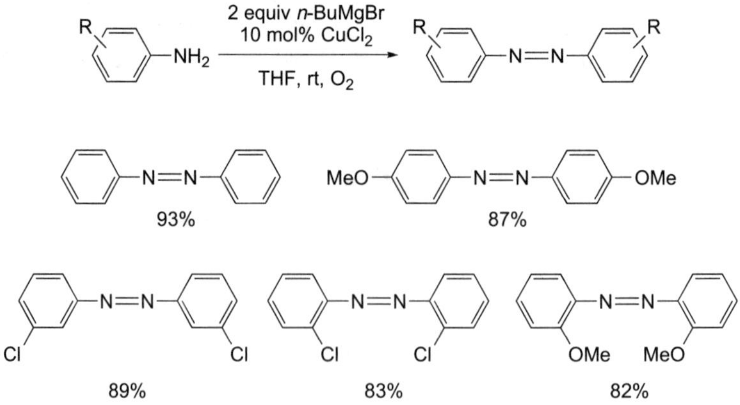

SCHEME 11.50. Oxidative dimerization of anilines to azobenzenes.

11.2.6. Reactions of Arenes

Oxidative C—H insertion reactions of arenes have attracted a great deal of attention, especially in the arena of palladium catalysis. However, the combination of inexpensive base metal catalysts, (e.g., Cu), with oxygen, an inexpensive and green oxidant, has much merit. Recent discoveries have provided a strong foundation for the utility of this combination. While there is considerable evidence supporting the viability of copper C—H insertion to yield an aryl copper intermediate,[167,168] the various arene oxidative C—H insertion reactions reported in the literature encompass a range of different mechanisms that can be broadly divided into three categories: (1) nucleophilic attack of the arene onto an electrophilic metal or generation of a radical cation intermediate, (2) reaction of the arene via a radical cation with copper coordinated to a substituent, or (3) C—H insertion. For related sections in this chapter, see Sections 11.3.3 (Sections 11.3.3.4, alkane, and 11.3.3.5, arene, oxidation) and 11.2.4.6 (halogenation of phenols).

11.2.6.1. Nucleophilic. One type of oxidative cross-coupling of arenes with arylboronic acids involves the use of the highly electrophilic $Cu(OCOCF_3)_2$ reagent (Scheme 11.51).[169] In this transformation, nucleophilic attack of an electron-rich arene onto the $Cu(OCOCF_3)_2$ may generate a new $ArCu(OCOCF_3)$ accompanied by released of CF_3CO_2H. Transmetalation followed by an oxidatively induced reductive elimination (see Scheme 11.20) would then provide the biaryl product. The reaction is selective for cross-coupling (cf., arylboronic acid homocoupling in Section 11.2.3.4)

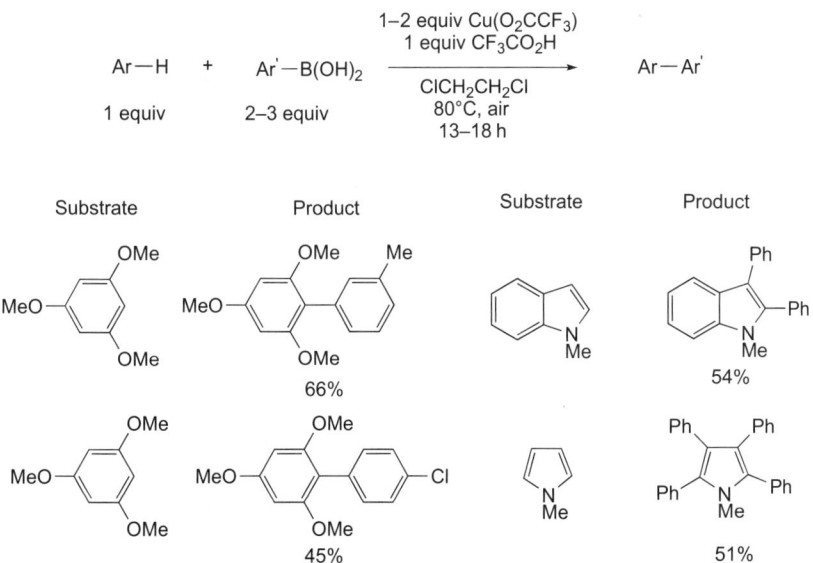

SCHEME 11.51. Oxidative coupling of nucleophilic arenes with aryl boronic acids.

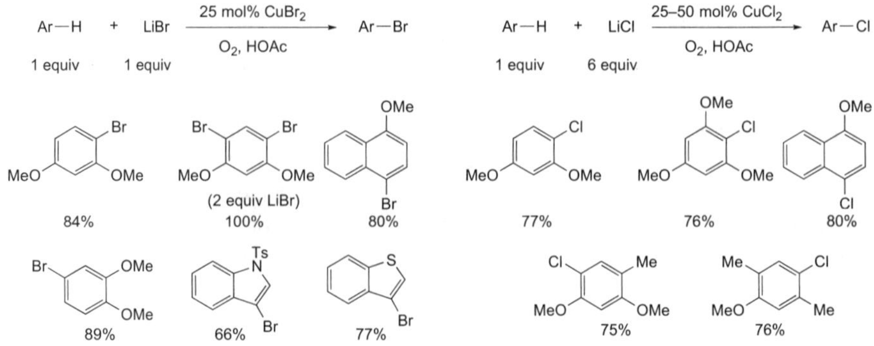

SCHEME 11.52. Copper-catalyzed oxidative halogenation of non-phenols.

and produces a range of biaryls in good yields. Notably, multiple C–H bond arylation is possible with indoles and pyrroles (right half of Scheme 11.51).

Substitution of electron-rich arenes with halides has been reported by Stahl and co-workers[153] using catalytic Cu(II) halides and oxygen (Scheme 11.52). Regioselective chlorination and bromination is observed in all cases. Preliminary mechanistic insights suggest that the bromination and chlorination reactions proceed by different pathways. For bromination, the copper catalysts generate bromine from bromide giving way to an electrophilic aromatic substitution pathway. For chlorination, the copper catalyst likely causes a single ET generating a radical cation from the arene that then reacts with a chlorine source.

11.2.6.2. Directed Insertion. Coordinating groups have proven remarkably effective in directing the position of substitution in palladium-catalyzed oxidative arene insertion chemistry. A similar effect has been observed in copper-catalyzed oxidative insertion chemistry. For example, Buchwald and Brasche[170] discovered that aryl amidines gave rise to benzimidazoles in high yields when Cu(OAc)$_2$ was employed in the presence of oxygen (Scheme 11.53). Notably, a number of additional examples of directed copper-catalyzed oxidative arene insertions have appeared providing an avenue to a range of heterocyclic systems.[171–173]

As of yet, the mechanism of these types of transformations remains unclear. Scheme 11.54 outlines three possible pathways, which are all united by the formation of an initial copper adduct accompanied by deprotonation of the amidines. Pathways A and B both involve oxidation of the electron-rich aromatic substrate by single ET to form a radical cation. In pathway A, the radical cation attacks the copper; after deprotonation a reductive elimination generates the product along with concurrent release of a reduced copper species. In pathway B, the radical cation displaces the copper amidine causing a release of the reduced Cu$^{(I)}$OAc; rearomatization then generates the benzimidazole product. Pathway C involves a copper nitrene that can then undergo a concerted insertion of the nitrogen into a C–H bond or an electrocyclic ring closure and a [1,3]-shift of a hydrogen to the product. The reactivities of amidines substituted with electron-donating or

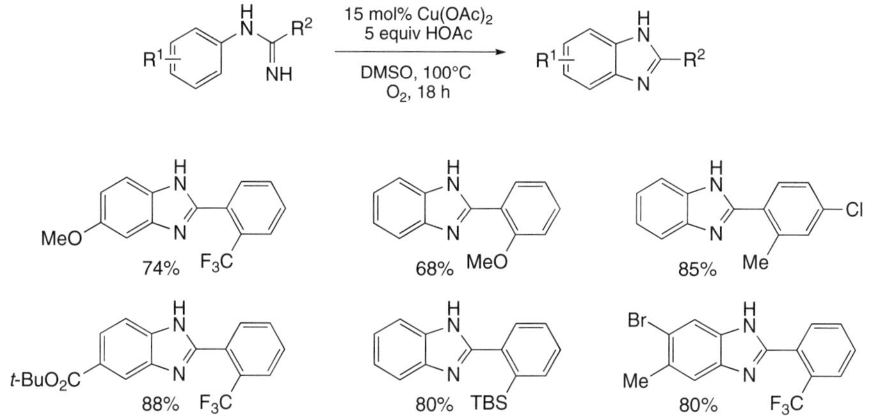

SCHEME 11.53. Directed oxidative intramolecular C—H insertion into arenes.

electron-withdrawing groups are consistent in direction, but not in magnitude, with the ability of these substituents to stabilize the cationic intermediates in paths A or B. Thus, a thorough mechanistic study is needed to unravel the mechanistic intricacies of this process.

Yu and co-workers[174] described the oxidative chlorination of aromatics containing a directing 2-pyridyl substituent using catalytic $CuCl_2$ with oxygen. Typically, bischlorination occurred except when one ortho-position was blocked or steric hindrance was large. Similar to the directed C—H insertion in Scheme 11.54, a radical cation intermediate is proposed for this transformation. Subsequent trapping of this radical cation with $CuCl_2$ coordinated to the o-pyridyl group provides an explanation for the regioselection, as well as the fact that biphenyl does not react under these conditions.

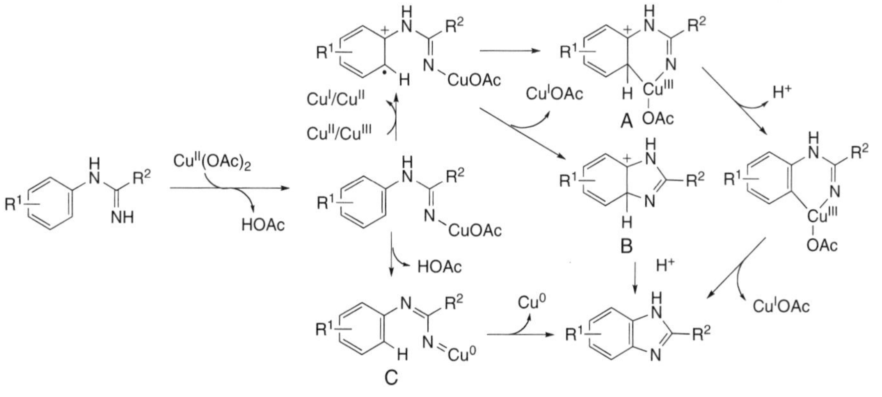

SCHEME 11.54. Possible mechanisms for the directed oxidative insertion.

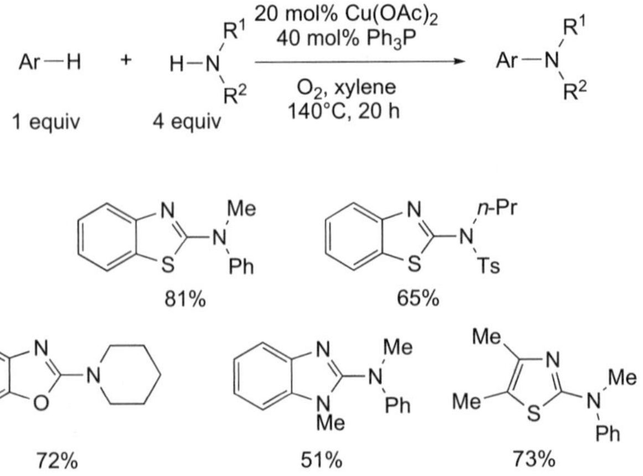

SCHEME 11.55. Copper-catalyzed C—H insertion into azoles.

11.2.6.3. Functionalization of Acidic Sites. The oxidative amination of azoles at the 2-position is outlined in Scheme 11.55. For example, the reaction of benzothiazole with N-methylaniline in the presence of sodium acetate and 20 mol% Cu(OAc)₂ in xylene under an oxygen atmosphere afforded the aminated product in 81% yield.[175]

Although the mechanism of copper-catalyzed oxidative C—H, N—H coupling is still unclear, the reaction may proceed under a similar pathway to the reaction of terminal alkynes reported by Stahl and co-workers[25] (Scheme 11.7, see Section 11.2.1.2). The acidic C—H bond of the azole (pK_a DMSO ~ 27)[176] would be replaced with Cu(II) to form an organocopper (Scheme 11.56). This step may either precede or follow coordination of the amine and deprotonation. Either order would lead to an aryl amido Cu(II) species. Reductive elimination accompanied by oxygen mediated oxidation would then provide the product and regenerate the Cu(II) catalyst.

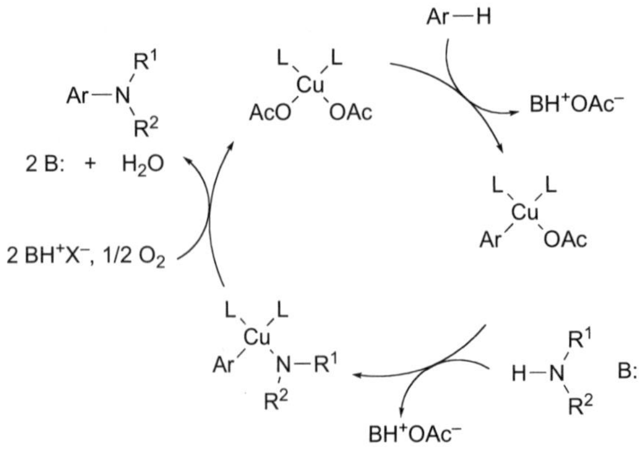

SCHEME 11.56. Possible mechanism of copper-catalyzed C—H insertion into azoles.

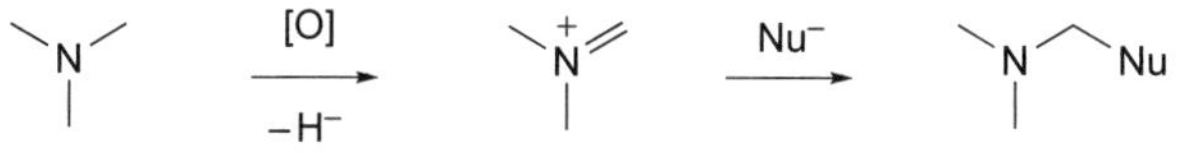

SCHEME 11.57. Oxidative coupling of amines with nucleophiles.

In addition to the considerable synthetic potential of these reactions, these results indicate that the role of stoichiometric Cu salts used in palladium-catalyzed oxidative C−H bond transformations may not always be limited to that of a reoxidant for Pd(0).

11.2.7. Amines to Iminiums

The oxidative coupling of amines has received much attention recently.[177,178] In these reactions, amines are generally believed to be oxidized to iminium ions (Scheme 11.57), similar to the reaction of amine oxidases (see Section 11.3.4). Subsequent reaction with nucleophiles (Nu) can occur via a Mannich-like process. While many metals have been employed in conjunction with a variety of oxidants, only the copper-catalyzed processes using oxygen as an oxidant are summarized here.

Seminal work on this type of reaction was described by Miura and co-workers[179,180] who showed that N,N-dimethylanilines couple with alkynes in the presence of oxygen. While formation of other byproducts was problematic (Scheme 11.58), this work firmly established that copper and oxygen are a functional pair for this type of transformation.

Later work on copper-catalyzed amine to iminium oxidation predominantly focused on *tert*-butyl hydroperoxide (TBHP) as the oxidant with a variety of nucleophiles, but has also recently shown high selectivities with oxygen.[177] The reaction of tetrahydroisoquinolines and anilines with nitroalkanes is illustrative (Scheme 11.59).[181] Other nucleophiles including malonates have proven effective in this system.[181] The use of less acidic nucleophiles (e.g., ketones) by means of an acetic acid additive also has been documented.[182]

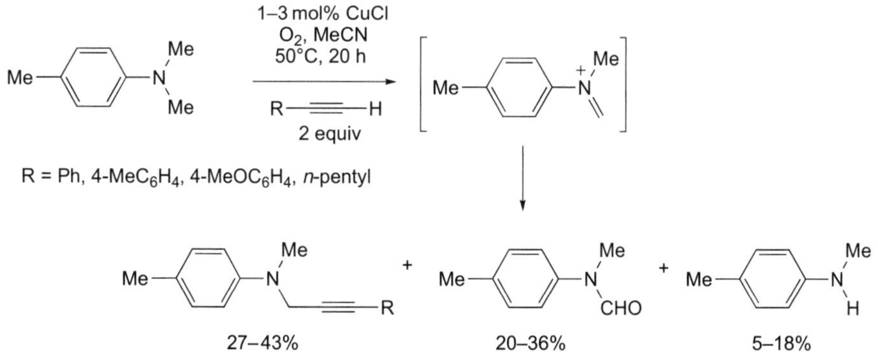

SCHEME 11.58. Oxidative coupling of amines with alkynes.

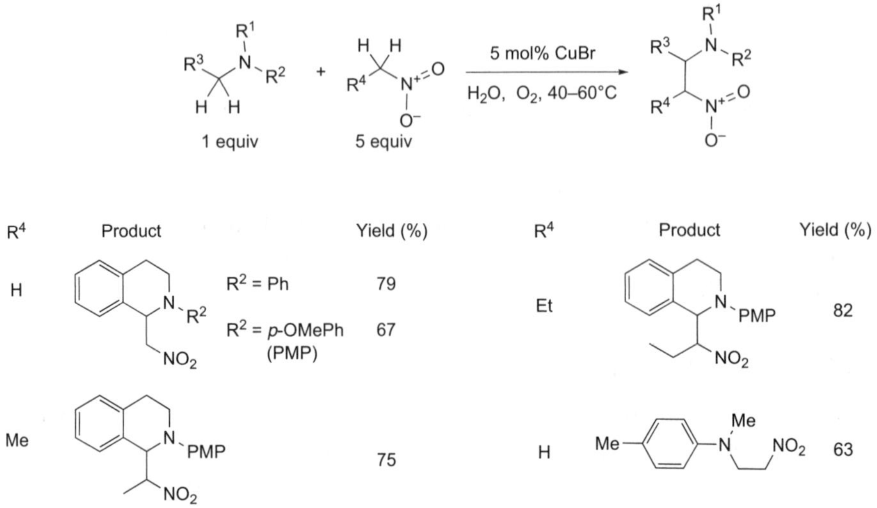

SCHEME 11.59. Oxidative coupling of amines and nitroalkanes.

Slightly higher yields were obtained using O_2 versus TBHP with acetone, but only O_2 was effective with butanone. The addition of 3 equiv of acetic acid and molecular sieves increased yields further. For the methyl alkyl ketones, there was a general downward trend in the product yield as the alkyl chain became longer. Even with oxygen, diethyl ketone gave poor yield (24%) and poor diastereoselection (1.1:1).

Highly nucleophilic silyl ketene acetals, for which *in situ* deprotonation was not required, were also highly effective nucleophiles in this transformation.[183] In this case, the reduction product from O_2 must be a silanol instead of water. Good results have also been reported with the silyl ketene acetals and the less nucleophilic silyl enol ethers using $CuCl_2$ as the catalyst.[178]

Aryl boronic acids have also been successfully employed as nucleophiles in the reaction of copper-generated iminiums.[184] Higher reactivity under O_2 in the presence of water was observed that is consistent with activation of the boronic acid. It is likely that copper is playing more than one role in this transformation including oxidizing the aniline to an iminium and transmetaling the aryl boronic acid in a manner similar to that described in Section 11.2.3 on boronic acid couplings.

Molecular oxygen uptake measurement during coupling with nitroalkanes[181] (see Scheme 11.59) and ketones[182] showed that one-half of equivalent of oxygen was consumed. The reactions proceeded in the presence of 2 equiv of BHT (BHT = Butylated hydroxytoluene), a free radical inhibitor, strongly suggesting that reaction does not involve a radical process.[178,183–185] A $2e^-$ oxidation of the amine by the copper is suggested producing an iminium-type intermediate that may still interact with the copper center (Scheme 11.60). Additional Cu(II) may act as a Lewis acid to faciliate deprotonation with the nitroalkane, ketone, and malonate substrates.

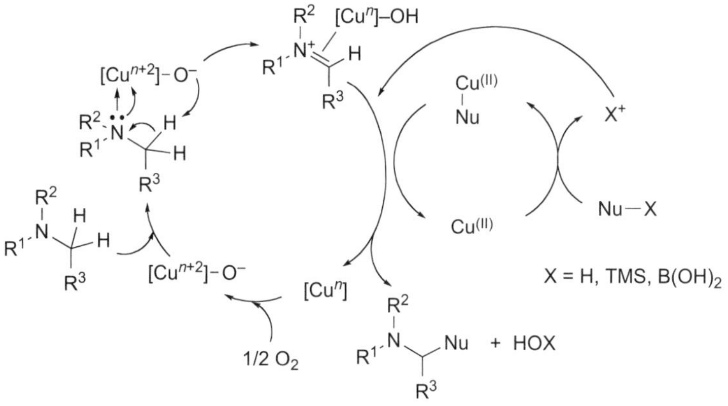

SCHEME 11.60. Copper-catalyzed oxidation of an amine to an iminium using oxygen followed by nucleophilic trapping.

11.2.8. Ethers to Oxocarbeniums

In general, the radical abstraction at the position alpha to oxygen is less favorable than alpha to an amine (see Section 11.2.7). For this reason, stronger oxidants than oxygen are typically required in the copper-catalyzed processes.[177,186] However, this problem can be overcome by using a catalytic oxidant, such as N-hydroxyphthalimide (NHPI) (Scheme 11.61).[187] Under these conditions, cyclic benzyl ethers can be oxidatively

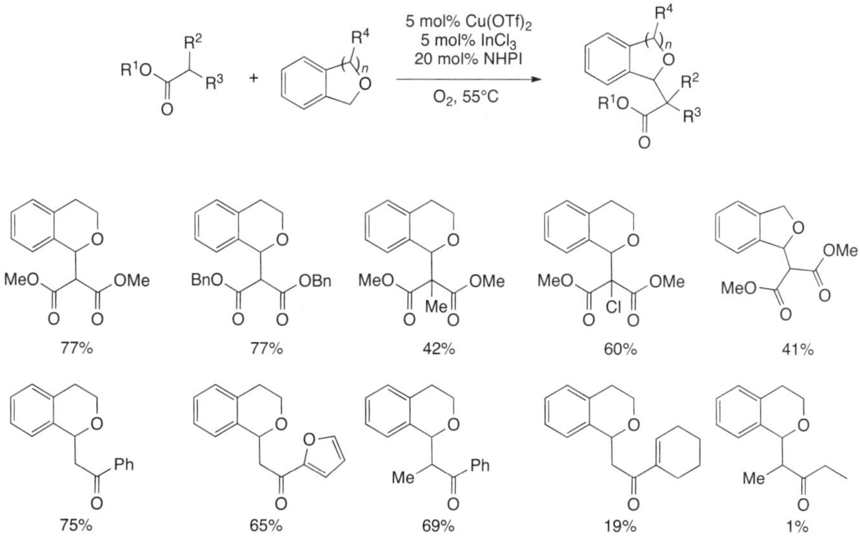

SCHEME 11.61. Oxidative coupling of ethers and ketones using the cooxidant, NHPI.

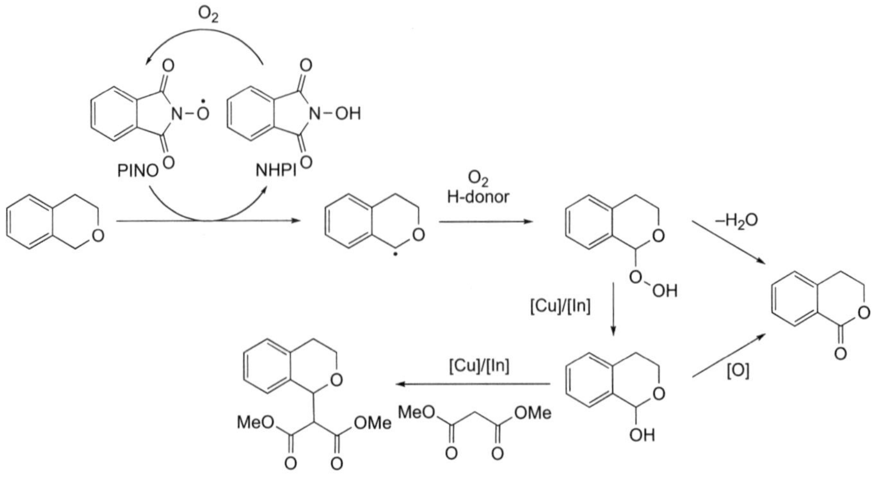

(PINO = photodimide-N-oxyl)

SCHEME 11.62. Mechansim of the oxidative coupling of ethers and ketones.

coupled with a range of malonate and ketone nucleophiles. With the exception of ethyl phenyl ketone, couplings to the ethyl group of ketones do not proceed well.

The mechanism proposed for this transformation (Scheme 11.62) is similar to that proposed for the amine oxidation (see Scheme 11.60). Key differences include a single-electron oxidation catalyzed by NHPI, which is in turn reoxidized by O_2. The resultant radical traps oxygen to form a peroxyketal that is then reduced by the copper or indium to provide a hemiketal. The oxocarbenium formed from the hemiketal can then trap a nucleophile. The copper–indium plays an additional role as a Lewis acid to facilitate deprotonation of the nucleophile.

Another example of two oxidants working in tandem can be found in Scheme 11.63, where both TBHP and O_2 are required to achieve the transformation.[188] This tandem process is comprised of a hydroxyalkylation of an alkene

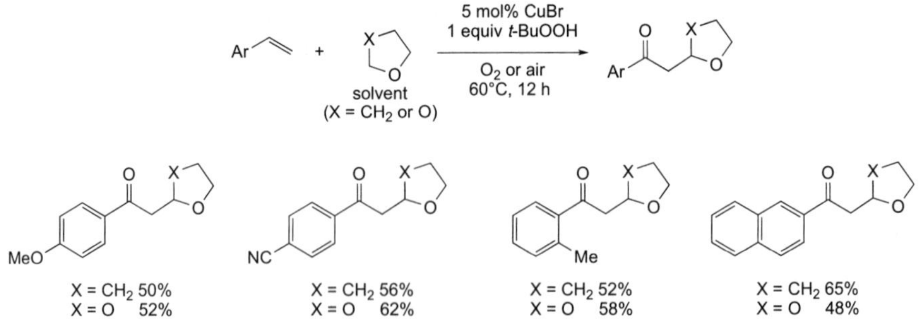

SCHEME 11.63. Oxidative coupling of ethers and alkenes.

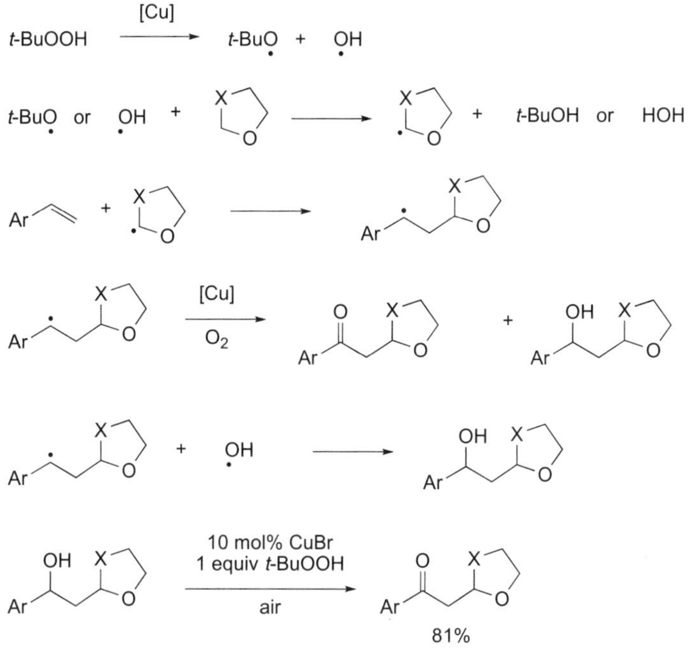

SCHEME 11.64. Possible mechanism for the copper-catalyzed coupling of ethers and alkenes.

followed by further oxidation to the ketone. A putative mechanism for this reaction is outlined in Scheme 11.64.

11.2.9. Thioamides to Nitriliums

A copper-catalyzed oxidative cyclization reaction of thioamides results in desulfurization (Scheme 11.65).[189] Molecular oxygen is used as the stoichiometric oxidant and a wide range of five- to seven-membered nitrogen-containing heterocycles can be synthesized. The reactions do not take place in the absence of either oxygen or CuCl, and elemental sulfur is the sole coproduct.

A possible reaction pathway for the oxidative cyclization involves oxidation by *in situ* generated Cu(II) to give a sulfenylium intermediate (Scheme 11.66). Because the corresponding alkylations of this sterically hindered intermediate do not occur, it is proposed that elemental sulfur is then spontaneously eliminated to give the nitrilium ion. Subsequent cyclization by the internal alcohol gives the corresponding heterocyclic product.

11.2.10. Alcohol Oxidation

Copper-containing enzymes (e.g., galactose oxidase, GO) that catalyzes the 2e$^-$ oxidation of alcohols to aldehydes, have been studied extensively.[1–8] These systems

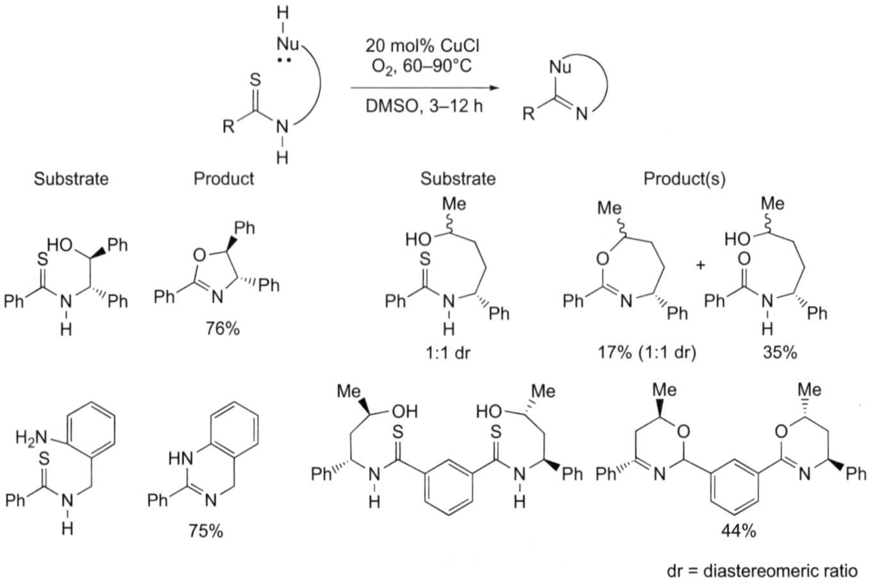

SCHEME 11.65. Copper-catalyzed oxidation of thioamides to oxazolines and amidines.

have served as the inspiration for numerous small molecule copper catalysts that can effect similar transformations.[9–15] In developing such small molecule catalysts, the greatest challenges have been (1) substrate scope, especially primary alcohols and hindered alkanols; (2) identification of safe and inexpensive cooxidants; and (3) identification of simple, inexpensive copper catalysts. Many of these challenges have been met and in the next sections, a selection of the most practical methods is presented.

11.2.10.1. Simple Oxidation to Aldehydes.

Galactose oxidase functions by means of a copper catalytic site in which a ligand-stabilized radical is key to the oxidation.

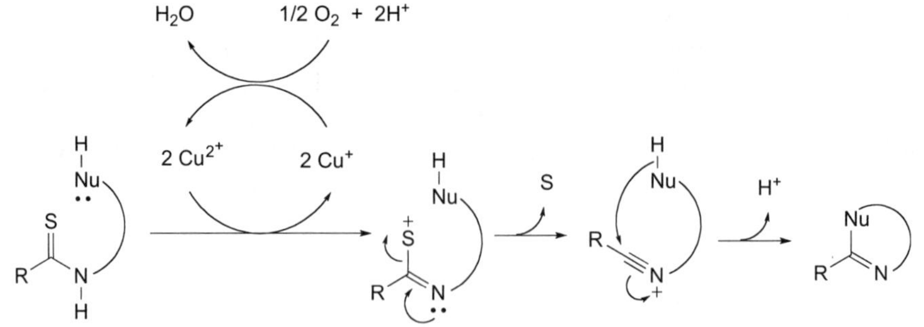

SCHEME 11.66. Proposed mechanism for the oxidation of thioamides to oxazolines and amidines.

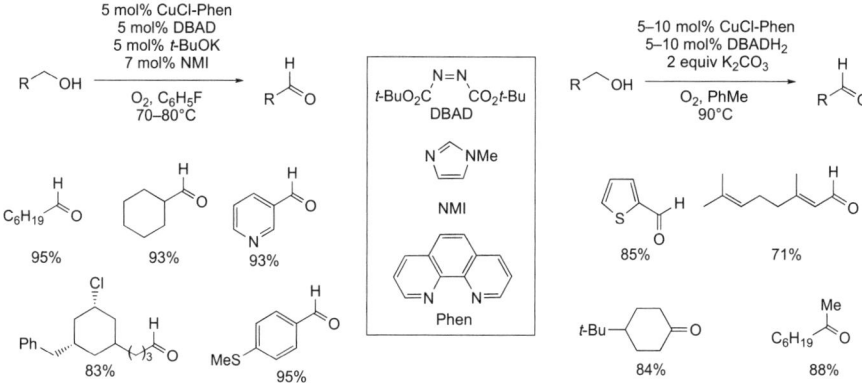

SCHEME 11.67. Aerobic copper-catalyzed alcohol oxidations using an azodicarboxylate as a cocatalyst.

Small molecule copper catalysts have mimicked this activity in three ways: (1) by using a hydride anion acceptor as a cofactor; (2) by using a hydrogen radical acceptor as a cofactor; or (3) by using a ligand that is a functional analogue of that found in GO.[9]

Copper-catalyzed aerobic oxidation reactions have emerged as a powerful methodology for the transformation of alcohols into carbonyl compounds. Among these methods, Marko et al.[190,191] have developed the most general and efficient catalytic oxidation. Using CuCl with a 1,10-phenanthroline (phen) ligand and di-*tert*-butyl azodicarboxylate (DBAD) as a cocatalyst, very efficient oxidations are possible (Scheme 11.67). Furthermore, the presence of a catalytic amount of either *t*-BuOK or *N*-methylimidazole (NMI) does not inhibit the reaction, permitting development of protocols for hindered alcohols, as well as the much less reactive primary alcohols. Thus, these systems function well for allylic, benzylic, and aliphatic alcohols, including primary and secondary alcohols.

The mechanism of the aerobic copper-catalyzed alcohol oxidation employing an azodicarboxylate as a cocatalyst proceeds via a Cu(I)/Cu(II) redox cycle in which the azodicarboxylate acts as a hydride acceptor thereby, oxidizing the alcohol coordinated to the copper center (Scheme 11.68). Subsequently, oxygen oxidizes the Cu(I) intermidate to a Cu(II) peroxo species. Homolysis of the peroxide bond and 1e⁻ reduction of the Cu(II) then accompany reoxidation of the hydrazine ligand to re-form the azodicarboxylate.

Another class of general and efficient aerobic copper-catalyzed oxidations employs TEMPO (2,2,6,6-tetramethylpiperidine-1-oxyl) as a cocatalyst. In this case, the stable TEMPO radical plays the role of the ligand sphere radical found in GO as outlined in the mechanism in Scheme 11.69.

Numerous examples of this type of system have been reported. One, comprised of a biphasic fluorous system, shows broad efficacy for primary, secondary, allylic, and benzylic alcohols (Scheme 11.70).[192] High chemoselectivities are observed in the oxidation of substituted cyclohexanols with axial cyclohexanols reacting six to eight times faster than the corresponding equatorial cyclohexanols.

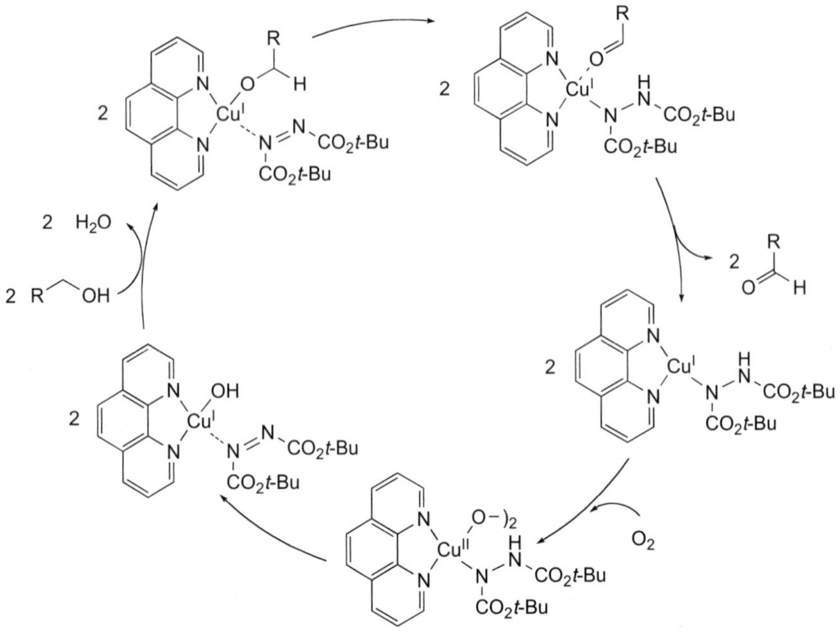

SCHEME 11.68. Mechanism of the copper–azodicarboxylate aerobic alcohol oxidation.

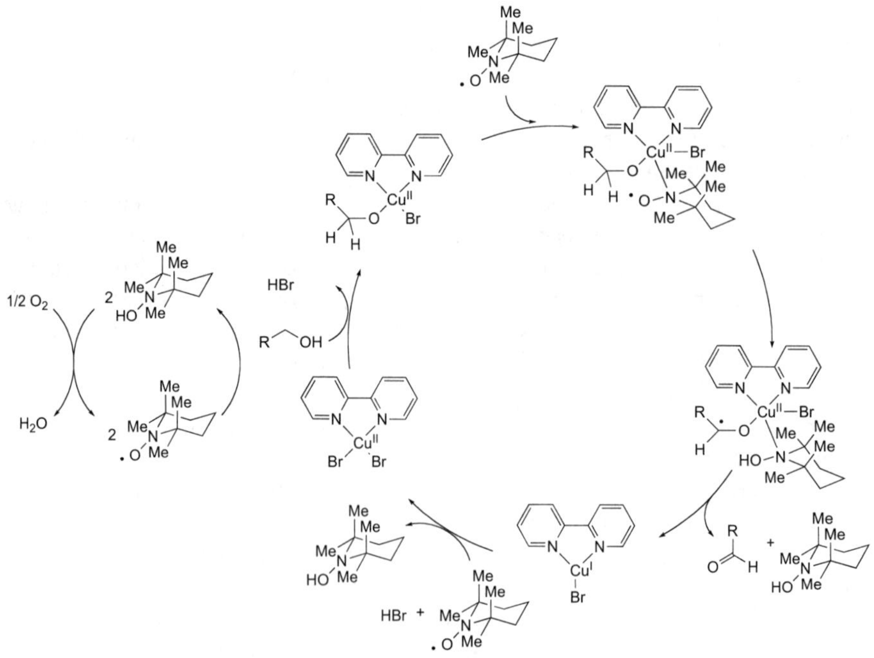

SCHEME 11.69. Mechanism of the Cu–TEMPO aerobic alcohol oxidation.

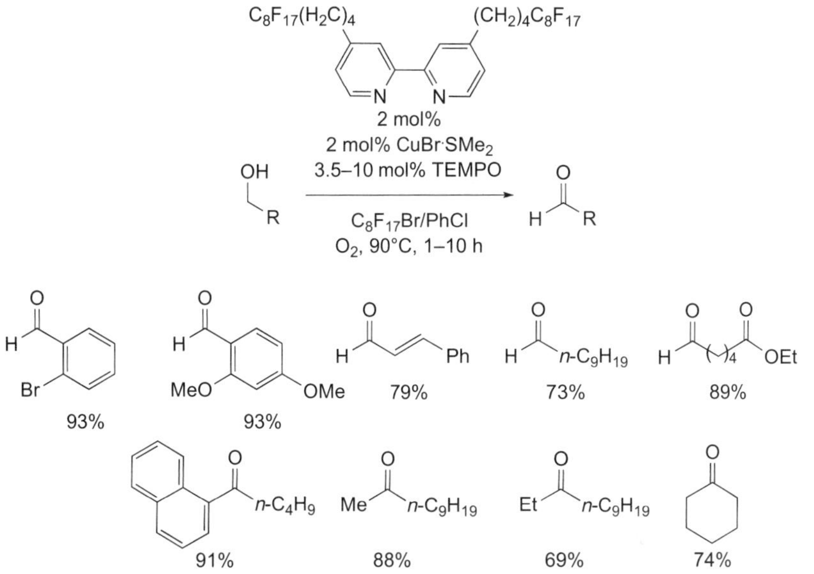

SCHEME 11.70. Fluorous aerobic copper-catalyzed alcohol oxidations using TEMPO as a cocatalyst.

The use of high temperatures (90°C) in the above system does present some limitations. One system that operates well at ambient temperature is outlined in Scheme 11.71.[193] However, the scope is limited because the method does not transform secondary benzylic or aliphatic alcohols.

Stack and co-workers[194] described the first small molecule catalyst able to mimic the mode of action of GO (Scheme 11.72). This report has inspired much activity[9]

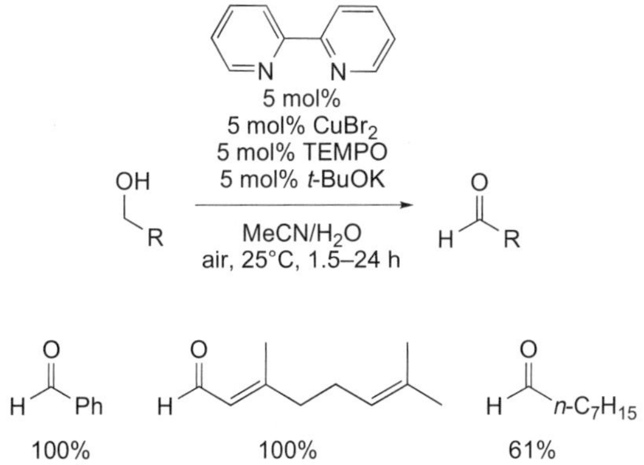

SCHEME 11.71. Room temperature copper-catalyzed alcohol oxidation using TEMPO as a cocatalyst.

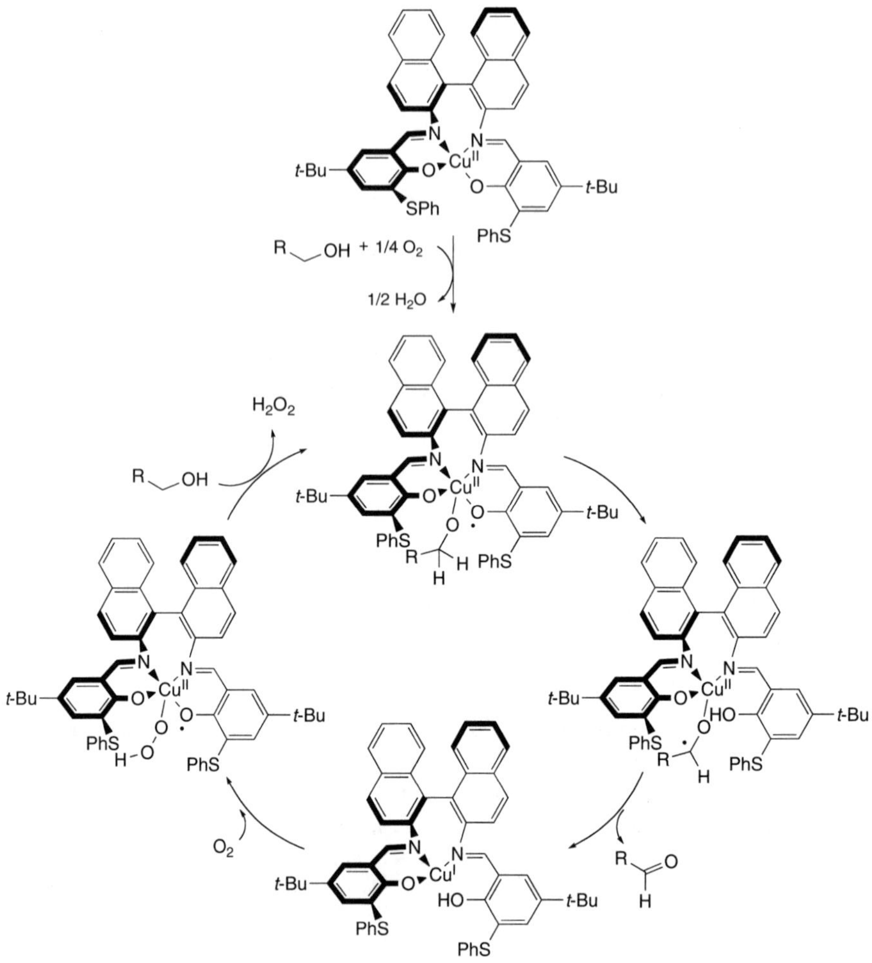

SCHEME 11.72. Mechanism of a small molecule GO mimic.

including theoretical studies that support the proposed mechanism.[195] Since hydrogen-atom abstraction from the α-position of the alcohol is rate determining, the most challenging substrate is methanol with its very strong C—H bonds. Even so, catalyst systems mimicking GO have been reported that can oxidize methanol.[196]

Tandem reactions are extremely powerful in that complexity can be built up in one reaction flask thereby reducing the effort spent purifying materials. Even more efficient is when one catalyst can be used to catalyze more than one transformation. A novel catalytic sequence employing a tandem aerobic oxidation olefination is outlined in Scheme 11.73. A single and inexpensive copper catalyst provides a large range of olefins from both primary and secondary alcohols in good-to-excellent yields. The reaction exhibits excellent functional group compatibility, and the

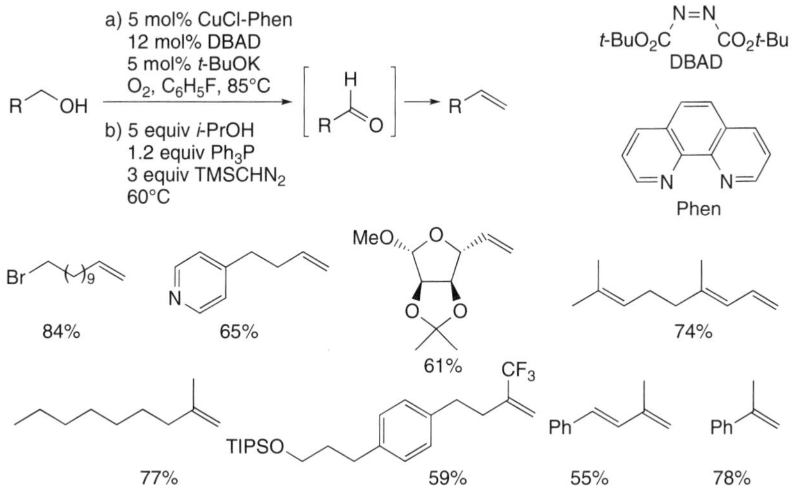

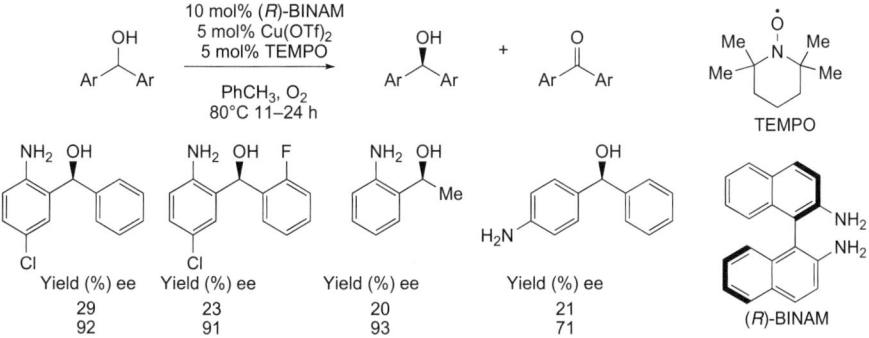

SCHEME 11.73. One-pot aerobic oxidation and methylenation of alcohols.

nonbasic reaction conditions allow the transformation of chiral substrates without racemization.[197]

11.2.10.2. Asymmetric Oxidation. An enantiopure GO enzyme model has been synthesized from readily available (R)-BINAM and Cu(OTf)$_2$ in which molecular oxygen is used as the terminal oxidant and TEMPO as the cooxidant.[198] This system permits the synthesis of chiral diarylmethanols through oxidative kinetic resolution[199] of the racemate (Scheme 11.74), where one enantiomer of the product is oxidized more rapidly by the chiral catalyst. In this case, the unreacted starting material is chiral and enantioenriched while the product is achiral and can be readily removed. Although k_{rel} values are not given, the enantioselectivies at the conversions given would indicate a moderate resolution ($k_{rel} \sim 5$). Under the reaction conditions, o-diarylmethanols were resolved with higher enantioselectivities when compared to

SCHEME 11.74. Asymmetric oxidative kinetic resolution of diarylmethanols.

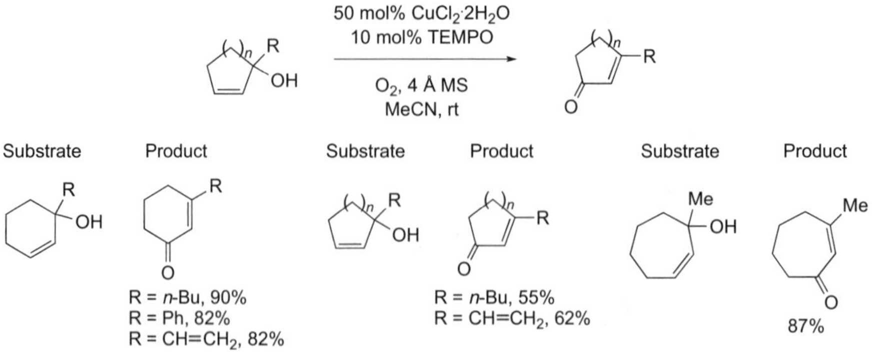

SCHEME 11.75. Oxidative rearrangement of tertiary alcohols.

the para-substrates, where the enantioselectivities were moderate at 70–80% conversion. Using this same catalyst system, Sekar and co-workers[200] also accomplished the asymmetric oxidative kinetic resolution of racemic benzoins.

11.2.10.3. Oxidation with Rearrangement. As is the case in chromium oxidations, tertiary alcohols also undergo rearrangement in copper-catalyzed oxidations. As a result, a mild method for the oxidative rearrangement of tertiary allylic alcohols to β-substituted enones has been developed (Scheme 11.75).[201] A range of substrates can be oxidized by using a TEMPO/CuCl$_2$/O$_2$ system in the presence of molecular sieves. Here, oxygen is the stoichiometric oxidant that together with copper serves to regenerate the TEMPO.

11.3. OXYGENASE-TYPE REACTIONS

11.3.1. Aldehyde Oxidation

11.3.1.1. Oxidation to Carboxylic Acids. The oxidation of aldehydes to carboxylic acids is an important transformation in organic synthesis and industry for which many reagents and catalysts have been developed. Nonetheless, inexpensive catalysts that use green oxidants, such as oxygen, and that can perform under mild conditions with high chemoselectivity are elusive. To this end, CuO was investigated as a catalyst and found to be useful in the the oxidation of aromatic aldehydes to their corresponding carboxylic acids by molecular oxygen (Scheme 11.76). Copper oxide is a good choice for industrial production since it can be easily collected and regenerated. The reaction can be conducted in water although some substrates required a small amount of a cosolvent.[202]

11.3.1.2. Oxidative Decarbonylation. Aliphatic aldehydes with an α-hydrogen can undergo oxidative decarbonylation to the ketone under mild conditions (Scheme 11.77).[203] This reaction was found to be first order in the aldehyde and

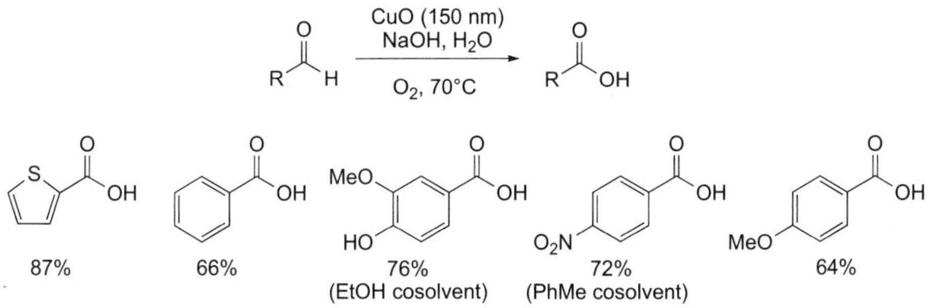

SCHEME 11.76. Copper-catalyzed oxidation of aldehydes to carboxylic acids.

diazabicyclo[2.2.2]octane (DABCO), but zero order in copper and oxygen. The rate-determining step is most likely enolization. Notably, the reaction with DABCO did not occur in the absence of copper even though ketones are known to oxygenate under strongly enolizing conditions. The reaction is proposed to proceed via Cu(II) induced oxidation of the enolate that then traps oxygen. Rearrangement to an acyl radical is then followed by decarbonylation accompanied by copper-catalyzed reduction of the intermediate peroxo species.

11.3.2. Phenols and Anilines

11.3.2.1. Quinone Formation. The copper containing enzyme tyrosinase under-takes both the hydroxylation of phenols and further oxidation to form *ortho-* or *para-*quinones. A number of very simple copper catalysts have been reported to possess this type of combined oxygenase and oxidase ability. The *p*-quinones are formed most readily from phenols, provided that the *p*-position is unsubstituted.

One *p*-quinone, trimethyl-1,4-benzoquinone, is a particularly important target because it is a key intermediate in the industrial production of vitamin E. A one-step direct preparation from 2,3,6-trimethylphenol using an inexpensive catalyst and

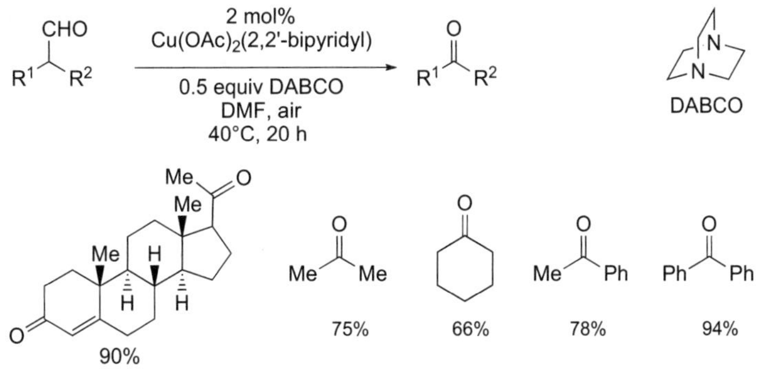

SCHEME 11.77. Oxidative decarbonylation of aldehydes to ketones.

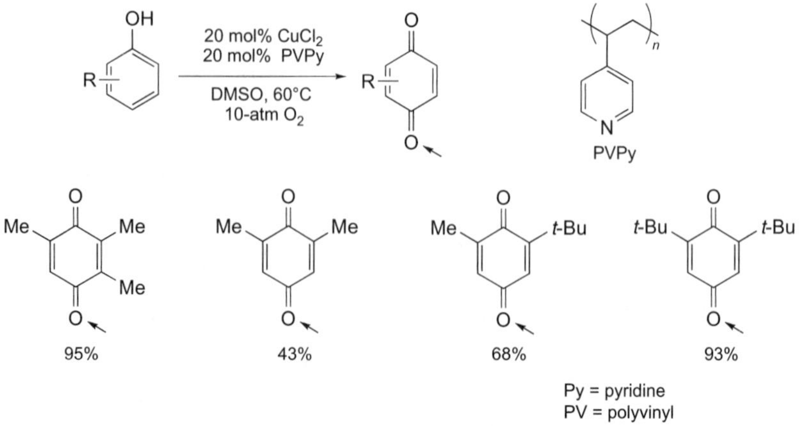

SCHEME 11.78. Copper-catalyzed oxygenation of phenols to *p*-quinones.

copper source would be highly attractive. Several efforts have been made toward this end including the use of 10 mol% CuCl$_2$·2H$_2$O−Et$_2$NH·HCl[204] and 2.5 mol% CuCl$_2$ in an ionic liquid medium[205] to produce trimethyl-1,4-benzoquinone with 100% conversion in 84 and 86% selectivity, respectively. Takehira and co-workers[148] also observed that a pyridine−CuCl$_2$ complex in DMSO was very effective in the oxidation of 2,3,6-trimethylphenol to the corresponding *p*-quinone. A solid supported pyridyl complex displayed similar behavior and was also very effective in the oxygenation of 2,6-dialkylphenols (Scheme 11.78). In this case, a pyridyl catalyst was more effective that a bipyridyl catalyst in contrast to related studies oxidizing *p*-methylphenols to *p*-quinone methides (see Scheme 11.43). Notably, little to no dimer or polyether were observed that is typically the major pathway for the oxidation of the less hindered 2,6-dialkylphenols (see Scheme 11.29a and Scheme 11.44).

11.3.2.2. Dearomatization. Inspired by reports from Stack and co-workers[206,207] of tyrosinase mimics[208–210] derived from bidentate nitrogen ligands, Porco and co-workers[211] constructed an asymmetric tyrosinase mimic from (−)-sparteine (Scheme 11.79) with the aim of applying it to the "tyrosine-like" substrate in Scheme 11.80. The biscopper(II)-peroxo species illustrated in Scheme 11.79 was determined to be the dominant species under the most favorable reaction conditions

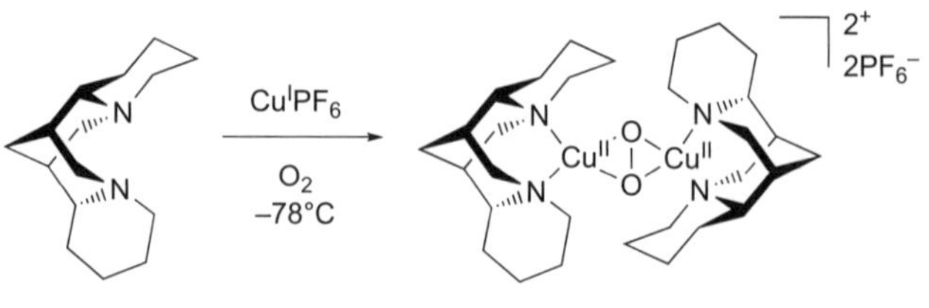

SCHEME 11.79. (−)-Sparteine derived tyrosinase mimic.

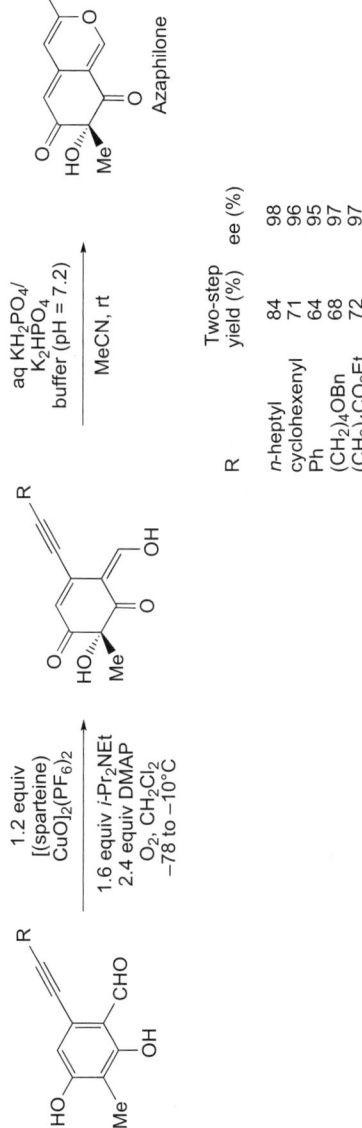

SCHEME 11.80. Asymmetric dearomatization of bisphenols en route to azaphilones.

R	Two-step yield (%)	ee (%)
n-heptyl	84	98
cyclohexenyl	71	96
Ph	64	95
(CH₂)₄OBn	68	97
(CH₂)₄CO₂Et	72	97

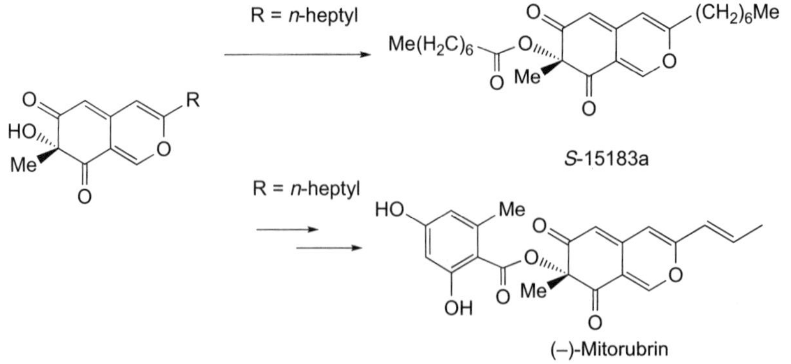

SCHEME 11.81. Total syntheses using asymmetric dearomatization as a key step.

and is postulated to be responsible for the oxygenation of the substrates in Scheme 11.80. The addition of a base was found to facilitate the reaction, presumably by formation of the phenolate anion that was subsequently oxygenated by the copper reagent. Direct cyclization of the dearomatized species provided the more stable azaphilone that could be isolated in good yields with very high enantioselectivities (95–98% ee).[211] This biomimetic strategy provided a facile entry to (S)-15183a[211] and (−)-mitorubrin (Scheme 11.81).[212]

Porco and co-workers[213] applied this same copper reagent (Scheme 11.80) to good effect to monophenol substrates, producing a dearomatized intermediate that underwent dimerization via a [4 + 2] cycloaddition to provide complex adducts (Scheme 11.82) with very high enantioselection (98–99% ee). This discovery led to a biomimetic synthesis of aquaticol, in which a chiral substrate was combined with the chiral reagent (Scheme 11.83). Notably, the chiral substrate gave rise to a product as a single diastereomer, suggesting that the chiral reagent was the dominant stereocontrol element.

The above examples of dearomatization all rely upon a stoichiometric copper reagent as the oxidant. Another report that moves a step closer to an enzymatic mimic uses catalytic Cu(OAc)$_2$ to effect the dearomatization of an aniline substrate (Scheme 11.84).[214]

A preliminary investigation revealed that molecular oxygen is a prerequisite for achieving the present catalytic cyclization and that one of the oxygen atoms of O$_2$ is

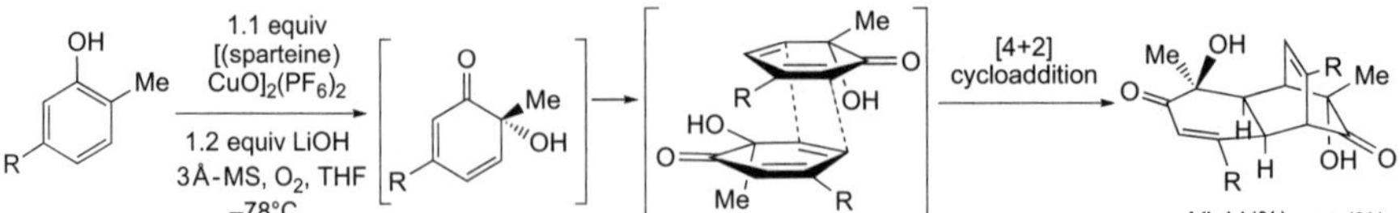

R	Yield (%)	ee (%)
i-Pr	80	99
Me	40	98
t-Bu	82	99
MeO	52	99

SCHEME 11.82. Tandem asymmetric dearomatization.

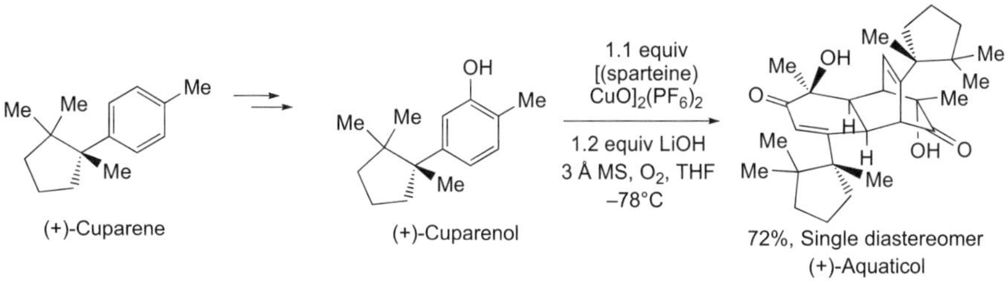

SCHEME 11.83. Total synthesis of aquaticol via asymmetric dearomatization.

incorporated into the cyclohexadienones (Scheme 11.85). A possible mechanism commences with denitrogenative formation of iminyl copper followed by its oxidation with O_2 to form a peroxycopper(III) species. The reaction of *p*-tolylamide (product **A** Scheme 11.84) suggests that an intramolecular imino-cupration of the aniline ring might form C−N and C−Cu bonds concurrently at the ipso and its para positions. Subsequent isomerization to a peroxydiene followed by elimination to regenerate the Cu(II) species would deliver the azaspirodienone.

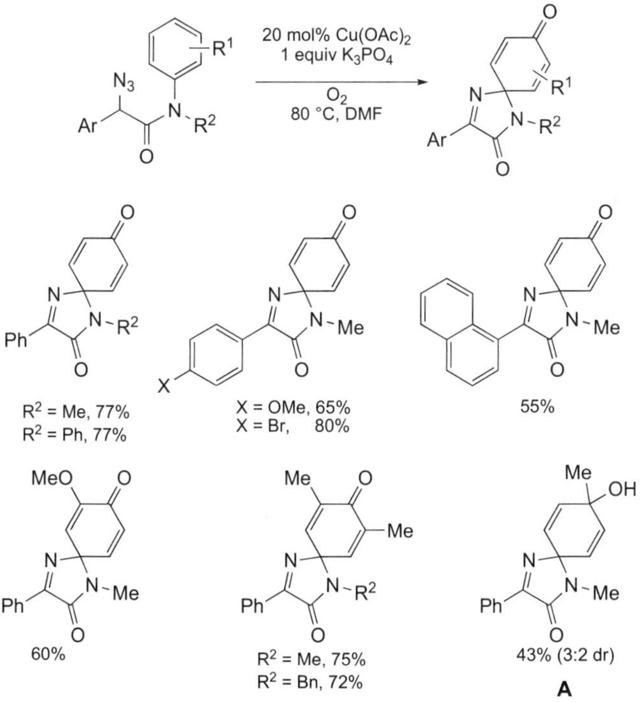

DMF = dimethylfoumamide

SCHEME 11.84. Copper-catalyzed synthesis of azaspirocyclohexadienones.

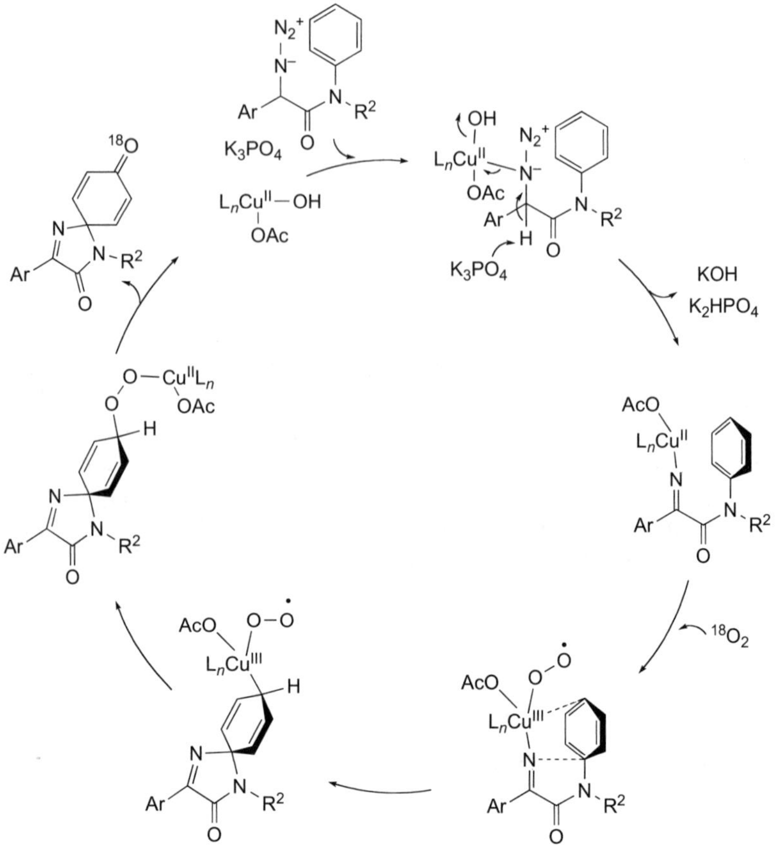

SCHEME 11.85. Proposed mechanism of copper-catalyzed azaspirocyclohexadienone formation.

Dearomatization as a synthetic strategy is very powerful because even complex aromatic systems are easily constructed from readily available materials. By generating a quaternary stereocenter via dearomatization it is then possible to rapidly create highly complex structures with many other useful functional groups. Since nature clearly exploits similar reactions,[82–85,87] the development of small-molecule analogues permits the use of biomimetic routes to complex natural products and natural product analogues. Remaining challenges in this area include control of regiochemistry, which is dictated to a large extent by the electronics of the substrates.

11.3.3. Hydrocarbon Oxidation

Enzymatic copper–dioxygen derived species also effect aliphatic and arene C–H bond hydroxylations, for example, in dopamine β-monooxygenase, peptidylglycine α-amidating monooxygenase, and particulate methane monooxygenase. Numerous attempts have been made to mimic the exquisite selectivity and reactivity of these enzymatic systems with small-molecule copper catalysts and oxygen that typically react via radical intermediates. While much progress has been made, the selective functionalization of alkane or arene C—H groups is not yet useful as a regular synthetic procedure for a range of structures. Selectivity remains the foremost issue.

11.3.3.1. Benzylic. The most success has been seen in benzylic oxidation that is aided by the much weaker C—H bond strength at benzylic centers, thereby aiding processes involving radical abstraction. Controlling the degree of oxygenation is also a problem since the acid, aldehyde, or alcohol can be produced. As of yet there is no general catalyst available for a broad range of substrates. However, a variety of different copper catalysts have been devised that operate on electron-poor or electron-rich substrates. For example, a copper phthalocyanine catalyst in an ionic solvent ([omin]BF$_4$) has been shown to be effective in converting nitrotoluenes to nitrobenzoic acids (Scheme 11.86a) and is suited to large scale industrial applications.[215] On the other hand, oxidation of electron-rich p-cresol to generate p-hydroxybenzaldehyde proceeds well in the presence of a basic additive using a mixed-manganese/copper catalyst (Scheme 11.86b).[216] With copper(II) chloride and an oxime ligand in ethanol, oxidation of 2,6-di-*tert*-butyl-4-methylphenol could be directed to the ether

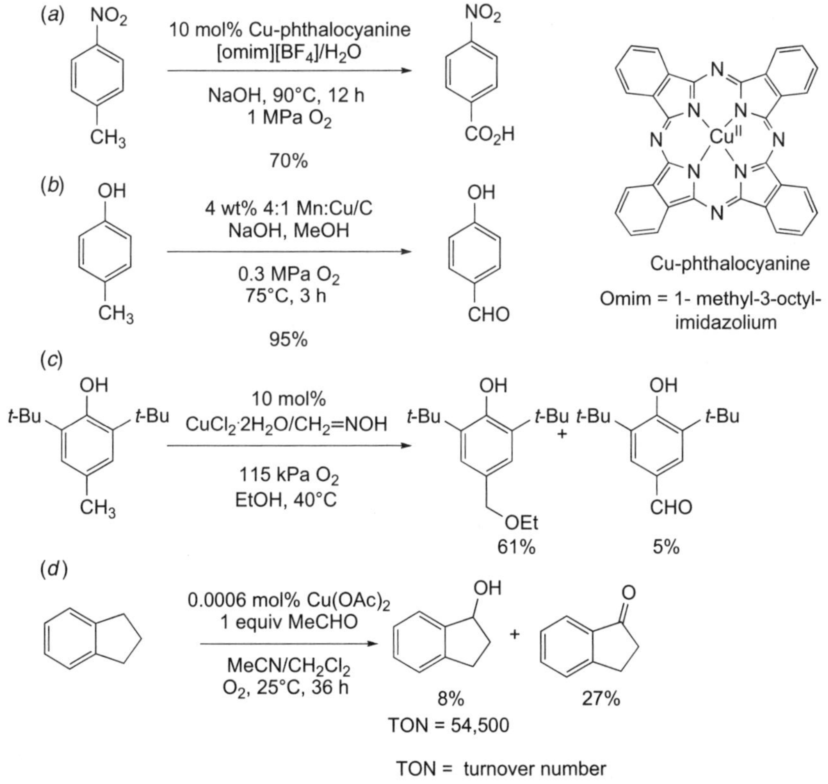

SCHEME 11.86. Selected oxidations of benzylic substrates.

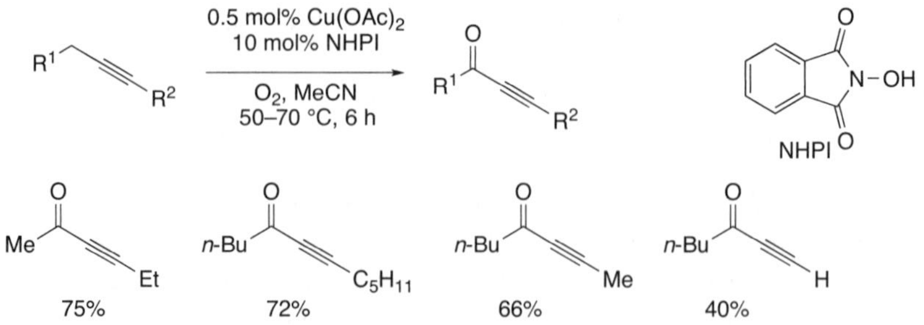

SCHEME 11.87. Oxygenation of alkynes to α,β-acetylenic ketones.

product with relatively little overoxidation (Scheme 11.86c).[217] Extremely high turnovers (TON = 54,500) have been observed in selected systems using copper(II) acetate (0.0006 mol%) in conjunction with acetonitrile, which can function as a ligand and acetaldehyde, which functions as a cooxidant (Scheme 11.86d).[218]

11.3.3.2. Propargylic. Aerobic oxygenation of propargylic systems is complicated by the potential for the alkyne to also react. This tendency has been overcome by the use of N-hydroxyphthalimide (NHPI) as a cocatalyst in conjuction with copper(II) acetate (Scheme 11.87).[219] While NHPI alone can catalyze the transformations of some substrates, the conversion is greatly enhanced by the addition of the copper catalyst. The metal catalyst allows lower temperatures, thereby preventing over-oxidation to the oxidatively cleaved carboxylic acids, and plays a key role in conversion of the initial proparygyl alcohol to the ketone. The mechanism presumably involves formation of the radical from NHPI (e.g., see Scheme 11.62).

11.3.3.3. Allylic. As was the case for propargylic systems (see Section 11.3.3.2), aerobic oxygenation of allylic systems is complicated by the potential for the alkene to also react. Nonetheless, allylic oxidation is a powerful synthetic transformation and a copper catalytic system using oxygen under milder conditions would possess many advantages over reported reagents and catalysts that rely on toxic metals or costly oxidants.[220]

An example of an oxyhemocyanin–oxytyrosinase model compound that converts alkenes to allyl alcohols and α,β-unsaturated ketones is outlined in Scheme 11.88.[221] The importance of the ligand sphere is particularly evident here when compared to other copper-catalysts systems that cause epoxidation of alkenes (see Section 11.3.6). Labeling experiments established that the oxygen atoms in the product arise from exogeneous oxygen and not from the peroxo ligand of the catalyst.

The potential for copper catalysts to undertake allylic oxidation in complex systems can be seen in the example in Scheme 11.89, where oxidation is accompanied by a double-bond migration. Thus, Δ^5-cholestenone was rapidly converted to Δ^4-cholestene-3,6-dione in 75% yield using a copper(II) acetate catalyst with a measured amount of oxygen.[222] This transformation could be applied to a number of α,β- and

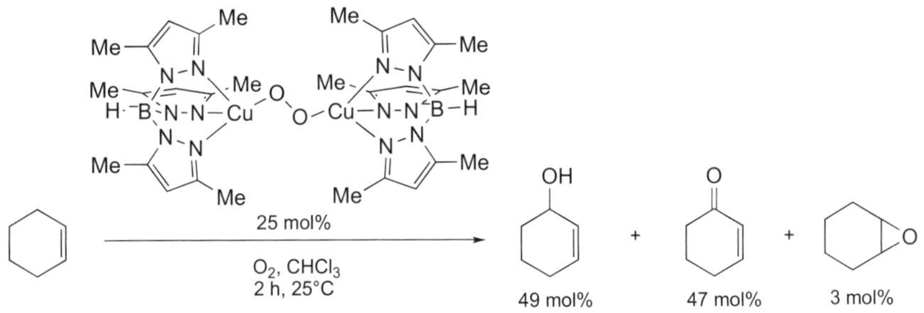

SCHEME 11.88. Allylic alkene oxidation with a oxyhemocyanin–oxytyrosinase model compound.

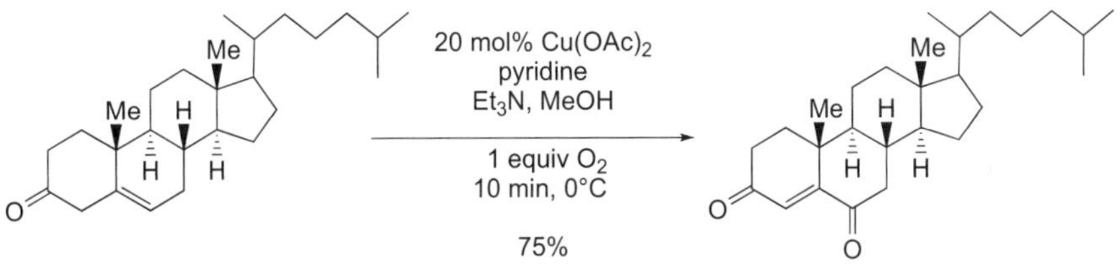

SCHEME 11.89. Allylic oxidation–rearrangement of a cholestenone.

β,γ-unsaturated ketones usually resulting in introduction of the oxygen atom at the γ-position.

11.3.3.4. Alkanes. Because alkane C—H bonds are strong and similar in strength compared to those at allylic or benzylic centers, selective alkane oxygenation via the radical processes typical with copper catalysts using oxygen suffers from low selectivity. Thus, most reports center on symmetrical compounds. Even so, yields are low and mixtures of alcohol and ketone are seen. For example, the combination of CuCl$_2$ with 18-C-6 (18-crown-6), acetaldehyde, and oxygen is proposed to generate peracetic acid intermediates and provides good yields of alkane oxgyenation products (Scheme 11.90).[223,224] With unsymmetrical substrates (e.g., hexane or decane), similar conversions are observed, but the comparable reactivity of all the secondary centers leads to a mixture of products. Light-induced copper-catalyzed oxygenations have also been reported that do not require the acetaldehyde cooxidant, but again yields are moderate and selectivities are low.[225–227]

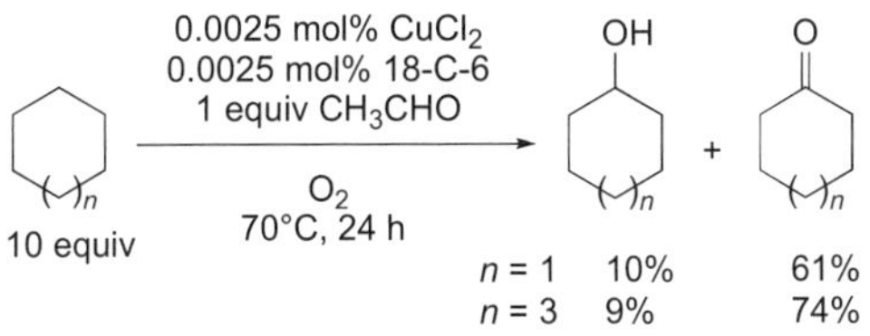

SCHEME 11.90. Copper–catalyzed oxygenation of alkanes (yields based on MeCHO).

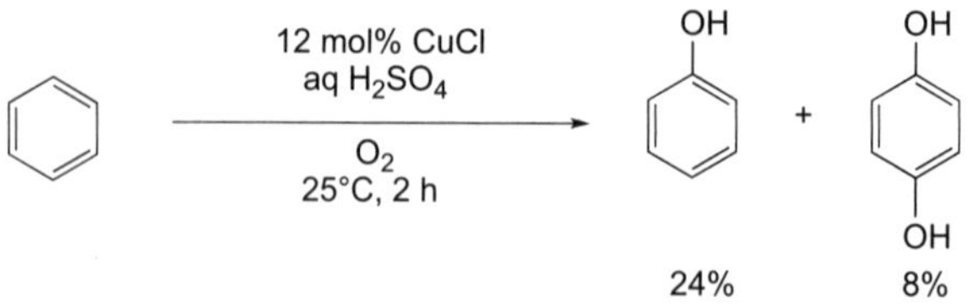

SCHEME 11.91. Oxidation of benzene to phenol.

11.3.3.5. Arenes. Most hydrocarbon oxygenative C−H functionalizations proceed via radicals. As such, selective functionalization of arenes is challenging since any other benzylic or alkyl center in the structure possesses weaker C−H bonds and will react first. Thus, the main target for arene oxygenation is benzene to form phenol. Phenol, an important commodity chemical for resins and fibers, is made industrially via a three-step process from cumene. An efficient preparation from benzene using inexpensive catalysts and oxygen as a green oxidant would be competitive. To this end, numerous catalysts have been screened including Cu[(I)]Cl (Scheme 11.91), which gives among the highest selectivities for homogenous catalysts.[228] Mechanistic evidence indicates that phenol is formed via the intermediacy of a hydroxyl radical. A number of heterogeneous and supported copper-containing catalysts have been reported for this transformation.[229–233] Sections 11.2.4 and 11.2.6 discuss other oxidative transformations of phenols and oxidative C−H insertions of arenes.

11.3.4. Amines to Carbonyls

The oxidative deamination of amines is a classic example of a reaction catalyzed by copper-containing enzymes, the amine oxidases. One of the many examples of amine oxidase mimics[130] is illustrated in Scheme 11.92. Here, a bridging biscopper(II) complex converts primary amines with at least one α-H atom into the aldehyde that subsequently condenses with additional substrate to provide the illustrated imine.[234] Notably, the reaction conditions are very mild (ambient temperature, under air).

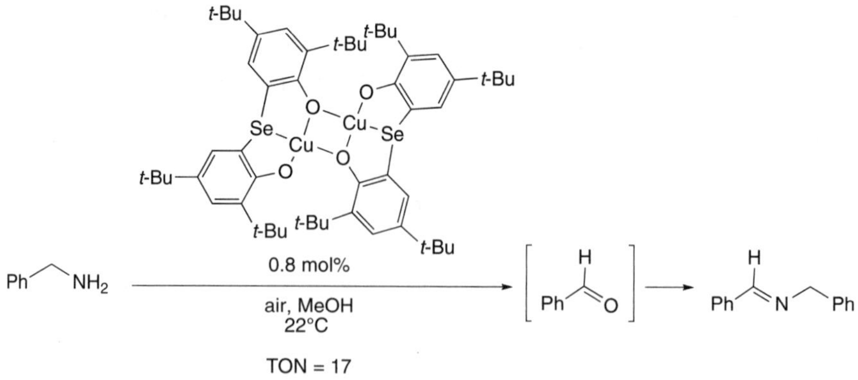

SCHEME 11.92. Oxidative deamination of amines.

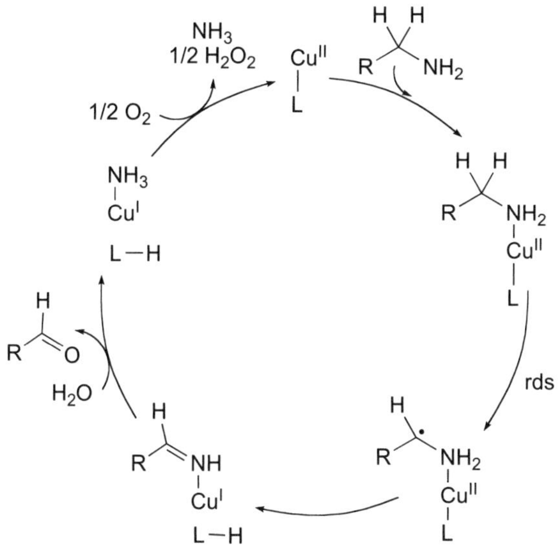

SCHEME 11.93. Mechanism of an amine oxidase mimic.

The development of a simple inexpensive copper catalyst with the same reactivity would have broad utility in organic synthesis.

By using benzylamine selectively deuterated at the α-carbon as substrate (PhCD$_2$NH$_2$), a KIE of 4 was observed. This result implies that hydrogen-atom abstraction from the α-carbon atom of the coordinated benzylamine represents the rate-determining step, consistent with the mechanistic proposal in Scheme 11.93. The imine–copper intermediate is similar to the reactive species proposed in the amine α-functionalizations described in Section 11.2.7 (see Scheme 11.60).

A copper-catalyzed system utilizing dehydroascorbic acid (dehydroAsc) functions by a very different mechanism. DehydroAsc is known to oxidize amines to the corresponding aldehydes easily. In this case, however, the dehydroAsc is generated *in situ* from a stoichiometric amount of ascorbic acid (Asc) via a copper-catalyzed aerobic oxidation. A variety of amines could be oxidized under very mild conditions (Scheme 11.94).[235]

Omission of either the Asc or the Cu catalyst from the reaction mixture abrogated the activity indicating that each plays a critical role as outlined in the proposed mechanism (Scheme 11.95). Either Asc or its derivative could be, in theory, reoxidized a number of times, thus rendering Asc catalytic. However, 2 equiv were needed to obtain the full conversion of amine into the desired carbonyl product. The likely culprit of this stoichiometry requirement is the reaction of the amine-containing byproduct with dehydroAsc.

In addition to proceeding under mild conditions, the dehydroAsc oxidation exhibits very high chemoselectivity. Since alcohols cannot form the requisite imine required by the mechanism, the selectivity for amines over alcohols is very high (Scheme 11.96).

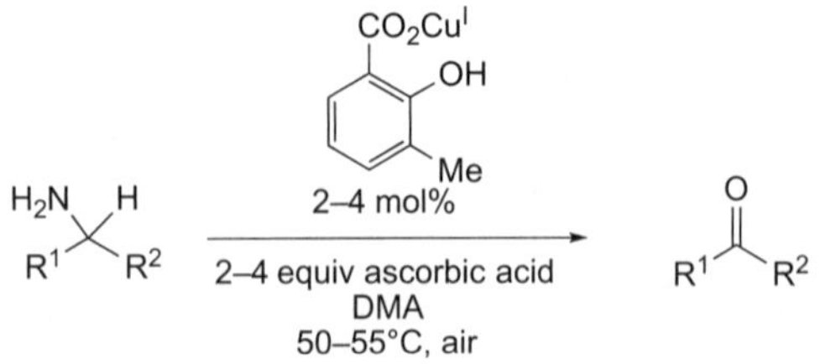

R¹	R²	Yield (%)	R¹	R²	Yield (%)
Ph	H	82	4-HO$_2$CPh	H	67
3-MePh	H	85	Ph	Ph	81
4-MePh	H	73	Ph	Me	53
4-t-BuPh	H	70	—(CH$_2$)$_6$—		58
3-MeOPh	H	78	—(CH$_2$)$_{11}$—		51
4-MeOPh	H	87	PhCH$_2$	CO$_2$Et	63
4-MeO$_2$CPh	H	89	C$_{12}$H$_{25}$	H	60

SCHEME 11.94. Dehydroascorbic acid mediated, copper-catalyzed oxidation of amines.

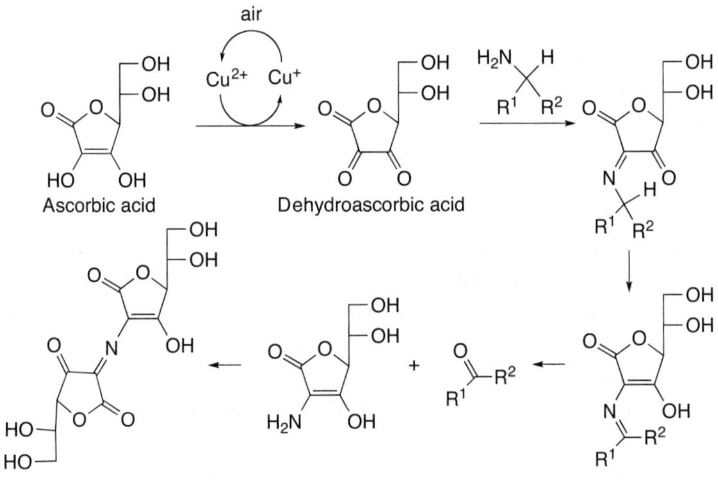

SCHEME 11.95. Mechanism of the dehydroascorbic acid mediated oxidation of amines.

11.3.5. Thioamides to Nitrilium Equivalents

An aerobic copper-catalyzed oxidation of thioamides and selenoamides to the corresponding amides has been reported (Scheme 11.97).[236] This transformation is related to the capture of an intramolecular nucleophile by a thioamide (Scheme 11.65, see Section 11.2.9) except that tertiary thioamides and thioureas can be employed in addition to secondary thioamides. Importantly, both a copper source and O$_2$ are

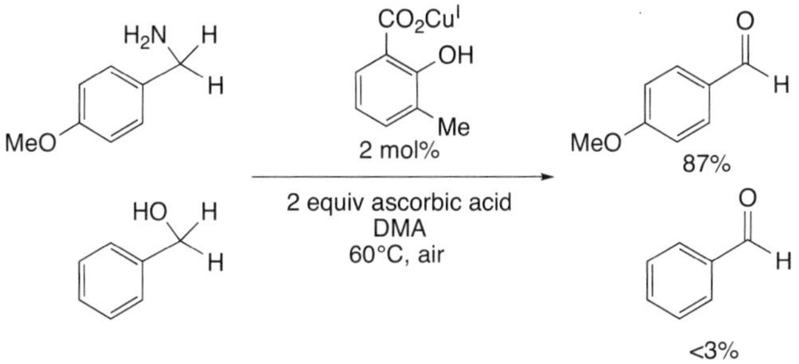

SCHEME 11.96. Chemoselectivity of the dehydroascorbic acid mediated oxidation.

necessary. Highly polar solvents (e.g., DMSO and DMF) are required; only trace amounts of product are formed when less polar solvents are employed.

The incorporation of $^{18}O_2$ in this process is optimized when using DMF as the solvent. The power of this novel oxygen isotope labeling method was demonstrated in the context of an ^{18}O labeled sialic acid derivative that contained several other ester groups that were unmodified (Scheme 11.98).

Although the intermediates involved in the desulfurization reaction have not yet been identified, a possible mechanism has been proposed involving an initial stepwise $2e^-$ oxidation of the thiocarbonyl sulfur to give a sulfenylium intermediate (Scheme 11.99). The reduced Cu(I) species is then reoxidized to Cu(II) by molecular oxygen. Collapse of the sulfenylium to the nitrilium intermediate, as proposed for a related transformation, discussed in Section 11.2.9 (see Scheme 11.66), is unlikely because tertiary thioamides (R^2 and $R^3 \neq H$) react well. Rather, attack by some nucleophilic oxygen species is proposed followed by the expulsion of elemental sulfur.

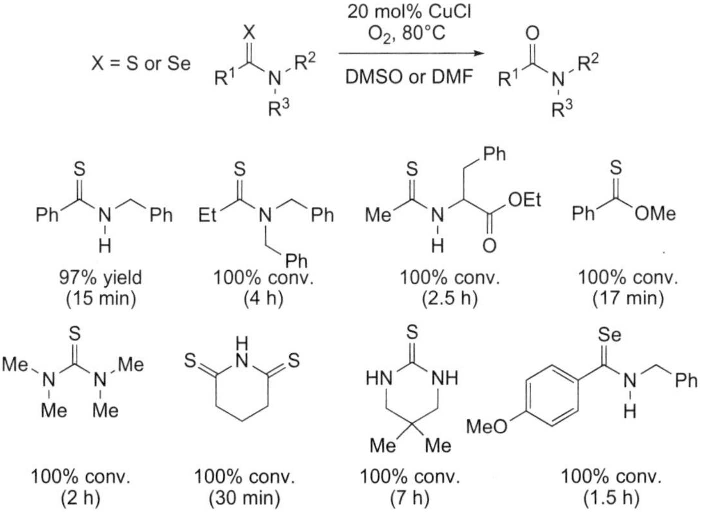

SCHEME 11.97. Oxidation of thioamides and selenoamides to amides.

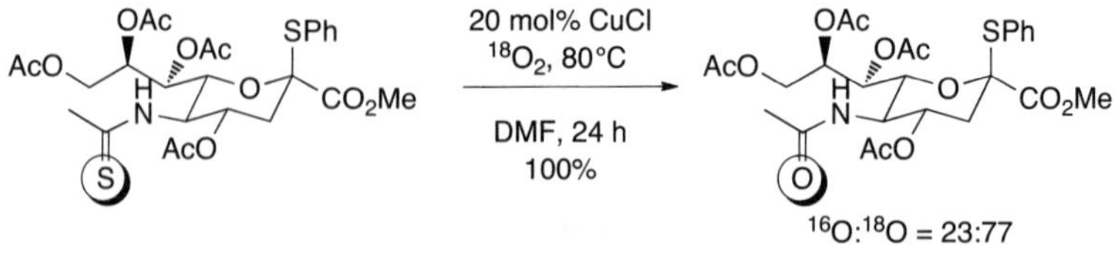

SCHEME 11.98. The ^{18}O labeling of amides by oxidation of the thioamide.

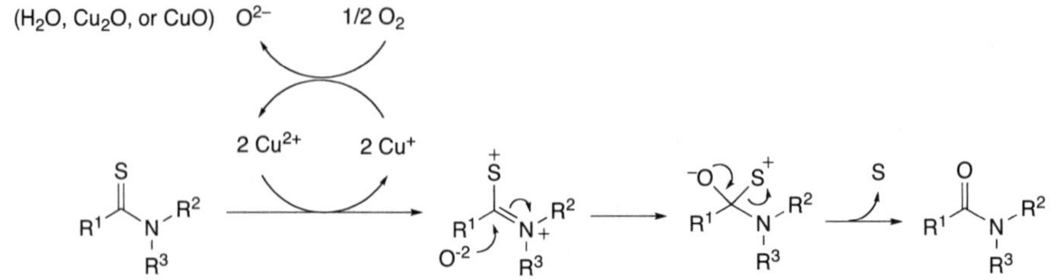

SCHEME 11.99. Proposed mechanism of the oxidation of thioamides to amides.

11.3.6. Epoxidation

Catalytic oxidations of hydrocarbons (e.g., alkanes and alkenes) with molecular oxygen under mild conditions are especially rewarding goals. Even though copper-

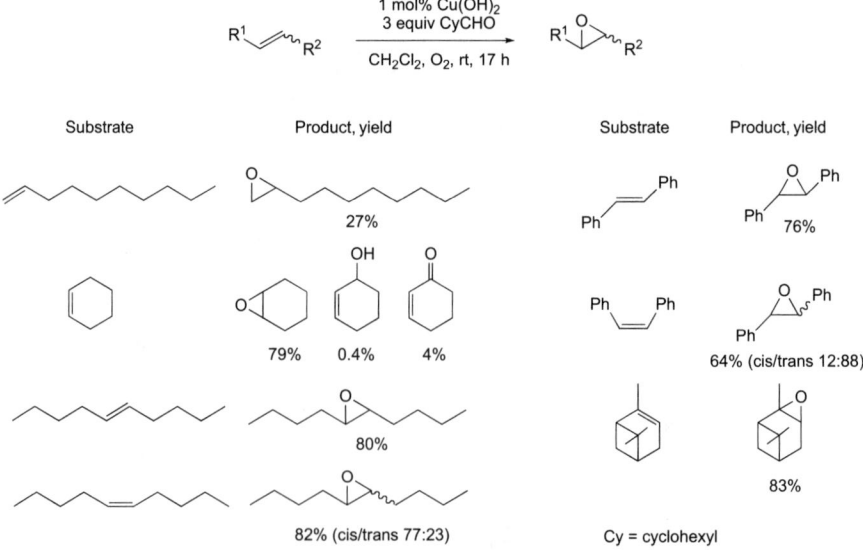

SCHEME 11.100. Epoxidation of alkenes with a copper catalyst.

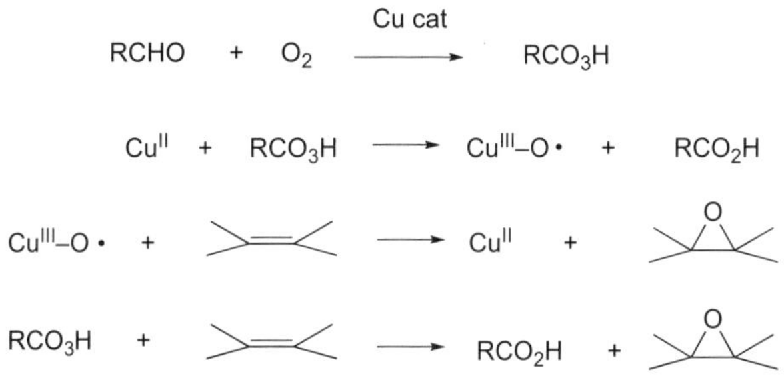

SCHEME 11.101. Mechanism of epoxidation of alkenes with a copper catalyst.

containing oxygenase enzymes can accomplish epoxidation,[1–8] synthetic copper catalysts have been much less explored in the context of epoxidation. An early example demonstrating the potential of copper in this context is outlined in Scheme 11.100.[237] The process works well for di- and trisubstituted alkenes, producing very little of the allylic C−H insertion products.

An aldehyde is employed under these conditions as a cooxidant and the mechanism is believed to involve both peroxy acids and radical species (Scheme 11.101). The latter is consistent with the formation of mixtures of epoxide diastereomers when beginning with the cis alkenes (Scheme 11.100). On the other hand, the high ratio of hydroxyl directed cis-epoxidation seen with the copper catalyst is very similar to that seen with MCPBA (*meta*-chloroperoxybenzoic acid) supporting the intervention of a peroxy acid (Scheme 11.102). Conversely, the acetate protected version give predominantly the trans-epoxide, again similar to MCPBA (Scheme 11.102). It appears that both peroxy acids and radical species form under the reaction conditions and that the former accounts for the products except when the substrates can stabilize radicals (i.e., the stilbenes, Scheme 11.100).

Other sources of Cu have been investigated in similar epoxidations of styrene, cylcohexene, and decene utilizing air–dioxygen and an aldehyde with limited

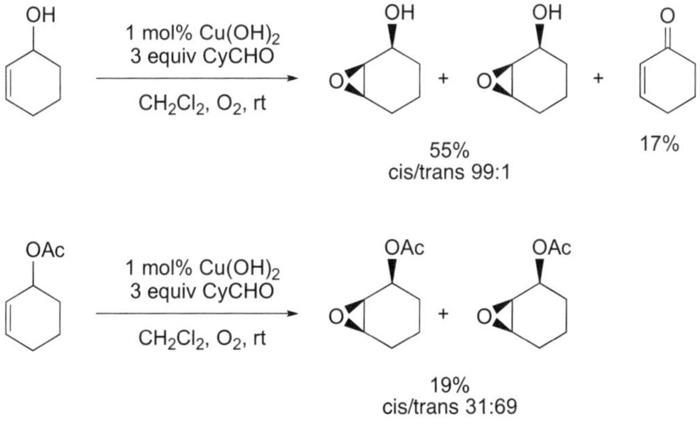

SCHEME 11.102. Substrate directed copper-catalyzed epoxidation.

success.[238] The most promising system at this time is comprised of a copper perchlorophthalocyanine ($CuCl_{16}Pc$) complex, the activity of which is greatly improved by placement in the channels of HSi-MCM-41 molecular sieves.[239]

11.3.7. Baeyer–Villager Reaction

The Baeyer–Villager insertion of oxygen into ketones to form esters has long been a useful reaction in organic synthesis complementing biosynthetic versions. With the success of copper and oxygen in epoxidation (see Section 11.3.6), application to the Baeyer–Villager reaction, which occurs with many of the same oxidants used in alkene epoxidation, proved fruitful. For example, Bolm et al.[240] demonsrated that $Cu(OAc)_2$ was an effective catalyst for the Baeyer–Villager reaction of several ketones (Scheme 11.103). In line with the typical Baeyer–Villager reaction, the more substituted ketone carbon undergoes the migration. An aldehyde additive was needed and most likely acted as a transfer oxidant, reacting first with the oxygen to form a peracid.[241]

By means of a copper catalyst containing chiral ligands, Baeyer–Villager reactions of racemic ketones could be undertaken resulting in a kinetic resolution.[242] With 1 mol% of Cu catalyst and pivaldehyde as a transfer oxidant, oxidation of 2-arylcyclohexanones afforded the corresponding lactones with up to 69% ee. A related chiral catalyst has also been reported for this transformation, but with poorer selectivity.[243]

A Baeyer–Villager reaction effecting parallel kinetic resolution,[244] where the two enantiomers of a racemate undergo different transformations, has also been reported (Scheme 11.104).[244] Oxidation of racemic cyclobutanones with 1 mol% of a chiral copper catalyst yielded optically active butyrolactones with up to 95% ee. The oxidation proceeded in an enantiodivergent manner and two isomeric lactones were obtained. Apparently, one enantiomer of the substrate reacts selectively with the catalyst to undergo the "normal" insertion at the more substituted ketone position, whereas the opposite enantiomer of the substrate interacts with the catalyst to undergo insertion at the less substituted position giving rise to the "abnormal" product.

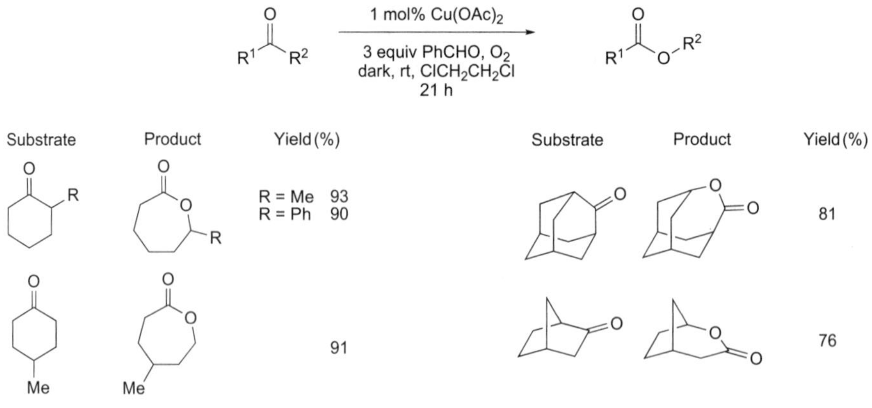

SCHEME 11.103. Copper-catalyzed Baeyer–Villager reaction.

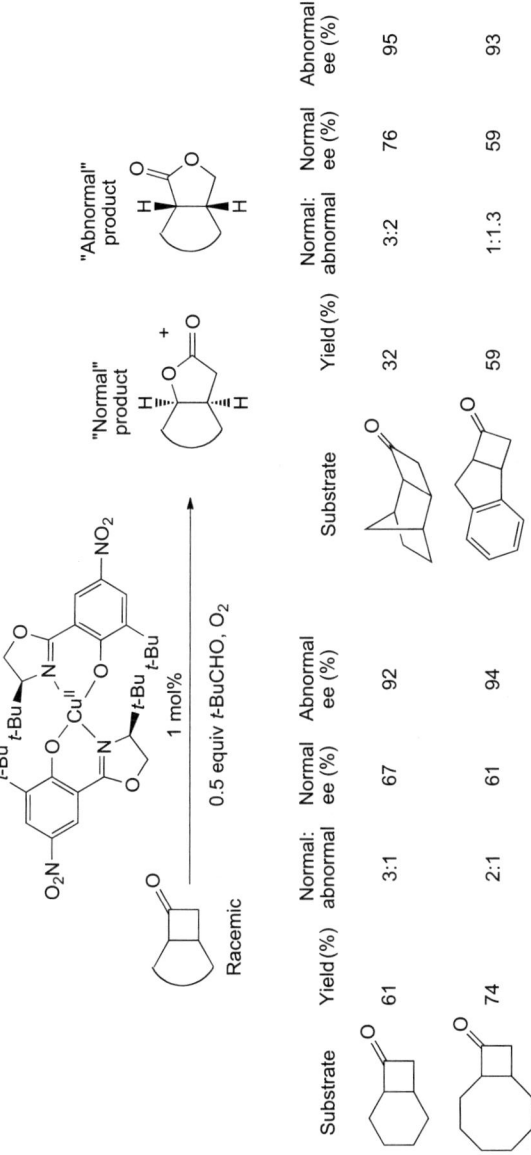

SCHEME 11.104. Parallel kinetic resolution in the copper-catalyzed Baeyer–Villager reaction.

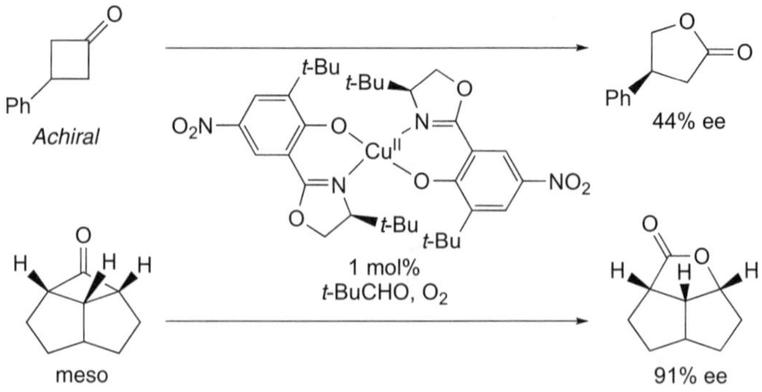

SCHEME 11.105. Enantioselective copper-catalyzed Baeyer–Villager reactions.

Achiral cyclobutanones have also been subjected to Baeyer–Villager oxidation with the same catalyst (Scheme 11.105).[245] With a simple cyclobutanone, the asymmetric induction remained low and the ee of butyrolactone was only 44%. On the other hand, the more complex meso substrate provided the product with high selectivity (91% ee). Apparently the very hindered butyrolactone undergoes greater interaction with the catalyst permitting better differentiation of the two prochiral ketone substituents.

11.3.8. Sulfoxidation

Sulfoxides are key intermediates in organic synthesis and medicinal chemistry and many methods have been reported for their synthesis. Even so, the majority use peroxide or peracid oxidants. For example, copper-catalyzed oxidations of sulfides to sulfoxides with TBHP have been reported as highly effective.[14] However, the corresponding reaction with oxygen has lagged due to overoxidation of the sulfides. Some progress has been made, as shown in Scheme 11.106 suggesting that a general aerobic copper-catalyzed oxidation is viable.[246,247]

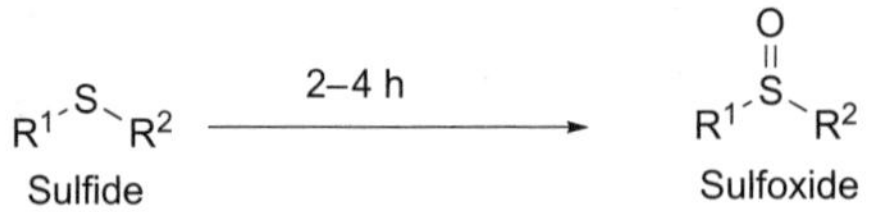

R^1	R^2	Catalyst	T (°C)	Solvent		Yield (%)
Et	Et	1 mol% Cu(OAc)$_2$, xs (Me)$_2$CHCH$_2$CHO	40	ClCH$_2$CH$_2$Cl	O$_2$	75
Ph	Ph	1 mol% Cu(OAc)$_2$, xs(Me)$_2$CHCH$_2$CHO	40	ClCH$_2$CH$_2$Cl	O$_2$	76
Ph	Me	10 mol% Cu(NO$_3$)$_2$, 5 mol% CuBr$_2$	25	MeCN	Air	82
		Xs = excess				

SCHEME 11.106. Copper-catalyzed oxidation of sulfides to sulfoxides.

11.4. CONCLUSION AND FUTURE PROSPECTS

Much is still unknown about Cu and, in particular, its reaction with O_2 under different sets of conditions. However, it is clear that Cu is highly versatile, with different ligand spheres and different reaction conditions giving rise to a large range of reaction chemistries useful in organic synthesis. Of particular importance are oxidative functionalization reactions that allow the forging of bonds by direct functionalization of C−H bonds. Such processes have the potential to significantly reduce the number of steps and waste products in a reaction sequence as no special leaving groups have to be introduced.

With the drive toward sustainable and environmentally benign synthetic methods, oxidation and oxygenation reactions with Cu catalysts and reagents that employ O_2 as the terminal oxidant are logical targets for further development. Molecular oxygen is readily available, highly efficient, atom-economical, and green with the byproduct from its use, (water) being nontoxic. Copper is an inexpensive metal that is readily available. Even with the successes achieved to date, considerable work remains to reach the levels of reactivity and selectivity seen in many of the enzymatic counterparts. While many of the reaction described in this chapter are biomimetic, a good number are not. Undoubtedly, many biological copper activities remain to be discovered, some of which may correspond to some of the above transformations, others of which will provide inspiration for new transformations.

ACKNOWLEDGMENTS

The support of the NSF (CHE-0911713) and the NIH (CA-109164) for research efforts in this area is gratefully acknowledged.

REFERENCES

1. Himes, R. A.; Karlin, K. D. Copper–dioxygen complex mediated C−H bond oxygenation: relevance for particulate methane monooxygenase (pMMO) *Curr. Opin. Chem. Biol.* **2009**, *13*, 119–131.

2. van der Vlugt, J. I.; Meryer, F. Homogeneous Copper-Catalyzed Oxidations *Top. Organomet. Chem.* **2007**, *22*, 191–240.

3. Cramer, C. J.; Tolman, W. B. Mononuclear Cu−O_2 Complexes: Geometries, Spectroscopic Properties, Electronic Structures, and Reactivity *Acc. Chem. Res.* **2007**, *40*, 601–608.

4. Chaudhuri, P.; Wieghardt, K.; Weyhermüller, T.; Paine, T. K.; Mukherjee, S.; Mukherjee, C. Biomimetic metal-radical reactivity: aerial oxidation of alcohols, amines, aminophenols and catechols catalyzed by transition metal complexes *Biol. Chem.* **2005**, *386*, 1023–1033.

5. Mirica, L. M.; Ottenwaelder, X.; Stack, T. D. P. Structure and Spectroscopy of Copper–Dioxygen Complexes *Chem. Rev.* **2004**, *104*, 1013–1045.

6. Kim, E.; Chufan, E. E.; Kamaraj, K.; Karlin, K. D. Synthetic Models for Heme–Copper Oxidases. *Chem. Rev.* **2004**, *104*, 1077–1133.

7. Zhang, C. X.; Liang, H.-C.; Humphreys, K. J.; Karlin, K. D. Copper–dioxygen complexes and their roles in biomimetic oxidation reactions *Catalysis by Metal Complexes* **2003**, *26*, 79–121.

8. Klinman, J. P. Mechanisms Whereby Mononuclear Copper Proteins Functionalize Organic Substrates. *Chem. Rev.* **1996**, *96*, 2541–2561.

9. *Advances in Organic Chemistry, Volume 58: Homogeneous Biomimetic Oxidation Catalysts*, R. van Eldik and J. Reedijk, Eds. Elsevier, Amsterdam, 2006.

10. *Organic Syntheses by Oxidation with Metal Compounds* W. J. Mijs and C. R. H. I. de Jonge, Eds., Plenum Press, New York, 1986.

11. Sheldon, R. A.; Kochi, J. K. *Metal-Catalyzed Oxidations of Organic Compounds*, Academic Press, New York, 1981.

12. *Oxidation in Organic Chemistry, Part B* W. S. Trahnovsky, Ed., Academic Press, New York, 1973.

13. Punniyamurthy, T.; Velusamy, S.; Iqbal, J. Recent Advances in Transition Metal Catalyzed Oxidation of Organic Substrates with Molecular Oxygen *Chem. Rev.* **2005**, *105*, 2329–2363.

14. Punniyamurthy, T.; Rout, L. Recent advances in copper-catalyzed oxidation of organic compounds *Coord. Chem. Rev.* **2008**, *252*, 134–154.

15. Hayashi, M.; Kawabata, H. Environmentally benign oxidation of alcohols using transition metal catalysts *Advan. Chem. Res.* **2006**, *1*, 45–62.

16. Siemsen, P. Livingston, R. C.; Diederich, F. Acetylenic Coupling: A Powerful Tool in Molecular Construction *Angew. Chem. Int. Ed.* **2000**, *39*, 2632–2657.

17. Glaser, C. Beiträge zur Kenntniss des Acetenylbenzols *Ber. Dtsch. Chem. Ges.* **1869**, *2*, 422–424.

18. Eglington, G.; Kocovsky Galbraith, R. Macrocyclic acetylenic compounds. Part I. Cyclotetradeca-1:3-diyne and related compounds *J. Chem. Soc.* **1959**, 889–896.

19. Hay, A. S. Oxidative Coupling of Acetylenes. I. *J. Org. Chem.* **1960**, *25*, 1275–1276. Hay, A. S. Oxidative Coupling of Acetylenes. II. *J. Org. Chem.* **1962**, *27*, 3320–3321.

20. Jones, G. E.; Kendrick, D. A.; Holmes, A. B. 1,4-Bis(trimethylsilyl)buta-1,3-diyne *Org. Syn. Coll. Vol.* **1993**, *8*, 63–67.

21. Yamaguchi, M.; Park, H.-J.; Hirama, M.; Torisu, K.; Nakamura, S.; Minami, T.; Nishihara, H.; Hiraoka, T. Synthesis and Reactions of Monosilylated 1,3,5-Hexatriyne and 1,3,5,7-Octatetrayne. Total Synthesis of Caryoynencins *Bull. Chem. Soc. Jpn.* **1994**, *67*, 1717–1725.

22. Bunz, U. H. F. A Soft Spot for Alkynes *Synlett* **1997**, 1117–1127.

23. Hoan, H. M.; Brailovskii, S. M.; Temkin, O. N. Kinetics Of Dialkyne Synthesis In Aqueous-Solutions Of Cu(I) And Cu(II) Chlorides *Kinet. Catal. (Engl. Transl.)* **1994**, *35*, 242–246.

24. Adimurthy, S.; Malakar, C. C.; Beifuss, U. Influence of Bases and Ligands on the Outcome of the Cu(I)-Catalyzed Oxidative Homocoupling of Terminal Alkynes to 1,4-Disubstituted 1,3-Diynes Using Oxygen as an Oxidant *J. Org. Chem.* **2009**, *74*, 5648–5651.

25. Hamada, T.; Ye, X.; Stahl, S. S. Copper-Catalyzed Aerobic Oxidative Amidation of Terminal Alkynes: Efficient Synthesis of Ynamides *J. Am. Chem. Soc.* **2008**, *130*, 833–835.

26. Chu, L.; Feng-Ling Qing, F.-L. Copper-Mediated Aerobic Oxidative Trifluoromethylation of Terminal Alkynes with Me_3SiCF_3 *J. Am. Chem. Soc.* **2010**, *132*, 7262–7263.

27. Whitesides, G. M.; San Filippo, J.; Casey, C. P.; Panek, E. J. Oxidative Coupling Using Copper(1) Ate Complexes *J. Am. Chem. Soc.* **1967**, *89*, 5302–5303.

28. Whitesides, G. M.; Fischer, W. F.; San Filippo, J.; Bashe, R. W.; House, H. O. Reaction of Lithium Dialkyl- and Diarylcuprates with Organic Halides *J. Am. Chem. Soc.* **1969**, *91*, 4871–4882.

29. Lipshutz, B. H.; Siegmann, K.; Garcia, E. "Kinetic" Higher Order Cyanocuprates: Applications to Biaryl Synthesis *J. Am. Chem. Soc.* **1991**, *113*, 8161–8162.

30. Lipshutz, B. H.; Siegmann, K.; Garcia, E., Kayser, F. Synthesis of Unsymmetrical Biaryls via "Kinetic" Higher Order Cyanocuprates: Scope, Limitations, and Spectroscopic Insights *J. Am. Chem. Soc.* **1993**, *115*, 9276–9282.

31. Lipshutz, B. H.; Kayser, F.; Maullin, N. Inter- and Intramolecular Biaryl Couplings via Cyanocuprate Intermediates *Tetrahedron Lett.* **1994**, *35*, 815–818.

32. Coleman, R. S.; Grant, E. B. Atropdiastereoselective Total Synthesis of Phleichrome and the Protein Kinase C Inhibitor Calphostin A *J. Am. Chem. Soc.* **1994**, *116*, 8795–8796.

33. Coleman, R. S.; Grant, E. B. Synthesis of Helically Chiral Molecules: Stereoselective Total Synthesis of the Perylenequinones Phleichrome and Calphostin A *J. Am. Chem. Soc.* **1995**, *117*, 10889–10904.

34. Su, X.; Fox, D. J.; Blackwell, D. T.; Tanaka, K.; Spring, D. R. Copper catalyzed oxidation of organozinc halides *Chem. Commun.* **2006**, 3883–3885.

35. Do, H.-Q.; Daugulis, O. An Aromatic Glaser–Hay Reaction *J. Am. Chem. Soc.* **2009**, *131*, 17052–17053.

36. De Jongh, H. A. P.; De Jongh, C. R. H.; Mijs, W. J. Oxidative Carbon–Carbon Coupling. I. The Oxidative Coupling of a-Substituted Benzyl Cyanides *J. Org. Chem.* **1971**, *36*, 3160–3168.

37. De Jongh, H. A. P.; De Jongh, C. R. H.; Sinnigew, J. M.; De Klein, J.; Huysmnasn, W. G. B.; Mijs, W. J.; van den Hoek, W. J.; Smidt, J. Oxidative Carbon–Carbon Coupling. II. The Effect of Ring Substituents on the Oxidative Carbon–Carbon Coupling of Arylmalonic Esters, Arylmalonodinitriles, and Arylcyanoacetic Esters *J. Org. Chem.* **1972**, *37*, 1960–1966.

38. Kozlowski, M. C.; DiVirgilio, E. S.; Malolanarasimhan, K.; Mulrooney, C. A. Oxidation of Chiral α-Phenylacetate Derivatives: Formation of Dimers with Contiguous Quaternary Stereocenters versus Tertiary Alcohols *Tetrahedron: Asym.* **2005**, *16*, 3599–3605.

39. Kronenburg, C. M. P.; Amijs, C. H. M.; Wijkens, P.; Jastrzebski, J. T. B. H.; van Koten, G. Unexpected formation of aryl-cyanides during the oxidative decomposition of aryl-cyanocuprates. Transfer of a non-transferable group? *Tetrahedron Lett.* **2002**, *43*, 1113–1115.

40. Chan, D. M. T.; Monaco, K. L.; Wang, R.-P.; Winters, M. P. New *N*- and *O*-Arylations with Phenylboronic Acids and Cupric Acetate *Tetrahedron Lett.* **1998**, *39*, 2933–2936.

41. Evans, D. A.; Katz, J. L.; West, T. R. Synthesis of Diaryl Ethers through the Copper-Promoted Arylation of Phenols with Arylboronic Acids. An Expedient Synthesis of Thyroxine *Tetrahedron Lett.* **1998**, *39*, 2937–2940.

42. Lam, P. Y. S.; Clark, C. G.; Saubern, S.; Adams, J.; Winters, M. P.; Chan, D. M. T.; Combs, A. New Aryl/Heteroaryl C—N Bond Cross-coupling Reactions via Arylboronic Acid/Cupric Acetate Arylation *Tetrahedron Lett.* **1998**, *39*, 2941–2944.

43. Lam, P. Y. S.; Deudon, S.; Averill, K. M.; Li, R.; He, M. Y.; DeShong, P.; Clark, C. G. Copper-Promoted C—N Bond Cross-Coupling with Hypervalent Aryl Siloxanes and Room-Temperature *N*-Arylation with Aryl Iodide *J. Am. Chem. Soc.* **2000**, *122*, 7600–7601.

44. Hassan, J.; Sévignon, M.; Gozzi, C.; Schulz, E.; Lemaire, M. Aryl–Aryl Bond Formation One Century after the Discovery of the Ullmann Reaction *Chem. Rev.* **2002**, *102*, 1359–1469.

45. Ley, S. V.; Thomas, A. W. Modern Synthetic Methods for Copper-Mediated C(aryl)—O, C(aryl)—N, and C(aryl)-S Bond Formation *Angew. Chem. Int. Ed.* **2003**, *42*, 5400–5449.

46. Kunz, K.; Scholz, U.; Ganzer, D. Renaissance of Ullmann and Goldberg Reactions - Progress in Copper Catalyzed C-N-, C-O- and C-S-Coupling *Synlett* **2003**, *15*, 2428–2439.

47. Beletskaya, I. P.; Cheprakov, A. V. Copper in cross-coupling reactions The post-Ullmann chemistry *Coord. Chem. Rev.* **2004**, *248*, 2337–2364.

48. Evano, G.; Blanchard, N.; Toumi, M. Copper-Mediated Coupling Reactions and Their Applications in Natural Products and Designed Biomolecules Synthesis *Chem. Rev.* **2008**, *108*, 3054–3131.

49. Collman, J. P.; Zhong, M. An Efficient Diamine, Copper Complex-Catalyzed Coupling of Arylboronic Acids with Imidazoles *Org. Lett.* **2000**, *2*, 1233–1236.

50. Noji, M.; Nakajima, M.; Koga, K. A New Catalytic System for Aerobic Oxidative Coupling of 2-Naphthol Derivatives by the Use of CuCl-Amine Complex: A Practical Synthesis of Binaphthol Derivatives *Tetrahedron Lett.* **1994**, *35*, 7983–7984.

51. Lopez-Alvarado, P.; Avendaño, C.; Menendez, J. C. New Synthetic Applications of Aryllead Triacetates. *N*-Arylation of Azoles *J. Org. Chem.* **1995**, *60*, 5678–5682.

52. Huffman, L. M.; Stahl, S. S. Carbon-Nitrogen Bond Formation Involving Well-Defined Aryl-Copper(III) Complexes *J. Am. Chem. Soc.* **2008**, *130*, 9196–9197.

53. Strieter, E. R.; Blackmond, D. G.; Buchwald, S. L. The Role of Chelating Diamine Ligands in the Goldberg Reaction: A Kinetic Study on the Copper-Catalyzed Amidation of Aryl Iodides *J. Am. Chem. Soc.* **2005**, *127*, 4120–4121.

54. Lam, P. Y. S.; Vincent, G.; Bonne, D.; Clark, C. G. Copper-promoted/catalyzed C—N and C—O bond cross-coupling with vinylboronic acid and its utilities *Tetrahedron Lett.* **2003**, *44*, 4927–4931.

55. Quach, T. D.; Robert A. Batey, R. A. Ligand- and Base-Free Copper(II)-Catalyzed C—N Bond Formation: Cross-Coupling Reactions of Organoboron Compounds with Aliphatic Amines and Anilines *Org. Lett.* **2003**, *5*, 4397–4400.

56. Zhong, W.; Liu, Z.; Yu, C.; Su, W. Copper(II) Acetate Catalyzed Cross-Coupling of Pentafluorophenylboronic Acid with Amines under an Atmosphere of Oxygen *Synlett* **2008**, 2888–2892.

57. Yu, X.-Q.; Yamamoto, Y.; Miyaura, N. Triolborates: Novel Reagent for Copper-Catalyzed N Arylation of Amines, Anilines, and Imidzaoles *Chem. Asian J.* **2008**, *3*, 1517–1522.

58. Nishiura, K.; Urawa, Y.; Soda, S. N-Arylation of Benzimidazole with Arylboronate, Boroxine and Boronic Acids. Acceleration with an Optimal Amount of Water *Adv. Synth. Catal.* **2004**, *346*, 1679–1684.

59. Kantam, M. L.; Neelima, B.; Reddy, C. V.; Neeraja, V. *N*-Arylation of imidazoles, imides, amines, amides and sulfonamides with boronic acids using a recyclable Cu(OAc)$_2$·H$_2$O/ [bmim][BF$_4$] system *J. Mol. Cat. A* **2006**, *249*, 201–206.

60. Kantam, M. L.; Venkanna, G. T.; Sridhar, C.; Sreedhar, B.; Choudary, B. M. An Efficient Base-Free *N*-Arylation and Amines with Arylboronic Acids Copper-Exchanged Fluorapatite *J. Org. Chem.* **2006**, *71*, 9522–9524.

61. Jon C. Antilla and Stephen L. Buchwald, S. L. Copper-Catalyzed Coupling of Arylboronic Acids and Amines *Org. Lett.* **2001**, *3*, 2077–2079.

62. Lan, J.-B.; Zhang, G.-L.; Yu, X.-Q.; You, J.-S.; Chen, L.; Yan, M.; Xie, R.-G. A simple copper salt N-Arylation of Amines, Amides, Imides, and Sulfonamides with Arylboronic Acids *Synlett* **2004**, 1095–1097.

63. Lam, P. Y. S.; Vincent, G.; Clark, C. G.; Deudon, S.; Jadhav, P. K. Copper-catalyzed general C−N and C−O bond cross-coupling with aryl boronic acid *Tetrahedron Lett.* **2001**, *42*, 3415–3418.

64. Lan, J.-B.; Chen, L.; Yu, X.-Q.; You, J.-S.; Xie, R.-G. A simple copper salt catalysed the coupling of imidazole with arylboronic acids in protic solvent *Chem. Comm.* **2004**, 188–189.

65. Yue, Y.; Zheng, Z. G.; Wu, B.; Xia, C. Q.; Yu, X. Q. Copper-Catalyzed Cross-Coupling Reactions of Nucleobases with Arylboronic Acids: An Efficient Access to *N*-Arylnucleobases *Eur. J. Org. Chem.* **2005**, 5154–5157.

66. Likhar, R. R.; Roy, S.; Roy, M.; Kantama, M. L.; De, R. L. Silica immobilized copper complexes: Efficient and reusable catalysts for *N*-arylation of N(H)-heterocycles and benzyl amines with aryl halides and arylboronic acids *J. Mol. Cat. A* **2007**, *271*, 57–62.

67. Hosseinzadeh, R.; Tajbakhsh, M.; Alikarami, M. Copper-catalyzed N-arylation of diazoles with aryl bromides using KF/Al$_2$O$_3$: an improved protocol *Tetrahedron Lett.* **2006**, *47*, 5203–5205.

68. Moessner, C.; Bolm, C. Cu(OAc)$_2$-catalyzed *N*-arylations of sulfoximines with aryl boronic acids *Org. Lett.* **2005**, *7*, 2667–2669.

69. Wentzel, M. T.; Kamble, R.; Wall, P.; Hewgley, J. B.; Kozlowski, M. C. Copper Catalyzed *N*-Arylation of Hindered Substrates Under Mild Conditions *Adv. Synth. Cat.* **2009**, *351*, 931–937.

70. Chan, D. M. T. Monaco, K. L.; Li, R.; Bonne, D.; Clark; C. G.; Lam, P. Y. S. Copper promoted C−N and C−O bond cross-coupling with phenyl and pyridylboronates *Tetrahedron Lett.* **2003**, *44*, 3863–3865.

71. Quach, T. D.; Batey, R. A. Copper(II)-Catalyzed Ether Synthesis from Aliphatic Alcohols and Potassium Organotrifluoroborate Salts *Org. Lett.* **2003**, *5*, 1381–1384.

72. King, A. E.; Brunold, T. C.; Stahl, S. S. Mechanistic Study of Copper-Catalyzed Aerobic Oxidative Coupling of Arylboronic Esters and Methanol: Insights into an Organometallic Oxidase Reaction *J. Am. Chem. Soc.* **2009**, *131*, 5044–5045.

73. Ikegai, K.; Fukumoto, K.; Mukaiyama, T. Copper(II)-catalyzed O-phenylation of tertiary alcohols with organobismuth(V) reagents *Chem. Lett.* **2006**, *35*, 612–613.

74. Mukaiyama, T.; Sakurai, N.; Ikegai, K. Copper(II)-catalyzed *O*-phenylation of alcohols with organobismuth(V) reagents: A convenient method for the synthesis of simple *tert*-alkyl phenyl ethers *Chem. Lett.* **2006**, *35*, 1140–1141.

75. Herradura, P. S.; Pendola, K. A.; Guy, R. K. Copper-Mediated Cross-Coupling of Aryl Boronic Acids and Alkyl Thiols *Org. Lett.* **2000**, *2*, 2019–2022.

76. Savarin, C.; Srogl, J.; Liebeskind, L. S. A Mild, Nonbasic Synthesis of Thioethers. The Copper-Catalyzed Coupling of Boronic Acids with N-Thio(alkyl, aryl, heteroaryl)imides *Org. Lett.* **2002**, *4*, 4309–4312.

77. Huang F.; Batey, R. A. Cross-coupling of organoboronic acids and sulfinate salts using catalytic copper(II) acetate and 1,10-phenanthroline: synthesis of aryl and alkenylsulfones *Tetrahedron* **2007**, *63*, 7667–7672.

78. Demir, A. S.; Reis, O.; Emrullahoglu, M. Role of Copper Species in the Oxidative Dimerization of Arylboronic Acids: Synthesis of Symmetrical Biaryls *J. Org. Chem.* **2003**, *68*, 10130–10134.

79. Kirai, N.; Yamamoto, Y. Homocoupling of Arylboronic Acids Catalyzed by 1,10-Phenanthroline-Ligated Copper Complexes in Air *Eur. J. Org. Chem.* **2009**, 1864–1867.

80. Villalobos, J. M.; Srogl, J.; Liebeskind, L. S. A New Paradigm for Carbon-Carbon Bond Formation: Aerobic, Copper-Templated Cross-Coupling *J. Am. Chem. Soc.* **2007**, *129*, 15734–15735.

81. For a general review: Whiting, D. A. Oxidative Coupling of Phenols and Phenol Ethers in *Comprehensive Organic Synthesis* Trost, B. M.; Fleming, I.; Pattenden, G.; Eds., Pergamon: Oxford, 1991; Vol 3, p 659.

82. Barton, D. H. R.; Cohen, T. Some Biogenic Aspects of Phenol Oxidation. In *Festschrift Arthur Stoll*; Birkhauser A.G.: Basel, 1956; pp 117–143.

83. Pal, T.; Pal, A. Oxidative Phenol Coupling: A Key Step for the Biomimetic Synthesis of Many Important Natural Products *Current Science* **1996**, *71*, 106–109.

84. Keseru, G. M.; Nogradi, M. Natural Products by Oxidative Phenolic Coupling: Phytochemistry, Biosynthesis and Synthesis, in *Studies in Natural Products Chemistry*; Atta-ur-Rahman, Ed.; Elsevier Science, New York: **1998**; Vol 20, pp 263–322.

85. *Tetrahedron* **2001**, 57, The entire issue No. 2. Quideau, S.; Feldman, K. S., Eds.

86. Kozlowski, M. C.; Morgan, B. J.; Elizabeth C. Linton, E. C. Total Synthesis of Chiral Biaryl Natural Products by Asymmetric Biaryl Coupling *Chem. Soc. Rev.* **2009**, *38*, 3193–3207.

87. Hudlicky, T.; Butora, G.; Fearnley, S. P.; Gum, A. G.; Stabile, M. R. A historical perspective of morphine syntheses *Studies Nat. Prod. Chem.* **1996**, *18*, 43–154.

88. Sainsbury, M. Modern Methods of Aryl-Aryl Bond Formation *Tetrahedron*, **1980**, *36*, 3327–3359.

89. Bringmann, G.; Walter, R.; Weirich, R. The Directed Synthesis of Biaryl Compounds: Modern Concepts and Strategies *Angew. Chem. Int. Ed. Engl.* **1990**, *29*, 977–991.

90. Hassan, J.; Sevignon, M.; Gozzi, C.; Schulz, E.; Lemaire, M. Aryl-Aryl Bond Formation One Century after the Discovery of the Ullmann Reaction *Chem. Rev.* **2002**, *102*, 1359–1469.

91. Armstrong, D. R.; Cameron, C.; Nonhebel, D. C.; Perkins, P. G. Oxidative Coupling of Phenols. Part 9. The Role of Steric Effects in the Oxidaiton of Methyl-substituted Phenols *J. Chem. Soc. Perkin Trans. II* **1983**, 581–585.

92. Pummerer, R.; Puttfarcken, H.; Schopflocher, P. Die Dehydrierung von p-Kresol *Chem. Ber.* **1925**, *58*, 1808–1820.

93. Correct structure of Pummerer's ketone: Barton, D. H. R.; Parekh, S. I. *Half a Century of Free Radical Chemistry* Cambridge University Press: Cambridge, 1993.

94. Cliffton, M. D.; Carter, S. J. Oxidative coupling of alkylphenols by copper catalysts US 4851589, **1989**, 6 pp.

95. Inui, N.; Kikuchi, T.; Tanaka, S. Regioselective oxidative coupling method and catalysts for producing 2,2'-dihydroxybiphenyls from phenols. WO 9946227, **1999**, 29 pp.

96. Takuma, Y.; Tanaka, Y.; Nakashima, I.; Kasuga, Y.; Urata, T. Preparation of 2,2'-dihydroxybiphenyls from phenols JP 2002069022, **2002**, 11 pp.

97. Feringa, B.; Wynberg, H. Biomimetic Asymmetric Oxidative Coupling of Phenols *Bioorg. Chem.* **1978**, *7*, 397–408.

98. Brussee, J.; Jansen, A. C. A. A Highly Stereoselective Synthesis of *S*(−)-[1,1'-Binaphthalene]-2,2'-Diol *Tetrahedron Lett.* **1983**, *31*, 3261–3262.

99. Yamamoto, K.; Fukushima, H.; Nakazaki, M. Stereoselective Oxidative Coupling and Asymmetric Hydride Reduction related to (−)-(*S*)-10,10'-Dihydroxy-9,9'-biphenanthryl *J. Chem. Soc. Chem. Commun.* **1984**, 1490–1491.

100. Brussee, J.; Groenendijk, J. L. G.; te Koppele, J. M.; Jansen, A. C. A. On The Mechanism of the Formation of *S*(−)-[1,1'-Binaphthalene]-2,2'-Diol via Copper(II)Amine Complexes *Tetrahedron* **1985**, *41*, 3313–3319.

101. Smrcina, M.; Lorenc, M.; Hanus, V.; Sedmera, P.; Kocovsky, P. Synthesis of Enantiomerically Pure 2,2'-Dihydroxy-1,1'-binaphthyl,2,2'-Diamino-1,1'-binaphthyl and 2-amino-2'-hydroxy-1,1'-binaphthyl. Comparison of Processes Operating as Diastereoselective Crystallization and as Second-Order Asymmetric Transformation *J. Org. Chem.* **1992**, *57*, 1917–1920.

102. Smrcina, M.; Polakova, J.; Vyskocil, S.; Kocovsky, P. Synthesis of Enantiomerically Pure Binaphthyl Derivatives. Mechanism of the Enantioselective, Oxidative Coupling of Naphthols and Designing a Catalytic Cycle *J. Org. Chem.* **1993**, *58*, 4534–4538.

103. Jaffres, P.-A.; Bar, N.; Villemin, D. Phosphonation of 1,1-binaphthalene-2,2'-diol (BINOL): synthesis of (*R*) and (*S*)-2,2'-dihydroxy-1,1'-binaphthalene-6,6'-diyldiphosphonic acid *J. Chem. Soc. Perkin Trans. 1* **1998**, 2083–2089.

104. Botman, P. N. M.; Postma, M.; Fraanje, J.; Goubitz, K.; Schenk, H.; van Maarseveen, J. H.; Hiemstra, H. Synthesis and Resolution of BICOL, a Carbazole Analogue of BINOL *Eur. J. Org. Chem.* **2002**, 1952–1955.

105. Majumder, P. L.; Rahaman, B.; Roychowdhury, M.; Dhara, K. P. Monomeric and dimeric stilbenoids from the orchid *Dendrobium amplum J. Indian Chem. Soc.* **2008**, *85*, 192–199.

106. Lipshutz, B.H.; James, B., Vance, S. I. Carrico A Potentially General Intramolecular Biaryl Coupling Approach to Optically Pure 2,2'-BINOL Analogs *Tetrahedron Lett.* **1997**, *38*, 753–756.

107. Lipshutz, B.H.; Young-Jun Shin, Y-J. A Modular Approach to Nonracemic *cyclo*-BINOLs. Preparation of Symmetrically & Unsymmetrically Substituted Ligands *Tetrahedron Lett.* **1998**, *39*, 7017–7020.

108. Xin, Z.-Q.; Da, C.-S.; Dong, S.-L.; Liu, D.-X.; Wei, J.; Wang, R. A novel method for the synthesis of (*R*)-2,2'-dihydroxy-1,1'-binaphthyl-3,3'-dicarboxylic acid by asymmetric

oxidative coupling of a chiral β-naphthol derivative catalyzed by CuCl *Tetrahedron: Asymm.* **2002**, *13*, 1937–1940.

109. Sridhar, M.; Vadivel, S. K.; Bhalerao, U. T. Novel Horseradish Peroxidase Catalysed Enantioselective Oxidation of 2-Naphthols to 1,1'-Binaphthyl-2,2'-diols *Tetrahedron Lett.* **1997**, *38*, 5695–5696.

110. Takemoto, M; Suzuki, Y.; Tanaka, K. Enantioselective oxidative coupling of 2-naphthol derivatives catalyzed by *Camellia sinensis* cell culture *Tetrahedron Lett.* **2002**, *43*, 8499–8501.

111. Nakajima, M.; Kanayama, K.; Miyoshi, I.; Hashimoto, S. Catalytic Asymmetric Synthesis of Binaphthol Derivatives by Aerobic Oxidative Coupling of 3-Hydroxy-2-naphthoates with Chiral Diamine-Copper Complex *Tetrahedron Lett.* **1995**, *36*, 9519–9520.

112. Nakajima, M.; Miyoshi, I.; Kanayama, K.; Hashimoto, S.-I.; Noji, M.; Koga, K. Enantioselective Synthesis of Binaphthol Derivatives by Oxidative Coupling of Naphthol Derivatives Catalyzed by Chiral Diamine·Copper Complexes *J. Org. Chem.* **1999**, *64*, 2264–2271.

113. Li, X.; Yang, J.; Kozlowski, M. C. Enantioselective Oxidative Biaryl Coupling Reactions Catalyzed by 1,5-Diazadecalin Metal Complexes *Org. Lett.* **2001**, *3*, 1137–1140.

114. Li, X.; Hewgley, J. B.; Mulrooney, C.; Yang, J.; Kozlowski, M. C. Enantioselective Oxidative Biaryl Coupling Reactions Catalyzed by 1,5-Diazadecalin Metal Complexes: Efficient Formation of Chiral Functionalized BINOL Derivatives *J. Org. Chem.* **2003**, *68*, 5500–5511.

115. DiVirgilio, E. S.; Dugan, E. C.; Mulrooney, C. A.; Kozlowski, M. C. Asymmetric Total Synthesis of Nigerone *Org. Lett.* **2007**, *9*, 385–388.

116. Kozlowski, M. C., Dugan, E. C.; DiVirgilio, E. S.; Maksimenka, K.; Bringmann, G. Asymmetric Total Synthesis of Nigerone and *ent*-Nigerone: Enantioselective Oxidative Biaryl Coupling of Highly Hindered Naphthols *Adv. Synth. Cat.* **2007**, *349*, 583–594.

117. Mulrooney, C. A.; Li, X.; DiVirgilio, E. S.; Kozlowski, M. C. General Approach for the Synthesis of Chiral Perylenequinones via Catalytic Enantioselective Oxidative Biaryl Coupling *J. Am. Chem. Soc.* **2003**, *125*, 6856–6857.

118. O'Brien, E. M.; Morgan; B. J.; Kozlowski, M. C. Dynamic Sterochemistry Transfer in a Transannular Aldol Reaction: Total Synthesis of Hypocrellin A *Angew. Chem., Int. Ed.* **2008**, *47*, 6877–6880.

119. Morgan, B. J.; Dey, S.; Johnson, S. W.; Kozlowski, M. C. Total Synthesis of Cercosporin and New Photodynamic Perylenequinones: Inhibition of the Protein Kinase C Regulatory Domain *J. Am. Chem. Soc.* **2009**, *131*, 9413–9425.

120. Mulrooney, C. A.; Morgan, B. J.; Li, X.; Kozlowski, M. C. Perylenequinone Natural Products: Asymmetric Synthesis of the Oxidized Pentacyclic Core *J. Org. Chem.* **2010**, *75*, 16–29.

121. Morgan, B. J.; Mulrooney, C. A.; O'Brien, E. M.; Kozlowski, M. C. Perylenequinone Natural Products: Total Syntheses of the Diastereomers (+)-Phleichrome and (+)-Calphostin D by Assembly of Centrochiral and Axial Chiral Fragments *J. Org. Chem.* **2010**, *75*, 30–43.

122. Morgan, B. J.; Mulrooney, C. A.; Kozlowski, M. C. Perylenequinone Natural Products: Evolution of the Total Synthesis of Cercosporin *J. Org. Chem.* **2010**, *75*, 44–56.

123. O'Brien, E. M.; Morgan, B. J.; Mulrooney, C. A.; Kozlowski, M. C. Perylenequinone Natural Products: Total Synthesis of Hypocrellin A *J. Org. Chem.* **2010**, *75*, 57–68.

124. Kozlowski, M. C.; Li, X.; Carroll, P. J.; Xu, Z. Copper(II) Complexes of Novel 1,5-Diaza-*cis*-decalin Diamine Ligands: An Investigation of Structure and Reactivity *Organometallics* **2002**, *21*, 4513–4522.

125. Hewgley, J. B.; Stahl, S. S.; Kozlowski, M. C. Mechanistic Study of Asymmetric Oxidative Biaryl Coupling: Evidence for Self-Processing of the Copper Catalyst to Achieve Control of Oxidase vs. Oxygenase Activity, *J. Am. Chem. Soc.* **2008**, *130*, 12232–12233.

126. Brazeau, B.J.; Johnson, B. J.; Wilmot, C. M. Copper-Containing Amine Oxidases: Biogenisis and Catalysis; A Structural Perspective *Arch. Biochem. Biophys.* **2004**, *428*, 22–31.

127. Mure, M. Tyrosine Derived Quinone Cofactors *Acc. Chem. Res.* **2004**, *37*, 131–139.

128. DuBois, J. L.; Klinman, J. P. Mechanism of Post-Translational Quinone Formation in Copper Amine Oxidases and Its Relationship to the Catalytic Turnover *Arch. Biochem. Biophys.* **2005**, *433*, 255–265.

129. Roithova, J.; Petr Milko, P. Naphthol Coupling Monitored by Infrared Spectroscopy in the Gas Phase *J. Am. Chem. Soc.* **2010**, *132*, 281–288.

130. Kim, K. H.; Lee, D.-W.; Lee, Y.-S.; Ko, D.-H.; Ha, D.-C. Enantioselective oxidative coupling of methyl 3-hydroxy-2-naphthoate using mono-*N*-alkylated octahydrobinaphthyl-2,2′-diamine ligand *Tetrahedron* **2004**, *60*, 9037–9042.

131. Gao, J.; Reibenspies, J. H.; Martell, A. E. Structurally Defined Catalysts for Enantioselective Oxidative Coupling Reactions *Angew. Chem. Int. Ed.* **2003**, *42*, 6008–6012.

132. Majumder, P. L.; Bandyopadhyay, S.; Pal, S. Rigidanthrin, a new dimeric phenanthrene derivative of the orchid *Bulbophyllum rigidum J. Indian Chem. Soc.* **2008**, *85*, 1116–1123.

133. Temma, T.; Habaue, S. Highly selective oxidative cross-coupling of 2-naphthol derivatives with chiral copper(I)–bisoxazoline catalysts *Tetrahedron Lett.* **2005**, *46*, 5655–5657.

134. Habaue, S.; Takahashi, Y.; Temma, T. Oxidative cross-coupling leading to 3-amido substituted 1,1'-bi-2-naphthol derivatives *Tetrahedron Lett.* **2007**, *48*, 7301–7304.

135. Temma, T.; Hatano, B.; Habaue, S. Cu(I)-catalyzed asymmetric oxidative cross-coupling of 2-naphthol derivatives *Tetrahedron* **2006**, *62*, 8559–8563.

136. Habaue, S.; Temma, T.; Sugiyama, Y.; Yan, P. Ytterbium triflate-assisted catalytic oxidative cross-coupling of 2-naphthol derivatives *Tetrahedron Lett.* **2007**, *48*, 8595–8598.

137. Habaue, S.; Seko, T.; Okamoto, Y. Asymmetric Oxidative Coupling Polymerization of Optically Active Tetrahydroxybinaphthalene Derivative *Macromolecules* **2002**, *35*, 2437–2439.

138. Habaue, S.; Seko, T.; Isonaga, M.; Ajiro, H.; Okamoto, Y. Stereoselective Synthesis of (*R,R*)-,(*S,S*)-, and (*R,S*)-Poly(2,3-dihydroxy-1,4-naphthylene) Derivatives by Asymmetric Oxidative Coupling Polymerization *Polym. J.* **2003**, *35*, 592–597.

139. Habaue, S.; Seko, T.; Okamoto, Y. Copper(I)-Catalyzed Asymmetric Oxidative Coupling Polymerization of 2,3-Dihydroxynaphthalene Using Bisoxazoline Ligands *Macromolecules* **2003**, *36*, 2604–2608.

140. Habaue, S.; Ajiro, H.; Yoshii, Y.; Hirasa, T. Asymmetric Oxidative Coupling Polymerizations Affording Polynaphthylene with 1,1'-bi-2-Naphthol Units *J. Polym. Sci: Part A: Polym. Chem.* **2004**, *42*, 4528–4534.

141. Yan, P.; Sugiyama, Y.; Takahashi, Y.; Kinemuchi, H.; Temma, T.; Habaue, S. Lewis acid-assisted oxidative cross-coupling of 2-naphthol derivatives with copper catalysts *Tetrahedron* **2008**, *64*, 4325–4331.

142. Yan, P.; Temma, T.; Habaue, S. Lewis-Acid-Assisted Highly Selective Oxidative Cross-Coupling Polymerization with Copper Catalysts *Polym. J.* **2008**, *40*, 710–715.

143. Xie, X.; Phuan, P.-W.; Kozlowski, M. C. Novel Pathways for the Formation of Chiral Binaphthyl Polymers: Oxidative Asymmetric Phenolic Coupling and Tandem Glaser/Oxidative Asymmetric Phenolic Coupling *Angew. Chem. Int. Ed.* **2003**, *42*, 2168–2170.

144. Morgan, B. J.; Xie, X.; Phuan, P-W.; Kozlowski, M. C. Enantioselective Synthesis of Binaphthyl Polymers using Chiral Asymmetric Phenolic Coupling Catalysts: Oxidative Coupling and Tandem Glaser/Oxidative Coupling *J. Org. Chem.* **2007**, *72*, 6171–6182.

145. Daniels, R. N.; Fadeyi, O. O.; Lindsley, C. W. A New Catalytic Cu(II)/Sparteine Oxidant System for β,β-Phenolic Couplings of Styrenyl Phenols: Synthesis of Carpanone and Unnatural Analogs *Org. Lett.* **2008**, *10*, 4097–4100.

146. Liron, F.; Fontana, F.; Zirimwabagabo, J. O.; Prestat, G.; Rajabi, J.; Rosa, C.; Poli, G. A New Cross-Coupling-Based Synthesis of Carpanone *Org. Lett.* **2009**, *11*, 4378–4381.

147. Prokofieva, A.; Prikhod'ko, A. I.; Dechert, S.; Meyer, F. Selective benzylic C–C coupling catalyzed by a bioinspired dicopper complex *Chem. Commun.* **2008**, 1005–1007.

148. Takaki, K.; Shimasaki, Y.; Shishido, T.; Takehira, K. Selective Oxidation of Phenols to Hydroxybenzaldehydes and Benzoquinones with Dioxygen Catalyzed by a Polymer Supported Copper *Bull. Chem. Soc. Jpn.* **2002**, *75*, 311–317.

149. Hay, A. S.; Stafford, H. S.; Endres, G. F.; Eustance, J. W. Polymerizaton by Oxidative Coupling *J. Am. Chem. Soc.* **1959**, *81*, 6335–6336.

150. Baesjou, P. J.; Driessen, W. L.; Challa, G.; Reedijk, J. *Ab Initio* Calculations on 2,6-Dimethylphenol and 4-(2,6-Dimethylphenoxy)-2,6-dimethylphenol. Evidence of an Important Role for the Phenoxonium Cation in the Copper-Catalyzed Oxidative Phenol Coupling Reaction *J. Am. Chem. Soc.* **1997**, *119*, 12590–12594.

151. Suzuki, Y.; Shibasaki, Y.; Ueda, M. Regio-controlled Oxidative Polymerization of 2,5-Dimethylphenol by Using CuCl–TMEDA Complex *Chem. Lett.* **2007**, *36*, 1234–1235.

152. Menini, L.; Parreira L. A.; Gusevskaya, E. V. A practical highly selective oxybromination of phenols with dioxygen *Tetrahedron Lett.* **2007**, *48*, 6401–6404.

153. Yang, L.; Lub, Z.; S. Stahl, S. S. Regioselective copper-catalyzed chlorination and bromination of arenes with O2 as the oxidant *Chem. Commun.* **2009**, 6460–6462.

154. Menini L.; Gusevskaya, E. V. Novel highly selective catalytic oxychlorination of phenols *Chem. Commun.* **2006**, 209–211.

155. Menini L.; Gusevskaya, E. V. Aerobic oxychlorination of phenols catalyzed by copper(II) chloride *Appl. Catal. A* **2006**, *309*, 122–128.

156. Kupán, A.; Kaizer, J.; Speier, G.; Giorgi, M.; Réglier, M.; Pollreisz, F. Molecular structure and catechol oxidase activity of a new copper(I) complex with sterically crowded monondentate *N*-donor ligand *J. Inorg. Biochem.* **2009**, *103*, 389–395.

157. Gichinga, M. G.; Striegler, S. Effect of Water on the Catalytic Oxidation of Catechols *J. Am. Chem. Soc.*, **2008**, *130*, 5150–5156.

158. Balla, J.; Kiss, T.; Jameson, R. F. Copper(II)-catalyzed oxidation of catechol by molecular oxygen in aqueous solution *Inorg. Chem.*, **1992**, *31*, 58–62.

159. Mimmi, M. C.; Gullotti, M.; Santagostini, L.; Battaini, G.; Monzani, E.; Pagliarin, R.; Zoppellaro, G.; Casella, L. Models for biological trinuclear copper clusters. Characterization and enantioselective catalytic oxidation of catechols by the copper(II) complexes of a chiral ligand derived from (S)-(−)-1,1′-binaphthyl-2,2′-diamine *Dalton Trans.* **2004**, 2192–2201.

160. Smrcina, M.; Vyskocil, S.; Maca, B.; Polasek, M.; Claxton, T. A.; Abbott, A. P.; Kocovsky, P. Selective Cross-Coupling of 2-Naphthol and 2-Naphthylamine Derivatives. A Facile Synthesis of 2,2′,3-Trisubstituted and 2,2′,3,3′-Tetrasubstituted 1,1′-Binaphthyls *J. Org. Chem.* **1994**, *59*, 2156–2163.

161. Vyskocil, S.; Smrcina, M.; Lorenc, M.; Tislerova, I. Brooks, R. D. Kulagowski, J. J.; Langer, V.; Farrugia, L. J.; Kocovsky, P. Copper(II)-Mediated Oxidative Coupling of 2-Aminonaphthalene Homologues. Competition between the Straight Dimerization and the Formation of Carbazoles *J. Org. Chem.* **2001**, *66*, 1359–1365.

162. Yusa, Y.; Kaito, I.; Akiyama, K.; Mikami, K. Asymmetric Catalysis of Homo-Coupling of 3-Substituted Naphthylamine and Hetero-Coupling with 3-Substituted Naphthol Leading to 3,3′-Dimethyl-2,2′-diaminobinaphthyl and -2-amino-2′-hydroxybinaphthyl *Chirality* **2010**, *22*, 224–228.

163. Chen, Z.; Pina, C. D.; Falletta, E.; Rossi, M. A green route to conducting polyaniline by copper catalysis *J. Catal.* **2009**, *267*, 93–96.

164. Zhang, M.; Zhang, R.; Zhang, A.-Q.; Li, X.; Liang, H. Cupric Chloride–Catalyzed Synthesis of Symmetrical Azo Compounds from Primary Aromatic Amines *Synth. Commun.* **2009**, *39*, 3428–3435.

165. Jahn, F. P. The preparation of azomethane *J. Am. Chem. Soc.* **1937**, *59*, 1761–1762.

166. Menini, L.; da Cruz Santos, J. C.; Gusevskaya, E. V. Copper-Catalyzed Oxybromination and Oxychlorination of Primary Aromatic Amines Using LiBr or LiCl and Molecular Oxygen *Adv. Synth. Catal.* **2008**, *350*, 2052–2058.

167. Ribas, X.; Jackson, D. A.; Donnadieu, B.; Mahía, J.; Parella, T.; Xifra, R.; Britt Hedman, B.; Hodgson, K. O.; Llobet, A.; Stack, T. D. P. Aryl C-H Activation by Cu^{II} To Form an Organometallic Aryl-Cu^{III} Species: A Novel Twist on Copper Disproportionation *Angew. Chem. Int. Ed.* **2002**, *41*, 2991–2994.

168. Xifra, R.; Ribas, X.; Llobet, A.; Poater, A.; Duran, M.; Solà, M.; Stack, T. D. P.; Benet-Buchholz, J.; Donnadieu, B.; Mahía, J.; Parella, T. Fine-Tuning the Electronic Properties of Highly Stable Organometallic Cu^{III} Complexes Containing Monoanionic Macrocyclic Ligands *Chem. Eur. J.* **2005**, *11*, 5146–5156.

169. Ban, I.; Sudo, T.; Taniguchi, T.; Kenichiro Itami, K. Copper-Mediated C−H Bond Arylation of Arenes with Arylboronic Acids *Org. Lett.* **2008**, *10*, 3607–3609.

170. Brasche, G.; Buchwald, S. L. C−H Functionalization/C−N Bond Formation: Copper-Catalyzed Synthesis of Benzimidazoles from Amidines *Angew. Chem. Int. Ed.* **2008**, *47*, 1932–1934.

171. Ueda, S.; Nagasawa, H. Synthesis of 2-Arylbenzoxazoles by Copper-Catalyzed Intramolecular Oxidative C–O Coupling of Benzanilides *Angew. Chem. Int. Ed.* **2008**, *47*, 6411–6413.

172. Ueda, S.; Nagasawa, H. Copper-Catalyzed Synthesis of Benzoxazoles via a Regioselective C−H Functionalization/C−O Bond Formation under an Air Atmosphere *J. Org. Chem.* **2009**, *74*, 4272–4277.

173. Mizuhara, T.; Inuki, S.; Oishi, S.; Fujii, N.; Ohno, H. Cu(II)-mediated oxidative intermolecular ortho C–H functionalisation using tetrahydropyrimidine as the directing group *Chem. Commun.* **2009**, 3413–3415.

174. Chen, X.; Hao, X.-S.; Goodhue, C. E.; Yu, J.-Q. Cu(II)-Catalyzed Functionalizations of Aryl C–H Bonds Using O_2 as an Oxidant *J. Am. Chem. Soc.* **2006**, *128*, 6790–6791.

175. Monguchi, D.; Fujiwara, T.; Furukawa, H.; Mori, A. Direct Amination of Azoles via Catalytic C–H, N–H Coupling *Org. Lett.* **2009**, *11*, 1607–1610.

176. Bordwell, F. G. Equilibrium acidities in dimethyl sulfoxide solution *Acc. Chem. Res.*, **1988**, *21*, 456–463.

177. Li, C.-J. Cross-Dehydrogenative Coupling Exploring C–C Bond Formations Functional Group Transformations *Acc. Chem. Res.* **2009**, *42*, 335–344.

178. Sureshkumar, D.; Sud, A.; Klussmann, M. Aerobic Oxidative Coupling of Tertiary Amines with Silyl Enolates and Ketene Acetals *Synlett* **2009**, 1558–1561.

179. Murata, S.; Teramoto, K.; Miura, M.; Nomura, M. Copper catalyzed oxidative coupling of 4-substituted *N,N*-dimethylanilines with terminal alkynes under molecular oxygen *J. Chem. Res. (S)* **1993**, 434.

180. For an initial description of the amine to iminium transformation with copper, see: Murata, S.; Suzuki, K.; Tamatani, A.; Miura, M.; Nomura, M. Oxidative Dealkylation of 4-Dubstituted *N,N*-Dialkylanilines with Molecular Oxygen in the Presence of Acetic Anhydride Promoted by Colbalt(II) or Copper(I) Chloride *J. Chem. Soc. Perkin Trans. 1* **1992**, 1387–1392.

181. Basle, O.; Li, C.-J. Copper catalyzed oxidative alkylation of sp^3 C–H bond adjacent to a nitrogen atom using molecular oxygen in water *Green Chem.* **2007**, *9*, 1047–1050.

182. Shen, Y.; Li, M.; Wang, S.; Zhan, T.; Tan, Z.; Guo, C.-C. An efficient copper-catalyzed oxidative Mannich reaction between tertiary amines and methyl ketones *Chem. Commun.* **2009**, 953–955.

183. Shen, Y.; Tan, Z.; Chen, D.; Feng, X.; Li, M.; Guo, C.-C.; Zhu, C. Highly efficient Cu-catalyzed oxidative coupling of tertiary amines and siloxyfurans *Tetrahedron* **2009**, *65*, 158–163.

184. Basle, O.; Li, C.-J. Copper-Catalyzed Oxidative sp^3 C–H Bond Arylation with Aryl Boronic Acids *Org. Lett.* **2008**, *10*, 3661–3663.

185. Li, Z.; Bohle, S.; Li, C.-J. Cu-catalyzed cross-dehydrogenative coupling: A versatile strategy for C–C bond formations via the oxidative activation of sp^3 C–H bonds *PNAS* **2006**, *103*, 8928–8933.

186. Huang, L.; Cheng, K.; Yao, B.; Zhao, J.; Zhang, Y. Copper-Catalyzed Hydroalkylation of Alkynes: Addition Across Carbon–Carbon Triple Bonds *Synthesis* **2009**, 3504–3510.

187. Yoo, W.-J.; Correia, C. A.; Zhang, Y.; Li, C.-J. Oxidative Alkylation of Cyclic Benzyl Ethers with Malonates and Ketones Using Oxygen as the Terminal Oxidant *Synlett* **2009**, 138–142.

188. Cheng, K.; Huang, L.; Zhang, Y. CuBr-Mediated Oxyalkylation of Vinylarenes under Aerobic Conditions via Cleavage of sp^3 C–H Bonds α to Oxygen *Org. Lett.* **2009**, *11*, 2908–2911.

189. Shibahara, F.; Yoshida, A; Murai, T. Copper-catalyzed Oxidative Desulfurization-promoted Intramolecular Cyclization of Thioamides Using Molecular Oxygen as an Oxidant: An Efficient Route to Five- to Seven-membered Nitrogen-containing Heterocycles *Chem. Lett.* **2008**, *37*, 646–647.

190. Marko, I. E.; Giles, P. R.; Tsukazaki, M.; Chellé-Regnaut, I.; Gautier, A.; Brown, S. M.; Urch, C. J. Efficient, Ecologically Benign, Aerobic Oxidation of Alcohols *J. Org. Chem.* **1999**, *64*, 2433–2439.

191. Marko, I. E.; Gautier, A.; Dumeunier, R.; Doda, K.; Philippart, F.; Brown, S. M.; Urch, C. J. Efficient, Copper-Catalyzed, Aerobic Oxidation of Primary Alcohols *Angew. Chem. Int. Ed.* **2004**, *43*, 1588–1591.

192. Ragagnin, G.; Betzemeier, B.; Quicib, S.; Knochel, P. Copper-catalysed aerobic oxidation of alcohols using fluorous biphasic catalysis *Tetrahedron* **2002**, *58*, 3985–3991.

193. Gamez, P.; Arends, I. W. C. E.; Sheldon, R. A.; Reedijk, J. Room Temperature Aerobic Copper-Catalysed Selective Oxidation of Primary Alcohols to Aldehydes *Adv. Synth. Catal.* **2004**, *346*, 805–811.

194. Wang, Y.; DuBois, J. L.; Hedman, B.; Hodgson, K. O.; Stack, T. D. P. Catalytic Galactose Oxidase Models: Biomimetic Cu(II)–Phenoxyl-Radical Reactivity *Science* **1998**, *279*, 537–540.

195. Cheng, L.; Wang, J.; Wanga, M.; Wu, Z. Theoretical studies on the reaction mechanism of oxidation of primary alcohols by Zn/Cu(II)-phenoxyl radical catalyst *Dalton Trans.* **2009**, 3286–3297.

196. Chaudhuri, P.; Hess, M.; Müller, J.; Hildenbrand, K.; Bill, E.; Weyhermüller, T.; Wieghardt, K. Aerobic Oxidation of Primary Alcohols (Including Methanol) by Copper(II)- and Zinc(II)–Phenoxyl Radical Catalysts *J. Am. Chem. Soc.* **1999**, *121*, 9599–9610.

197. Davi, M.; Lebel, H. Copper-Catalyzed Tandem Oxidation–Olefination Process *Org. Lett.* **2009**, *11*, 41–44.

198. Mannam, S.; Sekar, G. An enantiopure galactose oxidase model: synthesis of chiral amino alcohols through oxidative kinetic resolution catalyzed by a chiral copper complex *Tetrahedron Asymm.* **2009**, *20*, 497–502.

199. Walsh, P. J.; Kozlowski, M. C. *Fundamentals of Asymmetric Catalysis*, University Science Books, Sausalito, CA, 2009, Chapter 8.

200. Alamsetti, S. K.; Mannam, S.; Mutupandi, P.; Sekar, G. Galactose Oxidase Model: Biomimetic Enantiomer-Differentiating Oxidation of Alcohols by a Chiral Copper Complex *Chem. Eur. J.* **2009**, *15*, 1086–1090.

201. Vatèle, J. M. Copper-Catalyzed Aerobic Oxidative Rearrangement of Tertiary Allylic Alcohols Mediated by TEMPO Oxidative Rearrangement of Tertiary Allylic Alcohols *Synlett* **2009**, 2143–2145.

202. Tian, Q.; Shi, D.; Sha, Y. CuO and Ag$_2$O/CuO Catalyzed Oxidation of Aldehydes to the Corresponding Carboxylic Acids by Molecular Oxygen *Molecules* **2008**, *13*, 948–957.

203. van Rheenen, V. Copper-catalyzed oxidation of branched aldehydes – an efficient ketone synthesis *Tetrahedron Lett.* **1969**, 985–988.

204. Takehira, K.; Shimizu, M.; Watanabe, Y.; Orita, H.; Hayakawa, T. A novel oxygenation of 2,3,6-trimethylphenol to trimethyl-*p*-benzoquinone by dioxygen with copper chloride/amine hydrochloride catalyst *Tetrahedron Lett.* **1989**, *30*, 6691–6692.

205. Sun, H.; Harms, K.; Sundermeyer, J. Aerobic Oxidation of 2,3,6-Trimethylphenol to Trimethyl-1,4-benzoquinone with Copper(II) Chloride as Catalyst in Ionic Liquid and Structure of the Active Species *J. Am. Chem. Soc.* **2004**, *126*, 9550–9551.

206. Mirica, L. M.; Vance, M.; Rudd, D. J.; Hedman, B.; Hodgson, K. O.; Solomon, E. I.; Stack, T. D. P. A Stabilized μ-η^2: η^2-Peroxodicopper(II) Complex with a Secondary Diamine Ligand and Its Tyrosinase-like Reactivity *J. Am. Chem. Soc.* **2002**, *124*, 9332–9333.

207. Stack, T. D. P. Complexity with simplicity: a steric continuum of chelating diamines with copper(I) and dioxygen *J. Chem. Soc. Dalton Trans.* **2003**, 1881–1889.

208. Mirica, L. M.; Ottenwaelder, X.; Stack, T. D. P. Structure and Spectroscopy of Copper–Dioxygen Complexes *Chem. Rev.* **2004**, *104*, 1013–1046.

209. Lewis, E. A.; Tolman, W. B. Reactivity of Dioxygen–Copper Systems *Chem. Rev.* **2004**, *104*, 1047–1076.

210. Hatcher, L. Q.; Karlin, K. D. Oxidant types in copper–dioxygen chemistry: the ligand coordination defines the $Cu_n–O_2$ structure and subsequent reactivity *J. Biol. Inorg. Chem.* **2004**, *9*, 669–683.

211. Zhu, J.; Grigoriadis, N. P.; Lee, J. P.; Porco, J. A. Synthesis of the Azaphilones Using Copper-Mediated Enantioselective Oxidative Dearomatization *J. Am. Chem. Soc.* **2005**, *127*, 9342–9343.

212. Zhu, J.; Porco, J. A. Asymmetric Syntheses of (–)-Mitorubrin and Related Azaphilone Natural Products *Org. Lett.* **2006**, *8*, 5169–5171.

213. Dong; S.; Zhu, J.; Porco, J. A. Enantioselective Synthesis of Bicyclo[2.2.2]octenones Using a Copper-Mediated Oxidative Dearomatization/[4 + 2] Dimerization Cascade *J. Am. Chem. Soc.* **2008**, *130*, 2738–2739.

214. Chiba, S.; Zhang, L.; J.-Y. Copper-Catalyzed Synthesis of Azaspirocyclohexadienones from α-Azido-N-arylamides under an Oxygen Atmosphere *J. Am. Chem. Soc.* **2010**, *132*, 7266–7267.

215. Zhao, X.; Kong, A.; Shan, C.; Wang, P.; Zhang, X.; Shan, Y. Highly Efficient and Green Oxidation of Nitrotoluenes with Dioxygen as Oxidant in a Novel Homogeneous and Recyclable Catalytic System *Catal. Lett.* **2009**, *131*, 526–529.

216. Wang, F.; Yang, G.-Y.; Zhang, W.; Wu, W.-H.; Xu, J. Copper and manganese: two concordant partners in the catalytic oxidation of *p*-cresol to *p*-hydroxybenzaldehyde *Chem. Commun.* **2003**, 1172–1173.

217. Shimizu, M.; Watanabe, Y.; Orita, H.; Hayakawa, T.; Takehira, K. A Facile Synthesis of 4-Alkoxymethylphenols by a Copper(II)Acetoxime Catalyst/O_2 System *Tetrahedron Lett.* **1991**, *32*, 2053–2056.

218. Murahashi, S.-I.; Komiya, N.; Hayashi, Y. Method for oxidizing alkanes and cycloalkanes into the corresponding alcohols and ketones with oxygen in the presence of aldehydes, copper -based catalysts and nitrogen-containing compounds *Eur. Pat. Appl.* **2002**, EP1174410, 7 pp.

219. Sakaguchi, S.; Takase, T.; Iwahama, T.; Ishii, Y. Oxygenation of alkynes to α,β-acetylenic ketones with dioxygen catalyzed by *N*-hydroxyphthalimide combined with a transition metal *Chem. Commun.* **1998**, 2037–2038.

220. Larock, R. C. *Comprehensive Organic Transformations* VCH Publishers, New York, 1989.

221. Nobumasa Kitajima, N.; Koda, T.; Iwata, Y.; Moro-oka, Y. Reaction Aspects of a μ-Peroxo Binuclear Copper(II) Complex *J. Am. Chem. Soc.* **1990**, *112*, 8833–8839.

222. Volger, H. C.; Brackman, W. Copper-catalysed oxidation of unsaturated carbonyl compounds. i. Catalysed oxidation of Δ5-cholestenone to Δ4-cholestene-3,6-dione *Rec. Trav. Chim.* **1965**, *84*, 579.

223. Komiya, N.; Naota, T.; Murahashi, S.-I. Aerobic Oxidation of Alkanes in the Presence of Acetaldehyde Catalysed by Copper-Crown Ether *Tetrahedron Lett.* **1996**, *37*, 1633–1636.

224. Murahashi, S.-I.; Komiya, N.; Hayashi, Y.; Kumano, T. Copper complexes for catalytic, aerobic oxidation of hydrocarbons *Pure Appl. Chem.* **2001**, *73*, 311–314.

225. Shul'pin, G. B.; Nizova, G. V. Photo-oxidation of cyclohexane by atmospheric oxygen in acetonitrile, catalysed by chloride complexes of iron, copper, and gold *Petrol. Chem.* **1993**, *33*, 107–112.

226. Shul'pin, G. B.; Bochkova, M. M.; Nizova, G. V. Aerobic oxidation of saturated hydrocarbons into alkylhydroperoxides induced by visible light and catalysed by a 'quinone-copper acetate' *J. Chem. Soc., Perkin Trans.* **1995**, 2 1465–1469.

227. Takaki, K.; Yamamoto, J.; Komeyama, K.; Kawabata, T.; Takehira, K. Photocatalytic oxidation of alkanes with dioxygen by visible light and copper(II) and iron(III) chlorides: preference oxidation of alkanes over alcohols and ketones *Bull. Chem. Soc. Jpn.* **2004**, *77*, 2251–2255.

228. Ito, S.; Yamasaki, T; Okada, H.; Okino, S.; Sasaki, K. Oxidation of Benzene to Phenols with Molecular Oxygen Promoted by Copper(I) Chloride *J. Chem. Soc., Perkin Trans.* **1988**, 2, 285–293.

229. Kunai, A.; Wani, T.; Uehara, Y.; Iwasaki, F.; Kuroda, Y.; Ito, S.; Sasaki, K. Catalytic Oxgenation of Benzene. Catalyst Design and its Performance *Bull. Chem. Soc. Jpn.* **1989**, *62*, 2613–2617.

230. Hiromitsu Yamanaka, H.; Hamada, R.; Nibuta, H.; Nishiyama, S.; Tsuruya, S. Gas-phase catalytic oxidation of benzene over Cu-supported ZSM-5 catalysts: an attempt of one-step production of phenol *J. Mol. Cat. A* **2002**, *178*, 89–95.

231. Bahidsky, M.; Hronec, M. Direct hydroxylation of aromatics over copper–calcium–phosphates in the gas phase *Catalysis Today* **2005**, *99*, 187–192.

232. Liu, Y.; Murata, K.; Inaba, M. Liquid-phase oxidation of benzene to phenol by molecular oxygen over transition metal substituted polyoxometalate compounds *Catal. Commun.* **2005**, *6*, 679–683.

233. Ichihashi, Y.; Kamizaki, Y.-H.; Terai, N.; Taniya, K.; Tsuruya, S.; Nishiyama, S. One-Step Oxidation of Benzene to Phenol over Cu/Ti/HZSM-5 Catalysts *Catal. Lett.* **2009**, *134*, 324–329.

234. Paine, T. K.; Weyhermüller, T.; Wieghardt, K.; Chaudhuri, P. Aerial oxidation of primary alcohols and amines catalyzed by Cu(II) complexes of 2,2′-selenobis(4,6-di-*tert*-butylphenol) providing [O,Se,O]-donor atoms *J. Chem. Soc. Dalton Trans.* **2004**, 2092–2101.

235. Srogl, J.; Voltrova, S. Copper/Ascorbic Acid Dyad as a Catalytic System for Selective Aerobic Oxidation of Amines *Org. Lett.* **2009**, *11*, 843–845.

236. Shibahara, F.; Suenami, A.; Yoshida, A; Murai, T. Copper-catalyzed oxidative desulfurization–oxygenation of thiocarbonyl compounds using molecular oxygen: an efficient method for the preparation of oxygen isotopically labeled carbonyl compounds *Chem. Commun.* **200**, *37*, 2354–2356.

237. Komiya, N.; Naota, T.; Oda, Y.; Murahashi, S.-I. Aerobic oxidation of alkanes and alkenes in the presence of aldehydes catalyzed by copper salts and copper-crown ether *J. Mol. Cat. A* **1997**, *17*, 21–37. Murahashi, S.-I.; Oda, Y.; Naota, T.; Komiya, N. Aerobic Oxidations of Alkanes and Alkenes in the Presence of Aldehydes catalysed by Copper Salts *J. Chem. Soc. Chem. Commun.* **1993**, 139–140.

238. Meng, X.; Lin, K.; Yang, X.; Sun, Z.; Jiang, D.; Xiao, F.-S. Catalytic oxidation of olefins and alcohols by molecular oxygen under air pressure over $Cu_2(OH)PO_4$ and $Cu_4O(PO_4)_2$ catalysts *J. Catal.* **2003**, *218*, 460–464.

239. Karandikar, P.; Agashe, M.; Vijayamohanan, K.; Chandwadkar, A.J. Cu^{2+}-perchlorophthalocyanine immobilized MCM-41: catalyst for oxidation of alkenes *Appl. Catal. A* **2004**, *257*, 133–143.

240. Bolm, C.; Schlingloff, G.; Weickhardt, K. Use of Molecular Oxygen in the Baeyer–Villlger Oxidation The Influence of Metal Catalysts *Tetrahedron Lett.* **1993**, *34*, 3405–3408.

241. Phillips, B.; Frostick Jr., F. C.; Starcher, P. S. A New Synthesis of Peracetic Acid *J. Am. Chem. Soc.* **1957**, *79*, 5982–5986.

242. Bolm, C.; Schlingloff, G.; Weickhardt, K. Optically Active Lactones from a Baeyer–Villiger-Type Metal-Catalyzed Oxidation with Molecular Oxygen *Angew. Chem. Int. Ed.* **1994**, *33*, 1848–1849.

243. Peng, Y.; Feng, X.; Yu, K.; Li, Z.; Jiang, Y.; Yeung, C.-H. Synthesis and crystal structure of bis-[(4S,5S)-4,5-dihydro-4,5-diphenyl-2-(2%-oxidophenyl-χO)oxazole-χN] copper(II) and its application in the asymmetric Baeyer–Villiger reaction *J. Organomet. Chem.* **2001**, *619*, 204–208.

244. Bolm, C.; Schlingloff, G. Copper- and vanadium-catalyzed asymmetric oxidations *J. Chem. Soc., Chem. Commun.* **1995**, 1247–1248.

245. Bolm, C.; Schlingloff, G.; Bienewald, F. Copper- and vanadium-catalyzed asymmetric oxidations *J. Mol. Cat. A* **1997**, *117*, 337–350.

246. Song, G.; Wang, F.; Zhang, H.; Lu X.; Wang, C. Efficient Oxidation of Sulfides Catalyzed by Transition Metal Salts with Molecular Oxygen in The Presence of Aldehydes *Synth. Commun.* **1998**, *28*, 2783–2787.

247. Martin, S. E.; Rossi, L. I. An efficient and selective aerobic oxidation of sulfides to sulfoxides catalyzed by $Fe(NO_3)_3–FeBr_3$ *Tetrahedron Lett.* **2001**, *42*, 7147–7151.

INDEX

Copper-Oxygen Chemistry, First Edition. Edited by Kenneth D. Karlin and Shinobu Itoh.
© 2011 John Wiley & Sons, Inc. Published 2011 by John Wiley & Sons, Inc.